The UK Pesticide Guide 2010

Editor: M. A. Lainsbury, BSc(Hons)

BCPC

cabi
www.cabi.org

CABI is one of the world's foremost publishers of databases, books and compendia in agriculture and applied life sciences. It has a worldwide reputation for producing high quality, value-added information, drawing on its links with the scientific community. CABI produces CAB Abstracts – the leading agricultural A&I database; together with the Crop Protection Compendium which provides solutions for identifying, preventing and solving crop health problems. CABI is a not-for-profit organization. For further information, please contact CABI, Nosworthy Way, Wallingford, Oxon OX10 8DE, UK.

Telephone: (01491) 832111
Fax: (01491) 833508
e-mail: enquiries@cabi.org
Web: www.cabi.org

BCPC (British Crop Production Council) is a self-supporting limited company with charity status, which was formed in 1968 to promote the knowledge and understanding of the science and practice of crop protection. The corporate members include government departments and research councils; advisory services; associations concerned with the farming industry; agrochemical manufacturers; agricultural engineering, agricultural contracting and distribution services; universities; scientific societies; organisations concerned with the environment; and some experienced independent members. Further details available from BCPC, 7 Omni Business Centre, Omega Park, Alton, Hampshire GU34 2QD, UK.

Telephone: (01420) 593200
Fax: (01420) 593209
e-mail: md@bcpc.org
Web: www.bcpc.org

ISBN 978-1-84593-615-0

Typeset and printed by Page Bros, Norwich, UK

Contents

Disclaimer

Every effort has been made to ensure that the information in this book is correct at the time of going to press, but the Editor and the publishers do not accept liability for any error or omission in the content, or for any loss, damage or other accident arising from the use of the products listed herein. Omission of a product does not necessarily mean that it is not approved or that it is unavailable.

It is essential to follow the instructions on the approved label when handling, storing or using any crop protection product. Approved 'off-label' extensions of use are undertaken entirely at the risk of the user.

Editor's Note

This is the 23rd edition of this publication and, with the impending upheaval in the pesticide arena, it is becoming ever more important to know what is legal to use and what is no longer approved. The Pesticides Safety Directorate (PSD) is no longer part of the Department for Environment, Food and Rural Affairs (Defra), but has become a division of the Health and Safety Executive (HSE) and is rebranded as the Chemicals Regulation Directorate. EEC 91/414, the European Directive controlling pesticide approval, has been changed to a Regulation, which means it is no longer advice that member states can choose to adopt, but is now law which they must follow or face fines. As the UK has already adopted most of the recommendations, such as advisor training and sprayer testing, this country will perhaps be less affected than some other states. However, the amendments to the Regulation also bring a significant change of approach. The old system was based on risk management while the new approach is based on hazard criteria, and this is likely to result in the loss of some important products. This guide may prove useful in identifying alternatives to lost favourite products.

Also approaching is the new Water Directive, due to be implemented in 2012. Under this legislation, countries that fail to meet water quality parameters face large fines, and this has contributed to the loss of actives such as isoproturon and trifluralin. Many other actives, not least metaldehyde, are also under scrutiny and we must deploy them with extreme care or face the prospect of losing them.

EEC 91/414 saw the introduction of Annex 1 listing for active ingredients, and those products containing listed active ingredients are now approved under the Plant Protection Products Regulation (PPPR). Products containing those actives that have not been listed are still approved under the Control of Pesticides Regulations (COPR), and all these approvals will expire on 31 December 2013. Many will be withdrawn or expire before that date, and this guide can help identify which products you need to use up before you are faced with an expensive disposal bill.

This guide is now also available by subscription online, and this is proving very popular. The online version has easy-to-use search menus as well as simple fact sheets for individual products, and has one big benefit over the book – while the book is published just once a year, the online version is updated at least four times a year, and more frequently if events dictate. The book is the most portable version, but the online version is the most comprehensive and, with the spread of wireless internet, it can only become more popular.

This 23rd edition is no exception to our aim of continuous improvement. Hundreds of replacements or introductions have been made since the previous edition. Many of these are parallel or generic approvals of existing products, which may offer cheaper alternatives – but we should bear in mind that some of the profit from the originals is used to fund the development of new active ingredients. Some of these new formulations are purer, and therefore more effective and with greater compatibility than the originals.

The increasing problem of managing pesticide resistance has been addressed by the inclusion of mode of action codes, as published by the various resistance action committees. The Long Term Arrangements for Extension of Use have now been withdrawn, and are largely superseded by Specific Off-label Approvals (SOLAs). These are constantly being amended and updated, and are listed in this guide. We remain very grateful to those who provide the information, enabling this Guide to remain a valuable reference for all. As always, criticisms and suggestions are welcome and, in particular, notification of any errors or omissions will help to improve the Guide.

M. Lainsbury
Editor
ukpg@bcpc.org

The information in this publication has been selected from official sources, from the suppliers' product labels and from the product manuals of pesticides approved for use in the UK under the Control of Pesticides Regulations 1986 or the Plant Protection Products Regulations 1995. The content is based on information received by the Editor up to October 2009.

Changes Since 2009 Edition

Pesticides and adjuvants added or deleted since the 2009 edition are listed below. Every effort has been made to ensure the accuracy of this list, but late notifications to the Editor are not included.

Products Added

Products new to this edition are shown here. In addition, products that were listed in the previous edition whose PSD/HSE registration number and/or supplier have changed are included.

Product	Reg. No.	Supplier
A-Cyper 100EC	14175	Goldengrass
Afalon	14187	Makhteshim
Agate	14482	Makhteshim
Agrimec	14491	Syngenta
Agrotech Asulam	12243	AgChem Access
Agrotech Ethephon	12170	AgChem Access
Agrotech-Nicosulfron 40 SC	12938	AgChem Access
Agrovista 3 See	14181	Agrovista
Agrovista Fenamid	13733	Agrovista
Agrovista Radni	14404	Agrovista
Agrovista Reggae	13706	Agrovista
Alpha Linuron 50SC	14215	Makhteshim
Aspire	14303	AgriGuard
Asset Express	14378	AgriGuard
Asu-Flex	A0677	Greenaway
Auxiliary	14576	BASF
Avro SX	14313	DuPont
Azzox	14296	Goldengrass
Bentazone 480	14423	Goldengrass
Berelex	14123	Interfarm
Betamax	14371	AgriGuard
Betanal Maxxim	14186	Bayer CropScience
Betanal Quattro	12779	Bayer CropScience
Betasana SC	14209	United Phosphorus
Betasana Trio SC	14262	United Phosphorus
Bettaquat B	13965	Taminco
Blighter	14326	Nufarm UK
Blizzard	14363	AgriGuard
B-Nine SG	14435	Certis
Bosco WG	14455	Goldengrass
Bounty	14616	AgChem Access
Broad-Flex	A0680	Greenaway
Broadway Star	14319	Dow
Brogue	14465	AgriGuard
Brutus	14353	BASF
Bulldog	14565	Makhteshim
Bullseye	12278	AgriGuard
Carakol 3	14309	Makhteshim
Casino	14352	AgriGuard
Celest Extra	14050	Syngenta
Certis Metaldehyde 3	14337	Certis
Charger B	14269	AgChem Access
Charger C	14216	AgChem Access
Cheer 500	14418	Goldengrass
ChemSource Diquat	14533	ChemSource
Chikara Weed Control	14189	Belchim
Churchill	14139	Makhteshim
Cimarron	13408	Headland
Cinder	14526	Makhteshim
Clayton Delilah XL	14443	Clayton
Clayton Florin	14351	Clayton
Clayton IQ	14519	Clayton
Clayton Myth	14489	Clayton
Clayton Myth XL	14466	Clayton
Clayton Orleans	14452	Clayton
Clayton Pontoon 480EC	14555	Clayton
Clayton Tonto	14436	Clayton
Cleancrop GYR 2	14578	Makhteshim
Cleancrop Hyde 3	14310	Makhteshim
Clodinafop 240	14417	Goldengrass
Cloister	14358	BASF
Clone	14320	AgriGuard
Cobra	14561	Interfarm
Conjoin	A0676	AgriGuard
Corinth	14199	Bayer CropScience
Corzal	14192	AgriChem BV
Cruiser OSR	14496	Syngenta
CTL 500	14290	Goldengrass

Product	Reg. No.	Supplier
Curfew	14289	AgriGuard
Dazide Enhance	14433	Fine
Deacon	14270	Makhteshim
Dedicate	13612	Bayer Environ.
Delta-M 2.5 EC	12604	AgChem Access
Destiny	14159	AgChem Access
Diagor	A0671	AgChem Access
Dicurane	14276	Makhteshim
Dicurane 70SC	14277	Makhteshim
Dicurane 90WDG	14273	Makhteshim
Dicurane Surpass	14544	Makhteshim
Dicurane XL	14569	Makhteshim
Difcor 250 EC	13917	Q-Chem
Difenikan 500	14600	Goldengrass
Discman Biograde	12856	Barclay
Dithane 945	14547	Dow
Dithane 945	14621	Interfarm
Dovekie	14121	AgChem Access
Drum	13853	Belchim
Duo 400 SC	14232	AgriGuard
Duro	14347	Makhteshim
Ennobe	14340	BASF
Enzo	14306	Makhteshim
Epoxi	14292	Goldengrass
Epoxi 125 SC	14442	Goldengrass
Ethefon 480	14392	Goldengrass
Euroagkem Pace	A0562	Intracrop
Euroagkem Pen-e-trate	A0564	Intracrop
Falcon	14552	Makhteshim
Fiddle	14451	Goldengrass
Finalsan	13102	Certis
Firecrest	14486	European Ag
Flycatcher	14135	AgChem Access
Fulmar	14073	AgChem Access
Fumesate 500 SC	14321	Hermoo
Gadwall	14522	AgChem Access
Glyfosat 36	14297	Goldengrass
Glypho-Rapid 450	13882	Barclay
Grasp 400 SC	13594	Nufarm UK
Green Turf King	14348	Standon
Gringo	14467	AgChem Access
Gusto 3	14308	Makhteshim
Harpoon	14278	Headland
Headland Charge	14394	Headland
Headlite	13624	AgChem Access
Hekla	14298	Makhteshim
Holster	A0682	Greenaway
Icarus	14471	BASF
Instrata	14154	Syngenta

Product	Reg. No.	Supplier
Intracrop Dictate	A0673	Intracrop
Intracrop Era	A0664	Intracrop
Intracrop Mica	A0663	Intracrop
Intracrop Micotex	A0688	Intracrop
Intracrop Perm-E8	A0565	Intracrop
Intracrop Protocol	A0687	Intracrop
Intracrop Rapide Beta	A0672	Intracrop
Intracrop Salute	A0683	Intracrop
Intracrop Sapper	A0684	Intracrop
Intracrop Signal XL	A0659	Intracrop
Intracrop Zenith	A0665	Intracrop
Jenga	14357	DuPont
Jockey Solo	14281	BASF
Juliet	14557	ChemSource
Kalypstowe	13844	AgChem Access
Karate 2.5WG	14060	Syngenta
Kerb Granules	14213	Barclay
Knoxdoon	13987	AgChem Access
Kula	14299	Makhteshim
Legara	13888	AgChem Access
Legolas	14439	Nufarm UK
Lingo	14075	Belchim
Lotus	13744	Nufarm UK
Luas	14359	AgriGuard
Lynx H	14426	De Sangosse
Mac-Bentazone 480 SL	13598	AgChem Access
MAC-Napropamide 450 SC	13972	AgChem Access
MAC-Prochloraz Plus 490 EC	13616	AgChem Access
MAC-Propaquizafop 100 EC	13978	AgChem Access
Mainman	13123	Belchim
Martinet	14614	AgChem Access
Marvel	14508	ChemSource
Mascot Crossbar	14453	Rigby Taylor
Mascot Eland	14549	Rigby Taylor
Mascot Rayzor	14329	Rigby Taylor
Maxim XL	14136	Syngenta
Meco-Flex	A0678	Greenaway
Medax Top	13282	BASF
Metam 510	09796	Taminco
Metribuzin WG	14315	Goldengrass
Metric	14214	Belchim
Mextrol DP	11529	Nufarm UK
Minstrel	13745	United Phosphorus
Mobius	13395	Bayer CropScience

Product	Reg. No.	Supplier
Movento	14446	Bayer CropScience
MSS Sprout Nip	14184	Whyte Agrochemicals
Muntjac	14540	BASF
Nando 500 SC	14372	Nufarm UK
Naspar TDI	14584	Q-Chem
Nemo	14140	AgChem Access
New Estermone	13792	Vitax
Nico 4	14293	Goldengrass
Nirvana	14256	BASF
Nufosate Ace	13794	Nufarm UK
Oblix MT	14510	AgriChem BV
Octolan	11675	Syngenta
Omaha	14234	Makhteshim
Oracle	14470	ChemSource
Oreto	14258	Makhteshim
Osarex W	14428	De Sangosse
Output	A0644	Nufarm UK
Oxytril CM	14511	Bayer CropScience
Pan Glory	14487	Pan Amenity
Pan Pencycuron P	10494	Pan Agriculture
Pan Theta	14501	Pan Agriculture
Panto	14307	Pan Amenity
Paradise	14504	Pan Agriculture
Penncozeb 80 WP	14580	United Phosphorus
Permesso	14108	AgChem Access
Pesta H	14427	De Sangosse
Pharaoh	14731	Dow
Phemo	14193	AgriChem BV
Piccant	14294	Goldengrass
Pik-Flex	A0681	Greenaway
Pirimate	14539	European Ag
Pirimicarb 50	14295	Goldengrass
Pizamide	14684	ChemSource
Pizza 400 SC	14430	Goldengrass
Platform S	14380	Belchim
Prime-Turf	14617	ChemSource
Prompto	14391	Bayer CropScience
Propyzamide 400	14685	European Ag
Pryz-Flex	A0679	Greenaway
Punch 25	14497	DuPont
Quail	14506	ChemSource
Questor	A0495	United Phosphorus
Raptor	14566	Makhteshim
Redigo Twin TXC	14259	Bayer CropScience

Product	Reg. No.	Supplier
Redstar	14197	AgChem Access
Renegade	14507	ChemSource
Reynard	14523	ChemSource
Rhino	14311	Certis
Rizolex	14204	Interfarm
Rovral AquaFlo	14206	BASF
Samurai	12674	Monsanto
SAN 703	11676	Syngenta
Sarabi	13453	AgChem Access
Scarlett	14271	AgriGuard
Seal Z	14201	AgChem Access
Selon	13966	Taminco
Serenade ASO	14318	Fargro
Shogun	14567	Makhteshim
Skora	13422	AgChem Access
Slingshot	14505	ChemSource
Slugdown 3	14366	Unicrop
Sluxx	14462	Certis
Sprint	14272	AgriGuard
Squire Ultra	14401	Interfarm
Stamp 330 EC	13995	AgChem Access
Standon Girder 720	14251	Standon
Stomp Aqua	14664	BASF
Super 3	14370	Unicrop
Swallow	14078	AgChem Access
Swiftsure	14468	Sherriff Amenity
Tacet	14342	Goldengrass
Tebcon	14300	Goldengrass
Tempo	13799	Syngenta
Tepra	14160	Goldengrass
Thor	14240	Nufarm UK
TM 1088	A0675	Intracrop
Torrent	14374	AgriGuard
Totril	13100	Bayer CropScience
Trample	14173	AgChem Access
Trilogy	14464	United Phosphorus
Trojan	14224	Chiltern
Trotter	14236	Agrovista
Tuli	13302	AgChem Access
Vareon	14376	DuPont
Viscount	12642	Syngenta
Visor	14513	Makhteshim
Vixen	13621	BASF
Warrior	13857	Syngenta
Woody	14163	AgChem Access
Zeal	A0685	Interagro
Zigzag	A0692	Headland

Products Deleted

The appearance of a product name in the following list may not necessarily mean that it is no longer available. In cases of doubt, refer to Section 3 or the supplier.

Product	Reg. No.	Supplier
AAprotect	03784	Unicrop
Agate	12016	Makhteshim
Aligran 83 WDG	13698	Agform
Alpha IPU 500	10733	Makhteshim
Alpha Isoproturon 500	05882	Makhteshim
Alpha Trifluralin 48 EC	13433	Makhteshim
Amuse	14379	Pan Amenity
Aristo 5	13495	Certis
Aristo 6	13494	Certis
Aristo M	13278	De Sangosse
Arizona	12615	Makhteshim
Banco	13019	Makhteshim
Barclay Hurler	12425	Barclay
Bastion T	06011	Rigby Taylor
Betanal Expert	14034	Bayer CropScience
Bio-syl Extra	A0596	Intracrop
Blois	12376	Makhteshim
Bonzi	10517	Syngenta
Bristol Blues	11798	Doff Portland
Buckler	13080	Nufarm UK
Bulwark Flo	10161	Interfarm
Centium 360 CS	11607	Belchim
Centium 360 CS	13846	Belchim
Certis 9363	13486	Certis
Certis 9363 Blue	13487	Certis
Certis Metaldehyde	13489	Certis
Citation WG	11929	United Phosphorus
Clayton Fenican 550	11160	Clayton
Clayton Impasse	12520	Clayton
Clayton Lambada	13231	Clayton
Clayton Propel 80 WG	11216	Clayton
Clayton Squat	12271	Clayton
Clean Crop Jekyll D	12490	United Agri
Cleancrop Egret	12667	United Agri
Clovotox	09928	Bayer Environ.
Colosseo 500 SC	13897	Nuvaros
Condor	11791	Doff Portland
Coronet	12267	Bayer CropScience
Das-320	12375	Dow
Dexter	12551	AgChem Access
Dixie 6 M	13478	Clayton
Duplosan KV	12073	Nufarm UK
Flarepath	13156	Headland
Flux 200 EC	13944	Nuvaros
Fungus Attack + LC	13834	Westland Horticulture
Fungus Attack + RTU	13832	Westland Horticulture
Genfarm IPU 500	12977	Genfarm
Glyphosate 360	12670	Cardel
Greencrop Champion	11755	Greencrop
Greencrop Glenroe	09903	Greencrop
Greencrop Majestic 2	12781	Greencrop
Greencrop Saffron FL	10244	Greencrop
Gusto	13448	Makhteshim
Headland Javelin	12451	Headland
Herboxone	13958	Headland
IPU Minrinse	12457	Makhteshim
Isotron 500	10668	DAPT
IT Fluroxypyr	12914	Inter-Trade
Kerb 50 W	12296	Dow
Kerb Flo	10768	Dow
Koala	12556	Nufarm UK
Landgold Clomazone 360	13074	Teliton
Landgold Clopyralid 200	12359	Teliton
Landgold Lambda-Z	13533	Teliton
Landgold Mecoprop-p	12121	Teliton
Landgold Mecoprop-P 600	12122	Teliton
Landgold Pirimicarb 50	12126	Teliton
Landgold Propyzamide 400 SC	13614	Teliton
Leguma Plus	13913	Nuvaros
Luxan Deal	12934	Luxan
Luxan Pencycuron	12144	Luxan
Luxan Pencycuron 250 SC	12142	Luxan
Luxan Taurus	12442	Luxan
Lynx	11767	De Sangosse
Lynx M	13280	De Sangosse
Lynx S	12397	Lonza
MAC-Propiconazole 250 EC	13583	AgChem Access
Magdelin	13184	Nufarm UK
Matilda WG	13732	Nufarm UK
Metasec M	12396	De Sangosse
Metasec M	13279	De Sangosse
Mifaslug	10292	FCC

Product	Reg. No.	Supplier
Metasec M	13279	De Sangosse
Mifaslug	10292	FCC
Orius	12105	Makhteshim
Pan Ethephon	11020	Pan Agriculture
Pan Stiffen	11644	Pan Agriculture
Pastel M	12395	De Sangosse
Path Weedkiller	13195	Bayer CropScience
Pathfinder Excel	12501	Barclay
Phostrogen Mosskiller and Lawn Tonic	13112	Bayer CropScience
Pirimate	13047	AgChem Access
Platform S	11611	Belchim
Protugan	10043	Makhteshim
Protugan 80 WDG	10430	Makhteshim
Reliance	12676	Cardel
Roundup Biactive K	13248	Monsanto
RP 4169	13027	Makhteshim
Rubigan	12355	Gowan
Signal	A0461	Intracrop
Slug-Lentils	12533	Luxan
Snapper	13460	Nufarm UK
sodium chloride	-	various
Spread and Seal	A0449	Mandops

Product	Reg. No.	Supplier
Standon Diflufenican-IPU	11109	Standon
Supertox 30	09946	Bayer Environ.
Sword	13015	Makhteshim
Tandril 48	10643	DAPT
Tesco Complete Weedkiller Spray	13151	Tesco
Tolkan Liquid	11994	Bayer CropScience
Tolugan Extra	09393	Makhteshim
Triflurex 48 EC	13445	Makhteshim
Trigger	13492	Certis
Trilogy	08996	United Phosphorus
Trojan 500 SC	13946	Nuvaros
Turbair Super Flydown	H7225	Summit Agro
Uranus	12506	Makhteshim
Valiant	14456	Sherriff Amenity
Wombat	13028	Bayer CropScience
Wyevale Weed Killer	12687	Wyevale
Zebra	13155	Headland
Zimbali	12748	Makhteshim

Introduction

Purpose

The primary aim of this book is to provide a practical handbook of pesticides, plant growth regulators and adjuvants that the farmer or grower can realistically and legally obtain in the UK and to identify the purposes for which they can be used. It is designed to help in the identification of products appropriate to a particular problem. In addition to uses recommended on product labels, details are provided of those uses which do not appear on product labels but which have been granted specific off-label approval (SOLAs). As well as identifying the products available, the book provides guidance on how to use them safely and effectively but without giving details of doses, volumes, spray schedules or approved tank mixtures. Sections 5 and 6 provide essential background information on a wide range of pesticide-related issues including legislation, codes of practice, poisons and treatment of poisoning, products for use in special situations and weed & crop growth stage keys.

While we have tried to cover all other important factors, this book does not provide a full statement of product recommendations. Before using any pesticide product it is essential that the user should read the label carefully and comply strictly with the instructions it contains. **This *Guide* does not substitute for the product label.**

Scope

The *Guide* is confined to pesticides registered in the UK for use in arable agriculture, horticulture (including amenity use), forestry and areas in or near water. Within these fields of use there are about 2500 products in the UK with current approval. The *Guide* is a reference for virtually all of them given in two sections. Section 2 gives full details of the products notified to the editor as available on the market. Products are included in this section only if requested by the supplier and supported by evidence of approval. Section 3 gives brief details of all other registered products with extant approval.

Section 2 lists some 450 individual active ingredients and mixtures. For each entry a list is shown of the available approved products together with the name of the supplier from whom each may be obtained. All types of pesticide covered by Control of Pesticide Regulations or Plant Protection Products Regulations are included. This embraces acaricides, algicides, herbicides, fungicides, insecticides, lumbricides, molluscicides, nematacides and rodenticides together with plant growth regulators used as straw shorteners, sprout inhibitors and for various horticultural purposes. The total number of products included in section 2 (i.e. available in the market in 2010) is about 1460.

The *Guide* also gives information (in section 4) on around 130 authorised adjuvants which, although not active pesticides themselves, may be added to pesticide products to improve their effectiveness. The *Guide* does not include products solely approved for amateur, home & garden, domestic, food storage, public or animal health uses.

Sources of Information

The information in this edition has been drawn from these authoritative sources:

- approved labels and product manuals received from suppliers of pesticides up to October 2009.
- websites of the Chemicals Regulation Directorate (CRD, www.pesticides.gov.uk) and the Health and Safety Executive (HSE, www.hse.gov.uk).

Criteria for Inclusion

To be included, a pesticide must meet the following conditions:

- it must have extant approval under UK pesticides legislation
- information on the approved uses must have been provided by the supplier
- it must be expected to be on the UK market during the currency of this edition.

When a company changes its name, whether by merger, takeover or joint venture, it is obliged to re-register its products in the name of the new company, and new MAPP numbers are normally assigned. Where stocks of the previously registered product remain legally available, both old and new identities are included in the *Guide* and remain until approval of the former lapses or stocks are notified as exhausted.

Products that have been withdrawn from the market and whose approval will finally lapse during 2010 are identified in each profile. After the indicated date, sale, storage or use of the product bearing that approval number becomes illegal. Where there is a direct replacement product, this is indicated.

The Voluntary Initiative

The Voluntary Initiative (VI) is a programme of measures, agreed by the crop protection industry and related organisations with Government, to minimise the environmental impacts of crop protection products. The programme has provided a framework of practices and principles to help Government achieve its objective of protection of water quality and enhancement of farmland biodiversity. Many of the environmental protection schemes launched under the VI represent current best practice.

The first five-year phase of the programme formally concluded in March 2006, but it is continuing with a rolling two-year review of proposals and targets; current proposals include a new online crop protection management plan and greater involvement in catchment-sensitive farming (see below).

A key element of the VI has been the provision of environmental information on crop protection products. Members of the Crop Protection Association (CPA) have committed to do this by producing Environmental Information Sheets (EISs) for all their marketed professional products.

EISs reinforce and supplement information on a product label by giving specific environmental impact information in a standardised format. They highlight any situations where risk management is essential to ensure environmental protection. Their purpose is to provide user-friendly information to advisers (including those in the amenity sector), farmers and growers on the environmental impact of crop protection products, to allow planning with a better understanding of the practical implications.

Links are provided to give users rapid access to EISs on the VI website: www.voluntaryinitiative.org.uk.

Catchment-sensitive Farming

Catchment-sensitive Farming (CSF) is the government's response to climate change and the new European Water Directive. Rainfall events are likely to become heavier, with a greater risk of soil, nutrients and pesticides ending up in the waterways. The CSF programme is investigating the effects of specific targeted advice on 20 catchment areas perceived as being at risk, and new entries in this *Guide* can help identify the pesticides that pose the greatest risk of water contamination (see Environmental Safety).

Notes on Contents and Layout

The book consists of six main sections:

1 Crop/Pest Guide
2 Pesticide Profiles
3 Products Also Registered
4 Adjuvants
5 Useful Information
6 Appendices

1 Crop/Pest Guide

This section enables the user to identify which active ingredients are approved for a particular crop/pest combination. The crops are grouped as shown in the Crop Index. For convenience, some crops and pests have been grouped into generic units (for example, 'cereals' are covered as a group, not individually). Indications from the Crop/Pest Guide must always be checked against the specific entry in the pesticide profiles section because a product may not be approved on all crops in the group. Chemicals indicated as having uses in cereals, for example, may be approved for use only on winter wheat and winter barley, not for other cereals. Because of the difference in wording of product labels, it may sometimes be necessary to refer to broader categories of organism in addition to specific organisms (e.g. annual grasses and annual weeds in addition to annual meadowgrass).

2 Pesticide Profiles

Each active ingredient and mixture of ingredients has a separate numbered profile entry. The entries are arranged in alphabetical order using the common names approved by the British Standards Institution and used in *The Pesticide Manual* (15th Edition; BCPC, 2009). Where an active ingredient is available only in mixtures, this is stated. The ingredients of the mixtures are themselves ordered alphabetically, and the entries appear at the correct point for the first named active ingredient.

Below the profile title, a brief phrase describes the main use of the active ingredient(s). This is followed where appropriate by the mode of action code(s) as published by the Fungicide, Herbicide and Insecticide Resistance Action Committees.

Within each profile entry, a table lists the products approved and available on the market in the following style:

Product name	Main supplier	Active ingredient contents	Formulation type	Registration number
1 Broadsword	United Phosphorus	200:85:65 g/l	EC	09140
2 Greengard	SumiAgro Amenity	200:85:65 g/l	EC	11715
3 Nu-Shot	Nufarm UK	200:85:65 g/l	EC	11148

Many of the **product names** are registered trademarks, but no special indication of this is given. Individual products are indexed under their entry number in the Index of Proprietary Names at the back of the book. The **main supplier** indicates the marketing outlet through which the product may be purchased. Full addresses and contact details of suppliers are listed in Appendix 1. For mixtures, the **active ingredient contents** are given in the same order as they appear in the profile heading. The **formulation types** are listed in full in the Key to Abbreviations and Acronyms (Appendix 5). The **registration number** normally refers to the registration with the Chemicals Regulation Directorate (CRD). In cases where the products registered with the Health and Safety Executive are included, HSE numbers are quoted, e.g. H0462.

Below the product table, a **Uses** section lists all approved uses notified by suppliers to the Editor by October 2009, giving both the principal target organisms and the recommended crops or situations (identified in ***bold italics***). Where there is an important condition of use, or where the approval is off-label, this is shown in parentheses as (*off-label*). Numbers in square brackets refer to the numbered products in the table above. Thus, in the example shown above, a use approved for Nu-Shot (product 3) but not for Broadsword or Greengard (products 1 and 2) appears as:

Annual dicotyledons in ***rotational grassland*** [3]

Any **Specific Off-Label Approvals (SOLAs)** for products in the profile are detailed below the list of approved uses. Each SOLA has a separate entry and shows the crops to which it applies, the Notice of Approval number, the expiry date (if due to fall during the current edition), and the product reference number in square brackets. SOLAs do not appear on the product label and are undertaken entirely at the risk of the user.

Below the SOLA paragraph, **Notes** are listed under the following headings. Unless otherwise stated, any reference to dose is made in terms of product rather than active ingredient. Where a note refers to a particular product rather than to the entry generally, this is indicated by numbers in square brackets as described above.

Approval information	Information of a general nature about the approval status of the active ingredient or products in the profile is given here. Notes on approval for aerial or ULV application are given. Inclusion of the active ingredient in Annex I under EC Directive 91/414 is shown, as well as any acceptance from the British Beer and Pub Association (BBPA) for its use on barley or hops for brewing or malting. Where a product approval will finally expire in 2010, the expiry date is shown here.
Efficacy	This entry identifies factors important in making the most effective use of the treatment. Certain factors, such as the need to apply chemicals uniformly, the need to use appropriate volume rates, and the need to prevent settling out of the active ingredient in the spray tank, are assumed to be important for all treatments and are not emphasised in individual profiles.
Restrictions	Notes are included in this section where products are subject to the Poisons Law, and where the label warns about organophosphorus and/or anticholinesterase compounds. Factors important in optimising performance and minimising the risk of crop damage, including statutory conditions relating to the maximum permitted number of applications, are listed. Any restrictions on crop varieties that may not be sprayed are mentioned here.
Environmental safety	Where the label specifies an environmental hazard classification, it is noted together with the associated risk phrases. Any other special operator precautions and any conditions concerning withholding livestock from treated areas are specified. Where any of the products in the profile are subject to Category A, Category B or broadcast air-assisted buffer zone restrictions under the LERAP scheme, the relevant classification is shown. Other environmental hazards are also noted here, including potential dangers to livestock, game, wildlife, bees and fish. The need to avoid drift onto neighbouring crops and the need to wash out equipment thoroughly after use are important with all pesticide treatments, but may receive special mention here if of particular significance.
Crop-specific information	Instructions about the timing of application or cultivations that are specific to a particular crop, rather than generally applicable, are mentioned here. The latest permitted use and harvest intervals (if any) are shown for individual crops.
Following crops guidance	Any specific label instructions about what may be grown after a treated crop, whether harvested normally or after failure, are shown here,

Hazard classification and safety precautions	The label hazard classifications and precautions are shown using a series of letter and number codes which are explained in Appendix 4. Hazard warnings are now given in full, with the other precautions listed under the following subheadings:

- Risk phrases
- Operator protection
- Environmental protection
- Consumer protection
- Storage and disposal
- Treated seed
- Vertebrate/rodent control products
- Medical advice

This section is given for information only and should not be used for the purpose of making a COSHH assessment without reference to the actual label of the product to be used.

3 Products also Registered

Products with an extant approval for all or part of the year of the edition in which they appear are listed in this section if they have not been requested by their supplier or manufacturer for inclusion in Section 2. Details shown are the same (apart from formulation) as those in the product tables of Section 2 and, where relevant, an approval expiry date is shown. Not all products listed here will be available in the market, but this list, together with the products in section 2, comprises a comprehensive listing of all approved products in the UK for uses within the scope of the *Guide* for the year in question.

4 Adjuvants

Adjuvants are listed in a table ordered alphabetically by product name. For each product, details are shown of the main supplier, the authorisation number and the mode of action (e.g. wetter, sticker, etc.) as shown on the label. A brief statement of the uses of the adjuvant is given. Protective clothing requirements and label precautions are listed using the codes from Appendix 4.

5 Useful Information

This section summarises legislation covering approval, storage, sale and use of pesticides in the UK. There are brief articles on the broader issues concerning the use of crop protection chemicals, including resistance management. Lists are provided of products approved for use in or near water, in forestry, as seed treatments and for aerial application. Chemicals subject to the Poisons Laws are listed, and there is a summary of first aid measures if pesticide poisoning should be suspected. Finally, this section provides guidance on environmental protection issues and covers the protection of bees and the use of pesticides in or near water.

6 Appendices

Appendix 1	Gives names, addresses, telephone and fax numbers of all companies listed as main suppliers in the pesticide profiles. E-mail and website addresses are also listed where available. Where a supplier is no longer trading under that name (usually following a merger or takeover), but products under that name are still listed in the *Guide* because they are still available in the supply chain, a cross-reference indicates the new 'parent' company from which technical or commercial information can be obtained.
Appendix 2	Gives names, addresses, telephone and fax numbers of useful contacts, including the National Poisons Information Service. Website addresses are included where available.

Appendix 3	Gives details of the keys to crop and weed growth stages, including the publication reference for each. The numeric codes are used in the descriptive sections of the pesticide profiles (Section 2).
Appendix 4	Shows the full text for code letters and numbers used in the pesticide profiles (Section 2) to indicate personal protective equipment requirements and label precautions.
Appendix 5	Shows the full text for the formulation abbreviations used in the pesticide profiles (Section 2). Other abbreviations and acronyms used in the *Guide* are also explained here.
Appendix 6	Provides full definitions of officially agreed descriptive phrases for crops or situations used in the pesticide profiles (Section 2) where misunderstandings can occur.
Appendix 7	Shows a list of useful reference publications, which amplify the summarised information provided in this section.

SECTION 1
CROP/PEST GUIDE

Crop/Pest Guide Index

Important Note: The Crop/Pest Guide Index refers to pages on which the subject can be located.

Crop/Pest Guide

Important note: For convenience, some crops and pests or targets have been brought together into generic groups in this guide, e.g. 'cereals', 'annual grasses'. It is essential to check the profile entry in Section 2 *and* the label to ensure that a product is approved for a specific crop/pest combination, e.g. winter wheat/blackgrass.

Arable and vegetable crops

Agricultural herbage - Grass

Diseases	Crown rust	propiconazole
	Drechslera leaf spot	propiconazole
	Powdery mildew	propiconazole
	Rhynchosporium	propiconazole
Pests	Birds/mammals	aluminium ammonium sulphate, aluminium phosphide, strychnine hydrochloride (commodity substance) (*areas of restricted public access*)
	Flies	chlorpyrifos, cypermethrin
	Leatherjackets	chlorpyrifos, methiocarb (*seed admixture*)
	Slugs/snails	methiocarb (*seed admixture*)
Plant growth regulation	Quality/yield control	sulphur
Weeds	Aquatic weeds	MCPA
	Broad-leaved weeds	2,4-D, 2,4-D + dicamba, 2,4-D + dicamba + dichlorprop-P, 2,4-D + dicamba + triclopyr, 2,4-D + MCPA, 2,4-DB + linuron + MCPA, 2,4-DB + MCPA, amidosulfuron, aminopyralid + fluroxypyr, aminopyralid + triclopyr, asulam, bentazone + MCPA + MCPB, citronella oil, clopyralid, clopyralid (*off-label - spot treatment*), clopyralid + fluroxypyr + triclopyr, clopyralid + triclopyr, clopyralid + triclopyr (*off-label - via weed wiper*), dicamba + MCPA + mecoprop-P, dicamba + mecoprop-P, dichlorprop-P, fluroxypyr, fluroxypyr + triclopyr, glyphosate, glyphosate (*wiper application*), MCPA, MCPA + MCPB, MCPB, mecoprop-P, metsulfuron-methyl, thifensulfuron-methyl
	Crops as weeds	fluroxypyr
	Grass weeds	glyphosate
	Weeds, miscellaneous	aminopyralid + triclopyr, fluroxypyr, glufosinate-ammonium, glyphosate
	Woody weeds/scrub	asulam

Agricultural herbage - Herbage legumes

Crop control	Desiccation	diquat
Weeds	Broad-leaved weeds	carbetamide, diquat (*off-label*), isoxaben (*off-label - grown for game cover*), MCPA + MCPB, MCPB (*seed crops*), propyzamide
	Crops as weeds	carbetamide
	Grass weeds	carbetamide, diquat (*seed crop*), fluazifop-P-butyl (*off-label*), propyzamide, tri-allate

	Weeds, miscellaneous	clopyralid (*off-label - game cover*), diquat (*seed crop*)

Agricultural herbage - Herbage seed

Crop control	Desiccation	diquat
Diseases	Crown rust	propiconazole
	Damping off	thiram (*seed treatment*)
	Disease control/foliar feed	chlorothalonil (*off-label*), mancozeb (*off-label*), prochloraz (*off-label*), prochloraz + propiconazole (*off-label*), tebuconazole (*off-label*)
	Drechslera leaf spot	propiconazole
	Foliar diseases	azoxystrobin (*off-label*), azoxystrobin + chlorothalonil (*off-label*), azoxystrobin + cyproconazole (*off-label*), azoxystrobin + fenpropimorph (*off-label*), chlorothalonil (*off-label*), chlorothalonil + flutriafol (*off-label*), chlorothalonil + mancozeb (*off-label*), cyproconazole (*off-label*), cyproconazole + cyprodinil (*off-label*), cyproconazole + quinoxyfen (*off-label*), cyproconazole + trifloxystrobin (*off-label*), cyprodinil + picoxystrobin (*off-label*), difenoconazole (*off-label*), dimoxystrobin + epoxiconazole (*off-label*), epoxiconazole (*off-label*), epoxiconazole + fenpropimorph (*off-label*), epoxiconazole + fenpropimorph + kresoxim-methyl (*off-label*), epoxiconazole + kresoxim-methyl (*off-label*), epoxiconazole + kresoxim-methyl + pyraclostrobin (*off-label*), epoxiconazole + pyraclostrobin (*off-label*), famoxadone + flusilazole (*off-label*), fenpropimorph (*off-label*), fenpropimorph + flusilazole (*off-label*), fenpropimorph + kresoxim-methyl (*off-label*), fenpropimorph + pyraclostrobin (*off-label*), fluquinconazole (*off-label*), fluquinconazole + prochloraz (*off-label*), mancozeb (*off-label*), metconazole (*off-label*), prothioconazole + tebuconazole (*off-label*), pyraclostrobin (*off-label*), spiroxamine + tebuconazole (*off-label*), tebuconazole (*off-label*), tebuconazole + triadimenol (*off-label*), trifloxystrobin (*off-label*)
	Powdery mildew	cyflufenamid (*off-label*), epoxiconazole (*off-label*), fenpropimorph + quinoxyfen (*off-label*), propiconazole, quinoxyfen (*off-label*), sulphur (*off-label*)
	Rhynchosporium	propiconazole
Pests	Aphids	deltamethrin (*off-label*), dimethoate, esfenvalerate (*off-label*), lambda-cyhalothrin (*off-label*), lambda-cyhalothrin + pirimicarb (*off-label*), tau-fluvalinate (*off-label*), zeta-cypermethrin (*off-label*)
	Beetles	deltamethrin (*off-label*)
	Caterpillars	deltamethrin (*off-label*)
	Leafhoppers	deltamethrin (*off-label*)
	Pests, miscellaneous	alpha-cypermethrin (*off-label*), lambda-cyhalothrin (*off-label*), tau-fluvalinate (*off-label*), zeta-cypermethrin (*off-label*)
	Thrips	deltamethrin (*off-label*)
Plant growth regulation	Growth control	trinexapac-ethyl
Weeds	Broad-leaved weeds	amidosulfuron + iodosulfuron-methyl-sodium (*off-label*), asulam (*off-label*), bifenox (*off-label*), bromoxynil + diflufenican + ioxynil (*off-label*), bromoxynil + ioxynil (*off-*

		label), carfentrazone-ethyl (*off-label*), carfentrazone-ethyl + flupyrsulfuron-methyl (*off-label*), clopyralid (*off-label*), dicamba + MCPA + mecoprop-P, dichlorprop-P + ioxynil (*off-label*), dichlorprop-P + MCPA + mecoprop-P (*off-label*), diflufenican (*off-label*), diflufenican + flufenacet (*off-label*), diflufenican + flurtamone (*off-label*), ethofumesate, florasulam (*off-label*), florasulam + fluroxypyr (*off-label*), iodosulfuron-methyl-sodium (*off-label*), MCPA, MCPA + MCPB, MCPB, mecoprop-P, pendimethalin (*off-label*), propyzamide, sulfosulfuron (*off-label*)
	Crops as weeds	iodosulfuron-methyl-sodium (*off-label - from seed*)
	Grass weeds	clodinafop-propargyl (*off-label*), diflufenican (*off-label*), diflufenican + flufenacet (*off-label*), diflufenican + flurtamone (*off-label*), ethofumesate, fenoxaprop-P-ethyl (*off-label*), pendimethalin (*off-label*), propyzamide, sulfosulfuron (*off-label*), tralkoxydim (*off-label*)

Brassicas - Brassica seed crops

Diseases	Alternaria	iprodione (*pre-storage only*)
	Disease control/foliar feed	chlorothalonil (*off-label*)
Weeds	Broad-leaved weeds	carbetamide, propyzamide
	Crops as weeds	carbetamide
	Grass weeds	carbetamide, propyzamide

Brassicas - Brassicas, general

Diseases	Botrytis	Bacillus subtilis (*off-label*)
	Disease control/foliar feed	difenoconazole (*off-label*)
	Downy mildew	chlorothalonil, chlorothalonil (*moderate control*), copper oxychloride (*off-label*)
	Ring spot	chlorothalonil
Pests	Aphids	deltamethrin (*off-label*), pymetrozine (*off-label*), thiacloprid (*off-label*)
	Beetles	deltamethrin (*off-label*)
	Caterpillars	deltamethrin (*off-label*)
	Pests, miscellaneous	deltamethrin (*off-label*)
Plant growth regulation	Quality/yield control	sulphur
Weeds	Broad-leaved weeds	pyridate (*off-label*)
	Grass weeds	fluazifop-P-butyl (*off-label*)

Brassicas - Fodder brassicas

Diseases	Alternaria	azoxystrobin, fludioxonil + metalaxyl-M + thiamethoxam (*moderate control*), iprodione (*off-label*)
	Bacterial blight	copper oxychloride (*off-label*)
	Black rot	copper oxychloride (*off-label*)
	Bottom rot	copper oxychloride (*off-label*)
	Disease control/foliar feed	difenoconazole (*off-label*)
	Downy mildew	fludioxonil + metalaxyl-M + thiamethoxam, fosetyl-aluminium (*off-label*)

	Phytophthora	copper oxychloride (*off-label*)
	Powdery mildew	azoxystrobin + difenoconazole
	Pythium	fludioxonil + metalaxyl-M + thiamethoxam (*reduction*)
	Ring spot	azoxystrobin, difenoconazole (*off-label*)
	Seed-borne diseases	fludioxonil + metalaxyl-M + thiamethoxam
	Spear rot	copper oxychloride (*off-label*)
	Stem canker	copper oxychloride (*off-label*)
	White blister	azoxystrobin, azoxystrobin + difenoconazole, mancozeb + metalaxyl-M (*off-label*)
Pests	Aphids	cypermethrin, pirimicarb, pymetrozine (*off-label*), thiacloprid (*off-label*)
	Beetles	alpha-cypermethrin, deltamethrin, fludioxonil + metalaxyl-M + thiamethoxam (*reduction of damage*)
	Caterpillars	alpha-cypermethrin, Bacillus thuringiensis (*off-label*), cypermethrin, deltamethrin
	Flies	chlorpyrifos (*off-label*)
	Pests, miscellaneous	dimethoate (*off-label*)
Weeds	Broad-leaved weeds	chlorthal-dimethyl, chlorthal-dimethyl + propachlor, clomazone (*off-label*), clopyralid, metazachlor (*off-label*), napropamide, propachlor, pyridate (*off-label*)
	Crops as weeds	fluazifop-P-butyl (*stockfeed only*)
	Grass weeds	chlorthal-dimethyl + propachlor, fluazifop-P-butyl (*stockfeed only*), metazachlor (*off-label*), napropamide, propachlor

Brassicas - Leaf and flowerhead brassicas

Diseases	Alternaria	azoxystrobin, boscalid + pyraclostrobin, chlorothalonil + metalaxyl-M (*moderate control*), difenoconazole, iprodione, iprodione (*off-label*), prothioconazole, tebuconazole, tebuconazole + trifloxystrobin
	Bacterial blight	copper oxychloride (*off-label*)
	Black rot	copper oxychloride (*off-label*)
	Botrytis	chlorothalonil, chlorothalonil (*moderate control*), chlorothalonil (*off-label*), iprodione (*off-label*)
	Bottom rot	copper oxychloride (*off-label*)
	Damping off	thiram (*seed treatment*), tolclofos-methyl
	Disease control/foliar feed	chlorothalonil (*off-label*), difenoconazole (*off-label*)
	Downy mildew	chlorothalonil, chlorothalonil (*moderate control*), chlorothalonil (*off-label*), chlorothalonil + metalaxyl-M, fosetyl-aluminium (*off-label*), propamocarb hydrochloride
	Foliar diseases	flusilazole (*off-label*)
	Light leaf spot	prothioconazole, tebuconazole, tebuconazole + trifloxystrobin
	Phoma leaf spot	tebuconazole + trifloxystrobin
	Phytophthora	copper oxychloride (*off-label*), propamocarb hydrochloride
	Powdery mildew	azoxystrobin + difenoconazole, prothioconazole, tebuconazole, tebuconazole + trifloxystrobin
	Pythium	propamocarb hydrochloride
	Rhizoctonia	tolclofos-methyl (*off-label*)

	Ring spot	azoxystrobin, boscalid + pyraclostrobin, chlorothalonil, chlorothalonil (*off-label*), chlorothalonil + metalaxyl-M (*reduction*), difenoconazole, difenoconazole (*off-label*), prothioconazole, tebuconazole, tebuconazole (*off-label*), tebuconazole + trifloxystrobin
	Spear rot	copper oxychloride (*off-label*)
	Stem canker	copper oxychloride (*off-label*), flusilazole (*off-label*), prothioconazole
	Storage rots	metalaxyl-M (*off-label*)
	White blister	azoxystrobin, azoxystrobin + difenoconazole, boscalid + pyraclostrobin, boscalid + pyraclostrobin (*qualified minor use*), chlorothalonil + metalaxyl-M, mancozeb + metalaxyl-M (*off-label*), tebuconazole + trifloxystrobin
Pests	Aphids	acetamiprid (*off-label*), bifenthrin, chlorpyrifos, cypermethrin, deltamethrin (*off-label*), dimethoate, fatty acids, lambda-cyhalothrin + pirimicarb, pirimicarb, pymetrozine, pymetrozine (*off-label*), pymetrozine (*useful levels of control*), pyrethrins, spirotetramat, thiacloprid (*off-label*)
	Beetles	alpha-cypermethrin, deltamethrin, deltamethrin (*off-label*), lambda-cyhalothrin + pirimicarb, pyrethrins
	Birds/mammals	aluminium ammonium sulphate
	Caterpillars	alpha-cypermethrin, Bacillus thuringiensis, Bacillus thuringiensis (*off-label*), bifenthrin, chlorpyrifos, cypermethrin, deltamethrin, deltamethrin (*off-label*), diflubenzuron, indoxacarb, indoxacarb (*off-label*), lambda-cyhalothrin, lambda-cyhalothrin + pirimicarb, pyrethrins, spinosad, teflubenzuron (*off-label*)
	Cutworms	chlorpyrifos
	Flies	chlorpyrifos, chlorpyrifos (*off-label*), teflubenzuron (*off-label*)
	Leatherjackets	chlorpyrifos
	Mealybugs	fatty acids
	Pests, miscellaneous	alpha-cypermethrin (*off-label*), deltamethrin (*off-label*), dimethoate (*off-label*), teflubenzuron (*off-label*)
	Scale insects	fatty acids
	Slugs/snails	metaldehyde, methiocarb
	Spider mites	fatty acids
	Whiteflies	acetamiprid (*off-label*), bifenthrin, chlorpyrifos, cypermethrin, fatty acids, lambda-cyhalothrin, lambda-cyhalothrin + pirimicarb, pyrethrins, spirotetramat
Weeds	Broad-leaved weeds	carbetamide, chlorthal-dimethyl, chlorthal-dimethyl (*off-label*), chlorthal-dimethyl + propachlor, clomazone (*off-label*), clopyralid, clopyralid (*off-label*), metazachlor, metazachlor (*off-label*), napropamide, napropamide (*off-label*), pendimethalin, propachlor, propyzamide (*off-label*), pyridate, pyridate (*off-label*)
	Crops as weeds	carbetamide, cycloxydim, pendimethalin, tepraloxydim
	Grass weeds	carbetamide, chlorthal-dimethyl + propachlor, cycloxydim, fluazifop-P-butyl (*off-label*), metazachlor, metazachlor (*off-label*), napropamide, napropamide (*off-label*), pendimethalin, propachlor, propyzamide (*off-label*), tepraloxydim

Brassicas - Mustard

Crop control	Desiccation	diquat (*off-label*), glyphosate
Diseases	Alternaria	fludioxonil + metalaxyl-M + thiamethoxam (*moderate control*)
	Botrytis	boscalid (*off-label*), chlorothalonil (*off-label*)
	Disease control/foliar feed	tebuconazole (*off-label*)
	Downy mildew	fludioxonil + metalaxyl-M + thiamethoxam
	Foliar diseases	cyproconazole (*off-label*), difenoconazole (*off-label*), prothioconazole (*off-label*), tebuconazole (*off-label*)
	Pythium	fludioxonil + metalaxyl-M + thiamethoxam (*reduction*)
	Sclerotinia	boscalid (*off-label*)
	Seed-borne diseases	fludioxonil + metalaxyl-M + thiamethoxam
Pests	Aphids	lambda-cyhalothrin (*off-label*), lambda-cyhalothrin + pirimicarb (*off-label*), pirimicarb (*off-label*), tau-fluvalinate (*off-label*)
	Beetles	beta-cyfluthrin + imidacloprid (*off-label - seed treatment*), deltamethrin, deltamethrin (*off-label*), fludioxonil + metalaxyl-M + thiamethoxam (*reduction of damage*), thiacloprid
	Caterpillars	deltamethrin (*off-label*)
	Flies	tefluthrin (*off-label - seed treatment*)
	Pests, miscellaneous	lambda-cyhalothrin (*off-label*), tau-fluvalinate (*off-label*)
	Weevils	deltamethrin
Weeds	Broad-leaved weeds	bifenox (*off-label*), carbetamide (*off-label*), chlorthal-dimethyl + propachlor, clomazone (*off-label*), clopyralid (*off-label*), clopyralid + picloram (*off-label*), glyphosate, metazachlor (*off-label*), metazachlor + quinmerac (*off-label*), napropamide (*off-label*), phenmedipham (*off-label*), propachlor
	Crops as weeds	glyphosate, propaquizafop (*off-label*)
	Grass weeds	carbetamide (*off-label*), chlorthal-dimethyl + propachlor, cycloxydim (*off-label*), fluazifop-P-butyl (*off-label*), glyphosate, metazachlor (*off-label*), metazachlor + quinmerac (*off-label*), napropamide (*off-label*), propachlor, propaquizafop (*off-label*), tepraloxydim (*off-label*)
	Weeds, miscellaneous	glyphosate

Brassicas - Root brassicas

Diseases	Alternaria	azoxystrobin (*off-label*), azoxystrobin + difenoconazole (*off-label*), cyprodinil + fludioxonil (*off-label*), fenpropimorph (*off-label*), iprodione + thiophanate-methyl (*off-label*), tebuconazole
	Botrytis	propamocarb hydrochloride (*off-label*)
	Crown rot	fenpropimorph (*off-label*), iprodione + thiophanate-methyl (*off-label*)
	Damping off	propamocarb hydrochloride (*off-label*), thiram (*off-label - seed treatment*), thiram (*seed treatment*), tolclofos-methyl (*off-label*)
	Disease control/foliar feed	tebuconazole (*off-label*)

	Downy mildew	metalaxyl-M (*off-label*), propamocarb hydrochloride (*off-label*)
	Foliar diseases	azoxystrobin (*off-label*), flusilazole (*off-label*), tebuconazole (*off-label*), tebuconazole + trifloxystrobin (*off-label*)
	Fungus diseases	tebuconazole (*off-label*)
	Phytophthora	propamocarb hydrochloride (*off-label*)
	Powdery mildew	azoxystrobin (*off-label*), azoxystrobin + difenoconazole (*off-label*), fenpropimorph (*off-label*), sulphur, sulphur (*off-label*), tebuconazole
	Rhizoctonia	azoxystrobin (*off-label*), tolclofos-methyl (*with fleece or mesh covers*), tolclofos-methyl (*without covers*)
	Ring spot	tebuconazole (*off-label*)
	Sclerotinia	boscalid + pyraclostrobin (*off-label*), cyprodinil + fludioxonil (*off-label*), tebuconazole
	Seed-borne diseases	thiram (*off-label - Seed treatment*)
	Stem canker	flusilazole (*off-label*)
	White blister	mancozeb + metalaxyl-M (*off-label*), metalaxyl-M (*off-label*), propamocarb hydrochloride (*off-label*)
Pests	Aphids	deltamethrin (*off-label*), lambda-cyhalothrin + pirimicarb (*off-label*), pirimicarb, pirimicarb (*off-label*), thiacloprid (*off-label*)
	Beetles	deltamethrin, deltamethrin (*off-label*)
	Caterpillars	deltamethrin, deltamethrin (*off-label*)
	Cutworms	Bacillus thuringiensis (*off-label*), cypermethrin (*off-label*), lambda-cyhalothrin (*off-label*), lambda-cyhalothrin + pirimicarb (*off-label*)
	Flies	chlorpyrifos (*off-label*), lambda-cyhalothrin (*off-label*)
	Leafhoppers	deltamethrin (*off-label*)
	Pests, miscellaneous	deltamethrin (*off-label*), dimethoate (*off-label*), lambda-cyhalothrin (*off-label*)
	Thrips	deltamethrin (*off-label*)
	Weevils	lambda-cyhalothrin (*off-label*)
Plant growth regulation	Quality/yield control	sulphur
Weeds	Broad-leaved weeds	chlorthal-dimethyl, chlorthal-dimethyl + propachlor, clomazone (*off-label*), clopyralid, glyphosate, metamitron (*off-label*), metazachlor, propachlor, prosulfocarb (*off-label*)
	Crops as weeds	cycloxydim, fluazifop-P-butyl (*stockfeed only*), glyphosate, propaquizafop (*off-label*)
	Grass weeds	chlorthal-dimethyl + propachlor, cycloxydim, cycloxydim (*off-label*), fluazifop-P-butyl (*stockfeed only*), glyphosate, metamitron (*off-label*), metazachlor, propachlor, propaquizafop, propaquizafop (*off-label*), prosulfocarb (*off-label*), tepraloxydim (*off-label*)
	Weeds, miscellaneous	glyphosate, metamitron (*off-label*)

Brassicas - Salad greens

Diseases	Alternaria	azoxystrobin (*off-label - for baby leaf production*), iprodione (*off-label*)
	Bacterial blight	copper oxychloride (*off-label*)
	Black rot	copper oxychloride (*off-label*)

	Botrytis	fenhexamid (*off-label - baby leaf production*), propamocarb hydrochloride (*off-label*)
	Bottom rot	copper oxychloride (*off-label*)
	Damping off	cymoxanil + fludioxonil + metalaxyl-M (*off-label*), fosetyl-aluminium + propamocarb hydrochloride (*off-label - baby leaf production*), propamocarb hydrochloride (*off-label*), tolclofos-methyl
	Disease control/foliar feed	difenoconazole (*off-label*)
	Downy mildew	azoxystrobin (*off-label - for baby leaf production*), copper oxychloride (*off-label*), dimethomorph + mancozeb (*off-label*), dimethomorph + mancozeb (*off-label - for baby leaf production*), fosetyl-aluminium (*off-label*), fosetyl-aluminium (*off-label - for baby leaf production*), fosetyl-aluminium + propamocarb hydrochloride (*off-label - baby leaf production*), propamocarb hydrochloride, propamocarb hydrochloride (*off-label*)
	Foliar diseases	fosetyl-aluminium (*off-label*)
	Phytophthora	copper oxychloride (*off-label*), propamocarb hydrochloride, propamocarb hydrochloride (*off-label*)
	Pythium	propamocarb hydrochloride
	Rhizoctonia	tolclofos-methyl (*off-label - baby leaf production only*)
	Ring spot	difenoconazole (*off-label*), tebuconazole (*off-label*)
	Spear rot	copper oxychloride (*off-label*)
	Stem canker	copper oxychloride (*off-label*)
	White blister	boscalid + pyraclostrobin (*off-label*), mancozeb + metalaxyl-M (*off-label*)
Pests	Aphids	acetamiprid (*off-label*), chlorpyrifos, cypermethrin (*off-label*), cypermethrin (*off-label - outdoor and protected crops*), deltamethrin (*off-label*), lambda-cyhalothrin (*off-label - for baby leaf production*), pirimicarb, pirimicarb (*off-label*), pirimicarb (*off-label - baby leaf production only*), pirimicarb (*off-label - for baby leaf production*), pymetrozine (*off-label*), pymetrozine (*off-label - for baby leaf production*), thiacloprid (*off-label*), thiacloprid (*off-label - baby leaf production*)
	Beetles	deltamethrin (*off-label*), deltamethrin (*off-label - Baby leaf production*)
	Birds/mammals	aluminium ammonium sulphate
	Caterpillars	alpha-cypermethrin (*off-label - for baby leaf production*), Bacillus thuringiensis (*off-label*), chlorpyrifos, cypermethrin (*off-label - outdoor and protected crops*), deltamethrin (*off-label*), deltamethrin (*off-label - Baby leaf production*), diflubenzuron (*off-label - for baby leaf production*), teflubenzuron (*off-label*)
	Cutworms	chlorpyrifos
	Flies	chlorpyrifos, chlorpyrifos (*off-label*), teflubenzuron (*off-label*), tefluthrin (*off-label - seed treatment*)
	Leafhoppers	deltamethrin (*off-label*)
	Leatherjackets	chlorpyrifos
	Pests, miscellaneous	alpha-cypermethrin (*off-label*), deltamethrin (*off-label*), dimethoate (*off-label*), teflubenzuron (*off-label*)
	Thrips	deltamethrin (*off-label*)
	Whiteflies	chlorpyrifos

Weeds	Broad-leaved weeds	chlorthal-dimethyl (*off-label*), clomazone (*off-label*), clopyralid (*off-label*), metazachlor (*off-label*), napropamide (*off-label*), pendimethalin (*off-label*), pendimethalin (*off-label - for baby leaf production*), propyzamide (*off-label*)
	Grass weeds	fluazifop-P-butyl (*off-label*), metazachlor (*off-label*), napropamide (*off-label*), pendimethalin (*off-label*), propyzamide (*off-label*)

Cereals - Barley

Crop control	Desiccation	diquat, diquat (*stockfeed only*), glyphosate
Diseases	Covered smut	carboxin + thiram (*seed treatment*), clothianidin + prothioconazole (*seed treatment*), clothianidin + prothioconazole + tebuconazole + triazoxide (*seed treatment*), fludioxonil (*seed treatment*), fludioxonil + flutriafol (*seed treatment*), fludioxonil + tefluthrin, fluquinconazole + prochloraz (*seed treatment*), fuberidazole + imidacloprid + triadimenol (*seed treatment*), fuberidazole + triadimenol (*seed treatment*), guazatine + imazalil (*seed treatment*), prochloraz + triticonazole (*seed treatment*), prothioconazole (*seed treatment*), prothioconazole + tebuconazole + triazoxide (*seed treatment*)
	Ear blight	prothioconazole + tebuconazole
	Eyespot	azoxystrobin + cyproconazole (*reduction*), boscalid + epoxiconazole (*moderate*), boscalid + epoxiconazole (*moderate control*), bromuconazole, carbendazim, carbendazim + flusilazole, chlorothalonil + cyproconazole, cyproconazole (*useful reduction*), cyproconazole + cyprodinil, cyprodinil, cyprodinil + picoxystrobin, epoxiconazole (*reduction*), epoxiconazole + fenpropimorph (*reduction*), epoxiconazole + fenpropimorph + kresoxim-methyl (*reduction*), epoxiconazole + kresoxim-methyl (*reduction*), epoxiconazole + metconazole (*reduction*), epoxiconazole + prochloraz, fluoxastrobin + prothioconazole (*reduction*), fluoxastrobin + prothioconazole + trifloxystrobin (*reduction*), flusilazole, prochloraz, prochloraz + propiconazole, prochloraz + proquinazid + tebuconazole, prochloraz + tebuconazole, prothioconazole, prothioconazole + spiroxamine, prothioconazole + spiroxamine + tebuconazole (*reduction*), prothioconazole + tebuconazole, prothioconazole + tebuconazole (*reduction*), prothioconazole + trifloxystrobin (*reduction*), prothioconazole + trifloxystrobin (*reduction in severity*)
	Foot rot	carboxin + thiram (*seed treatment*), clothianidin + prothioconazole (*seed treatment*), clothianidin + prothioconazole + tebuconazole + triazoxide (*seed treatment*), fludioxonil (*seed treatment*), fludioxonil + flutriafol (*seed treatment*), fludioxonil + tefluthrin, fluquinconazole + prochloraz (*seed treatment*), fuberidazole + imidacloprid + triadimenol (*seed treatment*), fuberidazole + imidacloprid + triadimenol (*seed treatment - reduction*), fuberidazole + triadimenol (*seed treatment*), guazatine (*seed treatment*), guazatine (*seed treatment - reduction*), guazatine + imazalil (*seed treatment*), guazatine + imazalil (*seed treatment - moderate control*), prochloraz + triticonazole (*seed*

	treatment), prothioconazole (*seed treatment*), prothioconazole + tebuconazole + triazoxide (*seed treatment*)
Late ear diseases	fluoxastrobin + prothioconazole, fluoxastrobin + prothioconazole + trifloxystrobin, prothioconazole, prothioconazole + spiroxamine + tebuconazole, prothioconazole + tebuconazole
Leaf stripe	carboxin + thiram (*seed treatment*), carboxin + thiram (*seed treatment - reduction*), clothianidin + prothioconazole (*seed treatment*), clothianidin + prothioconazole + tebuconazole + triazoxide (*seed treatment*), fludioxonil (*seed treatment - reduction*), fludioxonil + flutriafol (*seed treatment*), fludioxonil + tefluthrin (*partial control*), fuberidazole + triadimenol (*seed treatment*), guazatine (*seed treatment*), guazatine + imazalil (*seed treatment*), imazalil, prochloraz + triticonazole (*seed treatment*), prothioconazole (*seed treatment*), prothioconazole + tebuconazole + triazoxide (*seed treatment*), tebuconazole
Loose smut	carboxin + thiram (*seed treatment - reduction*), clothianidin + prothioconazole (*seed treatment*), clothianidin + prothioconazole + tebuconazole + triazoxide (*seed treatment*), fludioxonil + flutriafol (*seed treatment*), fluquinconazole + prochloraz (*seed treatment*), fuberidazole + imidacloprid + triadimenol (*seed treatment*), fuberidazole + triadimenol (*seed treatment*), prochloraz + triticonazole (*seed treatment*), prothioconazole (*seed treatment*), prothioconazole + tebuconazole + triazoxide (*seed treatment*), tebuconazole
Net blotch	azoxystrobin, azoxystrobin + chlorothalonil, azoxystrobin + cyproconazole, azoxystrobin + fenpropimorph, boscalid + epoxiconazole, bromuconazole, carboxin + thiram (*seed treatment*), chlorothalonil + cyproconazole, chlorothalonil + cyproconazole + propiconazole (*reduction*), chlorothalonil + picoxystrobin, clothianidin + prothioconazole + tebuconazole + triazoxide (*seed treatment*), cyproconazole (*useful reduction*), cyproconazole + cyprodinil, cyproconazole + propiconazole, cyproconazole + trifloxystrobin, cyprodinil, cyprodinil + picoxystrobin, epoxiconazole, epoxiconazole + fenpropimorph, epoxiconazole + fenpropimorph + kresoxim-methyl, epoxiconazole + fenpropimorph + metrafenone, epoxiconazole + fenpropimorph + pyraclostrobin, epoxiconazole + kresoxim-methyl, epoxiconazole + kresoxim-methyl + pyraclostrobin, epoxiconazole + metconazole, epoxiconazole + metrafenone, epoxiconazole + prochloraz, epoxiconazole + pyraclostrobin, famoxadone + flusilazole, fenpropimorph + flusilazole, fenpropimorph + pyraclostrobin, fluoxastrobin + prothioconazole, fluoxastrobin + prothioconazole + trifloxystrobin, flusilazole, fuberidazole + imidacloprid + triadimenol (*seed treatment - seed-borne only*), guazatine (*seed treatment*), guazatine + imazalil (*seed treatment - moderate control*), imazalil, mancozeb, mancozeb (*moderate control*), metconazole (*reduction*), picoxystrobin, prochloraz, prochloraz + proquinazid + tebuconazole (*moderate control*), prochloraz +

	tebuconazole, propiconazole + tebuconazole, prothioconazole, prothioconazole + spiroxamine, prothioconazole + spiroxamine + tebuconazole, prothioconazole + tebuconazole, prothioconazole + tebuconazole + triazoxide (*seed treatment*), prothioconazole + trifloxystrobin, pyraclostrobin, spiroxamine + tebuconazole, tebuconazole, tebuconazole + triadimenol, trifloxystrobin
Powdery mildew	azoxystrobin, azoxystrobin + cyproconazole, azoxystrobin + fenpropimorph, boscalid + epoxiconazole, bromuconazole, carbendazim + flusilazole, chlorothalonil + cyproconazole, chlorothalonil + cyproconazole + propiconazole, cyflufenamid, cyproconazole, cyproconazole + cyprodinil, cyproconazole + propiconazole, cyproconazole + trifloxystrobin, cyprodinil, cyprodinil + picoxystrobin, epoxiconazole, epoxiconazole + fenpropimorph, epoxiconazole + fenpropimorph + kresoxim-methyl, epoxiconazole + fenpropimorph + metrafenone, epoxiconazole + fenpropimorph + pyraclostrobin, epoxiconazole + kresoxim-methyl, epoxiconazole + metconazole (*moderate control*), epoxiconazole + metrafenone, epoxiconazole + prochloraz, epoxiconazole + prochloraz (*moderate control*), fenpropidin, fenpropimorph, fenpropimorph + flusilazole, fenpropimorph + kresoxim-methyl, fenpropimorph + pyraclostrobin, fenpropimorph + quinoxyfen, fluoxastrobin + prothioconazole, fluoxastrobin + prothioconazole + trifloxystrobin, fluquinconazole + prochloraz, flusilazole, flusilazole (*moderate control*), flutriafol, fuberidazole + imidacloprid + triadimenol (*seed treatment*), metconazole, metconazole (*moderate control*), metrafenone, picoxystrobin, prochloraz, prochloraz + proquinazid + tebuconazole, prochloraz + tebuconazole, propiconazole, propiconazole + tebuconazole, proquinazid, prothioconazole, prothioconazole + spiroxamine, prothioconazole + spiroxamine + tebuconazole, prothioconazole + tebuconazole, prothioconazole + trifloxystrobin, quinoxyfen, spiroxamine, spiroxamine + tebuconazole, sulphur, tebuconazole, tebuconazole + triadimenol
Ramularia leaf spots	epoxiconazole + metconazole, epoxiconazole + metconazole (*moderate control*)
Rhynchosporium	azoxystrobin, azoxystrobin + chlorothalonil, azoxystrobin + chlorothalonil (*moderate control*), azoxystrobin + cyproconazole (*moderate control*), azoxystrobin + fenpropimorph, boscalid + epoxiconazole (*moderate*), boscalid + epoxiconazole (*moderate control*), bromuconazole, carbendazim, chlorothalonil, chlorothalonil (*moderate control*), chlorothalonil (*moderate control only*), chlorothalonil + cyproconazole, chlorothalonil + cyproconazole + propiconazole (*reduction*), chlorothalonil + flusilazole, chlorothalonil + picoxystrobin, chlorothalonil + propiconazole, cyproconazole, cyproconazole + cyprodinil, cyproconazole + propiconazole, cyproconazole + trifloxystrobin, cyprodinil, cyprodinil (*moderate control*), cyprodinil + picoxystrobin, epoxiconazole, epoxiconazole + fenpropimorph, epoxiconazole + fenpropimorph + kresoxim-methyl,

	epoxiconazole + fenpropimorph + metrafenone, epoxiconazole + fenpropimorph + pyraclostrobin, epoxiconazole + kresoxim-methyl, epoxiconazole + kresoxim-methyl + pyraclostrobin, epoxiconazole + metconazole, epoxiconazole + metrafenone, epoxiconazole + prochloraz, epoxiconazole + pyraclostrobin, famoxadone + flusilazole, fenpropidin (*moderate control*), fenpropimorph, fenpropimorph + flusilazole, fenpropimorph + kresoxim-methyl, fenpropimorph + pyraclostrobin, fluoxastrobin + prothioconazole, fluoxastrobin + prothioconazole + trifloxystrobin, fluquinconazole + prochloraz, flusilazole, flusilazole (*moderate control*), flutriafol, mancozeb, metconazole, picoxystrobin, prochloraz, prochloraz + propiconazole, prochloraz + proquinazid + tebuconazole (*Moderate control*), prochloraz + tebuconazole, propiconazole, propiconazole + tebuconazole, prothioconazole, prothioconazole + spiroxamine, prothioconazole + spiroxamine + tebuconazole, prothioconazole + tebuconazole, prothioconazole + trifloxystrobin, pyraclostrobin (*moderate*), spiroxamine (*reduction*), spiroxamine + tebuconazole, tebuconazole, tebuconazole + triadimenol, trifloxystrobin
Rust	azoxystrobin, azoxystrobin + chlorothalonil, azoxystrobin + cyproconazole, azoxystrobin + fenpropimorph, boscalid + epoxiconazole, bromuconazole, carbendazim + flusilazole, chlorothalonil + cyproconazole, chlorothalonil + cyproconazole + propiconazole, chlorothalonil + picoxystrobin, chlorothalonil + propiconazole, cyproconazole, cyproconazole + cyprodinil, cyproconazole + propiconazole, cyproconazole + trifloxystrobin, cyprodinil + picoxystrobin, epoxiconazole, epoxiconazole + fenpropimorph, epoxiconazole + fenpropimorph + kresoxim-methyl, epoxiconazole + fenpropimorph + metrafenone, epoxiconazole + fenpropimorph + pyraclostrobin, epoxiconazole + kresoxim-methyl, epoxiconazole + kresoxim-methyl + pyraclostrobin, epoxiconazole + metconazole, epoxiconazole + metrafenone, epoxiconazole + prochloraz, epoxiconazole + pyraclostrobin, famoxadone + flusilazole, fenpropidin, fenpropidin (*moderate control*), fenpropimorph, fenpropimorph + flusilazole, fenpropimorph + pyraclostrobin, fluoxastrobin + prothioconazole, fluoxastrobin + prothioconazole + trifloxystrobin, fluquinconazole + prochloraz, fluquinconazole + prochloraz (*seed treatment*), flusilazole, flutriafol, fuberidazole + imidacloprid + triadimenol (*seed treatment*), mancozeb, mancozeb (*reduction*), metconazole, picoxystrobin, prochloraz + proquinazid + tebuconazole, prochloraz + tebuconazole, propiconazole, propiconazole + tebuconazole, prothioconazole, prothioconazole + spiroxamine, prothioconazole + spiroxamine + tebuconazole, prothioconazole + tebuconazole, prothioconazole + trifloxystrobin, pyraclostrobin, spiroxamine, spiroxamine + tebuconazole, tebuconazole, tebuconazole + triadimenol, trifloxystrobin
Seed-borne diseases	fluquinconazole, ipconazole
Septoria	bromuconazole, flusilazole

	Snow mould	fludioxonil (*seed treatment*), fludioxonil + flutriafol (*seed treatment*), fludioxonil + tefluthrin, fuberidazole + imidacloprid + triadimenol (*seed treatment - reduction*), guazatine + imazalil (*seed treatment*)
	Soil-borne diseases	ipconazole
	Sooty moulds	bromuconazole, mancozeb, prothioconazole + tebuconazole
	Take-all	azoxystrobin (*reduction*), azoxystrobin + chlorothalonil (*reduction*), azoxystrobin + chlorothalonil (*reduction only*), azoxystrobin + cyproconazole (*reduction*), azoxystrobin + fenpropimorph (*reduction*), fluoxastrobin + prothioconazole (*reduction*), fluquinconazole + prochloraz (*seed treatment - reduction*), silthiofam (*seed treatment*)
Pests	Aphids	alpha-cypermethrin, bifenthrin, chlorpyrifos, clothianidin (*seed treatment*), clothianidin + prothioconazole (*seed treatment*), clothianidin + prothioconazole + tebuconazole + triazoxide (*seed treatment*), cypermethrin, cypermethrin (*autumn sown*), deltamethrin, esfenvalerate, fuberidazole + imidacloprid + triadimenol (*seed treatment*), lambda-cyhalothrin, lambda-cyhalothrin + pirimicarb, pirimicarb, tau-fluvalinate, zeta-cypermethrin
	Birds/mammals	aluminium ammonium sulphate
	Flies	alpha-cypermethrin, alpha-cypermethrin (*Some control if present at application*), chlorpyrifos, cypermethrin, cypermethrin (*autumn sown*), fludioxonil + tefluthrin
	Leafhoppers	clothianidin (*seed treatment*), clothianidin + prothioconazole (*seed treatment*), clothianidin + prothioconazole + tebuconazole + triazoxide (*seed treatment*)
	Leatherjackets	chlorpyrifos
	Midges	chlorpyrifos
	Slugs/snails	clothianidin (*seed treatment*), clothianidin + prothioconazole (*seed treatment*), clothianidin + prothioconazole + tebuconazole + triazoxide (*seed treatment - reduction*)
	Wireworms	clothianidin (*seed treatment*), clothianidin + prothioconazole (*seed treatment*), clothianidin + prothioconazole + tebuconazole + triazoxide (*seed treatment - reduction*), fludioxonil + tefluthrin, fuberidazole + imidacloprid + triadimenol (*reduction of damage*)
Plant growth regulation	Growth control	2-chloroethylphosphonic acid, 2-chloroethylphosphonic acid + mepiquat chloride, chlormequat, chlormequat + 2-chloroethylphosphonic acid, chlormequat + 2-chloroethylphosphonic acid + mepiquat chloride, mepiquat chloride + prohexadione-calcium, trinexapac-ethyl
	Quality/yield control	2-chloroethylphosphonic acid + mepiquat chloride (*low lodging situations*), chlormequat, mepiquat chloride + prohexadione-calcium, sulphur
Weeds	Broad-leaved weeds	2,4-D, 2,4-D + MCPA, 2,4-DB + MCPA, amidosulfuron, amidosulfuron + iodosulfuron-methyl-sodium, bifenox, bromoxynil, bromoxynil + diflufenican + ioxynil, bromoxynil + ioxynil, carfentrazone-ethyl, carfentrazone-ethyl + mecoprop-P, carfentrazone-ethyl + metsulfuron-

		methyl, chlorotoluron, chlorotoluron (*off-label*), chlorotoluron + diflufenican, cinidon-ethyl, clopyralid, clopyralid + florasulam + fluroxypyr, dicamba + MCPA + mecoprop-P, dicamba + mecoprop-P, dichlorprop-P, dichlorprop-P + ioxynil, dichlorprop-P + MCPA + mecoprop-P, diflufenican, diflufenican + flufenacet, diflufenican + flurtamone, diflufenican + mecoprop-P, diflufenican + pendimethalin, florasulam, florasulam + fluroxypyr, flufenacet + pendimethalin, flupyrsulfuron-methyl, fluroxypyr, fluroxypyr + thifensulfuron-methyl + tribenuron-methyl, glyphosate, iodosulfuron-methyl-sodium, MCPA, MCPA + MCPB, MCPB, mecoprop-P, metsulfuron-methyl, metsulfuron-methyl + thifensulfuron-methyl, metsulfuron-methyl + tribenuron-methyl, pendimethalin, pendimethalin + picolinafen, picolinafen, prosulfocarb, thifensulfuron-methyl + tribenuron-methyl, tribenuron-methyl
	Crops as weeds	amidosulfuron + iodosulfuron-methyl-sodium, diflufenican, diflufenican + flufenacet, diflufenican + flurtamone, florasulam, fluroxypyr, glyphosate, iodosulfuron-methyl-sodium, metsulfuron-methyl + tribenuron-methyl, pendimethalin
	Grass weeds	chlorotoluron, chlorotoluron (*off-label*), chlorotoluron + diflufenican, diflufenican, diflufenican + flufenacet, diflufenican + flurtamone, diflufenican + pendimethalin, diquat (*animal feed*), fenoxaprop-P-ethyl, flufenacet + pendimethalin, flupyrsulfuron-methyl, glyphosate, pendimethalin, pendimethalin + picolinafen, pinoxaden, prosulfocarb, prosulfocarb (*off-label*), tralkoxydim, tri-allate
	Weeds, miscellaneous	diquat (*animal feed only*), fluroxypyr, glyphosate

Cereals - Cereals, general

Crop control	Desiccation	glyphosate (*off-label - for wild bird seed production*)
Diseases	Foliar diseases	cyprodinil (*off-label*)
Pests	Aphids	pirimicarb
	Flies	chlorpyrifos
	Leatherjackets	methiocarb (*reduction*)
	Slugs/snails	methiocarb
Weeds	Broad-leaved weeds	2,4-DB + MCPA, bromoxynil + prosulfuron (*off-label*), fluroxypyr (*off-label - grown for wild bird seed production*), isoxaben (*off-label - grown for game cover*), MCPA + MCPB
	Weeds, miscellaneous	clopyralid (*off-label - game cover*), glyphosate (*off-label - grown for wild bird seed production*)

Cereals - Maize/sweetcorn

Diseases	Damping off	fludioxonil + metalaxyl M, thiram (*seed treatment*)
	Eyespot	flusilazole (*off-label*)
	Foliar diseases	flusilazole (*off-label*), pyraclostrobin (*off-label*)
	Fusarium	fludioxonil + metalaxyl M
	Pythium	fludioxonil + metalaxyl M
Pests	Aphids	pirimicarb, pirimicarb (*off-label*), pymetrozine (*off-label*)
	Caterpillars	Bacillus thuringiensis (*off-label*), indoxacarb (*off-label*)

	Flies	chlorpyrifos, clothianidin, lambda-cyhalothrin (*off-label*)
	Pests, miscellaneous	lambda-cyhalothrin (*off-label*)
	Slugs/snails	methiocarb
	Symphylids	clothianidin (*reduction*)
	Wireworms	clothianidin
Weeds	Broad-leaved weeds	bromoxynil, bromoxynil + prosulfuron, bromoxynil + terbuthylazine, bromoxynil + terbuthylazine (*off-label*), clopyralid, flufenacet + isoxaflutole, flufenacet + isoxaflutole (*off-label*), fluroxypyr, isoxaben (*off-label - grown for game cover*), mesotrione, mesotrione (*off-label*), mesotrione + terbuthylazine, mesotrione + terbuthylazine (*off-label*), nicosulfuron, nicosulfuron (*off-label*), pendimethalin, pendimethalin (*off-label*), pendimethalin (*off-label - under covers*), rimsulfuron
	Crops as weeds	fluroxypyr, fluroxypyr (*off-label*), mesotrione, nicosulfuron, pendimethalin, rimsulfuron
	Grass weeds	bromoxynil + terbuthylazine (*off-label*), flufenacet + isoxaflutole, flufenacet + isoxaflutole (*off-label*), mesotrione, mesotrione (*off-label*), mesotrione + terbuthylazine, mesotrione + terbuthylazine (*off-label*), nicosulfuron, nicosulfuron (*off-label*), pendimethalin, pendimethalin (*off-label*)
	Weeds, miscellaneous	bromoxynil + terbuthylazine (*off-label*), fluroxypyr

Cereals - Oats

Crop control	Desiccation	diquat, diquat (*stockfeed only*), glyphosate
Diseases	Bunt	fluoxastrobin + prothioconazole
	Covered smut	carboxin + thiram (*seed treatment*), fluoxastrobin + prothioconazole, fluoxastrobin + prothioconazole (*seed treatment*), prochloraz + triticonazole (*seed treatment*)
	Crown rust	azoxystrobin, azoxystrobin + cyproconazole, azoxystrobin + fenpropimorph, boscalid + epoxiconazole, cyproconazole, epoxiconazole + fenpropimorph + metrafenone, epoxiconazole + fenpropimorph + pyraclostrobin, epoxiconazole + kresoxim-methyl + pyraclostrobin, epoxiconazole + metrafenone, epoxiconazole + pyraclostrobin, epoxiconazole + pyraclostrobin (*qualified minor use*), fenpropimorph + pyraclostrobin, fluoxastrobin + prothioconazole, picoxystrobin, prochloraz + proquinazid + tebuconazole (*qualified minor use recommendation*), propiconazole, prothioconazole, prothioconazole + spiroxamine, prothioconazole + spiroxamine + tebuconazole, prothioconazole + tebuconazole, pyraclostrobin, tebuconazole, tebuconazole (*reduction*), tebuconazole + triadimenol
	Eyespot	boscalid + epoxiconazole (*moderate control*), epoxiconazole + fenpropimorph + kresoxim-methyl (*reduction*), epoxiconazole + kresoxim-methyl (*reduction*), fluoxastrobin + prothioconazole, prothioconazole, prothioconazole + spiroxamine, prothioconazole + spiroxamine + tebuconazole, prothioconazole + tebuconazole
	Foot rot	carboxin + thiram (*seed treatment*), clothianidin + prothioconazole (*seed treatment*), difenoconazole + fludioxonil, fludioxonil (*seed treatment*), fludioxonil +

	epoxiconazole (*off-label*), epoxiconazole (*off-label*), epoxiconazole + fenpropimorph (*off-label*), epoxiconazole + fenpropimorph + kresoxim-methyl (*off-label*), epoxiconazole + kresoxim-methyl (*off-label*), epoxiconazole + kresoxim-methyl + pyraclostrobin (*off-label*), epoxiconazole + pyraclostrobin (*off-label*), famoxadone + flusilazole (*off-label*), fenpropidin (*off-label*), fenpropimorph (*off-label*), fenpropimorph + flusilazole (*off-label*), fenpropimorph + kresoxim-methyl (*off-label*), fluquinconazole + prochloraz (*off-label*), mancozeb (*off-label*), metconazole (*off-label*), pyraclostrobin (*off-label*), trifloxystrobin (*off-label*)
Foot rot	carboxin + thiram (*seed treatment*), clothianidin + prothioconazole (*seed treatment*), difenoconazole + fludioxonil, fludioxonil (*seed treatment*), fludioxonil + flutriafol (*seed treatment*), fludioxonil + tefluthrin, fluoxastrobin + prothioconazole, fluoxastrobin + prothioconazole (*reduction*), fluoxastrobin + prothioconazole (*seed treatment*), fluquinconazole + prochloraz (*seed treatment*), fuberidazole + imidacloprid + triadimenol (*seed treatment - reduction*), fuberidazole + triadimenol (*seed treatment*), guazatine (*seed treatment - reduction*), prochloraz + triticonazole (*off-label - seed treatment*), prochloraz + triticonazole (*seed treatment*), prothioconazole (*seed treatment*)
Fusarium	thiophanate-methyl
Late ear diseases	azoxystrobin, azoxystrobin + fenpropimorph (*moderate control*), chlorothalonil + flutriafol, cyproconazole, cyproconazole + propiconazole, fenpropidin + tebuconazole, fluoxastrobin + prothioconazole, fluoxastrobin + prothioconazole + trifloxystrobin, picoxystrobin, prochloraz + proquinazid + tebuconazole (*reduction only*), prochloraz + tebuconazole, prothioconazole, prothioconazole + spiroxamine, prothioconazole + spiroxamine + tebuconazole, prothioconazole + tebuconazole, prothioconazole + trifloxystrobin
Loose smut	clothianidin + prothioconazole (*seed treatment*), fludioxonil + flutriafol (*seed treatment*), fluoxastrobin + prothioconazole, fluoxastrobin + prothioconazole (*seed treatment*), fuberidazole + imidacloprid + triadimenol (*seed treatment*), fuberidazole + triadimenol (*seed treatment*), prochloraz + triticonazole (*off-label - seed treatment*), prochloraz + triticonazole (*seed treatment*), prothioconazole (*seed treatment*)
Net blotch	bromuconazole
Powdery mildew	azoxystrobin, azoxystrobin + cyproconazole, boscalid + epoxiconazole, bromuconazole, carbendazim + flusilazole, chlorothalonil + cyproconazole, chlorothalonil + cyproconazole + propiconazole (*moderate*), chlorothalonil + flutriafol, cyflufenamid, cyproconazole, cyproconazole + cyprodinil, cyproconazole + propiconazole, cyproconazole + quinoxyfen, cyproconazole + trifloxystrobin, cyprodinil (*moderate control*), epoxiconazole, epoxiconazole (*off-label*), epoxiconazole + fenpropimorph, epoxiconazole + fenpropimorph + metrafenone, epoxiconazole + kresoxim-methyl, epoxiconazole + metconazole (*moderate control*), epoxiconazole + metrafenone, epoxiconazole + prochloraz (*moderate control*),

	fenpropidin, fenpropimorph, fenpropimorph + flusilazole, fenpropimorph + quinoxyfen, fluoxastrobin + prothioconazole, fluoxastrobin + prothioconazole + trifloxystrobin, fluquinconazole, fluquinconazole + prochloraz (*moderate control*), flusilazole, flutriafol, fuberidazole + imidacloprid + triadimenol (*seed treatment*), metconazole (*moderate control*), metrafenone, prochloraz, prochloraz + proquinazid + tebuconazole, prochloraz + tebuconazole, propiconazole, propiconazole + tebuconazole, proquinazid, proquinazid (*off-label*), prothioconazole, prothioconazole + spiroxamine, prothioconazole + spiroxamine + tebuconazole, prothioconazole + tebuconazole, prothioconazole + trifloxystrobin, quinoxyfen, spiroxamine, spiroxamine + tebuconazole, sulphur, tebuconazole, tebuconazole + triadimenol
Rhynchosporium	bromuconazole
Rust	azoxystrobin, azoxystrobin + chlorothalonil, azoxystrobin + cyproconazole, azoxystrobin + fenpropimorph, boscalid + epoxiconazole, bromuconazole, carbendazim + flusilazole, chlorothalonil + cyproconazole, chlorothalonil + cyproconazole + propiconazole, chlorothalonil + flutriafol, chlorothalonil + picoxystrobin, chlorothalonil + propiconazole, cyproconazole, cyproconazole + cyprodinil, cyproconazole + propiconazole, cyproconazole + quinoxyfen, cyproconazole + trifloxystrobin, cyprodinil + picoxystrobin, difenoconazole, dimoxystrobin + epoxiconazole, epoxiconazole, epoxiconazole + fenpropimorph, epoxiconazole + fenpropimorph + kresoxim-methyl, epoxiconazole + fenpropimorph + metrafenone, epoxiconazole + fenpropimorph + pyraclostrobin, epoxiconazole + kresoxim-methyl, epoxiconazole + kresoxim-methyl + pyraclostrobin, epoxiconazole + metconazole, epoxiconazole + metrafenone, epoxiconazole + prochloraz, epoxiconazole + pyraclostrobin, famoxadone + flusilazole, fenpropidin (*moderate control*), fenpropimorph, fenpropimorph + flusilazole, fenpropimorph + pyraclostrobin, fluoxastrobin + prothioconazole, fluoxastrobin + prothioconazole + trifloxystrobin, fluquinconazole, fluquinconazole (*seed treatment*), fluquinconazole + prochloraz, fluquinconazole + prochloraz (*seed treatment*), fluquinconazole + prochloraz (*seed treatment - moderate control*), flusilazole, flutriafol, fuberidazole + imidacloprid + triadimenol (*seed treatment*), mancozeb, mancozeb (*useful control*), metconazole, picoxystrobin, prochloraz + proquinazid + tebuconazole, prochloraz + tebuconazole, propiconazole, propiconazole + tebuconazole, prothioconazole, prothioconazole + spiroxamine, prothioconazole + spiroxamine + tebuconazole, prothioconazole + tebuconazole, prothioconazole + trifloxystrobin, pyraclostrobin, spiroxamine, spiroxamine + tebuconazole, tebuconazole, tebuconazole + triadimenol, trifloxystrobin
Seed-borne diseases	difenoconazole + fludioxonil, fludioxonil (*off-label*), fluquinconazole, fluquinconazole (*off-label*), fluquinconazole + prochloraz (*off-label*), ipconazole, silthiofam (*off-label*)

Septoria	azoxystrobin, azoxystrobin + chlorothalonil, azoxystrobin + cyproconazole, azoxystrobin + fenpropimorph, boscalid + epoxiconazole, bromuconazole, carbendazim + flusilazole, carboxin + thiram (*seed treatment*), chlorothalonil, chlorothalonil + cyproconazole, chlorothalonil + cyproconazole + propiconazole, chlorothalonil + flusilazole, chlorothalonil + flutriafol, chlorothalonil + mancozeb, chlorothalonil + picoxystrobin, chlorothalonil + propiconazole, chlorothalonil + tebuconazole, cyproconazole, cyproconazole + cyprodinil, cyproconazole + propiconazole, cyproconazole + quinoxyfen, cyproconazole + trifloxystrobin, cyprodinil + picoxystrobin, difenoconazole, difenoconazole + fludioxonil, dimoxystrobin + epoxiconazole, epoxiconazole, epoxiconazole + fenpropimorph, epoxiconazole + fenpropimorph + kresoxim-methyl, epoxiconazole + fenpropimorph + metrafenone, epoxiconazole + fenpropimorph + pyraclostrobin, epoxiconazole + kresoxim-methyl, epoxiconazole + kresoxim-methyl + pyraclostrobin, epoxiconazole + metconazole, epoxiconazole + metrafenone, epoxiconazole + prochloraz, epoxiconazole + pyraclostrobin, famoxadone + flusilazole, fenpropidin + tebuconazole, fenpropimorph + flusilazole, fenpropimorph + kresoxim-methyl (*reduction*), fenpropimorph + pyraclostrobin, fludioxonil (*seed treatment*), fludioxonil + flutriafol (*seed treatment*), fludioxonil + tefluthrin, fluoxastrobin + prothioconazole, fluoxastrobin + prothioconazole (*reduction*), fluoxastrobin + prothioconazole + trifloxystrobin, fluquinconazole, fluquinconazole (*seed treatment*), fluquinconazole + prochloraz, fluquinconazole + prochloraz (*seed treatment*), fluquinconazole + prochloraz (*seed treatment - moderate control*), flusilazole, flutriafol, fuberidazole + imidacloprid + triadimenol (*seed treatment*), fuberidazole + imidacloprid + triadimenol (*seed treatment - reduction*), guazatine (*seed treatment*), guazatine (*seed treatment - partial control*), mancozeb, mancozeb (*reduction*), metconazole, picoxystrobin, prochloraz, prochloraz + propiconazole, prochloraz + proquinazid + tebuconazole, prochloraz + proquinazid + tebuconazole (*Moderate control*), prochloraz + tebuconazole, prochloraz + triticonazole (*off-label - seed treatment*), prochloraz + triticonazole (*seed treatment*), propiconazole, propiconazole + tebuconazole, prothioconazole, prothioconazole + spiroxamine, prothioconazole + spiroxamine + tebuconazole, prothioconazole + tebuconazole, prothioconazole + trifloxystrobin, pyraclostrobin, spiroxamine + tebuconazole, tebuconazole, tebuconazole + triadimenol, trifloxystrobin
Sharp eyespot	fluoxastrobin + prothioconazole, fluoxastrobin + prothioconazole (*reduction*)
Snow mould	difenoconazole + fludioxonil, fludioxonil (*seed treatment*), fludioxonil + flutriafol (*seed treatment*), fludioxonil + tefluthrin
Soil-borne diseases	ipconazole

	Sooty moulds	boscalid + epoxiconazole, bromuconazole, epoxiconazole (*reduction*), epoxiconazole + fenpropimorph (*reduction*), epoxiconazole + fenpropimorph + kresoxim-methyl (*reduction*), epoxiconazole + fenpropimorph + metrafenone (*reduction*), epoxiconazole + kresoxim-methyl (*reduction*), epoxiconazole + metconazole (*good reduction*), epoxiconazole + metrafenone (*reduction*), epoxiconazole + prochloraz (*good reduction*), fluoxastrobin + prothioconazole (*reduction*), mancozeb, propiconazole, prothioconazole + tebuconazole, prothioconazole + tebuconazole (*reduction*), spiroxamine + tebuconazole, tebuconazole, tebuconazole + triadimenol
	Take-all	azoxystrobin (*reduction*), azoxystrobin + chlorothalonil (*reduction*), azoxystrobin + chlorothalonil (*reduction only*), azoxystrobin + cyproconazole (*reduction*), azoxystrobin + fenpropimorph (*reduction*), fluoxastrobin + prothioconazole (*reduction*), fluquinconazole (*seed treatment - reduction*), fluquinconazole + prochloraz (*seed treatment - reduction*), silthiofam (*seed treatment*)
	Tan spot	boscalid + epoxiconazole (*reduction - qualified minor use*), boscalid + epoxiconazole (*reduction only*), chlorothalonil + picoxystrobin, dimoxystrobin + epoxiconazole, epoxiconazole + metconazole (*moderate control*), epoxiconazole + pyraclostrobin, fluoxastrobin + prothioconazole, fluoxastrobin + prothioconazole + trifloxystrobin, prothioconazole, prothioconazole + spiroxamine, prothioconazole + tebuconazole
Pests	Aphids	alpha-cypermethrin, bifenthrin, chlorpyrifos, chlorpyrifos (*off-label*), clothianidin (*seed treatment*), clothianidin + prothioconazole (*seed treatment*), cypermethrin, cypermethrin (*autumn sown*), deltamethrin, deltamethrin (*off-label*), dimethoate, esfenvalerate, flonicamid, fuberidazole + imidacloprid + triadimenol (*seed treatment*), lambda-cyhalothrin, lambda-cyhalothrin + pirimicarb, pirimicarb, tau-fluvalinate, tau-fluvalinate (*off-label*), zeta-cypermethrin, zeta-cypermethrin (*off-label*)
	Beetles	lambda-cyhalothrin
	Birds/mammals	aluminium ammonium sulphate
	Caterpillars	lambda-cyhalothrin
	Flies	alpha-cypermethrin, alpha-cypermethrin (*Some control if present at application*), chlorpyrifos, chlorpyrifos (*off-label*), cypermethrin, cypermethrin (*autumn sown*), dimethoate, fludioxonil + tefluthrin, lambda-cyhalothrin
	Leafhoppers	clothianidin (*seed treatment*), clothianidin + prothioconazole (*seed treatment*)
	Leatherjackets	chlorpyrifos, chlorpyrifos (*off-label*)
	Midges	chlorpyrifos, chlorpyrifos (*off-label*), lambda-cyhalothrin, thiacloprid
	Pests, miscellaneous	alpha-cypermethrin (*off-label*), chlorpyrifos (*off-label*), deltamethrin (*off-label*), dimethoate (*off-label*), zeta-cypermethrin (*off-label*)
	Slugs/snails	clothianidin (*seed treatment*), clothianidin + prothioconazole (*seed treatment*)
	Suckers	lambda-cyhalothrin

	Weeds, miscellaneous	diquat (*off-label*)

Leafy vegetables - Endives

Diseases	Botrytis	cyprodinil + fludioxonil (*off-label*), fenhexamid (*off-label*), propamocarb hydrochloride (*off-label*)
	Damping off	propamocarb hydrochloride (*off-label*), thiram (*off-label - seed treatment*)
	Downy mildew	copper oxychloride (*off-label*), dimethomorph + mancozeb (*off-label*), fosetyl-aluminium (*off-label*), fosetyl-aluminium + propamocarb hydrochloride (*off-label*), propamocarb hydrochloride (*off-label*)
	Foliar diseases	fosetyl-aluminium (*off-label*), mancozeb (*off-label*)
	Phytophthora	propamocarb hydrochloride (*off-label*)
Pests	Aphids	acetamiprid (*off-label*), cypermethrin (*off-label*), deltamethrin (*off-label*), lambda-cyhalothrin (*off-label*), pirimicarb (*off-label*), pymetrozine (*off-label*)
	Caterpillars	Bacillus thuringiensis (*off-label*), diflubenzuron (*off-label*)
	Pests, miscellaneous	deltamethrin (*off-label*), spinosad
	Thrips	pirimicarb (*off-label*)
Weeds	Broad-leaved weeds	pendimethalin (*off-label*), propyzamide (*off-label*)
	Grass weeds	pendimethalin (*off-label*), propyzamide (*off-label*)

Leafy vegetables - Lettuce

Diseases	Botrytis	cyprodinil + fludioxonil (*off-label*), fenhexamid (*off-label*), iprodione, propamocarb hydrochloride
	Bottom rot	boscalid + pyraclostrobin
	Damping off	propamocarb hydrochloride (*off-label*), thiram (*off-label - seed treatment*)
	Downy mildew	copper oxychloride (*off-label*), dimethomorph + mancozeb (*off-label*), fosetyl-aluminium (*off-label*), mancozeb, propamocarb hydrochloride
	Foliar diseases	fosetyl-aluminium (*off-label*), mancozeb (*off-label*)
	Fungus diseases	mancozeb + metalaxyl-M (*off-label*)
	Sclerotinia	azoxystrobin (*off-label*), boscalid + pyraclostrobin
Pests	Aphids	acetamiprid (*off-label*), cypermethrin, cypermethrin (*off-label*), deltamethrin (*off-label*), dimethoate, fatty acids, lambda-cyhalothrin (*off-label*), lambda-cyhalothrin + pirimicarb, pirimicarb, pirimicarb (*off-label*), pirimicarb (*outdoor crops*), pymetrozine (*off-label*), pyrethrins, spirotetramat
	Beetles	deltamethrin (*off-label*)
	Caterpillars	Bacillus thuringiensis (*off-label*), cypermethrin, deltamethrin (*off-label*), diflubenzuron (*off-label*), pyrethrins
	Cutworms	cypermethrin, deltamethrin, lambda-cyhalothrin, lambda-cyhalothrin + pirimicarb
	Leafhoppers	deltamethrin (*off-label*)
	Mealybugs	fatty acids
	Pests, miscellaneous	deltamethrin (*off-label*), spinosad
	Scale insects	fatty acids
	Slugs/snails	methiocarb

	Spider mites	fatty acids
	Thrips	deltamethrin (*off-label*)
	Whiteflies	fatty acids
Weeds	Broad-leaved weeds	pendimethalin (*off-label*), propyzamide, propyzamide (*off-label*), propyzamide (*outdoor crops*)
	Grass weeds	pendimethalin (*off-label*), propyzamide, propyzamide (*off-label*), propyzamide (*outdoor crops*)

Leafy vegetables - Spinach

Diseases	Damping off	cymoxanil + fludioxonil + metalaxyl-M (*off-label*), fosetyl-aluminium + propamocarb hydrochloride (*off-label*)
	Downy mildew	boscalid + pyraclostrobin (*off-label*), copper oxychloride (*off-label*), fosetyl-aluminium (*off-label*), fosetyl-aluminium + propamocarb hydrochloride (*off-label*), metalaxyl-M (*off-label*)
	Foliar diseases	fosetyl-aluminium (*off-label*)
Pests	Aphids	cypermethrin (*off-label*), pirimicarb (*off-label*)
	Caterpillars	Bacillus thuringiensis (*off-label*), cypermethrin (*off-label - outdoor and protected crops*), diflubenzuron (*off-label*)
	Flies	tefluthrin (*off-label - seed treatment*)
	Slugs/snails	methiocarb
Weeds	Broad-leaved weeds	chloridazon (*off-label*), chloridazon + quinmerac (*off-label*), clopyralid (*off-label*), lenacil (*off-label*), phenmedipham (*off-label*)
	Grass weeds	chloridazon (*off-label*), chloridazon + quinmerac (*off-label*), lenacil (*off-label*)

Leafy vegetables - Watercress

Diseases	Bacterial blight	copper oxychloride (*off-label*)
	Bottom rot	copper oxychloride (*off-label*)
	Damping off	metalaxyl-M (*off-label*), propamocarb hydrochloride (*off-label*)
	Downy mildew	fosetyl-aluminium (*off-label*), fosetyl-aluminium (*off-label - during propagation*), metalaxyl-M (*off-label*), propamocarb hydrochloride (*off-label - under protection*)
	Phytophthora	copper oxychloride (*off-label*), fosetyl-aluminium (*off-label - during propagation*), propamocarb hydrochloride (*off-label*), propamocarb hydrochloride (*off-label - under protection*)
	Pythium	copper oxychloride (*off-label - during propagation*), fosetyl-aluminium (*off-label - during propagation*), propamocarb hydrochloride (*off-label - under protection*)
	Rhizoctonia	copper oxychloride (*off-label - during propagation*)
	Spear rot	copper oxychloride (*off-label*)
	Stem canker	copper oxychloride (*off-label*)
Pests	Caterpillars	Bacillus thuringiensis (*off-label*)

Legumes - Beans (Phaseolus)

Crop control	Desiccation	diquat (*off-label*)
Diseases	Alternaria	iprodione (*off-label*)

	Botrytis	azoxystrobin (*off-label*), cyprodinil + fludioxonil (*moderate control*), iprodione (*off-label*)
	Damping off	thiram (*seed treatment*)
	Rust	tebuconazole (*off-label*)
	Sclerotinia	cyprodinil + fludioxonil
Pests	Aphids	fatty acids, pirimicarb
	Caterpillars	Bacillus thuringiensis (*off-label*), lambda-cyhalothrin (*off-label*)
	Flies	chlorpyrifos (*off-label*), chlorpyrifos (*off-label - harvested as a dry pulse*), chlorpyrifos (*off-label - harvested dry as a pulse*)
	Mealybugs	fatty acids
	Pests, miscellaneous	lambda-cyhalothrin (*off-label*)
	Scale insects	fatty acids
	Spider mites	fatty acids
	Whiteflies	fatty acids
Weeds	Broad-leaved weeds	bentazone, chlorthal-dimethyl, chlorthal-dimethyl (*off-label*), linuron, linuron (*off-label*), pendimethalin (*off-label*)
	Crops as weeds	chlorthal-dimethyl (*off-label*), cycloxydim
	Grass weeds	cycloxydim, cycloxydim (*off-label*), fluazifop-P-butyl (*off-label*), linuron, linuron (*off-label*), pendimethalin (*off-label*)

Legumes - Beans (Vicia)

Crop control	Desiccation	diquat, glufosinate-ammonium, glyphosate
Diseases	Ascochyta	azoxystrobin (*off-label*)
	Botrytis	azoxystrobin (*off-label*), cyprodinil + fludioxonil (*moderate control*)
	Chocolate spot	boscalid + pyraclostrobin (*moderate*), chlorothalonil, chlorothalonil (*moderate control*), chlorothalonil + cyproconazole, chlorothalonil + pyrimethanil, cyproconazole, iprodione + thiophanate-methyl, tebuconazole
	Damping off	thiram (*seed treatment*)
	Downy mildew	chlorothalonil + metalaxyl-M, cymoxanil + fludioxonil + metalaxyl-M (*off-label - seed treatment*), fosetyl-aluminium
	Rust	azoxystrobin, boscalid + pyraclostrobin, chlorothalonil + cyproconazole, cyproconazole, metconazole, metconazole (*off-label*), tebuconazole, tebuconazole (*off-label*)
	Sclerotinia	cyprodinil + fludioxonil
Pests	Aphids	cypermethrin, cypermethrin (*off-label*), fatty acids, lambda-cyhalothrin, lambda-cyhalothrin + pirimicarb, pirimicarb, zeta-cypermethrin (*off-label*)
	Beetles	lambda-cyhalothrin, lambda-cyhalothrin (*off-label*)
	Birds/mammals	aluminium ammonium sulphate
	Caterpillars	Bacillus thuringiensis (*off-label*), lambda-cyhalothrin
	Mealybugs	fatty acids
	Pests, miscellaneous	lambda-cyhalothrin (*off-label*), zeta-cypermethrin (*off-label*)
	Scale insects	fatty acids

	Spider mites	fatty acids
	Thrips	lambda-cyhalothrin (*off-label*)
	Weevils	alpha-cypermethrin, cypermethrin, deltamethrin, lambda-cyhalothrin, lambda-cyhalothrin + pirimicarb, zeta-cypermethrin
	Whiteflies	fatty acids
Weeds	Broad-leaved weeds	bentazone, carbetamide, clomazone, clomazone (*off-label*), clomazone + linuron, glyphosate, imazamox + pendimethalin, isoxaben + terbuthylazine, isoxaben + terbuthylazine (*off-label*), linuron, pendimethalin (*off-label*), propyzamide, prosulfocarb (*off-label*)
	Crops as weeds	carbetamide, cycloxydim, fluazifop-P-butyl, glyphosate, pendimethalin (*off-label*), propyzamide, quizalofop-P-ethyl, quizalofop-P-tefuryl, tepraloxydim
	Grass weeds	carbetamide, clomazone + linuron, cycloxydim, cycloxydim (*off-label*), diquat, fluazifop-P-butyl, fluazifop-P-butyl (*off-label*), glyphosate, isoxaben + terbuthylazine (*off-label*), linuron, pendimethalin (*off-label*), propaquizafop, propyzamide, prosulfocarb (*off-label*), quizalofop-P-ethyl, quizalofop-P-tefuryl, tepraloxydim, tri-allate, tri-allate (*off-label*)
	Weeds, miscellaneous	diquat, glyphosate

Legumes - Lupins

Crop control	Desiccation	diquat (*off-label*), glyphosate (*off-label*)
Diseases	Ascochyta	azoxystrobin (*off-label*), metconazole (*qualified minor use*)
	Botrytis	chlorothalonil (*off-label*), cyprodinil + fludioxonil (*off-label*), metconazole (*qualified minor use*)
	Damping off	cymoxanil + fludioxonil + metalaxyl-M (*off-label*), thiram (*off-label - seed treatment*)
	Disease control/foliar feed	chlorothalonil (*off-label*)
	Downy mildew	cymoxanil + fludioxonil + metalaxyl-M (*off-label*), fosetyl-aluminium (*off-label*)
	Foliar diseases	chlorothalonil (*off-label*)
	Pythium	cymoxanil + fludioxonil + metalaxyl-M (*off-label*)
	Rust	azoxystrobin (*off-label*), metconazole (*qualified minor use*)
	Seed-borne diseases	fosetyl-aluminium (*off-label*), thiram (*off-label - Seed treatment*)
Pests	Aphids	deltamethrin (*off-label*), lambda-cyhalothrin (*off-label*), lambda-cyhalothrin + pirimicarb (*off-label*), zeta-cypermethrin (*off-label*)
	Pests, miscellaneous	alpha-cypermethrin (*off-label*), deltamethrin (*off-label*), zeta-cypermethrin (*off-label*)
Weeds	Broad-leaved weeds	carbetamide (*off-label*), clomazone (*off-label*), isoxaben + terbuthylazine (*off-label*), pendimethalin (*off-label*), pyridate (*off-label - (yellow)*)
	Crops as weeds	propaquizafop (*off-label*)
	Grass weeds	carbetamide (*off-label*), cycloxydim (*off-label*), fluazifop-P-butyl (*off-label*), pendimethalin (*off-label*), propaquizafop (*off-label*), tepraloxydim (*off-label*), tri-allate (*off-label*)

	Grass weeds	diquat (*around only*), ethofumesate (*off-label*), flufenacet + isoxaflutole (*off-label*), flufenacet + pendimethalin (*off-label*), linuron (*off-label*), mesotrione (*off-label*), pendimethalin (*off-label*), pendimethalin + picolinafen (*off-label*), pinoxaden (*off-label*), propyzamide (*off-label*), quizalofop-P-tefuryl, sulfosulfuron (*off-label*), tepraloxydim (*off-label*), tri-allate (*off-label - for wild bird seed production*)
	Weeds, miscellaneous	clopyralid (*off-label - game cover*), diquat (*around only*), glyphosate, glyphosate (*before planting*), glyphosate (*off-label*), glyphosate (*off-label - grown for wild bird seed production*), glyphosate (*pre-sowing/planting*)

Miscellaneous arable - Miscellaneous arable situations

Diseases	Soil-borne diseases	dazomet (*soil fumigation*)
Pests	Free-living nematodes	dazomet (*soil fumigation*)
	Slugs/snails	metaldehyde
	Soil pests	dazomet (*soil fumigation*)
Weeds	Broad-leaved weeds	amitrole, citronella oil, glufosinate-ammonium, glufosinate-ammonium (*uncropped*), glyphosate
	Crops as weeds	amitrole (*barley stubble*), fluazifop-P-butyl, glyphosate, tepraloxydim
	Grass weeds	amitrole, fluazifop-P-butyl, glufosinate-ammonium, glufosinate-ammonium (*uncropped*), glyphosate, tepraloxydim
	Weeds, miscellaneous	2,4-D + dicamba + triclopyr, cycloxydim, dazomet (*soil fumigation*), fluazifop-P-butyl, glufosinate-ammonium, glyphosate, glyphosate (*wiper application*), metsulfuron-methyl, tepraloxydim, thifensulfuron-methyl

Miscellaneous field vegetables - All vegetables

Diseases	Soil-borne diseases	dazomet (*soil fumigation*)
Pests	Aphids	physical pest control
	Free-living nematodes	dazomet (*soil fumigation*)
	Leafhoppers	physical pest control
	Mites	physical pest control
	Soil pests	dazomet (*soil fumigation*)
	Thrips	physical pest control
	Whiteflies	physical pest control
Plant growth regulation	Quality/yield control	sulphur
Weeds	Broad-leaved weeds	glufosinate-ammonium
	Grass weeds	glufosinate-ammonium
	Weeds, miscellaneous	dazomet (*soil fumigation*), glufosinate-ammonium

Oilseed crops - Linseed/flax

Crop control	Desiccation	diquat, glufosinate-ammonium, glyphosate
Diseases	Botrytis	boscalid (*off-label*), chlorothalonil (*off-label*), metconazole (*off-label*), tebuconazole (*reduction*)
	Disease control/foliar feed	prochloraz (*off-label*)

	Foliar diseases	cyproconazole (*off-label*), difenoconazole (*off-label*)
	Powdery mildew	tebuconazole
	Sclerotinia	boscalid (*off-label*)
	Seed-borne diseases	prochloraz (*seed treatment*)
Pests	Aphids	cypermethrin (*off-label*), lambda-cyhalothrin (*off-label*), lambda-cyhalothrin + pirimicarb (*off-label*), pirimicarb (*off-label*), tau-fluvalinate (*off-label*)
	Beetles	beta-cyfluthrin + imidacloprid (*off-label - seed treatment*), bifenthrin, zeta-cypermethrin
	Pests, miscellaneous	alpha-cypermethrin (*off-label*), lambda-cyhalothrin (*off-label*)
Weeds	Broad-leaved weeds	amidosulfuron, bentazone, bifenox (*off-label*), bromoxynil, carbetamide (*off-label*), clomazone (*off-label*), clopyralid, clopyralid (*off-label*), clopyralid + picloram (*off-label*), flupyrsulfuron-methyl (*off-label*), glyphosate, MCPA, mesotrione (*off-label*), metazachlor (*off-label*), metazachlor + quinmerac (*off-label*), metsulfuron-methyl, napropamide (*off-label*), prosulfocarb (*off-label*)
	Crops as weeds	cycloxydim, fluazifop-P-butyl, glyphosate, mesotrione (*off-label*), propaquizafop (*off-label*), quizalofop-P-ethyl, quizalofop-P-tefuryl, tepraloxydim
	Grass weeds	carbetamide (*off-label*), cycloxydim, diquat, fluazifop-P-butyl, flupyrsulfuron-methyl (*off-label*), glyphosate, mesotrione (*off-label*), metazachlor (*off-label*), metazachlor + quinmerac (*off-label*), napropamide (*off-label*), propaquizafop, propaquizafop (*off-label*), prosulfocarb (*off-label*), quizalofop-P-ethyl, quizalofop-P-tefuryl, tepraloxydim
	Weeds, miscellaneous	diquat, glyphosate

Oilseed crops - Miscellaneous oilseeds

Crop control	Desiccation	diquat (*off-label*), glyphosate (*off-label*)
Diseases	Botrytis	boscalid (*off-label*), chlorothalonil (*off-label*)
	Damping off	cymoxanil + fludioxonil + metalaxyl-M (*off-label - seed treatment*), thiram (*off-label - seed treatment*)
	Disease control/foliar feed	chlorothalonil + metalaxyl-M (*off-label*), propiconazole (*off-label*), tebuconazole (*off-label*)
	Downy mildew	chlorothalonil (*off-label*), cymoxanil + fludioxonil + metalaxyl-M (*off-label - seed treatment*), dimethomorph + mancozeb (*off-label*), mancozeb (*off-label*)
	Foliar diseases	azoxystrobin (*off-label*), cyproconazole (*off-label*), difenoconazole (*off-label*), mancozeb (*off-label*), prothioconazole (*off-label*), tebuconazole (*off-label*)
	Powdery mildew	azoxystrobin (*off-label*), sulphur (*off-label*)
	Pythium	cymoxanil + fludioxonil + metalaxyl-M (*off-label - seed treatment*), thiram (*off-label - seed treatment*)
	Sclerotinia	boscalid (*off-label*)
	Seed-borne diseases	prochloraz (*off-label - seed treatment*)
Pests	Aphids	deltamethrin (*off-label*), lambda-cyhalothrin (*off-label*), lambda-cyhalothrin + pirimicarb (*off-label*), pirimicarb (*off-label*), tau-fluvalinate (*off-label*)

	Beetles	beta-cyfluthrin + clothianidin (*off-label*), beta-cyfluthrin + imidacloprid (*off-label - seed treatment*), deltamethrin (*off-label*), lambda-cyhalothrin (*off-label*)
	Caterpillars	deltamethrin (*off-label*)
	Leafhoppers	deltamethrin (*off-label*)
	Pests, miscellaneous	alpha-cypermethrin (*off-label*), lambda-cyhalothrin (*off-label*)
	Thrips	deltamethrin (*off-label*)
Weeds	Broad-leaved weeds	asulam (*off-label*), bifenox (*off-label*), carbetamide (*off-label*), chlorthal-dimethyl (*off-label*), clomazone (*off-label*), clopyralid (*off-label*), clopyralid + picloram (*off-label*), diquat (*off-label*), ethofumesate (*off-label*), fluroxypyr (*off-label*), metazachlor (*off-label*), metazachlor + quinmerac (*off-label*), napropamide (*off-label*), pendimethalin (*off-label*), prosulfocarb (*off-label*)
	Crops as weeds	chlorthal-dimethyl (*off-label*), fluroxypyr (*off-label*), propaquizafop (*off-label*)
	Grass weeds	asulam (*off-label*), carbetamide (*off-label*), cycloxydim (*off-label*), ethofumesate (*off-label*), fluazifop-P-butyl (*off-label*), metazachlor (*off-label*), metazachlor + quinmerac (*off-label*), napropamide (*off-label*), pendimethalin (*off-label*), propaquizafop (*off-label*), prosulfocarb (*off-label*), tepraloxydim (*off-label*)
	Weeds, miscellaneous	diquat (*off-label*)

Oilseed crops - Oilseed rape

Crop control	Desiccation	diquat, glufosinate-ammonium, glyphosate
Diseases	Alternaria	azoxystrobin, azoxystrobin + cyproconazole, boscalid, difenoconazole, fludioxonil + metalaxyl-M + thiamethoxam (*moderate control*), iprodione, iprodione + thiophanate-methyl, metconazole, prochloraz, tebuconazole
	Black scurf and stem canker	difenoconazole, iprodione + thiophanate-methyl, prothioconazole + tebuconazole, tebuconazole
	Botrytis	chlorothalonil, chlorothalonil (*moderate control*), iprodione, iprodione + thiophanate-methyl, prochloraz
	Damping off	thiram (*seed treatment*)
	Downy mildew	chlorothalonil, chlorothalonil (*moderate control*), fludioxonil + metalaxyl-M + thiamethoxam, mancozeb
	Ear blight	thiophanate-methyl (*reduction only*)
	Fusarium	thiophanate-methyl
	Light leaf spot	carbendazim, carbendazim + flusilazole, cyproconazole, difenoconazole, famoxadone + flusilazole, flusilazole, iprodione + thiophanate-methyl, metconazole, prochloraz, prochloraz + propiconazole, propiconazole (*reduction*), prothioconazole + tebuconazole, prothioconazole + tebuconazole (*moderate control*), prothioconazole + tebuconazole (*moderate control only*), tebuconazole
	Phoma leaf spot	carbendazim + flusilazole, cyproconazole, prothioconazole + tebuconazole
	Powdery mildew	sulphur
	Pythium	fludioxonil + metalaxyl-M + thiamethoxam (*reduction*)
	Ring spot	tebuconazole (*reduction*)

	Sclerotinia	azoxystrobin, azoxystrobin + cyproconazole, boscalid, iprodione + thiophanate-methyl, prochloraz, prochloraz + tebuconazole, prothioconazole, prothioconazole + tebuconazole, tebuconazole
	Seed-borne diseases	fludioxonil + metalaxyl-M + thiamethoxam
	Stem canker	carbendazim + flusilazole (*reduction*), famoxadone + flusilazole (*suppression*), famoxadone + flusilazole (*Suppression*), metconazole (*reduction*), prochloraz, prochloraz + propiconazole, prochloraz + thiram (*seed treatment*), prothioconazole + tebuconazole, tebuconazole
	White leaf spot	prochloraz
Pests	Aphids	acetamiprid, alpha-cypermethrin, beta-cyfluthrin + clothianidin, cypermethrin, deltamethrin, lambda-cyhalothrin, lambda-cyhalothrin + pirimicarb, pirimicarb, tau-fluvalinate
	Beetles	alpha-cypermethrin, beta-cyfluthrin + clothianidin, beta-cyfluthrin + imidacloprid (*seed treatment*), bifenthrin, cypermethrin, deltamethrin, fludioxonil + metalaxyl-M + thiamethoxam (*reduction of damage*), lambda-cyhalothrin, lambda-cyhalothrin + pirimicarb, tau-fluvalinate, thiacloprid, zeta-cypermethrin
	Birds/mammals	aluminium ammonium sulphate
	Caterpillars	lambda-cyhalothrin
	Flies	beta-cyfluthrin + clothianidin, beta-cyfluthrin + clothianidin (*reduction of feeding activity only*)
	Midges	alpha-cypermethrin, bifenthrin, cypermethrin, cypermethrin (*some coincidental control only*), lambda-cyhalothrin, lambda-cyhalothrin + pirimicarb, zeta-cypermethrin
	Slugs/snails	methiocarb
	Weevils	alpha-cypermethrin, bifenthrin, cypermethrin, deltamethrin, lambda-cyhalothrin, lambda-cyhalothrin + pirimicarb, zeta-cypermethrin
Plant growth regulation	Growth control	tebuconazole
	Quality/yield control	sulphur
Weeds	Broad-leaved weeds	bifenox (*off-label*), carbetamide, clomazone, clomazone + metazachlor, clopyralid, clopyralid + picloram, clopyralid + picloram (*off-label*), dimethenamid-p + metazachlor, dimethenamid-p + metazachlor + quinmerac, glyphosate, metazachlor, metazachlor + quinmerac, napropamide, propachlor, propyzamide, pyridate (*off-label*)
	Crops as weeds	carbetamide, cycloxydim, fluazifop-P-butyl, glyphosate, propyzamide, quizalofop-P-ethyl, quizalofop-P-tefuryl, tepraloxydim
	Grass weeds	carbetamide, clomazone + metazachlor, cycloxydim, dimethenamid-p + metazachlor + quinmerac, diquat, fluazifop-P-butyl, glyphosate, metazachlor, metazachlor + quinmerac, napropamide, propachlor, propaquizafop, propyzamide, quizalofop-P-ethyl, quizalofop-P-tefuryl, tepraloxydim
	Weeds, miscellaneous	diquat, glyphosate

Oilseed crops - Soya

Diseases	Damping off	thiram (*seed treatment - qualified minor use*), thiram (*seed treatment - qualified minor use*))
Weeds	Broad-leaved weeds	bentazone (*off-label*)

Oilseed crops - Sunflowers

Crop control	Desiccation	diquat (*off-label*)
Pests	Slugs/snails	methiocarb
Plant growth regulation	Growth control	daminozide
Weeds	Broad-leaved weeds	isoxaben (*off-label - grown for game cover*), pendimethalin
	Crops as weeds	pendimethalin
	Grass weeds	pendimethalin

Root and tuber crops - Beet crops

Diseases	Alternaria	cyprodinil + fludioxonil (*off-label*)
	Aphanomyces	hymexazol (*off-label - seed treatment*)
	Black leg	hymexazol (*seed treatment*)
	Botrytis	iprodione (*off-label*)
	Cercospora leaf spot	azoxystrobin + cyproconazole, cyproconazole + trifloxystrobin, epoxiconazole + pyraclostrobin
	Damping off	cymoxanil + fludioxonil + metalaxyl-M (*off-label - seed treatment*)
	Downy mildew	copper oxychloride (*off-label*), fosetyl-aluminium (*off-label*), fosetyl-aluminium + propamocarb hydrochloride, mancozeb + metalaxyl-M (*off-label*), metalaxyl-M (*off-label*)
	Foliar diseases	cyproconazole (*off-label*), epoxiconazole + pyraclostrobin (*off-label*), flusilazole (*off-label*), fosetyl-aluminium (*off-label*)
	Powdery mildew	azoxystrobin + cyproconazole, carbendazim + flusilazole, cyproconazole, cyproconazole + trifloxystrobin, difenoconazole + fenpropidin, epoxiconazole + pyraclostrobin, flusilazole, quinoxyfen, quinoxyfen (*off-label*), sulphur, sulphur (*off-label*)
	Ramularia leaf spots	azoxystrobin + cyproconazole, cyproconazole, cyproconazole + trifloxystrobin, difenoconazole + fenpropidin, epoxiconazole + pyraclostrobin, propiconazole (*reduction*)
	Root malformation disorder	azoxystrobin (*off-label*), metalaxyl-M (*off-label*)
	Rust	azoxystrobin + cyproconazole, carbendazim + flusilazole, cyproconazole + trifloxystrobin, difenoconazole + fenpropidin, epoxiconazole + pyraclostrobin, fenpropimorph (*off-label*), flusilazole, propiconazole
	Sclerotinia	cyprodinil + fludioxonil (*off-label*)
	Seed-borne diseases	thiram (*seed soak*)
Pests	Aphids	beta-cyfluthrin + clothianidin (*seed treatment*), chlorpyrifos (*off-label*), cypermethrin, cypermethrin (*off-label*), cypermethrin (*off-label - outdoor and protected crops*), deltamethrin (*off-label*), dimethoate, dimethoate (*excluding Myzus persicae*), imidacloprid, lambda-

		cyhalothrin, lambda-cyhalothrin + pirimicarb, lambda-cyhalothrin + pirimicarb (*off-label*), oxamyl, oxamyl (*off-label*), pirimicarb, pirimicarb (*off-label*), thiacloprid (*off-label*), zeta-cypermethrin (*off-label*)
	Beetles	chlorpyrifos, chlorpyrifos (*off-label*), deltamethrin, imidacloprid, lambda-cyhalothrin, lambda-cyhalothrin (*off-label*), lambda-cyhalothrin + pirimicarb, lambda-cyhalothrin + pirimicarb (*off-label*), oxamyl, oxamyl (*off-label*), tefluthrin (*seed treatment*), thiamethoxam (*seed treatment*)
	Birds/mammals	aluminium ammonium sulphate
	Caterpillars	Bacillus thuringiensis (*off-label*), cypermethrin, cypermethrin (*off-label - outdoor and protected crops*), lambda-cyhalothrin, lambda-cyhalothrin (*off-label*)
	Cutworms	Bacillus thuringiensis (*off-label*), cypermethrin, lambda-cyhalothrin, lambda-cyhalothrin (*off-label*), lambda-cyhalothrin + pirimicarb, lambda-cyhalothrin + pirimicarb (*off-label*), methiocarb (*reduction*), zeta-cypermethrin
	Flies	lambda-cyhalothrin, oxamyl, oxamyl (*off-label*)
	Free-living nematodes	oxamyl, oxamyl (*off-label*)
	Leaf miners	dimethoate, imidacloprid, lambda-cyhalothrin, lambda-cyhalothrin (*off-label*), lambda-cyhalothrin + pirimicarb, lambda-cyhalothrin + pirimicarb (*off-label*), thiamethoxam (*seed treatment*)
	Leatherjackets	chlorpyrifos, chlorpyrifos (*off-label*), methiocarb (*reduction*)
	Millipedes	imidacloprid, methiocarb (*reduction*), oxamyl, oxamyl (*off-label*), tefluthrin (*seed treatment*), thiamethoxam (*seed treatment*)
	Pests, miscellaneous	chlorpyrifos (*off-label*), deltamethrin (*off-label*), lambda-cyhalothrin (*off-label*), zeta-cypermethrin (*off-label*)
	Slugs/snails	methiocarb
	Springtails	imidacloprid, tefluthrin (*seed treatment*), thiamethoxam (*seed treatment*)
	Symphylids	imidacloprid, tefluthrin (*seed treatment*), thiamethoxam (*seed treatment*)
	Weevils	lambda-cyhalothrin
	Wireworms	thiamethoxam (*seed treatment - reduction*)
Plant growth regulation	Quality/yield control	sulphur
Weeds	Broad-leaved weeds	carbetamide, chloridazon, chloridazon + ethofumesate, chloridazon + metamitron, chloridazon + metamitron (*off-label*), chloridazon + quinmerac, chloridazon + quinmerac (*off-label*), chlorpropham + metamitron, clopyralid, desmedipham + ethofumesate + metamitron + phenmedipham, desmedipham + ethofumesate + phenmedipham, desmedipham + phenmedipham, diquat, ethofumesate, ethofumesate + metamitron, ethofumesate + metamitron + phenmedipham, ethofumesate + phenmedipham, glufosinate-ammonium, glyphosate, lenacil, lenacil (*off-label*), lenacil + triflusulfuron-methyl, lenacil + triflusulfuron-methyl (*off-label*), metamitron, phenmedipham, phenmedipham (*off-label*), propyzamide, triflusulfuron-methyl, triflusulfuron-methyl (*off-label*)
	Crops as weeds	carbetamide, cycloxydim, fluazifop-P-butyl, glyphosate, glyphosate (*wiper application*), lenacil + triflusulfuron-

		methyl, propaquizafop (*off-label*), propyzamide, quizalofop-P-ethyl, quizalofop-P-tefuryl, tepraloxydim
	Grass weeds	carbetamide, chloridazon, chloridazon + ethofumesate, chloridazon + metamitron, chloridazon + metamitron (*off-label*), chloridazon + quinmerac, chloridazon + quinmerac (*off-label*), chlorpropham + metamitron, cycloxydim, desmedipham + ethofumesate + metamitron + phenmedipham, desmedipham + ethofumesate + phenmedipham, diquat, ethofumesate, ethofumesate + metamitron, ethofumesate + metamitron + phenmedipham, ethofumesate + phenmedipham, fluazifop-P-butyl, fluazifop-P-butyl (*off-label*), glufosinate-ammonium, glyphosate, lenacil, lenacil (*off-label*), lenacil + triflusulfuron-methyl (*off-label*), metamitron, propaquizafop, propaquizafop (*off-label*), propyzamide, quizalofop-P-ethyl, quizalofop-P-tefuryl, tepraloxydim, tepraloxydim (*off-label*), tri-allate
	Weeds, miscellaneous	diquat, glufosinate-ammonium, glyphosate

Root and tuber crops - Carrots/parsnips

Diseases	Alternaria	azoxystrobin, azoxystrobin (*off-label*), azoxystrobin + difenoconazole, azoxystrobin + difenoconazole (*off-label*), boscalid + pyraclostrobin (*moderate*), cyprodinil + fludioxonil (*off-label*), fenpropimorph (*off-label*), iprodione + thiophanate-methyl (*off-label*), tebuconazole, tebuconazole + trifloxystrobin
	Cavity spot	metalaxyl-M, metalaxyl-M (*off-label*), metalaxyl-M (*reduction only*)
	Crown rot	fenpropimorph (*off-label*), iprodione + thiophanate-methyl (*off-label*)
	Foliar diseases	tebuconazole + trifloxystrobin (*off-label*)
	Powdery mildew	azoxystrobin, azoxystrobin (*off-label*), azoxystrobin + difenoconazole, azoxystrobin + difenoconazole (*off-label*), boscalid + pyraclostrobin, fenpropimorph (*off-label*), sulphur (*off-label*), tebuconazole, tebuconazole + trifloxystrobin
	Pythium	cymoxanil + fludioxonil + metalaxyl-M (*off-label - seed treatment*)
	Sclerotinia	boscalid + pyraclostrobin (*moderate*), boscalid + pyraclostrobin (*off-label*), cyprodinil + fludioxonil (*off-label*), tebuconazole, tebuconazole + trifloxystrobin
	Seed-borne diseases	thiram (*seed soak*)
Pests	Aphids	deltamethrin (*off-label*), lambda-cyhalothrin + pirimicarb, lambda-cyhalothrin + pirimicarb (*off-label*), pirimicarb, thiacloprid (*off-label*)
	Beetles	deltamethrin (*off-label*)
	Birds/mammals	aluminium ammonium sulphate
	Caterpillars	deltamethrin (*off-label*)
	Cutworms	Bacillus thuringiensis (*off-label*), cypermethrin (*off-label*), lambda-cyhalothrin, lambda-cyhalothrin + pirimicarb, lambda-cyhalothrin + pirimicarb (*off-label*)
	Flies	lambda-cyhalothrin, lambda-cyhalothrin (*off-label*), tefluthrin (*off-label - seed treatment*)
	Leafhoppers	deltamethrin (*off-label*)
	Pests, miscellaneous	deltamethrin (*off-label*), lambda-cyhalothrin (*off-label*)

	Stem nematodes	oxamyl
	Thrips	deltamethrin (*off-label*)
Plant growth regulation	Growth control	maleic hydrazide (*off-label*)
Weeds	Broad-leaved weeds	clomazone, flumioxazine (*off-label*), isoxaben (*off-label*), linuron, metamitron (*off-label*), metribuzin (*off-label*), pendimethalin, prosulfocarb (*off-label*)
	Crops as weeds	cycloxydim, fluazifop-P-butyl, flumioxazine (*off-label*), pendimethalin, tepraloxydim
	Grass weeds	cycloxydim, fluazifop-P-butyl, fluazifop-P-butyl (*off-label*), linuron, metamitron (*off-label*), pendimethalin, propaquizafop, prosulfocarb (*off-label*), tepraloxydim, tepraloxydim (*off-label*)
	Weeds, miscellaneous	metamitron (*off-label*)

Root and tuber crops - Miscellaneous root crops

Crop control	Desiccation	diquat (*off-label*)
Diseases	Alternaria	azoxystrobin (*off-label*), azoxystrobin + difenoconazole (*off-label*), cyprodinil + fludioxonil (*off-label*), fenpropimorph (*off-label*)
	Crown rot	fenpropimorph (*off-label*)
	Disease control/foliar feed	tebuconazole (*off-label*)
	Foliar diseases	tebuconazole (*off-label*), tebuconazole + trifloxystrobin (*off-label*)
	Leaf spot	chlorothalonil (*off-label*)
	Powdery mildew	azoxystrobin (*off-label*), azoxystrobin + difenoconazole (*off-label*), fenpropimorph (*off-label*)
	Rust	azoxystrobin (*off-label*)
	Sclerotinia	azoxystrobin (*off-label*), cyprodinil + fludioxonil (*off-label*)
	Septoria	difenoconazole (*off-label*)
Pests	Aphids	deltamethrin (*off-label*), pirimicarb (*off-label*), pymetrozine (*off-label*), thiacloprid (*off-label*)
	Cutworms	Bacillus thuringiensis (*off-label*), cypermethrin (*off-label*), lambda-cyhalothrin (*off-label*)
	Flies	lambda-cyhalothrin (*off-label*)
	Pests, miscellaneous	deltamethrin (*off-label*), dimethoate (*off-label*), lambda-cyhalothrin (*off-label*)
	Weevils	lambda-cyhalothrin (*off-label*)
Weeds	Broad-leaved weeds	asulam (*off-label*), clomazone (*off-label*), linuron (*off-label*), metribuzin (*off-label*), pendimethalin (*off-label*), propyzamide (*off-label*), prosulfocarb (*off-label*)
	Grass weeds	asulam (*off-label*), fluazifop-P-butyl (*off-label*), linuron (*off-label*), metribuzin (*off-label*), pendimethalin (*off-label*), propyzamide (*off-label*), prosulfocarb (*off-label*)
	Weeds, miscellaneous	metribuzin (*off-label*)

Root and tuber crops - Potatoes

Crop control	Desiccation	carfentrazone-ethyl, diquat, glufosinate-ammonium, sulphuric acid (commodity substance)
Diseases	Alternaria	azoxystrobin + chlorothalonil (*off-label*)

	Black dot	azoxystrobin
	Black scurf and stem canker	azoxystrobin, flutolanil (*tuber treatment*), imazalil + pencycuron (*tuber treatment*), imazalil + pencycuron (*tuber treatment - reduction*), pencycuron (*tuber treatment*), tolclofos-methyl (*tuber treatment*), tolclofos-methyl (*tuber treatment only to be used with automatic planters*)
	Dry rot	imazalil, thiabendazole, thiabendazole (*tuber treatment - post-harvest*)
	Fungus diseases	peroxyacetic acid (commodity substance) (*tuber treatment*)
	Fusarium	imazalil
	Gangrene	imazalil, thiabendazole, thiabendazole (*tuber treatment - post-harvest*)
	Phytophthora	amisulbrom, benalaxyl + mancozeb, benthiavalicarb-isopropyl + mancozeb, Bordeaux mixture, chlorothalonil, chlorothalonil + mancozeb, chlorothalonil + propamocarb hydrochloride, copper oxychloride, cyazofamid, cymoxanil, cymoxanil + famoxadone, cymoxanil + mancozeb, cymoxanil + propamocarb, dimethomorph + mancozeb, fenamidone + propamocarb hydrochloride, fluazinam, fluazinam (*off-label*), fluazinam + metalaxyl-M, fluopicolide + propamocarb hydrochloride, mancozeb, mancozeb + metalaxyl-M, mancozeb + zoxamide, mandipropamid
	Powdery scab	fluazinam (*off-label*)
	Rhizoctonia	flutolanil (*off-label*), tolclofos-methyl (*off-label - chitted seed treatment*)
	Scab	fluazinam (*off-label*)
	Seed-borne diseases	iprodione
	Silver scurf	imazalil, imazalil + pencycuron (*tuber treatment - reduction*), thiabendazole, thiabendazole (*tuber treatment - post-harvest*)
	Skin spot	imazalil, thiabendazole, thiabendazole (*tuber treatment - post-harvest*)
Pests	Aphids	acetamiprid, cypermethrin, flonicamid, lambda-cyhalothrin, lambda-cyhalothrin + pirimicarb, nicotine, oxamyl, pirimicarb, pymetrozine, thiacloprid, thiamethoxam
	Beetles	lambda-cyhalothrin
	Caterpillars	cypermethrin, lambda-cyhalothrin
	Cutworms	chlorpyrifos, cypermethrin, lambda-cyhalothrin + pirimicarb, zeta-cypermethrin
	Cyst nematodes	ethoprophos, fosthiazate, oxamyl
	Free-living nematodes	fosthiazate (*reduction*), oxamyl, peroxyacetic acid (commodity substance) (*tuber treatment*)
	Leatherjackets	methiocarb (*reduction*)
	Slugs/snails	metaldehyde, methiocarb
	Weevils	lambda-cyhalothrin
	Wireworms	ethoprophos, fosthiazate (*reduction*)
Plant growth regulation	Growth control	chlorpropham, chlorpropham (*thermal fog*), ethylene, ethylene (*in store*), maleic hydrazide
	Quality/yield control	sulphur
Weeds	Broad-leaved weeds	bentazone, carfentrazone-ethyl, clomazone, clomazone + linuron, clomazone + metribuzin, diquat, flufenacet +

		metribuzin, glufosinate-ammonium, linuron, metribuzin, pendimethalin, prosulfocarb, rimsulfuron
	Crops as weeds	carfentrazone-ethyl, cycloxydim, metribuzin, pendimethalin, quizalofop-P-tefuryl, rimsulfuron
	Grass weeds	clomazone + linuron, clomazone + metribuzin, cycloxydim, dichlobenil, diquat (*weed control and dessication*), flufenacet + metribuzin, glufosinate-ammonium, linuron, metribuzin, pendimethalin, propaquizafop, prosulfocarb, quizalofop-P-tefuryl
	Weeds, miscellaneous	dichlobenil, diquat, diquat (*weed control or dessication*), glufosinate-ammonium (*not for seed*)

Stem and bulb vegetables - Asparagus

Diseases	Botrytis	azoxystrobin + chlorothalonil (*qualified minor use recommendation*), cyprodinil + fludioxonil (*off-label*)
	Downy mildew	metalaxyl-M (*off-label*)
	Foliar diseases	boscalid + pyraclostrobin (*off-label*)
	Rust	azoxystrobin, azoxystrobin + chlorothalonil (*moderate control only*), azoxystrobin + difenoconazole (*off-label*), difenoconazole (*off-label*)
	Stemphylium	azoxystrobin, azoxystrobin + chlorothalonil (*qualified minor use recommendation*), azoxystrobin + difenoconazole (*off-label*), cyprodinil + fludioxonil (*off-label*)
Pests	Beetles	cypermethrin (*off-label*)
Weeds	Broad-leaved weeds	clomazone (*off-label*), clopyralid (*off-label*), isoxaben (*off-label*), metamitron (*off-label*), metribuzin (*off-label*), pendimethalin (*off-label*), pyridate (*off-label*), triclopyr (*off-label - directed treatment*)
	Grass weeds	fluazifop-P-butyl (*off-label*), metamitron (*off-label*), metribuzin (*off-label*), pendimethalin (*off-label*), triclopyr (*off-label - directed treatment*)
	Weeds, miscellaneous	glyphosate, glyphosate (*off-label*), metamitron (*off-label*), metribuzin (*off-label*)

Stem and bulb vegetables - Celery/chicory

Diseases	Alternaria	iprodione (*off-label*)
	Botrytis	azoxystrobin (*off-label*), cyprodinil + fludioxonil (*off-label*), iprodione (*off-label*)
	Damping off	thiram (*off-label - seed treatment*)
	Downy mildew	copper oxychloride (*off-label*), dimethomorph + mancozeb (*off-label*)
	Foliar diseases	fosetyl-aluminium (*off-label*), mancozeb (*off-label*)
	Phytophthora	difenoconazole (*off-label*), fosetyl-aluminium (*off-label - for forcing*)
	Rhizoctonia	azoxystrobin (*off-label*)
	Sclerotinia	azoxystrobin (*off-label*)
	Seed-borne diseases	thiram (*seed soak*)
	Septoria	Bordeaux mixture, chlorothalonil (*qualified minor use*), copper ammonium carbonate, copper oxychloride
Pests	Aphids	acetamiprid (*off-label*), cypermethrin (*off-label*), deltamethrin (*off-label*), pirimicarb (*off-label*), pirimicarb (*off-label - for forcing*), pymetrozine (*off-label*)
	Beetles	deltamethrin (*off-label*)

	Caterpillars	Bacillus thuringiensis (*off-label*), deltamethrin (*off-label*), diflubenzuron (*off-label*), lambda-cyhalothrin (*off-label*)
	Cutworms	lambda-cyhalothrin (*off-label - for forcing*)
	Flies	lambda-cyhalothrin (*off-label*)
	Leafhoppers	deltamethrin (*off-label*)
	Pests, miscellaneous	deltamethrin (*off-label*), lambda-cyhalothrin (*off-label*), spinosad
	Thrips	deltamethrin (*off-label*)
Plant growth regulation	Quality/yield control	gibberellins
Weeds	Broad-leaved weeds	asulam (*off-label*), linuron (*off-label*), metamitron (*off-label*), pendimethalin (*off-label*), propyzamide (*off-label*), prosulfocarb (*off-label*), triflusulfuron-methyl (*off-label*)
	Crops as weeds	propaquizafop (*off-label*)
	Grass weeds	asulam (*off-label*), fluazifop-P-butyl (*off-label*), linuron (*off-label*), metamitron (*off-label*), pendimethalin (*off-label*), propaquizafop (*off-label*), propyzamide (*off-label*), prosulfocarb (*off-label*)

Stem and bulb vegetables - Globe artichokes/cardoons

Diseases	Downy mildew	azoxystrobin (*off-label*)
	Powdery mildew	azoxystrobin (*off-label*), myclobutanil (*off-label*)
Weeds	Broad-leaved weeds	metamitron (*off-label*)
	Grass weeds	fluazifop-P-butyl (*off-label*), metamitron (*off-label*)

Stem and bulb vegetables - Onions/leeks/garlic

Diseases	Alternaria	iprodione (*off-label*)
	Bacterial blight	copper oxychloride (*off-label*)
	Botrytis	azoxystrobin + chlorothalonil, chlorothalonil, chlorothalonil (*moderate control*), chlorothalonil (*off-label*), cyprodinil + fludioxonil (*off-label*), iprodione, iprodione (*off-label*)
	Bottom rot	copper oxychloride (*off-label*)
	Collar rot	iprodione, iprodione (*off-label*)
	Damping off	thiram (*off-label - seed treatment*), thiram (*seed treatment*)
	Disease control/foliar feed	chlorothalonil (*off-label*)
	Downy mildew	azoxystrobin, azoxystrobin (*off-label*), azoxystrobin + chlorothalonil, benthiavalicarb-isopropyl + mancozeb (*off-label*), chlorothalonil + metalaxyl-M (*qualified minor use*), dimethomorph + mancozeb (*off-label*), fosetyl-aluminium (*off-label*), mancozeb (*off-label*), mancozeb + metalaxyl-M (*off-label*), metalaxyl-M (*off-label*), propiconazole (*off-label*)
	Foliar diseases	chlorothalonil (*off-label*), fosetyl-aluminium (*off-label*), mancozeb (*off-label*)
	Fusarium	thiophanate-methyl (*off-label*)
	Phytophthora	copper oxychloride (*off-label*), propamocarb hydrochloride
	Purple blotch	azoxystrobin, azoxystrobin + difenoconazole (*moderate control*), prothioconazole, tebuconazole + trifloxystrobin
	Pythium	propamocarb hydrochloride

	Rust	azoxystrobin, azoxystrobin + difenoconazole, fenpropimorph, prothioconazole, tebuconazole, tebuconazole + trifloxystrobin
	Seed-borne diseases	thiram (*off-label - Seed treatment*)
	Spear rot	copper oxychloride (*off-label*)
	Stem canker	copper oxychloride (*off-label*)
	Storage rots	copper oxychloride (*off-label*)
	White rot	boscalid + pyraclostrobin (*off-label*), tebuconazole (*off-label*)
	White tip	azoxystrobin + difenoconazole (*qualified minor use*), boscalid + pyraclostrobin (*off-label*), chlorothalonil + metalaxyl-M (*qualified minor use*), dimethomorph + mancozeb (*off-label*), mancozeb (*off-label*), tebuconazole + trifloxystrobin
Pests	Aphids	deltamethrin (*off-label*)
	Beetles	deltamethrin (*off-label*)
	Caterpillars	deltamethrin (*off-label*)
	Cutworms	Bacillus thuringiensis (*off-label*), chlorpyrifos
	Flies	tefluthrin (*off-label - seed treatment*)
	Free-living nematodes	oxamyl (*off-label*)
	Leafhoppers	deltamethrin (*off-label*)
	Pests, miscellaneous	chlorpyrifos (*off-label*), deltamethrin (*off-label*), lambda-cyhalothrin (*off-label*), oxamyl (*off-label*)
	Stem nematodes	oxamyl (*off-label*)
	Thrips	chlorpyrifos (*off-label*), deltamethrin (*off-label*), lambda-cyhalothrin (*off-label*), spinosad
Plant growth regulation	Growth control	maleic hydrazide, maleic hydrazide (*off-label*)
Weeds	Broad-leaved weeds	bentazone (*off-label*), chloridazon (*off-label*), chlorthal-dimethyl, chlorthal-dimethyl (*off-label*), chlorthal-dimethyl + propachlor, clopyralid, clopyralid (*off-label*), flumioxazine (*off-label*), fluroxypyr (*off-label*), glyphosate, ioxynil, linuron (*off-label*), metazachlor (*off-label*), pendimethalin, pendimethalin (*off-label*), pendimethalin (*off-label - pre + post emergence treatment*), propachlor, prosulfocarb (*off-label*), pyridate, pyridate (*off-label*)
	Crops as weeds	cycloxydim, fluazifop-P-butyl, flumioxazine (*off-label*), fluroxypyr (*off-label*), glyphosate, pendimethalin, propaquizafop (*off-label*), prosulfocarb (*off-label*), tepraloxydim
	Grass weeds	chloridazon (*off-label*), chlorthal-dimethyl + propachlor, cycloxydim, cycloxydim (*off-label*), fluazifop-P-butyl, fluazifop-P-butyl (*off-label*), glyphosate, linuron (*off-label*), metazachlor (*off-label*), pendimethalin, pendimethalin (*off-label*), propachlor, propaquizafop, propaquizafop (*off-label*), prosulfocarb (*off-label*), tepraloxydim, tepraloxydim (*off-label*)
	Weeds, miscellaneous	glyphosate

Stem and bulb vegetables - Vegetables

Diseases	Botrytis	Bacillus subtilis (*off-label*)

Flowers and ornamentals

Flowers - Bedding plants, general

Diseases		thiram
	Petal blight	mancozeb
	Phytophthora	propamocarb hydrochloride
	Powdery mildew	bupirimate, copper ammonium carbonate
	Pythium	propamocarb hydrochloride
	Rust	azoxystrobin (*off-label - in pots*), mancozeb, thiram
Pests	Aphids	imidacloprid, thiacloprid
	Beetles	thiacloprid
	Flies	imidacloprid
	Weevils	imidacloprid, thiacloprid
	Whiteflies	imidacloprid, thiacloprid
Plant growth regulation	Flowering control	paclobutrazol
	Growth control	chlormequat, daminozide, paclobutrazol
Weeds	Broad-leaved weeds	chlorthal-dimethyl

Flowers - Bulbs/corms

Crop control	Desiccation	sulphuric acid (commodity substance)
Diseases	Basal stem rot	chlorothalonil (*off-label*)
	Botrytis	Bacillus subtilis (*off-label*), chlorothalonil (*off-label*), tebuconazole (*off-label*), tebuconazole (*off-label - for galanthamine production*), thiram
	Crown rot	metalaxyl-M (*off-label*)
	Downy mildew	metalaxyl-M (*off-label*)
	Fire	mancozeb
	Fungus diseases	peroxyacetic acid (commodity substance) (*dip*)
	Fusarium	thiabendazole, thiabendazole (*off-label - post-lifting or pre-planting dip*), thiabendazole (*post-lifting spray*)
	Ink disease	chlorothalonil (*qualified minor use*)
	Phytophthora	propamocarb hydrochloride
	Powdery mildew	bupirimate
	Pythium	propamocarb hydrochloride
	White mould	mancozeb (*off-label*)
Pests	Flies	chlorpyrifos (*off-label*)
	Free-living nematodes	peroxyacetic acid (commodity substance) (*dip*)
Plant growth regulation	Flowering control	paclobutrazol
	Growth control	chlormequat, paclobutrazol
Weeds	Broad-leaved weeds	bentazone, carfentrazone-ethyl (*off-label*), pendimethalin (*off-label*)
	Crops as weeds	cycloxydim, pendimethalin (*off-label*)
	Grass weeds	cycloxydim, pendimethalin (*off-label*)

Flowers - Miscellaneous flowers

Diseases	Black spot	captan, kresoxim-methyl, mancozeb, myclobutanil

	Disease control/foliar feed	tebuconazole (*off-label*)
	Foliar diseases	tebuconazole (*off-label*)
	Powdery mildew	bupirimate, kresoxim-methyl, myclobutanil
	Rust	mancozeb, myclobutanil, penconazole
Pests	Aphids	cypermethrin (*off-label*), deltamethrin (*off-label*), imidacloprid, lambda-cyhalothrin + pirimicarb (*off-label*), pirimicarb (*off-label*)
	Beetles	deltamethrin (*off-label*)
	Caterpillars	deltamethrin (*off-label*)
	Cutworms	lambda-cyhalothrin + pirimicarb (*off-label*)
	Flies	imidacloprid, lambda-cyhalothrin (*off-label*)
	Leafhoppers	deltamethrin (*off-label*)
	Pests, miscellaneous	deltamethrin (*off-label*), lambda-cyhalothrin (*off-label*)
	Thrips	deltamethrin (*off-label*)
	Weevils	imidacloprid
	Whiteflies	imidacloprid
Plant growth regulation	Flowering control	paclobutrazol
	Growth control	paclobutrazol
Weeds	Broad-leaved weeds	chlorthal-dimethyl, dichlobenil, lenacil, linuron (*off-label*), metribuzin (*off-label*), pendimethalin (*off-label*), propyzamide
	Crops as weeds	propaquizafop (*off-label*)
	Grass weeds	cycloxydim (*off-label*), dichlobenil, fluazifop-P-butyl (*off-label*), lenacil, linuron (*off-label*), pendimethalin (*off-label*), propaquizafop (*off-label*), propyzamide, tepraloxydim (*off-label*)

Flowers - Pot plants

Diseases	Phytophthora	fosetyl-aluminium, propamocarb hydrochloride
	Pythium	propamocarb hydrochloride
	Root rot	fosetyl-aluminium
	Rust	mancozeb
Pests	Aphids	deltamethrin, imidacloprid, thiacloprid
	Beetles	thiacloprid
	Caterpillars	deltamethrin
	Flies	imidacloprid
	Mealybugs	deltamethrin, petroleum oil
	Scale insects	deltamethrin, petroleum oil
	Spider mites	petroleum oil
	Weevils	imidacloprid, thiacloprid
	Whiteflies	deltamethrin, imidacloprid, thiacloprid
Plant growth regulation	Flowering control	paclobutrazol
	Fruiting control	paclobutrazol
	Growth control	chlormequat, daminozide, paclobutrazol

Flowers - Protected bulbs/corms

Pests	Aphids	pirimicarb

Flowers - Protected flowers

Diseases	Black spot	kresoxim-methyl
	Crown rot	metalaxyl-M (*off-label*)
	Downy mildew	metalaxyl-M (*off-label*)
	Powdery mildew	kresoxim-methyl
	Ring spot	difenoconazole (*off-label - ground grown (or pot grown - SOLA 20050159)*), difenoconazole (*off-label - ground grown (or pot grown - SOLA 20051488)*)
	Rust	azoxystrobin (*off-label*), difenoconazole (*off-label*), difenoconazole (*off-label - ground grown (or pot grown - SOLA 20050159)*), difenoconazole (*off-label - ground grown (or pot grown - SOLA 20051488)*)
Pests	Aphids	pirimicarb, Verticillium lecanii
	Spider mites	tebufenpyrad

Miscellaneous flowers and ornamentals - Miscellaneous uses

Diseases	Fungus diseases	potassium bicarbonate (commodity substance)
	Powdery mildew	physical pest control
	Sclerotinia	Coniothyrium minitans
Pests	Aphids	physical pest control, pyrethrins, thiacloprid (*off-label*)
	Caterpillars	diflubenzuron, pyrethrins
	Mealybugs	physical pest control
	Scale insects	physical pest control
	Slugs/snails	ferric phosphate, metaldehyde, metaldehyde (*around*), metaldehyde (*do not apply to cauliflowers*), metaldehyde (*Excluding potato and cauliflower*), methiocarb (*outdoor only*)
	Spider mites	physical pest control, pyrethrins
	Suckers	physical pest control
	Thrips	pyrethrins
	Whiteflies	physical pest control
Weeds	Broad-leaved weeds	carfentrazone-ethyl (*before planting*), diquat, diquat (*around*), propyzamide
	Crops as weeds	carfentrazone-ethyl (*before planting*)
	Grass weeds	diquat (*around only*), glyphosate, propyzamide
	Weeds, miscellaneous	diquat (*around only*), glyphosate, glyphosate (*before planting*), glyphosate (*pre-sowing/planting*)

Miscellaneous flowers and ornamentals - Soils

Diseases	Soil-borne diseases	metam-sodium
Pests	Free-living nematodes	metam-sodium
	Soil pests	metam-sodium
Weeds	Weeds, miscellaneous	metam-sodium

Ornamentals - Nursery stock

Diseases	Black spot	myclobutanil
	Botrytis	Bacillus subtilis (*off-label*), boscalid + pyraclostrobin (*off-label*), chlorothalonil, chlorothalonil (*moderate control*), cyprodinil + fludioxonil, iprodione, thiram, thiram (*except Hydrangea*)

	Damping off	copper ammonium carbonate, fosetyl-aluminium + propamocarb hydrochloride (*off-label*), tolclofos-methyl
	Disease control/foliar feed	azoxystrobin (*off-label*), mandipropamid (*off-label*)
	Downy mildew	benthiavalicarb-isopropyl + mancozeb (*off-label*)
	Foliar diseases	picoxystrobin (*off-label*), pyraclostrobin (*off-label*), thiamethoxam (*off-label*), trifloxystrobin (*off-label*)
	Foot rot	thiophanate-methyl (*off-label*), tolclofos-methyl
	Fungus diseases	prochloraz
	Leaf curl	thiophanate-methyl (*off-label*)
	Phytophthora	fosetyl-aluminium, metalaxyl-M, propamocarb hydrochloride
	Powdery mildew	cyflufenamid (*off-label*), imazalil (*off-label - only ornamentals grown on synthetic/inert material*), metrafenone, myclobutanil, physical pest control, quinoxyfen (*off-label*)
	Pruning wounds	tebuconazole
	Pythium	metalaxyl-M, propamocarb hydrochloride
	Root rot	tolclofos-methyl
	Rust	myclobutanil, propiconazole (*off-label*)
	Seed-borne diseases	iprodione
	Stem canker	tebuconazole, thiophanate-methyl (*off-label*)
Pests	Aphids	acetamiprid, cypermethrin, deltamethrin, dimethoate, flonicamid (*off-label*), imidacloprid, imidacloprid (*container grown*), lambda-cyhalothrin (*off-label*), nicotine, physical pest control, pirimicarb, pymetrozine, pyrethrins, thiacloprid, thiacloprid (*off-label*), Verticillium lecanii
	Beetles	teflubenzuron (*off-label*), thiacloprid
	Capsid bugs	cypermethrin, deltamethrin
	Caterpillars	Bacillus thuringiensis, cypermethrin, deltamethrin, diflubenzuron, indoxacarb, indoxacarb (*off-label*), pyrethrins, spinosad (*off-label*), teflubenzuron, teflubenzuron (*off-label*)
	Cutworms	cypermethrin
	Flies	chlorpyrifos (*off-label*), imidacloprid, imidacloprid (*container grown*)
	Free-living nematodes	fosthiazate (*off-label*)
	Leaf miners	dimethoate, oxamyl (*off-label*), thiacloprid (*off-label*)
	Leafhoppers	nicotine, physical pest control
	Mealybugs	deltamethrin, petroleum oil, physical pest control
	Mites	physical pest control
	Pests, miscellaneous	lambda-cyhalothrin (*off-label*), methoxyfenozide (*off-label*), spinosad (*off-label*), teflubenzuron (*off-label*), thiacloprid (*off-label*)
	Scale insects	deltamethrin, petroleum oil, physical pest control
	Slugs/snails	metaldehyde
	Spider mites	abamectin, bifenthrin, dimethoate, etoxazole (*off-label*), petroleum oil, physical pest control, spirodiclofen (*off-label*), spiromesifen (*off-label*)
	Suckers	physical pest control
	Thrips	abamectin, cypermethrin, deltamethrin, deltamethrin (*off-label*), nicotine, physical pest control, spinosad,

		teflubenzuron (*off-label*), thiacloprid (*off-label*), thiamethoxam (*off-label*)
	Weevils	chlorpyrifos, imidacloprid, imidacloprid (*container grown*), thiacloprid
	Whiteflies	acetamiprid, cypermethrin, deltamethrin, imidacloprid, imidacloprid (*container grown*), nicotine, physical pest control, spiromesifen (*off-label*), teflubenzuron, thiacloprid, thiacloprid (*off-label*), Verticillium lecanii
Plant growth regulation	Growth control	1-naphthylacetic acid, 2-chloroethylphosphonic acid (*off-label*), 4-indol-3-ylbutyric acid, 4-indol-3-ylbutyric acid + 1-naphthylacetic acid, chlormequat, daminozide, gibberellins (*off-label*), indol-3-ylacetic acid, prohexadione-calcium (*off-label*)
Weeds	Broad-leaved weeds	2,4-D (*off-label*), bentazone (*off-label*), carfentrazone-ethyl (*off-label*), clopyralid, desmedipham + ethofumesate + phenmedipham (*off-label*), dichlobenil, diquat, ethofumesate (*off-label*), florasulam (*off-label*), florasulam + fluroxypyr (*off-label*), flufenacet + isoxaflutole (*off-label*), flumioxazine (*off-label*), fluroxypyr (*off-label*), glufosinate-ammonium, isoxaben, isoxaben (*off-label*), linuron (*off-label*), metazachlor, metsulfuron-methyl (*off-label*), napropamide, oxadiazon, pendimethalin (*off-label*), phenmedipham (*off-label*), propachlor, propyzamide, thifensulfuron-methyl + tribenuron-methyl (*off-label*)
	Crops as weeds	flumioxazine (*off-label*)
	Grass weeds	dichlobenil, diquat, ethofumesate (*off-label*), fluazifop-P-butyl (*off-label*), flufenacet + isoxaflutole (*off-label*), glufosinate-ammonium, linuron (*off-label*), maleic hydrazide (*off-label*), metazachlor, napropamide, oxadiazon, pendimethalin (*off-label*), propachlor, propyzamide, quizalofop-P-tefuryl (*off-label*), tepraloxydim (*off-label*)
	Mosses	fatty acids
	Weeds, miscellaneous	diquat, fatty acids, glyphosate, glyphosate (*off-label*)

Ornamentals - Trees and shrubs

Diseases	Downy mildew	benthiavalicarb-isopropyl + mancozeb (*off-label*)
	Fungus diseases	prochloraz
	Powdery mildew	cyflufenamid (*off-label*), penconazole
	Rhizoctonia	chloropicrin (*soil fumigation*)
	Scab	penconazole
Pests	Aphids	deltamethrin, imidacloprid (*off-label*)
	Beetles	teflubenzuron (*off-label*)
	Birds/mammals	aluminium ammonium sulphate, warfarin (*nuts*)
	Capsid bugs	deltamethrin
	Caterpillars	Bacillus thuringiensis, deltamethrin, diflubenzuron, teflubenzuron (*off-label*)
	Mealybugs	deltamethrin
	Pests, miscellaneous	teflubenzuron (*off-label*)
	Scale insects	deltamethrin
	Slugs/snails	ferric phosphate
	Thrips	deltamethrin, teflubenzuron (*off-label*)
	Whiteflies	deltamethrin

Weeds	Broad-leaved weeds	asulam, chlorthal-dimethyl, dichlobenil, isoxaben, metazachlor, propyzamide
	Grass weeds	cycloxydim (*off-label*), dichlobenil, metazachlor, propyzamide, quizalofop-P-tefuryl (*off-label*)
	Mosses	fatty acids
	Weeds, miscellaneous	dichlobenil, fatty acids, glyphosate, glyphosate (*wiper application*), glyphosate + sulfosulfuron
	Woody weeds/scrub	asulam, glyphosate

Ornamentals - Woody ornamentals

Diseases	Fungus diseases	prochloraz
Pests	Aphids	fatty acids
	Mealybugs	fatty acids
	Scale insects	fatty acids
	Spider mites	fatty acids
	Whiteflies	fatty acids
Weeds	Bindweeds	oxadiazon
	Broad-leaved weeds	dichlobenil, glufosinate-ammonium, lenacil, napropamide, oxadiazon, propyzamide
	Grass weeds	dichlobenil, glufosinate-ammonium, lenacil, napropamide, oxadiazon, propyzamide
	Weeds, miscellaneous	glyphosate

Forestry

Forest nurseries, general - Forest nurseries

Diseases	Botrytis	cyprodinil + fludioxonil, fenhexamid (*off-label*), iprodione (*off-label*), mepanipyrim (*off-label*)
	Disease control/foliar feed	azoxystrobin (*off-label*)
	Foliar diseases	fluoxastrobin + prothioconazole (*off-label*)
	Powdery mildew	cyflufenamid (*off-label*), metrafenone (*off-label*), proquinazid (*off-label*), quinoxyfen (*off-label*)
	Pruning wounds	tebuconazole
	Stem canker	tebuconazole
Pests	Aphids	lambda-cyhalothrin (*off-label*), pirimicarb, pymetrozine (*off-label*)
	Beetles	teflubenzuron (*off-label*)
	Birds/mammals	aluminium ammonium sulphate
	Caterpillars	indoxacarb (*off-label*), teflubenzuron (*off-label*)
	Pests, miscellaneous	lambda-cyhalothrin (*off-label*), spiroxamine (*off-label*), teflubenzuron (*off-label*), thiacloprid (*off-label*)
	Spider mites	bifenazate (*off-label*)
	Thrips	teflubenzuron (*off-label*)
	Weevils	alpha-cypermethrin (*off-label*)
Weeds	Broad-leaved weeds	2,4-D (*off-label*), carfentrazone-ethyl (*off-label*), ethofumesate (*off-label*), florasulam (*off-label*), flumioxazine (*off-label*), fluroxypyr (*off-label*), linuron (*off-label*), metazachlor, metsulfuron-methyl (*off-label*), napropamide, pendimethalin (*off-label*), pendimethalin +

		picolinafen (*off-label*), propyzamide, propyzamide (*off-label*)
	Crops as weeds	flumioxazine (*off-label*)
	Grass weeds	ethofumesate (*off-label*), linuron (*off-label*), metazachlor, napropamide, pendimethalin (*off-label*), pendimethalin + picolinafen (*off-label*), propaquizafop, propyzamide, propyzamide (*off-label*), tepraloxydim (*off-label*)
	Weeds, miscellaneous	diquat (*off-label*)

Forestry plantations, general - Forestry plantations

Crop control	Chemical stripping/ thinning	glyphosate, glyphosate (*stump treatment*)
Diseases	Pruning wounds	tebuconazole
	Stem canker	tebuconazole
Pests	Birds/mammals	aluminium ammonium sulphate, warfarin
	Caterpillars	diflubenzuron
	Weevils	alpha-cypermethrin (*off-label*), cypermethrin
Plant growth regulation	Growth control	gibberellins (*off-label*)
Weeds	Aquatic weeds	glyphosate, propyzamide
	Broad-leaved weeds	2,4-D + dicamba + triclopyr, carfentrazone-ethyl (*off-label*), clopyralid (*off-label*), glufosinate-ammonium, glyphosate, glyphosate (*wiper application*), isoxaben, metamitron (*off-label*), metazachlor, napropamide (*off-label*), pendimethalin (*off-label*), propyzamide, triclopyr
	Grass weeds	cycloxydim, glufosinate-ammonium, glyphosate, metamitron (*off-label*), metazachlor, napropamide (*off-label*), propaquizafop, propyzamide
	Weeds, miscellaneous	glufosinate-ammonium, glyphosate, metamitron (*off-label*), propyzamide
	Woody weeds/scrub	2,4-D + dicamba + triclopyr, asulam, asulam (*off-label*), glyphosate, triclopyr

Miscellaneous forestry situations - Cut logs/timber

Pests	Beetles	chlorpyrifos, cypermethrin
Weeds	Grass weeds	propaquizafop

Woodland on farms - Woodland

Diseases	Pruning wounds	tebuconazole
	Stem canker	tebuconazole
Pests	Aphids	lambda-cyhalothrin (*off-label*)
	Beetles	lambda-cyhalothrin (*off-label*)
	Birds/mammals	aluminium phosphide
	Pests, miscellaneous	lambda-cyhalothrin (*off-label*)
	Sawflies	lambda-cyhalothrin (*off-label*)
Plant growth regulation	Growth control	gibberellins (*do not exceed 2.5 g per 5 litres when used as a seed soak*), gibberellins (*off-label*)
	Plant growth regulation, miscellaneous	gibberellins (*qualified minor use*)
Weeds	Broad-leaved weeds	2,4-D (*off-label*), 2,4-D + dicamba + triclopyr, florasulam (*off-label*), flumioxazine (*off-label*), fluroxypyr (*off-label*),

		lenacil (*off-label*), metazachlor, metsulfuron-methyl (*off-label*), napropamide (*off-label*), pendimethalin (*off-label*), propyzamide
	Crops as weeds	fluazifop-P-butyl, flumioxazine (*off-label*)
	Grass weeds	cycloxydim, fluazifop-P-butyl, lenacil (*off-label*), metazachlor, napropamide (*off-label*), pendimethalin (*off-label*), propaquizafop, propyzamide
	Weeds, miscellaneous	amitrole (*off-label*), glyphosate, glyphosate (*off-label*)
	Woody weeds/scrub	2,4-D + dicamba + triclopyr

Fruit and hops

All bush fruit - All currants

Diseases	Botrytis	Bacillus subtilis (*off-label*), boscalid + pyraclostrobin (*off-label*), chlorothalonil, chlorothalonil (*qualified minor use*), cyprodinil + fludioxonil (*qualified minor use recommendation*), fenhexamid, pyrimethanil (*off-label*)
	Disease control/foliar feed	sulphur (*off-label*)
	Downy mildew	copper oxychloride (*off-label*)
	Foliar diseases	mancozeb (*off-label*)
	Leaf spot	Bordeaux mixture, chlorothalonil, chlorothalonil (*qualified minor use*), copper ammonium carbonate, dodine, dodine (*off-label*), mancozeb
	Powdery mildew	boscalid + pyraclostrobin (*off-label*), bupirimate, bupirimate (*off-label*), chlorothalonil, chlorothalonil (*qualified minor use*), fenpropimorph (*off-label*), kresoxim-methyl, kresoxim-methyl (*off-label*), myclobutanil, myclobutanil (*off-label*), penconazole, penconazole (*off-label*)
	Rust	copper oxychloride
	Septoria	boscalid + pyraclostrobin (*off-label*)
	Verticillium wilt	chloropicrin (*off-label*)
Pests	Aphids	chlorpyrifos, pirimicarb, pymetrozine (*off-label*)
	Capsid bugs	chlorpyrifos
	Caterpillars	Bacillus thuringiensis (*off-label*), chlorpyrifos, diflubenzuron, diflubenzuron (*off-label*)
	Flies	diflubenzuron (*off-label*)
	Gall, rust and leaf & bud mites	sulphur, tebufenpyrad (*off-label*)
	Midges	lambda-cyhalothrin (*off-label*)
	Pests, miscellaneous	lambda-cyhalothrin (*off-label*), thiacloprid (*off-label*)
	Sawflies	lambda-cyhalothrin (*off-label*)
	Scale insects	thiacloprid (*off-label*)
	Spider mites	bifenthrin (*off-label*), chlorpyrifos, chlorpyrifos (*non OP resistant strains only*), clofentezine (*off-label*)
Weeds	Bindweeds	oxadiazon, oxadiazon (*off-label*)
	Broad-leaved weeds	asulam, asulam (*off-label*), chlorthal-dimethyl, chlorthal-dimethyl (*off-label*), flufenacet + metribuzin (*off-label*), glufosinate-ammonium, isoxaben, isoxaben (*off-label*), lenacil, lenacil (*off-label*), MCPB, MCPB (*off-label*), napropamide, napropamide (*off-label*), oxadiazon,

		oxadiazon (*off-label*), pendimethalin, pendimethalin (*off-label*), propyzamide
	Crops as weeds	fluazifop-P-butyl, pendimethalin
	Grass weeds	dichlobenil, fluazifop-P-butyl, fluazifop-P-butyl (*off-label*), flufenacet + metribuzin (*off-label*), glufosinate-ammonium, lenacil, lenacil (*off-label*), napropamide, napropamide (*off-label*), oxadiazon, oxadiazon (*off-label*), pendimethalin, pendimethalin (*off-label*), propyzamide
	Weeds, miscellaneous	dichlobenil, glufosinate-ammonium, glyphosate (*off-label*)

All bush fruit - All protected bush fruit

Pests	Aphids	pymetrozine (*off-label*), pyrethrins
	Caterpillars	pyrethrins
	Pests, miscellaneous	thiacloprid (*off-label*)
	Spider mites	bifenthrin (*off-label*)

All bush fruit - All vines

Diseases	Botrytis	Bacillus subtilis (*off-label*), fenhexamid, pyrimethanil (*off-label*)
	Downy mildew	copper oxychloride, fosetyl-aluminium (*off-label*)
	Foliar diseases	mancozeb (*off-label*)
	Powdery mildew	fenbuconazole (*off-label*), meptyldinocap, myclobutanil (*off-label*), sulphur, tebuconazole + trifloxystrobin (*off-label*)
Pests	Caterpillars	lambda-cyhalothrin (*off-label*)
	Mealybugs	petroleum oil
	Scale insects	petroleum oil
	Spider mites	petroleum oil
Weeds	Bindweeds	oxadiazon
	Broad-leaved weeds	glufosinate-ammonium, isoxaben, oxadiazon, propyzamide (*off-label*)
	Grass weeds	glufosinate-ammonium, oxadiazon, propyzamide (*off-label*)
	Weeds, miscellaneous	glufosinate-ammonium, glyphosate (*off-label*)

All bush fruit - Bilberries/blueberries/cranberries

Diseases	Botrytis	Bacillus subtilis (*off-label*), cyprodinil + fludioxonil (*qualified minor use recommendation*), fenhexamid (*off-label*), pyrimethanil (*off-label*)
	Disease control/foliar feed	sulphur (*off-label*)
	Downy mildew	copper oxychloride (*off-label*)
	Leaf spot	dodine (*off-label*)
	Powdery mildew	fenpropimorph (*off-label*), penconazole (*off-label*)
Pests	Aphids	pirimicarb (*off-label*), pymetrozine (*off-label*)
	Caterpillars	Bacillus thuringiensis (*off-label*), diflubenzuron (*off-label*)
	Flies	diflubenzuron (*off-label*)
	Gall, rust and leaf & bud mites	tebufenpyrad (*off-label*)
	Pests, miscellaneous	thiacloprid (*off-label*)

	Scale insects	thiacloprid (*off-label*)
Weeds	Bindweeds	oxadiazon (*off-label*)
	Broad-leaved weeds	asulam (*off-label*), chlorthal-dimethyl (*off-label*), flufenacet + metribuzin (*off-label*), isoxaben (*off-label*), lenacil (*off-label*), MCPB (*off-label*), napropamide (*off-label*), oxadiazon (*off-label*), pendimethalin (*off-label*), propyzamide (*off-label*)
	Grass weeds	fluazifop-P-butyl (*off-label*), flufenacet + metribuzin (*off-label*), lenacil (*off-label*), napropamide (*off-label*), oxadiazon (*off-label*), pendimethalin (*off-label*), propyzamide (*off-label*)
	Weeds, miscellaneous	glufosinate-ammonium, glyphosate (*off-label*)

All bush fruit - Bush fruit, general

Diseases	Phytophthora	fluazinam (*off-label*)
Pests	Aphids	pyrethrins
	Birds/mammals	aluminium ammonium sulphate
	Caterpillars	pyrethrins
	Mealybugs	petroleum oil
	Scale insects	petroleum oil
	Spider mites	petroleum oil
Plant growth regulation	Quality/yield control	sulphur

All bush fruit - Gooseberries

Diseases	Botrytis	Bacillus subtilis (*off-label*), chlorothalonil, chlorothalonil (*qualified minor use*), cyprodinil + fludioxonil (*qualified minor use recommendation*), fenhexamid, pyrimethanil (*off-label*)
	Downy mildew	copper oxychloride (*off-label*)
	Leaf spot	chlorothalonil (*qualified minor use*), dodine (*off-label*), mancozeb
	Powdery mildew	bupirimate, chlorothalonil, chlorothalonil (*qualified minor use*), fenpropimorph (*off-label*), kresoxim-methyl (*off-label*), myclobutanil, penconazole (*off-label*), sulphur
Pests	Aphids	chlorpyrifos, pirimicarb, pymetrozine (*off-label*)
	Capsid bugs	chlorpyrifos
	Caterpillars	Bacillus thuringiensis (*off-label*), chlorpyrifos, diflubenzuron (*off-label*)
	Flies	diflubenzuron (*off-label*)
	Gall, rust and leaf & bud mites	tebufenpyrad (*off-label*)
	Midges	lambda-cyhalothrin (*off-label*)
	Pests, miscellaneous	lambda-cyhalothrin (*off-label*), thiacloprid (*off-label*)
	Sawflies	lambda-cyhalothrin (*off-label*)
	Scale insects	thiacloprid (*off-label*)
	Spider mites	chlorpyrifos, chlorpyrifos (*non OP resistant strains only*)
Weeds	Bindweeds	oxadiazon
	Broad-leaved weeds	asulam (*off-label*), chlorthal-dimethyl, flufenacet + metribuzin (*off-label*), isoxaben, lenacil, lenacil (*off-label*),

		MCPB, MCPB (*off-label*), napropamide, oxadiazon, pendimethalin, propyzamide
	Crops as weeds	fluazifop-P-butyl, pendimethalin
	Grass weeds	dichlobenil, fluazifop-P-butyl, flufenacet + metribuzin (*off-label*), lenacil, lenacil (*off-label*), napropamide, oxadiazon, pendimethalin, propyzamide
	Weeds, miscellaneous	dichlobenil, glyphosate (*off-label*)

All bush fruit - Miscellaneous bush fruit

Diseases	Alternaria	iprodione (*off-label*)
	Botrytis	cyprodinil + fludioxonil (*off-label*), fenhexamid, iprodione (*off-label*)
	Downy mildew	metalaxyl-M (*off-label*)
	Foliar diseases	fosetyl-aluminium (*off-label*), mancozeb (*off-label*)
	Powdery mildew	sulphur
Pests	Capsid bugs	lambda-cyhalothrin (*off-label*)
	Caterpillars	indoxacarb (*off-label*)
	Wasps	lambda-cyhalothrin (*off-label*)
Weeds	Grass weeds	fluazifop-P-butyl (*off-label*)
	Weeds, miscellaneous	glyphosate (*off-label*)

Cane fruit - All outdoor cane fruit

Crop control	Desiccation	carfentrazone-ethyl (*off-label*)
Diseases	Botrytis	Bacillus subtilis (*off-label*), fenhexamid, iprodione, pyrimethanil (*off-label*)
	Botrytis fruit rot	cyprodinil + fludioxonil
	Cane blight	tebuconazole (*off-label*)
	Cane spot	Bordeaux mixture, copper ammonium carbonate, copper oxychloride, copper oxychloride (*off-label*)
	Downy mildew	metalaxyl-M (*off-label*)
	Foliar diseases	boscalid + pyraclostrobin (*off-label*)
	Phytophthora	fluazinam (*off-label*), tebuconazole (*off-label*)
	Powdery mildew	azoxystrobin (*off-label*), bupirimate (*outdoor only*), fenpropimorph (*off-label*), myclobutanil
	Purple blotch	copper oxychloride
	Root rot	fluazinam (*off-label*)
	Spur blight	Bordeaux mixture
	Verticillium wilt	chloropicrin (*off-label*)
Pests	Aphids	chlorpyrifos, chlorpyrifos (*off-label*), deltamethrin (*off-label*), pirimicarb, pirimicarb (*off-label*), pymetrozine (*off-label*), pyrethrins
	Beetles	chlorpyrifos, deltamethrin
	Birds/mammals	aluminium ammonium sulphate
	Capsid bugs	lambda-cyhalothrin (*off-label*), thiacloprid (*off-label*)
	Caterpillars	Bacillus thuringiensis, Bacillus thuringiensis (*off-label*), diflubenzuron (*off-label*), pyrethrins
	Mealybugs	petroleum oil
	Midges	chlorpyrifos

	Pests, miscellaneous	chlorpyrifos (*off-label*), deltamethrin (*off-label*), lambda-cyhalothrin (*off-label*), spinosad (*off-label*), thiacloprid (*off-label*)
	Scale insects	petroleum oil
	Spider mites	bifenthrin (*off-label*), chlorpyrifos, chlorpyrifos (*non OP resistant strains only*), chlorpyrifos (*off-label*), clofentezine (*off-label*), petroleum oil
	Thrips	spinosad (*off-label*)
	Weevils	bifenthrin (*off-label*), lambda-cyhalothrin (*off-label*)
Plant growth regulation	Quality/yield control	sulphur
Weeds	Bindweeds	oxadiazon, oxadiazon (*off-label*)
	Broad-leaved weeds	asulam (*off-label*), chlorthal-dimethyl, chlorthal-dimethyl (*off-label*), glufosinate-ammonium, isoxaben, isoxaben (*off-label*), lenacil, lenacil (*off-label*), MCPB, MCPB (*off-label*), napropamide, napropamide (*off-label*), oxadiazon, oxadiazon (*off-label*), pendimethalin, propyzamide, propyzamide (*England only*)
	Crops as weeds	fluazifop-P-butyl, pendimethalin
	Grass weeds	dichlobenil, fluazifop-P-butyl, fluazifop-P-butyl (*off-label*), glufosinate-ammonium, lenacil, lenacil (*off-label*), napropamide, napropamide (*off-label*), oxadiazon, oxadiazon (*off-label*), pendimethalin, propyzamide, propyzamide (*England only*)
	Weeds, miscellaneous	dichlobenil, glufosinate-ammonium

Cane fruit - All protected cane fruit

Crop control	Desiccation	carfentrazone-ethyl (*off-label*)
Diseases	Botrytis	pyrimethanil (*off-label*)
	Phytophthora	fluazinam (*off-label*)
	Powdery mildew	azoxystrobin (*off-label*), myclobutanil (*off-label*)
	Root rot	fluazinam (*off-label*)
	Verticillium wilt	chloropicrin (*off-label*)
Pests	Aphids	pymetrozine (*off-label*), pyrethrins
	Beetles	thiacloprid (*off-label*)
	Capsid bugs	thiacloprid (*off-label*)
	Caterpillars	pyrethrins
	Gall, rust and leaf & bud mites	abamectin (*off-label*)
	Pests, miscellaneous	thiacloprid (*off-label*)
	Spider mites	abamectin (*off-label*), bifenthrin (*off-label*), clofentezine (*off-label*)
	Weevils	bifenthrin (*off-label*)

Hops, general - Hops

Crop control	Chemical stripping/ thinning	diquat
Diseases	Botrytis	Bacillus subtilis (*off-label*), fenhexamid (*off-label*), iprodione (*off-label*)
	Disease control	boscalid + pyraclostrobin (*off-label*), cyazofamid (*off-label*), thiophanate-methyl (*off-label - do not harvest for*

		human or animal consumption (including idling) within 12 months of treatment.)
	Downy mildew	Bordeaux mixture, chlorothalonil, copper oxychloride, fosetyl-aluminium, metalaxyl-M (*off-label*)
	Powdery mildew	bupirimate, cyflufenamid (*off-label*), fenpropimorph (*off-label*), kresoxim-methyl (*off-label*), myclobutanil (*off-label*), penconazole, quinoxyfen (*off-label*), sulphur
Pests	Aphids	acetamiprid (*off-label*), bifenthrin, cypermethrin, deltamethrin, flonicamid (*off-label*), imidacloprid, lambda-cyhalothrin (*off-label*), pymetrozine (*off-label*), tebufenpyrad, thiamethoxam (*off-label*)
	Caterpillars	indoxacarb (*off-label*), spinosad (*off-label*)
	Free-living nematodes	fosthiazate (*off-label*)
	Mealybugs	petroleum oil
	Pests, miscellaneous	lambda-cyhalothrin (*off-label*), methoxyfenozide (*off-label*), spinosad (*off-label*), thiacloprid (*off-label*)
	Scale insects	petroleum oil
	Spider mites	abamectin (*off-label*), bifenthrin, petroleum oil, spirodiclofen (*off-label*), tebufenpyrad
Plant growth regulation	Growth control	diquat, prohexadione-calcium (*off-label*)
Weeds	Bindweeds	oxadiazon
	Broad-leaved weeds	asulam, bentazone (*off-label*), flufenacet + isoxaflutole (*off-label*), flumioxazine (*off-label*), isoxaben, oxadiazon, pendimethalin, pendimethalin (*off-label*), propyzamide (*off-label*)
	Crops as weeds	fluazifop-P-butyl, flumioxazine (*off-label*), pendimethalin
	Grass weeds	diquat, fluazifop-P-butyl, flufenacet + isoxaflutole (*off-label*), maleic hydrazide (*off-label*), oxadiazon, pendimethalin, pendimethalin (*off-label*), propyzamide (*off-label*), tepraloxydim (*off-label*)
	Weeds, miscellaneous	diquat, glyphosate (*off-label*)

Miscellaneous fruit situations - Fruit crops, general

Diseases	Botrytis	Bacillus subtilis (*off-label*), fenhexamid (*off-label*), iprodione (*off-label*)
	Disease control	azoxystrobin (*off-label*), boscalid + pyraclostrobin (*off-label*), thiophanate-methyl (*off-label - any fruit harvested within 12 months of treatment must be destroyed.*)
	Downy mildew	metalaxyl-M (*off-label*)
	Powdery mildew	cyflufenamid (*off-label*), imazalil (*off-label*), kresoxim-methyl (*off-label*), proquinazid (*off-label*)
Pests	Aphids	acetamiprid (*off-label*), fatty acids, lambda-cyhalothrin (*off-label*), physical pest control, pymetrozine (*off-label*), thiamethoxam (*off-label*)
	Caterpillars	indoxacarb (*off-label*), spinosad (*off-label*)
	Free-living nematodes	fosthiazate (*off-label*)
	Leafhoppers	physical pest control
	Mealybugs	fatty acids, petroleum oil
	Mites	physical pest control
	Pests, miscellaneous	lambda-cyhalothrin (*off-label*), methoxyfenozide (*off-label*), spinosad (*off-label*), thiacloprid (*off-label*)
	Scale insects	fatty acids, petroleum oil

	Spider mites	fatty acids, petroleum oil, spirodiclofen (*off-label*)
	Thrips	physical pest control
	Whiteflies	fatty acids, physical pest control
Plant growth regulation	Fruiting control	ethylene (*in store*)
	Growth control	2-chloroethylphosphonic acid (*off-label*), prohexadione-calcium (*off-label*)
Weeds	Broad-leaved weeds	carfentrazone-ethyl (*off-label*), flufenacet + isoxaflutole (*off-label*), flumioxazine (*off-label*), pendimethalin (*off-label*), propyzamide (*off-label*)
	Crops as weeds	flumioxazine (*off-label*)
	Grass weeds	fluazifop-P-butyl (*off-label*), flufenacet + isoxaflutole (*off-label*), maleic hydrazide (*off-label*), pendimethalin (*off-label*), propyzamide (*off-label*), tepraloxydim (*off-label*)
	Weeds, miscellaneous	glyphosate (*off-label*)

Miscellaneous fruit situations - Fruit nursery stock

Weeds	Broad-leaved weeds	metazachlor
	Grass weeds	metazachlor

Other fruit - All outdoor strawberries

Diseases	Black spot	azoxystrobin (*off-label*), boscalid + pyraclostrobin, cyprodinil + fludioxonil (*qualified minor use*)
	Botrytis	Bacillus subtilis (*off-label*), boscalid + pyraclostrobin, captan, chlorothalonil, chlorothalonil (*moderate control*), chlorothalonil (*outdoor crops only*), cyprodinil + fludioxonil, fenhexamid, iprodione, mepanipyrim, pyrimethanil, thiram
	Crown rot	chloropicrin (*soil fumigation*), fosetyl-aluminium (*off-label*)
	Foliar diseases	fosetyl-aluminium (*off-label*)
	Leaf spot	chlorothalonil (*off-label*)
	Powdery mildew	boscalid + pyraclostrobin, bupirimate, fenpropimorph (*off-label*), kresoxim-methyl, myclobutanil, penconazole (*off-label*), quinoxyfen (*off-label*), sulphur
	Red core	chloropicrin (*soil fumigation*), fosetyl-aluminium, fosetyl-aluminium (*off-label*)
	Verticillium wilt	chloropicrin (*soil fumigation*), thiophanate-methyl (*off-label*)
Pests	Aphids	chlorpyrifos, pirimicarb, pymetrozine (*off-label*)
	Beetles	methiocarb
	Birds/mammals	aluminium ammonium sulphate
	Capsid bugs	thiacloprid (*off-label*)
	Caterpillars	Bacillus thuringiensis, chlorpyrifos
	Free-living nematodes	chloropicrin (*soil fumigation*)
	Slugs/snails	methiocarb
	Spider mites	abamectin (*off-label*), bifenthrin, chlorpyrifos, chlorpyrifos (*non OP resistant strains only*), clofentezine (*off-label*), spirodiclofen (*off-label*), tebufenpyrad
	Tarsonemid mites	abamectin (*off-label*), abamectin (*off-label - in propagation*), fenpyroximate (*off-label*)
	Weevils	chlorpyrifos, chlorpyrifos (*non-protected crops only*)

Weeds	Broad-leaved weeds	asulam (*off-label*), chlorthal-dimethyl, clopyralid, glufosinate-ammonium, isoxaben, lenacil, metamitron (*off-label*), napropamide, pendimethalin, phenmedipham, propyzamide
	Crops as weeds	cycloxydim, fluazifop-P-butyl, pendimethalin
	Grass weeds	cycloxydim, fluazifop-P-butyl, glufosinate-ammonium, lenacil, metamitron (*off-label*), napropamide, pendimethalin, propyzamide
	Weeds, miscellaneous	glufosinate-ammonium

Other fruit - Protected miscellaneous fruit

Diseases	Black spot	azoxystrobin (*off-label*), boscalid + pyraclostrobin, cyprodinil + fludioxonil (*qualified minor use*)
	Botrytis	Bacillus subtilis (*reduction of damage to fruit*), boscalid + pyraclostrobin, cyprodinil + fludioxonil, iprodione, iprodione (*off-label*), mepanipyrim, thiram
	Botrytis fruit rot	Bacillus subtilis (*reduction of damage to fruit*)
	Crown rot	fosetyl-aluminium (*off-label*)
	Disease control	boscalid + pyraclostrobin (*off-label*)
	Foliar diseases	fosetyl-aluminium (*off-label*)
	Powdery mildew	boscalid + pyraclostrobin, bupirimate, imazalil (*off-label*), kresoxim-methyl, penconazole (*off-label*), quinoxyfen (*off-label*)
	Red core	fosetyl-aluminium (*off-label*)
Pests	Aphids	acetamiprid (*off-label*), deltamethrin (*off-label*), physical pest control, pymetrozine (*off-label*)
	Beetles	deltamethrin (*off-label*)
	Capsid bugs	thiacloprid (*off-label*)
	Caterpillars	Bacillus thuringiensis (*off-label*), deltamethrin (*off-label*), indoxacarb (*off-label*)
	Leafhoppers	deltamethrin (*off-label*), physical pest control
	Mites	physical pest control
	Pests, miscellaneous	deltamethrin (*off-label*)
	Spider mites	abamectin (*off-label*), bifenazate, bifenazate (*off-label*), clofentezine (*off-label*), etoxazole (*off-label*), spirodiclofen (*off-label*), spiromesifen (*off-label - on inert media or by NFT*)
	Tarsonemid mites	abamectin (*off-label*), fenpyroximate (*off-label*)
	Thrips	deltamethrin (*off-label*), physical pest control
	Whiteflies	physical pest control, spiromesifen (*off-label - on inert media or by NFT*)
Plant growth regulation	Quality/yield control	gibberellins

Other fruit - Rhubarb

Diseases	Bacterial blight	copper oxychloride (*off-label*)
	Damping off	thiram (*off-label - seed treatment*)
	Downy mildew	mancozeb + metalaxyl-M (*off-label*)
	Foliar diseases	copper oxychloride (*off-label*)
	Phytophthora	difenoconazole (*off-label*)

Pests	Aphids	deltamethrin (*off-label*), pirimicarb (*off-label*)
	Beetles	deltamethrin (*off-label*)
	Caterpillars	Bacillus thuringiensis (*off-label*), deltamethrin (*off-label*)
	Leafhoppers	deltamethrin (*off-label*)
	Pests, miscellaneous	deltamethrin (*off-label*)
	Thrips	deltamethrin (*off-label*)
Plant growth regulation	Quality/yield control	gibberellins
Weeds	Broad-leaved weeds	metamitron (*off-label*), pendimethalin (*off-label*), propyzamide, propyzamide (*outdoor*)
	Crops as weeds	propaquizafop (*off-label*)
	Grass weeds	metamitron (*off-label*), pendimethalin (*off-label*), propaquizafop (*off-label*), propyzamide, propyzamide (*outdoor*)
	Weeds, miscellaneous	glyphosate (*off-label*)

Tree fruit - All nuts

Diseases	Bacterial blight	copper oxychloride (*off-label*)
	Bacterial canker	copper oxychloride (*off-label*)
	Bottom rot	copper oxychloride (*off-label*)
	Phytophthora	copper oxychloride (*off-label*)
	Pruning wounds	tebuconazole
	Spear rot	copper oxychloride (*off-label*)
	Stem canker	copper oxychloride (*off-label*), tebuconazole, tebuconazole (*off-label*)
Pests	Aphids	lambda-cyhalothrin (*off-label*)
	Caterpillars	diflubenzuron (*off-label*), lambda-cyhalothrin (*off-label*)
	Flies	diflubenzuron (*off-label*)
	Mites	tebufenpyrad (*off-label*), thiacloprid (*off-label*)
	Pests, miscellaneous	lambda-cyhalothrin (*off-label*)
Weeds	Bindweeds	oxadiazon (*off-label*)
	Broad-leaved weeds	2,4-D + dichlorprop-P + MCPA + mecoprop-P (*off-label*), asulam (*off-label*), dicamba + MCPA + mecoprop-P (*off-label*), dicamba + MCPA + mecoprop-P (*off-label - undersown with rotational grass*), glufosinate-ammonium, isoxaben (*off-label*), metazachlor, oxadiazon (*off-label*), pendimethalin (*off-label*), propyzamide (*off-label*)
	Grass weeds	fluazifop-P-butyl (*off-label*), glufosinate-ammonium, metazachlor, oxadiazon (*off-label*), pendimethalin (*off-label*), propyzamide (*off-label*)
	Weeds, miscellaneous	glufosinate-ammonium, glyphosate (*off-label*)

Tree fruit - All pome fruit

Crop control	Sucker/shoot control	glyphosate
Diseases	Alternaria	cyprodinil + fludioxonil
	Botrytis	iprodione (*off-label*), pyrimethanil (*off-label*), thiram
	Botrytis fruit rot	cyprodinil + fludioxonil, thiram
	Collar rot	copper oxychloride (*off-label*), fosetyl-aluminium

	Crown rot	fosetyl-aluminium
	Fusarium	cyprodinil + fludioxonil
	Leaf curl	copper oxychloride (*off-label*)
	Penicillium rot	cyprodinil + fludioxonil
	Phytophthora	mancozeb + metalaxyl-M (*off-label - applied to orchard floor*)
	Powdery mildew	boscalid + pyraclostrobin, bupirimate, dithianon + pyraclostrobin, fenbuconazole (*reduction*), kresoxim-methyl (*off-label*), kresoxim-methyl (*reduction*), meptyldinocap (*off-label*), myclobutanil, myclobutanil (*off-label*), penconazole, sulphur, sulphur (*off-label*)
	Pruning wounds	tebuconazole
	Scab	Bordeaux mixture, boscalid + pyraclostrobin, captan, captan (*off-label*), dithianon, dithianon (*off-label*), dithianon + pyraclostrobin, dodine, dodine (*off-label*), fenbuconazole, kresoxim-methyl, mancozeb, myclobutanil, pyrimethanil, pyrimethanil (*off-label*), sulphur, sulphur (*off-label*), thiram
	Stem canker	Bordeaux mixture, copper oxychloride, tebuconazole, tebuconazole (*off-label*), thiophanate-methyl (*off-label*)
	Storage rots	captan, cyprodinil + fludioxonil, thiophanate-methyl (*off-label*), thiram
Pests	Aphids	acetamiprid, chlorpyrifos, chlorpyrifos (*off-label*), cypermethrin, deltamethrin, flonicamid, pirimicarb, thiacloprid, thiamethoxam
	Capsid bugs	chlorpyrifos, chlorpyrifos (*off-label*), cypermethrin, deltamethrin
	Caterpillars	Bacillus thuringiensis (*off-label*), chlorpyrifos, chlorpyrifos (*off-label*), Cydia pomonella GV, cypermethrin, deltamethrin, diflubenzuron, diflubenzuron (*off-label*), fenoxycarb, indoxacarb, methoxyfenozide, spinosad
	Flies	diflubenzuron (*off-label*)
	Gall, rust and leaf & bud mites	diflubenzuron, spirodiclofen
	Midges	thiacloprid (*off-label*)
	Pests, miscellaneous	chlorpyrifos (*off-label*), thiacloprid (*off-label*)
	Sawflies	chlorpyrifos, cypermethrin, deltamethrin
	Scale insects	spirodiclofen
	Spider mites	bifenthrin, chlorpyrifos, chlorpyrifos (*non OP resistant strains only*), clofentezine, fenpyroximate, spirodiclofen, tebufenpyrad
	Suckers	abamectin (*effective against larval stages but little activity against adults*), chlorpyrifos, cypermethrin, deltamethrin, diflubenzuron, lambda-cyhalothrin, spirodiclofen, thiamethoxam
	Weevils	chlorpyrifos
	Whiteflies	acetamiprid
Plant growth regulation	Fruiting control	1-methylcyclopropene (*post-harvest use*), gibberellins, paclobutrazol, paclobutrazol (*off-label*)
	Growth control	gibberellins, gibberellins (*off-label*), paclobutrazol, paclobutrazol (*off-label*), prohexadione-calcium
	Quality/yield control	gibberellins, gibberellins (*off-label*)
Weeds	Bindweeds	oxadiazon

	Broad-leaved weeds	2,4-D, 2,4-D + dichlorprop-P + MCPA + mecoprop-P, amitrole, asulam, asulam (*off-label*), clopyralid (*off-label*), dicamba + MCPA + mecoprop-P, fluroxypyr (*off-label*), glufosinate-ammonium, isoxaben, metazachlor, oxadiazon, pendimethalin, pendimethalin (*off-label*), propyzamide, propyzamide (*off-label*)
	Crops as weeds	pendimethalin
	Grass weeds	amitrole, dichlobenil, fluazifop-P-butyl (*off-label*), glufosinate-ammonium, glyphosate, metazachlor, oxadiazon, pendimethalin, pendimethalin (*off-label*), propyzamide, propyzamide (*off-label*)
	Weeds, miscellaneous	amitrole, amitrole (*off-label*), dichlobenil, glufosinate-ammonium, glyphosate, glyphosate (*off-label*)

Tree fruit - All stone fruit

Crop control	Sucker/shoot control	glyphosate
Diseases	Bacterial canker	Bordeaux mixture, copper oxychloride
	Blossom wilt	boscalid + pyraclostrobin (*off-label*), cyprodinil + fludioxonil (*off-label*), fenbuconazole (*off-label*), myclobutanil (*off-label*)
	Botrytis	fenhexamid (*off-label*)
	Leaf curl	Bordeaux mixture, copper ammonium carbonate, copper oxychloride, copper oxychloride (*off-label*)
	Powdery mildew	myclobutanil (*off-label*)
	Pruning wounds	tebuconazole
	Rust	myclobutanil (*off-label*)
	Sclerotinia	fenbuconazole (*off-label*), myclobutanil (*off-label*)
	Stem canker	tebuconazole
Pests	Aphids	acetamiprid, chlorpyrifos, chlorpyrifos (*Non OP resistant strains only*), cypermethrin, deltamethrin, pirimicarb, pirimicarb (*off-label*), tebufenpyrad (*off-label*), thiacloprid (*off-label*), thiacloprid (*off-label - under temporary rain covers*)
	Capsid bugs	chlorpyrifos
	Caterpillars	Bacillus thuringiensis (*off-label*), chlorpyrifos, cypermethrin, deltamethrin, diflubenzuron, indoxacarb (*off-label*)
	Gall, rust and leaf & bud mites	diflubenzuron
	Pests, miscellaneous	thiacloprid (*off-label - under temporary protective rain covers*)
	Sawflies	deltamethrin
	Spider mites	chlorpyrifos, chlorpyrifos (*non OP resistant strains only*), clofentezine, fenpyroximate (*off-label*)
	Suckers	chlorpyrifos
	Weevils	chlorpyrifos
	Whiteflies	acetamiprid
Plant growth regulation	Growth control	paclobutrazol (*off-label*)
	Quality/yield control	gibberellins (*off-label*)
Weeds	Broad-leaved weeds	asulam, asulam (*off-label*), glufosinate-ammonium, isoxaben, metazachlor, pendimethalin, propyzamide, propyzamide (*off-label*)

	Crops as weeds	pendimethalin
	Grass weeds	fluazifop-P-butyl (*off-label*), glufosinate-ammonium, metazachlor, pendimethalin, propyzamide, propyzamide (*off-label*)
	Weeds, miscellaneous	amitrole (*off-label*), glufosinate-ammonium, glyphosate, glyphosate (*off-label*)

Tree fruit - Miscellaneous top fruit

Diseases	Rhizoctonia	chloropicrin (*soil fumigation*)
Pests	Birds/mammals	aluminium ammonium sulphate
Plant growth regulation	Quality/yield control	sulphur

Grain/crop store uses

Stored seed - Stored grain/rape/linseed

Pests	Birds/mammals	aluminium ammonium sulphate
	Food/grain storage pests	chlorpyrifos-methyl, d-phenothrin + tetramethrin, pirimiphos-methyl
	Mites	physical pest control
	Pests, miscellaneous	aluminium phosphide, magnesium phosphide

Grass

Turf/amenity grass - Amenity grassland

Diseases	Anthracnose	azoxystrobin, chlorothalonil + fludioxonil + propiconazole (*reduction*), propiconazole (*qualified minor use*), tebuconazole + trifloxystrobin
	Brown patch	azoxystrobin, chlorothalonil + fludioxonil + propiconazole, iprodione, propiconazole (*qualified minor use*)
	Crown rust	azoxystrobin
	Dollar spot	chlorothalonil + fludioxonil + propiconazole, iprodione, propiconazole, tebuconazole + trifloxystrobin
	Fairy rings	azoxystrobin
	Foliar diseases	fosetyl-aluminium (*off-label*)
	Fusarium	azoxystrobin, chlorothalonil + fludioxonil + propiconazole, iprodione, propiconazole, tebuconazole + trifloxystrobin, trifloxystrobin
	Melting out	azoxystrobin, iprodione
	Red thread	iprodione, tebuconazole + trifloxystrobin, trifloxystrobin
	Snow mould	iprodione
	Take-all patch	azoxystrobin
Pests	Flies	chlorpyrifos
	Leatherjackets	chlorpyrifos
	Slugs/snails	metaldehyde
Plant growth regulation	Growth control	trinexapac-ethyl
Weeds	Broad-leaved weeds	2,4-D, 2,4-D + dicamba, 2,4-D + dicamba + dichlorprop-P, 2,4-D + dicamba + triclopyr, asulam, asulam (*not fine*

		turf), clopyralid + florasulam + fluroxypyr, clopyralid + triclopyr, dicamba + MCPA + mecoprop-P, ethofumesate, florasulam + fluroxypyr, MCPA, mecoprop-P
	Crops as weeds	florasulam + fluroxypyr
	Grass weeds	ethofumesate, maleic hydrazide (*off-label*)
	Mosses	carfentrazone-ethyl + mecoprop-P, fatty acids, ferrous sulphate
	Weeds, miscellaneous	carfentrazone-ethyl + mecoprop-P, fatty acids, florasulam + fluroxypyr, glyphosate, glyphosate (*wiper application*)
	Woody weeds/scrub	clopyralid + triclopyr

Turf/amenity grass - Managed amenity turf

Crop control	Miscellaneous non-selective situations	glufosinate-ammonium
Diseases	Anthracnose	azoxystrobin, chlorothalonil, chlorothalonil + fludioxonil + propiconazole (*reduction*), propiconazole (*qualified minor use*), tebuconazole + trifloxystrobin
	Brown patch	azoxystrobin, chlorothalonil + fludioxonil + propiconazole, iprodione, propiconazole (*qualified minor use*)
	Crown rust	azoxystrobin
	Disease control/foliar feed	iprodione
	Dollar spot	carbendazim, chlorothalonil, chlorothalonil + fludioxonil + propiconazole, iprodione, propiconazole, pyraclostrobin (*reduction*), pyraclostrobin (*useful reduction*), tebuconazole + trifloxystrobin
	Fairy rings	azoxystrobin
	Foliar diseases	fosetyl-aluminium (*off-label*)
	Fusarium	azoxystrobin, carbendazim, chlorothalonil, chlorothalonil + fludioxonil + propiconazole, iprodione, myclobutanil, prochloraz + tebuconazole, propiconazole, pyraclostrobin, pyraclostrobin (*moderate control*), pyraclostrobin (*moderate control only*), tebuconazole + trifloxystrobin, trifloxystrobin
	Melting out	azoxystrobin, iprodione
	Pythium	fosetyl-aluminium (*off-label*)
	Red thread	chlorothalonil, iprodione, pyraclostrobin, tebuconazole + trifloxystrobin, trifloxystrobin
	Snow mould	iprodione
	Take-all patch	azoxystrobin
Pests	Birds/mammals	aluminium phosphide
	Chafers	imidacloprid
	Earthworms	carbendazim
	Flies	chlorpyrifos
	Leatherjackets	chlorpyrifos, imidacloprid
	Slugs/snails	metaldehyde
Plant growth regulation	Growth control	trinexapac-ethyl
	Quality/yield control	sulphur
Weeds	Broad-leaved weeds	2,4-D, 2,4-D + dicamba + fluroxypyr, 2,4-D + florasulam, chlorthal-dimethyl, clopyralid + 2,4-D + MCPA,

		clopyralid + diflufenican + MCPA, clopyralid + florasulam + fluroxypyr, clopyralid + fluroxypyr + MCPA, dicamba + dichlorprop-P + MCPA, dicamba + MCPA + mecoprop-P, dichlorprop-P + ferrous sulphate + MCPA, dichlorprop-P + MCPA, ethofumesate (*off-label*), ferrous sulphate + MCPA + mecoprop-P, florasulam + fluroxypyr, MCPA, MCPA + mecoprop-P, mecoprop-P
	Crops as weeds	2,4-D + dicamba + fluroxypyr, 2,4-D + florasulam, dichlorprop-P + MCPA, florasulam + fluroxypyr
	Grass weeds	amitrole (*off-label - on amitrole resistant turf*), ethofumesate (*off-label*), tepraloxydim (*off-label*)
	Mosses	carfentrazone-ethyl + mecoprop-P, dichlorprop-P + ferrous sulphate + MCPA, fatty acids, ferrous sulphate, ferrous sulphate + MCPA + mecoprop-P
	Weeds, miscellaneous	2,4-D + dicamba, carfentrazone-ethyl + mecoprop-P, fatty acids, florasulam + fluroxypyr, glyphosate (*pre-establishment*), glyphosate (*pre-establishment only*)

Non-crop pest control

Farm buildings/yards - Farm buildings

Pests	Birds/mammals	aluminium ammonium sulphate, brodifacoum (*indoor use only*), bromadiolone, chlorophacinone, difenacoum, warfarin
	Flies	diflubenzuron, d-phenothrin + tetramethrin, tetramethrin
	Food/grain storage pests	alpha-cypermethrin
	Pests, miscellaneous	alpha-cypermethrin, cypermethrin, pyrethrins
	Wasps	d-phenothrin + tetramethrin
Weeds	Weeds, miscellaneous	glyphosate

Farm buildings/yards - Farmyards

Pests	Birds/mammals	bromadiolone, difenacoum

Farmland pest control - Farmland situations

Pests		aluminium phosphide

Miscellaneous non-crop pest control - Manure/rubbish

Pests	Flies	diflubenzuron

Miscellaneous non-crop pest control - Miscellaneous pest control situations

Pests	Birds/mammals	carbon dioxide (commodity substance), paraffin oil (commodity substance) (*egg treatment*), strychnine hydrochloride (commodity substance) (*areas of restricted public access*)

Protected salad and vegetable crops

Protected brassicas - Protected brassica vegetables

Diseases	Bacterial blight	copper oxychloride (*off-label*)
	Black rot	copper oxychloride (*off-label*)
	Bottom rot	copper oxychloride (*off-label*)
	Damping off	fosetyl-aluminium + propamocarb hydrochloride (*off-label*)
	Downy mildew	fosetyl-aluminium (*off-label*)
	Phytophthora	copper oxychloride (*off-label*)
	Spear rot	copper oxychloride (*off-label*)
	Stem canker	copper oxychloride (*off-label*)
Pests	Aphids	fatty acids (*off-label - baby leaf production*), pyrethrins
	Beetles	deltamethrin (*off-label*), pyrethrins
	Caterpillars	Bacillus thuringiensis (*off-label*), deltamethrin (*off-label*), pyrethrins
	Pests, miscellaneous	dimethoate (*off-label*), fatty acids (*off-label*)
	Whiteflies	pyrethrins

Protected brassicas - Protected salad brassicas

Diseases	Bacterial blight	copper oxychloride (*off-label*)
	Bottom rot	copper oxychloride (*off-label*)
	Damping off	fosetyl-aluminium + propamocarb hydrochloride (*off-label - baby leaf production*)
	Downy mildew	copper oxychloride (*off-label*), fosetyl-aluminium (*off-label*), fosetyl-aluminium (*off-label - for baby leaf production*), fosetyl-aluminium + propamocarb hydrochloride (*off-label - baby leaf production*)
	Foliar diseases	fosetyl-aluminium (*off-label*)
	Phytophthora	copper oxychloride (*off-label*)
	Sclerotinia	boscalid + pyraclostrobin (*off-label*)
	Spear rot	copper oxychloride (*off-label*)
	Stem canker	copper oxychloride (*off-label*)
Pests	Aphids	acetamiprid (*off-label*), deltamethrin (*off-label*), lambda-cyhalothrin (*off-label*), lambda-cyhalothrin + pirimicarb (*off-label - for baby leaf production*), pirimicarb (*off-label - for baby leaf production*), pymetrozine (*off-label - for baby leaf production*), thiacloprid (*off-label - for baby leaf production*)
	Beetles	deltamethrin (*off-label*)
	Caterpillars	Bacillus thuringiensis (*off-label*), deltamethrin (*off-label*)
	Leafhoppers	deltamethrin (*off-label*)
	Pests, miscellaneous	deltamethrin (*off-label*), dimethoate (*off-label*), lambda-cyhalothrin (*off-label*)
	Thrips	deltamethrin (*off-label*)
	Whiteflies	Verticillium lecanii (*off-label*)
Weeds	Broad-leaved weeds	pendimethalin (*off-label*)
	Grass weeds	pendimethalin (*off-label*)

Protected crops, general - All protected crops

Diseases	Disease control	boscalid + pyraclostrobin (*off-label - seed crops only*)
	Fungus diseases	potassium bicarbonate (commodity substance)
	Sclerotinia	Coniothyrium minitans
	Soil-borne diseases	dazomet (*soil fumigation*)
Pests	Aphids	pyrethrins
	Beetles	pyrethrins
	Caterpillars	pyrethrins
	Free-living nematodes	dazomet (*soil fumigation*)
	Mealybugs	pyrethrins
	Scale insects	pyrethrins
	Slugs/snails	ferric phosphate, metaldehyde
	Soil pests	dazomet (*soil fumigation*)
	Spider mites	pyrethrins
	Thrips	pyrethrins
	Whiteflies	pyrethrins
Weeds	Weeds, miscellaneous	dazomet (*soil fumigation*)

Protected crops, general - Glasshouses

Diseases	Phytophthora	fosetyl-aluminium
	Root rot	fosetyl-aluminium
	Soil-borne diseases	metam-sodium
Pests	Free-living nematodes	metam-sodium
	Pests, miscellaneous	pyrethrins
	Soil pests	metam-sodium
Weeds	Weeds, miscellaneous	metam-sodium

Protected crops, general - Protected vegetables, general

Diseases	Botrytis	propamocarb hydrochloride (*off-label*)
	Damping off	propamocarb hydrochloride (*off-label*)
	Downy mildew	propamocarb hydrochloride (*off-label*)
	Phytophthora	propamocarb hydrochloride (*off-label*)
	White blister	propamocarb hydrochloride (*off-label*)
Pests	Aphids	physical pest control, pirimicarb (*off-label*)
	Leafhoppers	physical pest control
	Mites	physical pest control
	Thrips	physical pest control
	Whiteflies	physical pest control

Protected crops, general - Soils and compost

Diseases	Root rot	metam-sodium
	Sclerotinia	metam-sodium
	Soil-borne diseases	metam-sodium
Pests	Cyst nematodes	metam-sodium
	Free-living nematodes	metam-sodium
	Millipedes	metam-sodium

	Pests, miscellaneous	dimethoate (*off-label - for watercress progagation*), dimethoate (*off-label - for watercress propagation*)
	Root-knot nematodes	metam-sodium
	Soil pests	metam-sodium
	Symphylids	metam-sodium
	Wireworms	metam-sodium
Weeds	Weeds, miscellaneous	metam-sodium

Protected fruiting vegetables - Protected aubergines

Diseases	Botrytis	cyprodinil + fludioxonil (*off-label*), fenhexamid (*off-label*), iprodione (*off-label*), thiophanate-methyl (*off-label*)
	Damping off	fosetyl-aluminium + propamocarb hydrochloride (*off-label*)
	Powdery mildew	myclobutanil (*off-label*), sulphur (*off-label*)
	Sclerotinia	thiophanate-methyl (*off-label*)
Pests	Aphids	acetamiprid, deltamethrin (*off-label*), fatty acids (*off-label*), pirimicarb (*off-label*), pymetrozine (*off-label*)
	Caterpillars	Bacillus thuringiensis (*off-label*), indoxacarb, teflubenzuron (*off-label*)
	Leaf miners	thiacloprid (*off-label*)
	Mites	etoxazole
	Pests, miscellaneous	deltamethrin (*off-label*), fatty acids (*off-label*), teflubenzuron (*off-label*)
	Spider mites	abamectin (*off-label*), spiromesifen (*off-label - on inert media or by NFT*)
	Thrips	abamectin (*off-label*), deltamethrin (*off-label*), spinosad, teflubenzuron (*off-label*), thiacloprid (*off-label*)
	Whiteflies	acetamiprid, spiromesifen (*off-label - on inert media or by NFT*), teflubenzuron (*off-label*), thiacloprid (*off-label*)

Protected fruiting vegetables - Protected cucurbits

Diseases	Botrytis	chlorothalonil, cyprodinil + fludioxonil (*off-label*), fenhexamid (*off-label*), thiophanate-methyl (*off-label*)
	Damping off	fosetyl-aluminium + propamocarb hydrochloride (*off-label*), propamocarb hydrochloride (*off-label*)
	Downy mildew	azoxystrobin (*off-label*), metalaxyl-M (*off-label*)
	Fusarium	cyprodinil + fludioxonil (*off-label*)
	Mycosphaerella	cyprodinil + fludioxonil (*off-label*)
	Phytophthora	propamocarb hydrochloride, propamocarb hydrochloride (*off-label*)
	Powdery mildew	azoxystrobin (*off-label*), bupirimate, imazalil, myclobutanil (*off-label*), sulphur (*off-label*)
	Pythium	fosetyl-aluminium + propamocarb hydrochloride, propamocarb hydrochloride
	Sclerotinia	thiophanate-methyl (*off-label*)
Pests	Aphids	acetamiprid, deltamethrin, deltamethrin (*off-label*), fatty acids (*off-label*), nicotine, pirimicarb (*off-label*), pymetrozine, pymetrozine (*off-label*), Verticillium lecanii, Verticillium lecanii (*off-label*)
	Caterpillars	Bacillus thuringiensis, Bacillus thuringiensis (*off-label*), deltamethrin, diflubenzuron (*off-label*), indoxacarb, teflubenzuron (*off-label*)

	Leaf miners	oxamyl (*off-label*), thiacloprid (*off-label*)
	Leafhoppers	nicotine
	Mealybugs	deltamethrin, petroleum oil
	Pests, miscellaneous	deltamethrin (*off-label*), fatty acids (*off-label*), teflubenzuron (*off-label*)
	Scale insects	deltamethrin, petroleum oil
	Spider mites	abamectin, abamectin (*off-label*), petroleum oil, spiromesifen (*off-label - on inert media or by NFT*)
	Thrips	abamectin, deltamethrin (*off-label*), nicotine, spinosad, teflubenzuron (*off-label*), thiacloprid (*off-label*)
	Whiteflies	acetamiprid, cypermethrin, deltamethrin, nicotine, spiromesifen (*off-label - on inert media or by NFT*), teflubenzuron (*off-label*), thiacloprid (*off-label*), Verticillium lecanii, Verticillium lecanii (*off-label*)
Weeds	Broad-leaved weeds	isoxaben (*off-label*)

Protected fruiting vegetables - Protected tomatoes

Diseases	Botrytis	azoxystrobin (*off-label*), chlorothalonil, cyprodinil + fludioxonil (*off-label*), fenhexamid (*off-label*), iprodione, pyrimethanil (*off-label*)
	Damping off	copper oxychloride, propamocarb hydrochloride (*off-label*)
	Didymella stem rot	azoxystrobin (*off-label*)
	Foot rot	copper oxychloride
	Leaf mould	chlorothalonil, copper ammonium carbonate
	Phytophthora	azoxystrobin (*off-label*), chlorothalonil, copper oxychloride, propamocarb hydrochloride, propamocarb hydrochloride (*off-label*)
	Powdery mildew	azoxystrobin (*off-label*), bupirimate (*off-label*), imazalil (*off-label - only tomatoes grown on synthetic/inert material*), myclobutanil (*off-label*), sulphur (*off-label*)
	Pythium	fosetyl-aluminium + propamocarb hydrochloride, propamocarb hydrochloride
	Verticillium wilt	thiophanate-methyl (*off-label*)
Pests	Aphids	acetamiprid, deltamethrin, fatty acids, nicotine, pirimicarb, pyrethrins, Verticillium lecanii
	Beetles	pyrethrins
	Caterpillars	Bacillus thuringiensis, deltamethrin, indoxacarb, pyrethrins, teflubenzuron (*off-label*)
	Leaf miners	abamectin, abamectin (*off-label*), oxamyl (*off-label*), thiacloprid (*off-label*)
	Leafhoppers	nicotine
	Mealybugs	deltamethrin, fatty acids, petroleum oil, pyrethrins (*off-label*)
	Mirid bugs	pyrethrins (*off-label*)
	Mites	etoxazole
	Pests, miscellaneous	teflubenzuron (*off-label*)
	Scale insects	deltamethrin, fatty acids, petroleum oil
	Spider mites	abamectin, fatty acids, petroleum oil
	Thrips	deltamethrin (*off-label*), nicotine, spinosad, teflubenzuron (*off-label*), thiacloprid (*off-label*)
	Whiteflies	acetamiprid, deltamethrin, fatty acids, nicotine, pymetrozine (*off-label*), pyrethrins, spiromesifen,

		teflubenzuron (*off-label*), thiacloprid (*off-label*), Verticillium lecanii
Plant growth regulation	Growth control	2-chloroethylphosphonic acid (*off-label*)

Protected herb crops - Protected herbs

Diseases	Alternaria	iprodione (*off-label*)
	Botrytis	cyprodinil + fludioxonil (*off-label*), fenhexamid (*off-label*), iprodione (*off-label*), propamocarb hydrochloride (*off-label*), pyrimethanil (*off-label*)
	Damping off	fosetyl-aluminium + propamocarb hydrochloride (*off-label*), propamocarb hydrochloride (*off-label*)
	Disease control/foliar feed	prochloraz (*off-label*)
	Downy mildew	copper oxychloride (*off-label*), fosetyl-aluminium (*off-label*), fosetyl-aluminium + propamocarb hydrochloride (*off-label*), metalaxyl-M (*off-label*), propamocarb hydrochloride (*off-label*)
	Foliar diseases	mancozeb (*off-label*)
	Fungus diseases	mancozeb + metalaxyl-M (*off-label*)
	Powdery mildew	sulphur (*off-label*)
	Rhizoctonia	azoxystrobin (*off-label*)
	Sclerotinia	boscalid + pyraclostrobin (*off-label*)
	Seed-borne diseases	fosetyl-aluminium (*off-label*)
Pests	Aphids	acetamiprid (*off-label*), deltamethrin (*off-label*), fatty acids (*off-label*), lambda-cyhalothrin + pirimicarb (*off-label*), nicotine, pirimicarb (*off-label*), pymetrozine (*off-label*), thiacloprid (*off-label*), Verticillium lecanii (*off-label*)
	Beetles	deltamethrin (*off-label*)
	Caterpillars	Bacillus thuringiensis (*off-label*), deltamethrin (*off-label*)
	Leafhoppers	deltamethrin (*off-label*), nicotine
	Pests, miscellaneous	abamectin (*off-label*), fatty acids (*off-label*), spinosad (*off-label*)
	Thrips	deltamethrin (*off-label*), nicotine, pirimicarb (*off-label*)
	Whiteflies	nicotine, Verticillium lecanii (*off-label*)
Weeds	Broad-leaved weeds	pendimethalin (*off-label*), propyzamide (*off-label*)
	Grass weeds	pendimethalin (*off-label*), propyzamide (*off-label*)

Protected leafy vegetables - Mustard and cress

Diseases	Botrytis	cyprodinil + fludioxonil (*off-label*), propamocarb hydrochloride (*off-label*)
	Damping off	thiram (*off-label - seed treatment*)
	Downy mildew	copper oxychloride (*off-label*), dimethomorph + mancozeb (*off-label*), propamocarb hydrochloride (*off-label*)
Pests	Pests, miscellaneous	spinosad
Weeds	Broad-leaved weeds	pendimethalin (*off-label*), propyzamide (*off-label*)
	Grass weeds	pendimethalin (*off-label*), propyzamide (*off-label*)

Protected leafy vegetables - Protected leafy vegetables

Diseases	Botrytis	cyprodinil + fludioxonil (*off-label*), iprodione, propamocarb hydrochloride

	Bottom rot	boscalid + pyraclostrobin, tolclofos-methyl
	Damping off	propamocarb hydrochloride (*off-label*)
	Downy mildew	copper oxychloride (*off-label*), fosetyl-aluminium, fosetyl-aluminium (*off-label*), fosetyl-aluminium + propamocarb hydrochloride, mancozeb, propamocarb hydrochloride
	Foliar diseases	fosetyl-aluminium (*off-label*)
	Fungus diseases	mancozeb + metalaxyl-M (*off-label*)
	Pythium	fosetyl-aluminium + propamocarb hydrochloride
	Rhizoctonia	azoxystrobin (*off-label*)
	Sclerotinia	boscalid + pyraclostrobin
Pests	Aphids	acetamiprid (*off-label*), deltamethrin (*off-label*), fatty acids (*off-label*), lambda-cyhalothrin (*off-label*), lambda-cyhalothrin + pirimicarb (*off-label*), nicotine, pirimicarb, pirimicarb (*off-label*), pymetrozine (*off-label*), pyrethrins, thiacloprid (*off-label*), Verticillium lecanii
	Beetles	deltamethrin (*off-label*)
	Caterpillars	Bacillus thuringiensis (*off-label*), deltamethrin (*off-label*), pyrethrins
	Leafhoppers	deltamethrin (*off-label*), nicotine
	Pests, miscellaneous	abamectin (*off-label*), fatty acids (*off-label*), lambda-cyhalothrin (*off-label*)
	Thrips	deltamethrin (*off-label*), nicotine, pirimicarb (*off-label*)
	Whiteflies	cypermethrin, nicotine, Verticillium lecanii
Weeds	Broad-leaved weeds	pendimethalin (*off-label*), propyzamide (*off-label*)
	Grass weeds	pendimethalin (*off-label*), propyzamide (*off-label*)

Protected leafy vegetables - Protected spinach

Diseases	Downy mildew	copper oxychloride (*off-label*), metalaxyl-M (*off-label*)
	Foliar diseases	fosetyl-aluminium (*off-label*)
Pests	Aphids	deltamethrin (*off-label*), fatty acids (*off-label*), pirimicarb (*off-label*)
	Beetles	deltamethrin (*off-label*)
	Caterpillars	Bacillus thuringiensis (*off-label*), deltamethrin (*off-label*)
	Leafhoppers	deltamethrin (*off-label*)
	Pests, miscellaneous	dimethoate (*off-label*), fatty acids (*off-label*)
	Thrips	deltamethrin (*off-label*)

Protected legumes - Protected peas and beans

Pests	Aphids	Verticillium lecanii, Verticillium lecanii (*off-label*)
	Leaf miners	oxamyl (*off-label*)
	Whiteflies	Verticillium lecanii, Verticillium lecanii (*off-label*)

Protected root and tuber vegetables - Protected carrots/parsnips/celeriac

Pests	Pests, miscellaneous	dimethoate (*off-label*)
Weeds	Broad-leaved weeds	isoxaben (*off-label - temporary protection*), metribuzin (*off-label*)

Protected root and tuber vegetables - Protected potatoes

Diseases	Phytophthora	benalaxyl + mancozeb
Pests	Aphids	nicotine
	Leafhoppers	nicotine
	Thrips	nicotine
	Whiteflies	nicotine

Protected root and tuber vegetables - Protected root brassicas

Diseases	Botrytis	propamocarb hydrochloride (*off-label*)
	Damping off	propamocarb hydrochloride (*off-label*)
	Downy mildew	propamocarb hydrochloride (*off-label*)
	Phytophthora	propamocarb hydrochloride (*off-label*)
	Rhizoctonia	tolclofos-methyl (*off-label*)
	White blister	mancozeb + metalaxyl-M (*off-label*), propamocarb hydrochloride (*off-label*)
Pests	Aphids	pirimicarb (*off-label*)

Protected stem and bulb vegetables - Protected celery/chicory

Diseases	Botrytis	azoxystrobin (*off-label*), cyprodinil + fludioxonil (*off-label*)
	Downy mildew	copper oxychloride (*off-label*), fosetyl-aluminium + propamocarb hydrochloride (*off-label*)
	Foliar diseases	fosetyl-aluminium (*off-label*)
	Phytophthora	azoxystrobin (*off-label - for forcing*)
	Rhizoctonia	azoxystrobin (*off-label*), tolclofos-methyl (*off-label*)
	Sclerotinia	azoxystrobin (*off-label*)
Pests	Aphids	acetamiprid (*off-label*), deltamethrin (*off-label*), fatty acids (*off-label*), lambda-cyhalothrin + pirimicarb (*off-label*), pirimicarb (*off-label*), pymetrozine (*off-label*)
	Beetles	deltamethrin (*off-label*)
	Caterpillars	Bacillus thuringiensis (*off-label*), deltamethrin (*off-label*)
	Leafhoppers	deltamethrin (*off-label*)
	Pests, miscellaneous	abamectin (*off-label*), deltamethrin (*off-label*), fatty acids (*off-label*)
	Thrips	deltamethrin (*off-label*)

Protected stem and bulb vegetables - Protected onions/leeks/ garlic

Pests	Aphids	deltamethrin (*off-label*), nicotine
	Beetles	deltamethrin (*off-label*)
	Caterpillars	deltamethrin (*off-label*)
	Leafhoppers	deltamethrin (*off-label*), nicotine
	Pests, miscellaneous	deltamethrin (*off-label*)
	Stem nematodes	oxamyl (*off-label*)
	Thrips	deltamethrin (*off-label*), nicotine
	Whiteflies	nicotine

Total vegetation control

Aquatic situations, general - Aquatic situations

Weeds	Aquatic weeds	2,4-D, dichlobenil, glyphosate
	Grass weeds	dichlobenil, glyphosate
	Weeds, miscellaneous	dichlobenil, glyphosate

Non-crop areas, general - Miscellaneous non-crop situations

Weeds	Broad-leaved weeds	asulam, MCPA, triclopyr
	Weeds, miscellaneous	flazasulfuron, glyphosate
	Woody weeds/scrub	glyphosate, triclopyr

Non-crop areas, general - Non-crop farm areas

Weeds	Broad-leaved weeds	2,4-D + dicamba + triclopyr, dicamba + MCPA + mecoprop-P, glufosinate-ammonium, MCPA, metsulfuron-methyl, picloram, triclopyr
	Grass weeds	glufosinate-ammonium, glyphosate
	Weeds, miscellaneous	glufosinate-ammonium, glyphosate, picloram, sodium chlorate
	Woody weeds/scrub	asulam, glyphosate, picloram, triclopyr

Non-crop areas, general - Paths/roads etc

Pests	Slugs/snails	metaldehyde
Weeds	Broad-leaved weeds	2,4-D + dicamba + triclopyr, MCPA
	Grass weeds	dichlobenil, glyphosate, quizalofop-P-tefuryl (*off-label*)
	Mosses	fatty acids
	Weeds, miscellaneous	acetic acid, acetic acid (*Foliage kill only. Regrowth may occur from roots.*), dichlobenil, diflufenican + glyphosate, fatty acids, flazasulfuron, glyphosate, glyphosate + sulfosulfuron, sodium chlorate
	Woody weeds/scrub	2,4-D + dicamba + triclopyr, glyphosate

SECTION 2
PESTICIDE PROFILES

1 abamectin

A selective acaricide and insecticide for use in ornamentals and other protected crops
IRAC mode of action code: 6

Products

1	Agrimec	Syngenta	18 g/l	EC	14491
2	Clayton Abba	Clayton	18 g/l	EC	13808
3	Dynamec	Syngenta Bioline	18 g/l	EC	13331

Uses

- Insect control in ***protected chives*** *(off-label)*, ***protected cress*** *(off-label)*, ***protected frise*** *(off-label)*, ***protected herbs (see appendix 6)*** *(off-label)*, ***protected lamb's lettuce*** *(off-label)*, ***protected lettuce*** *(off-label)*, ***protected parsley*** *(off-label)*, ***protected radicchio*** *(off-label)*, ***protected scarole*** *(off-label)* [3]
- Leaf and bud mite in ***protected blackberries*** *(off-label)*, ***protected raspberries*** *(off-label)* [3]
- Leaf miner in ***protected cherry tomatoes*** *(off-label)* [3]; ***protected tomatoes*** [2, 3]
- Pear sucker in ***pears*** *(effective against larval stages but little activity against adults)* [1]
- Red spider mites in ***hops*** *(off-label)* [1]; ***protected blackberries*** *(off-label)*, ***protected cayenne peppers*** *(off-label)*, ***protected peppers*** *(off-label)*, ***protected raspberries*** *(off-label)* [3]; ***protected strawberries*** *(off-label)* [2, 3]; ***strawberries*** *(off-label)* [2]
- Spider mites in ***hops*** *(off-label)* [1]
- Tarsonemid mites in ***protected strawberries*** *(off-label)*, ***strawberries*** *(off-label)* [2]; ***strawberries*** *(off-label - in propagation)* [3]
- Two-spotted spider mite in ***ornamental plant production***, ***protected cucumbers***, ***protected ornamentals***, ***protected tomatoes*** [2, 3]; ***protected aubergines*** *(off-label)* [3]
- Western flower thrips in ***ornamental plant production***, ***protected cucumbers***, ***protected ornamentals*** [2, 3]; ***protected aubergines*** *(off-label)* [3]

Specific Off-Label Approvals (SOLAs)

- ***hops*** *20091581* [1]
- ***protected aubergines*** *20070421* [3]
- ***protected blackberries*** *20072290* [3]
- ***protected cayenne peppers*** *20070422* [3]
- ***protected cherry tomatoes*** *20070422* [3]
- ***protected chives*** *20070430* [3]
- ***protected cress*** *20070430* [3]
- ***protected frise*** *20070430* [3]
- ***protected herbs (see appendix 6)*** *20070430* [3]
- ***protected lamb's lettuce*** *20070430* [3]
- ***protected lettuce*** *20070430* [3]
- ***protected parsley*** *20070430* [3]
- ***protected peppers*** *20070422* [3]
- ***protected radicchio*** *20070430* [3]
- ***protected raspberries*** *20072290* [3]
- ***protected scarole*** *20070430* [3]
- ***protected strawberries*** *20090342* [2], *20070423* [3]
- ***strawberries*** *20090342* [2], *(in propagation) 20070423* [3]

Approval information

- Abamectin included in Annex I under EC Directive 91/414

Efficacy guidance

- Abamectin controls adults and immature stages of the two-spotted spidier mite, the larval stages of leaf miners and the nymphs of Western Flower Thrips, plus a useful reduction in adults
- Treat at first sign of infestation. Repeat sprays may be required
- For effective control total cover of all plant surfaces is essential, but avoid run-off
- Target pests quickly become immobilised but 3-5 d may be required for maximum mortality
- Indoor applications should be made through a hydraulic nozzle applicator or a knapsack applicator. Outdoors suitable high volume hydraulic nozzle applicators should be used
- Limited data shows abamectin only slightly harmful to Anthocorid bugs and so compatible with biological control systems in which Anthocorid bugs are important [1]

SEE SECTION 3 FOR PRODUCTS ALSO REGISTERED

Restrictions

- Number of treatments 6 on protected tomatoes and cucumbers (only 4 of which can be made when flowers or fruit present); not restricted on flowers but rotation with other products advised
- Maximum concentration must not exceed 50 ml per 100 l water
- Do not mix with wetters, stickers or other adjuvants
- Do not use on ferns (*Adiantum* spp) or Shasta daisies
- Do not treat protected crops which are in flower or have set fruit between 1 Nov and 28 Feb
- For use on all varieties of pears [1]
- Do not treat cherry tomatoes (but protected cherry tomatoes may be treated off-label)
- Consult manufacturer for list of plant varieties tested for safety
- There is insufficient evidence to support product compatibility with integrated and biological pest control programmes
- Unprotected persons must be kept out of treated areas until the spray has dried

Crop-specific information

- HI 3 d for protected edible crops
- On tomato or cucumber crops that are in flower, or have started to set fruit, treat only between 1 Mar and 31 Oct. Seedling tomatoes or cucumbers that have not started to flower or set fruit may be treated at any time
- Some spotting or staining may occur on carnation, kalanchoe and begonia foliage

Environmental safety

- Dangerous for the environment
- Very toxic to aquatic organisms
- High risk to bees. Do not apply to crops in flower or to those in which bees are actively foraging. Do not apply when flowering weeds are present
- Keep in original container, tightly closed, in a safe place, under lock and key
- Where bumble bees are used in tomatoes as pollinators, keep them out for 24 h after treatment

Hazard classification and safety precautions

Hazard Harmful, Dangerous for the environment
Transport code 9
Packaging group III
UN number 3082
Risk phrases R22a, R50, R53a
Operator protection A, C, D, H, K, M; U02a, U05a [1-3]; U20a [2, 3]; U20c [1]
Environmental protection E02a [2, 3] (until spray has dried); E12a, E15b, E34, E38 [1-3]; E12e [2, 3]; E12f, E17b, E22c [1]
Storage and disposal D01, D02, D05, D09b, D10c, D12a
Medical advice M04a

2 acetamiprid

A neonicotinoid insecticide for use in top fruit and horticulture
IRAC mode of action code: 4A

Products

1	Gazelle SG	Certis	20% w/w	SG	13725
2	Insyst	Certis	20% w/w	SP	13414

Uses

- Aphids in ***apples***, ***cherries***, ***frise*** *(off-label)*, ***hops*** *(off-label)*, ***lamb's lettuce*** *(off-label)*, ***lettuce*** *(off-label)*, ***ornamental plant production***, ***parsley*** *(off-label)*, ***pears***, ***plums***, ***protected aubergines***, ***protected frise*** *(off-label)*, ***protected hops*** *(off-label)*, ***protected lamb's lettuce*** *(off-label)*, ***protected lettuce*** *(off-label)*, ***protected ornamentals***, ***protected parsley*** *(off-label)*, ***protected peppers***, ***protected radicchio*** *(off-label)*, ***protected rocket*** *(off-label)*, ***protected salad brassicas*** *(off-label)*, ***protected scarole*** *(off-label)*, ***protected soft fruit*** *(off-label)*, ***protected tomatoes***, ***radicchio*** *(off-label)*, ***rocket*** *(off-label)*, ***salad brassicas*** *(off-label)*, ***scarole*** *(off-label)*, ***soft fruit*** *(off-label)*, ***top fruit*** *(off-label)* [1]; ***brussels sprouts*** *(off-label)*, ***seed potatoes***, ***spring oilseed rape***, ***ware potatoes***, ***winter oilseed rape*** [2]

- Whitefly in ***apples***, ***cherries***, ***ornamental plant production***, ***pears***, ***plums***, ***protected aubergines***, ***protected ornamentals***, ***protected peppers***, ***protected tomatoes*** [1]; ***brussels sprouts*** *(off-label)* [2]

Specific Off-Label Approvals (SOLAs)

- ***brussels sprouts*** *20072866* [2]
- ***frise*** *20082234* [1]
- ***hops*** *20082857* [1]
- ***lamb's lettuce*** *20082234* [1]
- ***lettuce*** *20082234* [1]
- ***parsley*** *20082234* [1]
- ***protected frise*** *20082234* [1]
- ***protected hops*** *20082857* [1]
- ***protected lamb's lettuce*** *20082234* [1]
- ***protected lettuce*** *20082234* [1]
- ***protected parsley*** *20082234* [1]
- ***protected radicchio*** *20082234* [1]
- ***protected rocket*** *20082234* [1]
- ***protected salad brassicas*** *20082234* [1]
- ***protected scarole*** *20082234* [1]
- ***protected soft fruit*** *20082857* [1]
- ***radicchio*** *20082234* [1]
- ***rocket*** *20082234* [1]
- ***salad brassicas*** *20082234* [1]
- ***scarole*** *20082234* [1]
- ***soft fruit*** *20082857* [1]
- ***top fruit*** *20082857* [1]

Approval information

- Acetamiprid included in Annex I under EC Directive 91/414

Efficacy guidance

- Best results obtained from application at the first sign of pest attack or when appropriate thresholds are reached
- Thorough coverage of foliage is essential to ensure best control. Acetamiprid has contact, systemic and translaminar activity

Restrictions

- Maximum number of treatments 1 per yr for cherries, ware potatoes; 2 per yr or crop for apples, pears, plums, ornamental plant production, protected crops, seed potatoes
- Do not use more than two applications of any neonicotinoid insecticide (e.g. acetamiprid, clothianidin, imidacloprid, thiacloprid) on any crop. Previous soil or seed treatment with a neonicotinoid counts as one such treatment
- 2 treatments are permitted on seed potatoes but must not be used consecutively [2]

Crop-specific information

- HI 3 d for protected crops; 14 d for apples, cherries, pears, plums, potatoes

Environmental safety

- Harmful to aquatic organisms
- LERAP Category B
- Acetamiprid is slightly toxic to predatory mites and generally slightly toxic to other beneficials

Hazard classification and safety precautions

Hazard Dangerous for the environment

Risk phrases R22a [1]; R52, R53a [1, 2]

Operator protection A [1, 2]; H [1]; U02a, U19a [1, 2]; U20a [2]; U20b [1]

Environmental protection E15b, E16a, E16b [1, 2]; E17b [1] (18 m); E38 [2]

Storage and disposal D05 [1]; D09a, D10b [1, 2]; D12a [2]

SEE SECTION 3 FOR PRODUCTS ALSO REGISTERED

3 acetic acid

A non-selective herbicide for non-crop situations

Products

1	Acetum	Unicrop	240 g/l	SL	13881
2	Natural Weed & Moss Spray No 1	Headland Amenity	240 g/l	SL	14509

Uses

- Annual and perennial weeds in ***hard surfaces*** *(Foliage kill only. Regrowth may occur from roots.)*, ***natural surfaces not intended to bear vegetation*** *(Foliage kill only. Regrowth may occur from roots.)*, ***permeable surfaces overlying soil*** *(Foliage kill only. Regrowth may occur from roots.)* [1]
- General weed control in ***hard surfaces***, ***natural surfaces not intended to bear vegetation***, ***permeable surfaces overlying soil*** [2]

Efficacy guidance

- Best results obtained from treatment of young tender weeds less than 10 cm high
- Treat in spring and repeat as necessary throughout the growing season
- Treat survivors as soon as fresh growth is seen
- Ensure complete coverage of foliage to the point of run-off
- Rainfall after treatment may reduce efficacy

Restrictions

- No restriction on number of treatments

Following crops guidance

- There is no residual activity in the soil and sowing or planting may take place as soon as treated weeds have died

Environmental safety

- Harmful to aquatic organisms
- High risk to bees
- Do not apply to crops in flower or to those in which bees are actively foraging. Do not apply when flowering weeds are present
- Keep people and animals off treated dense weed patches until spray has dried. This is not necessary for areas with occasional low growing prostrate weeds such as on pathways

Hazard classification and safety precautions

Hazard Irritant
UN number N/C [2]
Risk phrases R36, R37, R38, R52
Operator protection A, C, H [1, 2]; D [2]; U05a, U09a, U19a [1, 2]; U20b [2]; U20c [1]
Environmental protection E12a, E12e, E15a
Storage and disposal D01, D02, D09a, D12a

4 alpha-cypermethrin

A contact and ingested pyrethroid insecticide for use in arable crops and agricultural buildings
IRAC mode of action code: 3

Products

1	A-Cyper 100EC	Goldengrass	100 g/l	EC	14175
2	Antec Durakil 1.5 SC	Antec Biosentry	15 g/l	SC	H7560
3	Antec Durakil 6SC	Antec Biosentry	60 g/l	SC	H7559
4	Contest	BASF	15% w/w	WG	10216
5	Strength AC	Certis	15 g/l	SC	H7469

Uses

- Aphids in ***spring barley***, ***spring oilseed rape***, ***spring wheat***, ***winter barley***, ***winter oilseed rape***, ***winter wheat*** [1]
- Brassica pod midge in ***spring oilseed rape*** [1]; ***winter oilseed rape*** [1, 4]
- Cabbage seed weevil in ***spring oilseed rape***, ***winter oilseed rape*** [1, 4]
- Cabbage stem flea beetle in ***spring oilseed rape*** [1]; ***winter oilseed rape*** [1, 4]
- Cabbage stem weevil in ***spring oilseed rape***, ***winter oilseed rape*** [1]

- Cabbage white butterfly in ***salad brassicas*** *(off-label - for baby leaf production)* [4]
- Caterpillars in ***broccoli, brussels sprouts, cabbages, calabrese, cauliflowers, kale*** [1, 4]
- Cereal aphid in ***spring barley, spring wheat, winter barley, winter wheat*** [4]
- Diamond-back moth in ***salad brassicas*** *(off-label - for baby leaf production)* [4]
- Flea beetle in ***broccoli, brussels sprouts, cabbages, calabrese, cauliflowers, kale*** [1, 4]
- Insect pests in ***borage for oilseed production*** *(off-label)*, ***chinese cabbage*** *(off-label)*, ***choi sum*** *(off-label)*, ***collards*** *(off-label)*, ***durum wheat*** *(off-label)*, ***grass seed crops*** *(off-label)*, ***linseed*** *(off-label)*, ***lupins*** *(off-label)*, ***pak choi*** *(off-label)*, ***spring rye*** *(off-label)*, ***triticale*** *(off-label)*, ***winter rye*** *(off-label)* [4]; ***farm buildings, poultry houses*** [5]
- Lesser mealworm in ***poultry houses*** [2, 3]
- Pea and bean weevil in ***broad beans, combining peas, spring field beans, vining peas, winter field beans*** [1, 4]
- Pea aphid in ***combining peas, vining peas*** [1]; ***combining peas*** *(reduction)*, ***vining peas*** *(reduction)* [4]
- Pea moth in ***combining peas, vining peas*** [1, 4]
- Pine weevil in ***forest nurseries*** *(off-label)*, ***forestry transplants*** *(off-label)* [4]
- Pollen beetle in ***spring oilseed rape, winter oilseed rape*** [1, 4]
- Poultry red mite in ***poultry houses*** [2, 3]
- Rape winter stem weevil in ***winter oilseed rape*** [4]
- Small white butterfly in ***salad brassicas*** *(off-label - for baby leaf production)* [4]
- Yellow cereal fly in ***spring barley*** *(Some control if present at application)*, ***spring wheat*** *(Some control if present at application)*, ***winter barley*** *(Some control if present at application)*, ***winter wheat*** *(Some control if present at application)* [1]; ***winter barley, winter wheat*** [4]

Specific Off-Label Approvals (SOLAs)

- ***borage for oilseed production*** *20052268* [4]
- ***chinese cabbage*** *20052265* [4]
- ***choi sum*** *20052265* [4]
- ***collards*** *20052265* [4]
- ***durum wheat*** *20052269* [4]
- ***forest nurseries*** *20062391 expires 31 Dec 2010* [4]
- ***forestry transplants*** *20062391 expires 31 Dec 2010* [4]
- ***grass seed crops*** *20052269* [4]
- ***linseed*** *20052268* [4]
- ***lupins*** *20052267* [4]
- ***pak choi*** *20052265* [4]
- ***salad brassicas*** *(for baby leaf production) 20011221* [4]
- ***spring rye*** *20052269* [4]
- ***triticale*** *20052269* [4]
- ***winter rye*** *20052269* [4]

Approval information

- Alpha-cypermethrin included in Annex I under EC Directive 91/414
- Accepted by BBPA for use on malting barley

Efficacy guidance

- For cabbage stem flea beetle control spray oilseed rape when adult or larval damage first seen and about 1 mth later [4]
- For flowering pests on oilseed rape apply at any time during flowering, against pollen beetle best results achieved at green to yellow bud stage (GS 3.3-3.7), against seed weevil between 20 pods set and 80% petal fall (GS 4.7-5.8) [4]
- Spray cereals in autumn for control of cereal aphids, in spring/summer for grain aphids. (See label for details) [4]
- For flea beetle, caterpillar and cabbage aphid control on brassicas apply when the pest or damage first seen or as a preventive spray. Repeat if necessary [4]
- For pea and bean weevil control in peas and beans apply when pest attack first seen and repeat as necessary [4]
- For lesser mealworm control in poultry houses apply a coarse, low-pressure spray as routine treatment after clean-out and before each new crop. Spray vertical surfaces and ensure an overlap onto ceilings. It is not necessary to treat the floor [2, 3, 5]

- Use the highest recommended concentration in poultry houses where extreme residual action is required or surfaces are dirty or highly absorbent [2, 3, 5]

Restrictions

- Maximum number of treatments 4 per crop on edible brassicas, 3 per crop on winter oilseed rape, peas, 2 on beans, spring oilseed rape (only 1 after yellow bud stage - GS 3,7) [4]
- Maximum number of applications in animal husbandry use 2 when used in premises that are occupied by poultry [2, 3]
- Apply up to 2 sprays on cereals in autumn and spring, 1 in summer between 1 Apr and 31 Aug. See label for details of rates and maximum total dose [4]
- Only 1 aphicide treatment may be applied in cereals between 1 Apr and 31 Aug and spray volume must not be reduced in this period [4]
- Do not apply to a cereal crop if any product containing a pyrethroid or dimethoate has been applied after the start of ear emergence (GS 51) [4]

Crop-specific information

- Latest use: before the end of flowering for oilseed rape; before 31 Aug for cereals [4]
- HI vining peas 1 d; brassicas 7 d; combining peas, broad beans, field beans 11 d [4]
- For summer cereal application do not spray within 6 m from edge of crop and do not reduce volume when used after 31 Mar [4]

Environmental safety

- Dangerous for the environment [4]
- Very toxic to aquatic organisms [4]
- Extremely dangerous (harmful [5]) to fish or other aquatic life. Do not contaminate surface waters or ditches with chemical or used container [2, 3]
- Dangerous to bees [2, 3]
- LERAP Category A [1, 4]
- Where possible spray oilseed rape crops in the late evening or early morning or in dull weather [4]
- Do not spray within 6 m of the edge of a cereal crop after 31 Mar in yr of harvest [4]
- Do not apply directly to poultry; collect eggs before application [2, 3]

Hazard classification and safety precautions

Hazard Harmful, Dangerous for the environment [1, 4]; Flammable [1]
Transport code 3 [1]; 9 [2-4]
Packaging group III [1-4]
UN number 1993 [1]; 3077 [4]; 3082 [2, 3]
Risk phrases R20, R22b, R66 [1]; R22a, R48, R50, R53a [1, 4]
Operator protection A, H [1-5]; C [1]; U02a [1-3]; U02b, U09a [2, 3, 5]; U04a, U11, U20a [1]; U05a, U10 [1, 4]; U14 [1, 5]; U15 [5]; U19a [1-3, 5]; U20b [2-5]
Environmental protection E02a [2, 3] (until dry); E02d, E05c, E13c [5]; E05a [2, 3, 5]; E12c, E13a [2, 3]; E15a [1-4]; E16c, E16d [1, 4]; E34 [1]; E38 [4]
Consumer protection C04, C10 [5]; C05, C06, C07, C09, C11 [2, 3, 5]
Storage and disposal D01, D05 [1, 4, 5]; D02, D10b [1, 4]; D06a, D06c [5]; D09a, D12a [1-5]; D11a [2, 3]
Vertebrate/rodent control products V03b [5]
Medical advice M04a [1, 5]; M05a [4]; M05b [1]

5 aluminium ammonium sulphate

An inorganic bird and animal repellent

Products

1	Curb Crop Spray Powder	Sphere	88% w/w	WP	02480
2	Guardsman	Chiltern	83 g/l	SL	05494
3	Liquid Curb Crop Spray	Sphere	83 g/l	SC	03164

Uses

- Animal repellent in ***agricultural premises, all top fruit, broad beans, bush fruit, cane fruit, carrots, flowerhead brassicas, forest nursery beds, forestry plantations, grain stores, leaf brassicas, peas, permanent grassland, spring barley, spring field beans, spring oats, spring oilseed rape, spring wheat, strawberries, sugar beet, winter barley, winter field beans, winter oats, winter oilseed rape, winter wheat*** [1, 3]; ***amenity vegetation*** [1]

- Bird repellent in ***agricultural premises, all top fruit, broad beans, bush fruit, cane fruit, carrots, flowerhead brassicas, forest nursery beds, forestry plantations, grain stores, leaf brassicas, peas, permanent grassland, spring barley, spring field beans, spring oats, spring oilseed rape, spring wheat, strawberries, sugar beet, winter barley, winter field beans, winter oats, winter oilseed rape, winter wheat*** [1, 3]; ***amenity vegetation*** [1]
- Birds in ***all top fruit, broad beans, bush fruit, cane fruit, carrots, flowerhead brassicas, forest nursery beds, forestry plantations, leaf brassicas, peas, permanent grassland, spring field beans, spring oilseed rape, strawberries, sugar beet, winter barley, winter field beans, winter oats, winter oilseed rape, winter wheat*** [2]
- Damaging mammals in ***all top fruit, broad beans, bush fruit, cane fruit, carrots, flowerhead brassicas, forest nursery beds, forestry plantations, leaf brassicas, peas, permanent grassland, spring field beans, spring oilseed rape, strawberries, sugar beet, winter barley, winter field beans, winter oats, winter oilseed rape, winter wheat*** [2]

Efficacy guidance

- Apply as overall spray to growing crops before damage starts or mix powder with seed depending on type of protection required
- Spray deposit protects growth present at spraying but gives little protection to new growth
- Product must be sprayed onto dry foliage to be effective and must dry completely before dew or frost forms. In winter this may require some wind

Restrictions

- Maximum number of treatments: 1 per batch for corms, flower bulbs & seeds [2]

Crop-specific information

- Latest use: no restriction

Hazard classification and safety precautions

Operator protection U04b, U10 [2]; U05a [1, 3]; U20a [1-3]
Environmental protection E15a [1-3]; E19b [1, 3]
Storage and disposal D01, D02, D05, D12a [1, 3]; D09a, D10b [2]; D11a [1-3]
Treated seed S02, S05 [2]
Medical advice M03 [1, 3]; M05b [2]

6 aluminium phosphide

A phosphine generating compound used against vertebrates and grain store pests
IRAC mode of action code: 24A

Products

1	Degesch Fumigation Tablets	Rentokil	56% w/w	GE	09313
2	Phostoxin	Rentokil	56% w/w	GE	09315
3	Talunex	Certis	57% w/w	GE	13798

Uses

- Insect pests in ***stored grain*** [1]
- Moles in ***farm woodland, lawns, managed amenity turf, permanent grassland*** [3]; ***farmland*** [2, 3]
- Rabbits in ***farm woodland, lawns, managed amenity turf, permanent grassland*** [3]; ***farmland*** [2, 3]
- Rats in ***farm woodland, lawns, managed amenity turf, permanent grassland*** [3]; ***farmland*** [2, 3]

Approval information

- Accepted by BBPA for use in stores for malting barley

Efficacy guidance

- Product releases poisonous hydrogen phosphide gas in contact with moisture
- Place fumigation tablets in grain stores as directed [1]
- Place pellets in burrows or runs and seal hole by heeling in or covering with turf. Do not cover pellets with soil. Inspect daily and treat any new or re-opened holes [2, 3]
- Apply pellets by means of Luxan Topex Applicator [3]

Restrictions

- Aluminium phosphide is subject to the Poisons Rules 1982 and the Poisons Act 1972. See Section 5 for more information
- Only to be used by operators instructed or trained in the use of aluminium phosphide and familiar with the precautionary measures to be taken. See label and HSE Guidance Notes for full precautions
- Only open container outdoors [2, 3] and for immediate use. Keep away from liquid or water as this causes immediate release of gas. Do not use in wet weather
- Do not use within 3 m of human or animal habitation. Before application ensure that no humans or domestic animals are in adjacent buildings or structures. Allow a minimum airing-off period of 4 h before re-admission

Environmental safety

- Product liberates very toxic, highly flammable gas
- Dangerous for the environment [3]
- Very toxic to aquatic organisms [3]
- Dangerous to fish or other aquatic life. Do not contaminate surface waters or ditches with chemical or used container [1, 2]
- Prevent access by livestock, pets and other non-target mammals and birds to buildings under fumigation and ventilation [1]
- Pellets must never be placed or allowed to remain on ground surface
- Do not use adjacent to watercourses
- Take particular care to avoid gassing non-target animals, especially those protected under the Wildlife and Countryside Act (e.g. badgers, polecat, reptiles, natterjack toads, most birds). Do not use in burrows where there is evidence of badger or fox activity, or when burrows might be occupied by birds
- Dust remaining after decomposition is harmless and of no environmental hazard
- Keep in original container, tightly closed, in a safe place, under lock and key
- Dispose of empty containers as directed on label

Hazard classification and safety precautions

Hazard Very toxic, Highly flammable [1-3]; Harmful [2]; Dangerous for the environment [3]
Risk phrases R21, R26, R28 [1-3]; R36, R37, R50 [3]
Operator protection A, H [1-3]; D [1, 2]; G [3]; U01, U13, U19a, U20a [1-3]; U05a [2, 3]; U05b [1]; U07 [1, 2]; U14, U15 [3]; U18 [1, 3]
Environmental protection E02a [1] (4 h min); E02a [2] (4 h); E02b [1]; E13b [1, 2]; E15a [3]; E34 [1-3]
Storage and disposal D01, D02, D09b, D11b [1-3]; D07 [1, 2]
Vertebrate/rodent control products V04a [2, 3]
Medical advice M04a

7 amidosulfuron

A post-emergence sulfonylurea herbicide for cleavers and other broad-leaved weed control in cereals
HRAC mode of action code: B

Products

1	Eagle	Bayer CropScience	75% w/w	WG	07318
2	Squire Ultra	Interfarm	75% w/w	WG	14401

Uses

- Annual dicotyledons in ***durum wheat, linseed, spring barley, spring oats, spring rye, spring wheat, triticale, winter barley, winter oats, winter rye, winter wheat*** [1]; ***grassland*** [2]
- Charlock in ***durum wheat*** *(off-label)* [1]
- Cleavers in ***durum wheat, linseed, spring barley, spring oats, spring rye, spring wheat, triticale, winter barley, winter oats, winter rye, winter wheat*** [1]
- Docks in ***grassland*** [2]

Specific Off-Label Approvals (SOLAs)

- ***durum wheat*** *20061297* [1]

Approval information

- Accepted by BBPA for use on malting barley
- Amidosulfuron included in Annex I under EC Directive 91/414

Efficacy guidance

- For best results apply in spring (from 1 Feb) in warm weather when soil moist and weeds growing actively. When used in grassland following cutting or grazing, docks should be allowed to regrow before treatment
- Weed kill is slow, especially under cool, dry conditions. Weeds may sometimes only be stunted but will have little or no competitive effect on crop
- May be used on all soil types unless certain sequences are used on linseed. See label
- Spray is rainfast after 1 h
- Cleavers controlled from emergence to flower bud stage. If present at application charlock (up to flower bud), shepherds purse (up to flower bud) and field forget-me-not (up to 6 leaves) will also be controlled
- Amidosulfuron is a member of the ALS-inhibitor group of herbicides and products should be used in a planned Resistance Management strategy. See Section 5 for more information

Restrictions

- Maximum number of treatments 1 per crop [1]
- Use after 1 Feb and do not apply to rotational grass after 30 Jun, or to permanent grassland after 15 Oct [2]
- Do not apply to crops undersown or due to be undersown with clover or alfalfa [1]
- Do not spray crops under stress, suffering drought, waterlogged, grazed, lacking nutrients or if soil compacted
- Do not spray if frost expected
- Do not roll or harrow within 1 wk of spraying
- Specific restrictions apply to use in sequence or tank mixture with other sulfonylurea or ALS-inhibiting herbicides. See label for details. There are no recommendations for mixtures with metsulfuron-methyl products on linseed
- Certain mixtures with fungicides are expressly forbidden. See label for details

Crop-specific information

- Latest use: before first spikelets just visible (GS 51) for cereals; before flower buds visible for linseed; 15 Oct for grassland
- Broadcast cereal crops should be sprayed post-emergence after plants have a well established root system

Following crops guidance

- If a treated crop fails cereals may be sown after 15 d and thorough cultivation
- After normal harvest of a treated crop only cereals, winter oilseed rape, mustard, turnips, winter field beans or vetches may be sown in the same year as treatment and these must be preceded by ploughing or thorough cultivation
- Only cereals may be sown within 12 mth of application to grassland [2]
- Cereals or potatoes must be sown as the following crop after use of permitted mixtures or sequences with other sulfonylurea herbicides in cereals. Only cereals may be sown after the use of such sequences in linseed

Environmental safety

- Take care to wash out sprayers thoroughly. See label for details
- Avoid drift onto neighbouring broad-leaved plants or onto surface waters or ditches

Hazard classification and safety precautions

Hazard Dangerous for the environment [2]
Transport code 9
Packaging group III
UN number 3077
Risk phrases R50, R53a [2]
Operator protection U20a [1]; U20b [2]
Environmental protection E07b [2] (1 week); E15a [1, 2]; E38, E41 [2]
Storage and disposal D10a [1, 2]; D12a [2]

SEE SECTION 3 FOR PRODUCTS ALSO REGISTERED

8 amidosulfuron + iodosulfuron-methyl-sodium

A post-emergence sulfonylurea herbicide mixture for cereals
HRAC mode of action code: B + B

See also iodosulfuron-methyl-sodium

Products

1	Chekker	Bayer CropScience	13.4:1.3% w/w	WG	10955
2	Sekator	Bayer CropScience	5:1.25% w/w	WG	12634

Uses

- Annual dicotyledons in ***durum wheat*** *(off-label)*, ***grass seed crops*** *(off-label)* [1]; ***spring barley, spring rye, spring wheat, triticale, winter barley, winter rye, winter wheat*** [1, 2]
- Chickweed in ***spring barley, spring rye, spring wheat, triticale, winter barley, winter rye, winter wheat*** [1, 2]
- Cleavers in ***spring barley, spring rye, spring wheat, triticale, winter barley, winter rye, winter wheat*** [1, 2]
- Mayweeds in ***spring barley, spring rye, spring wheat, triticale, winter barley, winter rye, winter wheat*** [1, 2]
- Volunteer oilseed rape in ***spring barley, spring rye, spring wheat, triticale, winter barley, winter rye, winter wheat*** [1, 2]

Specific Off-Label Approvals (SOLAs)

- ***durum wheat*** *20060858* [1]
- ***grass seed crops*** *20060858* [1]

Approval information

- Amidosulfuron and iodosulfuron-methyl-sodium included in Annex I under EC Directive 91/414
- Accepted by BBPA for use on malting barley

Efficacy guidance

- Best results obtained from treatment in warm weather when soil is moist and the weeds are growing actively
- Weeds must be present at application to be controlled
- Dry conditions resulting in moisture stress may reduce effectiveness
- Weed control is slow especially under cool dry conditions
- Occasionally weeds may only be stunted but they will normally have little or no competitive effect on the crop
- Amidosulfuron and iodosulfuron are members of the ALS-inhibitor group of herbicides and products should be used in a planned Resistance Management strategy. See Section 5 for more information

Restrictions

- Maximum number of treatments 1 per crop
- Must only be applied between 1 Feb in yr of harvest and specified latest time of application
- Do not apply to crops undersown or to be undersown with grass, clover or alfalfa
- Do not roll or harrow within 1 wk of spraying
- Do not spray crops under stress from any cause or if the soil is compacted
- Do not spray if rain or frost expected
- Do not apply in mixture or in sequence with any other ALS inhibitor

Crop-specific information

- Latest use: before first spikelet of inflorescence just visible (GS 51)
- Treat drilled crops after the 2-leaf stage; treat broadcast crops after the plants have a well-established root system
- Applications to spring barley may cause transient crop yellowing

Following crops guidance

- Cereals, winter oilseed rape and winter field beans may be sown in the same yr as treatment provided they are preceded by ploughing or thorough cultivation. Any crop may be sown in the spring of the yr following treatment
- A minimum of 3 mth must elapse between treatment and sowing winter oilseed rape

Environmental safety

- Dangerous for the environment
- Toxic to aquatic organisms
- LERAP Category B
- Take extreme care to avoid damage by drift onto broad-leaved plants outside the target area or onto ponds, waterways and ditches
- Observe carefully label instructions for sprayer cleaning

Hazard classification and safety precautions

Hazard Irritant, Dangerous for the environment
Risk phrases R36, R51, R53a
Operator protection A, C, H; U05a, U08, U11, U14, U15, U20b
Environmental protection E15a, E16a, E16b, E38
Storage and disposal D01, D02, D10a, D12a

9 aminopyralid

A pyridine carboxylic acid herbicide available only in mixtures
HRAC mode of action code: O

10 aminopyralid + fluroxypyr

A foliar acting herbicide mixture for use in grassland
HRAC mode of action code: O + O

See also fluroxypyr

Products

Forefront	Dow	30:100 g/l	EO	14701

Uses

- Buttercups in ***permanent grassland, rotational grass***
- Chickweed in ***permanent grassland, rotational grass***
- Dandelions in ***permanent grassland, rotational grass***
- Docks in ***permanent grassland, rotational grass***
- Stinging nettle in ***permanent grassland, rotational grass***
- Thistles in ***permanent grassland, rotational grass***

Approval information

- Fluroxypyr included in Annex I under EC Directive 91/414

Efficacy guidance

- For best results and to avoid crop check, grass and weeds must be growing actively
- Allow 2-3 wk after cutting for hay or silage for sufficient regrowth to occur before spraying and leave 7 d afterwards to allow maximum translocation
- Where there is a high reservoir of weed seed or a historically high weed population a programmed approach may be needed involving a second treatment in the following yr
- Control may be reduced if rain falls within 1 h of spraying

Restrictions

- Maximum number of treatments 1 per yr
- Use of an antifoam is compulsory
- Do not use any treated plant material, or manure from animals fed on treated crops, for composting or mulching
- Do not apply to newly sown grass within 8 wk of sowing and preferably not until tillering
- Do not use on crops grown for seed
- Manure from animals fed on pasture or silage treated with [1] should not leave the farm

Crop-specific information

- Treatment will kill clover
- Treatment may occasionally cause transient yellowing of the sward which is quickly outgrown
- Late treatments may lead to a slight transient leaning of grass that does not affect yield

Following crops guidance

- Do not drill clover or other legumes within 4 mth of treatment, or potatoes in the spring following treatment in the previous autumn
- In the event of failure of newly seeded treated grassland, grass may be re-seeded immediately, or wheat may be sown provided 4 mth have elapsed since application
- Residues in plant tissues, including manure, may affect succeeding susceptible crops of peas, beans, other legumes, carrots, other Umbelliferae, potatoes, tomatoes, lettuce and other Compositae. These crops should not be sown within 3 mth of ploughing up treated grassland

Environmental safety

- Dangerous for the environment
- Toxic to aquatic organisms
- Keep livestock out of treated areas for at least 7 d following treatment and until poisonous weeds, such as ragwort, have died down and become unpalatable
- To protect groundwater do not apply between 1Sep and 28 Feb
- Avoid damage by drift onto susceptible crops, non-target plants or waterways

Hazard classification and safety precautions

Hazard Irritant, Dangerous for the environment
Risk phrases R38, R41, R51, R53a, R67
Operator protection A, C; U05a, U08, U11, U14, U23a
Environmental protection E07a, E15b, E38
Consumer protection C01
Storage and disposal D01, D02, D05, D09a, D10b, D12a

11 aminopyralid + triclopyr

A foliar acting herbicide mixture for broad-leaved weed control in grassland
HRAC mode of action code: O + O

See also *triclopyr*

Products

Pharaoh	Dow	30:240g/l	EW	14731

Uses

- Buttercups in ***grassland***
- Common nettle in ***grassland***
- Dandelions in ***grassland***
- Docks in ***grassland***
- Thistles in ***grassland***

Approval information

- Triclopyr included in Annex I under EC Directive 91/414

Efficacy guidance

- For best results and to avoid crop check, grass and weeds must be growing actively
- Use adeqate water volume to ensure good weed coverage. Increase water volume if necessary where the weed population is high and where the grass is dense
- Allow 2-3 wk after cutting for hay or silage for sufficient regrowth to occur before spraying and leave 7 d afterwards to allow maximum translocation
- Where there is a high reservoir of weed seed or a historically high weed population a programmed approach may be needed involving a second treatment in the following yr
- Control may be reduced if rain falls within 1 h of spraying

Restrictions

- Maximum number of treatments 1 per yr
- Do not use any treated plant material for composting or mulching
- Do not use manure or slurry from animals fed on treated grass on susceptible crops (see following crops guidance)
- Do not roll or harrow 10 d before or within 7 d after application
- Do not cut grass for 7 d after application
- Do not apply to newly sown grass within 8 wk of sowing and preferably not until tillering
- Do not use on crops grown for seed

- Do not drill clover within 4 mth of application. Clover may be re-introduced 4 mth after application provided treated plant remains have completely decayed

Crop-specific information

- Latest use: 7 d before grazing or harvest for grassland
- Late applications may lead to transient leaning of grass which does not affect final yield

Following crops guidance

- Ensure that all plant remains of a treated crop have completely decayed before planting susceptible crops such as peas, beans and other legumes, sugar beet, carrots and other Umbelliferae, potatoes and tomatoes, lettuce and other Compositae
- Do not plant potatoes, sugar beet, vegetables, beans or other leguminous crops in the calendar yr following application

Environmental safety

- Dangerous for the environment
- Toxic to aquatic organisms
- Keep livestock out of treated areas for at least 7 d after treatment or until foliage of any poisonous weeds such as ragwort has died and become unpalatable
- To protect groundwater do not apply to grass leys less than 1 yr old
- Take extreme care to avoid drift onto susceptible crops, non-target plants or waterways. All conifers, especially pine and larch, are very sensitive and may be damaged by vapour drift in hot conditions
- LERAP Category B

Hazard classification and safety precautions

Hazard Irritant, Dangerous for the environment
Risk phrases R43, R51, R53a
Operator protection A, H; U05a, U08, U14, U23a
Environmental protection E06a, E15b, E15c, E16a, E16c, E38
Consumer protection C01
Storage and disposal D01, D02, D09a, D10b, D12a

12 amisulbrom

A fungicide for use in potatoes
FRAC mode of action code: C4

Products

Shinkon	Nissan	200 g/l	SC	13722

Uses

- Late blight in ***potatoes***

Hazard classification and safety precautions

Hazard Dangerous for the environment
Risk phrases R50, R53a
Operator protection A, C; U05a, U15, U20b
Environmental protection E13b, E15a, E16a, E34, E38
Storage and disposal D01, D02, D05, D09a, D10b, D12a

13 amitrole

A translocated, foliar-acting, non-selective triazole herbicide
HRAC mode of action code: F3

Products

Weedazol-TL	Nufarm UK	225 g/l	SL	11968

Uses

- Annual and perennial weeds in ***apricots*** *(off-label)*, ***cherries*** *(off-label)*, ***peaches*** *(off-label)*, ***plums*** *(off-label)*, ***quinces*** *(off-label)*
- Annual dicotyledons in ***fallows***, ***headlands***, ***stubbles***
- Annual grasses in ***fallows***, ***headlands***, ***stubbles***

- Barren brome in ***apple orchards, pear orchards***
- Couch in ***apple orchards, fallows, headlands, pear orchards, stubbles***
- Creeping thistle in ***fallows, stubbles***
- Docks in ***fallows, headlands, stubbles***
- General weed control in ***apple orchards, farm woodland*** *(off-label)*, ***pear orchards***
- Grass weeds in ***managed amenity turf*** *(off-label - on amitrole resistant turf)*
- Perennial dicotyledons in ***apple orchards, fallows, headlands, pear orchards***
- Perennial grasses in ***apple orchards, fallows, headlands, pear orchards***
- Volunteer potatoes in ***stubbles*** *(barley stubble)*

Specific Off-Label Approvals (SOLAs)

- ***apricots*** *20051981*
- ***cherries*** *20051981*
- ***farm woodland*** *20051980*
- ***managed amenity turf*** *(on amitrole resistant turf) 20051979*
- ***peaches*** *20051981*
- ***plums*** *20051981*
- ***quinces*** *20051981*

Approval information

- Amitrole included in Annex I under EC Directive 91/414

Efficacy guidance

- In non-crop land may be applied at any time from Apr to Oct. Best results achieved in spring or early summer when weeds growing actively. For coltsfoot, hogweed and horsetail summer and autumn applications are preferred
- Uptake is via foliage and heavy rain immediately after application will reduce efficacy. Amitrole is less affected by drought than some residual herbicides and remains effective for up to 2 mth
- Applications made in summer may not give complete control of couch if past the shooting stage or not actively growing
- Effective crop competition and efficient ploughing are essential for good couch control

Restrictions

- Maximum number of treatments 1 per yr
- Keep off suckers or foliage of desirable trees or shrubs
- Do not spray areas into which the roots of adjacent trees or shrubs extend
- Application to land intended for spring barley should be in the preceding autumn, not the spring
- Do not spray on sloping ground when rain imminent and run-off may occur
- Do not spray if foliage is wet or rain imminent
- Do not use in low temperatures or in drought
- Do not mix product with acids

Crop-specific information

- Latest use: before end Jun or after harvest for apple and pear orchards; end Oct for headlands; end Oct and at least 2 wk before cultivation and drilling for stubbles and fallows

Following crops guidance

- Amitrole breaks down fairly quickly in medium and heavy soils and 3 wk should be allowed between application and sowing or planting. On sandy soils the interval should be 6 wk

Environmental safety

- Harmful to aquatic organisms
- Keep livestock out of treated areas for at least two weeks following treatment and until poisonous weeds, such as ragwort, have died down and become unpalatable
- Harmful to fish or other aquatic life. Do not contaminate surface waters or ditches with chemical or used container

Hazard classification and safety precautions

Hazard Harmful
Risk phrases R48, R52, R53a, R63
Operator protection A, C; U05a, U08, U19a, U20b
Environmental protection E07a, E15a
Storage and disposal D01, D02, D09a, D10b

14 ammonium sulphate

An inorganic salt used to condition water and to promote rapid growth in plants to increase herbicide activity. Last product approval expired in 2008

15 asulam

A translocated carbamate herbicide for control of docks and bracken
HRAC mode of action code: I

Products

1	Agrotech Asulam	AgChem Access	400 g/l	SL	12243
2	Asulox	United Phosphorus	400 g/l	SL	13175
3	Brack-N	AgriGuard	400 g/l	SL	13138
4	Formule 1	AgriChem BV	400 g/l	SL	13255
5	Greencrop Frond	Greencrop	400 g/l	SL	11912

Uses

- Annual grasses in ***celeriac*** *(off-label)*, ***chicory*** *(off-label)*, ***chicory root*** *(off-label)*, ***herbs (see appendix 6)*** *(off-label)* [2]
- Annual meadow grass in ***celeriac*** *(off-label)*, ***chicory*** *(off-label)*, ***chicory root*** *(off-label)*, ***herbs (see appendix 6)*** *(off-label)* [2]
- Bracken in ***amenity vegetation*** [4, 5]; ***forest*** [1-5]; ***forest*** *(off-label)* [2]; ***grassland*** [4]; ***non-crop areas*** [3]; ***permanent grassland*** [1-3, 5]
- Brome grasses in ***poppies for morphine production*** *(off-label)* [2]
- Docks in ***almonds*** *(off-label)*, ***chestnuts*** *(off-label)*, ***hazel nuts*** *(off-label)*, ***walnuts*** *(off-label)* [2, 5]; ***amenity grassland***, ***apples***, ***grassland***, ***pears*** [4]; ***amenity grassland*** *(not fine turf)*, ***apple orchards***, ***pear orchards***, ***permanent grassland*** [1-3, 5]; ***amenity vegetation*** [1-4]; ***blackberries*** *(off-label)*, ***blueberries*** *(off-label)*, ***celeriac*** *(off-label)*, ***chicory*** *(off-label)*, ***chicory root*** *(off-label)*, ***clover seed crops*** *(off-label)*, ***cranberries*** *(off-label)*, ***damsons*** *(off-label)*, ***gooseberries*** *(off-label)*, ***herbs (see appendix 6)*** *(off-label)*, ***loganberries*** *(off-label)*, ***mint*** *(off-label)*, ***nectarines*** *(off-label)*, ***parsley*** *(off-label)*, ***poppies for morphine production*** *(off-label)*, ***quinces*** *(off-label)*, ***raspberries*** *(off-label)*, ***redcurrants*** *(off-label)*, ***strawberries*** *(off-label)*, ***tarragon*** *(off-label)*, ***whitecurrants*** *(off-label)* [2]; ***blackcurrants***, ***cherries***, ***hops***, ***plums*** [1-5]; ***road verges***, ***waste ground*** [1-3]; ***rotational grass*** [1, 2]
- Meadow grasses in ***poppies for morphine production*** *(off-label)* [2]

Specific Off-Label Approvals (SOLAs)

- ***almonds*** *20070816* [2], *20060866* [5]
- ***blackberries*** *20070799* [2]
- ***blueberries*** *20070799* [2]
- ***celeriac*** *20082738* [2]
- ***chestnuts*** *20070816* [2], *20060866* [5]
- ***chicory*** *20082739* [2]
- ***chicory root*** *20082739* [2]
- ***clover seed crops*** *20070825* [2]
- ***cranberries*** *20070799* [2]
- ***damsons*** *20070799* [2]
- ***forest*** *20070823* [2]
- ***gooseberries*** *20070799* [2]
- ***hazel nuts*** *20070816* [2], *20060866* [5]
- ***herbs (see appendix 6)*** *20082740* [2]
- ***loganberries*** *20070799* [2]
- ***mint*** *20070822* [2]
- ***nectarines*** *20070799* [2]
- ***parsley*** *20070822* [2]
- ***poppies for morphine production*** *20070826* [2]
- ***quinces*** *20070799* [2]
- ***raspberries*** *20070799* [2]
- ***redcurrants*** *20070799* [2]
- ***strawberries*** *20070824* [2]

- ***tarragon*** *20070822* [2]
- ***walnuts*** *20070816* [2], *20060866* [5]
- ***whitecurrants*** *20070799* [2]

Approval information

- May be applied through CDA equipment
- Approved for aerial application on bracken in agricultural grassland, amenity grassland, forestry and rough upland intended for grazing [2]. See Section 5 for more information
- Approved for use near surface waters. See Section 5 for more information
- Accepted by BBPA for use on hops

Efficacy guidance

- Spray bracken when fronds fully expanded but not senescent, usually Jul-Aug; docks in full leaf before flower stem emergence
- Bracken fronds must not be damaged by stock, frost or cutting before treatment
- Uptake and reliability of bracken control may be improved by use of specified additives - see label. Additives not recommended on forestry land
- To allow adequate translocation do not cut or admit stock for 14 d after spraying bracken or 7 d after spraying docks. Preferably leave undisturbed until late autumn
- Complete bracken control rarely achieved by one treatment. Survivors should be sprayed when they recover to full green frond, which may be in the ensuing year but more likely in the second year following initial application

Restrictions

- Maximum number of treatments 1 per crop or 1 per yr
- Do not apply in drought or hot, dry conditions
- Do not use in pasture before mowing for hay

Crop-specific information

- Latest use: Aug-Sep for all situations. See label for details
- In forestry areas some young trees may be checked if sprayed directly (see label)
- In fruit crops apply as a directed spray
- Do not treat blackcurrant cuttings, hop sets or weak hills
- Some grasses and herbs will be damaged by full dose. Most sensitive are cocksfoot, Yorkshire fog, timothy, bents, annual meadow-grass, daisies, docks, plantains, saxifrage
- Apply as spot treatment in parsley, mint and tarragon, not directly to crop

Following crops guidance

- Allow at least 6 wk between spraying and planting any crop

Environmental safety

- Dangerous for the environment
- Very toxic to aquatic organisms
- Keep livestock out of treated areas for at least two weeks following treatment and until poisonous weeds, such as ragwort, have died down and become unpalatable
- The use of asulam near surface waters has been considered by CRD. Whilst every care should be taken to avoid contamination, any that does occur during the normal course of spraying should offer no harm to operators, to users and consumers of the water, to domestic and farm animals and to wildlife. Before spraying such areas the appropriate regulatory authority should be notified

Hazard classification and safety precautions

Hazard Irritant, Dangerous for the environment [1-3, 5]
Transport code 9 [2, 3, 5]
Packaging group III [2, 3, 5]
UN number 3082 [2, 3, 5]; N/C [4]
Risk phrases R43, R50 [1-3, 5]; R52 [4]; R53a [5]
Operator protection A, C, D, H [1-3, 5]; M [1, 2]; M [3] (for ULV application); M [5]; U05a [4]; U08 [5]; U14 [1-3, 5]; U19a, U20b [1-3] (ULV use); U19a, U20b [4, 5]; U20c [1-3]
Environmental protection E07a [1-3, 5]; E07d [4]; E15a [1-5]; E38 [1-4]
Storage and disposal D01, D02 [4]; D05 [5]; D09a [1-5]; D10a [3-5]; D10c [1, 2]; D12a [1-4]

16 azoxystrobin

A systemic translaminar and protectant strobilurin fungicide for a wide range of crops
FRAC mode of action code: 11

Products

1	Amicron	AgriGuard	250 g/l	SC	13336
2	Amistar	Syngenta	250 g/l	SC	10443
3	Aubrac	AgChem Access	250 g/l	SC	13483
4	Azzox	Goldengrass	250 g/l	SC	14296
5	Clayton Belfry	Clayton	250 g/l	SC	12886
6	Heritage	Syngenta	50% w/w	WG	13536
7	Landgold Azzox	Goldengrass	250 g/l	SC	14003
8	Panama	Pan Agriculture	50% w/w	SG	13950
9	Standon Azoxystrobin	Standon	250 g/l	SC	09515

Uses

- Alternaria in ***broccoli***, ***brussels sprouts***, ***cabbages***, ***calabrese***, ***carrots***, ***cauliflowers***, ***collards***, ***kale***, ***spring oilseed rape***, ***winter oilseed rape*** [1-5, 7]; ***chicory root*** *(off-label)*, ***horseradish*** *(off-label)*, ***parsnips*** *(off-label)*, ***salad brassicas*** *(off-label - for baby leaf production)* [2]
- Anthracnose in ***amenity grassland***, ***managed amenity turf*** [6, 8]
- Ascochyta in ***broad beans*** *(off-label)*, ***lupins*** *(off-label)* [2]; ***combining peas***, ***vining peas*** [1-5, 7]
- Black dot in ***potatoes*** [1-5, 7]
- Black scurf and stem canker in ***potatoes*** [1-5, 7]
- Black spot in ***protected strawberries*** *(off-label)*, ***strawberries*** *(off-label)* [2]
- Botrytis in ***broad beans*** *(off-label)*, ***celery (outdoor)*** *(off-label)*, ***dwarf beans*** *(off-label)*, ***navy beans*** *(off-label)*, ***protected celery*** *(off-label)*, ***runner beans*** *(off-label)* [2]
- Brown patch in ***amenity grassland***, ***managed amenity turf*** [6, 8]
- Brown rust in ***spring barley***, ***spring wheat***, ***winter barley***, ***winter wheat*** [1-5, 7, 9]; ***spring rye***, ***triticale***, ***winter rye*** [1-5, 7]
- Crown rust in ***amenity grassland***, ***managed amenity turf*** [6, 8]; ***spring oats***, ***winter oats*** [1-5, 7]
- Didymella in ***inert substrate aubergines*** *(off-label)*, ***inert substrate tomatoes*** *(off-label)* [2]
- Didymella stem rot in ***aubergines*** *(off-label)*, ***protected tomatoes*** *(off-label)* [2]
- Disease control in ***all edible seed crops grown outdoors*** *(off-label)*, ***all non-edible seed crops grown outdoors*** *(off-label)*, ***forest nurseries*** *(off-label)*, ***ornamental plant production*** *(off-label)*, ***protected forest nurseries*** *(off-label)*, ***protected ornamentals*** *(off-label)*, ***protected soft fruit*** *(off-label)*, ***soft fruit*** *(off-label)* [2]
- Downy mildew in ***artichokes*** *(off-label)*, ***garlic*** *(off-label)*, ***inert substrate courgettes*** *(off-label)*, ***inert substrate cucumbers*** *(off-label)*, ***inert substrate gherkins*** *(off-label)*, ***protected courgettes*** *(off-label)*, ***protected cucumbers*** *(off-label)*, ***protected gherkins*** *(off-label)*, ***protected marrows*** *(off-label)*, ***protected melons*** *(off-label)*, ***protected pumpkins*** *(off-label)*, ***protected squashes*** *(off-label)*, ***salad brassicas*** *(off-label - for baby leaf production)*, ***salad onions*** *(off-label)*, ***shallots*** *(off-label)* [2]; ***bulb onions*** [1-5, 7]
- Fairy rings in ***amenity grassland***, ***managed amenity turf*** [6, 8]
- Foliar disease control in ***crambe*** *(off-label)*, ***durum wheat*** *(off-label)*, ***radishes*** *(off-label)* [2]; ***grass seed crops*** *(off-label)* [2, 9]
- Fusarium patch in ***amenity grassland***, ***managed amenity turf*** [6, 8]
- Glume blotch in ***spring wheat***, ***winter wheat*** [1-5, 7, 9]
- Grey mould in ***aubergines*** *(off-label)*, ***inert substrate aubergines*** *(off-label)*, ***inert substrate tomatoes*** *(off-label)*, ***protected tomatoes*** *(off-label)* [2]
- Late blight in ***aubergines*** *(off-label)*, ***inert substrate aubergines*** *(off-label)*, ***inert substrate tomatoes*** *(off-label)*, ***protected tomatoes*** *(off-label)* [2]
- Late ear diseases in ***spring wheat***, ***winter wheat*** [1-5, 7, 9]
- Melting out in ***amenity grassland***, ***managed amenity turf*** [6, 8]
- Net blotch in ***spring barley***, ***winter barley*** [1-5, 7, 9]
- Phytophthora in ***protected chicory*** *(off-label - for forcing)* [2]
- Powdery mildew in ***artichokes*** *(off-label)*, ***aubergines*** *(off-label)*, ***blackberries*** *(off-label)*, ***chicory root*** *(off-label)*, ***chives*** *(off-label)*, ***courgettes*** *(off-label)*, ***herbs (see appendix 6)*** *(off-label)*, ***horseradish*** *(off-label)*, ***inert substrate aubergines*** *(off-label)*, ***inert substrate courgettes*** *(off-label)*, ***inert substrate cucumbers*** *(off-label)*, ***inert substrate gherkins*** *(off-label)*, ***inert***

substrate tomatoes *(off-label)*, ***parsley*** *(off-label)*, ***parsnips*** *(off-label)*, ***poppies for morphine production*** *(off-label)*, ***protected blackberries*** *(off-label)*, ***protected cayenne peppers*** *(off-label)*, ***protected courgettes*** *(off-label)*, ***protected cucumbers*** *(off-label)*, ***protected gherkins*** *(off-label)*, ***protected marrows*** *(off-label)*, ***protected melons*** *(off-label)*, ***protected peppers*** *(off-label)*, ***protected pumpkins*** *(off-label)*, ***protected raspberries*** *(off-label)*, ***protected squashes*** *(off-label)*, ***protected tomatoes*** *(off-label)*, ***raspberries*** *(off-label)* [2]; ***carrots***, ***spring oats***, ***triticale***, ***winter oats***, ***winter rye*** [1-5, 7]; ***spring barley***, ***winter barley*** [1-5, 7, 9]; ***spring wheat***, ***winter wheat*** [9]

- Purple blotch in ***leeks*** [1-5, 7]
- Rhizoctonia in ***celery (outdoor)*** *(off-label)*, ***protected celery*** *(off-label)*, ***protected chives*** *(off-label)*, ***protected herbs (see appendix 6)*** *(off-label)*, ***protected lettuce*** *(off-label)*, ***protected parsley*** *(off-label)*, ***swedes*** *(off-label)*, ***turnips*** *(off-label)* [2]
- Rhynchosporium in ***spring barley***, ***winter barley*** [1-5, 7, 9]; ***spring rye***, ***triticale***, ***winter rye*** [1-5, 7]
- Ring spot in ***broccoli***, ***brussels sprouts***, ***cabbages***, ***calabrese***, ***cauliflowers***, ***collards***, ***kale*** [1-5, 7]; ***chives*** *(off-label)*, ***herbs (see appendix 6)*** *(off-label)*, ***parsley*** *(off-label)* [2]
- Root malformation disorder in ***red beet*** *(off-label)* [2]
- Rust in ***asparagus***, ***leeks***, ***spring field beans***, ***winter field beans*** [1-5, 7]; ***chicory root*** *(off-label)*, ***chives*** *(off-label)*, ***herbs (see appendix 6)*** *(off-label)*, ***lupins*** *(off-label)*, ***parsley*** *(off-label)* [2]
- Sclerotinia in ***celeriac*** *(off-label)*, ***celery (outdoor)*** *(off-label)*, ***lettuce*** *(off-label)*, ***protected celery*** *(off-label)* [2]
- Sclerotinia stem rot in ***spring oilseed rape***, ***winter oilseed rape*** [1-5, 7]
- Septoria leaf blotch in ***spring wheat***, ***winter wheat*** [1-5, 7, 9]
- Stemphylium in ***asparagus*** [1-5, 7]
- Take-all in ***spring barley*** *(reduction)*, ***spring wheat*** *(reduction)*, ***winter barley*** *(reduction)*, ***winter wheat*** *(reduction)* [1-5, 7]
- Take-all patch in ***amenity grassland***, ***managed amenity turf*** [6, 8]
- White blister in ***broccoli***, ***brussels sprouts***, ***cabbages***, ***calabrese***, ***cauliflowers***, ***collards***, ***kale*** [1-5, 7]
- White rust in ***chrysanthemums*** *(off-label - in pots)*, ***protected chrysanthemums*** *(off-label)* [2]
- Yellow rust in ***spring wheat***, ***winter wheat*** [1-5, 7, 9]

Specific Off-Label Approvals (SOLAs)

- ***all edible seed crops grown outdoors*** *20090443* [2]
- ***all non-edible seed crops grown outdoors*** *20090443* [2]
- ***artichokes*** *20051814* [2]
- ***aubergines*** *20021533* [2]
- ***blackberries*** *20030365* [2]
- ***broad beans*** *20032311* [2]
- ***celeriac*** *20031862* [2]
- ***celery (outdoor)*** *20011041* [2]
- ***chicory root*** *20051813* [2]
- ***chives*** *20021293* [2]
- ***chrysanthemums*** *(in pots) 20011684* [2]
- ***courgettes*** *20051985* [2]
- ***crambe*** *20081341* [2]
- ***durum wheat*** *20061722* [2]
- ***dwarf beans*** *20032311* [2]
- ***forest nurseries*** *20090443* [2]
- ***garlic*** *20061724* [2]
- ***grass seed crops*** *20061722* [2], *20061095* [9]
- ***herbs (see appendix 6)*** *20021293* [2]
- ***horseradish*** *20061721* [2]
- ***inert substrate aubergines*** *20011685* [2]
- ***inert substrate courgettes*** *20011685* [2]
- ***inert substrate cucumbers*** *20011685* [2]
- ***inert substrate gherkins*** *20011685* [2]
- ***inert substrate tomatoes*** *20011685* [2]
- ***lettuce*** *20011465* [2]
- ***lupins*** *20061723* [2]

- ***navy beans*** *20032311* [2]
- ***ornamental plant production*** *20090443* [2]
- ***parsley*** *20021293* [2]
- ***parsnips*** *20061721* [2]
- ***poppies for morphine production*** *20031137* [2]
- ***protected blackberries*** *20051194* [2]
- ***protected cayenne peppers*** *20021295* [2]
- ***protected celery*** *20011041* [2]
- ***protected chicory*** *(for forcing) 20051813* [2]
- ***protected chives*** *20030659* [2]
- ***protected chrysanthemums*** *20011684* [2]
- ***protected courgettes*** *20021533* [2]
- ***protected cucumbers*** *20021533* [2]
- ***protected forest nurseries*** *20090443* [2]
- ***protected gherkins*** *20021533* [2]
- ***protected herbs (see appendix 6)*** *20030659* [2]
- ***protected lettuce*** *20030659* [2]
- ***protected marrows*** *20070371* [2]
- ***protected melons*** *20070371* [2]
- ***protected ornamentals*** *20090443* [2]
- ***protected parsley*** *20030659* [2]
- ***protected peppers*** *20021295* [2]
- ***protected pumpkins*** *20070371* [2]
- ***protected raspberries*** *20051194* [2]
- ***protected soft fruit*** *20090443* [2]
- ***protected squashes*** *20070371* [2]
- ***protected strawberries*** *20021294* [2]
- ***protected tomatoes*** *20021533* [2]
- ***radishes*** *20081448* [2]
- ***raspberries*** *20030365* [2]
- ***red beet*** *20040614* [2]
- ***runner beans*** *20032311* [2]
- ***salad brassicas*** *(for baby leaf production) 20011465* [2]
- ***salad onions*** *20021687* [2]
- ***shallots*** *20061724* [2]
- ***soft fruit*** *20090443* [2]
- ***strawberries*** *20021294* [2]
- ***swedes*** *20040614* [2]
- ***turnips*** *20040614* [2]

Approval information

- Azoxystrobin included in Annex I under EC Directive 91/414
- Accepted by BBPA for use on malting barley

Efficacy guidance

- Best results obtained from use as a protectant or during early stages of disease establishment or when a predictive assessment indicates a risk of disease development
- Azoxystrobin inhibits fungal respiration and should always be used in mixture with fungicides with other modes of action
- Treatment under poor growing conditions may give less reliable results
- For good control of *Fusarium* patch in amenity turf and grass repeat treatment at minimum intervals of 2 wk
- Azoxystrobin is a member of the QoI cross resistance group. Product should be used preventatively and not relied on for its curative potential
- Use product in cereals as part of an Integrated Crop Management strategy incorporating other methods of control, including where appropriate other fungicides with a different mode of action. Do not apply more than two foliar applications of QoI containing products to any cereal crop
- There is a significant risk of widespread resistance occurring in *Septoria tritici* populations in UK. Failure to follow resistance management action may result in reduced levels of disease control

- On cereal crops product must always be used in mixture with another product, recommended for control of the same target disease, that contains a fungicide from a different cross resistance group and is applied at a dose that will give robust control
- Strains of barley powdery mildew resistant to QoIs are common in the UK

Restrictions

- Maximum number of treatments 1 per crop for potatoes; 2 per crop for brassicas, peas, cereals, oilseed rape; 4 per crop or yr for onions, carrots, leeks, amenity turf [1, 2, 5]
- Maximum total dose ranges from 2-4 times the single full dose depending on crop and product. See labels for details
- On turf the maximum number of treatments is 4 per yr but they must not exceed one third of the total number of fungicide treatments applied
- Do not use where there is risk of spray drift onto neighbouring apple crops
- The same spray equipment should not be used to treat apples

Crop-specific information

- Latest use: at planting for potatoes; grain watery ripe (GS 71) for cereals; before senescence for asparagus
- HI: 10 d for carrots;14 d for broccoli, Brussels sprouts, bulb onions, cabbages, calabrese, cauliflowers, collards, kale, vining peas; 21 d for leeks, spring oilseed rape, winter oilseed rape; 36 d for combining peas, 35 d for field beans [1, 2, 5]
- In cereals control of established infections can be improved by appropriate tank mixtures or application as part of a programme. Always use in mixture with another product from a different cross-resistance group
- In turf use product at full dose rate in a disease control programme, alternating with fungicides of different modes of action
- In potatoes when used as incorporated treatment apply overall to the entire area to be planted, incorporate to 15 cm and plant on the same day. In-furrow spray should be directed at the furrow and not the seed tubers
- Applications to brassica crops must only be made to a developed leaf canopy and not before growth stages specified on the label
- Heavy disease pressure in brassicae and oilseed rape may require a second treatment
- All crops should be treated when not under stress. Check leaf wax on peas if necessary
- Consult processor before treating any crops for processing
- Treat asparagus after the harvest season. Where a new bed is established do not treat within 3 wk of transplanting out the crowns
- Do not apply to turf when ground is frozen or during drought

Environmental safety

- Dangerous for the environment
- Very toxic to aquatic organisms
- Avoid spray drift onto surrounding areas or crops, especially apples, plums or privet
- LERAP Category B (potatoes only) [1, 2, 5]

Hazard classification and safety precautions

Hazard Dangerous for the environment [1-7, 9]

Transport code 9 [1, 2, 4-9]

Packaging group III [1, 2, 4-9]

UN number 3077 [6, 8]; 3082 [1, 2, 4, 5, 7, 9]

Risk phrases R50, R53a [1-7, 9]

Operator protection A [1-4]; A [5] (during treatment of potatoes); A [7]; U05a [1-7]; U09a, U19a [1-5, 7, 9]; U20b [1-7, 9]

Environmental protection E15a [1-5, 7, 9]; E15b [6]; E16a [1-5, 7] (potatoes only); E16b [2-5, 7] (potatoes only); E38 [1-7]

Storage and disposal D01, D02 [2-7]; D03 [6]; D05, D09a [1-7, 9]; D07 [1]; D10b [5, 6, 9]; D10c [1-4, 7]; D12a [1-7]

Medical advice M05a [6]

17 azoxystrobin + chlorothalonil

A preventative and systemic fungicide mixture for cereals
FRAC mode of action code: 11 + M5

See also chlorothalonil

Products

1	Amistar Opti	Syngenta	100:500 g/l	SC	12515
2	Curator	Syngenta	80:400 g/l	SC	13936
3	Olympus	Syngenta	80:400 g/l	SC	13797

Uses

- Alternaria in ***potatoes*** *(off-label)* [3]
- Botrytis in ***asparagus*** *(qualified minor use recommendation)*, ***bulb onions***, ***garlic***, ***shallots*** [3]
- Brown rust in ***spring barley***, ***spring wheat***, ***winter barley***, ***winter wheat*** [1, 2]
- Downy mildew in ***bulb onions***, ***garlic***, ***shallots*** [3]
- Foliar disease control in ***durum wheat*** *(off-label)*, ***grass seed crops*** *(off-label)*, ***spring rye*** *(off-label)*, ***triticale*** *(off-label)*, ***winter rye*** *(off-label)* [1]
- Glume blotch in ***spring wheat***, ***winter wheat*** [1, 2]
- Net blotch in ***spring barley***, ***winter barley*** [1, 2]
- Rhynchosporium in ***spring barley***, ***winter barley*** [2]; ***spring barley*** *(moderate control)*, ***winter barley*** *(moderate control)* [1]
- Rust in ***asparagus*** *(moderate control only)* [3]
- Septoria leaf blotch in ***spring wheat***, ***winter wheat*** [1, 2]
- Stemphylium in ***asparagus*** *(qualified minor use recommendation)* [3]
- Take-all in ***spring barley*** *(reduction)*, ***spring wheat*** *(reduction)*, ***winter barley*** *(reduction)*, ***winter wheat*** *(reduction)* [1]; ***spring barley*** *(reduction only)*, ***spring wheat*** *(reduction only)*, ***winter barley*** *(reduction only)*, ***winter wheat*** *(reduction only)* [2]
- Yellow rust in ***spring barley***, ***winter barley*** [2]; ***spring wheat***, ***winter wheat*** [1, 2]

Specific Off-Label Approvals (SOLAs)

- ***durum wheat*** *20092463* [1]
- ***grass seed crops*** *20092463* [1]
- ***potatoes*** *20090711* [3]
- ***spring rye*** *20092463* [1]
- ***triticale*** *20092463* [1]
- ***winter rye*** *20092463* [1]

Approval information

- Azoxystrobin and chlorothalonil included in Annex I under EC Directive 91/414
- Accepted by BBPA for use on malting barley

Efficacy guidance

- Best results obtained from applications made as a protectant treatment or in earliest stages of disease development. Further applications may be needed if disease attack is prolonged
- Reduction of barley spotting occurs when used as part of a programme with other fungicides
- Control of *Septoria* and rust diseases may be improved by mixture with a triazole fungicide
- Azoxystrobin is a member of the QoI cross resistance group. Product should be used preventatively and not relied on for its curative potential
- Use product in cereals as part of an Integrated Crop Management strategy incorporating other methods of control, including where appropriate other fungicides with a different mode of action. Do not apply more than two foliar applications of QoI containing products to any cereal crop
- There is a significant risk of widespread resistance occurring in *Septoria tritici* populations in UK. Failure to follow resistance management action may result in reduced levels of disease control

Restrictions

- Maximum number of treatments 2 per crop
- Maximum total dose on barley equivalent to one full dose treatment
- Do not use where there is risk of spray drift onto neighbouring apple crops
- The same spray equipment should not be used to treat apples

SEE SECTION 3 FOR PRODUCTS ALSO REGISTERED

Specific Off-Label Approvals (SOLAs)

- ***asparagus*** *20070831*
- ***horseradish*** *20061476*
- ***parsley root*** *20061476*
- ***parsnips*** *20061476*
- ***salsify*** *20061476*

Approval information

- Azoxystrobin and difenoconazole included in Annex I under EC Directive 91/414

Efficacy guidance

- Best results obtained from applications made in the earliest stages of disease development or as a protectant treatment following a disease risk assessment
- Ensure the crop is free from any stress caused by environmental or agronomic effects
- Azoxystrobin is a member of the QoI cross resistance group. Product should be used preventatively and not relied on for its curative potential
- Use as part of an Integrated Crop Management strategy incorporating other methods of control, including where appropriate other fungicides with a different mode of action. Do not apply more than two foliar applications of QoI containing products

Restrictions

- Maximum number of treatments 2 per crop
- Do not apply where there is a risk of spray drift onto neighbouring apple crops
- Consult processors before treating a crop destined for processing

Crop-specific information

- HI 14 d for carrots; 21 d for brassicas, leeks
- Minimum spray interval of 14 d must be observed on brassicas

Environmental safety

- Dangerous for the environment
- Very toxic to aquatic organisms

Hazard classification and safety precautions

Hazard Irritant, Dangerous for the environment
Transport code 9
Packaging group III
UN number 3082
Risk phrases R20, R43, R50, R53a
Operator protection A, H; U05a, U09a, U19a, U20b
Environmental protection E15b, E38
Storage and disposal D01, D02, D05, D09a, D10c, D12a

20 azoxystrobin + fenpropimorph

A protectant and eradicant fungicide mixture for cereals
FRAC mode of action code: 11 + 5

See also fenpropimorph

Products

1	Amistar Pro	Syngenta	100:280 g/l	SE	10513
2	Aspect	Syngenta	100:280 g/l	SE	10516

Uses

- Brown rust in ***spring barley***, ***spring rye***, ***spring wheat***, ***triticale***, ***winter barley***, ***winter rye***, ***winter wheat*** [1, 2]
- Crown rust in ***spring oats***, ***winter oats*** [1, 2]
- Foliar disease control in ***durum wheat*** *(off-label)*, ***grass seed crops*** *(off-label)* [1]
- Glume blotch in ***spring wheat***, ***winter wheat*** [1, 2]
- Late ear diseases in ***spring wheat*** *(moderate control)*, ***winter wheat*** *(moderate control)* [1, 2]
- Net blotch in ***spring barley***, ***winter barley*** [1, 2]
- Powdery mildew in ***spring barley***, ***spring oats***, ***spring rye***, ***triticale***, ***winter barley***, ***winter oats***, ***winter rye*** [1, 2]

- Rhynchosporium in ***spring barley***, ***spring rye***, ***triticale*** *(reduction)*, ***winter barley***, ***winter rye*** [1, 2]
- Septoria leaf blotch in ***spring wheat***, ***winter wheat*** [1, 2]
- Take-all in ***spring barley*** *(reduction)*, ***spring rye*** *(reduction)*, ***spring wheat*** *(reduction)*, ***triticale*** *(reduction)*, ***winter barley*** *(reduction)*, ***winter rye*** *(reduction)*, ***winter wheat*** *(reduction)* [1, 2]
- Yellow rust in ***spring wheat***, ***winter wheat*** [1, 2]

Specific Off-Label Approvals (SOLAs)

- ***durum wheat*** *20060962* [1]
- ***grass seed crops*** *20060962* [1]

Approval information

- Azoxystrobin and fenpropimorph included in Annex I under EC Directive 91/414
- Accepted by BBPA for use on malting barley

Efficacy guidance

- Best results obtained from application before infection following a disease risk assessment, or when disease first seen in crop
- Results may be less reliable when used on crops under stress
- Treatments for protection against ear disease should be made at ear emergence
- Azoxystrobin is a member of the QoI cross resistance group. Product should be used preventatively and not relied on for its curative potential
- Use product as part of an Integrated Crop Management strategy incorporating other methods of control, including where appropriate other fungicides with a different mode of action. Do not apply more than two foliar applications of QoI containing products to any cereal crop
- There is a significant risk of widespread resistance occurring in *Septoria tritici* populations in UK. Failure to follow resistance management action may result in reduced levels of disease control
- Strains of barley powdery mildew resistant to QoIs are common in the UK
- In wheat product must always be used in mixture with another product, recommended for control of the same target disease, that contains a fungicide from a different cross resistance group and is applied at a dose that will give robust control

Restrictions

- Maximum total dose equivalent to 2 full dose treatments
- Do not use where there is risk of spray drift onto neighbouring apple crops
- The same spray equipment should not be used to treat apples

Crop-specific information

- Latest use: before early milk stage (GS 73) for wheat, barley; up to and including grain watery ripe stage (GS 71) for oats, rye, triticale
- HI 5 wk

Environmental safety

- Dangerous for the environment
- Very toxic to aquatic organisms

Hazard classification and safety precautions

Hazard Harmful, Dangerous for the environment
Transport code 9
Packaging group III
UN number 3082
Risk phrases R20, R38, R43, R50, R53a, R63
Operator protection A, H; U02a, U05a, U09a, U14, U15, U19a, U20b
Environmental protection E15a, E38
Storage and disposal D01, D02, D05, D09a, D10c, D12a
Medical advice M03

21 Bacillus subtilis

A bacterial fungicide for the control of Botrytis cinerea
FRAC mode of action code: F6

Products

Serenade ASO	Fargro	13.96 g/l	SC	14318

SEE SECTION 3 FOR PRODUCTS ALSO REGISTERED

Uses

- Botrytis in ***bilberries*** *(off-label)*, ***blackcurrants*** *(off-label)*, ***blueberries*** *(off-label)*, ***bulb vegetables*** *(off-label)*, ***canary grass*** *(off-label)*, ***cane fruit*** *(off-label)*, ***cranberries*** *(off-label)*, ***figs*** *(off-label)*, ***fruiting vegetables*** *(off-label)*, ***gooseberries*** *(off-label)*, ***herbs (see appendix 6)*** *(off-label)*, ***hops*** *(off-label)*, ***leafy vegetables*** *(off-label)*, ***ornamental plant production*** *(off-label)*, ***redcurrants*** *(off-label)*, ***ribes hybrids*** *(off-label)*, ***root & tuber crops*** *(off-label)*, ***stem vegetables*** *(off-label)*, ***strawberries*** *(off-label)*, ***table grapes*** *(off-label)*, ***top fruit*** *(off-label)*, ***vegetable brassicas*** *(off-label)*, ***wine grapes*** *(off-label)*
- Botrytis fruit rot in ***protected strawberries*** *(reduction of damage to fruit)*
- Grey mould in ***protected strawberries*** *(reduction of damage to fruit)*

Specific Off-Label Approvals (SOLAs)

- ***bilberries*** *20090246*
- ***blackcurrants*** *20090246*
- ***blueberries*** *20090246*
- ***bulb vegetables*** *20090246*
- ***canary grass*** *20090246*
- ***cane fruit*** *20090246*
- ***cranberries*** *20090246*
- ***figs*** *20090246*
- ***fruiting vegetables*** *20090246*
- ***gooseberries*** *20090246*
- ***herbs (see appendix 6)*** *20090246*
- ***hops*** *20090246*
- ***leafy vegetables*** *20090246*
- ***ornamental plant production*** *20090246*
- ***redcurrants*** *20090246*
- ***ribes hybrids*** *20090246*
- ***root & tuber crops*** *20090246*
- ***stem vegetables*** *20090246*
- ***strawberries*** *20090246*
- ***table grapes*** *20090246*
- ***top fruit*** *20090246*
- ***vegetable brassicas*** *20090246*
- ***wine grapes*** *20090246*

Approval information

- Bacillus subtilis included in Annex I under EC Directive 91/414

Efficacy guidance

- Altenating applications with fungicides using a different mode of action is recommended for resistance management.
- Do not apply using irrigation equipment.
- For maximum effectiveness, start applications before disease development.
- Apply in a minimum water volume of 400 l/ha.

Restrictions

- Consult processor before using on crops grown for processing

Hazard classification and safety precautions

Hazard Irritant
Risk phrases R43
Operator protection A, H; U05a, U11, U14, U20b
Environmental protection E15b
Storage and disposal D01, D02, D05, D10a, D16

22 Bacillus thuringiensis

A bacterial insecticide for control of caterpillars
IRAC mode of action code: 11

Products

Dipel DF	Interfarm	32000 IU/mg	WG	14119

Uses

- Caterpillars in ***amenity vegetation***, ***apples*** *(off-label)*, ***blackberries*** *(off-label)*, ***borage*** *(off-label)*, ***broad beans*** *(off-label)*, ***broccoli***, ***brussels sprouts***, ***cabbages***, ***calabrese*** *(off-label)*, ***cauliflowers***, ***celery (outdoor)*** *(off-label)*, ***cherries*** *(off-label)*, ***chinese cabbage*** *(off-label)*, ***chives*** *(off-label)*, ***choi sum*** *(off-label)*, ***collards*** *(off-label)*, ***crab apples*** *(off-label)*, ***dwarf beans*** *(off-label)*, ***french beans*** *(off-label)*, ***frise*** *(off-label)*, ***herbs (see appendix 6)*** *(off-label)*, ***kale*** *(off-label)*, ***leaf spinach*** *(off-label)*, ***lettuce*** *(off-label)*, ***ornamental plant production***, ***pak choi*** *(off-label)*, ***parsley*** *(off-label)*, ***pears*** *(off-label)*, ***protected aubergines*** *(off-label)*, ***protected broccoli*** *(off-label)*, ***protected brussels sprouts*** *(off-label)*, ***protected cabbages*** *(off-label)*, ***protected calabrese*** *(off-label)*, ***protected cauliflowers*** *(off-label)*, ***protected cayenne peppers*** *(off-label)*, ***protected celery*** *(off-label)*, ***protected chinese cabbage*** *(off-label)*, ***protected chives*** *(off-label)*, ***protected choi sum*** *(off-label)*, ***protected collards*** *(off-label)*, ***protected cucumbers***, ***protected frise*** *(off-label)*, ***protected herbs (see appendix 6)*** *(off-label)*, ***protected kale*** *(off-label)*, ***protected lettuce*** *(off-label)*, ***protected ornamentals***, ***protected parsley*** *(off-label)*, ***protected peppers***, ***protected rhubarb*** *(off-label)*, ***protected salad brassicas*** *(off-label)*, ***protected scarole*** *(off-label)*, ***protected spinach*** *(off-label)*, ***protected tomatoes***, ***protected watercress*** *(off-label)*, ***radicchio*** *(off-label)*, ***raspberries***, ***rhubarb*** *(off-label)*, ***rubus hybrids*** *(off-label)*, ***runner beans*** *(off-label)*, ***salad brassicas*** *(off-label)*, ***scarole*** *(off-label)*, ***spinach*** *(off-label)*, ***spring cabbage*** *(off-label)*, ***strawberries***, ***sweetcorn*** *(off-label)*, ***watercress*** *(off-label)*
- Cutworms in ***bulb onions*** *(off-label)*, ***carrots*** *(off-label)*, ***garlic*** *(off-label)*, ***horseradish*** *(off-label)*, ***leeks*** *(off-label)*, ***parsley root*** *(off-label)*, ***parsnips*** *(off-label)*, ***red beet*** *(off-label)*, ***salsify*** *(off-label)*, ***shallots*** *(off-label)*
- Silver Y moth in ***broad beans*** *(off-label)*, ***dwarf beans*** *(off-label)*, ***french beans*** *(off-label)*, ***runner beans*** *(off-label)*, ***sweetcorn*** *(off-label)*
- Winter moth in ***bilberries*** *(off-label)*, ***blackcurrants*** *(off-label)*, ***blueberries*** *(off-label)*, ***cranberries*** *(off-label)*, ***gooseberries*** *(off-label)*, ***redcurrants*** *(off-label)*, ***vaccinium spp.*** *(off-label)*, ***whitecurrants*** *(off-label)*

Specific Off-Label Approvals (SOLAs)

- ***apples*** *20091070*
- ***bilberries*** *20091069*
- ***blackberries*** *20091058*
- ***blackcurrants*** *20091069*
- ***blueberries*** *20091069*
- ***borage*** *20091070*
- ***broad beans*** *20091067*
- ***bulb onions*** *20091065*
- ***calabrese*** *20091064*
- ***carrots*** *20091068*
- ***celery (outdoor)*** *20091064*
- ***cherries*** *20091070*
- ***chinese cabbage*** *20091064*
- ***chives*** *20091070*
- ***choi sum*** *20091064*
- ***collards*** *20091064*
- ***crab apples*** *20091070*
- ***cranberries*** *20091069*
- ***dwarf beans*** *20091067*
- ***french beans*** *20091067*
- ***frise*** *20091070*
- ***garlic*** *20091065*
- ***gooseberries*** *20091069*
- ***herbs (see appendix 6)*** *20091070*
- ***horseradish*** *20091068*
- ***kale*** *20091064*
- ***leaf spinach*** *20091070*
- ***leeks*** *20091062*
- ***lettuce*** *20091070*
- ***pak choi*** *20091064*
- ***parsley*** *20091070*

- ***parsley root*** *20091068*
- ***parsnips*** *20091068*
- ***pears*** *20091070*
- ***protected aubergines*** *20091059*
- ***protected broccoli*** *20091064*
- ***protected brussels sprouts*** *20091064*
- ***protected cabbages*** *20091064*
- ***protected calabrese*** *20091064*
- ***protected cauliflowers*** *20091064*
- ***protected cayenne peppers*** *20091060*
- ***protected celery*** *20091064*
- ***protected chinese cabbage*** *20091064*
- ***protected chives*** *20091070*
- ***protected choi sum*** *20091064*
- ***protected collards*** *20091064*
- ***protected frise*** *20091070*
- ***protected herbs (see appendix 6)*** *20091070*
- ***protected kale*** *20091064*
- ***protected lettuce*** *20091070*
- ***protected parsley*** *20091070*
- ***protected rhubarb*** *20091064*
- ***protected salad brassicas*** *20091070*
- ***protected scarole*** *20091070*
- ***protected spinach*** *20091070*
- ***protected watercress*** *20091063*
- ***radicchio*** *20091070*
- ***red beet*** *20091061*
- ***redcurrants*** *20091069*
- ***rhubarb*** *20091064*
- ***rubus hybrids*** *20091058*
- ***runner beans*** *20091067*
- ***salad brassicas*** *20091070*
- ***salsify*** *20091068*
- ***scarole*** *20091070*
- ***shallots*** *20091065*
- ***spinach*** *20091070*
- ***spring cabbage*** *20091064*
- ***sweetcorn*** *20091057*
- ***vaccinium spp.*** *20091069*
- ***watercress*** *20091063*
- ***whitecurrants*** *20091069*

Approval information

- Bacillus thuringiensis included in Annex I under EC Directive 91/414

Efficacy guidance

- Pest control achieved by ingestion by caterpillars of the treated plant vegetation. Caterpillars cease feeding and die in 1-3 d
- Apply as soon as larvae appear on crop and repeat every 7-10 d until the end of the hatching period
- Good coverage is essential, especially of undersides of leaves. Spray onto dry foliage and do not apply if rain expected within 6 h

Restrictions

- No restriction on number of treatments on edible crops
- Apply spray mixture as soon as possible after preparation

Crop-specific information

- HI zero

Environmental safety

- Store out of direct sunlight

Hazard classification and safety precautions

UN number N/C

Operator protection A, C, D; U05a, U15, U19a, U20c
Environmental protection E15a
Storage and disposal D01, D02, D05, D09a, D11a

23 benalaxyl

A phenylamide (acylalanine) fungicide available only in mixtures
FRAC mode of action code: 4

24 benalaxyl + mancozeb

A systemic and protectant fungicide mixture
FRAC mode of action code: 4 + M3

See also *mancozeb*

Products

Intro Plus	Interfarm	8:65% w/w	WP	11630

Uses
- Blight in ***potatoes***, ***protected potatoes***

Approval information
- Benalaxyl and mancozeb included in Annex I under EC Directive 91/414

Efficacy guidance
- Apply to potatoes at blight warning prior to crop becoming infected and repeat at 10-21 d intervals depending on risk of infection
- Spray irrigated potatoes after irrigation and at 14 d intervals, crops in polythene tunnels at 10 d intervals
- When active potato growth ceases use a non-systemic fungicide to end of season, starting not more than 10 d after last application
- For reduction of downy mildew in oilseed rape apply at seedling to 7-leaf stage, before end Nov as soon as infection seen

Restrictions
- Maximum number of treatments 5 per crop (including other phenylamide-based fungicides) for potatoes
- Do not treat potatoes showing active blight infection

Crop-specific information
- HI 7 d

Environmental safety
- Dangerous for the environment
- Very toxic to aquatic organisms

Hazard classification and safety precautions
Hazard Irritant, Dangerous for the environment
Transport code 9
Packaging group III
UN number 3077
Risk phrases R37, R43, R50, R53a
Operator protection A; U05a, U14, U15, U19a
Environmental protection E15a
Storage and disposal D01, D02, D12b
Medical advice M04a

25 bentazone

A post-emergence contact benzothiadiazinone herbicide
HRAC mode of action code: C3

Products

1	Basagran SG	BASF	87% w/w	SG	08360
2	Bentazone 480	Goldengrass	480 g/l	SL	14423
3	Mac-Bentazone 480 SL	AgChem Access	480 g/l	SL	13598
4	Troy 480	AgriChem BV	480 g/l	SL	12341
5	Zone 48	AgriGuard	480 g/l	SL	13165

Uses

- Annual dicotyledons in ***broad beans***, ***linseed***, ***narcissi***, ***navy beans***, ***potatoes***, ***runner beans***, ***spring field beans***, ***winter field beans*** [1-5]; ***bulb onions*** *(off-label)*, ***garlic*** *(off-label)*, ***leeks*** *(off-label)*, ***shallots*** *(off-label)*, ***soya beans*** *(off-label)* [1]; ***combining peas***, ***dwarf beans***, ***vining peas*** [4]; ***french beans***, ***peas*** [1-3, 5]
- Chickweed in ***chives*** *(off-label)*, ***game cover*** *(off-label)*, ***hops*** *(off-label)*, ***ornamental plant production*** *(off-label)*, ***salad onions*** *(off-label)* [1]
- Cleavers in ***chives*** *(off-label)*, ***game cover*** *(off-label)*, ***hops*** *(off-label)*, ***ornamental plant production*** *(off-label)*, ***salad onions*** *(off-label)* [1]
- Common storksbill in ***leeks*** *(off-label)* [1]
- Fool's parsley in ***leeks*** *(off-label)* [1]
- Groundsel in ***chives*** *(off-label)*, ***game cover*** *(off-label)*, ***hops*** *(off-label)*, ***ornamental plant production*** *(off-label)*, ***salad onions*** *(off-label)* [1]
- Mayweeds in ***bulb onions*** *(off-label)*, ***chives*** *(off-label)*, ***game cover*** *(off-label)*, ***garlic*** *(off-label)*, ***hops*** *(off-label)*, ***ornamental plant production*** *(off-label)*, ***salad onions*** *(off-label)*, ***shallots*** *(off-label)* [1]

Specific Off-Label Approvals (SOLAs)

- ***bulb onions*** *20061631* [1]
- ***chives*** *20090793* [1]
- ***game cover*** *20082819* [1]
- ***garlic*** *20061631* [1]
- ***hops*** *20082819* [1]
- ***leeks*** *20061630* [1]
- ***ornamental plant production*** *20082819* [1]
- ***salad onions*** *20090793* [1]
- ***shallots*** *20061631* [1]
- ***soya beans*** *20061629* [1]

Approval information

- Bentazone included in Annex I under EC Directive 91/414

Efficacy guidance

- Most effective control obtained when weeds are growing actively and less than 5 cm high or across. Good spray cover is essential
- The addition of specified adjuvant oils is recommended for use on some crops to improve fat hen control. Do not use under hot or humid conditions. See label for details
- Split dose application may be made in all recommended crops except peas and generally gives better weed control. See label for details

Restrictions

- Maximum number of treatments normally 2 per crop but check label
- Crops must be treated at correct stage of growth to avoid danger of scorch. See label for details
- Not all varieties of recommended crops are fully tolerant. Use only on tolerant varieties named in label. Do not use on forage or mange-tout varieties of peas
- Do not use on crops which have been affected by drought, waterlogging, frost or other stress conditions
- Do not apply insecticides within 7 d of treatment
- Leave 14 d after using a post-emergence grass herbicide and carry out a leaf wax test where relevant or 7 d where treatment precedes the grass herbicide
- Do not spray at temperatures above 21 °C. Delay spraying until evening if necessary

- Do not apply if rain or frost expected, if unseasonably cold, if foliage wet or in drought
- A minimum of 6 h (preferably 12 h) free from rain is required after application
- May be used on selected varieties of maincrop and second early potatoes (see label for details), not on seed crops or first earlies

Crop-specific information

- Latest use: before shoots exceed 15 cm high for potatoes and spring field beans (or 6-7 leaf pairs); 4 leaf pairs (6 pairs or 15 cm high with split dose) for broad beans; before flower buds visible for French, navy, runner and winter field beans, and linseed; before flower buds can be found enclosed in terminal shoot for peas
- Best results in narcissi obtained by using a suitable pre-emergence herbicide first
- Consult processor before using on crops for processing
- A satisfactory wax test must be carried out before use on peas
- Do not treat narcissi during flower bud initiation

Environmental safety

- Dangerous for the environment
- Harmful to aquatic organisms

Hazard classification and safety precautions

Hazard Harmful [1-3, 5]; Irritant [4]; Dangerous for the environment [5]
Packaging group III [1]
UN number 2588 [1]; N/C [2-5]
Risk phrases R22a [1-3, 5]; R36 [4, 5]; R41 [1, 3]; R43 [1-5]; R52, R53a [1-3]
Operator protection A, C [1-5]; H [4]; U05a, U08, U19a, U20b [1-5]; U11 [1-4]; U14 [1-3]
Environmental protection E15b, E34
Storage and disposal D01, D02, D09a [1-5]; D05, D07 [5]; D10b [1-3]; D10c [4, 5]
Medical advice M03 [1-5]; M05a [1-3]

26 bentazone + MCPA + MCPB

A post-emergence herbicide for undersown spring cereals and grass
HRAC mode of action code: C3 + O + O

See also *MCPA*
MCPB

Products

Acumen	BASF	200:80:200 g/l	SL	00028

Uses

- Annual dicotyledons in ***rotational grass***, ***undersown spring cereals***

Approval information

- Bentazone, MCPA and MCPB included in Annex I under EC Directive 91/414

Efficacy guidance

- Best results when weeds small and actively growing provided crop is at correct growth stage
- A minimum of 6 h free from rain is required after treatment

Restrictions

- Maximum number of treatments 1 per crop on undersown cereals; 1 per yr on grassland
- Do not treat red clover after 3-trifoliate leaf stage
- Do not use on crops suffering from herbicide damage or physical stress or in temperatures above 21 °C
- Do not use on seed crops or on cereals undersown with lucerne or seed mixtures containing lucerne
- Do not roll or harrow for 7 d before or after spraying
- Clovers may be scorched and undersown crop checked but effects likely to be outgrown. Clover stand in grassland may be reduced
- Do not apply if frost expected, if crop wet or when temperatures at or above 21 °C

Crop-specific information

- Latest use: before 1st node detectable for undersown cereals (GS 31), 2 wk before grazing for grassland

SEE SECTION 3 FOR PRODUCTS ALSO REGISTERED

- Apply to cereals from 2-fully expanded leaf stage but before first node detectable (GS 12-30) provided clover has reached 1-trifoliate leaf stage
- On first year grass leys use on seedling stage weeds before end of Sep
- Apply to newly sown leys after grass has reached 2-leaf stage provided clovers have at least 1 trifoliate leaf and red clover has not passed 3-trifoliate leaf stage

Environmental safety
- Dangerous for the environment
- Toxic to aquatic organisms
- Keep livestock out of treated areas for at least 2 wk and until foliage of poisonous weeds such as ragwort has died and become unpalatable

Hazard classification and safety precautions

Hazard Harmful, Dangerous for the environment
Transport code 9
Packaging group III
UN number 3082
Risk phrases R22a, R36, R38, R43, R51, R53a
Operator protection A, C; U05a, U08, U14, U19a, U20b
Environmental protection E07a, E15a, E34, E38
Storage and disposal D01, D02, D08, D09a, D10c, D12a
Medical advice M03, M05a

27 benthiavalicarb-isopropyl

An amino acid amide carbamate fungicide available only in mixtures
FRAC mode of action code: 40

28 benthiavalicarb-isopropyl + mancozeb

A fungicide mixture for potatoes
FRAC mode of action code: 40 + M3

See also mancozeb

Products

1	En-Garde	Certis	1.75:70% w/w	WG	12985
2	Valbon	Certis	1.75:70% w/w	WG	12603

Uses
- Blight in ***potatoes*** [1, 2]
- Downy mildew in ***amenity vegetation*** *(off-label)*, ***bulb onions*** *(off-label)*, ***ornamental plant production*** *(off-label)*, ***shallots*** *(off-label)* [2]

Specific Off-Label Approvals (SOLAs)
- ***amenity vegetation*** *20070240* [2]
- ***bulb onions*** *20081306* [2]
- ***ornamental plant production*** *20070240* [2]
- ***shallots*** *20081306* [2]

Approval information
- Benthiavalicarb-isopropyl and mancozeb included in Annex I under EC Directive 91/414

Efficacy guidance
- Apply as a protectant spray commencing before blight enters the crop irrespective of growth stage following a blight warning or where there is a local source of infection
- In the absence of a warning commence the spray programme before the foliage meets in the rows
- Repeat treatment every 7-10 d depending on disease pressure
- Increase water volume as necessary to ensure thorough coverage of the plant
- Use of air assisted sprayers or drop legs may help to improve coverage

Restrictions
- Maximum number of treatments 6 per crop

- Do not use more than three consecutive treatments. Include products with a different mode of action in the programme

Crop-specific information

- HI 7 d for potatoes

Environmental safety

- Very toxic to aquatic organisms
- Risk to non-target insects or other arthropods. Avoid spraying within 6 m of field boundary
- LERAP Category B

Hazard classification and safety precautions

Hazard Irritant
Transport code 9
Packaging group III
UN number 3077
Risk phrases R40, R43, R50, R53a
Operator protection A, H; U14
Environmental protection E15a, E16a, E22c
Storage and disposal D01, D02, D05, D09a, D11a, D12b

29 beta-cyfluthrin

A non-systemic pyrethroid insecticide available only in mixtures
IRAC mode of action code: 3

30 beta-cyfluthrin + clothianidin

An insecticidal seed treatment mixture for beet
IRAC mode of action code: 3 + 4A

See also *clothianidin*

Products

1	Modesto	Bayer CropScience	80:400 g/l	FS	14029
2	Poncho Beta	Bayer CropScience	53:400 g/l	FS	12076

Uses

- Aphids in ***winter oilseed rape*** [1]
- Beet virus yellows vectors in ***fodder beet*** *(seed treatment)*, ***sugar beet*** *(seed treatment)* [2]
- Cabbage root fly in ***winter oilseed rape*** *(reduction of feeding activity only)* [1]
- Cabbage stem flea beetle in ***winter oilseed rape*** [1]
- Flea beetle in ***poppies for morphine production*** *(off-label)* [1]
- Turnip sawfly larvae in ***winter oilseed rape*** [1]

Specific Off-Label Approvals (SOLAs)

- ***poppies for morphine production*** *20090479* [1]

Approval information

- Beta-cyfluthrin and clothianidin included in Annex I under EC Directive 91/414

Efficacy guidance

- In addition to control of aphid virus vectors, product improves crop establishment by reducing damage caused by symphylids, springtails, millipedes, wireworms and leatherjackets
- Additional control measures should be taken where very high populations of soil pests are present
- Product reduces direct feeding damage by foliar pests such as pygmy mangold beetle, beet flea beetle, mangold fly
- Product is not active against nematodes
- Treatment does not alter physical characteristics of pelleted seed and no change to standard drill settings should be necessary

Restrictions

- Maximum number of treatments: 1 per seed batch
- Product must be co-applied with a colouring dye
- Product must only be applied as a coating to pelleted seed using special treatment machinery

Crop-specific information

- Latest use: pre-drilling

Environmental safety

- Extremely dangerous to fish or other aquatic life. Do not contaminate surface waters or ditches with chemical or used container

Hazard classification and safety precautions

Hazard Harmful [1, 2]; Dangerous for the environment [1]
Transport code 9
Packaging group III
UN number 3082
Risk phrases R22a [1, 2]; R50, R53a [1]
Operator protection A, H; U04a, U05a, U07, U13, U14 [1, 2]; U20b, U24 [2]; U20c [1]
Environmental protection E03, E15a, E38 [1]; E13a [2]; E34, E36a [1, 2]
Storage and disposal D01, D02, D09a, D14 [1, 2]; D05 [2]; D12a [1]
Treated seed S02, S04b, S07 [1, 2]; S03, S04a, S05, S06a, S06b [2]
Medical advice M03

31 beta-cyfluthrin + imidacloprid

An insecticide mixture for seed treatment of oilseed rape
IRAC mode of action code: 3 + 4A

See also imidacloprid

Products

1	Chinook Blue	Bayer CropScience	100:100 g/l	LS	11262
2	Chinook Colourless	Bayer CropScience	100:100 g/l	LS	11206

Uses

- Cabbage stem flea beetle in ***evening primrose*** *(off-label - seed treatment)*, ***honesty*** *(off-label - seed treatment)*, ***linseed*** *(off-label - seed treatment)*, ***mustard*** *(off-label - seed treatment)*, ***winter oilseed rape*** *(seed treatment)* [1, 2]; ***spring oilseed rape*** *(seed treatment)* [1]
- Flea beetle in ***evening primrose*** *(off-label - seed treatment)*, ***honesty*** *(off-label - seed treatment)*, ***linseed*** *(off-label - seed treatment)*, ***mustard*** *(off-label - seed treatment)* [1, 2]; ***spring oilseed rape*** *(seed treatment)*, ***winter oilseed rape*** *(seed treatment)* [1]

Specific Off-Label Approvals (SOLAs)

- ***evening primrose*** *(seed treatment) 20052262* [1], *(seed treatment) 20052263* [2]
- ***honesty*** *(seed treatment) 20052262* [1], *(seed treatment) 20052263* [2]
- ***linseed*** *(seed treatment) 20052262* [1], *(seed treatment) 20052263* [2]
- ***mustard*** *(seed treatment) 20052262* [1], *(seed treatment) 20052263* [2]

Approval information

- Beta-cyfluthrin and imidacloprid included in Annex I under EC Directive 91/414

Efficacy guidance

- Drill treated seed as soon as possible after treatment
- Product will reduce damage by early attacks of flea beetle and may also reduce subsequent larval damage. However a follow-up spray treatment may be required if pest activity is heavy and prolonged

Restrictions

- Maximum number of treatments 1 per batch of seed
- Do not use on seed with more than 9% moisture content, on sprouted seed or on cracked, split or otherwise damaged seed
- Product must be co-applied with a colouring agent [2]
- Must be applied using seed treatment machine recommended by manufacturer
- Allow treated seed to dry before packaging. Drying can be assisted by subsequent application of talc
- Product can cause a transient tingling or numbing sensation to exposed skin. Avoid skin contact with the product, treated seed or dust throughout all operations in the treatment plant and during drilling

Crop-specific information

- Latest use: before drilling

Environmental safety

- Dangerous for the environment
- Toxic to aquatic organisms
- Extremely dangerous to fish or other aquatic life. Do not contaminate surface waters or ditches with chemical or used container
- Do not use treated seed as food or feed
- Dangerous to birds, game and other wildlife. Treated seed should not be left on the soil surface. Bury spillages
- Do not broadcast treated seed

Hazard classification and safety precautions

Hazard Harmful, Dangerous for the environment
Transport code 6.1
Packaging group III
UN number 3352
Risk phrases R22a, R51, R53a
Operator protection A, H; U04a, U05a, U07, U20b
Environmental protection E13a, E34, E36a, E38
Storage and disposal D01, D02, D05, D09a, D12a, D14
Treated seed S01, S02, S03, S04c, S05, S06a, S06b, S07, S08
Medical advice M03

32 bifenazate

A bifenazate acaricide for use on protected strawberries
IRAC mode of action code: 25

Products

Floramite 240 SC	Certis	240 g/l	SC	13686

Uses

- Spider mites in ***protected forest nurseries*** *(off-label)*, ***protected hops*** *(off-label)*
- Two-spotted spider mite in ***protected strawberries***

Specific Off-Label Approvals (SOLAs)

- ***protected forest nurseries*** *20082851*
- ***protected hops*** *20082851*

Approval information

- Bifenazate included in Annex I under EC Directive 91/414

Efficacy guidance

- Bifenazate acts by contact knockdown after approximately 4 d and a subsequent period of residual control
- All mobile stages of mites are controlled with occasionally some ovicidal activity
- Best results obtained from a programme of two treatments started as soon as first spider mites are seen
- To minimise possible development of resistance use in a planned Resistance Management strategy. Alternate at least two products with different modes of action between treatment programmes of bifenazate. See Section 5 for more information

Restrictions

- Maximum number of treatments 2 per yr for protected strawberries
- After use in protected environments avoid re-entry for at least 2 h while full ventilation is carried out allowing spray to dry on leaves
- Personnel working among treated crops should wear suitable long-sleeved garments and gloves for 14 d after treatment
- Do not apply as a low volume spray

Crop-specific information

- HI: 7 d for protected strawberries
- Carry out small scale tolerance test on the strawberry cultivar before large scale use

Environmental safety

- Dangerous for the environment
- Toxic to aquatic organisms
- Harmful to non-target predatory mites. Avoid spray drift onto field margins, hedges, ditches, surface water and neighbouring crops

Hazard classification and safety precautions

Hazard Irritant, Dangerous for the environment
Transport code 9
Packaging group III
UN number 3082
Risk phrases R43, R51, R53a
Operator protection A, H; U02a, U04a, U05a, U10, U14, U15, U19a, U20b
Environmental protection E15b, E38
Storage and disposal D01, D02, D05, D09a, D10b, D12b
Medical advice M03

33 bifenox

A diphenyl ether herbicide for cereals
HRAC mode of action code: E

Products

1	Aspire	AgriGuard	480 g/l	SC	14303
2	Clayton Belstone	Clayton	480 g/l	SC	14033
3	Fox	Makhteshim	480 g/l	SC	11981
4	Oram	Makhteshim	480 g/l	SC	13410
5	Sabine	Makhteshim	480 g/l	SC	13439

Uses

- Annual dicotyledons in ***canary flower (echium spp.)*** *(off-label)*, ***evening primrose*** *(off-label)*, ***honesty*** *(off-label)*, ***linseed*** *(off-label)*, ***mustard*** *(off-label)*, ***oilseed rape*** *(off-label)* [2, 4, 5]; ***durum wheat*** *(off-label)*, ***grass seed crops*** *(off-label)* [2-5]; ***triticale***, ***winter barley***, ***winter rye***, ***winter wheat*** [1, 2]
- Cleavers in ***triticale*** [3, 4]; ***winter barley***, ***winter rye***, ***winter wheat*** [3-5]
- Field pansy in ***triticale*** [3, 4]; ***winter barley***, ***winter rye***, ***winter wheat*** [3-5]
- Field speedwell in ***triticale*** [3, 4]; ***winter barley***, ***winter rye***, ***winter wheat*** [3-5]
- Forget-me-not in ***triticale*** [3, 4]; ***winter barley***, ***winter rye***, ***winter wheat*** [3-5]
- Geranium species in ***spring oilseed rape*** *(off-label)*, ***winter oilseed rape*** *(off-label)* [3]
- Ivy-leaved speedwell in ***triticale*** [3, 4]; ***winter barley***, ***winter rye***, ***winter wheat*** [3-5]
- Poppies in ***triticale*** [3, 4]; ***winter barley***, ***winter rye***, ***winter wheat*** [3-5]
- Red dead-nettle in ***triticale*** [3, 4]; ***winter barley***, ***winter rye***, ***winter wheat*** [3-5]

Specific Off-Label Approvals (SOLAs)

- ***canary flower (echium spp.)*** *20081839* [2], *20073141* [4], *20073143* [5]
- ***durum wheat*** *20081840* [2], *20072364* [3], *20073142* [4], *20073144* [5]
- ***evening primrose*** *20081839* [2], *20073141* [4], *20073143* [5]
- ***grass seed crops*** *20081840* [2], *20072364* [3], *20073142* [4], *20073144* [5]
- ***honesty*** *20081839* [2], *20073141* [4], *20073143* [5]
- ***linseed*** *20081839* [2], *20073141* [4], *20073143* [5]
- ***mustard*** *20081839* [2], *20073141* [4], *20073143* [5]
- ***oilseed rape*** *20081839* [2], *20073141* [4], *20073143* [5]
- ***spring oilseed rape*** *20060321* [3]
- ***winter oilseed rape*** *20060321* [3]

Approval information

- Accepted by BBPA for use on malting barley
- Bifenox included in Annex I under EC Directive 91/414

Efficacy guidance

- Bifenox is absorbed by foliage and emerging roots of susceptible species
- Best results obtained when weeds are growing actively with adequate soil moisture

Restrictions

- Maximum number of treatments: one per crop
- Do not apply to crops suffering from stress from whatever cause
- Do not apply if the crop is wet or if rain or frost is expected
- Avoid drift onto broad-leaved plants outside the target area

Crop-specific information

- Latest use: before 2nd node detectable (GS 32) for all crops

Environmental safety

- Dangerous for the environment
- Very toxic to aquatic organisms
- Do not empty into drains

Hazard classification and safety precautions

Hazard Dangerous for the environment
Transport code 9
Packaging group III
UN number 3082
Risk phrases R50, R53a
Operator protection A; U05a, U08, U20b
Environmental protection E15a, E38 [1-5]; E19b, E34 [2-5]
Storage and disposal D01, D02, D09a, D12a
Medical advice M03

34 bifenthrin

A contact and residual pyrethroid acaricide/insecticide for use in agricultural and horticultural crops
IRAC mode of action code: 3

Products

1	Brigade 80 SC	Belchim	80 g/l	SC	12853
2	Starion Flo	Belchim	80 g/l	SC	12455
3	Talstar 80 Flo	Belchim	80 g/l	SC	12352

Uses

- Aphids in ***broccoli***, ***brussels sprouts***, ***cabbages***, ***calabrese***, ***cauliflowers*** [1-3]; ***combining peas***, ***spring barley***, ***spring oats***, ***spring wheat***, ***vining peas***, ***winter barley***, ***winter oats***, ***winter wheat*** [1, 2]
- Cabbage seed weevil in ***spring oilseed rape***, ***winter oilseed rape*** [1, 2]
- Cabbage stem flea beetle in ***winter oilseed rape*** [1, 2]
- Caterpillars in ***broccoli***, ***brussels sprouts***, ***cabbages***, ***calabrese***, ***cauliflowers*** [1-3]
- Clay-coloured weevil in ***blackberries*** *(off-label)*, ***protected blackberries*** *(off-label)*, ***protected raspberries*** *(off-label)*, ***raspberries*** *(off-label)* [1, 2]
- Damson-hop aphid in ***hops*** [1-3]
- Flea beetle in ***linseed***, ***spring oilseed rape***, ***winter oilseed rape*** [1, 2]
- Fruit tree red spider mite in ***apples***, ***pears*** [1-3]
- Pod midge in ***spring oilseed rape***, ***winter oilseed rape*** [1, 2]
- Pollen beetle in ***spring oilseed rape***, ***winter oilseed rape*** [1, 2]
- Rape winter stem weevil in ***winter oilseed rape*** [1, 2]
- Two-spotted spider mite in ***blackberries*** *(off-label)*, ***hops***, ***ornamental plant production***, ***raspberries*** *(off-label)*, ***strawberries*** [1-3]; ***blackcurrants*** *(off-label)*, ***protected blackcurrants*** *(off-label)* [2, 3]; ***protected blackberries*** *(off-label)*, ***protected raspberries*** *(off-label)* [1, 2]
- Virus vectors in ***winter barley***, ***winter oats***, ***winter wheat*** [1, 2]
- Whitefly in ***broccoli***, ***brussels sprouts***, ***cabbages***, ***calabrese***, ***cauliflowers*** [1-3]

Specific Off-Label Approvals (SOLAs)

- ***blackberries*** *20061628* [1], *20060109* [2], *20053056* [3]
- ***blackcurrants*** *20063125* [2], *20063033* [3]
- ***protected blackberries*** *20061628* [1], *20060109* [2]
- ***protected blackcurrants*** *20063125* [2], *20063033* [3]
- ***protected raspberries*** *20061628* [1], *20060109* [2]
- ***raspberries*** *20061628* [1], *20060109* [2], *20053056* [3]

SEE SECTION 3 FOR PRODUCTS ALSO REGISTERED

Approval information

- Accepted by BBPA for use on malting barley and hops. Bifenthrin has not been included in Annex 1 and all approvals will lapse 31/12/2013.

Efficacy guidance

- Timing of application varies with crop and pest. See label for details
- Good spray cover of upper and lower plant surfaces essential to achieve effective pest control. Spray volumes should be increased where crop is dense

Restrictions

- Maximum number of treatments 5 per yr for hops; 4 per crop on brassicas; 2 per yr for apples, pears, strawberries. Other crops either 2 per yr or a maximum total dose is stipulated. See label
- Do not apply to a cereal crop if any product containing a pyrethroid insecticide or dimethoate has been applied after the start of ear emergence (GS 51)
- Consult processor before use on crops for processing
- Do not treat ornamental specimens in blossom or under stress

Crop-specific information

- Latest use: before 31 Mar in yr of harvest for cereals; before late milk stage (GS 77) for wheat, before early milk stage (GS 73) for barley, oats [1, 2];
- HI 2 d for brassicas, 3 d for peas; 8 wk for linseed [1, 2], oilseed rape [1, 2]

Environmental safety

- Dangerous for the environment
- Very toxic to aquatic organisms
- High risk to bees. Do not apply to crops in flower, or to those in which bees are actively foraging, except as directed on named crops. Do not apply when flowering weeds are present [1-3]
- Extremely dangerous to bees. Do not apply to crops in flower or to those in which bees are actively foraging. Do not apply when flowering weeds are present
- Extremely dangerous to fish or other aquatic life. Do not contaminate surface waters or ditches with chemical or used container
- High risk to non-target insects or other arthropods. Not compatible with IPM systems
- Do not spray cereal crops within 6 m of the field boundary or after 31 Mar in the yr of harvest
- Keep in original container, tightly closed, in a safe place, under lock and key [1-3]
- LERAP Category A and broadcast air-assisted LERAP (30 m [1-3])

Hazard classification and safety precautions

Hazard Harmful, Dangerous for the environment
Transport code 9
Packaging group III
UN number 3082
Risk phrases R20, R22a, R43, R50, R53a
Operator protection A, H; U02a, U04a, U05a, U08, U19a, U20b
Environmental protection E12a, E13a, E15b, E16c, E16d, E34, E38; E12f (named crops - see label); E17b (30 m); E22a (or after 31 Mar)
Storage and disposal D01, D02, D05, D09b, D10b, D12a
Medical advice M03, M05b

35 Bordeaux mixture

A protectant copper sulphate/lime complex fungicide
FRAC mode of action code: M1

Products

Wetcol 3 Copper Fungicide	Law Fertilisers	30 g/l (copper)	SC	13369

Uses

- Bacterial canker in ***cherries***
- Blight in ***potatoes***, ***tomatoes (outdoor)***
- Cane spot in ***loganberries***, ***raspberries***
- Canker in ***apples***, ***pears***
- Celery leaf spot in ***celery (outdoor)***
- Currant leaf spot in ***blackcurrants***

- Downy mildew in ***hops***
- Leaf curl in ***apricots, nectarines, peaches***
- Scab in ***apples, pears***
- Spur blight in ***raspberries***

Efficacy guidance

- Spray interval normally 7-14 d but varies with disease and crop. See label for details
- Commence spraying potatoes before crop meets in row or immediately first blight period occurs
- For canker control spray monthly from Aug to Oct
- For peach leaf curl control spray at leaf fall in autumn and again in Feb
- Spray when crop foliage dry. Do not spray if rain imminent

Restrictions

- Do not use on copper sensitive cultivars, including Doyenne du Comice pears

Environmental safety

- Harmful to fish or other aquatic life. Do not contaminate surface waters or ditches with chemical or used container
- Harmful to livestock. Keep all livestock out of treated areas for at least 3 wk

Hazard classification and safety precautions

Hazard Harmful, Irritant
Risk phrases R22a, R36, R37, R38
Operator protection U13, U15, U20a
Environmental protection E06c (3 wk); E13c
Storage and disposal D09a, D11a

SECTION 2

36 boscalid

A translocated and translaminar anilide fungicide
FRAC mode of action code: 7

Products

1	Bosco WG	Goldengrass	50% w/w	WG	14455
2	Filan	BASF	50% w/w	WG	11449
3	Fulmar	AgChem Access	50% w/w	WG	14073

Uses

- Alternaria in ***spring oilseed rape, winter oilseed rape*** [1-3]
- Botrytis in ***all edible seed crops grown outdoors*** *(off-label)*, ***all non-edible seed crops grown outdoors*** *(off-label)*, ***borage for oilseed production*** *(off-label)*, ***canary flower (echium spp.)*** *(off-label)*, ***evening primrose*** *(off-label)*, ***honesty*** *(off-label)*, ***linseed*** *(off-label)*, ***mustard*** *(off-label)*, ***poppies for morphine production*** *(off-label)* [2]
- Sclerotinia in ***all edible seed crops grown outdoors*** *(off-label)*, ***all non-edible seed crops grown outdoors*** *(off-label)*, ***borage for oilseed production*** *(off-label)*, ***canary flower (echium spp.)*** *(off-label)*, ***evening primrose*** *(off-label)*, ***honesty*** *(off-label)*, ***linseed*** *(off-label)*, ***mustard*** *(off-label)*, ***poppies for morphine production*** *(off-label)* [2]
- Sclerotinia stem rot in ***spring oilseed rape, winter oilseed rape*** [1-3]

Specific Off-Label Approvals (SOLAs)

- ***all edible seed crops grown outdoors*** *20082849* [2]
- ***all non-edible seed crops grown outdoors*** *20082849* [2]
- ***borage for oilseed production*** *20063642* [2]
- ***canary flower (echium spp.)*** *20063642* [2]
- ***evening primrose*** *20063642* [2]
- ***honesty*** *20063642* [2]
- ***linseed*** *20063642* [2]
- ***mustard*** *20063642* [2]
- ***poppies for morphine production*** *20040163* [2]

Approval information

- Boscalid included in Annex I under EC Directive 91/414
- Accepted by BBPA on malting barley before ear emergence only

Efficacy guidance

- For best results apply as a protectant spray before symptoms are visible
- Applications against *Sclerotinia* should be made in high disease risk situations at early to full flower
- *Sclerotinia* control may be reduced when high risk conditions occur after flowering, leading to secondary disease spread
- Applications against *Alternaria* should be made at full flowering. Later applications may result in reduced levels of control
- Ensure adequate spray penetration and good coverage

Restrictions

- Maximum total dose equivalent to two full dose treatments
- Do no treat crops to be used for seed production later than full flowering stage
- Avoid spray drift onto neighbouring crops

Crop-specific information

- Latest use: up to and including when 50% of pods have reached final size (GS 75) for oilseed rape

Environmental safety

- Dangerous for the environment
- Toxic to aquatic organisms
- Product represents minimal hazard to bees when used as directed. However local bee-keepers should be notified if crops are to be sprayed when in flower

Hazard classification and safety precautions

Hazard Dangerous for the environment
Transport code 9
Packaging group III
UN number 3077
Risk phrases R51, R53a
Operator protection U05a, U20c
Environmental protection E15a, E38
Storage and disposal D01, D02, D05, D09a, D10c, D12a

37 boscalid + epoxiconazole

A broad spectrum fungicide mixture for cereals
FRAC mode of action code: 7 + 3

See also epoxiconazole

Products

1	Chord	BASF	210:75 g/l	SC	13928
2	Deuce	BASF	233:67 g/l	SC	13240
3	Splice	BASF	233:67 g/l	SC	12315
4	Totem	BASF	233:67 g/l	SC	12692
5	Tracker	BASF	233:67 g/l	SC	12295
6	Venture	BASF	233:67 g/l	SC	12316

Uses

- Brown rust in ***spring barley, spring wheat, winter barley, winter wheat*** [1-6]
- Crown rust in ***spring oats, winter oats*** [1-6]
- Eyespot in ***spring barley*** *(moderate)*, ***spring wheat*** *(moderate)*, ***winter barley*** *(moderate)*, ***winter wheat*** *(moderate)* [2, 3, 5, 6]; ***spring barley*** *(moderate control)*, ***spring wheat*** *(moderate control)*, ***winter barley*** *(moderate control)*, ***winter wheat*** *(moderate control)* [1, 4]; ***spring oats*** *(moderate control)*, ***winter oats*** *(moderate control)* [1]
- Foliar disease control in ***durum wheat*** *(off-label)* [3-6]
- Fusarium ear blight in ***spring wheat*** *(good reduction)*, ***winter wheat*** *(good reduction)* [1]
- Glume blotch in ***spring wheat, winter wheat*** [1]
- Net blotch in ***spring barley, winter barley*** [1-6]
- Powdery mildew in ***spring barley, spring oats, spring wheat, winter barley, winter oats, winter wheat*** [1]
- Rhynchosporium in ***spring barley*** *(moderate)*, ***winter barley*** *(moderate)* [2, 3, 5, 6]; ***spring barley*** *(moderate control)*, ***winter barley*** *(moderate control)* [1, 4]
- Septoria leaf blotch in ***spring wheat, winter wheat*** [1-6]

- Sooty moulds in ***spring wheat*, *winter wheat*** [1]
- Tan spot in ***spring wheat*** *(reduction - qualified minor use)*, ***winter wheat*** *(reduction - qualified minor use)* [4]; ***spring wheat*** *(reduction only)*, ***winter wheat*** *(reduction only)* [1]
- Yellow rust in ***spring barley*, *spring wheat*, *winter barley*, *winter wheat*** [1]

Specific Off-Label Approvals (SOLAs)

- ***durum wheat*** *20070460* [3], *20070458* [4], *20070462* [5], *20070461* [6]

Approval information

- Boscalid and epoxiconazole included in Annex I under EC Directive 91/414
- Accepted by BBPA for use on malting barley (before ear emergence only)

Efficacy guidance

- Apply at the start of foliar or stem based disease attack
- Optimum effect against eyespot achieved by spraying between leaf-sheath erect and second node detectable stages (GS 30-32)
- Epoxiconazole is a DMI fungicide. Resistance to some DMI fungicides has been identified in Septoria leaf blotch which may seriously affect performance of some products. For further advice contact a specialist advisor and visit the Fungicide Resistance Action Group (FRAG)-UK website

Restrictions

- Maximum number of treatments 1 per crop [4]
- Maximum total dose equivalent to two full dose treatments [2, 3, 5, 6]

Crop-specific information

- Latest use: before cereal ear emergence

Environmental safety

- Dangerous for the environment
- Toxic to aquatic organisms
- LERAP Category B
- Avoid drift onto neighbouring crops. May cause damage to broad-leaved plant species

Hazard classification and safety precautions

Hazard Harmful, Dangerous for the environment
Transport code 9 [2-6]
Packaging group III [2-6]
UN number 3082 [2-6]; N/C [1]
Risk phrases R36, R40, R51, R53a, R62, R63
Operator protection A, C, H; U05a, U20b
Environmental protection E15a, E16a, E38 [1-6]; E34 [1]
Storage and disposal D01, D02, D08, D10c, D12a [1-6]; D09a [1]
Medical advice M03

38 boscalid + pyraclostrobin

A protectant and systemic fungicide mixture
FRAC mode of action code: 7 + 11

See also pyraclostrobin

Products

1	Bellis	BASF	25.2:12.8% w/w	WG	12522
2	Signum	BASF	26.7:6.7% w/w	WG	11450

Uses

- Alternaria in ***cabbages*, *carrots*** *(moderate)*, ***cauliflowers*** [2]
- Black spot in ***protected strawberries*, *strawberries*** [2]
- Blossom wilt in ***cherries*** *(off-label)*, ***mirabelles*** *(off-label)*, ***plums*** *(off-label)*, ***protected cherries*** *(off-label)*, ***protected mirabelles*** *(off-label)*, ***protected plums*** *(off-label)* [2]
- Botrytis in ***blackcurrants*** *(off-label)*, ***ornamental plant production*** *(off-label)*, ***protected ornamentals*** *(off-label)* [2]
- Chocolate spot in ***spring field beans*** *(moderate)*, ***winter field beans*** *(moderate)* [2]
- Disease control in ***all edible seed crops grown outdoors*** *(off-label)*, ***all non-edible seed crops grown outdoors*** *(off-label)*, ***protected edible crops*** *(off-label - seed crops only)*, ***protected hops***

(off-label), ***protected soft fruit*** *(off-label)*, ***protected top fruit*** *(off-label)*, ***soft fruit*** *(off-label)*, ***top fruit*** *(off-label)* [2]; ***hops*** *(off-label)* [1, 2]

- Downy mildew in ***bulls blood*** *(off-label)*, ***chard*** *(off-label)*, ***spinach beet*** *(off-label)* [2]
- Foliar disease control in ***asparagus*** *(off-label)*, ***blackberries*** *(off-label)*, ***raspberries*** *(off-label)* [2]
- Grey mould in ***protected strawberries***, ***strawberries*** [2]
- Powdery mildew in ***apples*** [1]; ***blackcurrants*** *(off-label)*, ***carrots***, ***protected strawberries***, ***strawberries*** [2]
- Rhizoctonia rot in ***lettuce***, ***protected lettuce*** [2]
- Ring spot in ***brussels sprouts***, ***cabbages***, ***cauliflowers*** [2]
- Rust in ***spring field beans***, ***winter field beans*** [2]
- Scab in ***apples***, ***pears*** [1]
- Sclerotinia in ***carrots*** *(moderate)*, ***herbs (see appendix 6)*** *(off-label)*, ***horseradish*** *(off-label)*, ***lettuce***, ***parsnips*** *(off-label)*, ***protected herbs (see appendix 6)*** *(off-label)*, ***protected lettuce***, ***protected salad brassicas*** *(off-label)* [2]
- Septoria leaf blotch in ***blackcurrants*** *(off-label)* [2]
- White blister in ***brussels sprouts***, ***cabbages*** *(qualified minor use)*, ***leaf brassicas*** *(off-label)*, ***salad brassicas*** *(off-label)* [2]
- White rot in ***bulb onions*** *(off-label)*, ***garlic*** *(off-label)*, ***salad onions*** *(off-label)*, ***shallots*** *(off-label)* [2]
- White tip in ***leeks*** *(off-label)* [2]

Specific Off-Label Approvals (SOLAs)

- ***all edible seed crops grown outdoors*** *20082924* [2]
- ***all non-edible seed crops grown outdoors*** *20082924* [2]
- ***asparagus*** *20080262* [2]
- ***blackberries*** *20080992* [2]
- ***blackcurrants*** *20071257* [2]
- ***bulb onions*** *20080790* [2]
- ***bulls blood*** *20073277* [2]
- ***chard*** *20073277* [2]
- ***cherries*** *20052089* [2]
- ***garlic*** *20080790* [2]
- ***herbs (see appendix 6)*** *20041984* [2]
- ***hops*** *20082822* [1], *20082924* [2]
- ***horseradish*** *20051317* [2]
- ***leaf brassicas*** *20081356* [2]
- ***leeks*** *20062848* [2]
- ***mirabelles*** *20052089* [2]
- ***ornamental plant production*** *20091842* [2]
- ***parsnips*** *20051317* [2]
- ***plums*** *20052089* [2]
- ***protected cherries*** *20052089* [2]
- ***protected edible crops*** *(seed crops only) 20082924* [2]
- ***protected herbs (see appendix 6)*** *20041984* [2]
- ***protected hops*** *20082924* [2]
- ***protected mirabelles*** *20052089* [2]
- ***protected ornamentals*** *20091842* [2]
- ***protected plums*** *20052089* [2]
- ***protected salad brassicas*** *20041984* [2]
- ***protected soft fruit*** *20082924* [2]
- ***protected top fruit*** *20082924* [2]
- ***raspberries*** *20080992* [2]
- ***salad brassicas*** *20081356* [2]
- ***salad onions*** *20080790* [2]
- ***shallots*** *20080790* [2]
- ***soft fruit*** *20082924* [2]
- ***spinach beet*** *20073277* [2]
- ***top fruit*** *20082924* [2]

Approval information

- Boscalid and pyraclostrobin included in Annex I under EC Directive 91/414
- Accepted by BBPA for use on malting barley (before ear emergence only)

Efficacy guidance

- On brassicas apply as a protectant spray or at the first sign of disease and repeat at 3-4 wk intervals depending on disease pressure [2]
- Ensure adequate spray penetration and coverage by increasing water volume in dense crops [2]
- For best results on strawberries apply as a protectant spray at the white bud stage. Applications should be made in sequence with other products as part of a fungicide spray programme during flowering at 7-10 day intervals [2]
- On carrots and field beans apply as a protectant spray or at the first sign of disease with a repeat treatment if needed, as directed on the label [2]
- On lettuce apply as a protectant spray 1-2 wk after planting [2]
- Optimum results on apples and pears obtained from a protectant treatment from bud burst [1]
- Application as the final 2 sprays on apples and pears gives reduction in storage rots [1]
- Pyraclostrobin is a member of the QoI cross-resistance group of fungicides and should be used in programmes with fungicides with a different mode of action

Restrictions

- Maximum total dose equivalent to three full dose treatments on brassicas; two full dose treatments on all other field crops [2]
- Maximum number of treatments (including other QoI treatments) on apples and pears 4 per yr if total number of applications is 12 or more, or 3 per yr if the total number is fewer than 12 [1]
- Do not use more than 2 consecutive treatments on apples and pears, and these must be separated by a minimum of 2 applications of a fungicide with a different mode of action [1]
- Use a maximum of three applications per yr on brassicas and no more than two per yr on other field crops [2]
- Do not use consecutive treatments; apply in alternation with fungicides from a different cross resistance group and effective against the target diseases [2]
- Do not apply more than 6 kg/ha to the same area of land per yr [2]
- Consult processor before use on crops for processing
- Applications to lettuce and protected salad crops may only be made between 1 Apr and 31 Oct [2]

Crop-specific information

- HI 21 d for field beans; 14 d for brassica crops, carrots, lettuce; 7 d for apples, pears; 3 d for strawberries

Environmental safety

- Dangerous for the environment
- Very toxic to aquatic organisms
- LERAP Category B [2]
- Broadcast air-assisted LERAP (40 m) [1]

Hazard classification and safety precautions

Hazard Harmful, Dangerous for the environment
Transport code 9
Packaging group III
UN number 3077
Risk phrases R22a, R50, R53a
Operator protection A; U05a, U20c [1, 2]; U13 [1]
Environmental protection E15a, E38 [1, 2]; E16a [2]; E17b [1] (40 m); E34 [1]
Storage and disposal D01, D02, D10c, D12a [1, 2]; D08 [1]; D09a [2]
Medical advice M03 [1]; M05a [1, 2]

39 brodifacoum

An anticoagulant coumarin rodenticide

Products

Sorexa Checkatube	Sorex	0.2% w/w	XX	H6702

SEE SECTION 3 FOR PRODUCTS ALSO REGISTERED

Uses

- Mice in ***farm buildings*** *(indoor use only)*

Efficacy guidance

- Product is a self-contained control device which requires no bait handling
- Best results obtained by placing tubes where mice are active and will readily enter the device
- To acquire a lethal dose mice must pass through the tube a number of times
- Where a continuous source of infestation is present, or where sustained protection is needed, tubes can be installed permanently and replaced at monthly intervals
- For complete eradication other control methods should also be used bearing in mind the resistance status of the target population

Restrictions

- For use only by professional pest contractors
- Do not use outdoors. Products must be used in situations where baits are placed within a building or other enclosed structure, and the target is living or feeding predominantly within that building or structure
- Do not attempt to open or recharge the tubes

Environmental safety

- Prevent access to baits by children, birds and other animals
- Do not place tubes where food, feed or water could become contaminated
- Remove all remains of tubes after use and burn or bury
- Search for and burn or bury all rodent bodies. Do not place in refuse bins or on open rubbish tips
- Keep in original container, tightly closed, in a safe place, under lock and key

Hazard classification and safety precautions

Operator protection U13, U20b
Storage and disposal D02, D09b
Vertebrate/rodent control products V01b, V02, V03a, V04a
Medical advice M03

40 bromadiolone

An anti-coagulant coumarin-derivative rodenticide

Products

1	Bromag	Killgerm	0.005% w/w	RB	H8285
2	Bromag Fresh Bait	Killgerm	0.005% w/w	RB	H8557
3	Endorats Premium Rat Killer	Irish Drugs	0.005% w/w	GB	H6725
4	Sakarat Bromabait	Killgerm	0.005% w/w	RB	H7902
5	Tomcat 2	Antec Biosentry	0.005% w/w	PT	H6736
6	Tomcat 2 Blox	Antec Biosentry	0.005% w/w	BB	H6731
7	Vermigon Bait	Certis	0.005% w/w	RB	H7925
8	Vermigon Blocks	Certis	0.005% w/w	BB	H7926
9	Vermigon Pasta	Certis	0.005% w/w	RB	H8191
10	Vermigon Pellets	Certis	0.005% w/w	PT	H7941
11	Vermigon Sewer Blocks	Certis	0.005% w/w	BB	H7927

Uses

- Mice in ***farm buildings*** [4-11]; ***farm buildings/yards*** [1, 2]; ***farmyards*** [4, 7-11]
- Rats in ***farm buildings*** [3-11]; ***farm buildings/yards*** [1, 2]; ***farmyards*** [4, 7-11]

Efficacy guidance

- Ready-to-use baits are formulated on a mould-resistant, whole-wheat base
- Use in baiting programme. Place baits in protected situations, sufficient for continuous feeding between treatments
- Chemical is effective against warfarin- and coumatetralyl-resistant rats and mice and does not induce bait shyness
- Use bait bags where loose baiting inconvenient (eg behind ricks, silage clamps etc)
- The resistance status of the rodent population should be assessed when considering the choice of product to use

Restrictions

- For use only by professional operators

Environmental safety

- Access to baits by children, birds and animals, particularly cats, dogs, pigs and poultry, must be prevented
- Baits must not be placed where food, feed or water could become contaminated
- Remains of bait and bait containers must be removed after treatment and burned or buried
- Rodent bodies must be searched for and burned or buried. They must not be placed in refuse bins or on rubbish tips
- Take extreme care to prevent domestic animals having access to the bait

Hazard classification and safety precautions

Operator protection A [1]; U13 [1-10]; U20b [1, 2, 4-10]
Environmental protection E10b [7-10]; E15a [5, 6]
Storage and disposal D05 [1, 2, 4, 7-9]; D06a, D06c, D09c [7, 9]; D07 [1, 2, 4, 7-10]; D09a [1-10]; D11a [1, 2, 4-10]
Treated seed S06a [3]
Vertebrate/rodent control products V01a, V03a, V04a [3, 5, 6]; V01b, V03b, V04b [1, 2, 4, 7-10]; V02 [1-10]
Medical advice M03 [1, 2, 4-10]

SECTION 2

41 bromoxynil

A contact acting HBN herbicide
HRAC mode of action code: C3

Products

1	Alpha Bromolin 225 EC	Makhteshim	225 g/l	EC	13431
2	Alpha Bromotril P	Makhteshim	240 g/l	SC	13326
3	Butryflow	Nufarm UK	401.58 g/l	SC	14056
4	Flagon 400 EC	Makhteshim	400 g/l	EC	13470

Uses

- Annual dicotyledons in ***forage maize*** [2, 3]; ***grain maize***, ***sweetcorn*** [3]; ***linseed***, ***spring barley***, ***spring oats***, ***spring wheat***, ***winter barley***, ***winter oats***, ***winter wheat*** [1, 4]

Approval information

- Bromoxynil included in Annex I under EC Directive 91/414
- Accepted by BBPA for use on malting barley
- Approval expires 31/8/2010 [1, 2, 4]
- Approval expiry 31 Aug 2010 [4]

Efficacy guidance

- Spray when main weed flush has germinated and the largest are at the 4 leaf stage
- Weed control can be enhanced by using a split treatment spraying each application when the weeds are seedling to 2 true leaves. Apply the second treatment before the crop canopy covers the ground [2]

Restrictions

- Maximum number of treatments 1 per crop or yr or maximum total dose equivalent to one full dose treatment
- Do not apply with oils or other adjuvants
- Do not apply using hand-held equipment or at concentrations higher than those recommended
- Do not apply during frosty weather, drought, when soil is waterlogged, when rain expected within 4 h or to crops under any stress
- Take particular care to avoid drift onto neighbouring susceptible crops or open water surfaces

Crop-specific information

- Latest use: before 10 fully expanded leaf stage of crop for maize, sweetcorn; before crop 20 cm tall and before flower buds visible for linseed; before 2nd node detectable (GS 32) for cereals
- Apply to spring sown maize from 2 fully expanded leaves up to 9 fully expanded leaves [2]
- Apply to spring sown linseed from 1 fully expanded leaf to 20 cm tall [1, 4]
- Foliar scorch, which rapidly disappears without affecting growth, will occur if treatment made in hot weather or during rapid growth

Environmental safety

- Dangerous for the environment [1, 4]
- Very toxic to aquatic organisms
- Flammable [4]
- Keep livestock out of treated areas for at least 6 wk after treatment
- High risk to bees. Do not apply to crops in flower or to those in which bees are actively foraging
- Do not contaminate surface waters or ditches with chemical or used container
- LERAP Category B [3]

Hazard classification and safety precautions

Hazard Harmful [1-4]; Flammable [4]; Dangerous for the environment [1, 3, 4]
Risk phrases R20, R22a, R53a, R63 [1-4]; R22b, R38 [1, 4]; R36, R37 [4]; R43 [3, 4]; R50 [1, 3, 4]; R52 [2]
Operator protection A [1-4]; C [1, 2, 4]; H [1, 3, 4]; K [3]; U02a, U20a [2, 4]; U03, U11 [4]; U05a, U14, U15 [1-4]; U08, U13, U19a [1, 2, 4]; U10 [1, 4]; U20b [1]; U23a [1, 3, 4]
Environmental protection E06a [1, 4] (6 wk); E12d [1, 2, 4]; E12e [1, 2]; E13b [2]; E15a, E34 [1, 3, 4]; E16a [3]; E38 [1-4]
Consumer protection C02a [1, 4] (6 wk)
Storage and disposal D01, D02, D05, D09a, D12a [1-4]; D10b [1, 2, 4]; D10c [3]
Medical advice M03 [2, 3]; M04a [2, 4]; M04b [1]; M05b [4]

42 bromoxynil + diflufenican + ioxynil

A selective contact and translocated herbicide for cereals
HRAC mode of action code: C3 + F1 + C3

See also *diflufenican*
ioxynil

Products

Capture	Nufarm UK	300:50:200 g/l	SC	12514

Uses

- Annual dicotyledons in ***grass seed crops*** *(off-label)*, ***spring barley***, ***spring wheat***, ***triticale***, ***winter barley***, ***winter rye***, ***winter wheat***
- Chickweed in ***spring barley***, ***spring wheat***, ***triticale***, ***winter barley***, ***winter rye***, ***winter wheat***
- Field speedwell in ***spring barley***, ***spring wheat***, ***triticale***, ***winter barley***, ***winter rye***, ***winter wheat***
- Ivy-leaved speedwell in ***spring barley***, ***spring wheat***, ***triticale***, ***winter barley***, ***winter rye***, ***winter wheat***
- Knotgrass in ***spring barley***, ***spring wheat***, ***triticale***, ***winter barley***, ***winter rye***, ***winter wheat***
- Mayweeds in ***spring barley***, ***spring wheat***, ***triticale***, ***winter barley***, ***winter rye***, ***winter wheat***

Specific Off-Label Approvals (SOLAs)

- ***grass seed crops*** *20052240*

Approval information

- Bromoxynil, diflufenican and ioxynil included in Annex I under EC Directive 91/414
- Accepted by BBPA for use on malting barley

Efficacy guidance

- Best results obtained on small weeds and when competition removed early
- Good spray coverage essential for good activity

Restrictions

- Maximum number of treatments 1 per crop
- Do not treat crops undersown or to be undersown
- Do not treat frosted crops or those that are under stress from any cause
- Do not apply by hand-held equipment or at concentrations higher than those recommended

Crop-specific information

- Latest use: before 2nd node detectable (GS 32)

SECTION 2

Following crops guidance

- In the event of crop failure winter wheat may be drilled immediately after normal cultivations. Winter barley may be re-drilled after ploughing
- Land must be ploughed and an interval of 12 wk elapse after treatment before planting spring crops of wheat, barley, oilseed rape, peas, field beans, sugar beet, potatoes, carrots, edible brassicas or onions
- Successive treatments with any products containing diflufenican can lead to soil build-up and inversion ploughing to 150 mm must precede sowing any following non-cereal crop. Even where ploughing occurs some crops may be damaged

Environmental safety

- Dangerous for the environment
- Toxic to aquatic organisms
- Keep livestock out of treated areas for at least 2 wk after treatment
- Harmful to bees. Do not apply to crops in flower or to those in which bees are actively foraging
- LERAP Category B

Hazard classification and safety precautions

Hazard Harmful, Dangerous for the environment
Transport code 9
Packaging group III
UN number 3082
Risk phrases R22a, R51, R53a, R63
Operator protection A, C, H; U05a, U08, U13, U19a, U20b
Environmental protection E06a (2 wk); E12d, E12e, E15a, E16a, E34, E38
Consumer protection C02a (2 wk)
Storage and disposal D01, D02, D05, D09a, D10b, D12a
Medical advice M03

43 bromoxynil + ethofumesate + ioxynil

A post-emergence herbicide for new grass leys; product approval expired 2008
HRAC mode of action code: C3 + N + C3

See also *ethofumesate*
ioxynil

44 bromoxynil + ioxynil

A contact acting post-emergence HBN herbicide for cereals
HRAC mode of action code: C3 + C3

See also *ioxynil*

Products

1	Alpha Briotril	Makhteshim	240:160 g/l	EC	13580
2	Alpha Briotril Plus 19/19	Makhteshim	190:190 g/l	EC	13579
3	Oxytril CM	Bayer CropScience	200:200 g/l	EC	14511

Uses

- Annual dicotyledons in ***durum wheat*** *(off-label)*, ***grass seed crops*** *(off-label)*, ***undersown barley***, ***undersown oats***, ***undersown rye***, ***undersown triticale***, ***undersown wheat*** [1, 2]; ***spring barley***, ***spring oats***, ***spring rye***, ***spring wheat***, ***triticale***, ***winter barley***, ***winter oats***, ***winter rye***, ***winter wheat*** [1-3]

Specific Off-Label Approvals (SOLAs)

- ***durum wheat*** *20081259* [1], *20081260* [2]
- ***grass seed crops*** *20081259* [1], *20081260* [2]

Approval information

- Bromoxynil and ioxynil included in Annex I under EC Directive 91/414
- Accepted by BBPA for use on malting barley

SEE SECTION 3 FOR PRODUCTS ALSO REGISTERED

- Approval expires 31/8/2010 [1, 2]
- Approval expiry 31 Aug 2010 [2]

Efficacy guidance

- Best results achieved on young weeds growing actively in a highly competitive crop
- Do not apply during periods of drought or when rain imminent (some labels say 'if likely within 4 or 6 h')
- Recommended for tank mixture with hormone herbicides to extend weed spectrum. See labels for details

Restrictions

- Maximum number of treatments 1 per crop
- Do not spray crops stressed by drought, waterlogging or other factors
- Do not roll or harrow for several days before or after spraying. Number of days specified varies with product. See label for details
- Do not apply by hand-held equipment or at concentrations higher than those recommended

Crop-specific information

- Latest use: before 2nd node detectable stage (GS 31)
- HI (animal consumption) 6 wk
- Apply to winter or spring cereals from 1-2 fully expanded leaf stage, but before second node detectable (GS 32)
- Spray oats in spring when danger of frost past. Do not spray winter oats in autumn
- Apply to undersown cereals pre-sowing or pre-emergence of legume provided cover crop is at correct stage. Only spray trefoil pre-sowing
- On crops undersown with grasses alone or on direct re-seeds apply from 2-leaf stage of grass

Environmental safety

- Dangerous for the environment
- Very toxic to aquatic organisms
- Flammable [1, 2]
- Keep livestock out of treated areas for at least 6 wk after treatment and until poisonous weeds, such as ragwort, have died down and become unpalatable

Hazard classification and safety precautions

Hazard Harmful, Dangerous for the environment [1-3]; Flammable [1, 2]
Transport code 3 [1, 2]; 9 [3]
Packaging group III
UN number 1993 [1, 2]; 3082 [3]
Risk phrases R22a, R22b, R43, R50, R53a, R63, R67 [1-3]; R36, R37, R38 [1, 2]; R66 [3]
Operator protection A, C [1-3]; H [3]; U02a, U14, U15, U20a [1, 2]; U05a, U20b [3]; U08, U13, U19a, U23a [1-3]
Environmental protection E06a [1, 2] (6 wk); E07a, E12d, E12e [3]; E13b, E34, E38 [1-3]
Consumer protection C02a [1, 2] (6 wk); C02a [3] (14 d)
Storage and disposal D01, D02, D12a [1-3]; D05 [1, 2]; D09a, D11a [3]
Medical advice M03, M05b

45 bromoxynil + prosulfuron

A contact and residual herbicide mixture for maize
HRAC mode of action code: C3 + B

See also *prosulfuron*

Products

Jester	Syngenta	60:3% w/w	WG	13500

Uses

- Annual dicotyledons in ***forage maize, grain maize, millet*** *(off-label)*
- Black bindweed in ***forage maize, grain maize***
- Chickweed in ***forage maize, grain maize***
- Fat hen in ***millet*** *(off-label)*
- Hemp-nettle in ***forage maize, grain maize***

- Knotgrass in ***forage maize*, *grain maize***
- Mayweeds in ***forage maize*, *grain maize***

Specific Off-Label Approvals (SOLAs)

- ***millet*** *20082960 expires 01 May 2010*

Approval information

- Bromoxynil and prosulfuron included in Annex I under EC Directive 91/414

Efficacy guidance

- Prosulfuron is a member of the ALS-inhibitor group of herbicides and products should be used in a planned Resistance Management strategy. See Section 5 for more information

Restrictions

- Maximum total dose equivalent to one full dose treatment
- Do not treat crops grown for seed production
- Consult before use on crops intended for processing
- Do not use in frosty weather or on crops under stress
- Do not tank mix with organophosphate insecticides or apply in tank mix or sequence with any other sulfonylurea
- Do not apply by knapsack sprayer or in volumes less than those recommended
- Take care to wash out sprayers thoroughly. See label for details

Crop-specific information

- Latest use: before 7 crop leaves unfolded for maize
- Apply post-emergence up to when the crop has six unfolded leaves
- Product must be used with a non-ionic wetter that is not an organosilicone
- Apply in cool conditions, or during the evening, to avoid scorch

Following crops guidance

- Winter or spring cereals, winter or spring beans, spring sown peas or oilseed rape may be sown following normal harvest of a treated crop. Mould-board ploughing to 20 cm is recommended in some cases

Environmental safety

- Dangerous for the environment
- Very toxic to aquatic organisms
- Take special care to avoid drift outside the target area

Hazard classification and safety precautions

Hazard Toxic, Dangerous for the environment
Transport code 6.1
Packaging group III
UN number 2588
Risk phrases R22a, R23, R36, R43, R50, R53a, R63
Operator protection A, D, H; U05a, U11, U15, U19a, U20c, U23a
Environmental protection E15b, E34, E38
Storage and disposal D01, D02, D05, D09a, D10c, D12a
Medical advice M04a

46 bromoxynil + terbuthylazine

A post-emergence contact and residual herbicide for maize
HRAC mode of action code: C3 + C1

See also *terbuthylazine*

Products

Templar	Makhteshim	200:300 g/l	SC	10254

Uses

- Annual and perennial weeds in ***forage maize (under plastic mulches)*** *(off-label)*, ***grain maize under plastic mulches*** *(off-label)*, ***sweetcorn under plastic mulches*** *(off-label)*
- Annual dicotyledons in ***forage maize*, *sweetcorn*** *(off-label)*
- Annual grasses in ***forage maize (under plastic mulches)*** *(off-label)*, ***grain maize under plastic mulches*** *(off-label)*, ***sweetcorn under plastic mulches*** *(off-label)*

SEE SECTION 3 FOR PRODUCTS ALSO REGISTERED

Specific Off-Label Approvals (SOLAs)

- ***forage maize (under plastic mulches)*** *20080279*
- ***grain maize under plastic mulches*** *20080279*
- ***sweetcorn*** *20072179*
- ***sweetcorn under plastic mulches*** *20080279*

Approval information

- Bromoxynil included in Annex I under EC Directive 91/414
- Terbuthylazine has not been included in Annex 1 but the date of withdrawal has yet to be set.

Efficacy guidance

- Best results obtained from treatment when main flush of weeds has germinated and largest at 4 leaf stage
- Control can be enhanced by use of split dose treatment but second spray must be applied before crop canopy covers ground
- Soils should be moist at time of treatment and should not be disturbed afterwards
- Residual activity may be reduced on soils with high OM content, or where high amounts of slurry or manure have been applied

Restrictions

- Maximum total dose equivalent to one full dose treatment
- Foliar scorch may occur after treatment in hot conditions or when growth particularly rapid. In such conditions spray in the evening
- Do not apply during frosty weather, drought, when soil is waterlogged or when rain expected within 12 hr
- Do not treat crops under any stress whatever
- Product should be used only where damaging populations of competitive weeds have emerged. Yield may be reduced when used in absence of significant weed populations
- Succeeding crops may not be planted until the spring following application. Mould board plough to 150 mm before doing so

Crop-specific information

- Apply before 9 fully expanded leaf stage

Following crops guidance

- Keep livestock out of treated areas for at least 6 wk

Environmental safety

- Dangerous for the environment
- Very toxic to aquatic organisms
- Keep livestock out of treated areas for at least 6 wk after treatment
- Harmful to bees. Do not apply to crops in flower or to those in which bees are actively foraging. Do not apply when flowering weeds are present
- Do not apply by hand-held equipment or at concentrations higher than those recommended

Hazard classification and safety precautions

Hazard Harmful, Dangerous for the environment
Transport code 9
Packaging group III
UN number 3082
Risk phrases R20, R22a, R50, R53a
Operator protection A, C, H; U02a, U05a, U08, U13, U14, U15, U19a, U20a, U23a
Environmental protection E06a (6 wk); E12d, E12e, E13b, E38
Storage and disposal D01, D02, D12a
Medical advice M04a

47 bromuconazole

A systemic triazole fungicide for cereals
FRAC mode of action code: 3

Products

Jazz	Interfarm	200 g/l	SC	13941

Uses

- Brown rust in ***spring barley, spring wheat, winter barley, winter wheat***
- Eyespot in ***spring barley, spring wheat, winter barley, winter wheat***
- Mildew in ***spring barley, spring wheat, winter barley, winter wheat***
- Net blotch in ***spring barley, spring wheat, winter barley, winter wheat***
- Rhynchosporium in ***spring barley, spring wheat, winter barley, winter wheat***
- Septoria leaf blotch in ***spring barley, spring wheat, winter barley, winter wheat***
- Sooty moulds in ***spring barley, spring wheat, winter barley, winter wheat***
- Yellow rust in ***spring barley, spring wheat, winter barley, winter wheat***

Approval information

- Bromuconazole has not been included in Annex 1.
- Approval expiry 03 May 2010 [1]
- Accepted by BBPA on malting barley before ear emergence only

Efficacy guidance

- Best results achieved from applications before disease becomes established
- If disease pressure persists a second treatment may be required to prevent late season attacks
- Best control of ear diseases in wheat obtained from application at or after the ear completely emerged stage
- More consistent control of powdery mildew obtained from tank mixture with fenpropimorph. See label
- Bromuconazole is a DMI fungicide. Resistance to some DMI fungicides has been identified in Septoria leaf blotch which may seriously affect performance of some products. For further advice contact a specialist advisor and visit the Fungicide Resistance Action Group (FRAG)-UK website

Restrictions

- Maximum number of treatments 2 per crop

Crop-specific information

- Latest use varies with dose used and crop - see label
- Apply at any time from the tillering stage up to the latest times indicated for each crop

Following crops guidance

- Only cereals, oilseed rape, field beans, peas, potatoes, linseed or Italian ryegrass may be sown as following crops

Environmental safety

- Dangerous to fish or other aquatic life. Do not contaminate surface waters or ditches with chemical or used container
- LERAP Category B

Hazard classification and safety precautions

Hazard Dangerous for the environment
Transport code 9
Packaging group III
UN number 3082
Risk phrases R53a
Operator protection A, C, H; U08, U19a, U20b
Environmental protection E15a, E16b, E38
Storage and disposal D01, D02, D09a, D10b, D12a
Medical advice M05a

48 bupirimate

A systemic aminopyrimidinol fungicide active against powdery mildew
FRAC mode of action code: 8

Products

Nimrod	Makhteshim	250 g/l	EC	13046

Uses

- Powdery mildew in ***apples, begonias, blackcurrants, chrysanthemums, courgettes*** *(outdoor only)*, ***gooseberries, hops, marrows*** *(outdoor only)*, ***pears, protected cucumbers, protected***

SEE SECTION 3 FOR PRODUCTS ALSO REGISTERED

strawberries, ***protected tomatoes*** *(off-label)*, ***pumpkins*** *(off-label)*, ***raspberries*** *(outdoor only)*, ***redcurrants*** *(off-label)*, ***roses***, ***squashes*** *(off-label)*, ***strawberries***, ***whitecurrants*** *(off-label)*

Specific Off-Label Approvals (SOLAs)

- ***protected tomatoes*** *20070997*
- ***pumpkins*** *20070994*
- ***redcurrants*** *20082082*
- ***squashes*** *20070994*
- ***whitecurrants*** *20082082*

Approval information

- Accepted by BBPA for use on hops

Efficacy guidance

- On apples during periods that favour disease development lower doses applied weekly give better results than higher rates fortnightly
- Not effective in protected crops against strains of mildew resistant to bupirimate

Restrictions

- Maximum number of treatments or maximum total dose depends on crop and dose (see label for details)

Crop-specific information

- HI: 1 d for apples, pears, strawberries; 2 d for cucurbits; 7 d for blackcurrants; 8 d for raspberries; 14 d for gooseberries, hops
- Apply before or at first signs of disease and repeat at 7-14 d intervals. Timing and maximum dose vary with crop. See label for details
- With apples, hops and ornamentals cultivars may vary in sensitivity to spray. See label for details
- If necessary to spray cucurbits in winter or early spring spray a few plants 10-14 d before spraying whole crop to test for likelihood of leaf spotting problem
- On roses some leaf puckering may occur on young soft growth in early spring or under low light intensity. Avoid use of high rates or wetter on such growth
- Never spray flowering begonias (or buds showing colour) as this can scorch petals
- Do not mix with other chemicals for application to begonias, cucumbers or gerberas

Environmental safety

- Dangerous for the environment
- Toxic to aquatic organisms
- Flammable
- Product has negligible effect on *Phytoseiulus* and *Encarsia* and may be used in conjunction with biological control of red spider mite

Hazard classification and safety precautions

Hazard Harmful, Flammable, Dangerous for the environment
Transport code 3
Packaging group III
UN number 1993
Risk phrases R22b, R38, R51, R53a
Operator protection A, C; U05a, U20b
Environmental protection E15a
Storage and disposal D01, D02, D09a, D10c
Medical advice M05b

49 captan

A protectant phthalimide fungicide with horticultural uses
FRAC mode of action code: M4

Products

1	Alpha Captan 80 WDG	Makhteshim	80% w/w	WG	07096
2	Alpha Captan 83 WP	Makhteshim	83% w/w	WP	04806

Uses

- Black spot in ***roses*** [1, 2]
- Botrytis in ***strawberries*** [1, 2]

- Gloeosporium rot in ***apples*** [1, 2]
- Scab in ***apples, pears*** [1, 2]; ***quinces*** *(off-label)* [1]

Specific Off-Label Approvals (SOLAs)

- ***quinces*** *20091388* [1]

Approval information

- Captan included in Annex I under EC Directive 91/414

Restrictions

- Maximum number of treatments 12 per yr on apples and pears as pre-harvest sprays
- Product must not be used as dip or drench on apples or pears [2]
- Do not use on apple cultivars Bramley, Monarch, Winston, King Edward, Spartan, Kidd's Orange or Red Delicious or on pear cultivar D'Anjou
- Do not mix with alkaline materials or oils
- Do not use on fruit for processing
- Powered visor respirator with hood and neck cape must be used when handling concentrate

Crop-specific information

- HI apples, pears 14 d; strawberries 7 d
- For control of scab apply at bud burst and repeat at 10-14 d intervals until danger of scab infection ceased
- For suppression of fruit storage rots apply from late Jul and repeat at 2-3 wk intervals
- For black spot control in roses apply after pruning with 3 further applications at 14 d intervals or spray when spots appear and repeat at 7-10 d intervals
- For grey mould in strawberries spray at first open flower and repeat every 7-10 d [1]
- Do not leave diluted material for more than 2 h. Agitate well before and during spraying

Environmental safety

- Dangerous for the environment
- Very toxic to aquatic organisms

Hazard classification and safety precautions

Hazard Toxic [2]; Harmful [1]; Dangerous for the environment [1, 2]
Transport code 6.1 [2]; 9 [1]
Packaging group III
UN number 2588 [2]; 3077 [1]
Risk phrases R23, R37, R38 [2]; R36, R40, R43, R50, R53a [1, 2]; R41 [1]
Operator protection A [1, 2]; D, E, H [1]; U05a, U11, U19a [1, 2]; U09a, U20c [1]; U14, U15, U20b [2]
Environmental protection E15a
Storage and disposal D01, D02, D09a, D11a, D12a

50 carbendazim

A systemic benzimidazole fungicide with curative and protectant activity
FRAC mode of action code: 1

Products

1	AgriGuard Pro-Turf	AgriGuard	500 g/l	SC	11837
2	Delsene 50 Flo	Nufarm UK	500 g/l	SC	11452
3	Ringer	Barclay	500 g/l	SC	13289

Uses

- Dollar spot in ***managed amenity turf*** [1]
- Eyespot in ***spring barley, winter barley, winter wheat*** [2]
- Foliar disease control in ***durum wheat*** *(off-label)*, ***spring rye*** *(off-label)*, ***triticale*** *(off-label)*, ***winter rye*** *(off-label)* [2]
- Fusarium patch in ***managed amenity turf*** [1]
- Light leaf spot in ***spring oilseed rape, winter oilseed rape*** [2]
- Rhynchosporium in ***spring barley, winter barley*** [2]
- Wormcast formation in ***managed amenity turf*** [1, 3]

Specific Off-Label Approvals (SOLAs)

- ***durum wheat*** *20073118 expires 31 Dec 2010* [2]

SEE SECTION 3 FOR PRODUCTS ALSO REGISTERED

- ***spring rye*** *20073118 expires 31 Dec 2010* [2]
- ***triticale*** *20073118 expires 31 Dec 2010* [2]
- ***winter rye*** *20073118 expires 31 Dec 2010* [2]

Approval information

- Carbendazim has been included in Annex I under Directive 91/414.
- Accepted by BBPA for use on malting barley
- Approval expiry 31 Dec 2010 [2]

Efficacy guidance

- Products vary in the diseases listed as controlled for several crops. Labels must be consulted for full details and for rates and timings
- Mostly applied as spray or drench. Spray treatments normally applied at first sign of disease and repeated after 1 mth if required
- Apply as drench rather than spray where red spider mite predators are being used
- Do not apply during drought conditions. Rain or irrigation after treatment may improve control
- Leave 4 mth intervals where used only for worm cast control [1, 3]
- To delay appearance of resistant strains alternate treatment with non-MBC fungicide. Eyespot in cereals and *Botrytis cinerea* in many crops are now widely resistant. To avoid the development of resistnce, a maximum of 2 applications of any MBC product (thiophanate-methyl or carbendazim) are allowed in any one crop. Avoid using MBC fungicides alone.

Restrictions

- Maximum number of treatments (including applications of any product containing benomyl, carbendazim or thiophanate-methyl) varies with crop treated and product used - see labels for details
- Do not treat crops or turf suffering from drought or other physical or chemical stress
- Consult processors before using on crops for processing
- Not compatible with alkaline products such as lime sulphur

Crop-specific information

- Latest use: before 2nd node detectable (GS32) for cereals
- HI 2 d for inert substrate cucumbers, protected tomatoes; 7-14 d (depends on dose) for apples, pears; 14 d for mushrooms, protected celery; 21 d for oilseed rape, dwarf beans, lettuce, 6 mth for cane fruit
- On turf apply as preventative treatment in spring or autumn during periods of high disease risk [1, 3]
- Apply as a drench to control soil-borne diseases in cucumbers and tomatoes and as a pre-planting dip treatment for bulbs

Environmental safety

- Dangerous for the environment
- Very toxic to aquatic organisms
- After use dipping suspension must not be discharged directly into ditches or drains

Hazard classification and safety precautions

Hazard Toxic [2, 3]; Harmful [1]; Dangerous for the environment [1-3]
Transport code 9
Packaging group III
UN number 3082
Risk phrases R22a [1]; R46, R50, R53a, R60, R61 [2, 3]; R68 [3]
Operator protection A, C, H, M; U14, U19a [3]; U20a [1]; U20b [2, 3]
Environmental protection E13c [2]; E14a, E34, E38 [3]; E15a [1]
Storage and disposal D01, D02, D11a [2]; D05, D09a [1-3]; D07, D10b [1]; D08, D10a, D12a [3]
Medical advice M04a [3]

51 carbendazim + epoxiconazole

A systemic fungicide mixture for managed amenity turf but product withdrawn in 2008
FRAC mode of action code: 1 + 3

See also epoxiconazole

52 carbendazim + flusilazole

A broad-spectrum systemic and protectant fungicide for cereals, oilseed rape and sugar beet
FRAC mode of action code: 1 + 3

See also flusilazole

Products

1	Contrast	DuPont	125:250 g/l	SC	13107
2	Harvesan	DuPont	125:250 g/l	SC	13106
3	Punch C	DuPont	125:250 g/l	SC	13023

Uses

- Brown rust in ***spring barley***, ***spring wheat***, ***winter barley***, ***winter wheat***
- Canker in ***spring oilseed rape*** *(reduction)*, ***winter oilseed rape*** *(reduction)*
- Eyespot in ***spring barley***, ***spring wheat***, ***winter barley***, ***winter wheat***
- Glume blotch in ***spring wheat***, ***winter wheat***
- Light leaf spot in ***spring oilseed rape***, ***winter oilseed rape***
- Phoma leaf spot in ***spring oilseed rape***, ***winter oilseed rape***
- Powdery mildew in ***spring barley***, ***spring wheat***, ***sugar beet***, ***winter barley***, ***winter wheat***
- Rust in ***sugar beet***
- Septoria leaf blotch in ***spring wheat***, ***winter wheat***
- Yellow rust in ***spring barley***, ***spring wheat***, ***winter barley***, ***winter wheat***

Approval information

- Carbendazim included in Annex I under Directive 91/414 for 3 years from January 2007 and all approvals for products containing this active ingredient will expire in December 2009
- Flusilazole has been reinstated in Annex 1 under Directive 91/414 following an appeal
- Accepted by BBPA for use on malting barley
- Approval expiry 31 Dec 2010 [3]

Efficacy guidance

- Apply at early stage of disease development or in routine preventive programme
- Most effective timing of treatment on cereals varies with disease. See label for details
- Higher rate active against both MBC-sensitive and MBC-resistant eyespot
- Rain occurring within 2 h of spraying may reduce effectiveness
- To prevent build-up of resistant strains of cereal mildew tank mix with approved morpholine fungicide
- Flusilazole is a DMI fungicide. Resistance to some DMI fungicides has been identified in Septoria leaf blotch which may seriously affect performance of some products. For further advice contact a specialist advisor and visit the Fungicide Resistance Action Group (FRAG)-UK website

Restrictions

- Maximum number of treatments 1 per crop on sugar beet; 2 per crop on cereals and oilseed rape
- Do not apply to crops under stress or during frosty weather

Crop-specific information

- Latest use on cereals varies with crop and dose used - see label for details; before first flower opened stage for oilseed rape
- HI 7 wk for sugar beet
- Treat oilseed rape in autumn when leaf lesions first appear and spring from the start of stem extension when disease appears

Environmental safety

- Dangerous for the environment
- Toxic to aquatic organisms
- Dangerous to fish or other aquatic life. Do not contaminate surface waters or ditches with chemical or used container

Hazard classification and safety precautions

Hazard Toxic, Dangerous for the environment
Transport code 9
Packaging group III
UN number 3082
Risk phrases R22a, R40, R41, R43, R46, R51, R53a, R60, R61

Operator protection A, C, H, M; U05a, U11, U19a, U20b
Environmental protection E13b, E34, E38
Storage and disposal D01, D02, D05, D09a, D10b, D12a
Medical advice M04a

53 carbetamide

A residual pre- and post-emergence carbamate herbicide for a range of field crops
HRAC mode of action code: K2

Products

1	Crawler	Makhteshim	60% w/w	WG	11840
2	Riot	Makhteshim	60% w/w	WG	13380
3	Scrum	Makhteshim	60% w/w	WG	13386

Uses

- Annual dicotyledons in ***cabbage seed crops, collards, fodder rape seed crops, kale seed crops, lucerne, red clover, sainfoin, spring cabbage, sugar beet seed crops, swede seed crops, turnip seed crops, white clover, winter field beans, winter oilseed rape*** [1-3]; ***evening primrose*** *(off-label)*, ***honesty*** *(off-label)*, ***linseed*** *(off-label)*, ***lupins*** *(off-label)*, ***mustard*** *(off-label)* [1]
- Annual grasses in ***cabbage seed crops, collards, fodder rape seed crops, kale seed crops, lucerne, red clover, sainfoin, spring cabbage, sugar beet seed crops, swede seed crops, turnip seed crops, white clover, winter field beans, winter oilseed rape*** [1-3]; ***evening primrose*** *(off-label)*, ***honesty*** *(off-label)*, ***linseed*** *(off-label)*, ***lupins*** *(off-label)*, ***mustard*** *(off-label)* [1]
- Volunteer cereals in ***cabbage seed crops, collards, fodder rape seed crops, kale seed crops, lucerne, red clover, sainfoin, spring cabbage, sugar beet seed crops, swede seed crops, turnip seed crops, white clover, winter field beans, winter oilseed rape*** [1-3]

Specific Off-Label Approvals (SOLAs)

- ***evening primrose*** *20061236* [1]
- ***honesty*** *20061236* [1]
- ***linseed*** *20061236* [1]
- ***lupins*** *20061237* [1]
- ***mustard*** *20061236* [1]

Efficacy guidance

- Best results pre- or early post-emergence of weeds under cool, moist conditions. Adequate soil moisture is essential
- Dicotyledons controlled include chickweed, cleavers and speedwell
- Weed growth stops rapidly after treatment but full effects may take 6-8 wk to develop
- Various tank mixes are recommended to broaden the weed spectrum. See label for details
- Always follow WRAG guidelines for preventing and managing herbicide resistant weeds. See Section 5 for more information

Restrictions

- Maximum number of treatments 1 per crop for all crops
- Do not treat any crop on waterlogged soil
- Do not use on soils with more than 10% organic matter as residual activity is impaired
- Do not apply during prolonged periods of cold weather when weeds are dormant

Crop-specific information

- HI 6 wk for all crops
- Apply to brassicas from mid-Oct to end-Feb provided crop has at least 4 true leaves (spring cabbage, spring greens), 3-4 true leaves (seed crops, oilseed rape)
- Apply to established lucerne or sainfoin from Nov to end-Feb
- Apply to established red or white clover from Feb to mid-Mar

Following crops guidance

- Succeeding crops may be sown 2 wk after treatment for brassicas, field beans, 8 wk after treatment for peas, runner beans, 16 wk after treatment for cereals, maize
- Ploughing is not necessary before sowing subsequent crops

Environmental safety

- Do not graze crops for at least 6 wk after treatment
- Some pesticides pose a greater threat of contamination of water than others and carbetamide is one of these pesticides. Take special care when applying carbetamide near water and do not apply if heavy rain is forecast

Hazard classification and safety precautions

UN number n/a
Operator protection U20c
Environmental protection E15a
Storage and disposal D09a, D11a

54 carbon dioxide (commodity substance)

A gas for the control of trapped rodents and other vertebrates

Products

carbon dioxide	various	99.9%	GA	-

Uses

- Birds in ***traps***
- Mice in ***traps***
- Rats in ***traps***

Approval information

- Approval for the use of carbon dioxide as a commodity substance was granted on 8 October 1993 by Ministers under regulation 5 of the Control of Pesticides Regulations 1986
- Only to be used where a licence has been issued in accordance with Section 16(1) of the Wildlife and Countryside Act 1981
- Carbon dioxide is being supported for review in the fourth stage of the EC Review Programme under Directive 91/414. Whether or not it is included in Annex I, the existing commodity chemical approval will have to be revoked because substances listed in Annex I must be approved for marketing and use under the EC regime. If carbon dioxide is included in Annex I, products containing it will need to gain approval in the normal way if they carry label claims for pesticidal activity

Efficacy guidance

- Use to destroy trapped rodent pests
- Use to control birds covered by general licences issued by the Agriculture and Environment Departments under Section 16(1) of the Wildlife and Countryside Act (1981) for the control of opportunistic bird species, where birds have been trapped or stupefied with alphachloralose/ seconal

Restrictions

- Operators must wear self-contained breathing apparatus when carbon dioxide levels are greater than 0.5% v/v
- Operators must be suitably trained and competent

Environmental safety

- Unprotected persons and non-target animals must be excluded from the treatment enclosures and surrounding areas unless the carbon dioxide levels are below 0.5% v/v

Hazard classification and safety precautions

Operator protection G

55 carboxin

A carboxamide fungicide available only in mixtures
FRAC mode of action code: 7

SEE SECTION 3 FOR PRODUCTS ALSO REGISTERED

56 carboxin + thiram

A fungicide seed dressing for cereals
FRAC mode of action code: 7 + M3

See also thiram

Products

Anchor	Chemtura	200:200 g/l	FS	08684

Uses

- Bunt in ***winter wheat*** *(seed treatment)*
- Covered smut in ***spring barley*** *(seed treatment)*, ***spring oats*** *(seed treatment)*, ***winter barley*** *(seed treatment)*, ***winter oats*** *(seed treatment)*
- Fusarium foot rot and seedling blight in ***spring barley*** *(seed treatment)*, ***spring oats*** *(seed treatment)*, ***spring rye*** *(seed treatment)*, ***spring wheat*** *(seed treatment)*, ***triticale*** *(seed treatment)*, ***winter barley*** *(seed treatment)*, ***winter oats*** *(seed treatment)*, ***winter rye*** *(seed treatment)*, ***winter wheat*** *(seed treatment)*
- Leaf stripe in ***spring barley*** *(seed treatment)*, ***winter barley*** *(seed treatment - reduction)*
- Loose smut in ***spring barley*** *(seed treatment - reduction)*, ***spring oats*** *(seed treatment - reduction)*, ***winter barley*** *(seed treatment - reduction)*, ***winter oats*** *(seed treatment - reduction)*
- Net blotch in ***spring barley*** *(seed treatment)*
- Septoria seedling blight in ***spring wheat*** *(seed treatment)*, ***winter wheat*** *(seed treatment)*
- Stinking smut in ***spring wheat*** *(seed treatment)*

Approval information

- Thiram included in Annex I under EC Directive 91/414
- Accepted by BBPA for use on malting barley

Efficacy guidance

- Apply through suitable liquid flowable seed treating equipment of the batch treatment or continuous flow type where a secondary mixing auger is fitted
- Drill flow may be affected by treatment. Always re-calibrate seed drill before use

Restrictions

- Maximum number of treatments 1 per batch of seed
- Do not treat seed with moisture content above 16%
- Do not apply to cracked, split or sprouted seed
- Do not store treated seed from one season to the next

Crop-specific information

- Latest use: pre-drilling

Environmental safety

- Dangerous for the environment
- Very toxic to aquatic organisms
- Do not use treated seed as food or feed
- Treated seed harmful to game and wildlife

Hazard classification and safety precautions

Hazard Harmful, Dangerous for the environment
Transport code 9
Packaging group III
UN number 3082
Risk phrases R22a, R48, R50, R53a
Operator protection A, D, H; U05a, U09a, U20c
Environmental protection E15a, E34, E38
Storage and disposal D01, D02, D05, D06a, D09a, D10a
Treated seed S01, S02, S04b, S05, S06a, S07
Medical advice M03

57 carfentrazone-ethyl

A triazolinone contact herbicide
HRAC mode of action code: E

Products

1	Aurora 50 WG	Belchim	50% w/w	WG	11613
2	Carfen	AgriGuard	50% w/w	WG	13005
3	Headlite	AgChem Access	60 g/l	ME	13624
4	Shark	Belchim	60 g/l	ME	12762
5	Spotlight Plus	Belchim	60 g/l	ME	12436

Uses

- Annual dicotyledons in ***all edible crops (outdoor)*** *(before planting)*, ***all non-edible crops (outdoor)*** *(before planting)*, ***forest*** *(off-label)*, ***forest nurseries*** *(off-label)*, ***ornamental plant production*** *(off-label)*, ***potatoes***, ***protected forest nurseries*** *(off-label)*, ***protected ornamentals*** *(off-label)*, ***protected soft fruit*** *(off-label)*, ***soft fruit*** *(off-label)* [4]; ***grass seed crops*** *(off-label)*, ***spring rye*** *(off-label)*, ***winter rye*** *(off-label)* [1]; ***narcissi*** *(off-label)* [5]
- Black bindweed in ***potatoes*** [4]
- Cleavers in ***durum wheat***, ***spring barley***, ***spring oats***, ***spring wheat***, ***triticale***, ***winter barley***, ***winter oats***, ***winter wheat*** [1, 2]; ***potatoes*** [4]
- Desiccation in ***blackberries*** *(off-label)*, ***protected blackberries*** *(off-label)*, ***protected raspberries*** *(off-label)*, ***protected rubus hybrids*** *(off-label)*, ***raspberries*** *(off-label)*, ***rubus hybrids*** *(off-label)* [4]
- Fat hen in ***potatoes*** [4]
- Haulm destruction in ***seed potatoes***, ***ware potatoes*** [3, 5]
- Ivy-leaved speedwell in ***durum wheat***, ***spring barley***, ***spring oats***, ***spring wheat***, ***triticale***, ***winter barley***, ***winter oats***, ***winter wheat*** [1, 2]; ***potatoes*** [4]
- Knotgrass in ***potatoes*** [4]
- Redshank in ***potatoes*** [4]
- Volunteer oilseed rape in ***all edible crops (outdoor)*** *(before planting)*, ***all non-edible crops (outdoor)*** *(before planting)*, ***potatoes*** [4]

Specific Off-Label Approvals (SOLAs)

- ***forest nurseries*** *20082922* [4]
- ***grass seed crops*** *20060954* [1]
- ***narcissi*** *20091003* [5]
- ***protected forest nurseries*** *20082922* [4]
- ***protected soft fruit*** *20082922* [4]
- ***soft fruit*** *20082922* [4]
- ***spring rye*** *20060954* [1]
- ***winter rye*** *20060954* [1]

Approval information

- Carfentrazone-ethyl included in Annex I under EC Directive 91/414
- Accepted by BBPA for use on malting barley and hops

Efficacy guidance

- Best weed control results achieved from good spray cover applied to small actively growing weeds [1, 2, 4]
- Carfentrazone-ethyl acts by contact only; see label for optimum timing on specified weeds. Weeds emerging after application will not be controlled [1, 2, 4]
- For weed control in cereals use as two spray programme with one application in autumn and one in spring [1, 2]
- Efficacy of haulm destruction will be reduced where flailed haulm covers the stems at application [5]
- For potato crops with very dense vigorous haulm or where regrowth occurs following a single application a second application may be necessary to achieve satisfactory desiccation. A minimum interval between applications of 7 d should be observed to achieve optimum performance [5]

SEE SECTION 3 FOR PRODUCTS ALSO REGISTERED

Restrictions
- Maximum number of treatments 2 per crop for cereals (1 in Autumn and 1 in Spring) [1, 2]; 1 per crop for weed control in potatoes [4]; 1 per yr for pre-planting treatments [4]
- Maximum total dose for potato haulm destruction equivalent to 1.6 full dose treatments [5]
- Do not treat cereal crops under stress from drought, waterlogging, cold, pests, diseases, nutrient or lime deficiency or any factors reducing plant growth [1, 2]
- Do not treat cereals undersown with clover or other legumes [1, 2]
- Allow at least 2-3 wk between application and lifting potatoes to allow skins to set if potatoes are to be stored [5]
- Follow label instructions for sprayer cleaning
- Do not apply through knapsack sprayers
- Contact processor before using a split dose on potatoes for processing

Crop-specific information
- Latest use: 1 mth before planting edible and non-edible crops; before 3rd node detectable (GS 33) on cereals [1, 2]
- HI 7 d for potatoes [5]
- For weed control in cereals treat from 2 leaf stage [1, 2]
- When used as a treatment prior to planting a subsequent crop apply before weeds exceed maximum sizes indicated in the label [4]
- If treated potato tubers are to be stored then allow at least 2-3 wk between the final application and lifting to allow skins to set [5]

Following crops guidance
- No restrictions apply on the planting of succeeding crops 1 mth after application to potatoes for haulm destruction or as a pre-planting treatment, or 3 mth after application to cereals for weed control
- In the event of failure of a treated cereal crop, all cereals, ryegrass, maize, oilseed rape, peas, sunflowers, *Phacelia*, vetches, carrots or onions may be planted within 1 mth of treatment

Environmental safety
- Dangerous for the environment
- Very toxic to aquatic organisms
- Some non-target crops are sensitive. Avoid drift onto broad-leaved plants outside the treated area, or onto ponds waterways or ditches

Hazard classification and safety precautions

Hazard Irritant, Dangerous for the environment
Transport code 9
Packaging group III
UN number 3077 [1, 2]; 3082 [3-5]
Risk phrases R38 [3, 5]; R43, R50, R53a [1-5]
Operator protection A, H; U05a, U14 [1-5]; U08, U13 [1, 2]; U20a [3-5]
Environmental protection E15a [1, 4]; E15b [2, 3, 5]; E34 [1, 2]; E38 [1-5]
Storage and disposal D01, D02, D09a, D12a [1-5]; D10b [1, 2]; D10c [3-5]
Medical advice M05a [1]

58 carfentrazone-ethyl + flupyrsulfuron-methyl

A foliar and residual acting herbicide for cereals
HRAC mode of action code: E + B

See also flupyrsulfuron-methyl

Products

Lexus Class	DuPont	33.3:16.7% w/w	WG	10809

Uses
- Annual dicotyledons in ***durum wheat*** *(off-label)*, ***grass seed crops*** *(off-label)*, ***triticale***, ***winter oats***, ***winter rye***, ***winter wheat***
- Blackgrass in ***triticale***, ***winter oats***, ***winter rye***, ***winter wheat***

Specific Off-Label Approvals (SOLAs)

- ***durum wheat*** *20080895*
- ***grass seed crops*** *20080895*

Approval information

- Carfentrazone-ethyl and flupyrsulfuron-methyl included in Annex I under EC Directive 91/414

Efficacy guidance

- Best results obtained when applied to small actively growing weeds
- Good spray cover of weeds must be obtained
- Increased degradation of active ingredient in high soil temperatures reduces residual activity
- Weed control may be reduced in dry soil conditions but susceptible weeds germinating soon after treatment will be controlled if adequate soil moisture present
- Symptoms may not be apparent on some weeds for up to 4 wk, depending on weather conditions
- Blackgrass should be treated from 1 leaf but before first node
- Flupyrsulfuron-methyl is a member of the ALS-inhibitor group of herbicides. To avoid the build up of resistance do not use any product containing an ALS-inhibitor herbicide with claims for control of grass weeds more than once on any crop
- Use these products as part of a resistance management strategy that includes cultural methods of control and does not use ALS inhibitors as the sole chemical method of grass weed control

Restrictions

- Maximum number of treatments 1 per crop
- Do not use on barley, on crops undersown with grasses or legumes, or any other broad-leaved crop
- Do not apply within 7 d of rolling
- Do not treat any crop suffering from drought, waterlogging, pest or disease attack, nutrient deficiency, or any other stress factors
- Specific restrictions apply to use in sequence or tank mixture with other sulfonylurea or ALS-inhibiting herbicides. See label for details

Crop-specific information

- Latest use: before 31 Dec in yr of sowing for winter oats, winter rye, triticale; before 1st node detectable (GS 31) for winter wheat
- Slight chlorosis and stunting may occur in certain conditions. Recovery is rapid and yield not affected

Following crops guidance

- Only cereals, oilseed rape, field beans, clover or grass may be sown in the yr of harvest of a treated crop
- In the event of crop failure only winter or spring wheat may be sown 1-3 mth after treatment. Land should be ploughed and cultivated to 15 cm minimum before resowing

Environmental safety

- Dangerous for the environment
- Very toxic to aquatic organisms
- Take extreme care to avoid damage by drift outside the target area or onto surface waters or ditches
- Spraying equipment should not be drained or flushed onto land planted, or to be planted, with trees or crops other than cereals and should be thoroughly cleansed after use - see label for instructions

Hazard classification and safety precautions

Hazard Irritant, Dangerous for the environment
Transport code 9
Packaging group III
UN number 3077
Risk phrases R43, R50, R53a
Operator protection A, H; U05a, U08, U14, U19a, U20b
Environmental protection E15b, E38
Storage and disposal D01, D02, D09a, D10b, D11a, D12a

SEE SECTION 3 FOR PRODUCTS ALSO REGISTERED

59 carfentrazone-ethyl + mecoprop-P

A foliar applied herbicide for cereals
HRAC mode of action code: E + O

See also *mecoprop-P*

Products

1	Pan Glory	Pan Amenity	1.5:60% w/w	WG	14487
2	Platform S	Belchim	1.5:60% w/w	WG	14380

Uses

- Annual and perennial weeds in ***amenity grassland, managed amenity turf*** [1]
- Charlock in ***spring barley, spring oats, spring wheat, winter barley, winter oats, winter wheat*** [2]
- Chickweed in ***spring barley, spring oats, spring wheat, winter barley, winter oats, winter wheat*** [2]
- Cleavers in ***spring barley, spring oats, spring wheat, winter barley, winter oats, winter wheat*** [2]
- Field speedwell in ***spring barley, spring oats, spring wheat, winter barley, winter oats, winter wheat*** [2]
- Ivy-leaved speedwell in ***spring barley, spring oats, spring wheat, winter barley, winter oats, winter wheat*** [2]
- Moss in ***amenity grassland, managed amenity turf*** [1]
- Red dead-nettle in ***spring barley, spring oats, spring wheat, winter barley, winter oats, winter wheat*** [2]

Approval information

- Carfentrazone-ethyl and mecoprop-P included in Annex I under EC Directive 91/414
- Accepted by BBPA for use on malting barley

Efficacy guidance

- Best results obtained when weeds have germinated and growing vigorously in warm moist conditions
- Treatment of large weeds and poor spray coverage may result in reduced weed control

Restrictions

- Maximum number of treatments 2 per crop. The total amount of mecoprop-P applied in a single yr must not exceed the maximum total dose approved for any single product for the crop/situation
- Do not treat crops suffering from stress from any cause
- Do not treat crops undersown or to be undersown

Crop-specific information

- Latest use: before 3rd node detectable (GS 33)
- Can be used on all varieties of wheat and barley in autumn or spring from the beginning of tillering
- Early sown crops may be prone to damage if treated after period of rapid growth in autumn

Following crops guidance

- In the event of crop failure, any cereal, maize, oilseed rape, peas, vetches or sunflowers may be sown 1 mth after a spring treatment. Any crop may be planted 3 mth after treatment

Environmental safety

- Dangerous for the environment
- Very toxic to aquatic organisms
- Keep livestock out of treated areas for at least two weeks following treatment and until poisonous weeds, such as ragwort, have died down and become unpalatable
- Some pesticides pose a greater threat of contamination of water than others and mecoprop-P is one of these pesticides. Take special care when applying mecoprop-P near water and do not apply if heavy rain is forecast

Hazard classification and safety precautions

Hazard Harmful, Dangerous for the environment
Transport code 9
Packaging group III
UN number 3077

Risk phrases R22a, R41, R43, R50, R53a
Operator protection A, C, H [1, 2]; M [2]; U05a, U11, U13, U14, U20b [1, 2]; U08 [2]; U09a [1]
Environmental protection E07a, E34, E38 [1, 2]; E15a [2]; E15b [1]
Storage and disposal D01, D02, D09a, D10b [1, 2]; D12a [2]
Medical advice M05a

60 carfentrazone-ethyl + metsulfuron-methyl

A foliar applied herbicide mixture for cereals
HRAC mode of action code: E + B

See also metsulfuron-methyl

Products

Asset Express	AgriGuard	40:10% w/w	WG	14378

Uses

- Annual dicotyledons in ***durum wheat***, ***spring barley***, ***spring oats***, ***spring wheat***, ***triticale***, ***winter barley***, ***winter oats***, ***winter wheat***

Approval information

- Carfentrazone-ethyl and metsulfuron-methyl included in Annex I under EC Directive 91/414
- Accepted by BBPA for use on malting barley

Efficacy guidance

- Best results achieved from applications made in good growing conditions
- Good spray cover of weeds must be obtained
- Growth of weeds is inhibited within hours of treatment but the time taken for visible colour changes to appear will vary according to species and weather
- Product has short residual life in soil. Under normal moisture conditions susceptible weeds germinating soon after treatment will be controlled
- Metsulfuron-methyl is a member of the ALS-inhibitor group of herbicides and products should be used in a planned Resistance Management strategy. See Section 5 for more information

Restrictions

- Maximum number of treatments 1 per crop
- Product must only be used after 1 Feb
- Do not use on any crop suffering stress from drought, waterlogging, cold, pest or disease attack, or nutrient deficiency
- Do not use on crops undersown with grass or legumes or on any broad-leaved crop
- Do not apply within 7 d of rolling
- Specific restrictions apply to use in sequence or tank mixture with other sulfonylurea or ALS-inhibiting herbicides. See label for details
- Do not tank mix with products containing chlorpyrifos, diclofop-methyl, difenzoquat, flamprop-methyl, flutriafol, propiconazole or tralkoxydim

Crop-specific information

- Latest use: before 3rd node detectable (GS 33)
- Can be used on all soil types
- Apply in the spring from the 3-leaf stage on all crops
- Slight necrotic spotting of crops can occur under certain crop and soil conditions. Recovery is rapid and there is no effect on grain yield or quality

Following crops guidance

- Only cereals, oilseed rape, field beans or grass may be sown as a following crop in the same calendar yr. Any crop may follow in the next spring
- In the event of failure of a treated crop, only winter wheat may be sown 1-3 mth later after ploughing and cultivating to at least 15 cm

Environmental safety

- Dangerous for the environment
- Very toxic to aquatic organisms
- Take extreme care to avoid drift onto broad-leaved plants outside the target area or onto ponds, waterways or ditches, or onto land intended for cropping

- Spraying equipment should not be drained or flushed onto land planted, or to be planted, with trees or crops other than cereals and should be thoroughly cleansed after use - see label for instructions

Hazard classification and safety precautions

Hazard Irritant, Dangerous for the environment
Transport code 9
Packaging group III
UN number 3077
Risk phrases R43, R50, R53a
Operator protection A, H; U05a, U08, U14, U19a, U20b
Environmental protection E15b, E38
Storage and disposal D01, D02, D09a, D11a, D12a

61 chloridazon

A residual pyridazinone herbicide for beet crops
HRAC mode of action code: C1

Products

1	Better DF	Sipcam	65% w/w	SG	06250
2	Better Flowable	Sipcam	430 g/l	SC	04924
3	Burex 430 SC	United Phosphorus	430 g/l	SC	12619
4	Lidazone 65 WG	Globachem	65% w/w	WG	10847
5	Parador	United Phosphorus	430 g/l	SC	12372
6	Pyramin DF	BASF	65% w/w	WG	03438
7	Takron	BASF	430 g/l	SC	11627
8	Tempest	AgriGuard	430 g/l	SC	12688

Uses

- Annual dicotyledons in ***bulb onions*** *(off-label)*, ***leeks*** *(off-label)*, ***salad onions*** *(off-label)* [1, 4, 6]; ***chard*** *(off-label)*, ***chives*** *(off-label)*, ***herbs (see appendix 6)*** *(off-label)*, ***leaf spinach*** *(off-label)*, ***parsley*** *(off-label)*, ***spinach beet*** *(off-label)* [6]; ***fodder beet***, ***mangels***, ***sugar beet*** [1-8]; ***garlic*** *(off-label)*, ***shallots*** *(off-label)* [1, 4]
- Annual meadow grass in ***bulb onions*** *(off-label)*, ***leeks*** *(off-label)*, ***salad onions*** *(off-label)* [1, 6]; ***chard*** *(off-label)*, ***chives*** *(off-label)*, ***herbs (see appendix 6)*** *(off-label)*, ***leaf spinach*** *(off-label)*, ***parsley*** *(off-label)*, ***spinach beet*** *(off-label)* [6]; ***fodder beet***, ***mangels***, ***sugar beet*** [1, 2, 4-8]; ***garlic*** *(off-label)*, ***shallots*** *(off-label)* [1]

Specific Off-Label Approvals (SOLAs)

- ***bulb onions*** *20012570* [1], *20092188 expires 31 Dec 2010* [4], *970732* [6]
- ***chard*** *20062979* [6]
- ***chives*** *20062979* [6]
- ***garlic*** *20012570* [1], *20092188 expires 31 Dec 2010* [4]
- ***herbs (see appendix 6)*** *20062979* [6]
- ***leaf spinach*** *20062979* [6]
- ***leeks*** *20012570* [1], *20092188 expires 31 Dec 2010* [4], *970732* [6]
- ***parsley*** *20062979* [6]
- ***salad onions*** *20012570* [1], *20092188 expires 31 Dec 2010* [4], *970732* [6]
- ***shallots*** *20012570* [1], *20092188 expires 31 Dec 2010* [4]
- ***spinach beet*** *20062979* [6]

Approval information

- Chloridazon included in Annex I under EC Directive 91/414
- Approval expiry 31 Dec 2010 [4]

Efficacy guidance

- Absorbed by roots of germinating weeds and best results achieved pre-emergence of weeds or crop when soil moist and adequate rain falls after application
- Application rate for some products depends on soil type. Check label for details

Restrictions

- Maximum number of treatments generally 1 per crop (pre-emergence) for fodder beet and mangels; 1 (pre-emergence) + 3 (post-emergence) per crop for sugar beet, but labels vary slightly. Check labels for maximum total dose for the crop to be treated

- Maximum total dose for onions and leeks equivalent to one full dose recommended for these crops [6]
- Do not use on Coarse Sands, Sands or Fine Sands or where organic matter exceeds 5%
- A maximum total dose of 2.6 kg/ha chloridazon may only be applied every third year on the same field

Crop-specific information

- Latest use: pre-emergence for fodder beet and mangels; normally before leaves of crop meet in row for sugar beet, but labels vary slightly; up to and including second true leaf stage for onions and leeks [6]
- Where used pre-emergence spray as soon as possible after drilling in mid-Mar to mid-Apr on fine, firm, clod-free seedbed
- Where crop drilled after mid-Apr or soil dry apply pre-drilling and incorporate to 2.5 cm immediately afterwards
- Various tank mixes recommended on sugar beet for pre- and post-emergence use and as repeated low dose treatments. See label for details
- Crop vigour may be reduced by treatment of crops growing under unfavourable conditions including poor tilth, drilling at incorrect depth, soil capping, physical damage, pest or disease damage, excess seed dressing, trace-element deficiency or a sudden rise in temperature after a cold spell

Following crops guidance

- Winter cereals or any spring sown crop may follow a treated beet crop harvested at the normal time and after ploughing to at least 150 mm
- In the event of failure of a treated crop only a beet crop or maize may be drilled, after cultivation

Environmental safety

- Dangerous for the environment
- Very toxic to aquatic organisms

Hazard classification and safety precautions

Hazard Harmful [1, 6]; Irritant [2-5, 7, 8]; Dangerous for the environment [1-8]
Transport code 9 [3-8]
Packaging group III [3-8]
UN number 3077 [4, 6]; 3082 [3, 5, 7, 8]; N/C [1, 2]
Risk phrases R22a [1, 6]; R36, R37, R38 [8]; R43 [2-5, 7, 8]; R50 [1, 2, 4-8]; R51 [3]; R53a [1-8]
Operator protection A [3, 5, 7, 8]; U05a, U08 [1-8]; U14 [2, 4, 5, 7, 8]; U15 [8]; U19a [2, 5, 7, 8]; U20a [3, 4]; U20b [1, 2, 5-8]
Environmental protection E13b [3, 5]; E13c [4]; E15a [1, 2, 4-8]; E38 [1, 2, 5-7]
Storage and disposal D01, D02, D09a [1-8]; D05, D12b [3]; D10b [2, 4, 5, 7, 8]; D10c [3, 5]; D11a [1, 6]; D12a [1, 2, 5-7]
Medical advice M05a [1, 2, 5-8]

62 chloridazon + chlorpropham + metamitron

A residual herbicide mixture for beet crops but approval expired in 2008
HRAC mode of action code: C1 + K2 + C1

See also *chlorpropham*
metamitron

63 chloridazon + ethofumesate

A contact and residual herbicide for beet crops
HRAC mode of action code: C1 + N

See also *ethofumesate*

Products

Magnum	BASF	275:170 g/l	SC	11727

Uses

- Annual dicotyledons in ***fodder beet*, *sugar beet***
- Annual meadow grass in ***fodder beet*, *sugar beet***

Approval information

- Chloridazon and ethofumesate included in Annex I under EC Directive 91/414

Efficacy guidance

- Best results achieved on a fine firm clod-free seedbed when soil moist and adequate rain falls after spraying. Efficacy and crop safety may be reduced if heavy rain falls just after incorporation
- Effectiveness may be reduced under conditions of low pH
- May be applied by conventional or repeat low dose method. See label for details

Restrictions

- Maximum number of treatments 1 pre-emergence for fodder beet; 1 pre-emergence plus 3 post-emergence for sugar beet
- May be used on soil classes Loamy Sand - Silty Clay Loam. Additional restrictions apply for some tank mixtures
- Do not treat beet post-emergence with recommended tank mixtures when temperature is, or is likely to be, above 21 °C on day of spraying or under conditions of high light intensity
- If a mixture with phenmedipham is applied to a crop previously treated with a pre-emergence herbicide and the crop is suffering from stress from whatever cause, no further such post-emergence applications may be made
- A maximum total dose of 2.6 kg/ha chloridazon may only be applied every third year on the same field

Crop-specific information

- Latest use: pre-crop emergence for fodder beet; before crop leaves meet between rows for sugar beet
- Apply up to cotyledon stage of weeds
- Crop vigour may be reduced by treatment of crops growing under unfavourable conditions including poor tilth, drilling at incorrect depth, soil capping, physical damage, excess nitrogen, excess seed dressing, trace element deficiency or a sudden rise in temperature after a cold spell. Frost after pre-emergence treatment may check crop growth

Following crops guidance

- In the event of crop failure only sugar beet, fodder beet or mangels may be re-drilled
- Any crop may be sown 3 mth after spraying following ploughing to 15 cm

Environmental safety

- Dangerous for the environment
- Very toxic to aquatic organisms

Hazard classification and safety precautions

Hazard Harmful, Dangerous for the environment
Transport code 9
Packaging group III
UN number 3082
Risk phrases R22a, R50, R53a
Operator protection A, H; U05a, U08, U19a, U20a
Environmental protection E15a, E38
Storage and disposal D01, D02, D09a, D10b, D12a
Medical advice M05a

64 chloridazon + lenacil

A residual pre- and post-emergence herbicide for beet crops but product withdrawn in 2008
HRAC mode of action code: C1 + C1

See also *lenacil*

65 chloridazon + metamitron

A contact and residual herbicide mixture for sugar beet
HRAC mode of action code: C1 + C1

See also metamitron

Products

Volcan Combi	Sipcam	300:280 g/l	SC	10256

Uses
- Annual dicotyledons in ***fodder beet*** *(off-label)*, ***sugar beet***
- Annual meadow grass in ***fodder beet*** *(off-label)*, ***sugar beet***

Specific Off-Label Approvals (SOLAs)
- ***fodder beet*** *20061540*

Approval information
- Chloridazon included in Annex I under EC Directive 91/414

Efficacy guidance
- Best results achieved from a sequential programmme of sprays
- Pre-emergence use improves efficacy of post-emergence programme. Best results obtained from application to a moist seed bed
- First post-emergence treatment should be made when weeds at early cotyledon stage and subsequent applications made when new weed flushes reach this stage
- Weeds surviving an earlier treatment should be treated again after 7-10 d even if no new weeds have appeared

Restrictions
- Maximum number of treatments 1 pre-emergence followed by 3 post-emergence
- Take advice if light soils have a high proportion of stones

Crop-specific information
- Latest use: when leaves of crop meet in rows
- Tolerance of crops growing under stress from any cause may be reduced
- Crops treated pre-emergence and subsequently subjected to frost may be checked and recovery may not be complete

Following crops guidance
- After the last application only sugar beet or mangels may be sown within 4 mth; cereals may be sown after 16 wk. Land should be mouldboard ploughed to 15 cm and thoroughly cultivated before any succeeding crop

Environmental safety
- Harmful to fish or other aquatic life. Do not contaminate surface waters or ditches with chemical or used container

Hazard classification and safety precautions

Hazard Irritant
UN number N/C
Risk phrases R36, R38
Operator protection U08, U14, U15, U20a
Environmental protection E13c, E34
Storage and disposal D09a, D10b

66 chloridazon + quinmerac

A herbicide mixture for use in beet crops
HRAC mode of action code: C1 + O

See also quinmerac

Products

Fiesta T	BASF	360:60 g/l	SC	11734

Uses

- Annual dicotyledons in ***beet leaves*** *(off-label)*, ***chard*** *(off-label)*, ***chives*** *(off-label)*, ***fodder beet***, ***herbs (see appendix 6)*** *(off-label)*, ***leaf spinach*** *(off-label)*, ***mangels***, ***parsley*** *(off-label)*, ***spinach beet*** *(off-label)*, ***sugar beet***
- Annual grasses in ***beet leaves*** *(off-label)*, ***chard*** *(off-label)*, ***chives*** *(off-label)*, ***herbs (see appendix 6)*** *(off-label)*, ***leaf spinach*** *(off-label)*, ***parsley*** *(off-label)*, ***spinach beet*** *(off-label)*
- Annual meadow grass in ***fodder beet***, ***mangels***, ***sugar beet***
- Chickweed in ***fodder beet***, ***mangels***, ***sugar beet***
- Cleavers in ***fodder beet***, ***mangels***, ***sugar beet***
- Field speedwell in ***fodder beet***, ***mangels***, ***sugar beet***
- Ivy-leaved speedwell in ***fodder beet***, ***mangels***, ***sugar beet***
- Mayweeds in ***fodder beet***, ***mangels***, ***sugar beet***
- Poppies in ***fodder beet***, ***mangels***, ***sugar beet***

Specific Off-Label Approvals (SOLAs)

- ***beet leaves*** *20063164*
- ***chard*** *20063164*
- ***chives*** *20063164*
- ***herbs (see appendix 6)*** *20063164*
- ***leaf spinach*** *20063164*
- ***parsley*** *20063164*
- ***spinach beet*** *20063164*

Approval information

- Chloridazon included in Annex I under EC Directive 91/414

Efficacy guidance

- Best results obtained when adequate soil moisture is present at application and afterwards to form an active herbicidal layer in the soil
- A programme of pre-emergence treatment followed by post-emergence application(s) in mixture with a contact herbicide optimises weed control and is essential for some difficult weed species such as cleavers
- Treatment pre-emergence only may not provide sufficient residual activity to give season-long weed control
- Effectiveness may be reduced under conditions of low pH
- Always follow WRAG guidelines for preventing and managing herbicide resistant weeds. See Section 5 for more information

Restrictions

- Maximum total dose on sugar beet equivalent to one treatment pre-emergence plus two treatments post-emergence, all at maximum individual dose; maximum total dose on fodder beet and mangels equivalent to one full dose pre-emergence treatment
- Heavy rain shortly after treatment may check crop growth particularly when it leaves water standing in surface depressions
- Treatment of stressed crops or those growing in unfavourable conditions may depress crop vigour and possibly reduce stand
- Do not use on Sands or soils of high organic matter content
- Where rates of nitrogen higher than those generally recommended are considered necessary, apply at least 3 wk before drilling
- A maximum total dose of 2.6 kg/ha chloridazon may only be applied every third year on the same field

Crop-specific information

- Latest use: before plants meet between the rows for sugar beet; pre-emergence for fodder beet, mangels

Following crops guidance

- In the spring following normal harvest of a treated crop sow only cereals, beet or mangel crops, potatoes or field beans, after ploughing
- In the event of failure of a treated crop only a beet crop may be drilled, after cultivation

Environmental safety

- Dangerous for the environment

- Toxic to aquatic organisms
- To reduce movement to groundwater do not apply to dry soil or when heavy rain is forecast

Hazard classification and safety precautions

Hazard Irritant, Dangerous for the environment

Transport code 9

Packaging group III

UN number 3082

Risk phrases R43, R51, R53a

Operator protection A; U05a, U08, U14, U19a, U20b

Environmental protection E15a, E38

Storage and disposal D01, D02, D08, D09a, D10c, D12a

Medical advice M03, M05a

67 chlormequat

A plant-growth regulator for reducing stem growth and lodging

Products

1	Adjust	Taminco	620 g/l	SL	13961
2	Agriguard Chlormequat 700	AgriGuard	700 g/l	SL	09782
3	Agriguard Chlormequat 720	AgriGuard	720 g/l	SL	09919
4	Agrovista 3 See	Agrovista	720 g/l	SL	14181
5	Barclay Holdup	Barclay	700 g/l	SL	11365
6	Barclay Holdup 640	Barclay	640 g/l	SL	11374
7	Barclay Take 5	Barclay	645 g/l	SL	11368
8	Barleyquat B	Taminco	620 g/l	SL	13962
9	BASF 3C Chlormequat 720	BASF	720 g/l	SL	06514
10	Bettaquat B	Taminco	637 g/l	SL	13965
11	Clayton Coldstream	Clayton	400 g/l	SL	12724
12	Clayton Standup	Clayton	700 g/l	SL	11760
13	Fargro Chlormequat	Fargro	460 g/l	SL	02600
14	Fernpath Tangent	AgriGuard	460 g/l	SL	10341
15	Greencrop Carna	Greencrop	600 g/l	SL	09403
16	Hive	Nufarm UK	730 g/l	SL	11392
17	K2	Taminco	620 g/l	SL	13964
18	Mandops chlormequat	Taminco	700 g/l	SL	13970
19	Manipulator	Taminco	620 g/l	SL	13963
20	Mirquat	Nufarm UK	730 g/l	SL	11406
21	New 5C Cycocel	BASF	645 g/l	SL	01482
22	New 5C Quintacel	Nufarm UK	645 g/l	SL	12074
23	Selon	Taminco	620 g/l	SL	13966
24	Sigma PCT	Nufarm UK	460 g/l	SL	11209
25	Stabilan 700	Nufarm UK	700 g/l	SL	11393
26	Standon Girder 720	Standon	720 g/l	SL	14251
27	Terbine	Nufarm UK	730 g/l	SL	11407

Uses

- Growth regulation in ***durum wheat*** *(off-label)* [4, 9, 10, 16, 21, 25]; ***spring barley*** [19]; ***spring oats, spring wheat, winter barley, winter oats, winter wheat*** [19, 26]; ***spring rye*** *(off-label)*, ***winter rye*** *(off-label)* [4, 10, 18]; ***triticale*** *(off-label)* [4, 10, 16, 25]; ***winter rye*** [26]
- Increasing yield in ***spring barley*** [1, 8, 18, 23]; ***spring oats, winter oats*** [1, 8, 23]; ***winter barley*** [1, 2, 5, 6, 8, 9, 14, 15, 18, 21-23]
- Lodging control in ***durum wheat*** *(off-label)* [1, 10, 17, 19]; ***spring barley*** [1, 8, 17-19, 23]; ***spring oats, winter oats*** [1-9, 11, 12, 14-23, 25, 27]; ***spring rye*** [9, 21, 22]; ***spring rye*** *(off-label)*, ***winter rye*** *(off-label)* [1, 8, 10, 17-19]; ***spring wheat, winter wheat*** [1-7, 9-12, 14-25, 27]; ***triticale*** [7, 9, 14, 16, 20-22, 25]; ***triticale*** *(off-label)* [1, 8, 10, 17, 19]; ***winter barley*** [1-9, 11, 12, 16-20, 23-25, 27]; ***winter rye*** [2-4, 7, 9, 11, 12, 16, 20-22, 25, 27]
- Stem shortening in ***bedding plants, camellias, hibiscus trionum, lilies*** [12, 13, 16, 20, 25, 27]; ***geraniums, poinsettias*** [9, 12, 13, 16, 20-22, 25, 27]

Specific Off-Label Approvals (SOLAs)

- ***durum wheat*** *20081166* [1], *20090998* [4], *20062655* [9], *20081173 expires 31 Jul 2010* [10], *20063619* [16], *20081175* [17], *20081179* [19], *20061298* [21], *20063487* [25]

- ***spring rye*** *20081166* [1], *20090998* [4], *20081170 expires 31 Jul 2010* [8], *20081173 expires 31 Jul 2010* [10], *20081175* [17], *20081177* [18], *20081179* [19]
- ***triticale*** *20081166* [1], *20090998* [4], *20081170 expires 31 Jul 2010* [8], *20081173 expires 31 Jul 2010* [10], *20063619* [16], *20081175* [17], *20081179* [19], *20063487* [25]
- ***winter rye*** *20081166* [1], *20090998* [4], *20081170 expires 31 Jul 2010* [8], *20081173 expires 31 Jul 2010* [10], *20081175* [17], *20081177* [18], *20081179* [19]

Approval information

- Approved for aerial application on wheat and oats [2, 9, 15, 16, 20, 24]; on winter barley [9, 15]; on triticale, rye [9]. See Section 5 for more information
- Accepted by BBPA for use on malting barley

Efficacy guidance

- Most effective results on cereals normally achieved from application from Apr onwards, on wheat and rye from leaf sheath erect to first node detectable (GS 30-31), on oats at second node detectable (GS 32), on winter barley from mid-tillering to leaf sheath erect (GS 25-30). However, recommendations vary with product. See label for details (n.b [13] only approved on flowers)
- Influence on growth varies with crop and growth stage. Risk of lodging reduced by application at early stem extension. Root development and yield can be improved by earlier treatment
- Results on barley can be variable
- In tank mixes with other pesticides on cereals optimum timing for herbicide action may differ from that for growth reduction. See label for details of tank mix recommendations
- Most products recommended for use on oats require addition of approved non-ionic wetter. Check label
- Some products are formulated with trace elements to help compensate for increased demand during rapid growth

Restrictions

- Maximum number of treatments or maximum total dose varies with crop and product and whether split dose treatments are recommended. Check labels
- Do not use on very late sown spring wheat or oats or on crops under stress (n.b [13] only approved on flowers)
- Mixtures with liquid nitrogen fertilizers may cause scorch and are specifically excluded on some labels
- Do not use on soils of low fertility unless such crops regularly receive adequate dressings of nitrogen
- At least 6 h, preferably 24 h, required before rain for maximum effectiveness. Do not apply to wet crops
- Check labels for tank mixtures known to be incompatible
- Not to be used on food crops [13]

Crop-specific information

- Latest use varies with crop and product. See label for details
- May be used on cereals undersown with grass or clovers (n.b [13] only approved on flowers)
- Ornamentals to be treated must be well established and growing vigorously. Do not treat in strong sunlight or when temperatures are likely to fall below 10 °C
- Temporary yellow spotting may occur on poinsettias. It can be minimised by use of a non-ionic wetting agent - see label

Environmental safety

- Harmful to aquatic organisms [12, 13, 16, 20, 25]
- Wash equipment thoroughly with water and wetting agent immediately after use and spray out. Spray out again before storing or using for another product. Traces can cause harm to susceptible crops sprayed later
- Do not use straw from treated cereals as horticultural growth medium or mulch [9, 21]

Hazard classification and safety precautions

Hazard Harmful [1-4, 6-18, 20-27]; Irritant [5]
Transport code 8 [2, 3, 5, 7, 11-15, 26]
Packaging group III [2, 3, 5, 7, 11-15, 26]
UN number 1760 [2, 3, 5, 7, 11-15, 26]
Risk phrases R21 [2-4, 11, 26]; R22a [1-18, 20-27]; R36 [18]; R52 [1, 8, 12, 13, 16, 18, 20, 23, 25, 26]; R53a [1, 8, 12, 13, 16, 20, 23, 25, 26]

Operator protection A [1-18, 20-27]; C [18]; U05a, U19a [1-18, 20-27]; U08 [1-15, 17, 18, 21, 23, 24, 27]; U09a [16, 20, 22, 25, 26]; U13 [5, 13]; U20a [5, 22]; U20b [1-4, 6-18, 20, 21, 23-27]
Environmental protection E15a [1, 3-18, 20-27]; E34 [1-18, 20-27]
Consumer protection C01 [5]
Storage and disposal D01, D02, D09a [1-18, 20-27]; D05 [2-7, 11-13, 15, 16, 20, 22, 25]; D10a [5, 12, 25, 26]; D10b [1-4, 6-8, 10, 11, 13-18, 20, 23, 24, 27]; D10c [9, 21, 22]; D12a [11]
Medical advice M03 [1-18, 20-27]; M05a [7, 9, 12, 13, 16, 20-22, 25]

68 chlormequat + 2-chloroethylphosphonic acid

A plant growth regulator for use in cereals

See also 2-chloroethylphosphonic acid

Products

1	Greencrop Tycoon	Greencrop	305:155 g/l	SL	09571
2	Strate	Bayer CropScience	360:180 g/l	SL	10020
3	Upgrade	Bayer CropScience	360:180 g/l	SL	10029

Uses

- Growth regulation in ***durum wheat*** *(off-label)*, ***spring rye*** *(off-label)*, ***triticale*** *(off-label)*, ***winter rye*** *(off-label)* [2, 3]
- Lodging control in ***spring barley***, ***winter barley***, ***winter wheat*** [1-3]

Specific Off-Label Approvals (SOLAs)

- ***durum wheat*** *20061541* [2], *20063100* [3]
- ***spring rye*** *20061541* [2], *20063100* [3]
- ***triticale*** *20061541* [2], *20063100* [3]
- ***winter rye*** *20061541* [2], *20063100* [3]

Approval information

- 2-chloroethylphosphonic acid (ethephon) included in Annex I under EC Directive 91/414
- Accepted by BBPA for use on malting barley
- In 2006 PSD required that all products containing 2-chloroethylphosphonic acid should carry the following warning in the main area of the container label: "2-chloroethylphosphonic acid is an anticholinesterase organophosphate. Handle with care"

Efficacy guidance

- Best results obtained when crops growing vigorously
- Recommended dose varies with growth stage. See labels for details and recommendations for use of sequential treatments

Restrictions

- 2-chloroethylphosphonic acid is an anticholinesterase organophosphorus compound. Do not use if under medical advice not to work with such compounds
- Maximum number of treatments 1 per crop; maximum total dose equivalent to one full dose treatment
- Product must always be used with specified non-ionic wetter - see labels
- Do not use on any crop in sequence with any other product containing 2-chloroethylphosphonic acid
- Do not spray when crop wet or rain imminent
- Do not spray during cold weather or periods of night frost, when soil is very dry, when crop diseased or suffering pest damage, nutrient deficiency or herbicide stress
- If used on seed crops grown for certification inform seed merchant beforehand
- Do not use on wheat variety Moulin or on any winter varieties sown in spring [1]
- Do not use on spring barley variety Triumph [1]
- Do not treat barley on soils with more than 10% organic matter [1]
- Do not use in programme with any other product containing 2-chloroethylphosphonic acid [1]
- Only crops growing under conditions of high fertility should be treated

Crop-specific information

- Latest use: before flag leaf ligule/collar just visible (GS 39) or 1st spikelet visible (GS 51) for wheat or barley at top dose; or before flag leaf sheath opening (GS 47) for winter wheat at reduced dose
- Apply before lodging has started

Environmental safety

- Dangerous for the environment [2, 3]
- Harmful to aquatic organisms [1]
- Harmful to fish or other aquatic life. Do not contaminate surface waters or ditches with chemical or used container [2, 3]
- Do not use straw from treated cereals as a horticultural growth medium or as a mulch [1]

Hazard classification and safety precautions

Hazard Harmful [1-3]; Dangerous for the environment [2, 3]
Transport code 8
Packaging group III
UN number 3265
Risk phrases R22a, R37 [1-3]; R41 [2, 3]; R52 [1]
Operator protection A [1-3]; C [2, 3]; U05a, U08, U19a, U20b [1-3]; U11, U13 [2, 3]
Environmental protection E13c [2, 3]; E15a [1]; E34 [1-3]
Storage and disposal D01, D02, D09a, D10b [1-3]; D05 [2, 3]
Medical advice M01, M03 [1-3]; M05a [1]

69 chlormequat + 2-chloroethylphosphonic acid + mepiquat chloride

A plant growth regulator for reducing lodging in cereals

See also *2-chloroethylphosphonic acid*
mepiquat chloride

Products

Cyclade	BASF	230:155:75 g/l	SL	08958

Uses

- Growth regulation in ***durum wheat*** *(off-label)*, ***spring rye*** *(off-label)*, ***triticale*** *(off-label)*, ***winter rye*** *(off-label)*
- Lodging control in ***spring barley***, ***winter barley***, ***winter wheat***

Specific Off-Label Approvals (SOLAs)

- ***durum wheat*** *20063098*
- ***spring rye*** *20063098*
- ***triticale*** *20063098*
- ***winter rye*** *20063098*

Approval information

- 2-chloroethylphosphonic acid (ethephon) and mepiquat chloride included in Annex I under EC Directive 91/414
- Accepted by BBPA for use on malting barley
- In 2006 PSD required that all products containing 2-chloroethylphosphonic acid should carry the following warning in the main area of the container label: "2-chloroethylphosphonic acid is an anticholinesterase organophosphate. Handle with care"

Efficacy guidance

- Best results achieved in a vigorous, actively growing crop with adequate fertility and moisture
- Optimum timing on all crops is from second node detectable stage (GS 32)
- Recommended for use as part of an intensive growing system which includes provision for optimum fertilizer treatment and disease control

Restrictions

- 2-chloroethylphosphonic acid is an anticholinesterase organophosphorus compound. Do not use if under medical advice not to work with such compounds
- Maximum number of treatments 1 per crop
- Maximum total dose depends on spraying regime adopted. See label
- Must be used with a non-ionic wetting agent
- Do not apply to stressed crops or those on soils of low fertility unless receiving adequate dressings of fertilizer
- Do not apply in temperatures above 21 °C or if crop is wet or if rain expected

- Do not treat variety Moulin nor any winter varieties sown in spring
- Do not use in a programme with any other product containing 2-chloroethylphosphonic acid
- Do not apply to barley on soils with more than 10% organic matter (winter wheat may be treated)
- Notify seed merchant in advance if use on a seed crop is proposed

Crop-specific information

- Latest use: before first spikelet of ear visible (GS 51) using reduced dose on winter barley; before flag leaf sheath opening (GS 47) using reduced dose on winter wheat; before flag leaf just visible on spring barley
- May be applied to crops undersown with grasses or clovers
- Treatment may cause some delay in ear emergence

Environmental safety

- Harmful to aquatic organisms
- Do not use straw from treated crops as a horticultural growth medium

Hazard classification and safety precautions

Hazard Harmful
Transport code 8
Packaging group III
UN number 3265
Risk phrases R22a, R37, R52
Operator protection A; U05a, U08, U19a, U20b
Environmental protection E15a, E34
Storage and disposal D01, D02, D09a, D10c
Medical advice M01, M03, M05a

70 chlormequat + imazaquin

A plant growth regulator mixture for winter wheat

See also *imazaquin*

Products

1	Meteor	BASF	368:0.8 g/l	SL	10403
2	Standon Imazaquin 5C	Standon	368:0.8 g/l	SL	08813
3	Upright	BASF	368:0.8 g/l	SL	10404

Uses

- Growth regulation in ***spring rye*** *(off-label)*, ***triticale*** *(off-label)*, ***winter rye*** *(off-label)* [1]
- Increasing yield in ***winter wheat*** [1-3]
- Lodging control in ***winter wheat*** [1-3]

Specific Off-Label Approvals (SOLAs)

- ***spring rye*** *20063099* [1]
- ***triticale*** *20063099* [1]
- ***winter rye*** *20063099* [1]

Approval information

- Imazaquin included in Annex I under EC Directive 91/414

Efficacy guidance

- Apply to crops during good growing conditions or to those at risk from lodging
- On soils of low fertility, best results obtained where adequate nitrogen fertilizer used

Restrictions

- Maximum number of applications 1 per crop (2 per crop at split dose)
- Do not treat durum wheat
- Do not apply to undersown crops
- Do not apply when crop wet or rain imminent

Crop-specific information

- Latest use: before second node detectable (GS 31)
- Apply as single dose from leaf sheath lengthening up to and including 1st node detectable or as split dose, the first from tillers formed to leaf sheath lengthening, the second from leaf sheath erect up to and including 1st node detectable

Environmental safety

- Dangerous for the environment
- Toxic to aquatic organisms
- Do not use straw from treated cereals as horticultural growth medium or mulch

Hazard classification and safety precautions

Hazard Harmful [1-3]; Dangerous for the environment [2]
Transport code 8
Packaging group III
UN number 1760
Risk phrases R20, R52 [1, 3]; R22a, R36, R53a [1-3]; R51 [2]
Operator protection A, C; U05a, U08, U13, U19a, U20b [1-3]; U11, U15 [1, 3]
Environmental protection E15a, E34
Storage and disposal D01, D02, D05, D09a, D10b [1-3]; D06c, D12a [1, 3]
Medical advice M03 [1-3]; M05a [1, 3]

71 chlormequat + mepiquat chloride

A plant growth regulator for reducing lodging in wheat

See also mepiquat chloride

Products

Stronghold	BASF	345:115 g/l	SL	09134

Uses

- Lodging control in ***winter wheat***

Approval information

- Mepiquat chloride included in Annex I under EC Directive 91/414

Efficacy guidance

- Optimum timing is when leaf sheaths erect (GS 30)
- Benefit will vary according to crop and stage of growth at application

Restrictions

- Maximum total dose equivalent to one full dose treatment
- Do not apply to stressed crops or those on soils of low fertility unless receiving adequate dressings of fertilizer
- Do not treat crops where significant foot diseases, especially take-all, are expected
- Do not treat crops on soils of low fertility
- Do not apply in temperatures above 21 °C or if crop is wet or if rain expected
- Do not treat any winter varieties sown in spring
- Notify seed merchant in advance if use on a seed crop is proposed

Crop-specific information

- Latest use: before 3rd node detectable (GS 33)
- Apply during good growing conditions at the correct timings - see label
- May be applied to crops undersown with grasses or clovers
- Treatment may cause some delay in ear emergence
- Mixtures with liquid fertilizers may cause scorching in some circumstances

Environmental safety

- Do not use straw from treated crops as a horticultural growth medium or mulch

Hazard classification and safety precautions

Hazard Harmful
Transport code 8
Packaging group III
UN number 1760
Risk phrases R22a
Operator protection A; U05a, U08, U19a, U20b
Environmental protection E15a, E34
Storage and disposal D01, D02, D09a, D10c
Medical advice M03, M05a

72 2-chloroethylphosphonic acid

A plant growth regulator for cereals and various horticultural crops

See also *chlormequat + 2-chloroethylphosphonic acid*
chlormequat + 2-chloroethylphosphonic acid + mepiquat chloride

Products

1	Agriguard Cerusite	AgriGuard	480 g/l	SL	11494
2	Agrotech Ethephon	AgChem Access	480 g/l	SL	12170
3	Cerone	Bayer CropScience	480 g/l	SL	09985
4	Ethefon 480	Goldengrass	480 g/l	SL	14392
5	Pan Ethephon	Pan Agriculture	480 g/l	SL	13957
6	Pan Stiffen	Pan Agriculture	480 g/l	SL	13956

Uses

- Growth regulation in ***apple for cider making*** *(off-label)*, ***durum wheat*** *(off-label)*, ***protected ornamentals*** *(off-label)*, ***protected tomatoes*** *(off-label)* [3]
- Lodging control in ***spring barley*** [1-4]; ***triticale***, ***winter barley***, ***winter rye***, ***winter wheat*** [1-6]

Specific Off-Label Approvals (SOLAs)

- ***apple for cider making*** *20091381* [3]
- ***durum wheat*** *20061195* [3]
- ***protected ornamentals*** *20091382* [3]
- ***protected tomatoes*** *20091383* [3]

Approval information

- 2-chloroethylphosphonic acid (ethephon) included in Annex I under EC Directive 91/414
- Approved for aerial application on winter barley [1, 3]. See Section 5 for more information
- In 2006 PSD required that all products containing this active ingredient should carry the following warning in the main area of the container label: "2-chloroethylphosphonic acid is an anticholinesterase organophosphate. Handle with care"

Efficacy guidance

- Best results achieved on crops growing vigorously under conditions of high fertility
- Optimum timing varies between crops and products. See labels for details
- Do not spray crops when wet or if rain imminent

Restrictions

- 2-chloroethylphosphonic acid is an anticholinesterase organophosphorus compound. Do not use if under medical advice not to work with such compounds
- Maximum number of treatments 1 per crop or yr
- Do not spray crops suffering from stress caused by any factor, during cold weather or period of night frost nor when soil very dry
- Do not apply to cereals within 10 d of herbicide or liquid fertilizer application
- Do not spray wheat or triticale where the leaf sheaths have split and the ear is visible

Crop-specific information

- Latest use: before 1st spikelet visible (GS 51) for spring barley, winter barley, winter rye; before flag leaf sheath opening (GS 47) for triticale, winter wheat
- HI cider apples, tomatoes 5 d

Environmental safety

- Dangerous for the environment [1, 5, 6]
- Harmful to aquatic organisms
- Avoid accidental deposits on painted objects such as cars, trucks, aircraft

Hazard classification and safety precautions

Hazard Irritant [1-6]; Dangerous for the environment [1, 5, 6]
Transport code 8
Packaging group III
UN number 3265
Risk phrases R36 [1]; R37, R41, R52, R53a [2-6]; R38 [1-6]
Operator protection A, C; U05a, U08, U13, U20b [1-6]; U11 [2-4]
Environmental protection E13c, E38 [2-4]; E15a [1, 5, 6]; E34 [1]

Storage and disposal D01, D02, D09a, D10b [1-6]; D12a [2-4]
Medical advice M01 [2-6]

73 2-chloroethylphosphonic acid + mepiquat chloride

A plant growth regulator for reducing lodging in cereals

See also *mepiquat chloride*

Products

1	Clayton Florin	Clayton	155:305 g/l	SL	14351
2	Clayton Mepiquat	Clayton	155:305 g/l	SL	12360
3	Standon Mepiquat Plus	Standon	155:305 g/l	SL	09373
4	Terpal	BASF	155:305 g/l	SL	02103

Uses

- Growth regulation in ***durum wheat*** *(off-label)* [4]; ***spring barley***, ***triticale***, ***winter barley***, ***winter rye***, ***winter wheat*** [1]
- Increasing yield in ***winter barley*** *(low lodging situations)* [2, 4]
- Lodging control in ***spring barley*** [2, 4]; ***triticale***, ***winter barley***, ***winter rye***, ***winter wheat*** [2-4]

Specific Off-Label Approvals (SOLAs)

- ***durum wheat*** *20063104* [4]

Approval information

- 2-chloroethylphosphonic acid (ethephon) and mepiquat chloride included in Annex I under EC Directive 91/414
- In 2006 PSD required that all products containing 2-chloroethylphosphonic acid should carry the following warning in the main area of the container label: "2-chloroethylphosphonic acid is an anticholinesterase organophosphate. Handle with care"

Efficacy guidance

- Best results achieved on crops growing vigorously under conditions of high fertility
- Recommended dose and timing vary with crop, cultivar, growing conditions, previous treatment and desired degree of lodging control. See label for details
- May be applied to crops undersown with grass or clovers
- Do not apply to crops if wet or rain expected as efficacy will be impaired

Restrictions

- 2-chloroethylphosphonic acid is an anticholinesterase organophosphorus compound. Do not use if under medical advice not to work with such compounds
- Maximum number of treatments 2 per crop
- Add an authorised non-ionic wetter to spray solution. See label for recommended product and rate
- Do not treat crops damaged by herbicides or stressed by drought, waterlogging etc
- Do not treat crops on soils of low fertility unless adequately fertilized
- Do not use in a programme with any other product containing 2-chloroethylphosphonic acid
- Do not apply to winter cultivars sown in spring or treat winter barley, triticale or winter rye on soils with more than 10% organic matter (winter wheat may be treated)
- Do not apply at temperatures above 21 °C

Crop-specific information

- Latest use: before ear visible (GS 49) for winter barley, spring barley, winter wheat and triticale; flag leaf just visible (GS 37) for winter rye
- Late tillering may be increased with crops subject to moisture stress and may reduce quality of malting barley

Environmental safety

- Do not use straw from treated cereals as a mulch or growing medium

Hazard classification and safety precautions

Hazard Harmful
Transport code 8
Packaging group III
UN number 3265
Risk phrases R22a, R37, R53a [1-4]; R36 [1, 3, 4]; R41, R53b [2]; R52 [1]
Operator protection A, C [1, 4]; U05a, U11 [1, 2]; U20b [2-4]; U20c [1]

Environmental protection E15a
Storage and disposal D01, D02, D09a [1-4]; D05, D12a [1, 2]; D08 [3, 4]; D10b [2]; D10c [1, 3, 4]
Medical advice M01 [1-4]; M05a [1, 3, 4]

74 chlorophacinone

An indandione anticoagulant rodenticide

Products

1	Drat Rat Bait	B H & B	0.005% w/w	RB	H6743
2	Endorats	Irish Drugs	0.005% w/w	RB	H6744

Uses
- Rats in ***farm buildings*** [1, 2]
- Voles in ***farm buildings*** [1]

Efficacy guidance
- Chemical formulated with oil, thus improving weather resistance of bait
- Use in baiting programme
- Bait stations should be sited where rats active, by rat holes, along runs or in harbourages. Place bait in suitable containers
- Replenish baits every few days and remove unused bait when take ceases or after 7-10 d

Restrictions
- For use only by professional operators
- Resistance status of target population should be taken into account when considering choice of rodenticide
- Bait stations may be sited conveniently but bait should be inaccessible to non-target animals and protected from prevailing weather

Environmental safety
- Prevent access to baits by children, domestic animals and birds; see label for other precautions required
- Harmful to game, wild birds and animals

Hazard classification and safety precautions
Operator protection U13 [1, 2]; U20a [2]; U20b [1]
Storage and disposal D09a [1, 2]; D10a [1]; D11a [2]
Vertebrate/rodent control products V01a, V02, V03a, V04a

75 chloropicrin

A highly toxic horticultural soil fumigant
IRAC mode of action code: 8B

Products

K & S Chlorofume	K & S Fumigation	99.3% w/w	VP	08722

Uses
- Crown rot in ***strawberries*** *(soil fumigation)*
- Nematodes in ***strawberries*** *(soil fumigation)*
- Red core in ***strawberries*** *(soil fumigation)*
- Replant disease in ***all top fruit*** *(soil fumigation)*, ***hardy ornamentals*** *(soil fumigation)*
- Verticillium wilt in ***blackcurrants*** *(off-label)*, ***protected blackberries*** *(off-label)*, ***protected raspberries*** *(off-label)*, ***protected rubus hybrids*** *(off-label)*, ***raspberries*** *(off-label)*, ***rubus hybrids*** *(off-label)*, ***strawberries*** *(soil fumigation)*

Specific Off-Label Approvals (SOLAs)
- ***blackcurrants*** *20072373*
- ***protected blackberries*** *20072373*
- ***protected raspberries*** *20072373*
- ***protected rubus hybrids*** *20072373*
- ***raspberries*** *20072373*
- ***rubus hybrids*** *20072373*

Efficacy guidance

- Treat pre-planting
- Apply with specialised injection equipment
- For treating small areas or re-planting a single tree, a hand-operated injector may be used. Mark the area to be treated and inject to 22 cm at intervals of 22 cm
- Polythene sheeting (150 gauge) should be progressively laid over soil as treatment proceeds. The margin of the sheeting around the treated area must be embedded or covered with treated soil. Remove progressively after at least 4 d provided good air movement conditions prevail. Aerate soil for 15 d before planting
- Carry out a cress test before planting

Restrictions

- Chloropicrin is subject to the Poisons Rules 1982 and the Poisons Act 1972. See Section 5 for more information
- Before use, consult the code of practice for the fumigation of soil with chloropicrin. 2 fumigators must be present
- Maximum number of treatments 1 per crop
- Remove contaminated gloves, boots or other clothing immediately and ventilate them in the open air until all odour is eliminated
- Keep unprotected persons out of treated areas and adjacent premises until advised otherwise by professional operator
- Keep treated areas covered with sheets of low gas permeability for at least 4 d after treatment
- Avoid releasing vapour into the air when persistent still air conditions prevail

Crop-specific information

- Latest use: at least 20 d before planting

Following crops guidance

- Carry out a cress test before replanting treated soil

Environmental safety

- Avoid treatment or vapour release when persistent still air conditions prevail
- Dangerous to game, wild birds and animals
- Dangerous to bees
- Dangerous to fish or other aquatic life. Do not contaminate surface waters or ditches with chemical or used container
- Dangerous to livestock. Keep all livestock out of treated areas until advised otherwise by fumigator
- Keep in original container, tightly closed, in a safe place, under lock and key

Hazard classification and safety precautions

Hazard Very toxic
Transport code 6.1
Packaging group I
UN number 1580
Risk phrases R26, R27, R28, R34, R37
Operator protection A, G, H, K, M; U01, U04a, U05a, U10, U19a, U19b, U20a
Environmental protection E02a, E06b (until advised); E15a, E34
Storage and disposal D01, D02, D09b, D11a, D14
Medical advice M04a

76 chlorothalonil

A protectant chlorophenyl fungicide for use in many crops and turf
FRAC mode of action code: M5

See also azoxystrobin + chlorothalonil

Products

1	Agriguard Chlorothalonil	AgriGuard	500 g/l	SC	12201
2	Bravo 500	Syngenta	500 g/l	SC	10518
3	Busa	AgChem Access	500 g/l	SC	13452
4	Cheer 500	Goldengrass	500 g/l	SC	14418
5	CTL 500	Goldengrass	500 g/l	SC	14290

Products – continued

6	Daconil Turf	Scotts	500 g/l	SC	09265
7	Daconil Weatherstik	Syngenta	500 g/l	SC	12636
8	Expert-Turf	AgriGuard	500 g/l	SC	12571
9	Greencrop Orchid 2	Greencrop	500 g/l	SC	12769
10	Greencrop Orchid B	Greencrop	500 g/l	SC	12251
11	Joules	Nufarm UK	500 g/l	SC	13773
12	Juliet	ChemSource	500 g/l	SC	14557
13	Landgold Chlorothalonil 50	Goldengrass	500 g/l	SC	14011
14	Mascot Contact	Rigby Taylor	500 g/l	SC	13718
15	Pan Trilby	Pan Agriculture	500 g/l	SC	13558
16	Repulse	Certis	500 g/l	SC	11328
17	Rezacur	Sherriff Amenity	500 g/l	SC	13717
18	Sanspor	Syngenta	500 g/l	SC	13313
19	Sonar	Syngenta	720 g/l	SC	13083
20	Supreme	AgriGuard	500 g/l	SC	12233

Uses

- American gooseberry mildew in ***gooseberries*** [2-5, 11, 12, 18]
- Anthracnose in ***managed amenity turf*** [6-8, 14, 15, 17]
- Ascochyta in ***combining peas*** [1-5, 9-13, 16, 18-20]
- Basal stem rot in ***narcissi*** *(off-label)* [2]
- Blight in ***potatoes***, ***protected tomatoes*** [1-5, 9-13, 16, 18-20]
- Botrytis in ***blackcurrants***, ***gooseberries*** *(qualified minor use)*, ***redcurrants*** *(qualified minor use)*, ***strawberries*** *(outdoor crops only)* [1, 13, 20]; ***borage for oilseed production*** *(off-label)*, ***daffodils grown for galanthamine production*** *(off-label)*, ***evening primrose*** *(off-label)*, ***honesty*** *(off-label)*, ***linseed*** *(off-label)*, ***mustard*** *(off-label)* [2]; ***broccoli***, ***brussels sprouts***, ***cabbages***, ***cauliflowers***, ***protected ornamentals*** [1, 9, 10, 13, 16, 18-20]; ***broccoli*** *(moderate control)*, ***brussels sprouts*** *(moderate control)*, ***bulb onions*** *(moderate control)*, ***cabbages*** *(moderate control)*, ***cauliflowers*** *(moderate control)*, ***combining peas*** *(moderate control)*, ***protected ornamentals*** *(moderate control)*, ***spring oilseed rape*** *(moderate control)*, ***strawberries*** *(moderate control)*, ***winter oilseed rape*** *(moderate control)* [2-5, 11, 12]; ***bulb onions***, ***combining peas*** [1, 13, 16, 18, 20]; ***calabrese*** *(off-label)* [16, 19]; ***garlic*** *(off-label)*, ***lupins*** *(off-label)*, ***shallots*** *(off-label)* [16]; ***gooseberries*** [2-5, 11, 12, 18]; ***protected cucumbers*** [1-5, 9-13, 16, 18-20]; ***protected tomatoes*** [2-5, 9-12, 16, 18, 19]; ***spring oilseed rape***, ***winter oilseed rape*** [1, 9, 10, 13, 18-20]; ***strawberries*** [9, 10, 16, 18, 19]
- Celery leaf spot in ***celery (outdoor)*** *(qualified minor use)* [1-5, 9-13, 16, 18-20]
- Chocolate spot in ***spring field beans***, ***winter field beans*** [1, 9, 10, 13, 18-20]; ***spring field beans*** *(moderate control)*, ***winter field beans*** *(moderate control)* [2-5, 11, 12]
- Currant leaf spot in ***blackcurrants***, ***gooseberries*** *(qualified minor use)*, ***redcurrants*** *(qualified minor use)* [1-5, 9-13, 16, 18-20]
- Disease control in ***calabrese*** *(off-label)*, ***durum wheat*** *(off-label)*, ***spring rye*** *(off-label)*, ***triticale*** *(off-label)*, ***winter rye*** *(off-label)* [2]; ***calabrese seedlings*** *(off-label)*, ***garlic*** *(off-label)*, ***grass seed crops*** *(off-label)*, ***lupins*** *(off-label)*, ***shallots*** *(off-label)* [1, 2]
- Dollar spot in ***managed amenity turf*** [6-8, 14, 15, 17]
- Downy mildew in ***brassica seed beds*** [9, 10, 13, 18, 19]; ***brassica seed beds*** *(moderate control)*, ***broccoli*** *(moderate control)*, ***brussels sprouts*** *(moderate control)*, ***cabbages*** *(moderate control)*, ***cauliflowers*** *(moderate control)*, ***spring oilseed rape*** *(moderate control)*, ***winter oilseed rape*** *(moderate control)* [2-5, 11, 12]; ***broccoli***, ***brussels sprouts***, ***cabbages***, ***cauliflowers*** [1, 9, 10, 13, 16, 18-20]; ***calabrese*** *(off-label)* [16, 19]; ***edible brassicas*** [16]; ***hops*** [1-5, 9-13, 16, 18-20]; ***poppies for morphine production*** *(off-label)* [2]; ***spring oilseed rape***, ***winter oilseed rape*** [1, 9, 10, 13, 18-20]
- Foliar disease control in ***durum wheat*** *(off-label)*, ***garlic*** *(off-label)*, ***grass seed crops*** *(off-label)*, ***lupins*** *(off-label)*, ***shallots*** *(off-label)*, ***spring rye*** *(off-label)*, ***winter rye*** *(off-label)* [19]
- Fusarium patch in ***managed amenity turf*** [6-8, 14, 15, 17]
- Glume blotch in ***spring wheat*** [1-5, 11-13, 18, 20]; ***winter wheat*** [1-5, 9-13, 18-20]
- Ink disease in ***irises*** *(qualified minor use)* [1-5, 9-13, 18-20]
- Leaf mould in ***protected tomatoes*** [1-5, 11-13, 16, 20]
- Leaf rot in ***bulb onions*** [1-5, 9-13, 16, 18-20]
- Leaf spot in ***celeriac*** *(off-label)*, ***strawberries*** *(off-label)* [2]
- Mycosphaerella in ***combining peas*** [1, 9, 10, 13, 16, 18-20]; ***combining peas*** *(moderate control)* [2-5, 11, 12]

- Neck rot in ***bulb onions*** [1-5, 9-13, 16, 18-20]; ***narcissi*** *(off-label)* [2]
- Powdery mildew in ***blackcurrants***, ***gooseberries*** *(qualified minor use)*, ***redcurrants*** *(qualified minor use)* [1, 13, 20]
- Red thread in ***managed amenity turf*** [6-8, 14, 15, 17]
- Rhynchosporium in ***spring barley***, ***winter barley*** [9, 10, 18]; ***spring barley*** *(moderate control)*, ***winter barley*** *(moderate control)* [2-5, 11, 12]; ***spring barley*** *(moderate control only)*, ***winter barley*** *(moderate control only)* [13]
- Ring spot in ***broccoli***, ***brussels sprouts***, ***cabbages***, ***cauliflowers*** [9, 10, 16, 19]; ***calabrese*** *(off-label)*, ***edible brassicas*** [16]
- Septoria leaf blotch in ***spring wheat*** [1-5, 11-13, 18, 20]; ***winter wheat*** [1-5, 9-13, 18-20]

Specific Off-Label Approvals (SOLAs)

- ***borage for oilseed production*** *20092486* [2]
- ***calabrese*** *20092488* [2], *20052888* [16], *20072053* [19]
- ***calabrese seedlings*** *20092480* [1], *20092488* [2]
- ***celeriac*** *20092482* [2]
- ***daffodils grown for galanthamine production*** *20092484* [2]
- ***durum wheat*** *20092486* [2], *20072050* [19]
- ***evening primrose*** *20092486* [2]
- ***garlic*** *20092479* [1], *20092489* [2], *20052887* [16], *20072052* [19]
- ***grass seed crops*** *20092478* [1], *20092486* [2], *20072050* [19]
- ***honesty*** *20092486* [2]
- ***linseed*** *20092486* [2]
- ***lupins*** *20092481* [1], *20092487* [2], *20052886* [16], *20072051* [19]
- ***mustard*** *20092486* [2]
- ***narcissi*** *20092491* [2]
- ***poppies for morphine production*** *20092490* [2]
- ***shallots*** *20092479* [1], *20092489* [2], *20052887* [16], *20072052* [19]
- ***spring rye*** *20092486* [2], *20072050* [19]
- ***strawberries*** *20092483* [2]
- ***triticale*** *20092486* [2]
- ***winter rye*** *20092486* [2], *20072050* [19]

Approval information

- Chlorothalonil included in Annex I under EC Directive 91/414. However a Standing Committee decision in July 2007 to reduce to the limit of determination the MRL set for chlorothalonil in blackberries and raspberries led to revocation of approvals for use on these crops
- Approval for aerial spraying on potatoes [1, 2, 9, 10, 13, 16, 18-20]. See Section 5 for more information
- Accepted by BBPA for use on malting barley and hops

Efficacy guidance

- For some crops products differ in diseases listed as controlled. See label for details and for application rates, timing and number of sprays
- Apply as protective spray or as soon as disease appears and repeat as directed
- On cereals activity against Septoria may be reduced where serious mildew or rust present. In such conditions mix with suitable mildew or rust fungicide
- May be used for preventive and curative treatment of turf but treatment at a late stage of disease development may not be so successful and can leave bare patches of soil requiring renovation. Products are not recommended for curative control of Fusarium patch or Anthracnose [6-8, 14, 15, 17]
- Optimum results on managed amenity turf obtained from treatment in spring or autumn when conditions most favourable for disease development [6-8, 14, 15, 17]

Restrictions

- Maximum number of treatments and maximum total doses vary with crop and product - see labels for details
- Operators of vehicle mounted equipment must use a vehicle fitted with a cab and a forced air filtration unit with a pesticide filter complying with HSE Guidance Note PM74 or to an equivalent or higher standard when making broadcast or air-assisted applications
- Must only be used on established turf [6-8, 14, 15, 17]

- On managed amenity turf must only be applied by a pedestrian controlled sprayer or vehicle mounted/drawn equipment [6-8, 14, 15, 17]
- Do not mow or water turf for 24 h after treatment. Do not add surfactant or mix with liquid fertilizer [6-8, 14, 15, 17]

Crop-specific information

- Latest time of application to cereals varies with product. Consult label
- Latest use for winter oilseed rape before flowering; end Aug in yr of harvest for blackcurrants, gooseberries, redcurrants,
- HI 8 wk for field beans; 6 wk for combining peas; 28 d or before 31 Aug for post-harvest treatment for blackcurrants, gooseberries, redcurrants; 14 d for onions, strawberries; 10 d for hops; 7-14 d for broccoli, Brussels sprouts, cabbages, cauliflowers, celery, onions, potatoes; 3 d for cane fruit; 12-48 h for protected cucumbers, protected tomatoes
- For Botrytis control in strawberries important to start spraying early in flowering period and repeat at least 3 times at 10 d intervals
- On strawberries some scorching of calyx may occur with protected crops

Environmental safety

- Dangerous for the environment
- Very toxic to aquatic organisms
- Do not spray from the air within 250 m horizontal distance of surface waters or ditches
- LERAP Category B
- Broadcast air-assisted LERAP (18 m) [1, 2, 9, 10, 13, 16, 18-20]

Hazard classification and safety precautions

Hazard Harmful, Dangerous for the environment
Transport code 9 [1, 2, 4, 5, 7-10, 13-15, 17-20]
Packaging group III [1, 2, 4, 5, 7-10, 13-15, 17-20]
UN number 3082 [1, 2, 4, 5, 7-10, 13-15, 17-20]
Risk phrases R20, R40, R43, R50, R53a [1-20]; R22a [19]; R36 [1-15, 17-20]; R37 [1-9, 11-20]; R38 [10, 16]; R41 [16]
Operator protection A, C, H, M [1-20]; J [1-5, 9-13, 18-20]; U02a [2-5, 7, 9-13, 18, 19]; U05a, U20a [1-20]; U09a, U19a [1-13, 16, 18-20]; U14 [16]
Environmental protection E15a [1-12, 14-20]; E15b [13]; E16a [1-20]; E16b [1-6, 9-20]; E17b [1-5, 9-13, 16, 18-20] (18 m); E18 [1-5, 9-13, 18-20]; E34 [1, 6-9, 13-15, 17, 20]; E38 [1-8, 10-12, 16, 18-20]
Storage and disposal D01, D02, D09a [1-20]; D05 [1-15, 17-20]; D07 [8]; D10a [6, 8, 14-17]; D10b [9]; D10c [1-5, 7, 10-13, 18-20]; D12a [1-7, 9-13, 16, 18-20]
Medical advice M03 [8]; M05a [9]

77 chlorothalonil + cyproconazole

A systemic protectant and curative fungicide mixture for cereals
FRAC mode of action code: M5 + 3

See also cyproconazole

Products

1	Alto Elite	Syngenta	375:40 g/l	SC	08467
2	Octolan	Syngenta	375:40 g/l	SC	11675
3	SAN 703	Syngenta	37.5:4	SC	11676

Uses

- Brown rust in ***spring barley***, ***winter barley***, ***winter wheat*** [1-3]
- Chocolate spot in ***spring field beans***, ***winter field beans*** [1-3]
- Eyespot in ***winter barley***, ***winter wheat*** [1-3]
- Glume blotch in ***winter wheat*** [1]
- Grey mould in ***combining peas*** [1-3]
- Net blotch in ***spring barley***, ***winter barley*** [1-3]
- Powdery mildew in ***spring barley***, ***winter barley***, ***winter wheat*** [1-3]
- Rhynchosporium in ***spring barley***, ***winter barley*** [1-3]
- Rust in ***spring field beans***, ***winter field beans*** [1]

- Septoria leaf blotch in ***winter wheat*** [1-3]
- Yellow rust in ***spring barley***, ***winter barley***, ***winter wheat*** [1-3]

Approval information

- Chlorothalonil included in Annex I under EC Directive 91/414
- Accepted by BBPA for use on malting barley

Efficacy guidance

- Apply at first signs of infection or as soon as disease becomes active
- A repeat application may be made if re-infection occurs
- For established mildew tank-mix with an approved mildewicide
- When applied prior to third node detectable (GS 33) a useful reduction of eyespot will be obtained
- Cyproconazole is a DMI fungicide. Resistance to some DMI fungicides has been identified in Septoria leaf blotch which may seriously affect performance of some products. For further advice contact a specialist advisor and visit the Fungicide Resistance Action Group (FRAG)-UK website

Restrictions

- Maximum total dose equivalent to 2 full dose treatments
- Do not apply at concentrations higher than recommended

Crop-specific information

- Latest use: before beginning of anthesis (GS 60) for barley; before caryopsis watery ripe (GS 71) for wheat
- HI peas, field beans 6 wk
- If applied to winter wheat in spring at GS 30-33 straw shortening may occur but yield is not reduced

Environmental safety

- Dangerous for the environment
- Very toxic to aquatic organisms
- LERAP Category B

Hazard classification and safety precautions

Hazard Harmful, Dangerous for the environment
Transport code 9
Packaging group III
UN number 3082
Risk phrases R20, R37, R40, R41, R43, R50, R53a
Operator protection A, C, H, M; U05a, U19a [1-3]; U11, U20a [1]; U12, U20b [2, 3]
Environmental protection E15a, E16b [1]; E15b [2, 3]; E16a, E38 [1-3]
Storage and disposal D01, D02, D09a, D10c, D12a [1-3]; D05 [1]
Medical advice M04a

78 chlorothalonil + cyproconazole + propiconazole

A broad-spectrum fungicide mixture for cereals
FRAC mode of action code: M5 + 3 + 3

See also *cyproconazole*
propiconazole

Products

Cherokee	Syngenta	375:50:62.5 g/l	SE	13251

Uses

- Brown rust in ***spring barley***, ***winter barley***, ***winter wheat***
- Net blotch in ***spring barley*** *(reduction)*, ***winter barley*** *(reduction)*
- Powdery mildew in ***spring barley***, ***winter barley***, ***winter wheat*** *(moderate)*
- Rhynchosporium in ***spring barley*** *(reduction)*, ***winter barley*** *(reduction)*
- Septoria leaf blotch in ***winter wheat***
- Yellow rust in ***winter wheat***

Approval information

- Chlorothalonil and propiconazole included in Annex I under EC Directive 91/414

Efficacy guidance

- Best results obtained from treatment at the early stages of disease development
- Cyproconazole and propiconazole are DMI fungicides. Resistance to some DMI fungicides has been identified in Septoria leaf blotch which may seriously affect performance of some products. For further advice contact a specialist advisor and visit the Fungicide Resistance Action Group (FRAG)-UK website
- Product should be used as part of an Integrated Crop Management strategy incorporating other methods of control and including, where appropriate, other fungicides with a different mode of action
- Product should be used preventatively and not relied on for its curative potential
- Always follow FRAG guidelines for resistance management. See Section 5 for more information

Restrictions

- Maximum total dose equivalent to two full dose treatments on wheat and barley
- Do not treat crops under stress
- Spray solution must be used as soon as possible after mixing

Crop-specific information

- Latest use: before caryopsis watery ripe (GS 71) for winter wheat; before first spikelet just visible (GS 51) for barley
- When applied to winter wheat in the spring straw shortening may occur which causes no loss of yield

Environmental safety

- Dangerous for the environment
- Very toxic to aquatic organisms
- LERAP Category B

Hazard classification and safety precautions

Hazard Harmful, Dangerous for the environment
Transport code 9
Packaging group III
UN number 3082
Risk phrases R20, R36, R37, R40, R43, R50, R53a
Operator protection A, C, H; U02a, U05a, U09a, U10, U11, U14, U15, U19a, U20b
Environmental protection E15b, E16a, E16b, E34, E38
Storage and disposal D01, D02, D05, D07, D09a, D10c, D12a
Medical advice M04a

79 chlorothalonil + fludioxonil + propiconazole

A fungicide treatment for amenity grassland
FRAC mode of action code: M5 + 12 + 3

See also *fludioxonil*
propiconazole

Products

Instrata	Syngenta	362:14.5:56.9 g/l	SE	14154

Uses

- Anthracnose in ***amenity grassland*** *(reduction)*, ***managed amenity turf*** *(reduction)*
- Brown patch in ***amenity grassland*, *managed amenity turf***
- Dollar spot in ***amenity grassland*, *managed amenity turf***
- Fusarium patch in ***amenity grassland*, *managed amenity turf***

Approval information

- Chlorothalonil, fludioxonil and propiconazole included in Annex I under EC Directive 91/414.

Hazard classification and safety precautions

Hazard Toxic, Dangerous for the environment
Transport code 9
Packaging group III
UN number 3082

Risk phrases R23, R36, R40, R43, R50, R53a
Operator protection A, C, H; U02a, U05a, U09a, U10, U11, U19a, U19c, U20b
Environmental protection E15b, E16a, E34, E38
Storage and disposal D01, D02, D09a, D12a
Medical advice M04a

80 chlorothalonil + flusilazole

A fungicide mixture for wheat and barley
FRAC mode of action code: M5 + 3

See also flusilazole

Products

Midas	DuPont	200:80 g/l	SC	13285

Uses
- Glume blotch in ***spring wheat, winter wheat***
- Rhynchosporium in ***spring barley, winter barley***
- Septoria leaf blotch in ***spring wheat, winter wheat***

Approval information
- Chlorothalonil included in Annex I under EC Directive 91/414.
- Flusilazole has been reinstated in Annex 1 under Directive 91/414 following an appeal

Efficacy guidance
- Best results achieved from treatment when diseases are at low levels and before they spread to new growth
- Under heavy disease pressure a second application may be made 3-4 wk later
- Flusilazole is a DMI fungicide. Resistance to some DMI fungicides has been identified in Septoria leaf blotch which may seriously affect performance of some products. For further advice contact a specialist advisor and visit the Fungicide Resistance Action Group (FRAG)-UK website

Restrictions
- Maximum number of treatments 2 per crop
- Do not apply to crops under stress
- Do not apply during frosty weather

Crop-specific information
- Latest use: end of flowering (GS 69) for wheat; end of heading (GS 59) for barley

Environmental safety
- Dangerous for the environment
- Very toxic to aquatic organisms
- LERAP Category B

Hazard classification and safety precautions

Hazard Toxic, Dangerous for the environment
Transport code 9
Packaging group III
UN number 3082
Risk phrases R37, R40, R41, R43, R50, R53a, R61
Operator protection A, C, H, M; U05a, U11, U20b, U23a
Environmental protection E15a, E16a, E34, E38
Storage and disposal D01, D02, D05, D09a, D10b, D12a
Medical advice M04a

81 chlorothalonil + flutriafol

A systemic eradicant and protectant fungicide for winter wheat
FRAC mode of action code: M5 + 3

See also flutriafol

Products

1	Impact Excel	Headland	300:47 g/l	SC	11547
2	Prospa	Headland	300:47 g/l	SC	11548

Uses

- Brown rust in ***winter wheat***
- Foliar disease control in ***durum wheat*** *(off-label)*, ***grass seed crops*** *(off-label)*, ***spring rye*** *(off-label)*, ***triticale*** *(off-label)*, ***winter rye*** *(off-label)*
- Late ear diseases in ***winter wheat***
- Powdery mildew in ***winter wheat***
- Septoria leaf blotch in ***winter wheat***
- Yellow rust in ***winter wheat***

Specific Off-Label Approvals (SOLAs)

- ***durum wheat*** *20052890* [1], *20052889* [2]
- ***grass seed crops*** *20052890* [1], *20052889* [2]
- ***spring rye*** *20052890* [1], *20052889* [2]
- ***triticale*** *20052890* [1], *20052889* [2]
- ***winter rye*** *20052890* [1], *20052889* [2]

Approval information

- Chlorothalonil included in Annex I under EC Directive 91/414

Efficacy guidance

- Generally disease control and yield benefit will be optimised when application is made at an early stage of disease development
- Apply as soon as disease is seen establishing in the crop

Restrictions

- Maximum number of treatments 2 per crop on winter wheat

Crop-specific information

- Latest use: before early milk stage (GS 73)
- On certain cultivars with erect leaves high transpiration can result in flag leaf tip scorch. This may be increased by treatment but does not affect yield

Environmental safety

- Dangerous for the environment [1, 2]
- Very toxic to aquatic organisms [1, 2]
- Dangerous to fish or other aquatic life. Do not contaminate surface waters or ditches with chemical or used container [2]
- LERAP Category B

Hazard classification and safety precautions

Hazard Toxic, Dangerous for the environment
Transport code 6.1
Packaging group II [1]; II [2]
UN number 2902
Risk phrases R23, R40, R41, R43, R50, R53a
Operator protection A, C, H; U02a, U05a, U09a, U11, U19a, U20b
Environmental protection E13b [2]; E15a [1]; E16a, E38 [1, 2]
Storage and disposal D01, D02, D09a, D10b [1, 2]; D05 [2]

82 chlorothalonil + mancozeb

A multi-site protectant fungicide mixture
FRAC mode of action code: M5 + M3

See also *mancozeb*

Products

1	Adagio	Interfarm	201:274 g/l	SC	10796
2	Guru	Interfarm	286:194 g/l	SC	10801

Uses

- Blight in ***potatoes*** [1]
- Foliar disease control in ***durum wheat*** *(off-label)*, ***grass seed crops*** *(off-label)*, ***spring rye*** *(off-label)*, ***winter rye*** *(off-label)* [2]
- Glume blotch in ***winter wheat*** [2]
- Septoria leaf blotch in ***winter wheat*** [2]

Specific Off-Label Approvals (SOLAs)

- ***durum wheat*** *20052892* [2]
- ***grass seed crops*** *20052892* [2]
- ***spring rye*** *20052892* [2]
- ***winter rye*** *20052892* [2]

Approval information

- Chlorothalonil and mancozeb included in Annex I under EC Directive 91/414

Efficacy guidance

- Start spray treatments on potatoes immediately after a blight warning or just before the haulm meets in the row [1]
- It is essential to start the blight spray programme before the disease appears in the crop [1]
- Repeat treatments for blight at 7, 10 or 14 d intervals depending on blight risk and continue until haulm is to be destroyed [1]
- Best results on winter wheat achieved from protective applications to the flag leaf. If disease already present on lower leaves treat as soon as flag leaf is just visible (GS 37) [2]
- Activity on Septoria may be reduced in presence of severe mildew infection [2]

Restrictions

- Maximum number of treatments 5 per crop for potatoes [1]
- Maximum total dose on winter wheat equivalent to 1.5 full dose treatments [2]
- Do not treat crops under stress from frost, drought, waterlogging, trace element deficiency or pest attack [2]
- Broadcast air assisted applications must only be made by equipment fitted with a cab with a forced air filtration unit plus a pesticide filter complying with HSE Guidance Note PM 74 or an equivalent or higher standard

Crop-specific information

- Latest use: before grain watery ripe (GS 71) for winter wheat
- HI potatoes 7 d
- Irrigated potato crops should be sprayed immediately after irrigation [1]

Environmental safety

- Dangerous for the environment
- Very toxic to aquatic organisms
- LERAP Category B

Hazard classification and safety precautions

Hazard Irritant, Dangerous for the environment
Transport code 9
Packaging group III
UN number 3082
Risk phrases R37, R43, R50, R53a [1, 2]; R40 [2]
Operator protection A, C, H, M; U02a, U05a, U08, U10, U11, U13, U14, U19a [1, 2]; U20a [1]; U20b [2]
Environmental protection E15a [1]; E16a, E16b, E38 [1, 2]

Storage and disposal D01, D02, D05, D09a, D10b, D12a
Medical advice M04a [1]

83 chlorothalonil + metalaxyl-M

A systemic and protectant fungicide for various crops
FRAC mode of action code: M5 + 4

See also metalaxyl-M

Products

Folio Gold	Syngenta	500:37.5 g/l	SC	10704

Uses
- Alternaria in ***brussels sprouts*** *(moderate control)*, ***calabrese*** *(moderate control)*, ***cauliflowers*** *(moderate control)*
- Disease control in ***poppies for morphine production*** *(off-label)*
- Downy mildew in ***broad beans***, ***brussels sprouts***, ***bulb onions*** *(qualified minor use)*, ***calabrese***, ***cauliflowers***, ***shallots*** *(qualified minor use)*, ***spring field beans***, ***winter field beans***
- Ring spot in ***brussels sprouts*** *(reduction)*, ***calabrese*** *(reduction)*, ***cauliflowers*** *(reduction)*
- White blister in ***brussels sprouts***, ***calabrese***, ***cauliflowers***
- White tip in ***leeks*** *(qualified minor use)*

Specific Off-Label Approvals (SOLAs)
- ***poppies for morphine production*** *20092464*

Approval information
- Chlorothalonil and metalaxyl-M included in Annex I under EC Directive 91/414

Efficacy guidance
- Apply at first signs of disease or when weather conditions favourable for disease pressure
- Best results obtained when used in a full and well-timed programme. Repeat treatment at 14-21 d intervals if necessary
- Treatment of established disease will be less effective
- Evidence of effectiveness in bulb onions, shallots and leeks is limited

Restrictions
- Maximum total dose equivalent to 2 full doses on broad beans, field beans, cauliflowers, calabrese; 3 full doses on Brussels sprouts, leeks, onions and shallots

Crop-specific information
- HI 14 d for all crops

Environmental safety
- Dangerous for the environment
- Very toxic to aquatic organisms
- LERAP Category B

Hazard classification and safety precautions

Hazard Harmful, Dangerous for the environment
Transport code 9
Packaging group III
UN number 3082
Risk phrases R20, R36, R37, R38, R40, R43, R50, R53a
Operator protection A, C, H; U05a, U09a, U15, U20a
Environmental protection E15a, E16a, E38
Consumer protection C02a (14 d)
Storage and disposal D01, D02, D05, D09a, D10c, D12a

84 chlorothalonil + picoxystrobin

A broad spectrum fungicide mixture for cereals
FRAC mode of action code: M5 + 11

See also *picoxystrobin*

Products

1	Credo	DuPont	500:100 g/l	SC	13042
2	Zimbrail	DuPont	500:100 g/l	SC	13172

Uses

- Brown rust in ***spring barley, spring wheat, winter barley, winter wheat***
- Glume blotch in ***spring wheat, winter wheat***
- Net blotch in ***spring barley, winter barley***
- Rhynchosporium in ***spring barley, winter barley***
- Septoria leaf blotch in ***spring wheat, winter wheat***
- Tan spot in ***spring wheat, winter wheat***
- Yellow rust in ***spring wheat, winter wheat***

Approval information

- Chlorothalonil and picoxystrobin included in Annex I under EC Directive 91/414
- Accepted by BBPA for use on malting barley

Efficacy guidance

- Best results obtained as a protectant treatment, or when disease first seen in crop, applied in good growing conditions with adequate soil moisture
- Disease control for 4-6 wk normally achieved during stem elongation
- Results may be less reliable when used on crops under stress
- Treatments for protection against ear disease should be made at ear emergence
- Picoxystrobin is a member of the QoI cross resistance group. Product should be used preventatively and not relied on for its curative potential
- Use product as part of an Integrated Crop Management strategy incorporating other methods of control, including where appropriate other fungicides with a different mode of action. Do not apply more than two foliar applications of QoI containing products to any cereal crop
- Control of many diseases may be improved by use in mixture with an appropriate triazole
- There is a significant risk of widespread resistance occurring in *Septoria tritici* populations in UK. Failure to follow resistance management action may result in reduced levels of disease control
- Strains of barley powdery mildew resistant to QoIs are common in the UK

Restrictions

- Maximum number of treatments 2 per crop of wheat or barley

Crop-specific information

- Latest use: before first spikelet just visible (GS 51) for barley; before grain watery ripe (GS 71) for wheat

Environmental safety

- Dangerous for the environment
- Very toxic to aquatic organisms
- LERAP Category B

Hazard classification and safety precautions

Hazard Harmful, Dangerous for the environment
Transport code 9
Packaging group III
UN number 3082
Risk phrases R20, R37, R40, R43, R50, R53a
Operator protection A, H; U05a, U09a, U14, U15, U19a, U20a
Environmental protection E15a, E16a, E16b, E38
Storage and disposal D01, D02, D05, D09a, D10c, D12a

85 chlorothalonil + propamocarb hydrochloride

A contact and systemic fungicide mixture for blight control in potatoes
FRAC mode of action code: M5 + 28

See also *propamocarb hydrochloride*

Products

Pan Magician	Pan Agriculture	375:375 g/l	SC	11992

Uses

- Blight in ***potatoes***

Approval information

- Chlorothalonil and propamocarb hydrochloride included in Annex I under EC Directive 91/414

Efficacy guidance

- Commence treatment early in the season as soon as there is risk of infection
- In the absence of a blight warning treatment should start just before potatoes meet along the row
- Use only as a protectant. Stop use when blight readily visible (1% leaf area destroyed)
- Repeat sprays at 10-14 d intervals depending on blight infection risk. See label for details
- Complete blight spray programme after end Aug up to haulm destruction with protectant fungicides

Restrictions

- Maximum total dose equivalent to 6 full dose treatments on potatoes
- Apply to dry foliage. Do not apply if rainfall or irrigation imminent

Crop-specific information

- HI 7 d for potatoes

Environmental safety

- Dangerous for the environment
- Very toxic to aquatic organisms
- LERAP Category B

Hazard classification and safety precautions

Hazard Harmful, Dangerous for the environment
Transport code 9
Packaging group III
UN number 3082
Risk phrases R20, R40, R41, R43, R50, R53a
Operator protection A, C, H; U05a, U08, U11, U14, U19a, U20a
Environmental protection E15a, E16a, E16b, E34, E38
Storage and disposal D01, D02, D05, D09a, D10b, D12a

86 chlorothalonil + propiconazole

A systemic and protectant fungicide for winter wheat and barley
FRAC mode of action code: M5 + 3

See also *propiconazole*

Products

Prairie	Syngenta	250:62.5 g/l	SC	13994

Uses

- Brown rust in ***spring barley***, ***spring wheat***, ***winter barley***, ***winter wheat***
- Glume blotch in ***spring wheat***, ***winter wheat***
- Rhynchosporium in ***spring barley***, ***winter barley***
- Septoria leaf blotch in ***spring wheat***, ***winter wheat***
- Yellow rust in ***spring wheat***, ***winter wheat***

Approval information

- Chlorothalonil and propiconazole included in Annex I under EC Directive 91/414
- Accepted by BBPA for use on malting barley

SEE SECTION 3 FOR PRODUCTS ALSO REGISTERED

Efficacy guidance

- On wheat apply from start of flag leaf emergence up and including when ears just fully emerged (GS 37-59), on barley at any time to ears fully emerged (GS 59)
- Best results achieved from early treatment, especially if weather wet, or as soon as disease develops

Restrictions

- Maximum number of treatments 2 per crop or 1 per crop if other propiconazole based fungicide used in programme

Crop-specific information

- Latest use: up to and including emergence of ear just complete (GS 59).
- HI 42 d

Environmental safety

- Irritating to eyes, skin and respiratory system
- Risk of serious damage to eyes
- Dangerous to fish or other aquatic life. Do not contaminate surface waters or ditches with chemical or used container
- LERAP Category B

Hazard classification and safety precautions

Hazard Harmful, Dangerous for the environment
Transport code 9
Packaging group III
UN number 3082
Risk phrases R20, R36, R37, R40, R43, R50, R53a
Operator protection A, C, H; U02a, U05a, U11, U20b
Environmental protection E15b, E16a, E16b, E38
Storage and disposal D01, D02, D09a, D10c, D12a

87 chlorothalonil + pyrimethanil

A fungicide mixture for use in combining peas
FRAC mode of action code: M5 + 9

See also pyrimethanil

Products

Walabi	BASF	375:150 g/l	SC	12265

Uses

- Ascochyta in ***combining peas***
- Botrytis in ***combining peas***
- Chocolate spot in ***spring field beans***, ***winter field beans***
- Mycosphaerella in ***combining peas***

Approval information

- Chlorothalonil and pyrimethanil included in Annex I under EC Directive 91/414.

Efficacy guidance

- Use only as a protectant spray in a two spray programme
- For best results ensure good foliar cover

Restrictions

- Maximum total dose equivalent to two full dose treatments
- Consult processors before use on crops for processing

Crop-specific information

- HI 6 wk for combining peas; 8 wk for field beans

Environmental safety

- Dangerous for the environment
- Very toxic to aquatic organisms
- LERAP Category B

Hazard classification and safety precautions

Hazard Harmful, Dangerous for the environment

Transport code 9

Packaging group III

UN number 3082

Risk phrases R36, R40, R43, R50, R53a

Operator protection A, C, H; U02a, U11

Environmental protection E15a, E16a

Storage and disposal D01, D02, D05, D09a, D10c, D12a

Medical advice M03

88 chlorothalonil + tebuconazole

A protectant and systemic fungicide mixture for winter wheat
FRAC mode of action code: M5 + 3

See also *tebuconazole*

Products

Pentangle	Nufarm UK	500:180 g/l	SC	13746

Uses

- Septoria leaf blotch in ***winter wheat***

Approval information

- Chlorothalonil and tebuconazole included in Annex I under EC Directive 91/414

Efficacy guidance

- Latest time of application before anthesis (GS61)
- Maximum individual dose 1.0 L/ha. Where disease pressure is severe, a second application may be required but applications after flag leaf visible (GS39) may only result in a reduction of disease severity

Environmental safety

- LERAP B [1]

Hazard classification and safety precautions

Hazard Very toxic, Dangerous for the environment

Transport code 9

Packaging group III

UN number 3082

Risk phrases R26, R37, R40, R41, R43, R50, R53a, R63

Operator protection A, C, H; U02a, U05a, U09a, U10, U11, U13, U14, U15, U19a, U19b

Environmental protection E15a, E16a, E16b, E34, E38

Storage and disposal D01, D02, D09a, D10b, D12a

Medical advice M04a

89 chlorotoluron

A contact and residual urea herbicide for cereals
HRAC mode of action code: C2

Products

1	Alpha Chlorotoluron 500	Makhteshim	500 g/l	SC	04848
2	Atol	Nufarm UK	700 g/l	SC	11138
3	Chute	Makhteshim	90% w/w	WG	13520
4	Clayton Chloron 500 FL	Clayton	500 g/l	SC	10791
5	Dicurane	Makhteshim	500 g/l	SC	14276
6	Dicurane 70SC	Makhteshim	700 g/l	SC	14277
7	Dicurane 90WDG	Makhteshim	90% w/w	WG	14273
8	Headland Tolerate	Headland	500 g/l	SC	10774
9	Lentipur CL 500	Nufarm UK	500 g/l	SC	08743
10	Rotor	Makhteshim	700 g/l	SC	13511
11	Tolugan 700	Makhteshim	700 g/l	SC	08064
12	Tolurex 90 WDG	Makhteshim	90% w/w	WG	11403

Uses

- Annual dicotyledons in ***durum wheat*** [3, 5-7, 9, 12]; ***spring barley*** *(off-label)* [7, 10]; ***spring rye, winter rye*** [9]; ***spring rye*** *(off-label)*, ***winter rye*** *(off-label)* [1, 7]; ***spring wheat*** [10]; ***spring wheat*** *(off-label)* [7]; ***triticale, winter barley, winter wheat*** [1-12]
- Annual grasses in ***durum wheat*** [3, 5, 7, 9, 12]; ***spring barley*** *(off-label)*, ***spring wheat*** *(off-label)* [3, 5-7, 11, 12]; ***spring rye, winter rye*** [9]; ***spring rye*** *(off-label)*, ***winter rye*** *(off-label)* [1, 5-7, 11]; ***spring wheat*** [10]; ***triticale, winter barley, winter wheat*** [1-5, 7-12]
- Annual meadow grass in ***durum wheat, triticale, winter barley, winter wheat*** [6, 7]; ***spring barley*** *(off-label)* [3, 5-7, 10-12]; ***spring rye*** *(off-label)*, ***winter rye*** *(off-label)* [5-7, 11]; ***spring wheat*** *(off-label)* [3, 5-7, 11, 12]
- Blackgrass in ***durum wheat*** [3, 6, 9, 12]; ***spring barley*** *(off-label)*, ***spring wheat*** [10]; ***triticale, winter barley, winter wheat*** [1-4, 6, 8-12]
- Italian ryegrass in ***durum wheat, triticale, winter barley, winter wheat*** [6]
- Perennial ryegrass in ***durum wheat, triticale, winter barley, winter wheat*** [6]
- Rough meadow grass in ***durum wheat*** [3, 6, 9, 12]; ***spring barley*** *(off-label)*, ***spring wheat*** [10]; ***triticale, winter barley, winter wheat*** [1-4, 6, 8-12]
- Wild oats in ***durum wheat*** [3, 6, 9, 12]; ***spring barley*** *(off-label)*, ***spring wheat*** [10]; ***triticale, winter barley, winter wheat*** [1-4, 6, 8-12]

Specific Off-Label Approvals (SOLAs)

- ***spring barley*** *20090215* [3], *20090290* [5], *20090292* [6], *20090291* [7], *20090218* [10], *20090207* [11], *20090204* [12]
- ***spring rye*** *20092465* [1], *20090290* [5], *20090292* [6], *20090291* [7], *20092467* [11]
- ***spring wheat*** *20090215* [3], *20090290* [5], *20090292* [6], *20090291* [7], *20090207* [11], *20090204* [12]
- ***winter rye*** *20092465* [1], *20090290* [5], *20090292* [6], *20090291* [7], *20092467* [11]

Approval information

- Chlorotoluron included in Annex I under EC Directive 91/414
- May be applied through CDA equipment. See labels for details [2, 9]
- Approved for aerial application on winter wheat, winter barley, durum wheat, triticale [9]. See Section 5 for more information
- Accepted by BBPA for use on malting barley

Efficacy guidance

- Best results achieved by application soon after drilling. Application in autumn controls most weeds germinating in early spring
- For wild oat control apply within 1 wk of drilling, not after 2-leaf stage. Blackgrass and meadow grasses controlled to 5 leaf, ryegrasses to 3 leaf stage
- Any trash or burnt straw should be buried and dispersed during seedbed preparation
- Control may be reduced if prolonged dry conditions follow application
- Harrowing after treatment may reduce weed control
- Always follow WRAG guidelines for preventing and managing herbicide resistant weeds. See Section 5 for more information

Restrictions

- Maximum number of treatments 1 per crop
- Use only on listed crop varieties. See label. Ensure seed well covered at drilling
- Apply only as pre-emergence spray in durum wheat, pre- or post-emergence in wheat, barley or triticale
- Do not apply pre-emergence to crops sown after 30 Nov
- Do not apply to crops severely checked by waterlogging, pests, frost or other factors
- Do not use on undersown crops or those due to be undersown
- Do not apply post-emergence in mixture with liquid fertilizers
- Do not roll for 7 d before or after application to an emerged crop
- Do not use on soils with more than 10% organic matter
- Crops on stony or gravelly soils may be damaged, especially after heavy rain

Crop-specific information

- Latest use: pre-emergence for durum wheat. Post emergence timings on other crops vary - see labels
- Early sown crops may be damaged if applied before or during a period of rapid autumn growth
- Autumn treated crops must be drilled by 1 Dec and spring treated crops must be drilled by 1 Feb

Environmental safety

- Dangerous for the environment
- Very toxic to aquatic organisms. May cause long-term adverse effects in the aquatic environment
- Some pesticides pose a greater threat of contamination of water than others and chlorotoluron is one of these pesticides. Take special care when applying chlorotoluron near water and do not apply if heavy rain is forecast

Hazard classification and safety precautions

Hazard Harmful [1, 2, 4-6, 8-10]; Irritant [1]; Dangerous for the environment [1-12]
Transport code 9
Packaging group III
UN number 3077 [3, 7, 12]; 3082 [1, 2, 4-6, 8-11]
Risk phrases R22a, R36, R38 [1, 4, 8]; R40 [4-6, 9, 10]; R50, R53a [1-12]; R63 [2, 4-6, 9, 10]; R68 [2]
Operator protection A [1-12]; C [1-5, 7-12]; H [6, 10]; U05a [1-5, 7, 8, 10-12]; U08 [1-9, 11, 12]; U09a [10]; U10 [8]; U11 [4, 8]; U19a [1-12]; U20a [2, 11]; U20b [1, 3-10, 12]
Environmental protection E15a [1-12]; E34 [1-8, 10-12]; E38 [5, 6, 10]
Storage and disposal D01, D02, D09a, D10b [1-12]; D05 [4, 5, 8]; D12a [2, 5, 6, 8-10]; D12b [1, 3, 4, 7, 11, 12]
Medical advice M03 [1-8, 10-12]; M04a [5, 6, 10]; M05a [4]

90 chlorotoluron + diflufenican

A soil and foliar acting herbicide mixture for winter cereals
HRAC mode of action code: C2 + F1

See also diflufenican

Products

1	Agena	Nufarm UK	600:25 g/l	SC	14051
2	Buckler	Nufarm UK	600:25 g/l	SC	13476
3	Dicurane Surpass	Makhteshim	400:25 g/l	SC	14544
4	Dicurane XL	Makhteshim	450:15 g/l	SC	14569
5	Gloster	Makhteshim	400:25 g/l	SC	14037
6	Hekla	Makhteshim	450:15 g/l	SC	14298
7	Kula	Makhteshim	450:10 g/l	SC	14299
8	Steel	Nufarm UK	620:22.5 g/l	SC	14111
9	Tawny	Makhteshim	400:25 g/l	SC	14038

Uses

- Annual dicotyledons in ***durum wheat, triticale*** [3-7, 9]; ***winter barley, winter wheat*** [1-9]
- Annual grasses in ***durum wheat, triticale*** [4, 6, 7, 9]; ***winter barley, winter wheat*** [4, 6-9]
- Annual meadow grass in ***durum wheat, triticale*** [4, 6, 7]; ***winter barley, winter wheat*** [1, 2, 4, 6, 7]

Approval information

- Chlorotoluron and diflufenican included in Annex I under EC Directive 91/414

Efficacy guidance

- Best results obtained in crops sown in a fine firm seedbed with clods no greater than fist size and adequate moisture present at spraying. Good even spray coverage of the soil is essential
- Good weed control depends on burial and dispersal of any trash or straw before drilling
- Weed control may be reduced in prolonged periods of below average rainfall
- In direct drilled crops soil surface should be cultivated and the drill slots closed before spraying
- Crops should be treated pre-emergence and before end Feb
- Spring germinating weeds will not be controlled
- Fluffy seedbeds should be rolled before treatment
- Harrowing a treated crop may reduce the level of weed control

Restrictions

- Maximum number of treatments 1 per crop for wheat, barley
- Do not apply if heavy rain is expected within 4 h or if crop is stressed
- Do not roll for 7 d before or after application. Do not roll autumn crops until spring
- Do not treat undersown crops, or those to be undersown

SEE SECTION 3 FOR PRODUCTS ALSO REGISTERED

- Do not apply on soils with more than 10% organic matter
- Treat only listed varieties of wheat and barley

Crop-specific information

- Latest use: before end Feb for winter barley, winter wheat
- Transient leaf discolouration occasionally occurs after treatment of wheat or barley
- Treatment of wheat or barley on stony or gravelly soils may cause damage especially if heavy rain falls after application
- Early sown crops of wheat or barley may be damaged if treated during a period of rapid autumn growth

Following crops guidance

- Any crop may be drilled or planted following normal harvest of a treated crop provided the soil is ploughed to 150 mm
- In the event of failure of a treated crop only winter wheat or winter barley may be drilled immediately after ploughing. A period of 12 wk must elapse after ploughing before spring sowing of wheat, barley, oilseed rape, peas, beans, sugar beet, potatoes, carrots, edible brassicas or onions
- Successive treatments with any products containing diflufenican can lead to soil build-up and inversion ploughing to 150 mm must precede sowing any following non-cereal crop. Even where ploughing occurs some crops may be damaged

Environmental safety

- Dangerous for the environment
- Very toxic to aquatic organisms
- LERAP Category B
- Do not apply to dry, cracked or waterlogged soils
- Some pesticides pose a greater threat of contamination of water than others and chlorotoluron is one of these pesticides. Take special care when applying chlorotoluron near water and do not apply if heavy rain is forecast

Hazard classification and safety precautions

Hazard Harmful, Dangerous for the environment
Transport code 9
Packaging group III
UN number 3082
Risk phrases R40, R63 [1-7, 9]; R43 [1, 2, 8]; R50, R53a [1-9]; R61 [8]
Operator protection A, H [1-9]; C [9]; U05a [1-9]; U08, U13, U19a [1, 2, 4, 6-8]; U14, U15 [3, 5, 9]; U20a [4, 6, 7]; U20b [1, 2, 8]
Environmental protection E15a [1, 2, 4, 6-8]; E16a, E38 [1-9]; E16b [3, 5, 9]; E34 [4, 6, 7]
Storage and disposal D01, D02, D12a [1-9]; D05 [1, 2, 4, 6, 7]; D09a [1, 2, 4, 6-8]; D10b [4, 6-8]; D10c [1, 2]
Medical advice M04a [3, 5, 9]

91 chlorpropham

A residual carbamate herbicide and potato sprout suppressant
HRAC mode of action code: K2

See also chloridazon + chlorpropham + metamitron

Products

1	BL 500	United Phosphorus	500 g/l	HN	14387
2	Gro-Stop 100	Certis	300 g/l	HN	14182
3	Gro-Stop Fog	Certis	300 g/l	HN	14183
4	Gro-Stop HN	Certis	300 g/l	HN	14146
5	Gro-Stop Solid	Certis	100% w/w	BR	14103
6	MSS CIPC 50 M	United Phosphorus	500 g/l	HN	14388
7	MSS Sprout Nip	Whyte Agrochemicals	100% w/w	HN	14184
8	Pro-Long	United Phosphorus	500 g/l	HN	14389
9	Warefog 25	Whyte Agrochemicals	600 g/l	HN	06776

Uses

- Sprout suppression in ***ware potatoes*** [2, 6, 7]; ***ware potatoes*** *(thermal fog)* [1, 3-5, 8, 9]

Approval information

- Chlorpropham included in Annex I under EC Directive 91/414
- Some products are formulated for application by thermal fogging. See labels for details [9]
- Approval expiry 31 Jul 2010 [9]

Efficacy guidance

- Apply weed control sprays to freshly cultivated soil. Adequate rainfall must occur after spraying. Activity is greater in cold, wet than warm, dry conditions
- For sprout suppression apply with suitable fogging or rotary atomiser equipment or sprinkle granules over dry tubers before sprouting commences. Repeat applications may be needed. See labels for details
- Best results on potatoes obtained in purpose-built box stores with suitable forced draft ventilation. Potatoes in bulk stores should not be stacked more than 3 m high
- Blockage of air spaces between tubers prevents circulation of vapour and consequent loss of efficacy
- It is important to treat potatoes before the eyes open to obtain best results
- Effectiveness of fogging reduced in non-dedicated stores without proper insulation and temperature controls. Best results obtained at 5-10 °C and 75-80% humidity

Restrictions

- Maximum number of treatments normally 1 per batch for sprout suppression of potatoes; 1 per crop or yr for other named crops or situations
- Excess rainfall after application may result in crop damage
- Do not use on Sands, Very Light soils or soils low in organic matter
- Poor conditions at drilling or planting, soil compaction, surface capping, waterlogging or attack by pests may result in crop damage
- On crops under glass high temperatures and poor ventilation may cause crop damage
- Only clean, mature, disease-free potatoes should be treated for sprout suppression. Use of chlorpropham can inhibit tuber wound healing and the severity of skin spot infection in store may be increased if damaged tubers are treated
- Do not fog potatoes with a high level of skin spot
- Do not use on potatoes for seed. Do not handle, dry or store seed potatoes or any other seed or bulbs in boxes or buildings in which potatoes are being or have been treated
- Do not remove treated potatoes from store for sale or processing for at least 21 d after application

Crop-specific information

- Latest use: 3 d before drilling carrots; 7d before planting out celery; 21 d before removal for sale or processing for potatoes [9]; before tulip leaves unfurl or flower bulbs 5 cm high
- Cure potatoes according to label instructions before treatment and allow 3 wk between completion of loading into store and first treatment
- Apply weed control sprays to seeded crops pre-emergence of crop or weeds, to onions as soon as first crop seedlings visible, to planted crops a few days before planting, to bulbs immediately after planting, to fruit crops in late autumn-early winter. See label for further details

Following crops guidance

- There is a risk of damage to seed potatoes which are handled or stored in boxes or buildings previously treated with chlorpropham

Environmental safety

- Dangerous for the environment
- Toxic to aquatic organisms
- Keep unprotected persons out of treated stores for at least 24 h after application
- Keep in original container, tightly closed, in a safe place, under lock and key

Hazard classification and safety precautions

Hazard Toxic, Highly flammable [1, 6-8]; Harmful [2-4, 9]; Dangerous for the environment [1-9]
Transport code 6.1 [1-4, 6, 8, 9]; 9 [5, 7]
Packaging group III
UN number 2902 [1-4, 6, 8, 9]; 3077 [5, 7]
Risk phrases R23, R24, R39 [1, 6-8]; R25 [1, 8]; R36 [1, 6-9]; R38, R48 [9]; R40, R42, R43 [2-4]; R51, R53a [1-6, 8, 9]
Operator protection A [1-8]; C [1, 6-8]; D, E, H, M [1-9]; J [1, 5-9]; U01, U19b [5]; U02a [2-5]; U04a, U20a [2-4]; U05a, U08, U19a [1-9]; U14, U15 [2-4, 9]; U20b [1, 5-8]; U20c [9]

Environmental protection E02a [1, 6-8] (24 h); E02a [9] (12 h); E15a [1-4, 6-9]; E15b, E38 [5]; E34 [1-9]
Consumer protection C02b [2-4, 7, 9]; C02c [1, 6, 8]; C12 [2-4]
Storage and disposal D01, D02 [1-9]; D05 [1-8]; D06b [1, 6, 7]; D07, D11a, D12a [2-5]; D09a [9]; D09b [1-4, 6-8]; D10a, D12b [1, 6-9]
Medical advice M04a [1-8]

92 chlorpropham with cetrimide

A soil-acting herbicide for lettuce under cold glass - Last product withdrawn prior to approval expiry 31/7/2010
HRAC mode of action code: K2

93 chlorpropham + metamitron

A contact and residual herbicide mixture for beet crops
HRAC mode of action code: K2 + C1

See also *metamitron*

Products

Newgold	Whyte Agrochemicals	24.5:280 g/l	SE	13937

Uses
- Annual dicotyledons in ***fodder beet***, ***mangels***, ***sugar beet***
- Grass weeds in ***fodder beet***, ***mangels***, ***sugar beet***

Approval information
- Chlorpropham included in Annex I under EC Directive 91/414

Efficacy guidance
- May be used pre-emergence alone or post-emergence in tank mixture or with an authorised adjuvant oil
- Use as part of a spray programme. For optimum efficacy a full programme of pre- and post-emergence sprays is recommended
- Ideally one pre-emergence application should be followed by a repeat overall low dose programme on mineral soils. On organic soils a programme of up to five post-emergence sprays is likely to be most effective
- Product combines residual and contact activity. Adequate soil moisture and a fine firm seedbed are essential for optimum residual activity

Restrictions
- Maximum number of treatments one pre-emergence plus three post-emergence or, where no pre-emergence treatment has been applied, up to five post-emergence treatments with an adjuvant oil

Crop-specific information
- Latest use: before crop foliage meets across the rows

Following crops guidance
- Only sugar beet, fodder beet or mangels should be sown as following crops within 4 mth of treatment. After normal harvest of a treated crop land should be mouldboard ploughed to 15 cm after which any spring crops may be sown or planted

Environmental safety
- Dangerous for the environment
- Very toxic to aquatic organisms

Hazard classification and safety precautions

Hazard Dangerous for the environment
Risk phrases R22a, R36, R38, R43, R50, R53a
Operator protection A, C; U05a, U08, U11, U14, U15, U19a, U20b
Environmental protection E15a, E34

Storage and disposal D01, D02, D09a, D10c
Medical advice M03

94 chlorpyrifos

A contact and ingested organophosphorus insecticide and acaricide
IRAC mode of action code: 1B

Products

1	Agriguard Chlorpyrifos	AgriGuard	480 g/l	EC	13298
2	Alpha Chlorpyrifos 48 EC	Makhteshim	480 g/l	EC	13532
3	Ballad	Headland	480 g/l	EC	11659
4	Chlobber	AgChem Access	480 g/l	EC	13723
5	Clayton Pontoon 480EC	Clayton	480 g/l	EC	14555
6	Crossfire 480	Dow	480 g/l	EC	12516
7	Cyren	Headland	480 g/l	EC	11028
8	Dursban WG	Dow	75% w/w	WG	09153
9	Equity	Dow	480 g/l	EC	12465
10	Govern	Dow	75% w/w	WG	13223
11	Parapet	Dow	75 % w/w	WG	12773
12	Spannit Granules	Barclay	6% w/w	GR	12950
13	Suscon Green Soil Insecticide	Fargro	10% w/w	CG	06312

Uses

- Ambrosia beetle in ***cut logs*** [1-5, 7-11]
- Aphids in ***apples, gooseberries, pears, plums, raspberries, strawberries*** [1-5, 7, 9-11]; ***blackberries*** *(off-label)* [9, 11]; ***blackcurrants*** [1, 2, 4, 5, 10, 11]; ***broccoli, cabbages, calabrese, cauliflowers, chinese cabbage*** [1-5, 7-11]; ***brussels sprouts*** [2-4, 7]; ***currants*** [3, 7, 9]; ***durum wheat*** *(off-label)*, ***fodder beet*** *(off-label)*, ***spring rye*** *(off-label)*, ***triticale*** *(off-label)*, ***winter rye*** *(off-label)* [10]; ***quinces*** *(off-label)* [9]; ***redcurrants, whitecurrants*** [1, 10, 11]
- Apple blossom weevil in ***apples*** [1-5, 7, 9-11]; ***pears, plums*** [1]
- Apple sucker in ***apples*** [1, 5, 10, 11]; ***pears, plums*** [1]
- Bean seed fly in ***green beans*** *(off-label - harvested as a dry pulse)* [7, 8]; ***green beans harvested dry as a pulse*** *(off-label)*, ***kohlrabi*** *(off-label)*, ***mooli*** *(off-label)*, ***radishes*** *(off-label)* [10]
- Cabbage root fly in ***broccoli, cabbages, calabrese, cauliflowers*** [1-5, 7-12]; ***brussels sprouts*** [1-5, 7-11]; ***chinese cabbage*** [1-4, 7-11]; ***choi sum*** *(off-label)*, ***collards*** *(off-label)*, ***kale*** *(off-label)*, ***kohlrabi*** *(off-label)*, ***mooli*** *(off-label)*, ***pak choi*** *(off-label)*, ***radishes*** *(off-label)* [2, 7, 8, 10, 11]; ***green beans*** *(off-label - harvested dry as a pulse)* [11]; ***green beans harvested dry as a pulse*** *(off-label)* [2, 10]; ***tatsoi*** *(off-label)* [2, 7]
- Capsids in ***apples, gooseberries, pears*** [1-5, 7, 9-11]; ***blackcurrants*** [1, 2, 4, 5, 10, 11]; ***currants*** [3, 7, 9]; ***plums*** [1]; ***quinces*** *(off-label)* [9]; ***redcurrants, whitecurrants*** [1, 10, 11]
- Caterpillars in ***apples, pears*** [2-4, 7, 9]; ***blackcurrants*** [1, 2, 4, 5, 10, 11]; ***broccoli, cabbages, calabrese, cauliflowers, chinese cabbage*** [1-5, 7-11]; ***brussels sprouts*** [2-4, 7]; ***currants*** [3, 7, 9]; ***gooseberries*** [1-5, 7, 9-11]; ***quinces*** *(off-label)* [9]; ***redcurrants, whitecurrants*** [1, 10, 11]
- Cereal aphid in ***spring barley, spring wheat, winter barley, winter wheat*** [5]
- Codling moth in ***apples, pears*** [2-5, 7, 9-11]
- Cutworms in ***broccoli, cabbages, calabrese, cauliflowers, chinese cabbage*** [1, 2, 4, 5, 8-11]; ***brussels sprouts*** [2, 4]; ***bulb onions*** [2-5, 7-11]; ***onions*** [1]; ***potatoes grown for seed*** [5]; ***seed potatoes*** [2-4, 8-11]
- Damson-hop aphid in ***plums*** [2, 4, 10, 11]; ***plums*** *(Non OP resistant strains only)* [5]
- Elm bark beetle in ***cut logs*** [3, 7]
- Frit fly in ***amenity grassland*** [1, 3, 6, 7]; ***durum wheat, spring rye, triticale, winter rye*** [8]; ***durum wheat*** *(off-label)*, ***spring rye*** *(off-label)*, ***triticale*** *(off-label)*, ***winter rye*** *(off-label)* [9, 11]; ***forage maize*** [2-5, 7-11]; ***grassland*** [1, 5]; ***maize*** [1]; ***managed amenity turf*** [2-7]; ***permanent grassland*** [2-4, 7-11]; ***rotational grass*** [2-4, 7, 8, 10, 11]; ***spring barley, winter barley*** [1, 5, 9]; ***spring oats, spring wheat, winter oats, winter wheat*** [1-5, 7-11]
- Insect pests in ***blackberries*** *(off-label)* [2, 7, 10]; ***durum wheat*** *(off-label)*, ***spring rye*** *(off-label)*, ***triticale*** *(off-label)*, ***winter rye*** *(off-label)* [2, 8, 10]; ***fodder beet*** *(off-label)* [2, 10]; ***leeks*** *(off-label)* [8]; ***quinces*** *(off-label)* [9]
- Larch shoot beetle in ***cut logs*** [1-5, 7-11]

SEE SECTION 3 FOR PRODUCTS ALSO REGISTERED

- Leatherjackets in ***amenity grassland*** [1, 3, 6, 7]; ***broccoli, cabbages, calabrese, cauliflowers, chinese cabbage*** [1, 3, 7-11]; ***brussels sprouts*** [3, 7]; ***durum wheat, spring rye, triticale, winter rye*** [8]; ***durum wheat*** *(off-label)*, ***spring rye*** *(off-label)*, ***triticale*** *(off-label)*, ***winter rye*** *(off-label)* [9, 11]; ***fodder beet*** *(off-label)* [8, 9, 11]; ***grassland*** [1, 5]; ***managed amenity turf*** [2-7]; ***permanent grassland*** [2-4, 7-11]; ***rotational grass*** [2-4, 7, 8, 10, 11]; ***spring barley, spring oats, spring wheat, sugar beet, winter barley, winter oats, winter wheat*** [1-5, 7-11]
- Mealy aphid in ***plums*** [2, 4]
- Mealy plum aphid in ***plums*** [5]
- Midges in ***raspberries*** [5]
- Narcissus fly in ***flower bulbs*** *(off-label)* [7]; ***non-edible ornamental bulbs*** *(off-label)* [2]
- Opomyza in ***spring barley, spring oats, spring wheat, winter barley, winter oats, winter wheat*** [1]
- Pear sucker in ***pears*** [5, 10, 11]
- Pine shoot beetle in ***cut logs*** [1-5, 7-11]
- Pygmy mangold beetle in ***fodder beet*** *(off-label)* [8, 9, 11]; ***sugar beet*** [1, 3, 7-11]
- Raspberry beetle in ***raspberries*** [1-5, 7, 9-11]
- Raspberry cane midge in ***raspberries*** [1-4, 7, 9-11]
- Red spider mites in ***apples, pears, plums, raspberries*** [2-4, 7, 9-11]; ***apples*** *(non OP resistant strains only)*, ***blackcurrants*** *(non OP resistant strains only)*, ***gooseberries*** *(non OP resistant strains only)*, ***pears*** *(non OP resistant strains only)*, ***plums*** *(non OP resistant strains only)*, ***raspberries*** *(non OP resistant strains only)*, ***strawberries*** *(non OP resistant strains only)* [5]; ***blackberries*** *(off-label)* [9]; ***blackcurrants*** [1, 2, 4, 10, 11]; ***currants*** [3, 7, 9]; ***gooseberries, strawberries*** [1-4, 7, 9-11]; ***redcurrants, whitecurrants*** [1, 10, 11]
- Sawflies in ***apples*** [2-5, 7, 9-11]; ***pears*** [9]
- Strawberry blossom weevil in ***strawberries*** [1, 9]
- Suckers in ***apples, pears*** [2-4, 7, 9]
- Summer-fruit tortrix moth in ***apples*** [5, 10, 11]
- Thrips in ***leeks*** *(off-label)* [2]; ***spring oats, spring wheat, winter oats, winter wheat*** [3, 7]
- Tortrix moths in ***apples, pears, plums, strawberries*** [1-5, 7, 9-11]
- Vine weevil in ***ornamental plant production*** [13]; ***strawberries*** [2-4, 7, 9-11]; ***strawberries*** *(non-protected crops only)* [5]
- Wheat bulb fly in ***durum wheat, spring rye, triticale, winter rye*** [8]; ***durum wheat*** *(off-label)*, ***spring rye*** *(off-label)*, ***triticale*** *(off-label)*, ***winter rye*** *(off-label)* [9, 11]; ***spring barley, winter barley*** [1, 5, 8-11]; ***spring oats, spring wheat, winter oats, winter wheat*** [1-5, 7-11]
- Wheat-blossom midge in ***durum wheat, spring rye, triticale, winter rye*** [8]; ***durum wheat*** *(off-label)*, ***spring rye*** *(off-label)*, ***triticale*** *(off-label)*, ***winter rye*** *(off-label)* [9, 11]; ***spring barley, winter barley*** [1, 3, 7]; ***spring oats, winter oats*** [1, 3, 7-11]; ***spring wheat, winter wheat*** [1, 3, 5, 7-11]
- Whitefly in ***broccoli, cabbages, calabrese, cauliflowers, chinese cabbage*** [1, 3, 7-11]; ***brussels sprouts*** [3, 7]
- Winter moth in ***apples, pears, plums*** [1-5, 7, 9-11]
- Woolly aphid in ***apples*** [2-5, 7, 9-11]; ***pears*** [3, 7, 9]

Specific Off-Label Approvals (SOLAs)

- ***blackberries*** *20091081* [2], *20052977* [7], *20052964* [9], *20080023* [10], *20061585* [11]
- ***choi sum*** *20091087* [2], *20050237* [7], *20031390* [8], *20080024* [10], *20061587* [11]
- ***collards*** *20091087* [2], *20050237* [7], *20031390* [8], *20080024* [10], *20061587* [11]
- ***durum wheat*** *20091086* [2], *20052978* [8], *20052966* [9], *20080025* [10], *20061583* [11]
- ***flower bulbs*** *20050236* [7]
- ***fodder beet*** *20091088* [2], *20052979* [8], *20052965* [9], *20080022* [10], *20061584* [11]
- ***green beans*** *(harvested as a dry pulse) 20050235* [7], *(harvested as a dry pulse) 20031391* [8], *(harvested dry as a pulse) 20061586* [11]
- ***green beans harvested dry as a pulse*** *20091082* [2], *20080026* [10]
- ***kale*** *20091087* [2], *20050237* [7], *20031390* [8], *20080024* [10], *20061587* [11]
- ***kohlrabi*** *20091082* [2], *20050235* [7], *20031391* [8], *20080026* [10], *20061586* [11]
- ***leeks*** *20091084* [2], *20081469* [8]
- ***mooli*** *20091082* [2], *20050235* [7], *20031391* [8], *20080026* [10], *20061586* [11]
- ***non-edible ornamental bulbs*** *20091085* [2]
- ***pak choi*** *20091087* [2], *20050237* [7], *20031390* [8], *20080024* [10], *20061587* [11]
- ***quinces*** *20052974* [9]

- ***radishes*** *20091082* [2], *20050235* [7], *20031391* [8], *20080026* [10], *20061586* [11]
- ***spring rye*** *20091086* [2], *20052978* [8], *20052966* [9], *20061481* [9], *20080025* [10], *20061583* [11]
- ***tatsoi*** *20091087* [2], *20050237* [7]
- ***triticale*** *20091086* [2], *20052978* [8], *20052966* [9], *20080025* [10], *20061583* [11]
- ***winter rye*** *20091086* [2], *20052978* [8], *20052966* [9], *20080025* [10], *20061583* [11]

Approval information

- Chlorpyrifos included in Annex I under EC Directive 91/414
- Accepted by BBPA for use on malting barley
- Following implementation of Directive 98/82/EC, approval for use of chlorpyrifos on numerous crops was revoked in 1999
- Following the environmental review of chlorpyrifos uses on ware potatoes, carrots, conifers (drench), forestry trees (dip) and cereals (for aphids and yellow cereal fly) were revoked in September 2005. In addition the maximum permitted number of treatments, or the maximum total dose, on several other crops was reduced.
- In 2006 PSD required that all products containing this active ingredient should carry the following warning in the main area of the container label: "Chlorpyrifos is an anticholinesterase organophosphate. Handle with care"

Efficacy guidance

- Brassicas raised in plant-raising beds may require retreatment at transplanting
- Activity on field crops may be reduced when soil temperature below 5 °C or on organic soils
- In dry conditions the effect of granules applied as a surface band may be reduced [12]
- Where pear suckers resistant to chlorpyrifos occur control is unlikely to be satisfactory
- Efficacy against frit fly and leatherjackets in grass may be reduced if applied during periods of frost
- Thorough incorporation by hand or mixing machine is essential when used in potting media for ornamental plant production
- When used for control of vine weevil in ornamental plant production the recently emerged and young larvae are controlled by incorporation into the growing media. A single application will control successive generations for the growing cycle of plants kept in the same containers for 2 yr, with partial control for a third yr. Where plants are potted-on into larger containers the fresh growing media must also be treated to obtain control [13]

Restrictions

- Contains an anticholinesterase organophosphate compound. Do not use if under medical advice not to work with such compounds
- Maximum number of treatments and timing vary with crop, product and pest. See label for details
- Do not apply to sugar beet under stress or within 4 d of applying a herbicide
- Once incorporated, treated growing media must be used within 30 d
- Do not use product in growing media used for aquatic or semi-aquatic plants, and not for propagation of any edible crop
- Do not graze lactating cows on treated pasture for 14 d after treatment

Crop-specific information

- Latest use 4d after transplanting edible brassica crops; end Jul for sugar beet; varies for other crops. See individual labels
- HI range from 7 d to 6 wk depending on crop. See labels
- For vine weevil control in ornamental nursery stock incorporate in growing medium when plants first potted from rooted cutting stage. Treat the fresh growing medium when plants are potted on into larger containers [13]
- For turf pests apply from Nov where high larval populations detected or damage seen
- On lettuce apply only to strong well developed plants when damage first seen, or on professional advice
- In apples use pre-blossom up to pink/white bud and post-blossom after petal fall
- Test tolerance of glasshouse ornamentals before widescale use for propagating unrooted cuttings, or potting unusual plants and new species, and when using any media with a high content of non-peat ingredients

Environmental safety

- Dangerous for the environment
- Very toxic to aquatic organisms
- Flammable [6, 9]

SEE SECTION 3 FOR PRODUCTS ALSO REGISTERED

- Keep livestock out of treated areas for at least 14 d after treatment [2, 3, 6-9]
- High risk to bees. Do not apply to crops in flower or to those in which bees are actively foraging. Do not apply when flowering weeds are present [2, 3, 6-11]
- Chlorpyrifos is dangerous to some beneficial arthropods, especially parasitoid wasps, ground beetles, rove beetles and hoverflies. Environmental protection advice is not to spray summer applications within 12 m of the edge of the growing crop
- LERAP Category A [2, 3, 6-11]
- Broadcast air-assisted LERAP (18 m) [2, 6, 8-11]

Hazard classification and safety precautions

Hazard Harmful, Dangerous for the environment [1-13]; Flammable [1, 5, 6, 9]
Transport code 3 [5]; 6.1 [1, 3, 6, 7, 9]; 9 [8, 10, 11, 13]
Packaging group III [1, 3, 5-11, 13]
UN number 1993 [5]; 3017 [1, 6, 9]; 3018 [3, 7]; 3077 [8, 10, 11, 13]
Risk phrases R20 [2-7, 9]; R21 [12]; R22a, R53a [1-13]; R22b, R38 [1-7, 9]; R36 [2, 4-6, 9]; R37 [1, 5, 6, 9]; R41, R43 [2, 4]; R50 [1-11]; R51 [12, 13]; R67 [1]
Operator protection A [1-13]; C [2-5, 7]; H [1-5, 7, 13]; P [9]; U02a [2, 4, 12, 13]; U05a [1-7, 9]; U08 [2-11]; U09b [1]; U11 [5, 6, 9]; U14 [2-4, 7]; U19a [1-11]; U20a [12]; U20b [1-11, 13]
Environmental protection E06a [1-9] (14 d); E12a [1, 3, 7, 10, 11]; E12c [2, 4-6, 8, 9]; E12e, E16c, E16d [1-11]; E13a [1-4, 7]; E15a [1, 5, 6, 8-13]; E17a [3, 7] (18 m); E17b [1, 2, 4-6, 8-11] (18 m); E22a [1]; E34 [2-5, 7-11]; E38 [2, 4-6, 8-13]
Consumer protection C01 [13]; C02a [12] (6 wk)
Storage and disposal D01 [2-13]; D02 [2, 4-6, 8, 9]; D05 [2-4, 7, 13]; D09a [2-9, 12, 13]; D09c [10, 11]; D10b [2-11]; D11a [12, 13]; D12a [2, 4-6, 8-13]; D12b [3, 7]
Medical advice M01 [1, 5, 6, 8-13]; M03 [1, 3, 5-11]; M04a [13]; M05b [1-7, 9]

95 chlorpyrifos-methyl

An organophosphorus insecticide and acaricide for grain store use
IRAC mode of action code: 1B

Products

Reldan 22	Dow	225 g/l	EC	12404

Uses

- Grain storage pests in ***stored grain***
- Pre-harvest hygiene in ***grain stores***

Approval information

- Chlorpyrifos-methyl included in Annex I under EC Directive 91/414
- Accepted by BBPA for use in stores for malting barley
- In 2006 PSD required that all products containing this active ingredient should carry the following warning in the main area of the container label: "Chlorpyrifos-methyl is an anticholinesterase organophosphate. Handle with care"

Efficacy guidance

- Apply to grain after drying to moisture content below 14%, cooling and cleaning
- Insecticide may become depleted at grain surface if grain is being cooled by continuous extraction of air from the base leading to reduced control of grain store pests especially mites
- Resistance to organophosphorus compounds sometimes occurs in insect and mite pests of stored products

Restrictions

- Contains an anticholinesterase organophosphorus compound. Do not use if under medical advice not to work with such compounds
- Maximum number of treatments 1 per batch or 1 per store, prior to storage
- Only treat grain in good condition
- Do not treat grain intended for sowing
- Minimum of 90 d must elapse between treatment of grain and removal from store for consumption or processing

Crop-specific information

- May be applied pre-harvest to surfaces of empty store and grain handling machinery and as admixture with grain
- May be used in wheat, barley, oats, rye or triticale stores

Environmental safety

- Dangerous for the environment
- Very toxic to aquatic organisms

Hazard classification and safety precautions

Hazard Harmful, Dangerous for the environment
Transport code m6
Packaging group III
UN number 3082
Risk phrases R22b, R41, R50, R53a
Operator protection A, C, H, M; U05a, U08, U19a, U20b
Environmental protection E15a, E34
Storage and disposal D01, D02, D05, D09a, D10b
Medical advice M01, M03, M05b

SECTION 2

96 chlorthal-dimethyl

A residual benzenedicarboxylic acid herbicide for use in horticulture
HRAC mode of action code: K1

Products

Dacthal W-75	Certis	75% w/w	WP	11323

Uses

- Annual dicotyledons in ***bedding plants***, ***bilberries*** *(off-label)*, ***blackberries*** *(off-label)*, ***blackcurrants***, ***blueberries*** *(off-label)*, ***broccoli***, ***brussels sprouts***, ***bulb onions***, ***cabbages***, ***calabrese***, ***cauliflowers***, ***chinese cabbage*** *(off-label)*, ***choi sum*** *(off-label)*, ***collards*** *(off-label)*, ***courgettes*** *(off-label)*, ***cranberries*** *(off-label)*,, ***dwarf beans*** *(off-label)*, ***edible podded peas*** *(off-label)*, ***fodder rape***, ***french beans*** *(off-label)*, ***garlic*** *(off-label)*, ***gooseberries***, ***herbs (see appendix 6)*** *(off-label)*, ***kale***, ***leeks***, ***marrows*** *(off-label)*, ***navy beans*** *(off-label)*, ***pak choi*** *(off-label)*, ***poppies for morphine production*** *(off-label)*, ***raspberries***, ***redcurrants*** *(off-label)*, ***roses***, ***rubus hybrids*** *(off-label)*, ***runner beans***, ***sage***, ***salad onions*** *(off-label)*, ***shallots*** *(off-label)*, ***shrubs***, ***strawberries***, ***swedes***, ***trees***, ***turnips***, ***whitecurrants*** *(off-label)*
- Fat hen in ***poppies for morphine production*** *(off-label)*
- Slender speedwell in ***managed amenity turf***
- Volunteer potatoes in ***dwarf beans*** *(off-label)*, ***french beans*** *(off-label)*, ***navy beans*** *(off-label)*, ***poppies for morphine production*** *(off-label)*

Specific Off-Label Approvals (SOLAs)

- ***bilberries*** *20092777*
- ***blackberries*** *20092779*
- ***blueberries*** *20092777*
- ***chinese cabbage*** *20092778*
- ***choi sum*** *20092778*
- ***collards*** *20092778*
- ***courgettes*** *20092775*
- ***cranberries*** *20092777*
- ***dwarf beans*** *20092773*
- ***edible podded peas*** *20092780*
- ***french beans*** *20092773*
- ***garlic*** *20092776*
- ***herbs (see appendix 6)*** *20092772*
- ***marrows*** *20092775*
- ***navy beans*** *20092773*
- ***pak choi*** *20092778*
- ***poppies for morphine production*** *20092774*
- ***redcurrants*** *20092777*

- ***rubus hybrids*** *20092779*
- ***salad onions*** *20092710*
- ***shallots*** *20092776*
- ***whitecurrants*** *20092777*

Approval information

- Chlorthal-dimethyl has not been included in Annex 1. Therefore approval has been given until 23 March 2011 for storage and use of the products or mixtures containing chlorthal-dimethyl.

Efficacy guidance

- Best results on fine firm weed-free soil when adequate rain or irrigation follows
- Apply after drilling or planting prior to weed emergence. Rates and timing vary with crop and soil type. See label for details
- For control of slender speedwell in turf apply when weeds growing actively. Do not mow for at least 3 d after treatment

Restrictions

- Maximum number of treatments 1 per crop or 1 per season for all crops
- Do not apply mixture with propachlor to newly planted strawberries after rolling or application of other herbicides
- Do not use on strawberries between flowering and harvest
- Many types of ornamental have been treated successfully. See label for details. For species of unknown susceptibility treat a small number of plants first
- Do not use on turf where bent grasses form a major constituent of sward
- Do not plant lettuce within 6 mth of application, seeded turf within 2 mth, other crops within 3 mth. In the event of crop failure deep plough before re-drilling or planting
- Do not use on dwarf French beans
- Do not use on organic soils

Crop-specific information

- Latest use: before crop emergence on runner beans, sage; end of flowering for strawberries. Some restrictions apply to other crops. See label for details
- Recommended alone on roses, runner beans, various ornamentals, strawberries and turf, in tank-mix with propachlor on brassicas, onions, leeks, sage, ornamentals, established strawberries and newly planted soft fruit

Following crops guidance

- The interval between treatment and planting a following crop should be 6 mth for lettuce, 2 mth for seeded turf, 3 mth for other crops
- In the event of failure of a treated crop the land must be deep mould-board ploughed before re-drilling or re-planting

Environmental safety

- May cause long-term adverse effects in the aquatic environment

Hazard classification and safety precautions

Hazard Harmful
UN number N/C
Risk phrases R40, R53a
Operator protection U19a, U20a
Environmental protection E15a
Storage and disposal D01, D02, D09a, D11a
Medical advice M04a

97 chlorthal-dimethyl + propachlor

A residual herbicide mixture for a range of vegetable crops
HRAC mode of action code: K1 + K3

See also propachlor

Products

Greencrop Vegex	Greencrop	225:216 g/l	SC	12439

Uses

- Annual dicotyledons in ***broccoli*, *brussels sprouts*, *bulb onions*, *cabbages*, *calabrese*, *cauliflowers*, *fodder rape*, *kale*, *leeks*, *mustard*, *salad onions*, *swedes*, *turnips***
- Annual meadow grass in ***broccoli*, *brussels sprouts*, *bulb onions*, *cabbages*, *calabrese*, *cauliflowers*, *fodder rape*, *kale*, *leeks*, *mustard*, *salad onions*, *swedes*, *turnips***

Approval information

- Since chlorthal-dimethyl has not been included in Annex 1, approval has been given until 23 March 2011 for storage and use of the mixtures containing chlorthal-dimethyl.
- Approval expiry 18 Mar 2010 [1]

Efficacy guidance

- Product may be applied pre- or post-emergence of the crop but pre-emergence of the weeds. Emerged weeds are not controlled
- Soils should be fine, firm and free of weed growth
- For best results apply to moist soil and before rain. In dry weather irrigate lightly after application and before seedling weeds appear
- Weed control may be reduced under very wet, dry or windy conditions
- Do not use on soils with more than 10% organic matter as weed control will be impaired

Restrictions

- Maximum number of treatments 1 per crop for all crops

Crop-specific information

- Use on brassica crops pre-crop emergence may result in a check to growth but this is usually outgrown within 8 wk without effect on yield or quality
- Post-emergence treatment on brassica crops should not be made until they have 3-4 true leaves. Post-emergence treatment on onions should be made post-crook stage
- Young transplanted plants, particularly cabbages and cauliflowers, raised under glass should be hardened off before treatment
- Care should be taken with modular transplants to ensure that roots are not exposed to spray

Following crops guidance

- The interval between treatment and planting succeeding crops should be 2 mth for seeded turf, 6 mth for lettuce, 3 mth for all other crops
- In the event of failure of a treated crop the land must be deep mouldboard ploughed before redrilling or replanting

Environmental safety

- Dangerous for the environment
- Toxic to aquatic organisms

Hazard classification and safety precautions

Hazard Harmful, Dangerous for the environment
Transport code 9
Packaging group III
UN number 3082
Risk phrases R38, R40, R43, R51, R53a
Operator protection A, C; U02a, U04a, U05a, U09b, U11, U14, U15, U19a, U20a
Environmental protection E15a
Storage and disposal D01, D02, D09a, D10a
Medical advice M04a

98 cinidon-ethyl

A novel phthalimide contact herbicide for cereals
HRAC mode of action code: E

Products

Lotus	Nufarm UK	200 g/l	EC	13744

Uses

- Annual dicotyledons in ***durum wheat*, *spring barley*, *winter barley*, *winter rye*, *winter wheat***

SEE SECTION 3 FOR PRODUCTS ALSO REGISTERED

Approval information

- Cinidon-ethyl included in Annex I under EC Directive 91/414

Efficacy guidance

- Controls sensitive species that have emerged at time of treatment
- Ensure good even spray cover for best results
- Weed spectrum can be widened by mixture with mecoprop-P (see label)

Restrictions

- Maximum total dose equivalent to one full dose application
- Do not treat crops suffering from herbicide damage or physical stress
- Do not roll or harrow for 7d before or after treatment

Crop-specific information

- HI 35 d for cutting for stock feed or grazing
- Latest use: before first node detectable (GS 31) for spring barley and winter rye; before third node detectable (GS 33) for winter wheat, durum wheat and winter barley
- Some necrotic spotting may occur after treatment but has no effect on final yield. Effect can be greater when mixture with mecoprop-P used, especially in spring, but yield not normally affected

Following crops guidance

- In the event of crop failure only winter and spring wheat and barley, oilseed rape, maize or sunflowers may be planted

Environmental safety

- Dangerous to fish or other aquatic life. Do not contaminate surface waters or ditches with chemical or used container
- Treated cereals must not be used as animal feed or bedding for at least 35 d after last application
- Do not allow animals to graze for 35 d after treatment

Hazard classification and safety precautions

Hazard Irritant, Dangerous for the environment
Transport code 9
Packaging group III
UN number 3082
Risk phrases R38, R41, R50, R53a
Operator protection A, C, H; U05a, U08, U11, U20b
Environmental protection E15b, E16a, E34, E38
Storage and disposal D01, D02, D05, D09a, D12a
Medical advice M03

99 citronella oil

A natural plant extract herbicide

Products

Barrier H	Barrier	22.9% w/w	EW	10136

Uses

- Ragwort in ***land temporarily removed from production***, ***permanent grassland***

Efficacy guidance

- Best results obtained from spot treatment of ragwort in the rosette stage, during dry still conditions
- Aerial growth of ragwort is rapidly destroyed. Longer term control depends on overall management strategy
- Check for regrowth after 28 d and re-apply as necessary

Crop-specific information

- Contact with grasses will result in transient scorch which is outgrown in good growing conditions

Environmental safety

- Apply away from bees
- Harmful to fish or other aquatic life. Do not contaminate surface waters or ditches with chemical or used container

- Keep livestock out of treated areas for at least 2 wk and until foliage of any poisonous weeds such as ragwort has died and become unpalatable

Hazard classification and safety precautions

Hazard Irritant
Risk phrases R36, R38
Operator protection A, C; U05a, U20c
Environmental protection E07b (2 wk); E12g, E13c
Storage and disposal D01, D02, D05, D09a, D11a

100 clodinafop-propargyl

A foliar acting herbicide for annual grass weed control in wheat, triticale and rye
HRAC mode of action code: A

Products

1	Clodinafop 240	Goldengrass	240 g/l	EC	14417
2	Topik	Syngenta	240 g/l	EC	12333
3	Tuli	AgChem Access	240 g/l	EC	13302
4	Viscount	Syngenta	240 g/l	EC	12642

Uses

- Blackgrass in ***durum wheat***, ***spring rye***, ***spring wheat***, ***triticale***, ***winter rye***, ***winter wheat*** [1-4]; ***grass seed crops*** *(off-label)* [2]
- Rough meadow grass in ***durum wheat***, ***spring rye***, ***spring wheat***, ***triticale***, ***winter rye***, ***winter wheat*** [1-4]; ***grass seed crops*** *(off-label)* [2]
- Wild oats in ***durum wheat***, ***spring rye***, ***spring wheat***, ***triticale***, ***winter rye***, ***winter wheat*** [1-4]; ***grass seed crops*** *(off-label)* [2]

Specific Off-Label Approvals (SOLAs)

- ***grass seed crops*** *20090210* [2]

Approval information

- Clodinafop-propargyl included in Annex I under EC Directive 91/414

Efficacy guidance

- Spray when majority of weeds have germinated but before competition reduces yield
- Products contain a herbicide safener (cloquintocet-mexyl) that improves crop tolerance to clodinafop-propargyl
- Optimum control achieved when all grass weeds emerged. Wait for delayed germination on dry or cloddy seedbed
- A mineral oil additive is recommended to give more consistent control of very high blackgrass populations or for late season treatments. See label for details
- Weed control not affected by soil type, organic matter or straw residues
- Control may be reduced if rain falls within 1 h of treatment
- Clodinafop-propargyl is an ACCase inhibitor herbicide. To avoid the build up of resistance do not apply products containing an ACCase inhibitor herbicide more than twice to any crop. In addition do not use any product containing clodinafop-propargyl in mixture or sequence with any other product containing the same ingredient
- Use these products as part of a resistance management strategy that includes cultural methods of control and does not use ACCase inhibitors as the sole chemical method of grass weed control
- Applying a second product containing an ACCase inhibitor to a crop will increase the risk of resistance development; only use a second ACCase inhibitor to control different weeds at a different timing
- Always follow WRAG guidelines for preventing and managing herbicide resistant weeds. See Section 5 for more information

Restrictions

- Maximum number of treatments 1 per crop
- Do not use on barley or oats
- Do not treat crops under stress or suffering from waterlogging, pest attack, disease or frost
- Do not treat crops undersown with grass mixtures
- Do not mix with products containing MCPA, mecoprop-P, 2,4-D or 2,4-DB

- MCPA, mecoprop, 2,4-D or 2,4-DB should not be applied within 21 d before, or 7 d after, treatment

Crop-specific information

- Latest use: before second node detectable stage (GS 32) for durum wheat, triticale, rye; before flag leaf sheath extending (GS 41) for wheat
- Spray in autumn, winter or spring from 1 true leaf stage (GS 11) to before second node detectable (GS 32) on durum, rye, triticale; before flag leaf sheath extends (GS 41) on wheat

Following crops guidance

- Any broad leaved crop or cereal (except oats) may be sown after failure of a treated crop provided that at least 3 wk have elapsed between application and drilling a cereal
- After normal harvest of a treated crop any broad leaved crop or wheat, durum wheat, rye, triticale or barley should be sown. Oats and grass should not be sown until the following spring

Environmental safety

- Dangerous for the environment
- Very toxic to aquatic organisms

Hazard classification and safety precautions

Hazard Dangerous for the environment
Transport code 9 [1, 2, 4]
Packaging group III [1, 2, 4]
UN number 3082 [1, 2, 4]
Risk phrases R50, R53a
Operator protection A, C, H, K; U02a, U05a, U09a, U20a
Environmental protection E15a, E38
Storage and disposal D01, D02, D05, D09a, D10c, D12a

101 clodinafop-propargyl + pinoxaden

A foliar acting herbicide mixture for winter wheat
HRAC mode of action code: A + A

See also *pinoxaden*

Products

Traxos	Syngenta	100:100 g/l	EC	12742

Uses

- Italian ryegrass in ***winter wheat***
- Perennial ryegrass in ***winter wheat*** *(from seed)*
- Wild oats in ***winter wheat***

Approval information

- Clodinafop-propargyl included in Annex I under EC Directive 91/414

Efficacy guidance

- Products contain a herbicide safener (cloquintocet-mexyl) that improves crop tolerance to clodinafop-propargyl
- Best results obtained from treatment when all grass weeds have emerged. There is no residual activity
- Broad-leaved weeds are not controlled
- Treat before emerged weed competition reduces yield
- For ryegrass control use as part of a programme including other products with activity against ryegrass
- For blackgrass control must be used as part of an integrated control strategy
- Grass weed control may be reduced if rain falls within 1 hr of application
- Clodinafop-propargyl and pinoxaden are ACCase inhibitor herbicides. To avoid the build up of resistance do not apply products containing an ACCase inhibitor herbicide more than twice to any crop. In addition do not use any product containing clodinafop-propargyl or pinoxaden in mixture or sequence with any other product containing the same ingredient
- Use as part of a resistance management strategy that includes cultural methods of control and does not use ACCase inhibitors as the sole chemical method of grass weed control

- Applying a second product containing an ACCase inhibitor to a crop will increase the risk of resistance development; only use a second ACCase inhibitor to control different weeds at a different timing
- Always follow WRAG guidelines for preventing and managing herbicide resistant weeds. See Section 5 for more information

Restrictions

- Maximum number of treatments 1 per crop
- Product must always be used with specified adjuvant. See label
- Do not use on other cereals
- Do not treat crops under stress from any cause
- Do not treat crops undersown with grass or grass mixtures
- Avoid use of hormone-containing herbicides in mixture or in sequence. See label for restrictions on tank mixes

Crop-specific information

- Latest use: before flag leaf sheath extended (before GS 41) for winter wheat
- All varieties of winter wheat may be treated

Following crops guidance

- There are no restrictions on succeeding crops in a normal rotation
- In the event of failure of a treated crop ryegrass, maize, oats or any broad-leaved crop may be planted after a minimum interval of 4 wk from application

Environmental safety

- Dangerous for the environment
- Toxic to aquatic organisms

Hazard classification and safety precautions

Hazard Irritant, Dangerous for the environment
Transport code 9
Packaging group III
UN number 3082
Risk phrases R36, R38, R43, R51, R53a
Operator protection A, C, H, K; U05a, U09a, U20b
Environmental protection E15b, E38
Storage and disposal D01, D02, D05, D09a, D10c, D11a, D12a

102 clodinafop-propargyl + prosulfocarb

A foliar acting herbicide mixture for grass and broad-leaved weed control in winter wheat
HRAC mode of action code: A + N

See also prosulfocarb

Products

Auxiliary	BASF	10:800 g/l	EC	14576

Uses

- Annual dicotyledons in ***winter wheat***
- Annual grasses in ***winter wheat***

Approval information

- Clodinafop-propargyl and prosulfocarb included in Annex I under EC Directive 91/414

Efficacy guidance

- To avoid the build up of resistance do not apply products containing an ACCase inhibitor herbicide more than twice to any crop. In addition, do not use this product in mixture or sequence with any other product containing clodinafop-propargyl.

Restrictions

- Do not use on barley or oats
- Do not spray crops undersown with grass mixtures
- When [1] is applied first, leave 7 days before applying hormone herbicides. If mecoprop-p or 2,4-DB containing products are applied first, leave 14 days before [1] is applied. If MCPA or 2,4-D containing products are applied first, leave 21 days before [1] is applied.

SEE SECTION 3 FOR PRODUCTS ALSO REGISTERED

- The cereal seed must be covered by 3cm of soil and for best results apply to a firm, moist seedbed free of clods.

Crop-specific information
- Only one application per crop
- Latest use before 5 tiller stage (GS25)

Hazard classification and safety precautions

Hazard Harmful, Dangerous for the environment
Transport code 9
Packaging group III
UN number 3082
Risk phrases R22b, R51, R53a
Operator protection A, H; U02a, U05a, U08, U20c
Environmental protection E15b, E16a, E16b, E38
Storage and disposal D01, D02, D05, D09a, D10c, D12a
Medical advice M05b

103 clofentezine

A selective ovicidal tetrazine acaricide for use in top fruit
IRAC mode of action code: 10A

Products

Apollo 50 SC	Makhteshim	500 g/l	SC	10590

Uses
- Red spider mites in ***apples***, ***cherries***, ***pears***, ***plums***
- Spider mites in ***blackcurrants*** *(off-label)*, ***protected blackberries*** *(off-label)*, ***protected raspberries*** *(off-label)*, ***protected strawberries*** *(off-label)*, ***raspberries*** *(off-label)*, ***strawberries*** *(off-label)*

Specific Off-Label Approvals (SOLAs)
- ***blackcurrants*** *20012269*
- ***protected blackberries*** *20012268*
- ***protected raspberries*** *20012268*
- ***protected strawberries*** *20012271*
- ***raspberries*** *20012269*
- ***strawberries*** *20012269*

Approval information
- Clofentezine included in Annex I under EC Directive 91/414

Efficacy guidance
- Acts on eggs and early motile stages of mites. For effective control total cover of plants is essential, particular care being needed to cover undersides of leaves

Restrictions
- Maximum number of treatments 1 per yr for apples, pears, cherries, plums

Crop-specific information
- HI apples, pears 28 d; cherries, plums 8 wk
- For red spider mite control spray apples and pears between bud burst and pink bud, plums and cherries between white bud and first flower. Rust mite is also suppressed
- On established infestations apply in conjunction with an adult acaricide

Environmental safety
- Harmful to aquatic organisms
- Product safe on predatory mites, bees and other predatory insects

Hazard classification and safety precautions

UN number N/C
Risk phrases R52
Operator protection U05a, U08, U20b
Environmental protection E15a
Storage and disposal D01, D02, D05, D09a, D11a

104 clomazone

An isoxazolidinone residual herbicide for oilseed rape, field beans, combining and vining peas

HRAC mode of action code: F3

Products

1	Centium 360 CS	Belchim	360 g/l	CS	13846
2	Cirrus CS	Belchim	360 g/l	CS	11751
3	Cirrus CS	Belchim	371 g/l	CS	13860
4	Clone	AgriGuard	360 g/l	CS	14320
5	Fiddle	Goldengrass	360 g/l	CS	14451
6	Gadwall	AgChem Access	360 g/l	CS	14522
7	Gamit 36 CS	Belchim	360 g/l	CS	12598
8	Gamit 36 CS	Belchim	360 g/l	CS	13861

Uses

- Annual dicotyledons in ***asparagus*** *(off-label)*, ***broad beans*** *(off-label)*, ***broccoli*** *(off-label)*, ***brussels sprouts*** *(off-label)*, ***cabbages*** *(off-label)*, ***calabrese*** *(off-label)*, ***cauliflowers*** *(off-label)*, ***chinese cabbage*** *(off-label)*, ***choi sum*** *(off-label)*, ***collards*** *(off-label)*, ***game cover*** *(off-label)*, ***kale*** *(off-label)*, ***lupins*** *(off-label)*, ***pak choi*** *(off-label)*, ***spring cabbage*** *(off-label)*, ***swedes*** *(off-label)*, ***sweet potato*** *(off-label)* [8]; ***poppies for morphine production*** *(off-label)* [1, 8]
- Chickweed in ***asparagus*** *(off-label)* [7]; ***borage*** *(off-label)*, ***canary flower (echium spp.)*** *(off-label)*, ***honesty*** *(off-label)*, ***lupins*** *(off-label)*, ***mustard*** *(off-label)*, ***spring linseed*** *(off-label)* [2]; ***carrots***, ***potatoes*** [7, 8]; ***combining peas***, ***spring field beans***, ***spring oilseed rape***, ***vining peas***, ***winter field beans***, ***winter oilseed rape*** [1-6]
- Cleavers in ***asparagus*** *(off-label)* [7]; ***borage*** *(off-label)*, ***canary flower (echium spp.)*** *(off-label)*, ***honesty*** *(off-label)*, ***lupins*** *(off-label)*, ***mustard*** *(off-label)*, ***spring linseed*** *(off-label)* [2]; ***carrots***, ***potatoes*** [7, 8]; ***combining peas***, ***spring field beans***, ***spring oilseed rape***, ***vining peas***, ***winter field beans***, ***winter oilseed rape*** [1-6]
- Fool's parsley in ***asparagus*** *(off-label)* [7]; ***borage*** *(off-label)*, ***canary flower (echium spp.)*** *(off-label)*, ***honesty*** *(off-label)*, ***lupins*** *(off-label)*, ***mustard*** *(off-label)*, ***spring linseed*** *(off-label)* [2]; ***carrots***, ***potatoes*** [7, 8]; ***combining peas***, ***spring field beans***, ***spring oilseed rape***, ***vining peas*** [1-6]
- Red dead-nettle in ***asparagus*** *(off-label)* [7]; ***borage*** *(off-label)*, ***canary flower (echium spp.)*** *(off-label)*, ***honesty*** *(off-label)*, ***lupins*** *(off-label)*, ***mustard*** *(off-label)*, ***spring linseed*** *(off-label)* [2]; ***carrots***, ***potatoes*** [7, 8]; ***combining peas***, ***spring field beans***, ***spring oilseed rape***, ***vining peas***, ***winter field beans***, ***winter oilseed rape*** [1-6]
- Shepherd's purse in ***asparagus*** *(off-label)* [7]; ***borage*** *(off-label)*, ***canary flower (echium spp.)*** *(off-label)*, ***honesty*** *(off-label)*, ***lupins*** *(off-label)*, ***mustard*** *(off-label)*, ***spring linseed*** *(off-label)* [2]; ***carrots***, ***potatoes*** [7, 8]; ***combining peas***, ***spring field beans***, ***spring oilseed rape***, ***vining peas***, ***winter field beans***, ***winter oilseed rape*** [1-6]

Specific Off-Label Approvals (SOLAs)

- ***asparagus*** *20060658* [7], *20091276* [8]
- ***borage*** *20061214* [2]
- ***broad beans*** *20091281* [8]
- ***broccoli*** *20091275* [8]
- ***brussels sprouts*** *20091275* [8]
- ***cabbages*** *20091275* [8]
- ***calabrese*** *20091275* [8]
- ***canary flower (echium spp.)*** *20061214* [2]
- ***cauliflowers*** *20091275* [8]
- ***chinese cabbage*** *20091275* [8]
- ***choi sum*** *20091275* [8]
- ***collards*** *20091275* [8]
- ***game cover*** *20091280* [8]
- ***honesty*** *20061214* [2]
- ***kale*** *20091275* [8]
- ***lupins*** *20061213* [2], *20091274* [8]
- ***mustard*** *20061214* [2]
- ***pak choi*** *20091275* [8]
- ***poppies for morphine production*** *20080593* [1], *20091283* [8]

- ***spring cabbage*** *20091275* [8]
- ***spring linseed*** *20061214* [2]
- ***swedes*** *20091282* [8]
- ***sweet potato*** *20091279* [8]

Approval information

- Clomazone included in Annex I under EC Directive 91/414
- Approval expiry 28 Feb 2010 [7]

Efficacy guidance

- Best results obtained from application as soon as possible after sowing crop and before emergence of crop or weeds
- Uptake is via roots and shoots. Seedbeds should be firm, level and free from clods. Loose puffy seedbeds should be consolidated before spraying
- Efficacy is reduced on organic soils, on dry cloddy seedbeds and if prolonged dry weather follows application
- Clomazone acts by inhibiting synthesis of chlorophyll pigments. Susceptible weeds emerge but are chlorotic and die shortly afterwards
- Season-long control of weeds may not be achieved
- Always follow WRAG guidelines for preventing and managing herbicide resistant weeds. See Section 5 for more information

Restrictions

- Maximum number of treatments one per crop
- Crops must be covered by a minimum of 20 mm settled soil. Do not apply to broadcast crops. Direct-drilled crops should be harrowed across the slits to cover seed before spraying
- Do not use on compacted soils or soils of poor structure that may be liable to waterlogging
- Do not use on Sands or Very Light soils or those with more than 10% organic matter
- Do not treat two consecutive crops of carrots with clomazone in one calendar yr [7]
- Consult manufacturer or your advisor before use on potato seed crops [7]
- Do not overlap spray swaths. Crop plants emerged at time of treatment may be severely damaged

Crop-specific information

- Latest use: pre-emergence of crop
- Severe, but normally transient, crop damage may occur in overlaps on field beans
- Some transient crop bleaching may occur under certain climatic conditions and can be severe where heavy rain follows application. This is normally rapidly outgrown and has no effect on final crop yield. Overlapping spray swaths may cause severe damage to field beans

Following crops guidance

- Following normal harvest of a spring or autumn treated crop, cereals, oilseed rape, field beans, combining peas, potatoes, maize, turnips, linseed or sugar beet may be sown
- In the event of failure of an autumn treated crop, winter cereals or winter beans may be sown in the autumn if 6 wk have elapsed since treatment. In the spring following crop failure combining peas, field beans or potatoes may be sown if 6 wk have elapsed since treatment, and spring cereals, maize, turnips, onions, carrots or linseed may be sown if 7 mth have elapsed since treatment
- In the event of a failure of a spring treated crop a wide range of crops may be sown provided intervals of 6-9 wk have elapsed since treatment. See label for details
- Prior to resowing any listed replacement crop the soil should be ploughed and cultivated to 15 cm

Environmental safety

- Take extreme care to avoid drift outside the target area, or on to ponds, waterways or ditches as considerable damage may occur. Apply using a coarse quality spray

Hazard classification and safety precautions

Hazard Irritant [4]
UN number N/C
Risk phrases R43 [4]
Operator protection A; U05a, U14, U20b
Environmental protection E15b [1-8]; E23, E34 [4]
Storage and disposal D01, D02, D09a, D10b

105 clomazone + linuron

An isoxazolidinone and urea residual herbicide for peas, field beans and potatoes
HRAC mode of action code: F3 + C2

See also linuron

Products

Lingo	Belchim	45:250 g/l	CC	14075

Uses

- Annual dicotyledons in ***combining peas, potatoes, spring field beans, vining peas, winter field beans***
- Annual meadow grass in ***combining peas, potatoes, spring field beans, vining peas, winter field beans***

Approval information

- Clomazone and linuron included in Annex I under EC Directive 91/414

Restrictions

- Do not apply to broadcast crops
- Do not apply on light soils if heavy rain is expected within 24 hours of application
- Transient crop bleaching may occur under some conditions. This is normally outgrown with no effect on final yield

Following crops guidance

- Following normal harvest of an autumn treated crop, cereals, oilseed rape, field beans, combining peas, potatoes, maize, turnip, linseed and sugar beet may be sown. Prior to planting, soil should be ploughed and cultivated to a minimum depth of 15cm.
- Following normal harvest of a spring treated crop, cereals, oilseed rape, field beans, combining peas, potatoes, maize, linseed, sugar beet and turnip may be sown. Prior to planting, soil should be ploughed and cultivated to a minimum depth of 15cm. Do not sow or plant lettuce in the same season as an application of [1]

Hazard classification and safety precautions

Hazard Toxic, Dangerous for the environment
Transport code 9
Packaging group III
UN number 3082
Risk phrases R22a, R40, R48, R50, R53a, R61, R62
Operator protection A, H; U02a, U05a, U13, U14, U15, U19a
Environmental protection E15a, E23, E34, E38
Storage and disposal D01, D02, D12a
Medical advice M04a

106 clomazone + metazachlor

A mixture of isoxazolidinone and residual anilide herbicides for weed control in oilseed rape
HRAC mode of action code: F3 + K3

See also metazachlor

Products

Nimbus CS	BASF	33.3:250	SC	14092

Uses

- Annual dicotyledons in ***winter oilseed rape***
- Annual grasses in ***winter oilseed rape***
- Cleavers in ***winter oilseed rape***

Approval information

- Clomazone and metazachlor included in Annex I under EC Directive 91/414

SEE SECTION 3 FOR PRODUCTS ALSO REGISTERED

Hazard classification and safety precautions

Hazard Irritant, Dangerous for the environment
Transport code 9
Packaging group III
UN number 3082
Risk phrases R43, R50, R53a
Operator protection A, H; U05a, U08, U14, U19a, U20b
Environmental protection E07a, E15b, E16a, E38
Storage and disposal D01, D02, D08, D09a, D10c, D12a
Medical advice M05a

107 clomazone + metribuzin

A mixture of isoxazolidinone and triazinone residual herbicides for use in potatoes
HRAC mode of action code: F3 + C1

See also metribuzin

Products

Metric	Belchim	60:233 g/l	CC	14214

Uses

- Annual dicotyledons in ***potatoes***
- Annual meadow grass in ***potatoes***

Approval information

- Clomazone and metribuzin included in Annex I under EC Directive 91/414

Restrictions

- For use on specified varieties of potato only
- Safety to daughter tubers has not been tested; Consult manufacturer before treating seed crops

Hazard classification and safety precautions

Hazard Dangerous for the environment
Transport code 9
Packaging group III
UN number 3082
Risk phrases R50, R53a
Operator protection A, H; U05a, U19a, U20b
Environmental protection E15b, E16a, E16b, E34, E38
Storage and disposal D01, D02, D09a, D10b, D12a
Medical advice M05a

108 clopyralid

A foliar translocated picolinic herbicide for a wide range of crops
HRAC mode of action code: O

Products

1	Dow Shield	Dow	200 g/l	SL	10988
2	Glopyr 200 SL	Globachem	200 g/l	SL	10979
3	Lontrel 200	Dow	18.02% w/w	SL	11558
4	Vivendi 200	AgriChem BV	200 g/l	SL	12782

Uses

- Annual dicotyledons in ***broccoli, brussels sprouts, bulb onions, cabbages, calabrese, fodder beet, fodder rape, forage maize, kale, linseed, mangels, ornamental plant production, red beet, salad onions, spring barley, spring oats, spring oilseed rape, spring wheat, strawberries, sugar beet, swedes, sweetcorn, turnips, winter barley, winter oats, winter oilseed rape, winter wheat*** [1-4]; ***cauliflowers*** [1, 3, 4]; ***conifers and broadleaved trees*** *(off-label)* [1, 3]; ***crambe*** *(off-label)* [1]; ***grassland*** [4]; ***permanent grassland*** [1-3]; ***rotational grass*** [3]

- Corn marigold in ***broccoli, brussels sprouts, bulb onions, cabbages, calabrese, cauliflowers, fodder beet, fodder rape, forage maize, kale, linseed, mangels, ornamental plant production, permanent grassland, red beet, salad onions, spring barley, spring oats, spring oilseed rape, spring wheat, strawberries, sugar beet, swedes, sweetcorn, turnips, winter barley, winter oats, winter oilseed rape, winter wheat*** [1-3]; ***rotational grass*** [3]
- Creeping thistle in ***broccoli, brussels sprouts, bulb onions, cabbages, calabrese, cauliflowers, fodder beet, fodder rape, forage maize, kale, linseed, mangels, ornamental plant production, permanent grassland, red beet, salad onions, spring barley, spring oats, spring oilseed rape, spring wheat, strawberries, sugar beet, swedes, sweetcorn, turnips, winter barley, winter oats, winter oilseed rape, winter wheat*** [1-3]; ***crambe*** *(off-label)* [1]; ***rotational grass*** [3]
- General weed control in ***canary grass*** *(off-label - game cover)*, ***quinoa*** *(off-label - game cover)*, ***tanka millet*** *(off-label - game cover)*, ***white millet*** *(off-label - game cover)* [1, 3]; ***sweet clover*** *(off-label - game cover)* [1]
- Groundsel in ***chives*** *(off-label)*, ***herbs (see appendix 6)*** *(off-label)*, ***leaf spinach*** *(off-label)*, ***parsley*** *(off-label)*, ***spinach beet*** *(off-label)* [1, 3]
- Mayweeds in ***borage for oilseed production*** *(off-label)*, ***chinese cabbage*** *(off-label)*, ***choi sum*** *(off-label)*, ***crambe*** *(off-label)*, ***durum wheat*** *(off-label)*, ***leeks*** *(off-label)*, ***pak choi*** *(off-label)*, ***spring rye*** *(off-label)*, ***winter rye*** *(off-label)* [1]; ***broccoli, brussels sprouts, bulb onions, cabbages, calabrese, cauliflowers, fodder beet, fodder rape, forage maize, kale, linseed, mangels, ornamental plant production, red beet, salad onions, spring barley, spring oats, spring oilseed rape, spring wheat, strawberries, sugar beet, swedes, sweetcorn, turnips, winter barley, winter oats, winter oilseed rape, winter wheat*** [1-4]; ***canary flower (echium spp.)*** *(off-label)*, ***chives*** *(off-label)*, ***evening primrose*** *(off-label)*, ***herbs (see appendix 6)*** *(off-label)*, ***leaf spinach*** *(off-label)*, ***parsley*** *(off-label)*, ***poppies for morphine production*** *(off-label)*, ***spinach beet*** *(off-label)* [1, 3]; ***collards*** *(off-label)*, ***garlic*** *(off-label)*, ***honesty*** *(off-label)*, ***mustard*** *(off-label)*, ***permanent grassland, shallots*** *(off-label)* [1-3]; ***grassland*** [4]; ***rotational grass, spring linseed*** *(off-label)* [3]; ***triticale*** *(off-label)* [1, 2]
- Thistles in ***apple orchards*** *(off-label)*, ***asparagus*** *(off-label)*, ***canary flower (echium spp.)*** *(off-label)*, ***evening primrose*** *(off-label)*, ***grass seed crops*** *(off-label)*, ***pear orchards*** *(off-label)*, ***permanent grassland*** *(off-label - spot treatment)*, ***poppies for morphine production*** *(off-label)* [1, 3]; ***borage for oilseed production*** *(off-label)*, ***chinese cabbage*** *(off-label)*, ***choi sum*** *(off-label)*, ***leeks*** *(off-label)*, ***pak choi*** *(off-label)* [1]; ***broccoli, brussels sprouts, bulb onions, cabbages, calabrese, cauliflowers, fodder beet, fodder rape, forage maize, grassland, kale, linseed, mangels, ornamental plant production, red beet, salad onions, spring barley, spring oats, spring oilseed rape, spring wheat, strawberries, sugar beet, swedes, sweetcorn, turnips, winter barley, winter oats, winter oilseed rape, winter wheat*** [4]; ***collards*** *(off-label)*, ***garlic*** *(off-label)*, ***honesty*** *(off-label)*, ***mustard*** *(off-label)*, ***shallots*** *(off-label)* [1-3]; ***spring linseed*** *(off-label)* [3]

Specific Off-Label Approvals (SOLAs)

- ***apple orchards*** *20050471* [1], *20062444* [3]
- ***asparagus*** *20050470* [1], *20062445* [3]
- ***borage for oilseed production*** *20062635* [1]
- ***canary flower (echium spp.)*** *20062635* [1], *20061933* [3]
- ***canary grass*** *(game cover) 20050474* [1], *(game cover) 20062450* [3]
- ***chinese cabbage*** *20062637* [1]
- ***chives*** *20050473* [1], *20062442* [3]
- ***choi sum*** *20062637* [1]
- ***collards*** *20062637* [1], *20061956* [2], *20061932* [3]
- ***conifers and broadleaved trees*** *20050468* [1], *20062449* [3]
- ***crambe*** *20081045* [1]
- ***durum wheat*** *20062634* [1]
- ***evening primrose*** *20062635* [1], *20061933* [3]
- ***garlic*** *20062636* [1], *20061954* [2], *20061936* [3]
- ***grass seed crops*** *20050467* [1], *20062448* [3]
- ***herbs (see appendix 6)*** *20050473* [1], *20062442* [3]
- ***honesty*** *20050469* [1], *20061953* [2], *20062446* [3]
- ***leaf spinach*** *20050472* [1], *20062443* [3]
- ***leeks*** *20062638* [1]

- ***mustard*** *20062635* [1], *20061953* [2], *20061933* [3]
- ***pak choi*** *20062637* [1]
- ***parsley*** *20050473* [1], *20062442* [3]
- ***pear orchards*** *20050471* [1], *20062444* [3]
- ***permanent grassland*** *(spot treatment) 20050467* [1], *(spot treatment) 20062448* [3]
- ***poppies for morphine production*** *20040161* [1], *20062447* [3]
- ***quinoa*** *(game cover) 20050474* [1], *(game cover) 20062450* [3]
- ***shallots*** *20062636* [1], *20061954* [2], *20061936* [3]
- ***spinach beet*** *20050472* [1], *20062443* [3]
- ***spring linseed*** *20061933* [3]
- ***spring rye*** *20062634* [1]
- ***sweet clover*** *(game cover) 20050474* [1]
- ***tanka millet*** *(game cover) 20050474* [1], *(game cover) 20062450* [3]
- ***triticale*** *20062634* [1], *20061955* [2]
- ***white millet*** *(game cover) 20050474* [1], *(game cover) 20062450* [3]
- ***winter rye*** *20062634* [1]

Approval information

- Clopyralid included in Annex I under EC Directive 91/414
- Accepted by BBPA for use on malting barley

Efficacy guidance

- Best results achieved by application to young actively growing weed seedlings. Treat creeping thistle at rosette stage and repeat 3-4 wk later as directed
- High activity on weeds of Compositae family. For most crops recommended for use in tank mixes. See label for details

Restrictions

- Maximum total dose varies between the equivalent of one and two full dose treatments, depending on the crop treated. See labels for details
- Do not apply to cereals later than the second node detectable stage (GS 32)
- Do not apply when crop damp or when rain expected within 6 h
- Do not use straw from treated cereals in compost or any other form for glasshouse crops. Straw may be used for strawing down strawberries
- Straw from treated grass seed crops or linseed should be baled and carted away. If incorporated do not plant winter beans in same year
- Do not use on onions at temperatures above 20 °C or when under stress
- Do not treat maiden strawberries or runner beds or apply to early leaf growth during blossom period or within 4 wk of picking. Aug or early Sep sprays may reduce yield

Crop-specific information

- Latest use: 7 d before cutting grass for hay or silage; before 3rd node detectable (GS 33) for cereals; before flower buds visible from above for oilseed rape, linseed
- HI grassland 7 d; apples, pears, strawberries 4 wk; maize, sweetcorn, onions, Brussels sprouts, broccoli, cabbage, cauliflowers, calabrese, kale, fodder rape, oilseed rape, swedes, turnips, sugar beet, red beet, fodder beet, mangels, sage, honesty 6 wk
- Timing of application varies with weed problem, crop and other ingredients of tank mixes. See labels for details
- Apply as directed spray in woody ornamentals, avoiding leaves, buds and green stems. Do not apply in root zone of families Compositae or Leguminosae

Following crops guidance

- Do not plant susceptible autumn-sown crops in same year as treatment. Do not apply later than Jul where susceptible crops are to be planted in spring. See label for details

Environmental safety

- Dangerous for the environment [2]
- Harmful to aquatic organisms [1-3]
- Wash spray equipment thoroughly with water and detergent immediately after use. Traces of product can damage susceptible plants sprayed later
- Keep livestock out of treated areas for at least 7 d and until foliage of any poisonous weeds such as ragwort has died and become unpalatable

- Some pesticides pose a greater threat of contamination of water than others and clopyralid is one of these pesticides. Take special care when applying clopyralid near water and do not apply if heavy rain is forecast

Hazard classification and safety precautions

Hazard Dangerous for the environment [2]
UN number N/C
Risk phrases R50 [2]; R52 [1, 3]; R53a [1-3]
Operator protection A, C; U05a [4]; U08, U19a, U20b [1-4]
Environmental protection E07a [1-3]; E15a [1-4]; E34 [4]
Storage and disposal D01 [1, 3, 4]; D02 [4]; D05, D09a, D10b [1-4]; D12a [1, 3]

109 clopyralid + 2,4-D + MCPA

A translocated herbicide mixture for managed amenity turf
HRAC mode of action code: O + O + O

See also *2,4-D*
MCPA

Products

Esteem	Vitax	35:150:175 g/l	SL	12555

Uses

- Annual dicotyledons in ***managed amenity turf***
- Perennial dicotyledons in ***managed amenity turf***

Approval information

- Clopyralid, 2,4-D and MCPA included in Annex I under EC Directive 91/414

Efficacy guidance

- Apply when soil is moist and weeds growing actively, normally between Apr and Oct
- Ensure sufficient leaf area for uptake and select appropriate water volume to achieve good spray coverage
- To allow maximum translocation do not cut grass for 3 d after treatment

Restrictions

- Maximum number of treatments on managed amenity turf 1 per yr
- Do not spray in periods of drought unless irrigation is applied
- Do not apply if night temrperatures are low, if ground frost is imminent or in periods of prolonged cold weather
- Do not treat turf less than 1 yr old
- Do not use any treated plant materials for composting or mulching
- Do not mow for 3 d before or after treatment

Crop-specific information

- Consult manufacturer or carry out small scale safety test on turf grass cultivars not previously treated

Following crops guidance

- If reseeding of treated turf is required allow at least 6 wk after treatment before drilling grasses into the sward
- Where treated land is subsequently to be sown with broad-leaved plants allow an interval of at least 6 mth

Environmental safety

- Dangerous for the environment
- Harmful to aquatic organisms
- Wash spray equipment thoroughly with water and detergent immediately after use. Traces of product can damage susceptible plants sprayed later
- Do not spray outside the target area and avoid drift onto non-target plants
- Some pesticides pose a greater threat of contamination of water than others and clopyralid is one of these pesticides. Take special care when applying clopyralid near water and do not apply if heavy rain is forecast

Hazard classification and safety precautions

Hazard Harmful, Dangerous for the environment
Transport code M6
Packaging group III
UN number 3082
Risk phrases R22a, R37, R41, R52, R53a
Operator protection A, C, H; U05a, U11, U15
Environmental protection E15a, E34
Storage and disposal D01, D02, D12a

110 clopyralid + diflufenican + MCPA

A selective herbicide for use in managed amenity turf
HRAC mode of action code: O + F1 + O

See also *diflufenican*
MCPA

Products

Spearhead	Bayer Environ.	20:15:300 g/l	SL	09941

Uses

- Annual dicotyledons in ***managed amenity turf***
- Perennial dicotyledons in ***managed amenity turf***

Approval information

- Clopyralid, diflufenican and MCPA included in Annex I under EC Directive 91/414

Efficacy guidance

- Best results achieved by application when grass and weeds are actively growing
- Treatment during early part of the season is recommended, but not during drought

Restrictions

- Maximum number of treatments 1 per yr
- Only use when sward is satisfactorily established and regular mowing has begun
- Turf sown in spring or early summer may be ready for treatment after 2 mth. Later sown turf should not be sprayed until growth is resumed in the following spring
- Avoid mowing within 3-4 d before or after treatment
- Do not use cuttings from treated area as a mulch for any crop

Environmental safety

- Extremely dangerous to fish or other aquatic life. Do not contaminate surface waters or ditches with chemical or used container
- Keep livestock out of treated areas/away from treated water for at least 1 wk and until foliage of any poisonous weeds, such as ragwort, has died and become unpalatable
- LERAP Category B
- Avoid drift. Small amounts of spray can cause serious injury to herbaceous plants, vegetables, fruit and glasshouse crops
- Keep livestock out of treated areas
- Some pesticides pose a greater threat of contamination of water than others and clopyralid is one of these pesticides. Take special care when applying clopyralid near water and do not apply if heavy rain is forecast

Hazard classification and safety precautions

Hazard Harmful, Irritant
Transport code 9
Packaging group III
UN number 3082
Risk phrases R20, R21, R22a, R36, R38
Operator protection A, C; U02a, U05a, U08, U19a, U20b
Environmental protection E06a (unstipulated period); E13a, E16a, E16b, E34
Storage and disposal D01, D02, D05, D09a, D10b
Medical advice M03

111 clopyralid + florasulam + fluroxypyr

A translocated herbicide mixture for cereals
HRAC mode of action code: O + B + O

See also *florasulam*
fluroxypyr

Products

1	Galaxy	Dow	80:2.5:100 g/l	EC	14085
2	Praxys	Dow	80:2.5;100 g/l	EC	13912

Uses

- Annual dicotyledons in ***amenity grassland, lawns, managed amenity turf*** [2]; ***spring barley, spring oats, spring wheat, winter barley, winter oats, winter wheat*** [1]
- Chickweed in ***spring barley, spring oats, spring wheat, winter barley, winter oats, winter wheat*** [1]
- Cleavers in ***spring barley, spring oats, spring wheat, winter barley, winter oats, winter wheat*** [1]

Approval information

- Clopyralid, florasulam and fluroxypyr are all included in Annex I under Directive 91/414

Efficacy guidance

- Best results obtained when weeds are small and actively growing
- Effectiveness may be reduced when soil is very dry
- Use adequate water volume to achieve complete spray coverage of the weeds
- Florasulam is a member of the ALS-inhibitor group of herbicides

Restrictions

- Maximum number of treatments 1 per yr for all crops
- Do not spray when crops are under stress from any cause
- Do not apply through CDA applicators
- Do not roll or harrow for 7 d before or after application
- Do not use any treated plant material for composting or mulching
- Do not use manure from animals fed on treated crops for composting
- Specific restrictions apply to use in sequence or tank mixture with other sulfonylurea or ALS-inhibiting herbicides. See label for details

Crop-specific information

- Latest use: before second node detectable for oats; before third node detectable for barley and wheat

Following crops guidance

- Where residues of a treated crop have not completely decayed by the time of planting a succeeding crop, avoid planting peas, beans and other legumes, carrots and other Umbelliferae, potatoes, lettuce and other Compositae, glasshouse and protected crops
- Where the product has been used in mixture with certain named products (see label) only cereals or grass may be sown in the autumn following harvest. Otherwise cereals, oilseed rape, grass or vegetable brassicas as transplants may be sown as a following crop in the same calendar yr as treatment. Oilseed rape may show some temporary reduction of vigour after a dry summer, but yields are not affected
- In addition to the above, field beans, linseed, peas, sugar beet, potatoes, maize, clover (for use in grass/clover mixtures) or carrots may be sown in the calendar yr following treatment
- In the event of failure of a treated crop in spring, only spring wheat, spring barley, spring oats, maize or ryegrass may be sown

Environmental safety

- Dangerous for the environment
- Very toxic to aquatic organisms
- Take extreme care to avoid drift outside the target area
- Some pesticides pose a greater threat of contamination of water than others and clopyralid is one of these pesticides. Take special care when applying clopyralid near water and do not apply if heavy rain is forecast

SEE SECTION 3 FOR PRODUCTS ALSO REGISTERED

Hazard classification and safety precautions

Hazard Harmful, Dangerous for the environment
Transport code 9
Packaging group III
UN number 3082
Risk phrases R20, R36, R38, R50, R53a
Operator protection A, H [1, 2]; C [1]; U05a, U11, U19a, U20a
Environmental protection E15b, E34, E38
Storage and disposal D01, D02, D09a, D10b, D12a [1, 2]; D05 [2]
Medical advice M03 [1, 2]; M05a [1]

112 clopyralid + fluroxypyr + MCPA

A translocated herbicide mixture for use in sports and amenity turf
HRAC mode of action code: O + O + O

See also *fluroxypyr*
MCPA

Products

Greenor	Rigby Taylor	20:40:200 g/l	ME	10909

Uses

- Annual dicotyledons in ***managed amenity turf***

Approval information

- Clopyralid, fluroxypyr and MCPA included in Annex I under EC Directive 91/414

Efficacy guidance

- Best results achieved when weeds actively growing and turf grass competitive
- Treatment should normally be between Apr-Sep when the soil is moist
- Do not apply during drought unless irrigation is applied
- Allow 3 d before or after mowing established turf to ensure sufficient weed leaf surface present to allow uptake and movement

Restrictions

- Maximum number of treatments 2 per yr
- Do not treat grass under stress from frost, drought, waterlogging, trace element deficiency, disease or pest attack
- Do not treat if night temperatures are low, when frost is imminent or during prolonged cold weather

Crop-specific information

- Treat young turf only in spring when at least 2 mth have elapsed since sowing
- Allow 5 d after mowing young turf before treatment
- Product selective on a number of turf grass species (see label) but consultation or testing recommended before treatment of any cultivar

Environmental safety

- Dangerous for the environment
- Very toxic to aquatic organisms
- Wash spray equipment thoroughly with water and detergent immediately after use. Traces of product can damage susceptible plants sprayed later
- Some pesticides pose a greater threat of contamination of water than others and clopyralid is one of these pesticides. Take special care when applying clopyralid near water and do not apply if heavy rain is forecast

Hazard classification and safety precautions

Hazard Irritant, Dangerous for the environment
Transport code 9
Packaging group III
UN number 3082
Risk phrases R36, R43, R50, R53a
Operator protection A, C; U05a, U08, U14, U19a, U20b

Environmental protection E15a, E38
Storage and disposal D01, D02, D05, D09a, D10b, D12a

113 clopyralid + fluroxypyr + triclopyr

A foliar acting herbicide mixture for grassland
HRAC mode of action code: O + O + O

See also *fluroxypyr*
triclopyr

Products

1	Pastor	Dow	50:75:100 g/l	EC	11168
2	Trinity	AgriGuard	50:75:100 g/l	EC	12738

Uses
- Annual dicotyledons in ***rotational grass***
- Docks in ***permanent grassland, rotational grass***
- Stinging nettle in ***permanent grassland***
- Thistles in ***permanent grassland***

Approval information
- Clopyralid, fluroxypyr and triclopyr included in Annex I under EC Directive 91/414

Efficacy guidance
- Apply in spring or autumn depending on weeds present or as a split treatment in spring followed by autumn
- Treatment must be made when weeds and grass are actively growing
- It is important to ensure sufficient leaf area is present for uptake especially on established docks and thistles
- On large well established docks and where there is a large soil seed reservoir further treatment in the following year may be needed
- To allow maximum translocation in the weeds do not cut grass for 4 wk after treatment

Restrictions
- Maximum number of treatments 1 per yr at full dose or 2 per yr at half dose
- Do not spray in drought, very hot or very cold weather
- Do not treat sports or amenity turf
- Product kills or severely checks clover and should not be used where clover is an important constituent of the sward
- Do not roll or harrow 10 d before or 7 d after treatment
- Do not allow spray or drift to reach other crops, amenity plantings, gardens, ponds, lakes or water courses

Crop-specific information
- HI 7 d
- Application during active growth ensures minimal check to grass. Newly sown grass may be treated from the third leaf visible stage
- Product may be used in established grassland which is under non-rotational setaside arrangements
- Very occasionally some yellowing of the sward may occur after treatment but is quickly outgrown

Following crops guidance
- Residues in incompletely decayed plant tissue may affect succeeding susceptible crops such as peas, beans and other legumes, carrots and related crops, potatoes, tomatoes, lettuce
- Do not plant susceptible autumn-sown crops in the same yr as treatment with product. Spring sown crops may follow if treatment was before end July in the previous yr

Environmental safety
- Dangerous for the environment
- Toxic to aquatic organisms
- Keep livestock out of treated areas for at least 7 d following treatment and until foliage of poisonous weeds such as ragwort has died and become unpalatable
- Wash spray equipment thoroughly with water and detergent immediately after use. Traces of product can damage susceptible plants sprayed later

- Some pesticides pose a greater threat of contamination of water than others and clopyralid is one of these pesticides. Take special care when applying clopyralid near water and do not apply if heavy rain is forecast

Hazard classification and safety precautions

Hazard Harmful, Flammable, Dangerous for the environment
Transport code 3
Packaging group III
UN number 1993
Risk phrases R22b, R37, R38, R41, R43, R51, R53a, R67
Operator protection A, C; U02a, U05a, U08, U11, U19a, U20b [1, 2]; U14 [1]
Environmental protection E07a, E15a, E34 [1, 2]; E38 [1]
Consumer protection C01 [1]
Storage and disposal D01, D02, D05, D09a, D10b, D12a [1, 2]; D07 [2]
Medical advice M05b

114 clopyralid + picloram

A post-emergence herbicide mixture for oilseed rape
HRAC mode of action code: O + O

See also picloram

Products

1	Galera	Dow	267:67 g/l	SL	11961
2	Landgold Piccant	Goldengrass	267:67 g/l	SL	13770
3	Legara	AgChem Access	267:67 g/l	SL	13888
4	Nugget	AgriGuard	267:67 g/l	SL	13121
5	Piccant	Goldengrass	267:67 g/l	SL	14294
6	Renegade	ChemSource	267:67 g/l	SL	14507

Uses

- Cleavers in ***evening primrose*** *(off-label)*, ***honesty*** *(off-label)*, ***linseed*** *(off-label)*, ***mustard*** *(off-label)*, ***spring oilseed rape*** *(off-label)* [1]; ***winter oilseed rape*** [1-6]
- Mayweeds in ***evening primrose*** *(off-label)*, ***honesty*** *(off-label)*, ***linseed*** *(off-label)*, ***mustard*** *(off-label)*, ***spring oilseed rape*** *(off-label)* [1]; ***winter oilseed rape*** [1-6]

Specific Off-Label Approvals (SOLAs)

- ***evening primrose*** *20060960* [1]
- ***honesty*** *20060960* [1]
- ***linseed*** *20060960* [1]
- ***mustard*** *20060960* [1]
- ***spring oilseed rape*** *20061437* [1]

Approval information

- Clopyralid and picloram included in Annex I under EC Directive 91/414

Efficacy guidance

- Best results obtained from treatment when weeds are small and actively growing
- Cleavers that germinate after treatment will not be controlled

Restrictions

- Maximum total dose equivalent to one full dose treatment
- Do not treat crops under stress from cold, drought, pest damage, nutrient deficiency or any other cause
- Do not roll or harrow for 7 d before or after spraying
- Do not apply through CDA applicators
- Do not use any treated plant material for composting or mulching
- Do not use manure from animals fed on treated crops for composting
- Chop and incorporate all treated plant remains in early autumn, or as soon as possible after harvest, to release any residues into the soil. Ensure that all treated plant remains have completely decayed before planting susceptible crops

Crop-specific information

- Latest Use: before flower buds visible above crop canopy for winter oilseed rape

Following crops guidance

- Any crop may be sown in the calendar yr following treatment
- Ploughing or thorough cultivation should be carried out before planting leguminous crops
- Do not attempt to plant peas, beans, other legumes, carrots, other umbelliferous crops, potatoes, lettuce, other Compositae, or any glasshouse or protected crops if treated crop remains have not fully decayed by the time of planting
- In the event of failure of an autumn treated crop only oilseed rape, wheat, barley, oats, maize or ryegrass may be sown in the spring and only after ploughing or thorough cultivation

Environmental safety

- Dangerous for the environment
- Toxic to aquatic organisms
- Take extreme care to avoid drift onto crops and non-target plants outside the target area
- Some pesticides pose a greater threat of contamination of water than others and clopyralid is one of these pesticides. Take special care when applying clopyralid near water and do not apply if heavy rain is forecast

Hazard classification and safety precautions

Hazard Dangerous for the environment
Transport code 9 [6]
Packaging group III [6]
UN number 3082 [6]; N/C [1-5]
Risk phrases R51, R53a
Operator protection A, C; U05a
Environmental protection E15a, E34
Storage and disposal D01, D02, D05, D07, D09a

115 clopyralid + triclopyr

A perennial and woody weed herbicide for use in grassland
HRAC mode of action code: O + O

See also *triclopyr*

Products

1	Blaster	Headland Amenity	60:240 g/l	EC	13267
2	Grazon 90	Dow	60:240 g/l	EC	13117
3	Thistlex	Dow	200:200 g/l	EC	11533

Uses

- Brambles in ***amenity grassland*** [1]
- Broom in ***amenity grassland*** [1]
- Creeping thistle in ***permanent grassland***, ***rotational grass*** [3]
- Docks in ***amenity grassland*** [1]
- Gorse in ***amenity grassland*** [1]
- Perennial dicotyledons in ***amenity grassland*** [1]; ***permanent grassland*** *(off-label - via weed wiper)* [2]
- Stinging nettle in ***amenity grassland*** [1]
- Thistles in ***amenity grassland*** [1]

Specific Off-Label Approvals (SOLAs)

- ***permanent grassland*** *(via weed wiper) 20063161* [2]

Approval information

- Clopyralid and triclopyr included in Annex I under EC Directive 91/414

Efficacy guidance

- Must be applied to actively growing weeds
- Correct timing crucial for good control. Spray stinging nettle before flowering, docks in rosette stage in spring, creeping thistle before flower stems 15-20 cm high, brambles, broom and gorse in Jun-Aug
- Allow 2-3 wk regrowth after grazing or mowing before spraying perennial weeds.
- Where there is a large reservoir of weed seed in the soil further treatment in the following yr may be needed

SEE SECTION 3 FOR PRODUCTS ALSO REGISTERED

Restrictions

- Maximum number of treatments 1 per yr
- Only use on permanent pasture or rotational grassland established for at least 1 yr
- Do not apply where clover is an important constituent of sward
- Do not roll or harrow within 10 d before or 7 d after spraying
- Do not cut grass for 21 d before or 28 d after spraying
- Do not use any treated plant material for composting or mulching, and do not use manure for composting from animals fed on treated crops
- Do not apply by hand-held rotary atomiser equipment
- Do not allow drift onto other crops, amenity plantings or gardens, ponds, lakes or water courses. All conifers, especially pine and larch, are very sensitive

Crop-specific information

- Latest use: 7 d before grazing or cutting grass
- Some transient yellowing of treated swards may occur but is quickly outgrown

Following crops guidance

- Residues in plant tissues which have not completely decayed may affect succeeding susceptible crops such as peas, beans, other legumes, carrots, parsnips, potatoes, tomatoes, lettuce, glasshouse and protected crops
- Do not plant susceptible autumn-sown crops (eg winter beans) in same year as treatment and allow at least 9 mth from treatment before planting a susceptible crop in the following yr
- Do not direct drill kale, swedes, turnips, grass or grass mixtures within 6 wk of spraying
- Do not spray after end Jul where susceptible crops are to be planted in the next spring

Environmental safety

- Dangerous for the environment
- Very toxic to aquatic organisms
- Keep livestock out of treated areas for at least 7 d after spraying and until foliage of any poisonous weeds such as ragwort or buttercup has died down and become unpalatable
- Some pesticides pose a greater threat of contamination of water than others and clopyralid is one of these pesticides. Take special care when applying clopyralid near water and do not apply if heavy rain is forecast

Hazard classification and safety precautions

Hazard Harmful [1, 2]; Irritant [3]; Dangerous for the environment [1-3]
Transport code 3 [1, 2]
Packaging group III [1, 2]
UN number 1993 [1, 2]; N/C [3]
Risk phrases R22a, R38, R43, R50 [1, 2]; R41, R53a [1-3]; R52 [3]
Operator protection A, C [1-3]; H, M [1, 2]; U02a, U05a, U11, U20b [1-3]; U08, U14, U19a, U23b [1, 2]; U15, U23a [3]
Environmental protection E07a, E15a [1-3]; E23, E34, E38 [1, 2]
Consumer protection C01 [1, 2]
Storage and disposal D01, D02, D05, D09a, D10b, D12a
Medical advice M03 [3]

116 clothianidin

A nitromethylene neonicotinoid insecticide
IRAC mode of action code: 4A

See also beta-cyfluthrin + clothianidin

Products

1	Deter	Bayer CropScience	250 g/l	FS	12411
2	Poncho	Bayer CropScience	600 g/l	SC	13910

Uses

- Frit fly in ***forage maize***, ***grain maize***, ***sweetcorn*** [2]
- Leafhoppers in ***durum wheat*** *(seed treatment)*, ***triticale*** *(seed treatment)*, ***winter barley*** *(seed treatment)*, ***winter oats*** *(seed treatment)*, ***winter rye*** *(seed treatment)*, ***winter wheat*** *(seed treatment)* [1]

- Slugs in ***durum wheat*** *(seed treatment)*, ***triticale*** *(seed treatment)*, ***winter barley*** *(seed treatment)*, ***winter oats*** *(seed treatment)*, ***winter rye*** *(seed treatment)*, ***winter wheat*** *(seed treatment)* [1]
- Symphylids in ***forage maize*** *(reduction)*, ***grain maize*** *(reduction)*, ***sweetcorn*** *(reduction)* [2]
- Virus vectors in ***durum wheat*** *(seed treatment)*, ***triticale*** *(seed treatment)*, ***winter barley*** *(seed treatment)*, ***winter oats*** *(seed treatment)*, ***winter rye*** *(seed treatment)*, ***winter wheat*** *(seed treatment)* [1]
- Wireworm in ***durum wheat*** *(seed treatment)*, ***triticale*** *(seed treatment)*, ***winter barley*** *(seed treatment)*, ***winter oats*** *(seed treatment)*, ***winter rye*** *(seed treatment)*, ***winter wheat*** *(seed treatment)* [1]; ***forage maize***, ***grain maize***, ***sweetcorn*** [2]

Approval information

- Clothianidin included in Annex I under EC Directive 91/414
- Accepted by BBPA for use on malting barley

Efficacy guidance

- May only be used in conjunction with manufacturer's approved seed treatment application equipment or by following procedures given in the operating instructions
- Treated seed should preferably be drilled in the same season
- Evenness of seed cover improved by simultaneous application with a small volume (1.5-3 litres/tonne) of water
- Calibrate drill for treated seed and drill at 2.5-4 cm into firm, well prepared seedbed
- Use minimum 125 kg treated seed per ha
- Seed should be drilled to a depth of 40 mm into a well prepared and firm seedbed. If seed is present on the surface, or if spills have occurred, the field should be harrowed and rolled if conditions are appropriate
- When aphid activity is unusually late, or is heavy and prolonged in areas of high risk, and mild weather predominates, follow-up foliar aphicide may be required
- Incidental suppression of leafhoppers in early spring (Mar/Apr) may be achieved but if a specific attack develops an additional foliar insecticide may be required

Restrictions

- Maximum number of seed treatments one per batch
- Product must be fully re-dispersed and homogeneous before use
- Do not use on seed with more than 16% moisture content, or on sprouted, cracked or skinned seed
- All seed batches should be tested to ensure they are suitable for treatment

Crop-specific information

- Latest use: pre-drilling

Environmental safety

- Harmful to game and wildlife. Treated seed should not be left on the surface. Bury spillages.

Hazard classification and safety precautions

Hazard Harmful, Dangerous for the environment [2]; Irritant [1]
UN number N/C
Risk phrases R22a, R50, R53a [2]; R43 [1]
Operator protection A, H; U04a, U05a, U13 [1, 2]; U07, U14, U20b [1]; U20c [2]
Environmental protection E03, E15b, E34 [1, 2]; E36a [1]; E38 [2]
Storage and disposal D01, D02, D09a, D14 [1, 2]; D05 [1]; D10d, D12a [2]
Treated seed S01, S02, S03, S04b, S05, S06b, S07, S08 [1]; S04d [2]; S06a [1, 2]
Medical advice M03

117 clothianidin + prothioconazole

A combined fungicide and insecticide seed dressing for cereals
IRAC mode of action code: 4A

See also *prothioconazole*

Products

Redigo Deter	Bayer CropScience	250:50 g/l	FS	12423

SEE SECTION 3 FOR PRODUCTS ALSO REGISTERED

Uses

- Bunt in ***durum wheat*** *(seed treatment)*, ***triticale*** *(seed treatment)*, ***winter rye*** *(seed treatment)*, ***winter wheat*** *(seed treatment)*
- Covered smut in ***winter barley*** *(seed treatment)*
- Fusarium foot rot and seedling blight in ***durum wheat*** *(seed treatment)*, ***triticale*** *(seed treatment)*, ***winter barley*** *(seed treatment)*, ***winter oats*** *(seed treatment)*, ***winter rye*** *(seed treatment)*, ***winter wheat*** *(seed treatment)*
- Leaf stripe in ***winter barley*** *(seed treatment)*
- Leafhoppers in ***durum wheat*** *(seed treatment)*, ***triticale*** *(seed treatment)*, ***winter barley*** *(seed treatment)*, ***winter oats*** *(seed treatment)*, ***winter rye*** *(seed treatment)*, ***winter wheat*** *(seed treatment)*
- Loose smut in ***durum wheat*** *(seed treatment)*, ***winter barley*** *(seed treatment)*, ***winter oats*** *(seed treatment)*, ***winter wheat*** *(seed treatment)*
- Slugs in ***durum wheat*** *(seed treatment)*, ***triticale*** *(seed treatment)*, ***winter barley*** *(seed treatment)*, ***winter oats*** *(seed treatment)*, ***winter rye*** *(seed treatment)*, ***winter wheat*** *(seed treatment)*
- Virus vectors in ***durum wheat*** *(seed treatment)*, ***triticale*** *(seed treatment)*, ***winter barley*** *(seed treatment)*, ***winter oats*** *(seed treatment)*, ***winter rye*** *(seed treatment)*, ***winter wheat*** *(seed treatment)*
- Wireworm in ***durum wheat*** *(seed treatment)*, ***triticale*** *(seed treatment)*, ***winter barley*** *(seed treatment)*, ***winter oats*** *(seed treatment)*, ***winter rye*** *(seed treatment)*, ***winter wheat*** *(seed treatment)*

Approval information

- Clothianidin and prothioconazole included in Annex I under EC Directive 91/414
- Accepted by BBPA for use on malting barley

Efficacy guidance

- May only be used in conjunction with manufacturer's approved seed treatment application equipment or by following procedures given in the operating instructions
- Treated wheat seed should preferably be drilled in the same season. Treated seed of other cereals must be used in the same season
- Evenness of seed cover improved by simultaneous application with a small volume (1.5-3 litres/ tonne) of water
- Calibrate drill for treated seed and drill at 2.5-4 cm into firm, well prepared seedbed
- Use minimum 125 kg treated seed per ha
- When aphid activity unusually late, or is heavy and prolonged in areas of high risk, and mild weather predominates, follow-up foliar aphicide may be required
- Incidental suppression of leafhoppers in early spring (Mar/Apr) may be achieved but if a specific attack develops an additional foliar insecticide may be required
- Protection against slugs only applies to germinating seeds and not to aerial parts after emergence
- Where very high populations of aphids, wireworms or slugs occur additional specific control measures may be needed

Restrictions

- Maximum number of seed treatments one per batch for all cereals
- Product must be fully re-dispersed and homogeneous before use
- Do not use on seed with more than 16% moisture content, or on sprouted, cracked or skinned seed
- All seed batches should be tested to ensure they are suitable for treatment

Crop-specific information

- Latest use: pre-drilling of all cereals

Environmental safety

- Harmful to aquatic organisms

Hazard classification and safety precautions

Hazard Irritant
UN number N/C
Risk phrases R43, R52, R53a
Operator protection A, H; U14, U19a, U20b
Environmental protection E15a, E34, E38

Storage and disposal D01, D02, D09a, D11a, D12a
Treated seed S01, S02, S03, S04b, S05, S06a, S07, S08

118 clothianidin + prothioconazole + tebuconazole + triazoxide

A fungicide and insecticide seed treatment mixture for winter barley
IRAC mode of action code: 4A

See also *prothioconazole*
tebuconazole
triazoxide

Products

Raxil Deter	Bayer CropScience	333.3:33.3:20:13.3 g/l	FS	13030

Uses
- Barley yellow dwarf virus vectors in ***winter barley*** *(seed treatment)*
- Covered smut in ***winter barley*** *(seed treatment)*
- Fusarium foot rot and seedling blight in ***winter barley*** *(seed treatment)*
- Leaf stripe in ***winter barley*** *(seed treatment)*
- Leafhoppers in ***winter barley*** *(seed treatment)*
- Loose smut in ***winter barley*** *(seed treatment)*
- Net blotch in ***winter barley*** *(seed treatment)*
- Slugs in ***winter barley*** *(seed treatment - reduction)*
- Wireworm in ***winter barley*** *(seed treatment - reduction)*

Approval information
- Clothianidin and prothioconazole included in Annex I under EC Directive 91/414
- Accepted by BBPA for use on malting barley

Efficacy guidance
- May only be used in conjunction with manufacturer's approved seed treatment application equipment or by following procedures given in the operating instructions
- Treated barley seed must be used in the same season
- Evenness of seed cover improved by simultaneous application with a small volume (1.5-3 litres/ tonne) of water
- Calibrate drill for treated seed and drill at 4 cm into firm, well prepared seedbed
- Use minimum 125 kg treated seed per ha
- When aphid activity unusually late, or is heavy and prolonged in areas of high risk, and mild weather predominates, follow-up foliar aphicide may be required
- Leafhoppers can be suppressed in early spring (Mar/Apr) but if a specific attack develops an additional foliar insecticide may be required
- Protection against slugs only applies to germinating seeds and not to aerial parts after emergence
- Where very high populations of aphids, wireworms or slugs occur additional specific control measures may be needed

Restrictions
- Maximum number of seed treatments one per batch
- Product must be fully re-dispersed and homogeneous before use
- Do not use on seed with more than 16% moisture content, or on sprouted, cracked or skinned seed
- All seed batches should be tested to ensure they are suitable for treatment

Crop-specific information
- Latest use: pre-drilling of winter barley

Environmental safety
- Harmful to aquatic organisms

Hazard classification and safety precautions

Hazard Irritant
Transport code 9
Packaging group III
UN number 3082

Risk phrases R43, R52, R53a
Operator protection A, D, H; U07, U14, U20a
Environmental protection E15a, E34, E36a, E38
Storage and disposal D01, D02, D09a, D12a, D14
Treated seed S01, S02, S03, S04b, S05, S07, S08

119 Coniothyrium minitans

A fungal parasite of sclerotia in soil

Products

Contans	Belchim	1000 IU/mg	WG	12616

Uses

- Sclerotinia in ***all edible crops (outdoor)***, ***all non-edible crops (outdoor)***, ***protected crops***

Efficacy guidance

- *C.minitans* is a soil acting biological fungicide with specific action against the resting bodies (sclerotia) of *Sclerotinia sclerotiorum* and *S.minor*
- Treat 3 mth before disease protection is required to allow time for the infective sclerotia in the soil to be reduced
- A post-harvest treatment of soil and debris prevents further contamination of the soil with sclerotia produced by the previous crop
- For best results the soil should be moist and the temperature between 12-20 °C
- Application should be followed by soil incorporation into the surface layer to 10 cm with a rotovator or rotary harrow. Thorough spray coverage of the soil is essential to ensure uniform distribution
- If soil temperature drops below 0 °C or rises above 27 °C fungicidal activity is suspended but restarts when soil temperature returns within this range
- In glasshouses untreated areas at the margins should be covered by a film to avoid spread of spores from untreated sclerotia

Restrictions

- Maximum number of treatments 1 per crop or situation
- Do not plough or cultivate after treatment
- Do not mix with other pesticides, acids, alkalines or any product that attacks organic material
- Store product under cool dry conditions away from direct sunlight and away from heat sources

Crop-specific information

- Latest use: pre-planting
- HI: zero

Following crops guidance

- There are no restrictions on following crops and no waiting interval is specified

Hazard classification and safety precautions

Hazard Harmful
UN number N/C
Risk phrases R42
Operator protection A, C, D; U05a, U19a
Environmental protection E15b, E34
Storage and disposal D01, D02, D12a
Medical advice M03

120 copper ammonium carbonate

A protectant copper fungicide
FRAC mode of action code: M1

Products

Croptex Fungex	Certis	8% w/w (copper)	SL	11049

Uses

- Blight in ***tomatoes (outdoor)***

- Cane spot in ***loganberries***, ***raspberries***
- Celery leaf spot in ***celery (outdoor)***
- Currant leaf spot in ***blackcurrants***
- Damping off in ***seedlings of ornamentals***
- Leaf curl in ***peaches***
- Leaf mould in ***protected tomatoes***
- Powdery mildew in ***chrysanthemums***, ***cucumbers***

Approval information

- Accepted by BBPA for use on hops

Efficacy guidance

- Apply spray to both sides of foliage
- With protected crops keep foliage dry before and after spraying
- Ventilate glasshouse immediately after spraying

Restrictions

- Maximum number of treatments 5 per crop for celery; 3 per yr for blackcurrants and loganberries; 2 per yr for peaches and raspberries; 1 per tray when used as a drench for ornamental seedlings
- Do not spray plants which are dry at the roots

Environmental safety

- Harmful to fish or other aquatic life. Do not contaminate surface waters or ditches with chemical or used container
- Keep all livestock out of treated areas for at least 3 wk. Bury or remove spillages

Hazard classification and safety precautions

Hazard Harmful
UN number N/C
Risk phrases R22a, R41
Operator protection U05a, U11, U15, U20c
Environmental protection E06c (3 wk); E13c, E34
Storage and disposal D01, D02, D09a, D10a, D12a
Medical advice M03

121 copper oxychloride

A protectant copper fungicide and bactericide
FRAC mode of action code: M1

Products

1	Cuprokylt	Unicrop	50% w/w (copper)	WP	00604
2	Cuprokylt FL	Unicrop	270 g/l (copper)	SC	08299
3	Headland Inorganic Liquid Copper	Headland	256 g/l (copper)	SC	13009

Uses

- Bacterial canker in ***cherries*** [1, 3]; ***cob nuts*** *(off-label)*, ***hazel nuts*** *(off-label)*, ***walnuts*** *(off-label)* [2]; ***plums*** [1, 2]
- Bacterial rot in ***broccoli*** *(off-label)*, ***brussels sprouts*** *(off-label)*, ***cabbages*** *(off-label)*, ***calabrese*** *(off-label)*, ***cauliflowers*** *(off-label)*, ***chinese cabbage*** *(off-label)*, ***choi sum*** *(off-label)*, ***cob nuts*** *(off-label)*, ***collards*** *(off-label)*, ***hazel nuts*** *(off-label)*, ***kale*** *(off-label)*, ***pak choi*** *(off-label)*, ***protected broccoli*** *(off-label)*, ***protected brussels sprouts*** *(off-label)*, ***protected cabbages*** *(off-label)*, ***protected calabrese*** *(off-label)*, ***protected cauliflowers*** *(off-label)*, ***protected chinese cabbage*** *(off-label)*, ***protected choi sum*** *(off-label)*, ***protected collards*** *(off-label)*, ***tatsoi*** *(off-label)*, ***walnuts*** *(off-label)*, ***watercress*** *(off-label)* [3]; ***bulb onions*** *(off-label)*, ***garlic*** *(off-label)*, ***leeks*** *(off-label)*, ***salad onions*** *(off-label)*, ***shallots*** *(off-label)* [1, 3]; ***rhubarb*** *(off-label)* [1, 2]
- Black rot in ***broccoli*** *(off-label)*, ***brussels sprouts*** *(off-label)*, ***cabbages*** *(off-label)*, ***calabrese*** *(off-label)*, ***cauliflowers*** *(off-label)*, ***chinese cabbage*** *(off-label)*, ***collards*** *(off-label)*, ***kale*** *(off-label)* [1]; ***protected brassica seedlings*** *(off-label)* [2]
- Blight in ***broccoli*** *(off-label)*, ***brussels sprouts*** *(off-label)*, ***bulb onions*** *(off-label)*, ***cabbages*** *(off-label)*, ***calabrese*** *(off-label)*, ***cauliflowers*** *(off-label)*, ***chinese cabbage*** *(off-label)*, ***choi sum*** *(off-label)*, ***collards*** *(off-label)*, ***garlic*** *(off-label)*, ***kale*** *(off-label)*, ***leeks*** *(off-label)*, ***pak choi*** *(off-label)*, ***protected broccoli*** *(off-label)*, ***protected brussels sprouts*** *(off-label)*, ***protected***

cabbages *(off-label)*, ***protected calabrese*** *(off-label)*, ***protected cauliflowers*** *(off-label)*, ***protected chinese cabbage*** *(off-label)*, ***protected choi sum*** *(off-label)*, ***protected collards*** *(off-label)*, ***salad onions*** *(off-label)*, ***shallots*** *(off-label)*, ***tatsoi*** *(off-label)*, ***watercress*** *(off-label)* [3]; ***cob nuts*** *(off-label)*, ***hazel nuts*** *(off-label)*, ***walnuts*** *(off-label)* [2, 3]; ***potatoes***, ***tomatoes (outdoor)*** [1-3]; ***protected tomatoes*** [1, 2]

- Buck-eye rot in ***protected tomatoes***, ***tomatoes (outdoor)*** [1-3]
- Cane spot in ***blackberries*** *(off-label)*, ***loganberries***, ***raspberries***, ***rubus hybrids*** *(off-label)* [1, 2]
- Canker in ***apples***, ***pears*** [1-3]; ***broccoli*** *(off-label)*, ***brussels sprouts*** *(off-label)*, ***bulb onions*** *(off-label)*, ***cabbages*** *(off-label)*, ***calabrese*** *(off-label)*, ***cauliflowers*** *(off-label)*, ***chinese cabbage*** *(off-label)*, ***choi sum*** *(off-label)*, ***cob nuts*** *(off-label)*, ***collards*** *(off-label)*, ***garlic*** *(off-label)*, ***hazel nuts*** *(off-label)*, ***kale*** *(off-label)*, ***leeks*** *(off-label)*, ***pak choi*** *(off-label)*, ***protected broccoli*** *(off-label)*, ***protected brussels sprouts*** *(off-label)*, ***protected cabbages*** *(off-label)*, ***protected calabrese*** *(off-label)*, ***protected cauliflowers*** *(off-label)*, ***protected chinese cabbage*** *(off-label)*, ***protected choi sum*** *(off-label)*, ***protected collards*** *(off-label)*, ***salad onions*** *(off-label)*, ***shallots*** *(off-label)*, ***tatsoi*** *(off-label)*, ***walnuts*** *(off-label)*, ***watercress*** *(off-label)* [3]
- Celery leaf spot in ***celery (outdoor)*** [1-3]
- Collar rot in ***apples*** *(off-label)* [2]
- Damping off in ***protected tomatoes***, ***tomatoes (outdoor)*** [1, 2]
- Downy mildew in ***bilberries*** *(off-label)*, ***blueberries*** *(off-label)*, ***cranberries*** *(off-label)*, ***gooseberries*** *(off-label)*, ***redcurrants*** *(off-label)*, ***whitecurrants*** *(off-label)* [1, 2]; ***chard*** *(off-label)*, ***cress*** *(off-label)*, ***frise*** *(off-label)*, ***herbs (see appendix 6)*** *(off-label)*, ***lamb's lettuce*** *(off-label)*, ***protected chard*** *(off-label)*, ***protected cress*** *(off-label)*, ***protected frise*** *(off-label)*, ***protected herbs (see appendix 6)*** *(off-label)*, ***protected lamb's lettuce*** *(off-label)*, ***protected radicchio*** *(off-label)*, ***protected salad brassicas*** *(off-label)*, ***protected scarole*** *(off-label)*, ***protected spinach*** *(off-label)*, ***protected spinach beet*** *(off-label)*, ***radicchio*** *(off-label)*, ***salad brassicas*** *(off-label)*, ***scarole*** *(off-label)*, ***spinach*** *(off-label)*, ***spinach beet*** *(off-label)*, ***wine grapes*** [3]; ***hops*** [1-3]
- Foliar disease control in ***rhubarb*** *(off-label)* [3]
- Foot rot in ***protected tomatoes***, ***tomatoes (outdoor)*** [1, 2]
- Leaf curl in ***apricots*** *(off-label)*, ***peaches*** [1, 2]; ***quinces*** *(off-label)* [1]
- Pseudomonas storage rots in ***bulb onions*** *(off-label)*, ***garlic*** *(off-label)*, ***leeks*** *(off-label)*, ***salad onions*** *(off-label)*, ***shallots*** *(off-label)* [1]
- Purple blotch in ***blackberries*** [3]
- Pythium in ***watercress*** *(off label - during propagation)* [2]
- Rhizoctonia in ***watercress*** *(off-label - during propagation)* [2]
- Rhizoctonia rot in ***broccoli*** *(off-label)*, ***brussels sprouts*** *(off-label)*, ***bulb onions*** *(off-label)*, ***cabbages*** *(off-label)*, ***calabrese*** *(off-label)*, ***cauliflowers*** *(off-label)*, ***chinese cabbage*** *(off-label)*, ***choi sum*** *(off-label)*, ***cob nuts*** *(off-label)*, ***collards*** *(off-label)*, ***garlic*** *(off-label)*, ***hazel nuts*** *(off-label)*, ***kale*** *(off-label)*, ***leeks*** *(off-label)*, ***pak choi*** *(off-label)*, ***protected broccoli*** *(off-label)*, ***protected brussels sprouts*** *(off-label)*, ***protected cabbages*** *(off-label)*, ***protected calabrese*** *(off-label)*, ***protected cauliflowers*** *(off-label)*, ***protected chinese cabbage*** *(off-label)*, ***protected choi sum*** *(off-label)*, ***protected collards*** *(off-label)*, ***salad onions*** *(off-label)*, ***shallots*** *(off-label)*, ***tatsoi*** *(off-label)*, ***walnuts*** *(off-label)*, ***watercress*** *(off-label)* [3]
- Rust in ***blackcurrants*** [1-3]
- Spear rot in ***broccoli*** *(off-label)*, ***brussels sprouts*** *(off-label)*, ***cabbages*** *(off-label)*, ***calabrese*** *(off-label)*, ***cauliflowers*** *(off-label)*, ***chinese cabbage*** *(off-label)*, ***collards*** *(off-label)*, ***kale*** *(off-label)* [1, 3]; ***bulb onions*** *(off-label)*, ***choi sum*** *(off-label)*, ***cob nuts*** *(off-label)*, ***garlic*** *(off-label)*, ***hazel nuts*** *(off-label)*, ***leeks*** *(off-label)*, ***pak choi*** *(off-label)*, ***protected broccoli*** *(off-label)*, ***protected brussels sprouts*** *(off-label)*, ***protected cabbages*** *(off-label)*, ***protected calabrese*** *(off-label)*, ***protected cauliflowers*** *(off-label)*, ***protected chinese cabbage*** *(off-label)*, ***protected choi sum*** *(off-label)*, ***protected collards*** *(off-label)*, ***salad onions*** *(off-label)*, ***shallots*** *(off-label)*, ***tatsoi*** *(off-label)*, ***walnuts*** *(off-label)*, ***watercress*** *(off-label)* [3]; ***protected brassica seedlings*** *(off-label)* [2]

Specific Off-Label Approvals (SOLAs)

- ***apples*** *982336* [2]
- ***apricots*** *20063134* [1], *20063140* [2]
- ***bilberries*** *20063131* [1], *20063137* [2]
- ***blackberries*** *20063132* [1], *20063139* [2]
- ***blueberries*** *20063131* [1], *20063137* [2]
- ***broccoli*** *20010115* [1], *20080156* [3]

- ***brussels sprouts*** *20010115* [1], *20080156* [3]
- ***bulb onions*** *991127* [1], *20080156* [3]
- ***cabbages*** *20010115* [1], *20080156* [3]
- ***calabrese*** *20010115* [1], *20080156* [3]
- ***cauliflowers*** *20010115* [1], *20080156* [3]
- ***chard*** *20080157* [3]
- ***chinese cabbage*** *20010115* [1], *20080156* [3]
- ***choi sum*** *20080156* [3]
- ***cob nuts*** *990385* [2], *20080156* [3]
- ***collards*** *20010115* [1], *20080156* [3]
- ***cranberries*** *20063131* [1], *20063137* [2]
- ***cress*** *20080157* [3]
- ***frise*** *20080157* [3]
- ***garlic*** *991127* [1], *20080156* [3]
- ***gooseberries*** *20063131* [1], *20063137* [2]
- ***hazel nuts*** *990385* [2], *20080156* [3]
- ***herbs (see appendix 6)*** *20080157* [3]
- ***kale*** *20010115* [1], *20080156* [3]
- ***lamb's lettuce*** *20080157* [3]
- ***leeks*** *991127* [1], *20080156* [3]
- ***pak choi*** *20080156* [3]
- ***protected brassica seedlings*** *20010117* [2]
- ***protected broccoli*** *20080156* [3]
- ***protected brussels sprouts*** *20080156* [3]
- ***protected cabbages*** *20080156* [3]
- ***protected calabrese*** *20080156* [3]
- ***protected cauliflowers*** *20080156* [3]
- ***protected chard*** *20080157* [3]
- ***protected chinese cabbage*** *20080156* [3]
- ***protected choi sum*** *20080156* [3]
- ***protected collards*** *20080156* [3]
- ***protected cress*** *20080157* [3]
- ***protected frise*** *20080157* [3]
- ***protected herbs (see appendix 6)*** *20080157* [3]
- ***protected lamb's lettuce*** *20080157* [3]
- ***protected radicchio*** *20080157* [3]
- ***protected salad brassicas*** *20080157* [3]
- ***protected scarole*** *20080157* [3]
- ***protected spinach*** *20080157* [3]
- ***protected spinach beet*** *20080157* [3]
- ***quinces*** *20063133* [1]
- ***radicchio*** *20080157* [3]
- ***redcurrants*** *20063131* [1], *20063137* [2]
- ***rhubarb*** *20063135* [1], *20063138* [2], *20080158* [3]
- ***rubus hybrids*** *20063132* [1], *20063139* [2]
- ***salad brassicas*** *20080157* [3]
- ***salad onions*** *991127* [1], *20080156* [3]
- ***scarole*** *20080157* [3]
- ***shallots*** *991127* [1], *20080156* [3]
- ***spinach*** *20080157* [3]
- ***spinach beet*** *20080157* [3]
- ***tatsoi*** *20080156* [3]
- ***walnuts*** *990385* [2], *20080156* [3]
- ***watercress*** *(during propagation) 20001538* [2], *20080156* [3]
- ***whitecurrants*** *20063131* [1], *20063137* [2]

Approval information

- Approved for aerial application on potatoes [1]. See Section 5 for more information
- Accepted by BBPA for use on hops

Efficacy guidance

- Spray crops at high volume when foliage dry but avoid run off. Do not spray if rain expected soon
- Spray interval commonly 10-14 d but varies with crop, see label for details
- If buck-eye rot occurs, spray soil surface and lower parts of tomato plants to protect unaffected fruit [1, 2]
- A follow-up spray in the following spring should be made to top fruit severely infected with bacterial canker

Restrictions

- Maximum number of treatments 3 per crop for apples, blackberries, grapevines, pears; 2 per crop for watercress. Not specified for other crops
- Some peach cultivars are sensitive to copper. Treat non-sensitive varieties only [1, 2]

Crop-specific information

- HI hops 7 d
- Slight damage may occur to leaves of cherries and plums

Environmental safety

- Dangerous for the environment
- Very toxic to aquatic organisms
- Keep all livestock out of treated areas for at least 3 wk

Hazard classification and safety precautions

Hazard Harmful [2]; Dangerous for the environment [1-3]
Transport code 6.1 [2]; 9 [1, 3]
Packaging group III
UN number 3010 [2]; 3077 [1]; 3082 [3]
Risk phrases R22a, R50 [2]; R51 [1, 3]; R53a [1-3]
Operator protection A [3]; U20a [1, 3]; U20c [2]
Environmental protection E06a [1-3] (3 wk); E13c, E34, E38 [1, 3]; E15a [2]
Storage and disposal D01, D09a [1-3]; D05, D10c, D12a [2]; D10b [1, 3]

122 copper sulphate

See Bordeaux Mixture
FRAC mode of action code: M1

123 cyazofamid

A cyanoimidazole sulfonamide protectant fungicide for potatoes
FRAC mode of action code: 21

Products

1	Linford	AgChem Access	400 g/l	SC	13824
2	Ranman Twinpack	Belchim	400 g/l	KL	11851
3	Swallow	AgChem Access	400 g/l	SC	14078

Uses

- Blight in ***potatoes*** [1-3]
- Disease control in ***hops*** *(off-label)* [2]

Specific Off-Label Approvals (SOLAs)

- ***hops*** *20082918* [2]

Approval information

- Cyazofamid included in Annex I under EC Directive 91/414

Efficacy guidance

- Apply as a protectant treatment before blight enters the crop and repeat every 7-10 d depending on severity of disease pressure
- Commence spray programme immediately the risk of blight in the locality occurs, usually when the crop meets along the rows
- Product must always be used with organosilicone adjuvant provided in the twin pack

- To minimise the chance of development of resistance no more than three applications should be made consecutively (out of a permissible total of six) in the blight control programme. For more information on Resistance Management see Section 5

Restrictions
- Maximum number of treatments 6 per crop (no more than three of which should be consecutive)
- Mixed product must not be allowed to stand overnight
- Consult processor before using on crops intended for processing

Crop-specific information
- HI 7 d

Environmental safety
- Dangerous for the environment
- Very toxic to aquatic organisms
- Do not empty into drains

Hazard classification and safety precautions

Hazard Harmful (adjuvant); Dangerous for the environment
Transport code 9 [2]
Packaging group III [2]
UN number 3082 [2]
Risk phrases R20, R36, R48 (adjuvant); R41, R50, R53a
Operator protection A, C; U02a, U05a, U11, U15, U20a; U19a (adjuvant)
Environmental protection E15a, E19b, E34, E38
Storage and disposal D01, D02, D09a, D10c, D12a
Medical advice M03

124 cycloxydim

A translocated post-emergence cyclohexanedione oxime herbicide for grass weed control
HRAC mode of action code: A

Products

Laser	BASF	200 g/l	EC	12930

Uses
- Annual grasses in ***amenity vegetation*** *(off-label)*, ***borage for oilseed production*** *(off-label)*, ***broad beans*** *(off-label)*, ***brussels sprouts*, *bulb onions*, *cabbages*, *canary flower (echium spp.)*** *(off-label)*, ***carrots*, *cauliflowers*, *combining peas*, *dwarf beans*, *early potatoes*, *edible podded peas*** *(off-label)*, ***evening primrose*** *(off-label)*, ***farm forestry*, *flower bulbs*, *fodder beet*, *forest*, *garlic*** *(off-label)*, ***honesty*** *(off-label)*, ***horseradish*** *(off-label)*, ***leeks*, *linseed*, *lupins*** *(off-label)*, ***maincrop potatoes*, *mallow (althaea spp.)*** *(off-label)*, ***mangels*, *mustard*** *(off-label)*, ***parsnips*, *runner beans*** *(off-label)*, ***salad onions*, *shallots*** *(off-label)*, ***spring field beans*, *spring oilseed rape*, *strawberries*, *sugar beet*, *swedes*, *turnips*** *(off-label)*, ***vining peas*, *winter field beans*, *winter oilseed rape***
- Black bent in ***brussels sprouts*, *bulb onions*, *cabbages*, *carrots*, *cauliflowers*, *combining peas*, *dwarf beans*, *early potatoes*, *flower bulbs*, *fodder beet*, *leeks*, *linseed*, *maincrop potatoes*, *mangels*, *parsnips*, *salad onions*, *spring field beans*, *spring oilseed rape*, *strawberries*, *sugar beet*, *swedes*, *vining peas*, *winter field beans*, *winter oilseed rape***
- Blackgrass in ***brussels sprouts*, *bulb onions*, *cabbages*, *carrots*, *cauliflowers*, *combining peas*, *dwarf beans*, *early potatoes*, *flower bulbs*, *fodder beet*, *leeks*, *linseed*, *maincrop potatoes*, *mangels*, *parsnips*, *salad onions*, *spring field beans*, *spring oilseed rape*, *strawberries*, *sugar beet*, *swedes*, *vining peas*, *winter field beans*, *winter oilseed rape***
- Couch in ***brussels sprouts*, *bulb onions*, *cabbages*, *carrots*, *cauliflowers*, *combining peas*, *dwarf beans*, *early potatoes*, *flower bulbs*, *fodder beet*, *leeks*, *linseed*, *maincrop potatoes*, *mangels*, *parsnips*, *salad onions*, *spring field beans*, *spring oilseed rape*, *strawberries*, *sugar beet*, *swedes*, *vining peas*, *winter field beans*, *winter oilseed rape***
- Creeping bent in ***brussels sprouts*, *bulb onions*, *cabbages*, *carrots*, *cauliflowers*, *combining peas*, *dwarf beans*, *early potatoes*, *flower bulbs*, *fodder beet*, *leeks*, *linseed*, *maincrop potatoes*, *mangels*, *parsnips*, *salad onions*, *spring field beans*, *spring oilseed rape*, *strawberries*, *sugar beet*, *swedes*, *vining peas*, *winter field beans*, *winter oilseed rape***
- Green cover in ***land temporarily removed from production***

- Onion couch in ***brussels sprouts, bulb onions, cabbages, carrots, cauliflowers, combining peas, dwarf beans, early potatoes, flower bulbs, fodder beet, leeks, linseed, maincrop potatoes, mangels, parsnips, salad onions, spring field beans, spring oilseed rape, strawberries, sugar beet, swedes, vining peas, winter field beans, winter oilseed rape***
- Perennial grasses in ***amenity vegetation*** *(off-label)*, ***borage for oilseed production*** *(off-label)*, ***broad beans*** *(off-label)*, ***canary flower (echium spp.)*** *(off-label)*, ***edible podded peas*** *(off-label)*, ***evening primrose*** *(off-label)*, ***farm forestry, forest, garlic*** *(off-label)*, ***honesty*** *(off-label)*, ***horseradish*** *(off-label)*, ***lupins*** *(off-label)*, ***mallow (althaea spp.)*** *(off-label)*, ***mustard*** *(off-label)*, ***runner beans*** *(off-label)*, ***shallots*** *(off-label)*, ***turnips*** *(off-label)*
- Volunteer cereals in ***brussels sprouts, bulb onions, cabbages, carrots, cauliflowers, combining peas, dwarf beans, early potatoes, flower bulbs, fodder beet, leeks, linseed, maincrop potatoes, mangels, parsnips, salad onions, spring field beans, spring oilseed rape, strawberries, sugar beet, swedes, vining peas, winter field beans, winter oilseed rape***
- Wild oats in ***brussels sprouts, bulb onions, cabbages, carrots, cauliflowers, combining peas, dwarf beans, early potatoes, flower bulbs, fodder beet, leeks, linseed, maincrop potatoes, mangels, parsnips, salad onions, spring field beans, spring oilseed rape, strawberries, sugar beet, swedes, vining peas, winter field beans, winter oilseed rape***

Specific Off-Label Approvals (SOLAs)

- ***amenity vegetation*** *20081092*
- ***borage for oilseed production*** *20081094*
- ***broad beans*** *20081136*
- ***canary flower (echium spp.)*** *20081094*
- ***edible podded peas*** *20081096*
- ***evening primrose*** *20081094*
- ***garlic*** *20081097*
- ***honesty*** *20081094*
- ***horseradish*** *20081098*
- ***lupins*** *20081093*
- ***mallow (althaea spp.)*** *20081098*
- ***mustard*** *20081094*
- ***runner beans*** *20081096*
- ***shallots*** *20081097*
- ***turnips*** *20081095*

Efficacy guidance

- Best results achieved when weeds small and have not begun to compete with crop. Effectiveness reduced by drought, cool conditions or stress. Weeds emerging after application are not controlled
- Foliage death usually complete after 3-4 wk but longer under cool conditions, especially late treatments to winter oilseed rape
- Perennial grasses should have sufficient foliage to absorb spray and should not be cultivated for at least 14 d after treatment
- On established couch pre-planting cultivation recommended to fragment rhizomes and encourage uniform emergence
- Split applications to volunteer wheat and barley at GS 12-14 will often give adequate control in winter oilseed rape. See label for details
- Apply to dry foliage when rain not expected for at least 2 h
- Cycloxydim is an ACCase inhibitor herbicide. To avoid the build up of resistance do not apply products containing an ACCase inhibitor herbicide more than twice to any crop. In addition do not use any product containing cycloxydim in mixture or sequence with any other product containing the same ingredient
- Use these products as part of a resistance management strategy that includes cultural methods of control and does not use ACCase inhibitors as the sole chemical method of grass weed control
- Applying a second product containing an ACCase inhibitor to a crop will increase the risk of resistance development; only use a second ACCase inhibitor to control different weeds at a different timing
- Always follow WRAG guidelines for preventing and managing herbicide resistant weeds. See Section 5 for more information

Restrictions

- Maximum number of treatments 1 per crop or yr in most situations. See label

- Must be used with authorised adjuvant oil. See label
- Do not apply to crops damaged or stressed by adverse weather, pest or disease attack or other pesticide treatment
- Prevent drift onto other crops, especially cereals and grass

Crop-specific information

- HI cabbage, cauliflower, calabrese, salad onions 4 wk; peas, dwarf beans 5 wk; bulb onions, carrots, parsnips, strawberries 6 wk; sugar and fodder beet, leeks, mangels, potatoes, field beans, swedes, Brussels sprouts, winter field beans 8 wk; oilseed rape, soya beans, linseed 12 wk
- Recommended time of application varies with crop. See label for details
- On peas a crystal violet wax test should be done if leaf wax likely to have been affected by weather conditions or other chemical treatment. The wax test is essential if other products are to be sprayed before or after treatment
- May be used on ornamental bulbs when crop 5-10 cm tall. Product has been used on tulips, narcissi, hyacinths and irises but some subjects may be more sensitive and growers advised to check tolerance on small number of plants before treating the rest of the crop
- May be applied to land temporarily removed from production where the green cover is made up predominantly of tolerant crops listed on label. Use on industrial crops of linseed and oilseed rape on land temporarily removed from production also permitted.

Following crops guidance

- Guideline intervals for sowing succeeding crops after failed treated crop: field beans, peas, sugar beet, rape, kale, swedes, radish, white clover, lucerne 1 wk; dwarf French beans 4 wk; wheat, barley, maize 8 wk
- Oats should not be sown after failure of a treated crop

Environmental safety

- Dangerous for the environment
- Toxic to aquatic organisms
- Harmful to fish or other aquatic life. Do not contaminate surface waters or ditches with chemical or used container

Hazard classification and safety precautions

Hazard Harmful, Dangerous for the environment
Transport code 9
Packaging group III
UN number 3082
Risk phrases R22b, R38, R51, R53a
Operator protection A, C; U05a, U08, U20b
Environmental protection E15a, E38
Storage and disposal D01, D02, D09a, D10c, D12a
Medical advice M05b

125 Cydia pomonella GV

An baculovirus insecticide for codling moth control

Products

Cyd-X	Certis	10 g/l granulovirus	SC	13535

Uses

- Codling moth in ***apples***, ***pears***

Approval information

- Cydia pomonella included in Annex I under EC Directive 91/414

Efficacy guidance

- Optimum results require a precise spraying schedule commencing soon after egg laying and before first hatch
- First generation of codling moth normally occurs in first 2 wk of Jun in UK
- Apply in sufficient water to achieve good coverage of whole tree
- Normally 3 applications at intervals of 8 sunny days are needed for each codling moth generation
- Fruit damage is reduced significantly in the first yr and will reduce the population of codling moth in the following yr

Restrictions

- Maximum number of treatments 3 per season
- Do not tank mix with copper or any products with a pH lower than 5 or higher than 8
- Do not use as the exclusive measure for codling moth control. Adopt an anti-resistance strategy. See Section 5 for more information
- Consult processor before use on crops for processing

Crop-specific information

- HI: 14 d for apples, pears
- Evidence of efficacy and crop safety on pears is limited

Environmental safety

- Product has no adverse effect on predatory insects and is fully compatible with IPM programmes

Hazard classification and safety precautions

Hazard Harmful
Risk phrases R42, R43
Operator protection A, D, H; U14, U19a
Environmental protection E15a
Storage and disposal D12a

126 cyflufenamid

An amidoxime fungicide for cereals
FRAC mode of action code: U6

Products

Cyflamid	Certis	50 g/l	EW	12403

Uses

- Powdery mildew in ***amenity vegetation*** *(off-label)*, ***durum wheat***, ***forest nurseries*** *(off-label)*, ***grass seed crops*** *(off-label)*, ***hops*** *(off-label)*, ***ornamental plant production*** *(off-label)*, ***soft fruit*** *(off-label)*, ***spring barley***, ***spring wheat***, ***triticale***, ***winter barley***, ***winter rye***, ***winter wheat***

Specific Off-Label Approvals (SOLAs)

- ***forest nurseries*** *20082915*
- ***hops*** *20082915*
- ***soft fruit*** *20082915*

Approval information

- Accepted by BBPA on malting barley

Efficacy guidance

- Best results obtained from treatment at the first visible signs of infection by powdery mildew
- Sustained disease presssure may require a second treatment
- Disease spectrum may be broadened by appropriate tank mixtures. See label
- Product is rainfast within 1 h
- Must be used as part of an Integrated Crop Management programme that includes alternating use, or mixture, with fungicides with a different mode of action effective against powdery mildew

Restrictions

- Maximum number of treatments 2 per crop on all recommended cereals
- Apply only in the spring
- Do not apply to crops under stress from drought, waterlogging, cold, pests or diseases, lime or nutrient deficiency or other factors affecting crop growth

Crop-specific information

- Latest use: before start of flowering of cereal crops

Following crops guidance

- No restrictions

Environmental safety

- Dangerous for the environment
- Very toxic to aquatic organisms
- Avoid drift onto ponds, waterways or ditches

Hazard classification and safety precautions

Hazard Dangerous for the environment
Transport code 9
Packaging group III
UN number 3082
Risk phrases R50, R53a
Operator protection A; U05a
Environmental protection E15b, E34
Storage and disposal D01, D02, D05, D10c, D12b

127 cyfluthrin

A non-systemic pyrethroid insecticide; product approval expired in 2008
IRAC mode of action code: 3

128 cymoxanil

A cyanoacetamide oxime fungicide for potatoes
FRAC mode of action code: 27

Products

1	Curzate 60 WG	Headland	60% w/w	WG	13904
2	Cymbal	Belchim	45% w/w	WG	13312
3	Drum	Belchim	45% w/w	WG	13853
4	Harpoon	Headland	60% w/w	WG	14278
5	Option	DuPont	60% w/w	WG	11834
6	Sipcam C 50	Sipcam	50% w/w	WP	10610

Uses

- Blight in ***potatoes***

Efficacy guidance

- Product to be used in mixture with specified mixture partners in order to combine systemic and protective activity. See labels for details
- Commence spray programme as soon as weather conditions favourable for disease development occur and before infection appears. At latest the first treatment should be made as the foliage meets along the rows
- Repeat treatments at 7-14 day intervals according to disease incidence and weather conditions
- Treat irrigated crops as soon as possible after irrigation and repeat at 10 d intervals

Restrictions

- Maximum total dose equivalent to four full dose treatments [6]. Not specified for other products
- Minimum spray interval 7 d
- Do not allow packs to become wet during storage and do not keep opened packs from one season to another
- Cymoxanil must be used in tank mixture with other fungicides

Crop-specific information

- HI for potatoes 7 d [6]; 14 d [2]
- Consult processor before treating potatoes grown for processing

Environmental safety

- Dangerous for the environment
- Very toxic to aquatic organisms

Hazard classification and safety precautions

Hazard Harmful [1, 4-6]; Irritant [2, 3]; Dangerous for the environment [1-5]
Transport code 9 [1-5]
Packaging group III [1-5]
UN number 3077 [1-5]; N/C [6]
Risk phrases R22a [1, 4-6]; R43 [2, 3]; R50, R53a [1-5]
Operator protection A [1, 2, 4-6]; H [2]; U05a [1-6]; U08 [1, 4-6]; U09a, U14 [2, 3]; U19a [1, 4, 5]; U20a [2]; U20b [1, 3-6]

Environmental protection E13a, E34 [1, 4, 5]; E13c [6]; E15a [2, 3]; E38 [2, 3, 5]
Storage and disposal D01, D02, D09a, D11a [1-6]; D12a [2, 3, 5]
Medical advice M03 [1, 4-6]

129 cymoxanil + famoxadone

A preventative and curative fungicide mixture for potatoes
FRAC mode of action code: 27 + 11

See also famoxadone

Products

Tanos	DuPont	25:25% w/w	WG	10677

Uses
- Blight in ***potatoes***

Approval information
- Famoxadone included in Annex I under EC Directive 91/414

Efficacy guidance
- Commence spray programme before infection appears as soon as weather conditions favourable for disease development to occur. At latest the first treatment should be made as the foliage meets along the rows
- Repeat treatments at 7-10 day intervals according to disease incidence and weather conditions
- Reduce spray interval if conditions are conducive to the spread of blight
- Spray as soon as possible after irrigation
- Increase water volume in dense crops
- Product combines active substances with different modes of action and is effective against strains of potato blight that are insensitive to phenylamide fungicides
- To minimise the likelihood of development of resistance to QoI fungicides these products should be used in a planned Resistance Management strategy. See Section 5 for more information

Restrictions
- Maximum number of treatments 6 per crop
- Consult processors before using on crops for processing
- Do not apply more than 3 consecutive treatments containing QoI active fungicides

Crop-specific information
- HI 14 d for potatoes

Environmental safety
- Dangerous for the environment
- Very toxic to aquatic organisms
- Dangerous to fish or other aquatic life. Do not contaminate surface waters or ditches with chemical or used container
- LERAP Category B

Hazard classification and safety precautions

Hazard Harmful, Dangerous for the environment
Transport code 9
Packaging group III
UN number 3077
Risk phrases R22a, R50, R53a
Operator protection A, D; U05a, U20b
Environmental protection E13b, E16a, E16b, E34, E38
Storage and disposal D01, D02, D09a, D10b, D12a
Medical advice M03

130 cymoxanil + fludioxonil + metalaxyl-M

A fungicide seed dressing for peas
FRAC mode of action code: 27 + 12 + 4

See also *fludioxonil*
metalaxyl-M

Products

Wakil XL	Syngenta	10:5:17.5% w/w	WS	10562

Uses

- Ascochyta in ***combining peas*** *(seed treatment)*, ***vining peas*** *(seed treatment)*
- Damping off in ***chard*** *(off-label)*, ***chives*** *(off-label)*, ***combining peas*** *(seed treatment)*, ***herbs (see appendix 6)*** *(off-label)*, ***leaf spinach*** *(off-label)*, ***lupins*** *(off-label)*, ***parsley*** *(off-label)*, ***poppies for morphine production*** *(off-label - seed treatment)*, ***red beet*** *(off-label - seed treatment)*, ***salad brassicas*** *(off-label)*, ***spinach beet*** *(off-label)*, ***vining peas*** *(seed treatment)*
- Downy mildew in ***broad beans*** *(off-label - seed treatment)*, ***combining peas*** *(seed treatment)*, ***lupins*** *(off-label)*, ***poppies for morphine production*** *(off-label - seed treatment)*, ***spring field beans*** *(off-label - seed treatment)*, ***vining peas*** *(seed treatment)*, ***winter field beans*** *(off-label - seed treatment)*
- Pythium in ***carrots*** *(off-label - seed treatment)*, ***combining peas*** *(seed treatment)*, ***lupins*** *(off-label)*, ***parsnips*** *(off-label - seed treatment)*, ***poppies for morphine production*** *(off-label - seed treatment)*, ***vining peas*** *(seed treatment)*

Specific Off-Label Approvals (SOLAs)

- ***broad beans*** *(seed treatment) 20023932*
- ***carrots*** *(seed treatment) 20021191*
- ***chard*** *20070791*
- ***chives*** *20070791*
- ***herbs (see appendix 6)*** *20070791*
- ***leaf spinach*** *20070791*
- ***lupins*** *20060614*
- ***parsley*** *20070791*
- ***parsnips*** *(seed treatment) 20021191*
- ***poppies for morphine production*** *(seed treatment) 20040683*
- ***red beet*** *(seed treatment) 20032313*
- ***salad brassicas*** *20070791*
- ***spinach beet*** *20070791*
- ***spring field beans*** *(seed treatment) 20023932*
- ***winter field beans*** *(seed treatment) 20023932*

Approval information

- Fludioxonil and metalaxyl-M included in Annex I under EC Directive 91/414

Efficacy guidance

- Apply through continuous flow seed treaters which should be calibrated before use

Restrictions

- Max number of treatments 1 per seed batch
- Ensure moisture content of treated seed satisfactory and store in a dry place
- Check calibration of seed drill with treated seed before drilling and sow as soon as possible after treatment
- Consult before using on crops for processing

Crop-specific information

- Latest use: pre-drilling

Environmental safety

- Harmful to aquatic organisms
- Do not use treated seed as food or feed

Hazard classification and safety precautions

UN number N/C
Risk phrases R52, R53a

Operator protection A, D, H; U02a, U05a, U20b
Environmental protection E03, E15a
Storage and disposal D01, D02, D05, D07, D09a, D11a, D12a
Treated seed S02, S04b, S05, S07

131 cymoxanil + mancozeb

A protectant and systemic fungicide for potato blight control
FRAC mode of action code: 27 + M3

See also mancozeb

Products

1	Agriguard Cymoxanil Plus	AgriGuard	4.5:68% w/w	WP	10893
2	Astral	AgChem Access	4.5:68% w/w	WP	14049
3	Besiege	DuPont	4.5:68% w/w	WP	08086
4	Blighter	Nufarm UK	5:68% w/w	WG	14326
5	Clayton Krypton	Clayton	4.5:68% w/w	WP	09398
6	Clayton Krypton MZ	Clayton	4.5:68% w/w	WP	14010
7	Curzate M WG	DuPont	4.5:68% w/w	WG	11901
8	Globe	Sipcam	6:70% w/w	WP	10339
9	Matilda	Nufarm UK	4.5:68% w/w	WP	12006
10	Rhapsody	DuPont	4.5:68% w/w	WG	11958
11	Rhythm	Interfarm	4.5:68% w/w	WP	09636
12	Solace	Nufarm UK	4.5:68% w/w	WP	11936

Uses

- Blight in ***potatoes***

Approval information

- Mancozeb included in Annex I under EC Directive 91/414

Efficacy guidance

- Apply immediately after blight warning or as soon as local conditions dictate and repeat at 7-14 d intervals until haulm dies down or is burnt off
- Spray interval should not be more than 7-10 d in irrigated crops (see product label). Apply treatment after irrigation
- To minimise the likelihood of development of resistance these products should be used in a planned Resistance Management strategy. See Section 5 for more information

Restrictions

- Maximum number of treatments 6 per crop [8]. Not specified for other products
- Do not apply at less than 7 d intervals
- Do not allow packs to become wet during storage

Crop-specific information

- HI 7 d [8]
- Destroy and remove any haulm that remains after harvest of early varieties to reduce blight pressure on neighbouring maincrop potatoes

Environmental safety

- Dangerous for the environment
- Very toxic to aquatic organisms
- Keep product away from fire or sparks

Hazard classification and safety precautions

Hazard Harmful [5]; Irritant [1-4, 6-12]; Dangerous for the environment [1-7, 10-12]
Transport code 9
Packaging group III
UN number 3077
Risk phrases R36 [1-3, 5-7, 9-12]; R37 [1-7, 9-12]; R38 [1, 6, 9, 12]; R42 [1-3, 5, 7, 10, 11]; R43 [1-12]; R50 [2-7, 10, 11]; R53a [2-7, 10-12]
Operator protection A, C [1-12]; H [8]; U02a [8]; U05a, U08 [1-12]; U11 [2-6, 9, 11]; U14 [2-4, 7, 10, 11]; U19a [1-7, 9-12]; U20a [1, 12]; U20b [2-11]
Environmental protection E13b [8]; E13c [2, 3, 7, 9, 10]; E15a [1-8, 10-12]; E38 [2-4, 7, 10, 11]

Storage and disposal D01, D02, D09a [1-12]; D06b [4]; D11a [1-3, 5-12]; D12a [2-4, 11]
Medical advice M04a, M04b [5, 6]

132 cymoxanil + propamocarb

A cyanoacetamide oxime fungicide + a carbamate fungicide for potatoes
FRAC mode of action code: 27 + 28

Products

Proxanil	Sipcam	50:400 g/l	SC	13945

Uses

- Blight in ***potatoes***

Approval information

- Propamocarb hydrochloride included in Annex I under EC Directive 91/414

Hazard classification and safety precautions

Hazard Irritant
UN number N/C
Risk phrases R43, R52, R53a
Operator protection A, H; U05a, U08, U14, U20b
Environmental protection E15a, E38
Storage and disposal D01, D02, D09a, D10a

133 cypermethrin

A contact and stomach acting pyrethroid insecticide
IRAC mode of action code: 3

Products

1	Curfew	AgriGuard	100 g/l	EC	14289
2	Cypermethrin Lacquer	Killgerm	7 g/l	LA	H3164
3	Forester	Fargro	100 g/l	EW	13164
4	Landgold Cyper 100	Goldengrass	100 g/l	EC	14105
5	Permasect C	Nufarm UK	100 g/l	EC	13158
6	Toppel 100 EC	United Phosphorus	100 g/l	EC	13704

Uses

- Aphids in ***apples***, ***vining peas*** [1, 5, 6]; ***beet leaves*** *(off-label - outdoor and protected crops)*, ***broad beans*** *(off-label)*, ***cherries***, ***chinese cabbage*** *(off-label - outdoor and protected crops)*, ***durum wheat***, ***fodder beet*** *(off-label)*, ***frise*** *(off-label)*, ***herbs (see appendix 6)*** *(off-label)*, ***hops***, ***lamb's lettuce*** *(off-label)*, ***leaf brassicas*** *(off-label)*, ***lettuce***, ***mallow (althaea spp.)*** *(off-label)*, ***ornamental plant production***, ***plums***, ***radicchio*** *(off-label)*, ***scarole*** *(off-label)*, ***spinach beet*** *(off-label)*, ***spring linseed*** *(off-label)*, ***winter linseed*** *(off-label)* [6]; ***broccoli***, ***brussels sprouts***, ***cabbages***, ***calabrese***, ***cauliflowers***, ***winter barley***, ***winter oats***, ***winter rye***, ***winter wheat*** [1, 4, 6]; ***fodder beet***, ***kale***, ***mangels***, ***potatoes***, ***red beet***, ***spring barley***, ***spring field beans***, ***spring oats***, ***spring oilseed rape***, ***spring wheat***, ***sugar beet***, ***winter field beans***, ***winter oilseed rape*** [4]; ***pears***, ***spring barley*** *(autumn sown)*, ***spring wheat*** *(autumn sown)* [1, 6]; ***triticale*** [4, 6]
- Asparagus beetle in ***asparagus*** *(off-label)* [6]
- Barley yellow dwarf virus vectors in ***durum wheat*** [6]; ***spring barley*** *(autumn sown)*, ***spring wheat*** *(autumn sown)*, ***triticale***, ***winter barley***, ***winter oats***, ***winter rye***, ***winter wheat*** [1, 5, 6]
- Beetles in ***cut logs*** [3]
- Bladder pod midge in ***spring oilseed rape***, ***winter oilseed rape*** [1, 6]
- Brassica pod midge in ***spring oilseed rape*** *(some coincidental control only)*, ***winter oilseed rape*** *(some coincidental control only)* [4]
- Cabbage seed weevil in ***spring oilseed rape***, ***winter oilseed rape*** [4, 5]
- Cabbage stem flea beetle in ***spring oilseed rape*** [4]; ***winter oilseed rape*** [1, 4-6]
- Capsids in ***apples*** [1, 6]; ***ornamental plant production*** [6]; ***pears*** [1, 5, 6]
- Caterpillars in ***apples***, ***pears*** [1, 6]; ***beet leaves*** *(off-label - outdoor and protected crops)*, ***cherries***, ***chinese cabbage*** *(off-label - outdoor and protected crops)*, ***lettuce***, ***ornamental plant production***, ***plums***, ***spinach beet*** *(off-label - outdoor and protected crops)* [6]; ***broccoli***, ***brussels***

sprouts, cabbages, calabrese, cauliflowers [1, 4-6]; ***fodder beet, kale, mangels, potatoes, red beet, sugar beet*** [4, 5]
- Codling moth in ***apples*** [1, 5, 6]
- Cutworms in ***carrots*** *(off-label)*, ***horseradish*** *(off-label)*, ***lettuce, ornamental plant production, parsley root*** *(off-label)*, ***parsnips*** *(off-label)* [6]; ***fodder beet, mangels, red beet*** [4, 5]; ***potatoes, sugar beet*** [1, 4-6]
- Frit fly in ***grassland*** [4]; ***permanent grassland*** [1, 6]; ***rotational grass*** [1, 5, 6]
- Insect pests in ***farm buildings*** [2]
- Large pine weevil in ***forest*** [3]
- Pea and bean weevil in ***spring field beans, vining peas, winter field beans*** [1, 4-6]
- Pea moth in ***vining peas*** [1, 4-6]
- Pod midge in ***spring oilseed rape, winter oilseed rape*** [5]
- Pollen beetle in ***spring oilseed rape, winter oilseed rape*** [1, 4-6]
- Rape winter stem weevil in ***winter oilseed rape*** [1, 6]
- Sawflies in ***apples*** [1, 6]
- Seed weevil in ***spring oilseed rape, winter oilseed rape*** [1, 6]
- Suckers in ***apples*** [5]; ***pears*** [1, 6]
- Thrips in ***ornamental plant production*** [6]
- Tortrix moths in ***apples*** [1, 5, 6]; ***pears*** [5]
- Whitefly in ***broccoli, brussels sprouts, cabbages, calabrese, cauliflowers*** [1, 6]; ***ornamental plant production, protected cucumbers, protected lettuce, protected ornamentals*** [6]
- Winter moth in ***apples, pears*** [5]
- Yellow cereal fly in ***durum wheat*** [6]; ***spring barley*** *(autumn sown)*, ***spring wheat*** *(autumn sown)*, ***triticale, winter barley, winter oats, winter rye*** [1, 5, 6]; ***winter wheat*** [1, 4-6]

Specific Off-Label Approvals (SOLAs)
- ***asparagus*** *20080639* [6]
- ***beet leaves*** *(outdoor and protected crops) 20080638* [6]
- ***broad beans*** *20080643* [6]
- ***carrots*** *20080637* [6]
- ***chinese cabbage*** *(outdoor and protected crops) 20080638* [6]
- ***fodder beet*** *20080640* [6]
- ***frise*** *20080642* [6]
- ***herbs (see appendix 6)*** *20080642* [6]
- ***horseradish*** *20080637* [6]
- ***lamb's lettuce*** *20080642* [6]
- ***leaf brassicas*** *20080642* [6]
- ***mallow (althaea spp.)*** *20080642* [6]
- ***parsley root*** *20080637* [6]
- ***parsnips*** *20080637* [6]
- ***radicchio*** *20080642* [6]
- ***scarole*** *20080642* [6]
- ***spinach beet*** *(outdoor and protected crops) 20080638* [6], *20080642* [6]
- ***spring linseed*** *20080641* [6]
- ***winter linseed*** *20080641* [6]

Approval information
- Cypermethrin included in Annex I under EC Directive 91/414
- Following implementation of Directive 98/82/EC, approval for use of cypermethrin on numerous crops was revoked in 1999
- Accepted by BBPA for use on malting barley and hops

Efficacy guidance
- Products combine rapid action, good persistence, and high activity on Lepidoptera.
- As effect is mainly via contact, good coverage is essential for effective action. Spray volume should be increased on dense crops [5, 6]
- A repeat spray after 10-14 d is needed for some pests of outdoor crops, several sprays at shorter intervals for whitefly and other glasshouse pests [5, 6]
- Rates and timing of sprays vary with crop and pest. See label for details [3, 5, 6]
- Where aphids in hops, pear suckers or glasshouse whitefly resistant to cypermethrin occur control is unlikely to be satisfactory [5, 6]

- Apply lacquer undiluted in a 5 cm wide band at strategic places. Ensure surfaces are clean and dry before application [2]
- Lacquer normally remains active for 12 mth [2]
- Temporary loss of activity for about 2 d will occur if lacquer surface is washed, until cypermethrin has migrated to the surface layer [2]

Restrictions

- Maximum number of treatments varies with crop and product. See label or approval notice for details [3, 5, 6]
- Apply by knapsack or hand-held sprayer [3]

Crop-specific information

- HI vining peas 7 d; other crops 0 d
- Test spray sample of new or unusual ornamentals or trees before committing whole batches [3, 6]
- Post-planting forestry applications should be made before damage is seen or at the onset of damage during the first 2 yrs after transplanting [3]

Environmental safety

- Dangerous for the environment [3, 5, 6]
- Very toxic to aquatic organisms
- High risk to bees. Do not apply to crops in flower, or to those in which bees are actively foraging, except as directed. Do not apply when flowering weeds are present
- Flammable [2, 5, 6]
- LERAP Category A [3, 5, 6]
- Broadcast air-assisted LERAP (18 m) [5, 6]
- Do not spray cereals after 31 Mar within 6 m of the edge of the growing crop [3, 5, 6]

Hazard classification and safety precautions

Hazard Harmful [1-5]; Irritant [6]; Flammable [1, 2, 4-6]; Dangerous for the environment [1, 3-6]
Transport code 3 [1, 4, 6]; 9 [3]
Packaging group III [1, 3, 4, 6]
UN number 1993 [1, 6]; 3082 [3]; 3295 [4]
Risk phrases R20 [1, 2, 6]; R21, R52 [2]; R22a, R50 [1, 3-6]; R22b, R37, R66 [1, 4-6]; R38, R43 [3]; R41 [1, 6]; R53a [1-6]; R67 [4, 5]
Operator protection A [1-6]; C [1, 2, 4-6]; H, M [3]; P [2]; U02a, U10 [3-5]; U04a [2-5]; U05a [1-6]; U08 [1, 6]; U09a, U15 [2]; U11 [4, 5]; U14 [3]; U16b [2, 3]; U19a [1, 3-6]; U20a [4]; U20b [1, 5, 6]
Environmental protection E02a [2] (48 h); E13a [1, 6]; E15a [2-5]; E16a [4]; E16c [1, 5, 6]; E16d, E34 [1, 3-6]; E17b [1, 4-6] (18 m); E38 [3]
Consumer protection C08, C09, C11, C12 [2]
Storage and disposal D01, D02, D09a [1-6]; D05, D12a [1, 3, 5, 6]; D10b [1, 3-6]; D12b [2]
Medical advice M03 [3]; M04a [2, 4, 5]; M05b [1, 4-6]

134 cyproconazole

A contact and systemic triazole fungicide
FRAC mode of action code: 3

See also *azoxystrobin + cyproconazole*
chlorothalonil + cyproconazole
chlorothalonil + cyproconazole + propiconazole

Products

Centaur	Bayer CropScience	200 g/l	EC	13852

Uses

- Brown rust in ***spring barley***, ***spring rye***, ***winter barley***, ***winter rye***, ***winter wheat***
- Chocolate spot in ***spring field beans***, ***winter field beans***
- Crown rust in ***spring oats***, ***winter oats***
- Ear diseases in ***winter wheat***
- Eyespot in ***winter barley*** *(useful reduction)*, ***winter wheat*** *(useful reduction)*
- Foliar disease control in ***borage for oilseed production*** *(off-label)*, ***canary flower (echium spp.)*** *(off-label)*, ***durum wheat*** *(off-label)*, ***fodder beet*** *(off-label)*, ***grass seed crops*** *(off-label)*, ***linseed*** *(off-label)*, ***mustard*** *(off-label)*, ***red beet*** *(off-label)*, ***triticale*** *(off-label)*

- Glume blotch in ***winter wheat***
- Light leaf spot in ***winter oilseed rape***
- Net blotch in ***spring barley*** *(useful reduction)*, ***winter barley*** *(useful reduction)*
- Phoma leaf spot in ***winter oilseed rape***
- Powdery mildew in ***spring barley***, ***spring oats***, ***spring rye***, ***sugar beet***, ***winter barley***, ***winter oats***, ***winter rye***, ***winter wheat***
- Ramularia leaf spots in ***sugar beet***
- Rhynchosporium in ***spring barley***, ***winter barley***
- Rust in ***spring field beans***, ***winter field beans***
- Septoria leaf blotch in ***winter wheat***
- Yellow rust in ***spring barley***, ***winter barley***, ***winter wheat***

Specific Off-Label Approvals (SOLAs)

- ***borage for oilseed production*** *20081099*
- ***canary flower (echium spp.)*** *20081099*
- ***durum wheat*** *20081101*
- ***fodder beet*** *20081100*
- ***grass seed crops*** *20081101*
- ***linseed*** *20081099*
- ***mustard*** *20081099*
- ***red beet*** *20081100*
- ***triticale*** *20081101*

Approval information

- Accepted by BBPA for use on malting barley
- Cyproconazole has not been included in Annex 1 but the date of withdrawal has yet to be set. Current approvals expire in 2013

Efficacy guidance

- Apply at start of disease development or as preventive treatment and repeat as necessary
- Most effective time of treatment varies with disease and use of tank mixes may be desirable. See label for details
- Product alone gives useful reduction of cereal eyespot. Where high infections probable a tank mix with prochloraz recommended
- On oilseed rape a two spray autumn/spring programme recommended for high risk situations and on susceptible varieties
- Cyproconazole is a DMI fungicide. Resistance to some DMI fungicides has been identified in Septoria leaf blotch which may seriously affect performance of some products. For further advice contact a specialist advisor and visit the Fungicide Resistance Action Group (FRAG)-UK website

Restrictions

- Maximum total dose equivalent to 1-3 full dose treatments depending on crop treated. See labels for details

Crop-specific information

- HI 6 wk for field beans; 14 d for sugar beet, red beet, leeks
- Latest use: completion of flowering for rye, wheat; emergence of ear complete (GS 59) for barley, oats; before lowest pods more than 2 cm long (GS 5,1) for winter oilseed rape
- Application to winter wheat in spring between the start of stem elongation and the third node detectable stage (GS 30-33) may cause straw shortening, but does not cause loss of yield

Environmental safety

- Dangerous for the environment
- Toxic to aquatic organisms
- Dangerous to fish or other aquatic life. Do not contaminate surface waters or ditches with chemical or used container

Hazard classification and safety precautions

Hazard Harmful, Dangerous for the environment
Transport code 9
Packaging group III
UN number 3082
Risk phrases R36, R37, R43, R51, R53a, R63
Operator protection A, C, H; U05a, U08, U20a

Environmental protection E15a, E38
Storage and disposal D01, D02, D09a, D10a, D12a
Medical advice M03

135 cyproconazole + cyprodinil

A broad spectrum fungicide mixture for cereals
FRAC mode of action code: 3 + 9

See also cyprodinil

Products

Radius	Syngenta	5.33:40 % w/w	WG	09387

Uses

- Brown rust in ***spring barley***, ***winter barley***, ***winter wheat***
- Eyespot in ***spring barley***, ***winter barley***, ***winter wheat***
- Foliar disease control in ***durum wheat*** *(off-label)*, ***grass seed crops*** *(off-label)*, ***spring rye*** *(off-label)*, ***triticale*** *(off-label)*, ***winter rye*** *(off-label)*
- Glume blotch in ***winter wheat***
- Net blotch in ***spring barley***, ***winter barley***
- Powdery mildew in ***spring barley***, ***winter barley***, ***winter wheat***
- Rhynchosporium in ***spring barley***, ***winter barley***
- Septoria leaf blotch in ***winter wheat***
- Yellow rust in ***spring barley***, ***winter barley***, ***winter wheat***

Specific Off-Label Approvals (SOLAs)

- ***durum wheat*** *20060613*
- ***grass seed crops*** *20060613*
- ***spring rye*** *20060613*
- ***triticale*** *20060613*
- ***winter rye*** *20060613*

Approval information

- Cyprodinil included in Annex I under EC Directive 91/414
- Accepted by BBPA for use on malting barley

Efficacy guidance

- Best results obtained from treatment at early stages of development. Later treatments may be required if disease pressure remains high
- Eyespot is fully controlled by treatment in spring during stem extension. Control may be reduced when very dry conditions follow application
- Cyproconazole is a DMI fungicide. Resistance to some DMI fungicides has been identified in Septoria leaf blotch which may seriously affect performance of some products. For further advice contact a specialist advisor and visit the Fungicide Resistance Action Group (FRAG)-UK website

Restrictions

- Maximum total dose equivalent to two full dose treatments

Crop-specific information

- Latest use: before first spikelet of inflorescence visible (GS 51) for barley; before caryopsis watery ripe (GS 71) for winter wheat

Environmental safety

- Dangerous for the environment
- Very toxic to aquatic organisms
- Risk to certain non-target insects or other arthropods. For advice on risk management and use in Integrated Pest Management (IPM) see directions for use

Hazard classification and safety precautions

Hazard Harmful, Dangerous for the environment
Transport code 9
Packaging group III
UN number 3077
Risk phrases R38, R43, R50, R53a, R63

SEE SECTION 3 FOR PRODUCTS ALSO REGISTERED

Operator protection A, H; U05a, U20a
Environmental protection E15a, E22b, E38
Storage and disposal D01, D02, D05, D07, D09a, D12a

136 cyproconazole + propiconazole

A broad-spectrum mixture of conazole fungicides for wheat and barley
FRAC mode of action code: 3 + 3

See also *propiconazole*

Products

1	Alto Xtra	Syngenta	160:250 g/l	EC	14437
2	Menara	Syngenta	160:250 g/l	EC	14398

Uses

- Brown rust in ***spring barley, winter barley, winter wheat*** [1, 2]
- Foliar disease control in ***durum wheat*** *(off-label)*, ***spring rye*** *(off-label)*, ***triticale*** *(off-label)*, ***winter rye*** *(off-label)* [2]
- Glume blotch in ***winter wheat*** [1, 2]
- Late ear diseases in ***winter wheat*** [1, 2]
- Net blotch in ***spring barley, winter barley*** [1, 2]
- Powdery mildew in ***spring barley, winter barley*** [1, 2]; ***winter wheat*** [1]
- Rhynchosporium in ***spring barley, winter barley*** [1, 2]
- Septoria leaf blotch in ***winter wheat*** [1, 2]
- Yellow rust in ***spring barley, winter barley*** [1]; ***winter wheat*** [1, 2]

Specific Off-Label Approvals (SOLAs)

- ***durum wheat*** *20090685* [2]
- ***spring rye*** *20090685* [2]
- ***triticale*** *20090685* [2]
- ***winter rye*** *20090685* [2]

Approval information

- Propiconazole included in Annex I under EC Directive 91/414
- Accepted by BBPA for use on malting barley

Efficacy guidance

- Treatment should be made at first sign of disease
- Useful reduction of eyespot when applied in spring during stem extension (GS 30-32)
- Cyproconazole and propiconazole are DMI fungicides. Resistance to some DMI fungicides has been identified in Septoria leaf blotch which may seriously affect performance of some products. For further advice contact a specialist advisor and visit the Fungicide Resistance Action Group (FRAG)-UK website

Restrictions

- Maximum total dose equivalent to one full dose treatment for barley; two full dose treatments for winter wheat

Crop-specific information

- Latest use: before beginning of anthesis (GS 60) for barley; before grain watery ripe (GS 71) for wheat
- Application to winter wheat in spring may cause straw shortening but does not cause loss of yield

Environmental safety

- Dangerous for the environment
- Very toxic to aquatic organisms

Hazard classification and safety precautions

Hazard Harmful, Dangerous for the environment
Transport code 9
Packaging group III
UN number 3082
Risk phrases R50, R53a, R63
Operator protection A, C, H; U05a, U08, U20b

Environmental protection E15a, E38
Storage and disposal D01, D02, D05, D07, D09a, D10c, D12a

137 cyproconazole + quinoxyfen

A contact and systemic fungicide mixture for cereals
FRAC mode of action code: 3 + 13

See also *quinoxyfen*

Products

Excelsior	Dow	80:75 g/l	SC	13378

Uses
- Brown rust in ***winter wheat***
- Eyespot in ***winter wheat*** *(reduction)*
- Foliar disease control in ***grass seed crops*** *(off-label)*
- Glume blotch in ***winter wheat***
- Powdery mildew in ***winter wheat***
- Septoria leaf blotch in ***winter wheat***
- Yellow rust in ***winter wheat***

Specific Off-Label Approvals (SOLAs)
- ***grass seed crops*** *20090887*

Approval information
- Quinoxyfen included in Annex I under Directive 91/414
- Accepted by BBPA for use on malting barley

Efficacy guidance
- Best results are achieved when disease is first detected and seen to be active. A second treatment may be needed under prolonged disease pressure
- Treatment for Septoria control should be made as soon as possible after weather that favours infection to prevent spread to the second and flag leaves
- Useful reduction of eyespot when applied between GS 30-34 to control other diseases
- Systemic activity may be reduced under conditions of severe drought stress

Restrictions
- Maximum number of treatments equivalent to a total of two full doses
- Application to winter wheat in spring may cause straw shortening but does not cause loss of yield

Crop-specific information
- Latest use: before ear emergence for wheat

Environmental safety
- Dangerous for the environment
- Very toxic to aquatic organisms
- LERAP Category B

Hazard classification and safety precautions
Hazard Harmful, Dangerous for the environment
Transport code 9
Packaging group III
UN number 3082
Risk phrases R43, R50, R53a, R63
Operator protection A, C, H; U05a, U20b
Environmental protection E15a, E16a, E16b, E38
Storage and disposal D01, D02, D09a, D10b, D12a

138 cyproconazole + trifloxystrobin

A conazole and strobilurin fungicide mixture for wheat and barley
FRAC mode of action code: 3 + 11

See also *trifloxystrobin*

Products

1	Escolta	Bayer CropScience	160:375 g/l	SC	13923
2	Sphere	Bayer CropScience	80:187.5 g/l	EC	11429

Uses

- Brown rust in ***spring barley***, ***winter barley***, ***winter wheat*** [2]
- Cercospora leaf spot in ***sugar beet*** [1]
- Foliar disease control in ***durum wheat*** *(off-label)*, ***grass seed crops*** *(off-label)*, ***spring rye*** *(off-label)*, ***triticale*** *(off-label)*, ***winter rye*** *(off-label)* [2]
- Glume blotch in ***winter wheat*** [2]
- Net blotch in ***spring barley***, ***winter barley*** [2]
- Powdery mildew in ***spring barley***, ***winter barley***, ***winter wheat*** [2]; ***sugar beet*** [1]
- Ramularia leaf spots in ***sugar beet*** [1]
- Rhynchosporium in ***spring barley***, ***winter barley*** [2]
- Rust in ***sugar beet*** [1]
- Septoria leaf blotch in ***winter wheat*** [2]
- Yellow rust in ***spring barley***, ***winter barley***, ***winter wheat*** [2]

Specific Off-Label Approvals (SOLAs)

- ***durum wheat*** *20060603* [2]
- ***grass seed crops*** *20060603* [2]
- ***spring rye*** *20060603* [2]
- ***triticale*** *20060603* [2]
- ***winter rye*** *20060603* [2]

Approval information

- Trifloxystrobin included in Annex I under EC Directive 91/414
- Accepted by BBPA for use on malting barley

Efficacy guidance

- Best results obtained from treatment at early stages of disease development. Further treatment may be needed if disease attack is prolonged
- Cyproconazole is a DMI fungicide. Resistance to some DMI fungicides has been identified in Septoria leaf blotch which may seriously affect performance of some products. For further advice contact a specialist advisor and visit the Fungicide Resistance Action Group (FRAG)-UK website
- Trifloxystrobin is a member of the QoI cross resistance group. Product should be used preventatively and not relied on for its curative potential
- Use product as part of an Integrated Crop Management strategy incorporating other methods of control, including where appropriate other fungicides with a different mode of action. Do not apply more than two foliar applications of QoI containing products to any cereal crop
- There is a significant risk of widespread resistance occurring in *Septoria tritici* populations in UK. Failure to follow resistance management action may result in reduced levels of disease control
- Strains of barley powdery mildew resistant to QoIs are common in the UK

Restrictions

- Maximum total dose per crop equivalent to two full dose treatments

Crop-specific information

- HI 35 d

Environmental safety

- Dangerous for the environment
- Very toxic to aquatic organisms

Hazard classification and safety precautions

Hazard Harmful, Dangerous for the environment
Transport code 9
Packaging group III

UN number 3082
Risk phrases R36 [2]; R50, R53a, R63 [1, 2]
Operator protection A [1, 2]; C, H [2]; U02a, U20a [1]; U05a, U09a, U19a [1, 2]; U20b [2]
Environmental protection E15a [2]; E15b [1]; E38 [1, 2]
Storage and disposal D01, D02, D09a, D10b, D12a [1, 2]; D05 [1]
Medical advice M03 [1]

139 cyprodinil

An anilinopyrimidine systemic broad spectrum fungicide for cereals
FRAC mode of action code: 9

See also cyproconazole + cyprodinil

Products

1	Kayak	Syngenta	300 g/l	EC	13119
2	Unix	Syngenta	75% w/w	WG	11512

Uses

- Eyespot in ***durum wheat*** *(off-label)*, ***spring rye*** *(off-label)*, ***spring wheat***, ***triticale*** *(off-label)*, ***winter rye*** *(off-label)*, ***winter wheat*** [2]; ***spring barley***, ***winter barley*** [1, 2]
- Foliar disease control in ***durum wheat*** *(off-label)*, ***millet*** *(off-label)*, ***spring rye*** *(off-label)*, ***triticale*** *(off-label)*, ***winter rye*** *(off-label)* [2]
- Net blotch in ***spring barley***, ***winter barley*** [1, 2]
- Powdery mildew in ***spring barley***, ***winter barley*** [1, 2]; ***spring wheat*** *(moderate control)*, ***winter wheat*** *(moderate control)* [2]
- Rhynchosporium in ***spring barley***, ***winter barley*** [1]; ***spring barley*** *(moderate control)*, ***winter barley*** *(moderate control)* [2]

Specific Off-Label Approvals (SOLAs)

- ***durum wheat*** *20060600* [2]
- ***millet*** *20091252* [2]
- ***spring rye*** *20060600* [2]
- ***triticale*** *20060600* [2]
- ***winter rye*** *20060600* [2]

Approval information

- Cyprodinil included in Annex I under EC Directive 91/414
- Accepted by BBPA for use on malting barley

Efficacy guidance

- Best results obtained from treatment at early stages of disease development
- For best control of eyespot spray before or during the period of stem extension in spring. Control may be reduced if very dry conditions follow treatment

Restrictions

- Maximum total dose equivalent to one and two thirds full dose on wheat and twice full dose on barley [2]
- Maximum number of treatments 2 per crop on wheat and barley [1]

Crop-specific information

- Latest use: up to and including first awns visible (GS 49) for spring barley, winter barley [1, 2]; up to and including grain watery ripe (GS 71) for spring wheat, winter wheat [2]

Environmental safety

- Dangerous for the environment
- Very toxic to aquatic organisms

Hazard classification and safety precautions

Hazard Irritant [1]; Dangerous for the environment [1, 2]
Transport code 9
Packaging group III
UN number 3077 [2]; 3082 [1]
Risk phrases R38, R43 [1]; R50, R53a [1, 2]
Operator protection A, H; U05a, U20a [1, 2]; U09a [1]

Environmental protection E15a [2]; E15b, E34 [1]; E38 [1, 2]
Storage and disposal D01, D02, D05, D07, D09a, D12a [1, 2]; D10c [1]

140 cyprodinil + fludioxonil

A broad-spectrum fungicide mixture for legumes and strawberries
FRAC mode of action code: 9 + 12

See also *fludioxonil*

Products

Switch	Syngenta	37.5:25 % w/w	WG	13185

Uses

- Alternaria in ***apples***, ***beetroot*** *(off-label)*, ***carrots*** *(off-label)*, ***crab apples***, ***ginger*** *(off-label)*, ***ginseng root*** *(off-label)*, ***horseradish*** *(off-label)*, ***liquorice*** *(off-label)*, ***parsley root*** *(off-label)*, ***parsnips*** *(off-label)*, ***pears***, ***quinces***, ***salsify*** *(off-label)*, ***tumeric*** *(off-label)*, ***valerian root*** *(off-label)*
- Ascochyta in ***mange-tout peas***, ***sugar snap peas***, ***vining peas***
- Black spot in ***protected strawberries*** *(qualified minor use)*, ***strawberries*** *(qualified minor use)*
- Blossom wilt in ***apricots*** *(off-label)*, ***cherries*** *(off-label)*, ***peaches*** *(off-label)*, ***plums*** *(off-label)*
- Botrytis in ***asparagus*** *(off-label)*, ***broad beans*** *(moderate control)*, ***cress*** *(off-label)*, ***forest nurseries***, ***frise*** *(off-label)*, ***grapevines*** *(off-label)*, ***green beans*** *(moderate control)*, ***lamb's lettuce*** *(off-label)*, ***lettuce*** *(off-label)*, ***lupins*** *(off-label)*, ***mange-tout peas*** *(moderate control)*, ***ornamental plant production***, ***protected aubergines*** *(off-label)*, ***protected cress*** *(off-label)*, ***protected cucumbers*** *(off-label)*, ***protected forest nurseries***, ***protected frise*** *(off-label)*, ***protected lamb's lettuce*** *(off-label)*, ***protected lettuce*** *(off-label)*, ***protected ornamentals***, ***protected peppers*** *(off-label)*, ***protected radicchio*** *(off-label)*, ***protected rocket*** *(off-label)*, ***protected scarole*** *(off-label)*, ***protected strawberries***, ***protected tomatoes*** *(off-label)*, ***radicchio*** *(off-label)*, ***rocket*** *(off-label)*, ***runner beans*** *(moderate control)*, ***salad onions*** *(off-label)*, ***scarole*** *(off-label)*, ***strawberries***, ***sugar snap peas*** *(moderate control)*, ***vining peas*** *(moderate control)*
- Botrytis fruit rot in ***apples***, ***blackberries***, ***crab apples***, ***pears***, ***quinces***, ***raspberries***
- Botrytis root rot in ***bilberries*** *(qualified minor use recommendation)*, ***blackcurrants*** *(qualified minor use recommendation)*, ***blueberries*** *(qualified minor use recommendation)*, ***cranberries*** *(qualified minor use recommendation)*, ***gooseberries*** *(qualified minor use recommendation)*, ***redcurrants*** *(qualified minor use recommendation)*, ***whitecurrants*** *(qualified minor use recommendation)*
- Fusarium in ***protected cucumbers*** *(off-label)*
- Fusarium diseases in ***apples***, ***crab apples***, ***pears***, ***quinces***
- Gloeosporium rot in ***apples***, ***crab apples***, ***pears***, ***quinces***
- Mycosphaerella in ***mange-tout peas***, ***protected cucumbers*** *(off-label)*, ***sugar snap peas***, ***vining peas***
- Penicillium rot in ***apples***, ***crab apples***, ***pears***, ***quinces***
- Sclerotinia in ***beetroot*** *(off-label)*, ***broad beans***, ***carrots*** *(off-label)*, ***ginger*** *(off-label)*, ***ginseng root*** *(off-label)*, ***green beans***, ***horseradish*** *(off-label)*, ***liquorice*** *(off-label)*, ***mange-tout peas***, ***parsley root*** *(off-label)*, ***parsnips*** *(off-label)*, ***runner beans***, ***salsify*** *(off-label)*, ***sugar snap peas***, ***tumeric*** *(off-label)*, ***valerian root*** *(off-label)*, ***vining peas***
- Stemphylium in ***asparagus*** *(off-label)*

Specific Off-Label Approvals (SOLAs)

- ***apricots*** *20070735*
- ***asparagus*** *20070235*
- ***beetroot*** *20092649*
- ***carrots*** *20092649*
- ***cherries*** *20070735*
- ***cress*** *20081931*
- ***frise*** *20081931*
- ***ginger*** *20092649*
- ***ginseng root*** *20092649*
- ***grapevines*** *20070236*
- ***horseradish*** *20092649*

- ***lamb's lettuce*** *20081931*
- ***lettuce*** *20081931*
- ***liquorice*** *20092649*
- ***lupins*** *20070237*
- ***parsley root*** *20092649*
- ***parsnips*** *20092649*
- ***peaches*** *20070735*
- ***plums*** *20070735*
- ***protected aubergines*** *20070238*
- ***protected cress*** *20081931*
- ***protected cucumbers*** *20070234*
- ***protected frise*** *20081931*
- ***protected lamb's lettuce*** *20081931*
- ***protected lettuce*** *20081931*
- ***protected peppers*** *20070238*
- ***protected radicchio*** *20081931*
- ***protected rocket*** *20081931*
- ***protected scarole*** *20081931*
- ***protected tomatoes*** *20071500*
- ***radicchio*** *20081931*
- ***rocket*** *20081931*
- ***salad onions*** *20070239*
- ***salsify*** *20092649*
- ***scarole*** *20081931*
- ***tumeric*** *20092649*
- ***valerian root*** *20092649*

Approval information

- Cyprodinil and fludioxonil included in Annex I under EC Directive 91/414

Efficacy guidance

- Best results obtained from application at the earliest stage of disease development or as a protective treatment following disease risk assessment
- Subsequent treatments may follow after a minimum of 10 d if disease pressure remains high
- To minimise the likelihood of development of resistance product should be used in a planned Resistance Management strategy. See Section 5 for more information

Restrictions

- Maximum number of treatments 2 per crop
- Consult processor before treating any crop for processing

Crop-specific information

- HI broad beans, green beans, mange-tout peas, runner beans, sugar snap peas, vining peas 14 d; protected strawberries, strawberries 3 d
- First signs of disease infection likely to be seen from early flowering for all crops
- Ensure peas are free from any stress before treatment. If necessary check wax with a crystal violet test

Environmental safety

- Dangerous for the environment
- Very toxic to aquatic organisms
- LERAP Category B

Hazard classification and safety precautions

Hazard Dangerous for the environment
Transport code 9
Packaging group III
UN number 3077
Risk phrases R50, R53a
Operator protection A, H; U05a, U20a
Environmental protection E15b, E16a, E16b, E38
Storage and disposal D01, D02, D05, D07, D09a, D10c, D12a

141 cyprodinil + picoxystrobin

A broad spectrum fungicide mixture for cereals
FRAC mode of action code: 9 + 11

See also *picoxystrobin*

Products

Acanto Prima	DuPont	30.8% w/w	WG	13041

Uses

- Brown rust in ***spring barley***, ***spring wheat***, ***winter barley***, ***winter wheat***
- Eyespot in ***spring barley***, ***spring wheat*** *(moderate control)*, ***winter barley***, ***winter wheat*** *(moderate control)*
- Foliar disease control in ***durum wheat*** *(off-label)*, ***grass seed crops*** *(off-label)*, ***spring rye*** *(off-label)*, ***triticale*** *(off-label)*, ***winter rye*** *(off-label)*
- Glume blotch in ***spring wheat***, ***winter wheat***
- Net blotch in ***spring barley***, ***winter barley***
- Powdery mildew in ***spring barley***, ***winter barley***
- Rhynchosporium in ***spring barley***, ***winter barley***
- Septoria leaf blotch in ***spring wheat***, ***winter wheat***
- Yellow rust in ***spring wheat***, ***winter wheat***

Specific Off-Label Approvals (SOLAs)

- ***durum wheat*** *20073510*
- ***grass seed crops*** *20073510*
- ***spring rye*** *20073510*
- ***triticale*** *20073510*
- ***winter rye*** *20073510*

Approval information

- Cyprodinil and picoxystrobin included in Annex I under EC Directive 91/414
- Accepted by BBPA for use on malting barley

Efficacy guidance

- Best results obtained from use as a protectant treatment or in the earliest stages of disease development. Further applications may be needed if disease attack is prolonged
- Picoxystrobin is a member of the QoI cross resistance group. Product should be used preventatively and not relied on for its curative potential
- Use product as part of an Integrated Crop Management strategy incorporating other methods of control, including where appropriate other fungicides with a different mode of action. Do not apply more than two foliar applications of QoI containing products to any cereal crop
- There is a significant risk of widespread resistance occurring in *Septoria tritici* populations in UK. Failure to follow resistance management action may result in reduced levels of disease control
- In wheat product must always be used in mixture with another product, recommended for control of the same target disease, that contains a fungicide from a different cross resistance group and is applied at a dose that will give robust control
- Strains of barley powdery mildew resistant to QoIs are common in the UK

Restrictions

- Maximum number of treatments 2 per crop

Crop-specific information

- Latest use: before early milk stage (GS 73) for wheat; before ear just visible (GS 51) for barley

Environmental safety

- Dangerous for the environment
- Very toxic to aquatic organisms

Hazard classification and safety precautions

Hazard Dangerous for the environment
Transport code 9
Packaging group III
UN number 3077
Risk phrases R50, R53a

Operator protection A, C, H; U05a, U09a, U11, U14, U15, U20a
Environmental protection E15a, E34, E38
Storage and disposal D01, D02, D09a, D10c, D12a

142 2,4-D

A translocated phenoxycarboxylic acid herbicide for cereals, grass and amenity use
HRAC mode of action code: O

See also clopyralid + 2,4-D + MCPA

Products

1	Depitox	Nufarm UK	500 g/l	SL	13258
2	Headland Staff 500	Headland	500 g/l	SL	13196
3	Herboxone	Headland	500 g/l	SL	13958
4	HY-D Super	Agrichem	500 g/l	SL	13198

Uses

- Annual dicotyledons in ***amenity grassland, managed amenity turf, permanent grassland, spring barley, spring wheat, winter barley, winter oats, winter rye, winter wheat*** [1-4]; ***apple orchards, farm forestry*** *(off-label)*, ***forest nurseries*** *(off-label)*, ***game cover*** *(off-label)*, ***ornamental plant production*** *(off-label)*, ***pear orchards, spring rye*** [1]; ***miscanthus*** *(off-label)*, ***undersown barley, undersown oats, undersown rye, undersown wheat*** [4]
- Aquatic weeds in ***enclosed waters, land immediately adjacent to aquatic areas, open waters*** [1]
- Perennial dicotyledons in ***amenity grassland, managed amenity turf, permanent grassland, spring barley, spring wheat, winter barley, winter oats, winter rye, winter wheat*** [1-4]; ***apple orchards, pear orchards, spring rye*** [1]; ***undersown barley, undersown oats, undersown rye, undersown wheat*** [4]

Specific Off-Label Approvals (SOLAs)

- ***farm forestry*** *20082843* [1]
- ***forest nurseries*** *20082843* [1]
- ***game cover*** *20082843* [1]
- ***miscanthus*** *20082914* [4]
- ***ornamental plant production*** *20082843* [1]

Approval information

- 2,4-D included in Annex I under EC Directive 91/414
- Approved for aquatic weed control [1]. See Section 5 for information on use of herbicides in or near water
- Accepted by BBPA for use on malting barley

Efficacy guidance

- Best results achieved by spraying weeds in seedling to young plant stage when growing actively in a strongly competing crop
- Most effective stage for spraying perennials varies with species. See label for details
- Spray aquatic weeds when in active growth between May and Sep

Restrictions

- Maximum number of treatments normally 1 per crop and in forestry, 1 per yr in grassland and 2 or 3 per yr in amenity turf. Check individual labels
- Do not use on newly sown leys containing clover
- Do not spray grass seed crops after ear emergence
- Do not spray within 6 mth of laying turf or sowing fine grass
- Do not dump surplus herbicide in water or ditch bottoms
- Do not plant conifers until at least 1 mth after treatment
- Do not spray crops stressed by cold weather or drought or if frost expected
- Do not roll or harrow within 7 d before or after spraying
- Do not spray if rain falling or imminent
- Do not mow or roll turf or amenity grass 4 d before or after spraying. The first 4 mowings after treatment must be composted for at least 6 mth before use
- Do not mow grassland or graze for at least 10 d after spraying

Crop-specific information

- Latest use: before 1st node detectable in cereals; end Aug for conifer plantations; before established grassland 25 cm high
- Spray winter cereals in spring when leaf-sheath erect but before first node detectable (GS 31), spring cereals from 5-leaf stage to before first node detectable (GS 15-31)
- Cereals undersown with grass and/or clover, but not with lucerne, may be treated
- Selective treatment of resistant conifers can be made in Aug when growth ceased and plants hardened off, spray must be directed if applied earlier. See label for details

Following crops guidance

- Do not use shortly before or after sowing any crop
- Do not plant succeeding crops within 3 mth [2]
- Do not direct drill brassicas or grass/clover mixtures within 3 wk of application

Environmental safety

- Dangerous for the environment
- Very toxic to aquatic organisms [2]
- Dangerous to aquatic higher plants. Do not contaminate surface waters or ditches with chemical or used container
- 2,4-D is active at low concentrations. Take extreme care to avoid drift onto neighbouring crops, especially beet crops, brassicas, most market garden crops including lettuce and tomatoes under glass, pears and vines
- May be used to control aquatic weeds in presence of fish if used in strict accordance with directions for waterweed control and precautions needed for aquatic use [1]
- Keep livestock out of treated areas for at least 2 wk following treatment and until poisonous weeds, such as ragwort, have died down and become unpalatable
- Water containing the herbicide must not be used for irrigation purposes within 3 wk of treatment or until the concentration in water is below 0.05 ppm

Hazard classification and safety precautions

Hazard Harmful, Dangerous for the environment [2-4]; Irritant [1]
Transport code 9
Packaging group III
UN number 3082
Risk phrases R22a [2-4]; R38, R52 [1]; R41, R53a [1-4]; R43, R51 [4], R50 [2, 3]
Operator protection A, C, H [1-4]; M [1]; U05a, U08, U11, U20b [1-4]; U14, U15 [4]
Environmental protection E07a, E34 [1-4]; E15a [1]; E15b, E38 [2-4]
Storage and disposal D01, D02, D05, D09a [1-4]; D10a, D12a [2-4]; D10b [1]
Medical advice M03 [1-4]; M05a [4]

143 2,4-D + dicamba

A translocated herbicide for use on turf
HRAC mode of action code: O + O

See also *dicamba*

Products

1	Magneto	Nufarm UK	344:120 g/l	SL	12339
2	New Estermone	Vitax	200:35 g/l	EC	13792
3	Thrust	Nufarm UK	344:120 g/l	SL	12230

Uses

- Annual and perennial weeds in ***managed amenity turf*** [2]
- Annual dicotyledons in ***amenity grassland***, ***permanent grassland*** [1, 3]
- Perennial dicotyledons in ***amenity grassland***, ***permanent grassland*** [1, 3]

Approval information

- 2,4-D and dicamba included in Annex I under EC Directive 91/414

Efficacy guidance

- Best results achieved by application when weeds growing actively in spring or early summer (later with irrigation and feeding)
- More resistant weeds may need repeat treatment after 3 wk

- Improved control of some weeds can be obtained by use of specifed oil adjuvant [1, 3]
- Do not use during drought conditions

Restrictions

- Maximum number of treatments 2 per yr on amenity grass and established grassland [1]
- Do not treat newly sown or turfed areas or grass less than 1 yr old
- Do not mow less than 3 d before treatment and at least 3-4 d afterwards [13076]
- Do not treat grass crops intended for seed production [1, 3]
- Do not treat grass suffering from drought, disease or other adverse factors [1, 3]
- Do not roll or harrow for 7 d before or after treatment [1, 3]
- Do not apply when grassland is flowering [1, 3]
- Do not re-seed for 6-8 wk after application [13076]
- Avoid spray drift onto cultivated crops or ornamentals
- Do not graze grass for at least seven days after spraying [3]
- Do not mow or roll four days before or after application. The first four mowings after treatment must be composted for at least six months before use [3]

Crop-specific information

- Latest use: before grass 25 cm high for amenity grass, established grassland [1, 3]
- The first four mowings after treatment must be composted for at least 6 mth before use

Environmental safety

- Dangerous for the environment [1, 3]
- Toxic to aquatic organisms
- Keep livestock out of treated areas for at least two weeks following treatment and until poisonous weeds, such as ragwort, have died down and become unpalatable [1, 3]

Hazard classification and safety precautions

Hazard Harmful, Dangerous for the environment
UN number N/C [2]
Risk phrases R22a, R43, R51, R53a [1-3]; R22b, R36, R66 [2]; R41 [1, 3]
Operator protection A, C, H, M; U05a [1, 3]; U08, U14, U15, U20b [2]; U11 [1-3]
Environmental protection E07a, E15a, E38 [1-3]; E34 [1, 3]
Storage and disposal D01, D02 [1, 3]; D05, D09a, D10a, D12a [1-3]
Medical advice M03 [1, 3]; M05b [2]

144 2,4-D + dicamba + dichlorprop-P

A translocated hormone mixture for the control of docks in grassland
HRAC mode of action code: O + O + O

See also *dicamba*
dichlorprop-P

Products

Legolas	Nufarm UK	300:120:400 g/l	SL	14439

Uses

- Creeping thistle in ***amenity grassland*, *grassland***
- Dandelions in ***amenity grassland*, *grassland***
- Docks in ***amenity grassland*, *grassland***
- Sowthistle in ***amenity grassland*, *grassland***

Approval information

- 2,4-D, dicamba and dichlorprop-P included in Annex I under EC Directive 91/414

Efficacy guidance

- Make applications when weed growth is vigorous due to warm, moist conditions

Restrictions

- Treatment will severely damage clover. Do not use where clover is an important component of the sward
- DO NOT mow or roll four days before or after application. The first four mowings after treatment must be composted for at least six months before use.

- DO NOT treat newly established grass less than one year old.
- DO NOT treat grass crops intended for seed production

Crop-specific information

- One application per year on established agricultural grassland, two applications per year on amenity grassland

Following crops guidance

- Do not re-seed within 3 months of treatment
- Avoid drift on to suceptible crops such as tomatoes, vegetables, sugar beet, glasshouse crops, fruit and ornamentals

Environmental safety

- Very toxic to aquatic organisms
- Keep livestock out of treated areas for at least 2 weeks following treatment and until poisonous weeds such as ragwort have died and become unpaletable

Hazard classification and safety precautions

Hazard Harmful, Dangerous for the environment
Transport code 9
Packaging group III
UN number 3082
Risk phrases R20, R21, R22a, R41, R43, R50, R53a
Operator protection A, C, H; U05a, U11, U14
Environmental protection E07a, E15a, E34, E38
Storage and disposal D01, D02, D05, D09a, D10a, D12a
Medical advice M03

145 2,4-D + dicamba + fluroxypyr

A translocated and contact herbicide mixture for amenity turf
HRAC mode of action code: O + O + O

See also *dicamba*
fluroxypyr

Products

1	Barclay Holster XL	Barclay	285:52.5:105 g/l	EC	13596
2	Mascot Crossbar	Rigby Taylor	285:52.5:105 g/l	EC	14453
3	Swiftsure	Sherriff Amenity	285:52.5:105 g/l	EC	14468

Uses

- Annual dicotyledons in ***managed amenity turf*** [1-3]
- Buttercups in ***managed amenity turf*** [1]
- Clover in ***managed amenity turf*** [1]
- Daisies in ***managed amenity turf*** [1]
- Dandelions in ***managed amenity turf*** [1]
- Yarrow in ***managed amenity turf*** [1]

Approval information

- 2,4-D, dicamba and fluroxypyr included in Annex I under EC Directive 91/414

Efficacy guidance

- Apply when weeds actively growing (normally between Apr and Sep) and when soil is moist
- Best results obtained from treatment in spring or early summer before weeds begin to flower
- Do not apply if turf is wet or if rainfall expected within 4 h of treatment. Both circumstances will reduce weed control

Restrictions

- Maximum number of treatments 2 per yr
- Do not mow for 3 d before or after treatment
- Avoid overlapping or overdosing, especially on newly sown turf
- Do not spray in drought conditions or if turf under stress from frost, waterlogging, trace element deficiency, pest or disease attack

Crop-specific information

- Latest use: normally Sep for managed amenity turf
- New turf may be treated in spring provided at least 2 mth have elapsed since sowing

Environmental safety

- Dangerous for the environment
- Toxic to aquatic organisms
- Keep livestock out of treated areas for at least two weeks following treatment and until poisonous weeds, such as ragwort, have died down and become unpalatable

Hazard classification and safety precautions

Hazard Harmful
Transport code 9
Packaging group III
UN number 3082
Risk phrases R22a, R22b, R36, R38
Operator protection A, C, H, M; U05a, U08, U19a, U20b
Environmental protection E07a, E15a, E34
Storage and disposal D01, D02, D09a, D10a
Medical advice M03

146 2,4-D + dicamba + triclopyr

A translocated herbicide for perennial and woody weed control
HRAC mode of action code: O + O + O

See also *dicamba*
triclopyr

Products

1	Broadsword	United Phosphorus	200:85:65 g/l	EC	09140
2	Nu-Shot	Nufarm UK	200:85:65 g/l	EC	13550

Uses

- Annual dicotyledons in ***amenity grassland*** [2]; ***land not intended to bear vegetation*** [1]; ***permanent grassland, rotational grass*** [1, 2]
- Brambles in ***farm forestry, forest, natural surfaces not intended to bear vegetation*** [1]
- Docks in ***amenity grassland*** [2]; ***land not intended to bear vegetation*** [1]; ***permanent grassland, rotational grass*** [1, 2]
- Gorse in ***farm forestry, forest, natural surfaces not intended to bear vegetation*** [1]
- Green cover in ***land temporarily removed from production*** [1]
- Japanese knotweed in ***farm forestry, forest, natural surfaces not intended to bear vegetation*** [1]
- Perennial dicotyledons in ***amenity grassland*** [2]; ***farm forestry, forest, land not intended to bear vegetation, natural surfaces not intended to bear vegetation*** [1]; ***permanent grassland, rotational grass*** [1, 2]
- Plantains in ***amenity grassland*** [2]; ***land not intended to bear vegetation*** [1]; ***permanent grassland, rotational grass*** [1, 2]
- Rhododendrons in ***farm forestry, forest, natural surfaces not intended to bear vegetation*** [1]
- Stinging nettle in ***amenity grassland*** [2]; ***permanent grassland, rotational grass*** [1, 2]
- Thistles in ***amenity grassland*** [2]; ***land not intended to bear vegetation*** [1]; ***permanent grassland, rotational grass*** [1, 2]
- Woody weeds in ***farm forestry, forest, natural surfaces not intended to bear vegetation*** [1]

Approval information

- 2,4-D, dicamba and triclopyr included in Annex I under EC Directive 91/414

Efficacy guidance

- Apply as foliar spray to herbaceous or woody weeds. Timing and growth stage for best results vary with species. See label for details
- Dilute with water for stump treatment and apply after felling up to the start of regrowth. Treat any regrowth with a spray to the growing foliage
- May be applied at 1/3 dilution in weed wipers or 1/8 dilution with ropewick applicators

Restrictions

- Maximum total dose per yr equivalent to two full dose treatments
- Do not use on pasture established less than 1 yr or on grass grown for seed
- Where clover a valued constituent of sward only use as a spot treatment
- Do not graze for 7 d or mow for 14 d after treatment
- Do not direct drill grass, clover or brassicas for at least 6 wk after grassland treatment
- Do not plant trees for 1-3 mth after spraying depending on dose applied. See label
- Avoid spray drift into greenhouses or onto crops or ornamentals. Vapour drift may occur in hot conditions

Crop-specific information

- Latest use: before weed flower buds open in grassland, non-cropped land, land temprorarily removed from production [1]; end of summer (spray); after felling up to the start of regrowth (cut stump spray) for farm forestry [1]
- Sprays may be applied in pines, spruce and fir providing drift is avoided. Optimum time is mid-autumn when tree growth ceased but weeds not yet senescent [1]

Environmental safety

- Dangerous for the environment
- Toxic (very toxic [1]) to aquatic organisms [2]
- Flammable
- Docks and other weeds may become increasingly palatable after treatment and may be preferentially grazed
- Keep livestock out of treated areas for at least two weeks following treatment and until poisonous weeds, such as ragwort, have died down and become unpalatable
- Take extreme care to avoid drift onto neighbouring crops, especially beet crops, brassicas, most market garden crops including lettuce and tomatoes under glass, pears and vines

Hazard classification and safety precautions

Hazard Harmful, Flammable, Dangerous for the environment
Transport code 3
Packaging group III
UN number 1993 [2]; 3295 [1]
Risk phrases R22a, R22b, R36, R37, R53a [1, 2]; R38, R50, R67 [1]; R43, R51 [2]
Operator protection A, C, H, M [1, 2]; D [2]; U02a, U04a, U13, U20a [1]; U05a, U08, U11, U19a [1, 2]; U14, U20b [2]
Environmental protection E07a, E34 [1, 2]; E13b [1]; E15a, E38 [2]
Consumer protection C01 [1]
Storage and disposal D01, D02, D05, D09a, D10b, D12a
Medical advice M03, M05b

147 2,4-D + dichlorprop-P + MCPA + mecoprop-P

A translocated herbicide for use in apple and pear orchards
HRAC mode of action code: O + O + O + O

See also *dichlorprop-P*
mecoprop-P

Products

UPL Camppex	United Phosphorus	34.5:66.6:52.8:82.1 g/l	SL	11661

Uses

- Annual dicotyledons in ***almonds*** *(off-label)*, ***apple orchards***, ***chestnuts*** *(off-label)*, ***hazel nuts*** *(off-label)*, ***pear orchards***, ***walnuts*** *(off-label)*
- Chickweed in ***apple orchards***, ***pear orchards***
- Cleavers in ***apple orchards***, ***pear orchards***
- Perennial dicotyledons in ***apple orchards***, ***pear orchards***

Specific Off-Label Approvals (SOLAs)

- ***almonds*** *20052244*
- ***chestnuts*** *20052244*

- ***hazel nuts*** *20052244*
- ***walnuts*** *20052244*

Approval information

- 2,4-D, dichlorprop-P, MCPA and mecoprop-P included in Annex I under EC Directive 91/414

Efficacy guidance

- Use on emerged weeds in established apple and pear orchards (from 1 yr after planting) as directed application
- Spray when weeds in active growth and at growth stage recommended on label
- Effectiveness may be reduced by rain within 12 h

Restrictions

- Maximum number of treatments 2 per yr
- Applications must be made around, and not directly to, trees. Do not allow drift onto trees
- Do not spray during blossom period. Do not spray to run-off
- Do not roll, harrow or cut grass crops on orchard floor within at least 3 d before or after spraying

Environmental safety

- Harmful to aquatic organisms
- Harmful to fish or other aquatic life. Do not contaminate surface waters or ditches with chemical or used container
- Keep livestock out of treated areas for at least two weeks following treatment and until poisonous weeds, such as ragwort, have died down and become unpalatable
- Take extreme care to avoid drift onto neighbouring sensitive crops
- Some pesticides pose a greater threat of contamination of water than others and mecoprop-p is one of these pesticides. Take special care when applying mecoprop-p near water and do not apply if heavy rain is forecast

Hazard classification and safety precautions

Hazard Harmful
UN number N/C
Risk phrases R20, R21, R22a, R43, R52, R53a
Operator protection A, C, H, M; U05a, U08, U20b
Environmental protection E07a, E13c, E34
Storage and disposal D01, D02, D05, D09a, D10c
Medical advice M03, M05a

148 2,4-D + florasulam

A post-emergence herbicide mixture for control of broad leaved weeds in managed amenity turf
HRAC mode of action code: O + B

See also florasulam

Products

Junction	Rigby Taylor	452:6.25 g/l	ME	12493

Uses

- Clover in ***managed amenity turf***
- Daisies in ***managed amenity turf***
- Dandelions in ***managed amenity turf***
- Plantains in ***managed amenity turf***
- Sticky mouse-ear in ***managed amenity turf***

Approval information

- 2,4-D and florasulam included in Annex I under EC Directive 91/414

Efficacy guidance

- Best results obtained from treatment of actively growing weeds between Mar and Oct when soil is moist
- Do not apply when rain is imminent or during periods of drought unless irrigation is applied

Restrictions

- Maximum number of treatments on managed amenity turf: 1 per yr

Crop-specific information

- Avoid mowing turf 3 d before and after spraying
- Ensure newly sown turf has established before treating. Turf sown in late summer or autumn should not be sprayed until growth is resumed in the following spring

Following crops guidance

- An interval of 4 wk must elapse between application and re-seeding turf

Environmental safety

- Dangerous for the environment
- Toxic to aquatic organisms

Hazard classification and safety precautions

Hazard Harmful, Dangerous for the environment
Transport code 9
Packaging group III
UN number 3082
Risk phrases R22a, R43, R51, R53a
Operator protection A, H; U02a, U05a, U08, U14, U20a
Environmental protection E15a, E34, E38
Storage and disposal D01, D02, D09a, D10b, D12a

149 2,4-D + MCPA

A translocated herbicide mixture for cereals and grass
HRAC mode of action code: O + O

See also *MCPA*

Products

1	Headland Polo	Headland	360:315 g/l	SL	10283
2	Lupo	Nufarm UK	345:345 g/l	SL	12725

Uses

- Annual dicotyledons in ***durum wheat*** *(off-label)*, ***spring barley***, ***spring rye*** *(off-label)*, ***spring wheat***, ***triticale*** *(off-label)*, ***winter barley***, ***winter oats***, ***winter rye*** *(off-label)*, ***winter wheat*** [1]; ***permanent grassland***, ***rotational grass*** [1, 2]
- Perennial dicotyledons in ***permanent grassland***, ***rotational grass*** [1, 2]; ***spring barley***, ***spring wheat***, ***winter barley***, ***winter oats***, ***winter wheat*** [1]

Specific Off-Label Approvals (SOLAs)

- ***durum wheat*** *20052243* [1]
- ***spring rye*** *20052243* [1]
- ***triticale*** *20052243* [1]
- ***winter rye*** *20052243* [1]

Approval information

- 2,4-D and MCPA included in Annex I under EC Directive 91/414
- Accepted by BBPA for use on malting barley

Efficacy guidance

- Best results achieved by spraying weeds in seedling to young plant stage when growing actively in a strongly competing crop
- Most effective stage for spraying perennials varies with species. See label for details

Restrictions

- Maximum number of treatments 1 per crop in cereals, permanent grassland; 2 per yr in rotational grassland [1]; 1 per crop in permanent and rotational grassland [2]
- Do not spray if rain falling or imminent
- Do not cut grass or graze for at least 10 d after spraying
- Do not use on newly sown leys containing clover or other legumes
- Do not spray crops stressed by cold weather or drought or if frost expected
- Do not use shortly before or after sowing any crop
- Do not roll or harrow within 7 d before or after spraying

Crop-specific information

- Spray winter cereals in spring when leaf-sheath erect but before first node detectable (GS 31), spring cereals from 5-leaf stage to before first node detectable (GS 15-31)
- Latest use: before first node detectable (GS 31) in cereals [1]; before crop is 25 cm high in grassland [2]

Environmental safety

- Dangerous for the environment
- Toxic to aquatic organisms
- Keep livestock out of treated areas for at least two weeks following treatment and until poisonous weeds, such as ragwort, have died down and become unpalatable
- 2,4-D and MCPA are active at low concentrations. Take extreme care to avoid drift onto neighbouring crops, especially beet crops, brassicas, most market garden crops including lettuce and tomatoes under glass, pears and vines

Hazard classification and safety precautions

Hazard Harmful, Dangerous for the environment
Transport code 9
Packaging group III
UN number 3082
Risk phrases R20, R21, R22a, R41, R43, R53a [1, 2]; R50 [2]; R51 [1]
Operator protection A, C, H, M; U05a, U08, U11, U14, U20b [1, 2]; U15 [1]
Environmental protection E07a, E15a, E34 [1, 2]; E38 [2]
Storage and disposal D01, D02, D05, D09a, D10a [1, 2]; D12a [2]
Medical advice M03 [1, 2]; M05a [1]

150 2,4-D + picloram

A persistent translocated herbicide for non-crop land; product approval expired 2008
HRAC mode of action code: O + O

See also picloram

151 daminozide

A hydrazide plant growth regulator for use in certain ornamentals

Products

1	B-Nine SG	Certis	85% w/w	SG	12734
2	B-Nine SG	Certis	85% w/w	SG	14435
3	Dazide Enhance	Fine	85% w/w	SG	11943
4	Dazide Enhance	Fine	85% w/w	SG	14433

Uses

- Growth regulation in ***azaleas***, ***bedding plants***, ***calibrachoa***, ***chrysanthemums***, ***hortensia***, ***kalanchoes***, ***ornamental specimens***, ***petunia***, ***sunflowers*** [4]; ***protected ornamentals*** [2]
- Internode reduction in ***azaleas***, ***bedding plants***, ***chrysanthemums***, ***hydrangeas*** [1, 3]

Approval information

- Daminozide included in Annex I under EC Directive 91/414

Efficacy guidance

- Best results obtained by application in late afternoon when glasshouse has cooled down
- Apply a fine spray using compressed air or power sprayers to give good coverage of dry foliage without run off
- Response to treatment differs widely depending on variety, stage of growth and physiological condition of the plant. It is recommended that any new variety is first tested on a small scale to observe if adverse effects occur
- For plants grown in modules or for seedlings prior to pricking out apply at reduced dose on a 7 d regime, starting at cotyledon stage

Restrictions

- Maximum number of treatments 2 per crop
- Apply only to turgid, well watered plants. Do not water for 24 h after spraying

SEE SECTION 3 FOR PRODUCTS ALSO REGISTERED

- Do not mix with other spray chemicals unless specifically recommended
- Do not store product in metal containers

Crop-specific information

- Do not use on chrysanthemum variety Fandango
- Evidence of effectiveness on Azaleas and Hydrangeas is limited
- See label for guidance on a range of bedding plant species

Hazard classification and safety precautions

UN number N/C
Operator protection A [1-4]; H [2-4]; U02a, U20c [4]; U05a [1-4]; U08, U19a [3, 4]; U20b [3]
Environmental protection E02a [2]; E15a [1, 3]; E15b [2, 4]
Storage and disposal D01, D02, D09a [1-4]; D08 [3]; D10b, D12a [3, 4]
Medical advice M05a [3, 4]

152 dazomet

A methyl isothiocyanate releasing soil fumigant
HRAC mode of action code: Z

Products

Basamid	Certis	97% w/w	GR	11324

Uses

- Nematodes in ***field crops*** *(soil fumigation)*, ***protected crops*** *(soil fumigation)*, ***vegetables*** *(soil fumigation)*
- Soil insects in ***field crops*** *(soil fumigation)*, ***protected crops*** *(soil fumigation)*, ***vegetables*** *(soil fumigation)*
- Soil-borne diseases in ***field crops*** *(soil fumigation)*, ***protected crops*** *(soil fumigation)*, ***vegetables*** *(soil fumigation)*
- Weed seeds in ***field crops*** *(soil fumigation)*, ***protected crops*** *(soil fumigation)*, ***vegetables*** *(soil fumigation)*

Efficacy guidance

- Dazomet acts by releasing methyl isothiocyanate in contact with moist soil
- Soil sterilization is carried out after harvesting one crop and before planting the next
- The soil must be of fine tilth, free of clods and evenly moist to the depth of sterilization
- Soil moisture must not be less than 50% of water-holding capacity or oversaturated. If too dry, water at least 7-14 d before treatment
- In order to obtain short treatment times it is recommended to treat soils when soil temperature is above 7 °C. Treatment should be used outdoors before winter rains make soil too wet to cultivate - usually early Nov
- For club root control treat only in summer when soil temperature above 10 °C
- Where onion white rot is a problem it is unlikely to give effective control where inoculum level high or crop under stress
- Apply granules with suitable applicators, mix into soil immediately to desired depth and seal surface with polythene sheeting, by flooding or heavy rolling. See label for suitable application and incorporation machinery
- With 'planting through' technique polythene seal is left in place to form mulch into which new crop can be planted

Restrictions

- Maximum number of treatments 1 per batch of soil for protected crops or field crop situation
- Do not treat ground where water table may rise into treated layer

Crop-specific information

- Latest use: pre-planting

Following crops guidance

- With 'planting through' technique no gas release cultivations are made and safety test with cress is particularly important. Conduct cress test on soil samples from centre as well as edges of bed. Observe minimum of 30 d from application to cress test in soils at 10 °C or above, at least 50 d in soils below 10 °C. See label for details

- In all other situations 14-28 d after treatment cultivate lightly to allow gas to disperse and conduct cress test after a further 14-28 d (timing depends on soil type and temperature). Do not treat structures containing live plants or any ground within 1 m of live plants

Environmental safety

- Dangerous for the environment
- Very toxic to aquatic organisms

Hazard classification and safety precautions

Hazard Harmful, Dangerous for the environment
Transport code 9
Packaging group III
UN number 3077
Risk phrases R22a, R50, R53a
Operator protection A, M; U04a, U05a, U09a, U19a, U20a
Environmental protection E15a, E34, E38
Storage and disposal D01, D02, D07, D09a, D11a, D12a
Medical advice M03, M05a

153 2,4-DB

A translocated phenoxycarboxylic acid herbicide available only in mixtures
HRAC mode of action code: O

154 2,4-DB + linuron + MCPA

A translocated herbicide for undersown cereals and grass
HRAC mode of action code: O + C2 + O

See also *linuron*
MCPA

Products

Alistell	United Phosphorus	220:30:30 g/l	EC	11053

Uses

- Annual dicotyledons in ***rotational grass***, ***undersown spring cereals*** *(undersown with clover)*, ***undersown winter cereals*** *(undersown with clover)*

Approval information

- 2,4-DB, linuron and MCPA included in Annex I under EC Directive 91/414
- Accepted by BBPA for use on malting barley

Efficacy guidance

- Best results achieved on young seedling weeds growing actively in warm, moist weather
- May be applied at any time of year provided crop at correct stage and weather suitable
- Avoid spraying if rain falling or imminent

Restrictions

- Maximum number of treatments 1 per crop or yr
- Spray winter cereals when fully tillered but before first node detectable (GS 29-30)
- Spray spring wheat from 5-fully expanded leaf stage (GS 15), barley and oats from 2-fully expanded leaves (GS 12)
- Do not spray cereals undersown with lucerne, peas or beans
- Apply to clovers after 1-trifoliate leaf, to grasses after 2-fully expanded leaf stage
- Do not spray in conditions of drought, waterlogging or extremes of temperature
- In frosty weather clover leaf scorch may occur but damage normally outgrown
- Do not use on sand or soils with more than 10% organic matter
- Do not roll or harrow within 7 d before or after spraying
- Avoid drift of spray or vapour onto susceptible crops
- Do not apply by hand-held sprayers

Crop-specific information

- Latest use: before first node detectable (GS 31) for undersown cereals; 2 wk before grazing for grassland

Environmental safety

- Dangerous for the environment
- Toxic to aquatic organisms
- Keep livestock out of treated areas for at least two weeks following treatment and until poisonous weeds, such as ragwort, have died down and become unpalatable
- LERAP Category B

Hazard classification and safety precautions

Hazard Harmful, Dangerous for the environment
Transport code 9
Packaging group III
UN number 3082
Risk phrases R22a, R22b, R40, R41, R51, R53a, R66
Operator protection A, C, H; U05a, U08, U11, U13, U19a, U20b
Environmental protection E07a, E13b, E16a, E34, E38
Storage and disposal D01, D02, D09a, D10c, D12a
Medical advice M03, M05a

155 2,4-DB + MCPA

A translocated herbicide for cereals, clovers and leys
HRAC mode of action code: O + O

See also MCPA

Products

1	Agrichem DB Plus	Agrichem	243:40 g/l	SL	00044
2	Headland Cedar	Headland	240:40 g/l	SL	12180
3	Redlegor	United Phosphorus	244:44 g/l	SL	07519

Uses

- Annual dicotyledons in ***all cereals*** [1]; ***durum wheat*** *(off-label)*, ***spring barley***, ***spring oats***, ***spring rye*** *(off-label)*, ***triticale*** *(off-label)*, ***undersown barley*** *(red or white clover)*, ***undersown oats*** *(red or white clover)*, ***undersown wheat*** *(red or white clover)*, ***winter barley***, ***winter oats***, ***winter rye*** *(off-label)* [2]; ***rotational grass*** [1-3]; ***spring wheat***, ***winter wheat*** [2, 3]; ***undersown spring cereals***, ***undersown winter cereals*** [1, 3]
- Perennial dicotyledons in ***all cereals*** [1]; ***rotational grass*** [1-3]; ***spring barley***, ***spring oats***, ***undersown barley*** *(red or white clover)*, ***undersown oats*** *(red or white clover)*, ***undersown wheat*** *(red or white clover)*, ***winter barley***, ***winter oats*** [2]; ***spring wheat***, ***winter wheat*** [2, 3]; ***undersown spring cereals***, ***undersown winter cereals*** [1, 3]
- Polygonums in ***rotational grass***, ***spring wheat***, ***undersown spring cereals***, ***undersown winter cereals***, ***winter wheat*** [3]

Specific Off-Label Approvals (SOLAs)

- ***durum wheat*** *20052247* [2]
- ***spring rye*** *20052247* [2]
- ***triticale*** *20052247* [2]
- ***winter rye*** *20052247* [2]

Approval information

- 2,4-DB and MCPA included in Annex I under EC Directive 91/414
- Accepted by BBPA for use on malting barley

Efficacy guidance

- Best results achieved on young seedling weeds under good growing conditions
- Spray thistles and other perennials when 10-20 cm high provided clover at correct stage
- Effectiveness may be reduced by rain within 12 h, by very cold conditions or drought

Restrictions

- Maximum number of treatments 1 per crop
- Do not spray established clover crops or lucerne

- Do not roll or harrow within 7 d before or after spraying
- Do not spray immediately before or after sowing any crop
- Avoid drift onto neighbouring sensitive crops

Crop-specific information

- Latest use: before first node detectable (GS 31) for cereals; 4th trifoliate leaf stage of clover for grass re-seeds
- Apply in spring to winter cereals from leaf sheath erect stage, to spring barley or oats from 2-leaf stage (GS 12), to spring wheat from 5-leaf stage (GS 15)
- Spray clovers as soon as possible after first trifoliate leaf, grasses after 2-3 leaf stage
- Red clover may suffer temporary distortion after treatment

Environmental safety

- Harmful to aquatic organisms
- Harmful to fish or other aquatic life. Do not contaminate surface waters or ditches with chemical or used container
- Keep livestock out of treated areas for at least two weeks following treatment and until poisonous weeds, such as ragwort, have died down and become unpalatable

Hazard classification and safety precautions

Hazard Harmful
UN number N/C
Risk phrases R22a, R41, R52 [1-3]; R38 [2, 3]; R53a [1, 3]
Operator protection A, C [1, 2]; U05a, U08, U11, U20b [1-3]; U13, U14 [3]; U15 [2, 3]
Environmental protection E07a, E13c [1-3]; E19b [1]; E34 [1, 2]; E38 [2]
Storage and disposal D01, D05, D09a [1-3]; D02, D10c [1, 3]; D10a [2]
Medical advice M03 [1-3]; M05a [1, 3]

SECTION 2

156 deltamethrin

A pyrethroid insecticide with contact and residual activity
IRAC mode of action code: 3

Products

1	Bandu	Headland	25 g/l	EC	10994
2	Decis	Bayer CropScience	25 g/l	EC	07172
3	Decis Protech	Bayer CropScience	15 g/l	EW	11502
4	Delta-M 2.5 EC	AgChem Access	25 g/l	EC	12604

Uses

- Aphids in ***amenity vegetation***, ***apples***, ***ornamental plant production***, ***protected cucumbers***, ***protected peppers***, ***protected pot plants***, ***protected tomatoes***, ***spring barley***, ***spring oats***, ***spring wheat***, ***winter barley***, ***winter oats***, ***winter wheat*** [1-4]; ***blackberries*** *(off-label)*, ***fodder beet*** *(off-label)*, ***frise*** *(off-label)*, ***herbs (see appendix 6)*** *(off-label)*, ***lamb's lettuce*** *(off-label)*, ***protected aubergines*** *(off-label)*, ***scarole*** *(off-label)*, ***spinach*** *(off-label)* [1, 2]; ***bulb onions*** *(off-label)*, ***garlic*** *(off-label)*, ***leaf brassicas*** *(off-label)*, ***leeks*** *(off-label)*, ***protected choi sum*** *(off-label)*, ***protected cress*** *(off-label)*, ***protected frise*** *(off-label)*, ***protected herbs (see appendix 6)*** *(off-label)*, ***protected lamb's lettuce*** *(off-label)*, ***protected lettuce*** *(off-label)*, ***protected ornamentals***, ***protected pak choi*** *(off-label)*, ***protected radicchio*** *(off-label)*, ***protected scarole*** *(off-label)*, ***protected spinach*** *(off-label)*, ***rocket*** *(off-label)*, ***shallots*** *(off-label)* [3]; ***carrots*** *(off-label)*, ***celery (outdoor)*** *(off-label)*, ***horseradish*** *(off-label)*, ***mallow (althaea spp.)*** *(off-label)*, ***parsnips*** *(off-label)*, ***protected celery*** *(off-label)*, ***protected chinese cabbage*** *(off-label)*, ***protected rhubarb*** *(off-label)*, ***protected spring onions*** *(off-label)*, ***rhubarb*** *(off-label)*, ***spring onions*** *(off-label)* [1, 3]; ***chinese cabbage*** *(off-label)*, ***durum wheat*** *(off-label)*, ***lettuce*** *(off-label)*, ***spring rye*** *(off-label)*, ***triticale*** *(off-label)*, ***winter rye*** *(off-label)* [1-3]; ***choi sum*** *(off-label)*, ***collards*** *(off-label)*, ***pak choi*** *(off-label)*, ***spring greens*** *(off-label)* [2]; ***evening primrose*** *(off-label)*, ***grass seed crops*** *(off-label)*, ***marjoram*** *(off-label)*, ***radicchio*** *(off-label)*, ***sorrel*** *(off-label)* [2, 3]; ***lupins*** *(off-label)*, ***parsley root*** *(off-label)*, ***protected cayenne peppers*** *(off-label)* [1]; ***nursery stock*** [2, 4]
- Apple sucker in ***apples*** [1-4]
- Barley yellow dwarf virus vectors in ***winter barley***, ***winter wheat*** [1-4]
- Beet virus yellows vectors in ***winter oilseed rape*** [1-4]
- Cabbage seed weevil in ***mustard***, ***spring oilseed rape***, ***winter oilseed rape*** [1-4]
- Cabbage stem flea beetle in ***winter oilseed rape*** [1-4]

- Cabbage stem weevil in ***mustard, spring oilseed rape, winter oilseed rape*** [1-4]
- Capsids in ***amenity vegetation, apples, ornamental plant production*** [1-4]; ***nursery stock*** [2, 4]
- Caterpillars in ***amenity vegetation, apples, broccoli, brussels sprouts, cabbages, cauliflowers, kale, ornamental plant production, plums, protected cucumbers, protected peppers, protected pot plants, protected tomatoes, swedes, turnips*** [1-4]; ***black cabbage*** *(off-label)*, ***borecole*** *(off-label)*, ***choi sum*** *(off-label)*, ***collards*** *(off-label)*, ***mustard*** *(off-label)*, ***orache*** *(off-label)*, ***pak choi*** *(off-label)*, ***spring greens*** *(off-label)*, ***tatsoi*** *(off-label)* [1]; ***bulb onions*** *(off-label)*, ***carrots*** *(off-label)*, ***celery (outdoor)*** *(off-label)*, ***chinese cabbage*** *(off-label)*, ***garlic*** *(off-label)*, ***horseradish*** *(off-label)*, ***leeks*** *(off-label)*, ***lettuce*** *(off-label)*, ***mallow (althaea spp.)*** *(off-label)*, ***parsnips*** *(off-label)*, ***protected celery*** *(off-label)*, ***protected chinese cabbage*** *(off-label)*, ***protected choi sum*** *(off-label)*, ***protected cress*** *(off-label)*, ***protected frise*** *(off-label)*, ***protected herbs (see appendix 6)*** *(off-label)*, ***protected lamb's lettuce*** *(off-label)*, ***protected lettuce*** *(off-label)*, ***protected ornamentals, protected pak choi*** *(off-label)*, ***protected radicchio*** *(off-label)*, ***protected rhubarb*** *(off-label)*, ***protected scarole*** *(off-label)*, ***protected spinach*** *(off-label)*, ***protected spring onions*** *(off-label)*, ***rhubarb*** *(off-label)*, ***rocket*** *(off-label)*, ***shallots*** *(off-label)*, ***spring onions*** *(off-label)* [3]; ***evening primrose*** *(off-label)*, ***grass seed crops*** *(off-label)*, ***marjoram*** *(off-label)*, ***radicchio*** *(off-label)*, ***sorrel*** *(off-label)* [2, 3]; ***leaf brassicas*** *(off-label)* [1, 3]; ***leaf brassicas*** *(off-label - Baby leaf production)*, ***protected brassica seedlings*** *(off-label)*, ***protected rocket*** *(off-label)*, ***rocket*** *(off-label - Baby leaf production)* [2]; ***nursery stock*** [2, 4]
- Codling moth in ***apples*** [1-4]
- Cutworms in ***lettuce*** [1, 3]
- Damson-hop aphid in ***hops, plums*** [1-4]
- Flea beetle in ***black cabbage*** *(off-label)*, ***borecole*** *(off-label)*, ***choi sum*** *(off-label)*, ***collards*** *(off-label)*, ***mustard*** *(off-label)*, ***orache*** *(off-label)*, ***pak choi*** *(off-label)*, ***spring greens*** *(off-label)*, ***tatsoi*** *(off-label)* [1]; ***broccoli, brussels sprouts, cabbages, cauliflowers, kale, swedes, turnips*** [2, 4]; ***bulb onions*** *(off-label)*, ***carrots*** *(off-label)*, ***celery (outdoor)*** *(off-label)*, ***chinese cabbage*** *(off-label)*, ***garlic*** *(off-label)*, ***horseradish*** *(off-label)*, ***leeks*** *(off-label)*, ***lettuce*** *(off-label)*, ***mallow (althaea spp.)*** *(off-label)*, ***parsnips*** *(off-label)*, ***protected celery*** *(off-label)*, ***protected chinese cabbage*** *(off-label)*, ***protected choi sum*** *(off-label)*, ***protected pak choi*** *(off-label)*, ***protected rhubarb*** *(off-label)*, ***protected spring onions*** *(off-label)*, ***rhubarb*** *(off-label)*, ***rocket*** *(off-label)*, ***shallots*** *(off-label)*, ***spring onions*** *(off-label)* [3]; ***evening primrose*** *(off-label)*, ***grass seed crops*** *(off-label)*, ***marjoram*** *(off-label)*, ***protected frise*** *(off-label)*, ***radicchio*** *(off label)*, ***sorrel*** *(off-label)* [2, 3]; ***leaf brassicas*** *(off-label)*, ***protected scarole*** *(off-label)* [1, 3]; ***leaf brassicas*** *(off-label - Baby leaf production)*, ***protected brassica seedlings*** *(off-label)*, ***protected rocket*** *(off-label)*, ***rocket*** *(off-label - Baby leaf production)* [2]; ***protected cress*** *(off-label)*, ***protected herbs (see appendix 6)*** *(off-label)*, ***protected lamb's lettuce*** *(off-label)*, ***protected lettuce*** *(off-label)*, ***protected radicchio*** *(off-label)*, ***protected spinach*** *(off-label)* [1-3]; ***sugar beet*** [1-4]
- Fruit tree tortrix moth in ***plums*** [1-4]
- Insect pests in ***blackberries*** *(off-label)*, ***frise*** *(off-label)*, ***herbs (see appendix 6)*** *(off-label)*, ***lamb's lettuce*** *(off-label)*, ***lupins*** *(off-label)*, ***protected aubergines*** *(off-label)*, ***scarole*** *(off-label)*, ***spinach*** *(off-label)* [1-3]; ***bulb onions*** *(off-label)*, ***choi sum*** *(off-label)*, ***collards*** *(off-label)*, ***garlic*** *(off-label)*, ***pak choi*** *(off-label)*, ***shallots*** *(off-label)*, ***spring greens*** *(off-label)* [2]; ***carrots*** *(off-label)*, ***celery (outdoor)*** *(off-label)*, ***fodder beet*** *(off-label)*, ***horseradish*** *(off-label)*, ***mallow (althaea spp.)*** *(off-label)*, ***parsley root*** *(off-label)*, ***parsnips*** *(off-label)*, ***protected celery*** *(off-label)*, ***protected chinese cabbage*** *(off-label)*, ***protected rhubarb*** *(off-label)*, ***protected spring onions*** *(off-label)*, ***rhubarb*** *(off-label)*, ***spring onions*** *(off-label)* [1, 2]; ***chinese cabbage*** *(off-label)* [1]; ***durum wheat*** *(off-label)*, ***protected cayenne peppers*** *(off-label)*, ***spring rye*** *(off-label)*, ***triticale*** *(off-label)*, ***winter rye*** *(off-label)* [1, 3]
- Leafhoppers in ***bulb onions*** *(off-label)*, ***carrots*** *(off-label)*, ***celery (outdoor)*** *(off-label)*, ***chinese cabbage*** *(off-label)*, ***garlic*** *(off-label)*, ***horseradish*** *(off-label)*, ***leaf brassicas*** *(off-label)*, ***leeks*** *(off-label)*, ***lettuce*** *(off-label)*, ***mallow (althaea spp.)*** *(off-label)*, ***parsnips*** *(off-label)*, ***protected celery*** *(off-label)*, ***protected chinese cabbage*** *(off-label)*, ***protected choi sum*** *(off-label)*, ***protected pak choi*** *(off-label)*, ***protected rhubarb*** *(off-label)*, ***protected spring onions*** *(off-label)*, ***rhubarb*** *(off-label)*, ***rocket*** *(off-label)*, ***shallots*** *(off-label)*, ***spring onions*** *(off-label)* [3]; ***evening primrose*** *(off-label)*, ***grass seed crops*** *(off-label)*, ***marjoram*** *(off-label)*, ***protected frise*** *(off-label)*, ***radicchio*** *(off-label)*, ***sorrel*** *(off-label)* [2, 3]; ***protected cress*** *(off-label)*, ***protected herbs (see appendix 6)*** *(off-label)*, ***protected lamb's lettuce*** *(off-label)*, ***protected lettuce*** *(off-label)*,

protected radicchio *(off-label)*, ***protected spinach*** *(off-label)* [1-3]; ***protected scarole*** *(off-label)* [1, 3]

- Mealybugs in ***amenity vegetation***, ***ornamental plant production***, ***protected cucumbers***, ***protected peppers***, ***protected pot plants***, ***protected tomatoes*** [1-4]; ***nursery stock*** [2, 4]; ***protected ornamentals*** [3]
- Pea and bean weevil in ***broad beans***, ***peas***, ***spring field beans***, ***winter field beans*** [1-4]
- Pea midge in ***peas*** [2-4]
- Pea moth in ***peas*** [1-4]
- Pear sucker in ***pears*** [1-4]
- Phorid flies in ***protected edible fungi*** *(off-label)* [1, 3]; ***protected mushrooms*** *(off-label)* [3]
- Pollen beetle in ***bulb onions*** *(off-label)*, ***carrots*** *(off-label)*, ***celery (outdoor)*** *(off-label)*, ***chinese cabbage*** *(off-label)*, ***garlic*** *(off-label)*, ***horseradish*** *(off-label)*, ***leaf brassicas*** *(off-label)*, ***leeks*** *(off-label)*, ***lettuce*** *(off-label)*, ***mallow (althaea spp.)*** *(off-label)*, ***parsnips*** *(off-label)*, ***protected celery*** *(off-label)*, ***protected chinese cabbage*** *(off-label)*, ***protected choi sum*** *(off-label)*, ***protected cress*** *(off-label)*, ***protected frise*** *(off-label)*, ***protected herbs (see appendix 6)*** *(off-label)*, ***protected lamb's lettuce*** *(off-label)*, ***protected lettuce*** *(off-label)*, ***protected pak choi*** *(off-label)*, ***protected radicchio*** *(off-label)*, ***protected rhubarb*** *(off-label)*, ***protected scarole*** *(off-label)*, ***protected spinach*** *(off-label)*, ***protected spring onions*** *(off-label)*, ***rhubarb*** *(off-label)*, ***rocket*** *(off-label)*, ***shallots*** *(off-label)*, ***spring onions*** *(off-label)* [3]; ***evening primrose*** *(off-label)*, ***grass seed crops*** *(off-label)*, ***marjoram*** *(off-label)*, ***radicchio*** *(off-label)*, ***sorrel*** *(off-label)* [2, 3]; ***mustard***, ***spring oilseed rape***, ***winter oilseed rape*** [1-4]
- Raspberry beetle in ***raspberries*** [1-4]
- Sawflies in ***apples***, ***plums*** [1-4]
- Scale insects in ***amenity vegetation***, ***ornamental plant production***, ***protected cucumbers***, ***protected peppers***, ***protected pot plants***, ***protected tomatoes*** [1-4]; ***nursery stock*** [2, 4]; ***protected ornamentals*** [3]
- Sciarid flies in ***protected edible fungi*** *(off-label)* [1-3]; ***protected mushrooms*** *(off-label)* [2, 3]
- Thrips in ***amenity vegetation***, ***ornamental plant production*** [1-4]; ***bulb onions*** *(off-label)*, ***garlic*** *(off-label)*, ***shallots*** *(off-label)* [1, 3]; ***carrots*** *(off-label)*, ***celery (outdoor)*** *(off-label)*, ***chinese cabbage*** *(off-label)*, ***evening primrose*** *(off-label)*, ***grass seed crops*** *(off-label)*, ***horseradish*** *(off-label)*, ***leaf brassicas*** *(off-label)*, ***lettuce*** *(off-label)*, ***mallow (althaea spp.)*** *(off-label)*, ***marjoram*** *(off-label)*, ***parsnips*** *(off-label)*, ***protected celery*** *(off-label)*, ***protected chinese cabbage*** *(off-label)*, ***protected choi sum*** *(off-label)*, ***protected cress*** *(off-label)*, ***protected frise*** *(off-label)*, ***protected herbs (see appendix 6)*** *(off-label)*, ***protected lamb's lettuce*** *(off-label)*, ***protected lettuce*** *(off-label)*, ***protected pak choi*** *(off-label)*, ***protected radicchio*** *(off-label)*, ***protected rhubarb*** *(off-label)*, ***protected scarole*** *(off-label)*, ***protected spinach*** *(off-label)*, ***protected spring onions*** *(off-label)*, ***radicchio*** *(off-label)*, ***rhubarb*** *(off-label)*, ***rocket*** *(off-label)*, ***sorrel*** *(off-label)*, ***spring onions*** *(off-label)* [3]; ***leeks*** *(off-label)*, ***protected courgettes*** *(off-label)*, ***protected cucumbers*** *(off-label)*, ***protected gherkins*** *(off-label)*, ***protected peppers*** *(off-label)*, ***protected tomatoes*** *(off-label)* [1-3]; ***nursery stock*** [2, 4]; ***protected aubergines*** *(off-label)*, ***protected cayenne peppers*** *(off-label)* [2]; ***protected non-edible flowers*** *(off-label)* [1]; ***protected ornamentals*** *(off-label)* [2, 3]
- Tortrix moths in ***apples*** [1-4]
- Western flower thrips in ***protected aubergines*** *(off-label)*, ***protected cayenne peppers*** *(off-label)* [2]; ***protected courgettes*** *(off-label)*, ***protected cucumbers*** *(off-label)*, ***protected gherkins*** *(off-label)*, ***protected peppers*** *(off-label)*, ***protected tomatoes*** *(off-label)* [1-3]; ***protected non-edible flowers*** *(off-label)* [1]; ***protected ornamentals*** *(off-label)* [2, 3]
- Whitefly in ***amenity vegetation***, ***ornamental plant production***, ***protected cucumbers***, ***protected peppers***, ***protected pot plants***, ***protected tomatoes*** [1-4]; ***nursery stock*** [2, 4]; ***protected ornamentals*** [3]

Specific Off-Label Approvals (SOLAs)

- ***black cabbage*** *20071582* [1]
- ***blackberries*** *20071605* [1], *20071705* [2], *20071654* [3]
- ***borecole*** *20071582* [1]
- ***bulb onions*** *20071613* [1], *20071698* [2], *20071158* [3]
- ***carrots*** *20071610* [1], *20071696* [2], *20071158* [3]
- ***celery (outdoor)*** *20071579* [1], *20071697* [2], *20071158* [3]
- ***chinese cabbage*** *20071606* [1], *20071695* [2], *20071158* [3]
- ***choi sum*** *20071582* [1], *20071704* [2]

- **collards** *20071582* [1], *20082458* [2]
- **durum wheat** *20071603* [1], *20071701* [2], *20071656* [3]
- **evening primrose** *20071694* [2], *20071158* [3]
- **fodder beet** *20071609* [1], *20071702* [2]
- **frise** *20071604* [1], *20071703* [2], *20071655* [3]
- **garlic** *20071613* [1], *20071698* [2], *20071158* [3]
- **grass seed crops** *20071694* [2], *20071158* [3]
- **herbs (see appendix 6)** *20071604* [1], *20071703* [2], *20071655* [3]
- **horseradish** *20071610* [1], *20071696* [2], *20071158* [3]
- **lamb's lettuce** *20071604* [1], *20071703* [2], *20071655* [3]
- **leaf brassicas** *20071582* [1], *(Baby leaf production) 20071708* [2], *20071158* [3]
- **leeks** *20071614* [1], *20071699* [2], *20071158* [3]
- **lettuce** *20071612* [1], *20071695* [2], *20071158* [3]
- **lupins** *20071583* [1], *20071700* [2], *20071653* [3]
- **mallow (althaea spp.)** *20071610* [1], *20071696* [2], *20071158* [3]
- **marjoram** *20071694* [2], *20071158* [3]
- **mustard** *20071582* [1]
- **orache** *20071582* [1]
- **pak choi** *20071582* [1], *20082458* [2]
- **parsley root** *20071610* [1], *20071696* [2]
- **parsnips** *20071610* [1], *20071696* [2], *20071158* [3]
- **protected aubergines** *20071608* [1], *20071693* [2], *20071707* [2], *20071649* [3]
- **protected brassica seedlings** *20071708* [2]
- **protected cayenne peppers** *20071607* [1], *20071693* [2], *20071657* [3]
- **protected celery** *20071579* [1], *20071697* [2], *20071158* [3]
- **protected chinese cabbage** *20071579* [1], *20071697* [2], *20071158* [3]
- **protected choi sum** *20071158* [3]
- **protected courgettes** *20071611* [1], *20071693* [2], *20071650* [3]
- **protected cress** *20080175* [1], *20071691* [2], *20071158* [3]
- **protected cucumbers** *20071611* [1], *20071693* [2], *20071650* [3]
- **protected edible fungi** *20071581* [1], *20071692* [2], *20071648* [3]
- **protected frise** *20071691* [2], *20071158* [3]
- **protected gherkins** *20071611* [1], *20071693* [2], *20071650* [3]
- **protected herbs (see appendix 6)** *20080175* [1], *20071691* [2], *20071158* [3]
- **protected lamb's lettuce** *20080175* [1], *20071691* [2], *20071158* [3]
- **protected lettuce** *20080175* [1], *20071691* [2], *20071158* [3]
- **protected mushrooms** *20071692* [2], *20071648* [3]
- **protected non-edible flowers** *20071611* [1]
- **protected ornamentals** *20071693* [2], *20071650* [3]
- **protected pak choi** *20071158* [3]
- **protected peppers** *20071611* [1], *20071693* [2], *20071650* [3]
- **protected radicchio** *20080175* [1], *20071691* [2], *20071158* [3]
- **protected rhubarb** *20071579* [1], *20071697* [2], *20071158* [3]
- **protected rocket** *20071708* [2]
- **protected scarole** *20080175* [1], *20071158* [3]
- **protected spinach** *20080175* [1], *20071691* [2], *20071158* [3]
- **protected spring onions** *20071579* [1], *20071697* [2], *20071158* [3]
- **protected tomatoes** *20071611* [1], *20071693* [2], *20071650* [3]
- **radicchio** *20071694* [2], *20071158* [3]
- **rhubarb** *20071579* [1], *20071697* [2], *20071158* [3]
- **rocket** *(Baby leaf production) 20071708* [2], *20071158* [3]
- **scarole** *20071604* [1], *20071703* [2], *20071655* [3]
- **shallots** *20071613* [1], *20071698* [2], *20071158* [3]
- **sorrel** *20071694* [2], *20071158* [3]
- **spinach** *20071604* [1], *20071703* [2], *20071655* [3]
- **spring greens** *20071582* [1], *20082458* [2]
- **spring onions** *20071579* [1], *20071697* [2], *20071158* [3]
- **spring rye** *20071603* [1], *20071701* [2], *20071656* [3]
- **tatsoi** *20071582* [1]

- ***triticale*** *20071603* [1], *20071701* [2], *20071656* [3]
- ***winter rye*** *20071603* [1], *20071701* [2], *20071656* [3]

Approval information

- Deltamethrin included in Annex I under EC Directive 91/414
- Accepted by BBPA for use on malting barley and hops

Efficacy guidance

- A contact and stomach poison with 3-4 wk persistence, particularly effective on caterpillars and sucking insects
- Normally applied at first signs of damage with follow-up treatments where necessary at 10-14 d intervals. Rates, timing and recommended combinations with other pesticides vary with crop and pest. See label for details
- Spray is rainfast within 1 h
- May be applied in frosty weather provided foliage not covered in ice
- Temperatures above 35 °C may reduce effectiveness or persistence

Restrictions

- Maximum number of treatments varies with crop and pest, 4 per crop for wheat and barley, only 1 application between 1 Apr and 31 Aug. See label for other crops
- Do not apply more than 1 aphicide treatment to cereals in summer
- Do not spray crops suffering from drought or other physical stress
- Consult processer before treating crops for processing
- Do not apply to a cereal crop if any product containing a pyrethroid insecticide or dimethoate has been applied to that crop after the start of ear emergence (GS 51)
- Do not spray cereals after 31 Mar in the year of harvest within 6 m of the outside edge of the crop
- Reduced volume spraying must not be used on cereals after 31 Mar in yr of harvest

Crop-specific information

- Latest use: early dough (GS 83) for barley, oats, wheat; before flowering for mustard, oilseed rape; before 31 Mar for grass seed crops [1-3]

Environmental safety

- Dangerous for the environment
- Toxic to aquatic organisms
- Flammable [1, 2]
- Extremely dangerous to fish or other aquatic life. Do not contaminate surface waters or ditches with chemical or used container
- High risk to bees. Do not apply to crops in flower or to those in which bees are actively foraging. Do not apply when flowering weeds are present
- Do not apply in tank-mixture with a triazole-containing fungicide when bees are likely to be actively foraging in the crop
- High risk to non-target insects or other arthropods. Do not spray within 6 m of the field boundary
- LERAP Category A
- Broadcast air-assisted LERAP (18 m) [2, 3]

Hazard classification and safety precautions

Hazard Harmful, Flammable [1, 2, 4]; Dangerous for the environment [1-4]
Transport code 3
Packaging group III
UN number 1993
Risk phrases R20, R22a, R22b, R38, R41, R51 [1, 2, 4]; R50 [3]; R53a [1-4]
Operator protection A, C, H; U04a, U05a, U08, U19a [1, 2, 4]; U10 [1]; U11, U14, U15 [3]; U20b [1-4]
Environmental protection E12a [2-4]; E12c [1]; E12f [1] (cereals, oilseed rape, peas, beans); E12f [2-4] (cereals, oilseed rape, peas, beans - see label for guidance); E15a, E16c, E16d, E38 [1-4]; E17b [2-4] (18 m); E22a [3]; E34 [1, 2, 4]
Storage and disposal D01, D02, D09a, D10c [1-4]; D05 [1, 2, 4]; D12a [2-4]
Medical advice M03, M05b [1, 2, 4]

SEE SECTION 3 FOR PRODUCTS ALSO REGISTERED

157 desmedipham

A contact phenyl carbamate herbicide available only in mixtures
HRAC mode of action code: C1

158 desmedipham + ethofumesate + metamitron + phenmedipham

A contact and residual herbicide mixture for beet
HRAC mode of action code: C1 + N + C1 + C1

See also *ethofumesate*
metamitron
phenmedipham

Products

Betanal Quattro	Bayer CropScience	20;100;200;60 g/l	SE	12779

Uses

- Annual dicotyledons in ***sugar beet***
- Annual meadow grass in ***sugar beet***

Approval information

- Desmedipham, ethofumesate and phenmedipham included in Annex I under EC Directive 91/414

Restrictions

- Avoid spraying within 6m of the field boundary to reduce the effects on non-target insects or arthropods

Crop-specific information

- A check to crop growth can occur if application is made to crops suffering stress from application of other treatments, manganese or lime deficiency, high temperatures or high light intensity, frost, wind or insect attack.

Following crops guidance

- Only sow sugar beet, fodder beet or mangels within 4 months of application. Ploughing to 15 cm followed by thorough cultivation is recommended before planting any crop. Winter cereals can be sown in the same season provided 16 weeks have elapsed since the last application. Any spring crop may be planted in the following year.

Hazard classification and safety precautions

Hazard Dangerous for the environment
Transport code 9
Packaging group III
UN number 3082
Risk phrases R51, R53a
Operator protection A; U08, U20a
Environmental protection E15b, E38, E40
Storage and disposal D05, D09a, D10b
Medical advice M03

159 desmedipham + ethofumesate + phenmedipham

A selective contact and residual herbicide for beet
HRAC mode of action code: C1 + N + C1

See also *ethofumesate*
phenmedipham

Products

1	Beetaweed	AgriGuard	25:151:75 g/l	EC	12260
2	Betanal Expert	Bayer CropScience	25:151:75 g/l	EC	14034
3	Betasana Trio SC	United Phosphorus	15:115:75 g/l	SC	14262
4	Trilogy	United Phosphorus	15:115:75 g/l	SC	14464

Uses

- Annual dicotyledons in ***fodder beet*, *mangels*** [3, 4]; ***game cover*** *(off-label)*, ***ornamental plant production*** *(off-label)* [2]; ***sugar beet*** [1-4]
- Annual meadow grass in ***sugar beet*** [1, 2]

Specific Off-Label Approvals (SOLAs)

- ***game cover*** *20082823* [2]
- ***ornamental plant production*** *20082823* [2]

Approval information

- Desmedipham, ethofumesate and phenmedipham included in Annex I under EC Directive 91/414
- Approval expiry 31 Aug 2010 [1]

Efficacy guidance

- Product recommended for low-volume overall application in a planned spray programme
- Product acts mainly by contact action. A full programme also gives some residual control but this may be reduced on soils with more than 5% organic matter
- Best results achieved from treatments applied at fully expanded cotyledon stage of largest weeds present. Occasional larger weeds will usually be controlled from a full programme of sprays
- Where a pre-emergence band spray has been applied treatment must be timed according to size of the untreated weeds between the rows
- Susceptible weeds may not all be killed by the first spray. Repeat applications as each flush of weeds reaches cotyledon size normally necessary for season long control
- Sequential treatments should be applied when the previous one is still showing an effect on the weeds
- Various mixtures with other beet herbicides are recommended. See label for details

Restrictions

- Maximum total dose equivalent to three full dose treatments [1], 7.0 l/ha [3]
- Do not spray crops stressed by nutrient deficiency, wind damage, pest or disease attack, or previous herbicide treatments. Stressed crops treated under conditions of high light intensity may be checked and not recover fully
- If temperature likely to exceed 21 °C spray after 5 pm
- Crystallisation may occur if spray volume exceeds that recommended or spray mixture not used within 2 h, especially if the water temperature is below 5 °C
- Before use, wash out sprayer to remove all traces of previous products, especially hormone and sulfonyl urea weedkillers

Crop-specific information

- Latest use: before crop meets between rows
- Apply first treatment when majority of crop plants have reached the fully expanded cotyledon stage
- Frost within 7 d of treatment may cause check from which the crop may not recover

Following crops guidance

- Beet crops may be sown at any time after treatment. Any other crop may be sown 3 mth after treatment following mouldboard ploughing to 15 cm minimum

Environmental safety

- Dangerous for the environment
- Toxic to aquatic organisms

Hazard classification and safety precautions

Hazard Irritant [3, 4]; Dangerous for the environment [1-4]
Transport code 9
Packaging group III
UN number 3082
Risk phrases R36, R50 [3, 4]; R51 [1, 2]; R53a [1-4]
Operator protection A, H [1, 2]; U08 [1-4]; U20a [1, 2]
Environmental protection E15a [1, 2]; E15b [3, 4]; E38 [1-4]
Storage and disposal D05, D09a, D12a [1-4]; D10b [1, 2]; D10c [3, 4]
Treated seed S06a [3, 4]
Medical advice M03

160 desmedipham + phenmedipham

A mixture of contact herbicides for use in sugar beet
HRAC mode of action code: C1 + C1

See also *phenmedipham*

Products

Betanal Maxxim	Bayer CropScience	160;160 g/l	EC	14186

Uses

- Annual dicotyledons in ***sugar beet***

Approval information

- Desmedipham and phenmedipham included in Annex I under EC Directive 91/414

Efficacy guidance

- Product is recommended for low volume overall spraying in a planned programme involving pre- and/or post-emergence treatments at doses recommended for low volume programmes
- Best results obtained from treatment when earliest germinating weeds have reached cotyledon stage
- Further treatments must be applied as each flush of weeds reaches cotyledon stage but allowing a minimum of 7 d between each spray
- Where a pre-emergence band spray has been applied, the first treatment should be timed according to the size of the weeds in the untreated area between the rows
- Product is absorbed by leaves of emerged weeds which are killed by scorching action in 2-10 d. Apply overall as a fine spray to optimise weed cover and spray retention
- Various tank mixtures and sequences recommended to widen weed spectrum and add residual activity - see label for details

Restrictions

- Maximum total dose equivalent to three full dose treatments
- Do not spray crops stressed by nutrient deficiency, frost, wind damage, pest or disease attack, or previous herbicide treatments. Stressed crops may be checked and not recover fully
- If temperature likely to exceed 21 °C spray after 5 pm
- Before use, wash out sprayer to remove all traces of previous products, especially hormone and sulfonyl urea weedkillers
- Crystallisation may occur if spray volume exceeds that recommended or spray mixture not used within 2 h, especially if the water temperature is below 5 °C
- Product may cause non-reinforced PVC pipes and hoses to soften and swell. Wherever possible, use reinforced PVC or synthetic rubber hoses

Crop-specific information

- Latest use: before crop leaves meet between rows
- Product safe to use on all soil types

Environmental safety

- Dangerous for the environment
- Toxic to aquatic organisms
- Risk to certain non-target insects or other arthropods. Avoid spraying within 6 m of field boundary

Hazard classification and safety precautions

Hazard Dangerous for the environment
Transport code 9
Packaging group III
UN number 3082
Risk phrases R50, R53a
Operator protection A; U05a, U08, U19a, U20a
Environmental protection E15b, E16b, E38, E39
Storage and disposal D01, D02, D09a, D10b, D12a

161 dicamba

A translocated benzoic herbicide available only in mixtures
HRAC mode of action code: O

See also *2,4-D + dicamba*
2,4-D + dicamba + dichlorprop-P
2,4-D + dicamba + fluroxypyr
2,4-D + dicamba + triclopyr

162 dicamba + dichlorprop-P + MCPA

A translocated herbicide mixture for turf
HRAC mode of action code: O + O + O

See also *dichlorprop-P*
MCPA

Products

Intrepid 2	Scotts	20.8:167:167 g/l	SL	11594

Uses
- Annual dicotyledons in ***managed amenity turf***
- Perennial dicotyledons in ***managed amenity turf***

Approval information
- Dicamba, dichlorprop-P and MCPA included in Annex I under EC Directive 91/414

Efficacy guidance
- Apply as directed on label between Apr and Sep when weeds growing actively
- Do not mow for at least 3 d before and 3-4 d after treatment so that there is sufficient leaf growth for spray uptake and sufficient time for translocation. See label for details
- If re-growth occurs or new weeds germinate re-treatment recommended

Restrictions
- Do not use during drought unless irrigation is carried out before and after treatment
- Do not apply during freezing conditions or when heavy rain is imminent
- Do not treat new turf until established for about 6 mth after seeding or turfing
- The first 4 mowings after treatment should not be used to mulch cultivated plants unless composted for at least 6 mth
- Do not re-seed turf within 8 wk of last treatment

Crop-specific information
- Latest use: end Sep for managed amenity turf

Environmental safety
- Prevent spray running or drifting onto cultivated plants, including shrubs and trees

Hazard classification and safety precautions

Hazard Harmful
Risk phrases R22a, R38, R41, R43
Operator protection A, C; U05a, U08, U11, U19a, U20a
Environmental protection E15a, E34
Storage and disposal D01, D02, D05, D09a, D10b
Medical advice M03, M05a

163 dicamba + MCPA + mecoprop-P

A translocated herbicide for cereals, grassland, amenity grass and orchards
HRAC mode of action code: O + O + O

See also *MCPA*
mecoprop-P

Products

1	Banlene Super	Certis	18:252:42 g/l	SL	14420
2	Field Marshal	United Phosphorus	18:360:80 g/l	SL	08956
3	Greencrop Triathlon	Greencrop	25:200:200 g/l	SL	10956
4	Headland Relay P	Headland	25:200:200 g/l	SL	08580
5	Headland Relay Turf	Headland Amenity	25:200:200 g/l	SL	08935
6	Headland Transfer	Headland	18:315:50 g/l	SL	11010
7	Headland Trinity	Headland	18:315:50 g/l	SL	10842
8	Hycamba Plus	Agrichem	20.4:125.9:240.2 g/l	SL	10180
9	Hyprone-P	Agrichem	16:101:92 g/l	SL	09125
10	Hysward-P	Agrichem	16:101:92 g/l	SL	09052
11	Mircam Plus	Nufarm UK	19.5:245:43.3 g/l	SL	11525
12	Nocweed	Bayer Environ.	18:252:42 g/l	SL	10609
13	Outrun	Barclay	20.4:125.9:240.2 g/l	SL	12838
14	Pasturol Plus	Headland	25:200:200 g/l	SL	10278
15	Pierce	Nufarm UK	31.2:256.2:237.5 g/l	SL	11924
16	Re-act	Scotts	31:256:237 g/l	SL	12231
17	Super Selective Plus	Rigby Taylor	31.25:256:238 g/l	SL	11928
18	T2 Green	Nufarm UK	31.25: 256: 237 g/l	SL	11925
19	Tribute	Nomix Enviro	18:252:42 g/l	SL	13864
20	Tribute Plus	Nomix Enviro	18:252:42 g/l	SL	13865
21	UPL Grassland Herbicide	United Phosphorus	25:200:200 g/l	SL	08934

Uses

- Annual dicotyledons in ***almonds*** *(off-label)*, ***chestnuts*** *(off-label)*, ***hazel nuts*** *(off-label)*, ***walnuts*** *(off-label)* [1, 6, 7]; ***almonds*** *(off-label - undersown with rotational grass)*, ***chestnuts*** *(off-label - undersown with rotational grass)*, ***durum wheat***, ***hazel nuts*** *(off-label - undersown with rotational grass)*, ***triticale***, ***walnuts*** *(off-label - undersown with rotational grass)* [11]; ***amenity grassland*** [3, 7, 8, 10, 13, 14, 20, 21]; ***apple orchards***, ***spring rye***, ***winter rye*** [1, 7, 11, 12]; ***durum wheat*** *(off-label)*, ***triticale*** *(off-label)* [7]; ***grass seed crops*** [1, 6, 7, 9, 11, 12, 15]; ***grassland***, ***green cover on land temporarily removed from production*** [20]; ***managed amenity turf*** [3, 5, 7, 8, 10-14, 16-20]; ***pear orchards*** [1, 7, 12]; ***permanent grassland*** [1, 3, 4, 6, 7, 10-12, 14, 15, 21]; ***rotational grass*** [1-4, 6, 7, 9-12, 14, 15, 21]; ***spring barley***, ***spring oats***, ***spring wheat***, ***winter barley***, ***winter oats***, ***winter wheat*** [1, 2, 6, 7, 9, 11, 12]; ***undersown barley***, ***undersown oats***, ***undersown wheat*** [9, 12]; ***undersown rye*** [12]; ***undersown spring cereals***, ***undersown winter cereals*** [1, 7]; ***undersown spring cereals*** *(grass only)*, ***undersown winter cereals*** *(grass only)* [2, 6]
- Docks in ***almonds*** *(off-label)*, ***chestnuts*** *(off-label)*, ***hazel nuts*** *(off-label)*, ***walnuts*** *(off-label)* [1]; ***amenity grassland*** [10, 21]; ***apple orchards***, ***pear orchards*** [1, 7, 12]; ***grass seed crops*** [1, 7, 9, 12, 15]; ***managed amenity turf*** [10]; ***permanent grassland***, ***rotational grass*** [1, 7, 10, 12, 15, 21]
- Perennial dicotyledons in ***almonds*** *(off-label)*, ***chestnuts*** *(off-label)*, ***hazel nuts*** *(off-label)*, ***walnuts*** *(off-label)* [6, 7]; ***almonds*** *(off-label - undersown with rotational grass)*, ***chestnuts*** *(off-label - undersown with rotational grass)*, ***durum wheat***, ***hazel nuts*** *(off-label - undersown with rotational grass)*, ***triticale***, ***walnuts*** *(off-label - undersown with rotational grass)* [11]; ***amenity grassland*** [3, 7, 8, 10, 13, 14, 20, 21]; ***apple orchards***, ***spring rye***, ***winter rye*** [1, 7, 11, 12]; ***grass seed crops*** [1, 6, 7, 9, 11, 12, 15]; ***grassland***, ***green cover on land temporarily removed from production*** [20]; ***managed amenity turf*** [3, 5, 7, 8, 10, 11, 13, 14, 16-20]; ***pear orchards*** [1, 7, 12]; ***permanent grassland*** [1, 3, 4, 6, 7, 10-12, 14, 15, 21]; ***rotational grass*** [1-4, 6, 7, 9-12, 14, 15, 21]; ***spring barley***, ***spring oats***, ***spring wheat***, ***winter barley***, ***winter oats***, ***winter wheat*** [1, 2, 6, 7, 9, 11, 12]; ***undersown barley***, ***undersown oats***, ***undersown wheat*** [9, 12]; ***undersown rye*** [12]; ***undersown spring cereals***, ***undersown winter cereals*** [1, 7]; ***undersown spring cereals*** *(grass only)*, ***undersown winter cereals*** *(grass only)* [2, 6]

Specific Off-Label Approvals (SOLAs)

- ***almonds*** *20090731* [1], *20060514* [6], *20060513* [7], *(undersown with rotational grass) 20061130* [11]
- ***chestnuts*** *20090731* [1], *20060514* [6], *20060513* [7], *(undersown with rotational grass) 20061130* [11]
- ***durum wheat*** *20060526* [7]
- ***hazel nuts*** *20090731* [1], *20060514* [6], *20060513* [7], *(undersown with rotational grass) 20061130* [11]
- ***triticale*** *20060526* [7]
- ***walnuts*** *20090731* [1], *20060514* [6], *20060513* [7], *(undersown with rotational grass) 20061130* [11]

Approval information

- Dicamba, MCPA and mecoprop-P included in Annex I under EC Directive 91/414
- Accepted by BBPA for use on malting barley
- Approval expiry 31 Dec 2010 [12]

Efficacy guidance

- Treatment should be made when weeds growing actively. Weeds hardened by winter weather may be less susceptible
- For best results apply in fine warm weather, preferably when soil is moist. Do not spray if rain expected within 6 h or in drought
- Application of fertilizer 1-2 wk before spraying aids weed control in turf
- Where a second treatment later in the season is needed in amenity situations and on grass allow 4-6 wk between applications to permit sufficient foliage regrowth for uptake

Restrictions

- Maximum number of treatments (including other mecoprop-P products) or maximum total dose varies with crop and product. See label for details. The total amount of mecoprop-P applied in a single yr must not exceed the maximum total dose approved for any single product for the crop/situation
- Do not apply to cereals after the first node is detectable (GS 31), or to grass under stress from drought or cold weather
- Do not spray cereals undersown with clovers or legumes, to be undersown with grass or legumes or grassland where clovers or other legumes are important
- Do not spray leys established less than 18 mth or orchards established less than 3 yr
- Do not roll or harrow within 7 d before or after treatment, or graze for at least 7 d afterwards (longer if poisonous weeds present)
- Do not use on turf or grass in year of establishment. Allow 6-8 wk after treatment before seeding bare patches
- The first mowings after use should not be used for mulching unless composted for 6 mth
- Turf should not be mown for 24 h before or after treatment (3-4 d for closely mown turf)
- Avoid drift onto all broad-leaved plants outside the target area

Crop-specific information

- Latest use: before first node detectable (GS 31) for cereals; 5-6 wk before head emergence for grass seed crops; mid-Oct for established grass
- HI 7-14 d before cutting or grazing for leys, permanent pasture
- Apply to winter cereals from the leaf sheath erect stage (GS 30), and to spring cereals from the 5 expanded leaf stage (GS 15)
- Spray grass seed crops 4-6 wk before flower heads begin to emerge (timothy 6 wk)
- Turf containing bulbs may be treated once the foliage has died down completely [4, 10]

Environmental safety

- Harmful to aquatic organisms
- Keep livestock out of treated areas for at least 2 wk following treatment and until poisonous weeds, such as ragwort, have died down and become unpalatable
- Harmful to fish or other aquatic life. Do not contaminate surface waters or ditches with chemical or used container
- Some pesticides pose a greater threat of contamination of water than others and mecoprop-P is one of these pesticides. Take special care when applying mecoprop-P near water and do not apply if heavy rain is forecast

SEE SECTION 3 FOR PRODUCTS ALSO REGISTERED

Hazard classification and safety precautions

Hazard Harmful [1-8, 11-20]; Irritant [9, 10, 21]

UN number N/C [2, 3, 5, 13, 16, 21]

Risk phrases R20 [4, 6, 7, 14]; R21 [1, 3-7, 11, 12, 14, 19, 20]; R22a [1-8, 11-18]; R36 [3, 4, 9-11, 14, 19-21]; R38 [6-8, 11, 13, 21]; R41 [1, 2, 4-8, 12-18]; R43 [21]; R52 [2, 4-11, 14-16, 18-21]; R53a [2, 6-11, 15, 16, 18, 21]

Operator protection A, C [1-21]; H, M [2-11, 13-21]; U05a [1, 3-21]; U08 [1-11, 13, 14, 19-21]; U09a [12]; U11 [1, 2, 4, 6-18, 21]; U15 [4, 14]; U19a [3, 8, 10, 13]; U20b [1-14, 19-21]

Environmental protection E07a [1-4, 6-20]; E13c [1-10, 12-18, 21]; E15a [2, 11, 15, 16, 18-20]; E23 [19, 20]; E34 [2-7, 11, 12, 14-20]; E38 [4, 14, 19, 20]

Storage and disposal D01, D02, D05, D09a [1-21]; D10a [6, 7, 17]; D10b [1, 3-5, 11, 12, 14-16, 18-20]; D10c [2, 8-10, 13, 21]; D12a [2, 19-21]

Medical advice M03 [2-7, 11, 14-18, 21]; M05a [2, 7-10, 13, 21]

164 dicamba + mecoprop-P

A translocated post-emergence herbicide for cereals and grassland
HRAC mode of action code: O + O

See also mecoprop-P

Products

1	Di-Farmon R	Headland	42:319 g/l	SL	08472
2	Dockmaster	Nufarm UK	18.7:150 g/l	SL	11651
3	Foundation	Headland	84:600 g/l	SL	11708
4	Headland Saxon	Headland	84:600 g/l	SL	11947
5	High Load Mircam	Nufarm UK	80:600 g/l	SL	11930
6	Hyban-P	Agrichem	18.7:150 g/l	SL	09129
7	Hygrass-P	Agrichem	18.7:150 g/l	SL	09130
8	Mircam	Nufarm UK	18.7:150 g/l	SL	11707
9	Prompt	Headland	84:600 g/l	SL	11948

Uses

- Annual dicotyledons in ***durum wheat*** *(off-label)*, ***spring rye*** *(off-label)*, ***triticale*** *(off-label)*, ***winter rye*** *(off-label)* [3, 4, 9]; ***permanent grassland*** [1, 3-5, 7-9]; ***rotational grass*** [5, 7, 8]; ***spring barley, spring oats, spring wheat, winter barley, winter oats, winter wheat*** [1-6, 8, 9]; ***spring rye, triticale, undersown rye, winter rye*** [6]; ***undersown barley, undersown oats, undersown wheat*** [2, 6]
- Buttercups in ***permanent grassland*** [2, 6]; ***rotational grass*** [6]
- Chickweed in ***permanent grassland*** [1, 2, 6]; ***rotational grass*** [5, 6]; ***spring barley, spring oats, spring wheat, winter barley, winter oats, winter wheat*** [1-6, 8, 9]; ***spring rye, triticale, undersown rye, winter rye*** [6]; ***undersown barley, undersown oats, undersown wheat*** [2, 6]
- Cleavers in ***permanent grassland*** [1]; ***rotational grass*** [5]; ***spring barley, spring oats, spring wheat, winter barley, winter oats, winter wheat*** [1-6, 8, 9]; ***spring rye, triticale, undersown rye, winter rye*** [6]; ***undersown barley, undersown oats, undersown wheat*** [2, 6]
- Creeping thistle in ***permanent grassland*** [2, 6]; ***rotational grass*** [6]
- Docks in ***permanent grassland*** [2, 6-8]; ***rotational grass*** [6-8]
- Mayweeds in ***permanent grassland*** [1]; ***rotational grass*** [5]; ***spring barley, spring oats, spring wheat, winter barley, winter oats, winter wheat*** [1-6, 8, 9]; ***spring rye, triticale, undersown rye, winter rye*** [6]; ***undersown barley, undersown oats, undersown wheat*** [2, 6]
- Perennial dicotyledons in ***permanent grassland*** [3-5, 7-9]; ***rotational grass*** [7, 8]; ***spring barley, spring oats, spring wheat, winter barley, winter oats, winter wheat*** [2, 6, 8]; ***spring rye, triticale, undersown rye, winter rye*** [6]; ***undersown barley, undersown oats, undersown wheat*** [2, 6]
- Plantains in ***spring barley, spring oats, spring rye, spring wheat, triticale, undersown barley, undersown oats, undersown rye, undersown wheat, winter barley, winter oats, winter rye, winter wheat*** [6]
- Polygonums in ***permanent grassland*** [1]; ***rotational grass*** [5]; ***spring barley, spring oats, spring wheat, winter barley, winter oats, winter wheat*** [1-6, 8, 9]; ***spring rye, triticale, undersown rye, winter rye*** [6]; ***undersown barley, undersown oats, undersown wheat*** [2, 6]
- Scentless mayweed in ***permanent grassland*** [2, 6]; ***rotational grass*** [6]

- Stinging nettle in ***permanent grassland*, *rotational grass*** [6]
- Thistles in ***permanent grassland*, *rotational grass*** [7, 8]

Specific Off-Label Approvals (SOLAs)

- ***durum wheat*** *20052953* [3], *20052951* [4], *20052952* [9]
- ***spring rye*** *20052953* [3], *20052951* [4], *20052952* [9]
- ***triticale*** *20052953* [3], *20052951* [4], *20052952* [9]
- ***winter rye*** *20052953* [3], *20052951* [4], *20052952* [9]

Approval information

- Dicamba and mecoprop-P included in Annex I under EC Directive 91/414
- Accepted by BBPA for use on malting barley

Efficacy guidance

- Best results by application in warm, moist weather when weeds are actively growing

Restrictions

- Maximum number of treatments 1 per crop for cereals and 1 or 2 per yr on grass depending on label. The total amount of mecoprop-P applied in a single yr must not exceed the maximum total dose approved for any single product for the crop/situation
- Do not spray in cold or frosty conditions
- Do not spray if rain expected within 6 h
- Do not treat undersown grass until tillering begins
- Do not spray cereals undersown with clover or legume mixtures
- Do not roll or harrow within 7 d before or after spraying
- Do not treat crops suffering from stress from any cause
- Use product immediately following dilution; do not allow diluted product to stand before use [4]
- Avoid treatment when drift may damage neighbouring susceptible crops

Crop-specific information

- Latest use: before 1st node detectable for cereals; 7 d before cutting or 14 d before grazing grass
- Apply to winter sown crops from 5 expanded leaf stage (GS 15)
- Apply to spring sown cereals from 5 expanded leaf stage but before first node is detectable (GS 15-31)
- Treat grassland just before perennial weeds flower
- Transient crop prostration may occur after spraying but recovery is rapid

Environmental safety

- Dangerous for the environment [1, 3-5, 9]
- Toxic to aquatic organisms
- Harmful to fish or other aquatic life. Do not contaminate surface waters or ditches with chemical or used container
- Keep livestock out of treated areas for at least 2 wk and until foliage of poisonous weeds such as ragwort has died and become unpalatable
- Some pesticides pose a greater threat of contamination of water than others and mecoprop-P is one of these pesticides. Take special care when applying mecoprop-P near water and do not apply if heavy rain is forecast

Hazard classification and safety precautions

Hazard Harmful [1-5, 8, 9]; Irritant [6, 7]; Dangerous for the environment [1, 3-5, 9]
UN number N/C
Risk phrases R21, R36 [2, 8]; R22a, R38 [1-5, 8, 9]; R41 [1, 3-7, 9]; R51 [1, 3-5, 9]; R52 [2, 6-8]; R53a [1-9]
Operator protection A, C [1-9]; H, M [1-4, 6-9]; U05a, U11 [1-9]; U08 [1, 3, 4, 6, 9]; U09a [2, 5, 7, 8]; U19a [7]; U20a [1]; U20b [2-9]
Environmental protection E07a [1-4, 6-9]; E13c [2, 5-8]; E15a [1, 3, 4, 9]; E34 [1-9]; E38 [1, 3]
Storage and disposal D01, D02, D05, D09a [1-9]; D10b [1-4, 8, 9]; D10c [6, 7]
Medical advice M03 [1-9]; M05a [6, 7]

165 dichlobenil

A residual benzonitrile herbicide for woody crops and non-crop uses
HRAC mode of action code: L

Products

1	Casoron G	Scotts	6.75% w/w	GR	10709
2	Casoron G4	Scotts	4% w/w	GR	10708
3	Dicore	AgriChem BV	6.75% w/w	GR	12542
4	Midstream GSR	Scotts	20% w/w	GR	11674
5	Scotts Dichlo G Macro	Scotts	6.75% w/w	GR	12011
6	Scotts Dichlo G Micro	Scotts	6.75% w/w	GR	11857

Uses

- Annual and perennial weeds in ***amenity vegetation, apples, blackberries, blackcurrants, enclosed waters, gooseberries, hard surfaces, loganberries, natural surfaces not intended to bear vegetation, open waters, pears, permeable surfaces overlying soil, potatoes, raspberries, redcurrants*** [3]
- Annual dicotyledons in ***amenity vegetation, ornamental plant production*** [1, 5, 6]; ***established woody ornamentals, ornamental trees*** [2]; ***roses*** [1, 2, 5, 6]
- Annual grasses in ***amenity vegetation, ornamental plant production*** [1, 5, 6]; ***established woody ornamentals, ornamental trees*** [2]; ***roses*** [1, 2, 5, 6]
- Aquatic weeds in ***enclosed waters, land immediately adjacent to aquatic areas, open waters*** [1, 4-6]
- Grass weeds in ***amenity vegetation, apples, blackberries, blackcurrants, enclosed waters, gooseberries, hard surfaces, loganberries, natural surfaces not intended to bear vegetation, open waters, pears, permeable surfaces overlying soil, potatoes, raspberries, redcurrants*** [3]
- Perennial dicotyledons in ***amenity vegetation, ornamental plant production*** [1, 5, 6]; ***established woody ornamentals, ornamental trees*** [2]; ***roses*** [1, 2, 5, 6]
- Perennial grasses in ***amenity vegetation, ornamental plant production*** [1, 5, 6]; ***established woody ornamentals, ornamental trees*** [2]; ***roses*** [1, 2, 5, 6]
- Total vegetation control in ***hard surfaces, natural surfaces not intended to bear vegetation, permeable surfaces overlying soil*** [1, 5, 6]

Approval information

- Dichlobenil has not been included in Annex 1 and the final use date for products containing dichlobenil is 18 March 2010
- Approval expiry 18 Mar 2010 [6]

Efficacy guidance

- Best results from soil applications achieved by application in winter to moist soil during cool weather, particularly if rain follows soon after. Do not disturb treated soil by hoeing
- In aquatic situations apply as soon as active growth commences to reduce the likelihood of subsequent water deoxygenation caused by die-back of heavy weed stands. Later partial treatment for rooted species may be effective [1, 4-6]
- Where mulching is practised, apply beforehand
- Lower rates control annuals, higher rates perennials, see label for details
- Residual activity in soil lasts 3-6 mth with selective, up to 12 mth with non-selective rates
- Do not use on fen peat or moss soils as efficacy will be severely impaired
- For control of emergent, floating and submerged aquatics apply to water surface in early spring. Intended for use in still or sluggish flowing water [1, 4-6]
- Control in water is effective where flow does not exceed 90 m/hr (2.5 cm/second). Treatment of faster flowing water will only be effective if the flow can be checked for a minimum of 7 d after application [1, 4-6]

Restrictions

- Maximum number of soil treatments 1 per yr in all situations
- Do not treat crops established for less than 2 yr. See label for lists of resistant and sensitive species. Do not treat stone fruit or Norway spruce (Christmas trees)
- Do not apply within 300 m of glasshouses or hops or near areas underplanted with bulbs, annuals or herbaceous stock
- Do not apply to frozen, snow-covered or waterlogged ground or when crop foliage wet

- Do not apply to sites less than 18 mth before replanting or sowing
- Store well away from corms, bulbs, tubers and seed

Crop-specific information

- Latest use: before end of spring for aquatic situations; 4 wk before budding for grapevines; before onset of spring growth for potato dumps and all other crops. See labels for precise details
- Apply to crops in dormant period. See label for details of timing and rates
- Apply evenly by hand applicator or suitable mechanical spreader
- Apply to potato dump sites before emergence of potatoes for blight prevention
- Aquatic treatments intended for use either in still waterbodies or in sluggishly flowing water courses [1, 4-6]

Environmental safety

- Dangerous for the environment [1]
- Toxic to aquatic organisms
- If fish are known to be present in a water body intended for treatment for aquatic weed control, partial treatment should be adopted. See label for details [1, 4-6]
- Do not dump surplus herbicide in water or ditch bottoms
- Use not permitted on reservoirs that form part of a water supply system [1, 4-6]
- Do not use treated water for irrigation purposes within 2 wk of treatment or until concentration of dichlobenil falls below 0.3 ppm [4-6]
- Use as an aquatic herbicide subject to requirements set out in *Guidelines for the use of herbicides in or near watercourses or lakes* available from Defra [1, 4-6]

Hazard classification and safety precautions

Hazard Harmful, Dangerous for the environment [1]; Irritant [3]
Risk phrases R21, R51 [1]; R43 [3]; R52 [2, 4-6]; R53a [1, 2, 4-6]
Operator protection A, H [1, 3]; U05a [1, 3]; U20b [3, 5, 6]; U20c [1, 2, 4]
Environmental protection E13c [1, 4]; E13e [3] (3 ppm); E15a [2, 3, 5, 6]; E19a [4-6]; E21 [4] (2 wk); E38 [2, 5, 6]
Storage and disposal D01, D02 [1, 3]; D04 [5, 6]; D07 [2, 4-6]; D09a, D11a [1-6]
Medical advice M05b [1]

166 1,3-dichloropropene

A halogenated hydrocarbon soil nematicide; product withdrawn 2008

167 dichlorprop-P

A translocated phenoxy carboxylic acid herbicide for use in cereals
HRAC mode of action code: O

See also *2,4-D + dicamba + dichlorprop-P*
2,4-D + dichlorprop-P + MCPA + mecoprop-P
dicamba + dichlorprop-P + MCPA

Products

Headland Link	Headland	500 g/l	SL	11091

Uses

- Annual dicotyledons in ***permanent grassland, rotational grass, spring barley, spring oats, spring wheat, undersown spring cereals, undersown winter cereals, winter barley, winter oats, winter wheat***
- Charlock in ***permanent grassland, rotational grass, spring barley, spring oats, spring wheat, undersown spring cereals, undersown winter cereals, winter barley, winter oats, winter wheat***
- Chickweed in ***permanent grassland, rotational grass, spring barley, spring oats, spring wheat, undersown spring cereals, undersown winter cereals, winter barley, winter oats, winter wheat***
- Fat hen in ***permanent grassland, rotational grass, spring barley, spring oats, spring wheat, undersown spring cereals, undersown winter cereals, winter barley, winter oats, winter wheat***

- Perennial dicotyledons in ***permanent grassland, rotational grass, spring barley, spring oats, spring wheat, undersown spring cereals, undersown winter cereals, winter barley, winter oats, winter wheat***
- Redshank in ***permanent grassland, rotational grass, spring barley, spring oats, spring wheat, undersown spring cereals, undersown winter cereals, winter barley, winter oats, winter wheat***

Approval information

- Dichlorprop-P included in Annex I under EC Directive 91/414
- Accepted by BBPA for use on malting barley

Efficacy guidance

- Best results achieved by application to young seedling weeds in good growing conditions in a strongly competing crop
- Where cleavers are to be controlled increase water volume if crop or weed is dense
- Control of redshank is best achieved in warm temperatures even if this means allowing the weed to grow beyond the seedling stage

Restrictions

- Maximum number of treatments 1 per crop
- Do not apply during cold weather, drought, rain or when rain is expected
- Do not roll or harrow within 7 d before or after spraying
- Do not apply immediately before sowing any crop

Crop-specific information

- Latest use: before second node detectable (GS 32)
- Spray winter cereals in the spring from the leaf sheath erect stage; spray spring cereals from the one leaf stage

Following crops guidance

- Keep livestock out of treated areas until foliage of any poisonous weeds such as ragwort has died and become unpalatable

Environmental safety

- Harmful to fish or other aquatic life. Do not contaminate surface waters or ditches with chemical or used container
- Keep livestock out of treated areas for at least 2 wk and until foliage of any poisonous weeds, such as ragwort, has died and become unpalatable
- Do not spray in windy conditions when drift may cause damage to neighbouring sensitive crops

Hazard classification and safety precautions

Hazard Harmful, Irritant
UN number N/C
Risk phrases R22a, R41, R43
Operator protection A, C, H; U05a, U08, U20b
Environmental protection E07a, E13c, E34
Storage and disposal D01, D02, D05, D09a, D10a
Medical advice M03

168 dichlorprop-P + ferrous sulphate + MCPA

A herbicide/fertilizer combination for moss and weed control in turf
HRAC mode of action code: O + O

See also *ferrous sulphate*
MCPA

Products

1	SHL Granular Feed, Weed & Mosskiller	Sinclair	0.2:10.9:0.3% w/w	GR	10972
2	SHL Turf Feed, Weed & Mosskiller	Sinclair	0.2:10.9:0.3% w/w	DP	10973

Uses

- Annual dicotyledons in ***managed amenity turf***
- Buttercups in ***managed amenity turf***

- Moss in ***managed amenity turf***
- Perennial dicotyledons in ***managed amenity turf***

Approval information

- Dichlorprop-P and MCPA included in Annex I under EC Directive 91/414

Efficacy guidance

- Apply between Mar and Sep when grass in active growth and soil moist
- A repeat treatment may be needed after 4-6 wk to control perennial weeds or if moss regrows
- Water in if rainfall does not occur within 48 h [2]

Restrictions

- Do not treat newly sown grass for 6 mth after establishment [1]
- Do not apply during drought or freezing conditions or when rain imminent
- Avoid walking on treated areas until it has rained or turf has been watered
- Do not mow for 3-4 d before or after application
- Do not use first 4 mowings after treatment for mulching. Mowings should be composted for 6 mth before use
- Do not treat areas of fine turf such as golf or bowling greens [1]
- Avoid contact with tarmac surfaces as staining may occur

Environmental safety

- Avoid drift onto nearby plants and borders [1]

Hazard classification and safety precautions

UN number N/C
Operator protection U20c
Storage and disposal D01, D09a, D11a

169 dichlorprop-P + ioxynil

A foliar applied translocated and contact herbicide mixture
HRAC mode of action code: O + C3

See also *ioxynil*

Products

Mextrol DP	Nufarm UK	500:116 g/l	SL	11529

Uses

- Annual dicotyledons in ***grass seed crops*** *(off-label)*, ***spring barley***, ***spring wheat***, ***triticale***, ***winter barley***, ***winter oats***, ***winter wheat***

Specific Off-Label Approvals (SOLAs)

- ***grass seed crops*** *20060525*

Approval information

- Dichlorprop-P and ioxynil included in Annex I under EC Directive 91/414
- Accepted by BBPA for use on malting barley

Hazard classification and safety precautions

Hazard Harmful
UN number N/C
Risk phrases R22a, R41, R52, R53a, R63
Operator protection A, C, H; U05a, U11, U15
Environmental protection E07a, E12a, E12e, E15a, E34
Storage and disposal D01, D02, D05, D12b
Medical advice M03, M05a

170 dichlorprop-P + MCPA

A translocated herbicide for use in managed amenity turf
HRAC mode of action code: O + O

See also *MCPA*

Products

1	SHL Granular Feed and Weed	Sinclair	0.2:0.3% w/w	GR	10970
2	SHL Turf Feed and Weed	Sinclair	0.2:0.3% w/w	DP	10963

Uses

- Annual dicotyledons in ***managed amenity turf*** [1, 2]
- Buttercups in ***managed amenity turf*** [1, 2]
- Clovers in ***managed amenity turf*** [2]
- Daisies in ***managed amenity turf*** [2]
- Dandelions in ***managed amenity turf*** [2]
- Perennial dicotyledons in ***managed amenity turf*** [1, 2]

Approval information

- Dichlorprop-P and MCPA included in Annex I under EC Directive 91/414

Efficacy guidance

- Best results achieved by application to young seedling weeds in active growth between Mar and Sep
- Water in if rainfall does not occur within 48 h [2]

Restrictions

- Do not use on newly sown turf for 6 mth after establishment
- Do not cut or mow for 2-3 d before and after treatment
- Do not spray during cold weather, if rain or frost expected, if crop wet or in drought
- The first four mowings should not be used for mulching and should be composted for 6 mth before use
- Avoid close mowing for 3-4 d before or after treatment
- Do not treat areas of fine turf such as golf or bowling greens [1]

Environmental safety

- Avoid drift onto nearby plants and borders [2]
- Avoid contact with tarmac surfaces as staining may occur

Hazard classification and safety precautions

UN number N/C
Operator protection U20c
Storage and disposal D01, D09a, D11a

171 dichlorprop-P + MCPA + mecoprop-P

A translocated herbicide mixture for winter and spring cereals
HRAC mode of action code: O + O + O

See also *MCPA*
mecoprop-P

Products

Hymec Triple	Agrichem	310:160:130 g/l	SL	09949

Uses

- Annual dicotyledons in ***durum wheat***, ***grass seed crops*** *(off-label)*, ***spring barley***, ***spring oats***, ***spring wheat***, ***winter barley***, ***winter oats***, ***winter wheat***
- Chickweed in ***durum wheat***, ***spring barley***, ***spring oats***, ***spring wheat***, ***winter barley***, ***winter oats***, ***winter wheat***
- Cleavers in ***durum wheat***, ***spring barley***, ***spring oats***, ***spring wheat***, ***winter barley***, ***winter oats***, ***winter wheat***

- Field pansy in ***durum wheat***, ***spring barley***, ***spring oats***, ***spring wheat***, ***winter barley***, ***winter oats***, ***winter wheat***
- Mayweeds in ***durum wheat***, ***spring barley***, ***spring oats***, ***spring wheat***, ***winter barley***, ***winter oats***, ***winter wheat***
- Poppies in ***durum wheat***, ***spring barley***, ***spring oats***, ***spring wheat***, ***winter barley***, ***winter oats***, ***winter wheat***

Specific Off-Label Approvals (SOLAs)

- ***grass seed crops*** *20060918*

Approval information

- Dichlorprop-P, MCPA and mecoprop-P included in Annex I under EC Directive 91/414
- Accepted by BBPA for use on malting barley

Efficacy guidance

- Best results obtained if application is made while majority of weeds are at seedling stage but not if temperatures are too low
- Optimum results achieved by spraying when temperature is above 10 °C. If temperatures are lower delay spraying until growth becomes more active

Restrictions

- Maximum number of treatments 1 per crop
- Do not spray in windy conditions where spray drift may cause damage to neighbouring crops, especially sugar beet, oilseed rape, peas, turnips and most horticultural crops including lettuce and tomatoes under glass

Crop-specific information

- Latest use: before second node detectable (GS 32) for all crops

Environmental safety

- Harmful to aquatic organisms
- Harmful to fish or other aquatic life. Do not contaminate surface waters or ditches with chemical or used container
- Some pesticides pose a greater threat of contamination of water than others and mecoprop-P is one of these pesticides. Take special care when applying mecoprop-P near water and do not apply if heavy rain is forecast

Hazard classification and safety precautions

Hazard Harmful
UN number N/C
Risk phrases R22a, R41, R52, R53a
Operator protection A, C, H, M; U05a, U08, U11, U20b
Environmental protection E13c, E34
Storage and disposal D01, D02, D05, D09a, D10c
Medical advice M03

172 difenacoum

An anticoagulant coumarin rodenticide

Products

1	Difenag	Killgerm	0.005% w/w	RB	H8492
2	Neokil	Sorex	0.005% w/w	RB	H6769
3	Neosorexa	Sorex	0.005% w/w	RB	H6773
4	Neosorexa Gold	Sorex	0.005% w/w	RB	H8301
5	Sakarat D (Whole Wheat)	Killgerm	0.005% w/w	RB	H7109
6	Sakarat D Wax Bait	Killgerm	0.005% w/w	RB	H7489
7	Sorexa Gel	Sorex	0.005% w/w	RB	H6790

Uses

- Mice in ***farm buildings***, ***farmyards*** [2-7]; ***farm buildings/yards*** [1]
- Rats in ***farm buildings***, ***farmyards*** [2-5]; ***farm buildings/yards*** [1]

Efficacy guidance

- Difenacoum is a chronic poison and rodents need to feed several times before accumulating a lethal dose. Effective against rodents resistant to other commonly used anticoagulants

SECTION 2

- Best results achieved by placing baits at points between nesting and feeding places, at entry points, in holes and where droppings are seen
- A minimum of five baiting points normally required for a small infestation; more than 40 for a large infestation
- Inspect bait sites frequently and top up as long as there is evidence of feeding
- Product formulated for application through a skeleton or caulking gun [7]
- Maintain a few baiting points to guard against reinfestation after a successful control campaign
- Products contain an insecticide and fungistat to prevent insect and mould attack [2, 3]

Restrictions

- Only for use by farmers, horticulturists and other professional users
- When working in rodent infested areas wear synthetic rubber/PVC gloves to protect against rodent-borne diseases

Environmental safety

- Harmful to wildlife
- Cover baits by placing in bait boxes, drain pipes or under boards to prevent access by children, animals or birds
- Products contain human taste deterrent

Hazard classification and safety precautions

Operator protection A [7]; U13, U20b
Environmental protection E10b [4]; E15a [7]
Storage and disposal D09a, D11a
Vertebrate/rodent control products V01a, V03a, V04a [2, 3, 7]; V01b, V04b [4, 6]; V02 [2-4, 6, 7]; V03b [6]; V04c [4]
Medical advice M03

173 difenoconazole

A diphenyl-ether triazole protectant and curative fungicide
FRAC mode of action code: 3

See also azoxystrobin + difenoconazole

Products

1	Difcor 250 EC	Nufarm UK	250 g/l	EC	12361
2	Difcor 250 EC	Q-Chem	250 g/l	EC	13917
3	Plover	Syngenta	250 g/l	EC	11763
4	Turnstone	AgChem Access	250 g/l	EC	14019

Uses

- Alternaria in ***broccoli***, ***brussels sprouts***, ***cabbages***, ***calabrese***, ***cauliflowers***, ***spring oilseed rape***, ***winter oilseed rape*** [1-4]
- Blight in ***celery (outdoor)*** *(off-label)*, ***rhubarb*** *(off-label)* [3]
- Brown rust in ***winter wheat*** [1-4]
- Celery leaf spot in ***celeriac*** *(off-label)* [3]
- Disease control in ***chinese cabbage*** *(off-label)*, ***choi sum*** *(off-label)*, ***collards*** *(off-label)*, ***kale*** *(off-label)*, ***pak choi*** *(off-label)*, ***spring greens*** *(off-label)* [2]
- Foliar disease control in ***borage*** *(off-label)*, ***canary flower (echium spp.)*** *(off-label)*, ***durum wheat*** *(off-label)*, ***evening primrose*** *(off-label)*, ***grass seed crops*** *(off-label)*, ***honesty*** *(off-label)*, ***linseed*** *(off-label)*, ***mustard*** *(off-label)*, ***spring rye*** *(off-label)*, ***triticale*** *(off-label)*, ***winter rye*** *(off-label)* [3]
- Late blight in ***celery (outdoor)*** *(off-label)*, ***rhubarb*** *(off-label)* [3]
- Light leaf spot in ***spring oilseed rape***, ***winter oilseed rape*** [1-4]
- Ring spot in ***broccoli***, ***brussels sprouts***, ***cabbages***, ***calabrese***, ***cauliflowers*** [1-4]; ***chinese cabbage*** *(off-label)*, ***choi sum*** *(off-label)*, ***collards*** *(off-label)*, ***pak choi*** *(off-label)*, ***protected pinks*** *(off-label - ground grown (or pot grown - SOLA 20050159))*, ***protected sweet williams*** *(off-label - ground grown (or pot grown - SOLA 20050159))* [3]; ***kale*** *(off-label)* [1, 3]; ***protected pinks*** *(off-label - ground grown (or pot grown - SOLA 20051488))*, ***protected sweet williams*** *(off-label - ground grown (or pot grown - SOLA 20051488))* [1]
- Rust in ***asparagus*** *(off-label)* [1-3]; ***protected hybrid pinks*** *(off-label)*, ***protected sweet williams*** *(off-label)* [2]; ***protected pinks*** *(off-label - ground grown (or pot grown - SOLA 20050159))*, ***protected sweet williams*** *(off-label - ground grown (or pot grown - SOLA 20050159))* [3];

protected pinks *(off-label - ground grown (or pot grown - SOLA 20051488)),* ***protected sweet williams*** *(off-label - ground grown (or pot grown - SOLA 20051488))* [1]
- Septoria leaf blotch in ***winter wheat*** [1-4]
- Stem canker in ***spring oilseed rape, winter oilseed rape*** [1-4]

Specific Off-Label Approvals (SOLAs)

- ***asparagus*** *20051493* [1], *20080617* [2], *20050158* [3]
- ***borage*** *20060559* [3]
- ***canary flower (echium spp.)*** *20060559* [3]
- ***celeriac*** *20060261* [3]
- ***celery (outdoor)*** *20091939* [3]
- ***chinese cabbage*** *20080615* [2], *20050558* [3]
- ***choi sum*** *20080615* [2], *20050558* [3]
- ***collards*** *20080615* [2], *20050558* [3]
- ***durum wheat*** *20060558* [3]
- ***evening primrose*** *20060559* [3]
- ***grass seed crops*** *20060558* [3]
- ***honesty*** *20060559* [3]
- ***kale*** *20051492* [1], *20080616* [2], *20050559* [3]
- ***linseed*** *20060559* [3]
- ***mustard*** *20060559* [3]
- ***pak choi*** *20080615* [2], *20050558* [3]
- ***protected hybrid pinks*** *20080613* [2]
- ***protected pinks*** *(ground grown (or pot grown - SOLA 20051488)) 20051486* [1], *(ground grown (or pot grown - SOLA 20050159)) 20050156* [3]
- ***protected sweet williams*** *(ground grown (or pot grown - SOLA 20051488)) 20051486* [1], *20080613* [2], *(ground grown (or pot grown - SOLA 20050159)) 20050156* [3]
- ***rhubarb*** *20091939* [3]
- ***spring greens*** *20080615* [2]
- ***spring rye*** *20060558* [3]
- ***triticale*** *20060558* [3]
- ***winter rye*** *20060558* [3]

Approval information

- Difenoconazole included in Annex I under EC Directive 91/414

Efficacy guidance

- Improved control of established infections on oilseed rape achieved by mixture with carbendazim. See label
- In brassicas a 3-spray programme should be used starting at the first sign of disease and repeated at 14-21 d intervals
- Product is fully rainfast 2 h after application
- For most effective control of Septoria, apply as part of a programme of sprays which includes a suitable flag leaf treatment
- Difenoconazole is a DMI fungicide. Resistance to some DMI fungicides has been identified in Septoria leaf blotch which may seriously affect performance of some products. For further advice contact a specialist advisor and visit the Fungicide Resistance Action Group (FRAG)-UK website

Restrictions

- Maximum number of treatments 3 per crop for brassicas; 2 per crop for oilseed rape; 1 per crop for wheat [1]
- Maximum total dose equivalent to 3 full dose treatments on brassicas; 2 full dose treatments on oilseed rape; 1 full dose treatment on wheat [3]
- Apply to wheat any time from ear fully emerged stage but before early milk-ripe stage (GS 59-73)

Crop-specific information

- Latest use: before grain early milk-ripe stage (GS 73) for cereals; end of flowering for oilseed rape
- HI brassicas 21 d
- Treat oilseed rape in autumn from 4 expanded true leaf stage (GS 1,4). A repeat spray may be made in spring at the beginning of stem extension (GS 2,0) if visible symptoms develop

SEE SECTION 3 FOR PRODUCTS ALSO REGISTERED

Environmental safety

- Dangerous for the environment
- Very toxic to aquatic organisms

Hazard classification and safety precautions

Hazard Irritant [1, 2]; Dangerous for the environment [1-4]
Transport code 9
Packaging group III
UN number 3082
Risk phrases R38, R41, R51 [1, 2]; R50 [3, 4]; R53a [1-4]
Operator protection A, C, H; U05a, U09a, U20b [1-4]; U11 [1, 2]
Environmental protection E15a, E38
Storage and disposal D01, D02, D09a, D12a [1-4]; D05, D07, D10c [3, 4]; D10b [1, 2]

174 difenoconazole + fenpropidin

A triazole morpholine fungicide mixture for beet crops
FRAC mode of action code: 3 + 5

See also *fenpropidin*

Products

Spyrale	Syngenta	375:100 g/l	EC	12566

Uses

- Powdery mildew in ***fodder beet***, ***sugar beet***
- Ramularia leaf spots in ***fodder beet***, ***sugar beet***
- Rust in ***fodder beet***, ***sugar beet***

Approval information

- Difenoconazole and fenpropidin included in Annex I under EC Directive 91/414

Efficacy guidance

- For best results apply as a preventative treatment or as soon as first symptoms of disease are seen
- Product gives prolonged protection against re-infection but a second application may be needed where crops remain under heavy disease pressure

Restrictions

- Maximum number of treatments 2 per crop of sugar beet or fodder beet

Crop-specific information

- HI: 28 d for fodder beet, sugar beet

Environmental safety

- Dangerous for the environment
- Very toxic to aquatic organisms

Hazard classification and safety precautions

Hazard Harmful, Dangerous for the environment
Transport code 9
Packaging group III
UN number 3082
Risk phrases R22a, R36, R37, R38, R50, R53a
Operator protection A, C, H; U02a, U04a, U05a, U09a, U10, U11, U14, U15, U20a
Environmental protection E15a, E34, E38
Storage and disposal D01, D02, D05, D07, D09a, D10c, D12a
Medical advice M03

SECTION 2

175 difenoconazole + fludioxonil

A triazole + phenylpyrrole seed treatment for use in cereals
FRAC mode of action code: 3 + 12

See also *fludioxonil*

Products

Celest Extra	Syngenta	25:25 g/l	FS	14050

Uses

- Bunt in ***winter wheat***
- Fusarium foot rot and seedling blight in ***winter oats***, ***winter wheat***
- Microdochium nivale in ***winter oats***, ***winter wheat***
- Pyrenophora leaf spot in ***winter oats***
- Seed-borne diseases in ***winter wheat***
- Seedling blight and foot rot in ***winter wheat***
- Septoria seedling blight in ***winter wheat***
- Snow mould in ***winter wheat***
- Stripe smut in ***winter rye***

Approval information

- Difenoconazole and fludioxonil included in Annex I under EC Directive 91/414

Efficacy guidance

- [1] is effective against benzimidazole-resistant and benzimidazole-sensitive strains of Microdochium nivale

Crop-specific information

- Under adverse environmental or soil conditions, seed rates should be increased to compensate for a slight drop in germination capacity. Flow rates of seed treated with [1] should be checked before drilling commences.

Hazard classification and safety precautions

Transport code 9
Packaging group III
UN number 3082
Risk phrases R52, R53a
Operator protection A, H; U05a, U20c
Environmental protection E15a, E38
Storage and disposal D01, D02, D09a, D11a, D12a
Treated seed S01, S02, S03, S04a, S04b, S05, S06a, S07, S08, S09

176 diflubenzuron

A selective, persistent, contact and stomach acting insecticide
IRAC mode of action code: 15

Products

1	Dimilin Flo	Chemtura	480 g/l	SC	08769
2	Dimilin Flo	Certis	480 g/l	SC	11056

Uses

- Browntail moth in ***amenity vegetation***, ***hedges***, ***nursery stock***, ***ornamental plant production*** [2]
- Bud moth in ***apples***, ***pears*** [2]
- Carnation tortrix moth in ***amenity vegetation***, ***hedges***, ***nursery stock***, ***ornamental plant production*** [2]
- Caterpillars in ***broccoli***, ***brussels sprouts***, ***cabbages***, ***calabrese***, ***cauliflowers***, ***chives*** *(off-label)*, ***endives*** *(off-label)*, ***frise*** *(off-label)*, ***herbs (see appendix 6)*** *(off-label)*, ***leaf spinach*** *(off-label)*, ***lettuce*** *(off-label)*, ***parsley*** *(off-label)*, ***radicchio*** *(off-label)*, ***salad brassicas*** *(off-label - for baby leaf production)*, ***scarole*** *(off-label)* [2]
- Clouded drab moth in ***apples***, ***pears*** [2]
- Codling moth in ***apples***, ***pears*** [2]

- Flies in ***almonds*** *(off-label)*, ***bilberries*** *(off-label)*, ***blueberries*** *(off-label)*, ***chestnuts*** *(off-label)*, ***crab apples*** *(off-label)*, ***cranberries*** *(off-label)*, ***edible fungi*** *(off-label)*, ***gooseberries*** *(off-label)*, ***hazel nuts*** *(off-label)*, ***quinces*** *(off-label)*, ***redcurrants*** *(off-label)*, ***walnuts*** *(off-label)*, ***whitecurrants*** *(off-label)* [1]
- Fruit tree tortrix moth in ***apples***, ***pears*** [2]
- Houseflies in ***livestock houses***, ***manure heaps***, ***refuse tips*** [2]
- Lackey moth in ***amenity vegetation***, ***hedges***, ***nursery stock***, ***ornamental plant production*** [2]
- Moths in ***almonds*** *(off-label)*, ***bilberries*** *(off-label)*, ***chestnuts*** *(off-label)*, ***crab apples*** *(off-label)*, ***cranberries*** *(off-label)*, ***gooseberries*** *(off-label)*, ***hazel nuts*** *(off-label)*, ***quinces*** *(off-label)*, ***redcurrants*** *(off-label)*, ***walnuts*** *(off-label)*, ***whitecurrants*** *(off-label)* [1, 2]; ***blackberries*** *(off-label)* [2]; ***blueberries*** *(off-label)*, ***edible fungi*** *(off-label)* [1]
- Oak leaf roller moth in ***forest*** [2]
- Pear sucker in ***pears*** [2]
- Phorid flies in ***edible fungi*** *(off-label - other than mushrooms)* [2]
- Pine beauty moth in ***forest*** [2]
- Pine looper in ***forest*** [2]
- Plum fruit moth in ***plums*** [2]
- Rust mite in ***apples***, ***pears***, ***plums*** [2]
- Sciarid flies in ***edible fungi*** *(off-label - other than mushrooms)* [2]; ***mushrooms*** [1]
- Small ermine moth in ***amenity vegetation***, ***hedges***, ***nursery stock***, ***ornamental plant production*** [2]
- Tomato moth in ***protected peppers*** *(off-label)* [1]
- Tortrix moths in ***plums*** [2]
- Winter moth in ***amenity vegetation***, ***apples***, ***blackcurrants***, ***forest***, ***hedges***, ***nursery stock***, ***ornamental plant production***, ***pears***, ***plums*** [2]

Specific Off-Label Approvals (SOLAs)

- ***almonds*** *20060571* [2], *20060575* [1]
- ***bilberries*** *20060573* [2], *20060577* [1]
- ***blackberries*** *20060573* [2]
- ***blueberries*** *20060577* [1]
- ***chestnuts*** *20060571* [2], *20060575* [1]
- ***chives*** *20051321* [2]
- ***crab apples*** *20060572* [2], *20060576* [1]
- ***cranberries*** *20060573* [2], *20060577* [1]
- ***edible fungi*** *(other than mushrooms) 20060574* [2], *20060578* [1]
- ***endives*** *20051321* [2]
- ***frise*** *20051321* [2]
- ***gooseberries*** *20060573* [2], *20060577* [1]
- ***hazel nuts*** *20060571* [2], *20060575* [1]
- ***herbs (see appendix 6)*** *20051321* [2]
- ***leaf spinach*** *20051321* [2]
- ***lettuce*** *20051321* [2]
- ***parsley*** *20051321* [2]
- ***protected peppers*** *992433* [1]
- ***quinces*** *20060572* [2], *20060576* [1]
- ***radicchio*** *20051321* [2]
- ***redcurrants*** *20060573* [2], *20060577* [1]
- ***salad brassicas*** *(for baby leaf production) 20051321* [2]
- ***scarole*** *20051321* [2]
- ***walnuts*** *20060571* [2], *20060575* [1]
- ***whitecurrants*** *20060573* [2], *20060577* [1]

Approval information

- Approved for aerial application in forestry when average wind velocity does not exceed 18 knots and gusts do not exceed 20 knots [2]. See Section 5 for more information
- Diflubenzuron included in Annex I under EC Directive 91/414

Efficacy guidance

- Most active on young caterpillars and most effective control achieved by spraying as eggs start to hatch

FOR FULL CONDITIONS OF USE ALWAYS READ THE PRODUCT LABEL

- Dose and timing of spray treatments vary with pest and crop. See label for details
- Addition of wetter recommended for use on brassicas and for pear sucker control in pears

Restrictions

- Maximum number of treatments 3 per yr for apples, pears; 2 per yr for plums, blackcurrants; 2 per crop for brassicas; 1 per yr for forest [2]
- Maximum number of treatments 1 per spawning for mushrooms [1]
- Before treating ornamentals check varietal tolerance on a small sample
- Do not use as a compost drench or incorporated treatment on ornamental crops
- Do not spray protected plants in flower or with flower buds showing colour
- For use only on the food crops specified on the label
- Do not apply directly to livestock/poultry

Crop-specific information

- HI apples, pears, plums, blackcurrants, brassicas 14 d
- Apply as casing mixing treatment (use immediately) on mushrooms or as post-casing drench [1]

Environmental safety

- Dangerous for the environment
- Very toxic to aquatic organisms
- LERAP Category B
- Broadcast air-assisted LERAP (20 m in orchards, 10 m in blackcurrants, forestry or ornamentals)

Hazard classification and safety precautions

Hazard Dangerous for the environment
Transport code 9
Packaging group III
UN number 3082
Risk phrases R50
Operator protection U20c
Environmental protection E05a, E16a, E16b, E18 [2]; E15a, E38 [1, 2]; E17b [2] (20 m when used in orchards; 10 m when used in blackcurrants, forestry or ornamentals)
Storage and disposal D05, D12b [2]; D09a, D11a [1, 2]

177 diflufenican

A shoot absorbed pyridinecarboxamide herbicide for winter cereals
HRAC mode of action code: F1

See also *bromoxynil + diflufenican + ioxynil*
chlorotoluron + diflufenican
clopyralid + diflufenican + MCPA

Products

1	Crater	Makhteshim	500 g/l	SC	13084
2	Dican	Agform	50% w/w	SG	13685
3	Difenikan 500	Goldengrass	500 g/l	SC	14600
4	Diflanil 500 SC	Q-Chem	500 g/l	SC	12489
5	Diflufenican GL 500	Globachem	500 g/l	SC	12531
6	Hurricane SC	Makhteshim	500 g/l	SC	12424
7	Overlord	Makhteshim	500 g/l	SC	13521
8	Sempra	AgriChem BV	500 g/l	SC	13525
9	Sipcam Solo-D	Sipcam	500 g/l	SC	12617

Uses

- Annual dicotyledons in ***durum wheat*** [4, 5]; ***grass seed crops*** *(off-label)* [4, 9]; ***spring barley*** [8]; ***spring oats*** *(off-label)*, ***winter oats*** *(off-label)* [1, 4, 6, 7, 9]; ***triticale***, ***winter rye*** [4, 5, 8]; ***winter barley***, ***winter wheat*** [4, 5, 8, 9]
- Annual grasses in ***grass seed crops*** *(off-label)*, ***spring oats*** *(off-label)*, ***winter oats*** *(off-label)* [9]; ***spring barley***, ***triticale***, ***winter barley***, ***winter rye***, ***winter wheat*** [8]
- Chickweed in ***durum wheat***, ***triticale***, ***winter rye*** [4, 5]; ***winter barley***, ***winter wheat*** [4, 5, 9]
- Cleavers in ***durum wheat***, ***triticale***, ***winter rye*** [1, 3-7]; ***spring barley***, ***spring wheat*** [1, 3, 6, 7]; ***winter barley***, ***winter wheat*** [1-7, 9]

SEE SECTION 3 FOR PRODUCTS ALSO REGISTERED

- Field pansy in ***durum wheat*, *spring barley*, *spring wheat*, *triticale*, *winter rye*** [1, 3, 6, 7]; ***winter barley*, *winter wheat*** [1-3, 6, 7]
- Field speedwell in ***durum wheat*, *triticale*, *winter rye*** [1, 3-7]; ***spring barley*, *spring wheat*** [1, 3, 6, 7]; ***winter barley*, *winter wheat*** [1-7, 9]
- Ivy-leaved speedwell in ***durum wheat*, *triticale*, *winter rye*** [1, 3-7]; ***spring barley*, *spring wheat*** [1, 3, 6, 7]; ***winter barley*, *winter wheat*** [1-7, 9]
- Mayweeds in ***durum wheat*, *spring barley*, *spring wheat*, *triticale*, *winter rye*** [1, 3, 6, 7]; ***winter barley*, *winter wheat*** [1-3, 6, 7]
- Poppies in ***durum wheat*, *spring barley*, *spring wheat*, *triticale*, *winter rye*** [1, 3, 6, 7]; ***winter barley*, *winter wheat*** [1-3, 6, 7]
- Red dead-nettle in ***durum wheat*, *spring barley*, *spring wheat*, *triticale*, *winter rye*** [1, 3, 6, 7]; ***winter barley*, *winter wheat*** [1-3, 6, 7]
- Volunteer oilseed rape in ***durum wheat*, *triticale*, *winter rye*** [4, 5]; ***winter barley*, *winter wheat*** [4, 5, 9]

Specific Off-Label Approvals (SOLAs)

- ***grass seed crops*** *20081935* [4], *20091625 expires 31 Dec 2010* [9]
- ***spring oats*** *20082172* [1], *20081936* [4], *20081320* [6], *20082170* [7], *20091625 expires 31 Dec 2010* [9]
- ***winter oats*** *20082172* [1], *20081936* [4], *20081320* [6], *20082170* [7], *20091625 expires 31 Dec 2010* [9]

Approval information

- Diflufenican included in Annex I under EC Directive 91/414
- Accepted by BBPA for use on malting barley
- Approval expiry 31 Dec 2010 [9]

Efficacy guidance

- Best results achieved from treatment of small actively growing weeds in early autumn or spring
- Good weed control depends on efficient burial of trash or straw before or during seedbed preparation
- Loose or fluffy seedbeds should be rolled before application
- The final seedbed should be moist, fine and firm with clods no bigger than fist size
- Ensure good even spray coverage and increase spray volume for post-emergence treatments where the crop or weed foliage is dense
- Activity may be slow under cool conditions and final level of weed control may take some time to appear
- Where cleavers are a particular problem a separate specific herbicide treatment may be required
- Efficacy may be impaired on soils with a Kd factor greater than 6
- Always follow WRAG guidelines for preventing and managing herbicide resistant weeds. See Section 5 for more information

Restrictions

- Maximum number of treatments 1 per crop
- Do not treat broadcast crops [4-6]
- Do not roll treated crops or harrow at any time after treatment
- Do apply to soils with more than 10% organic matter or on Sands, or very stony or gravelly soils
- Do not treat after a period of cold frosty weather

Crop-specific information

- Latest use: before end of tillering (GS 29) for wheat and barley [4-6], pre crop emergence for triticale and winter rye [4-6]
- Treat only named varieties of rye or triticale [4-6]

Following crops guidance

- Labels vary slightly but in general ploughing to 150 mm and thoroughly mixing the soil is recommended before drilling or planting any succeeding crops either after crop failure or after normal harvest
- In the event of crop failure only winter wheat or winter barley may be re-drilled immediately after ploughing. Spring crops of wheat, barley, oilseed rape, peas, field beans, sugar beet [4-6], potatoes, carrots, edible brassicas or onions may be sown provided an interval of 12 wk has elapsed after ploughing

- After normal harvest of a treated crop winter cereals, oilseed rape, field beans, leaf brassicas, sugar beet seed crops and winter onions may be drilled in the following autumn. Other crops listed above for crop failure may be sown in the spring after normal harvest
- Successive treatments with any products containing diflufenican can lead to soil build-up and inversion ploughing to 150 mm must precede sowing any following non-cereal crop. Even where ploughing occurs some crops may be damaged

Environmental safety

- Dangerous for the environment
- Very toxic to aquatic organisms
- LERAP Category B

Hazard classification and safety precautions

Hazard Dangerous for the environment
Transport code 9
Packaging group III
UN number 3077 [2]; 3082 [1, 3-9]
Risk phrases R50, R53a
Operator protection A [1-8]; C [1-7]; U05a [1-3, 6-9]; U08, U13, U19a [1-3, 6, 7]; U20a [1-3, 6-8]
Environmental protection E13a [8]; E13c, E34, E38 [1-3, 6, 7]; E15a [4, 5, 9]; E16a [1-9]
Storage and disposal D01, D02 [1-3, 6-9]; D05 [1-7, 9]; D09a, D10b [1-9]; D12a [4, 5, 8, 9]

178 diflufenican + flufenacet

A contact and residual herbicide mixture for cereals
HRAC mode of action code: F1 + K3

See also flufenacet

Products

1	Firebird	Bayer CropScience	200:400 g/l	SC	12421
2	Liberator	Bayer CropScience	100:400 g/l	SC	12032
3	Regatta	Bayer CropScience	100:400 g/l	SC	12054

Uses

- Annual dicotyledons in ***durum wheat*** *(off-label)*, ***triticale*** *(off-label)*, ***winter rye*** *(off-label)* [2]; ***grass seed crops*** *(off-label)* [2, 3]
- Annual grasses in ***durum wheat*** *(off-label)*, ***triticale*** *(off-label)*, ***winter rye*** *(off-label)* [2]; ***grass seed crops*** *(off-label)* [2, 3]
- Annual meadow grass in ***winter barley***, ***winter wheat*** [1-3]
- Blackgrass in ***winter barley***, ***winter wheat*** [1-3]
- Chickweed in ***winter barley***, ***winter wheat*** [1-3]
- Field pansy in ***winter barley***, ***winter wheat*** [1-3]
- Field speedwell in ***winter barley***, ***winter wheat*** [1-3]
- Ivy-leaved speedwell in ***winter barley***, ***winter wheat*** [1-3]
- Mayweeds in ***winter barley***, ***winter wheat*** [1-3]
- Red dead-nettle in ***winter barley***, ***winter wheat*** [2, 3]
- Shepherd's purse in ***winter barley***, ***winter wheat*** [1]
- Volunteer oilseed rape in ***winter barley***, ***winter wheat*** [1]

Specific Off-Label Approvals (SOLAs)

- ***durum wheat*** *20063650* [2]
- ***grass seed crops*** *20063650* [2], *20061029* [3]
- ***triticale*** *20063650* [2]
- ***winter rye*** *20063650* [2]

Approval information

- Diflufenican and flufenacet included in Annex I under EC Directive 91/414
- Accepted by BBPA for use on malting barley

Efficacy guidance

- Best results obtained when there is moist soil at and after application and rain falls within 7 d
- Residual control may be reduced under prolonged dry conditions

- Activity may be slow under cool conditions and final level of weed control may take some time to appear
- Good weed control depends on burying any trash or straw before or during seedbed preparation
- Established perennial grasses and broad-leaved weeds will not be controlled
- Do not use as a stand-alone treatment for blackgrass control. Always follow WRAG guidelines for preventing and managing herbicide resistant weeds. Section 5 for more information

Restrictions

- Maximum number of treatments 1 per crop
- Do not treat undersown cereals or those to be undersown
- Do not use on waterlogged soils or soils prone to waterlogging
- Do not use on Sands or Very Light soils, or very stony or gravelly soils, or on soils containing more than 10% organic matter
- Do not treat broadcast crops and treat shallow-drilled crops post-emergence only
- Do not incorporate into the soil or disturb the soil after application by rolling or harrowing
- Avoid treating crops under stress from whatever cause and avoid treating during periods of prolonged or severe frosts

Crop-specific information

- Latest use: before 31 Dec in yr of sowing and before 3rd tiller stage (GS 23) for wheat or 4th tiller stage (GS 24) for barley
- For pre-emergence treatments the seed should be covered with a minimum of 32 mm settled soil

Following crops guidance

- In the event of crop failure wheat, barley or potatoes may be sown provided the soil is ploughed to 15 cm, and a minimum of 12 weeks elapse between treatment and sowing spring wheat or spring barley
- After normal harvest wheat, barley or potatoes my be sown without special cultivations. Soil must be ploughed or cultivated to 15 cm before sowing oilseed rape, field beans, peas, sugar beet, carrots, onions or edible brassicae
- Successive treatments with any products containing diflufenican can lead to soil build-up and inversion ploughing to 150 mm must precede sowing any following non-cereal crop. Even where ploughing occurs some crops may be damaged

Environmental safety

- Dangerous for the environment
- Very toxic to aquatic organisms
- Risk to non-target insects or other arthropods. Avoid spraying within 6 m of the field boundary to reduce the effects on non-target insects or other arthropods
- LERAP Category B

Hazard classification and safety precautions

Hazard Harmful, Dangerous for the environment
Transport code 9
Packaging group III
UN number 3082
Risk phrases R22a, R43, R48, R50, R53a
Operator protection A, H; U05a, U14
Environmental protection E15a [1, 3]; E16a, E34, E38 [1-3]; E22c [1, 2]
Storage and disposal D01, D02, D09a, D10b, D12a
Medical advice M03, M05a

179 diflufenican + flupyrsulfuron-methyl

A pyridinecarboxamide and sulfonylurea herbicide mixture for wheat
HRAC mode of action code: F1 + B

See also flupyrsulfuron-methyl

Products

Absolute	DuPont	41.7:8.3% w/w	WG	12558

Uses
- Annual dicotyledons in ***winter oats*** *(off-label)*, ***winter wheat***
- Blackgrass in ***winter oats*** *(off-label)*, ***winter wheat***

Specific Off-Label Approvals (SOLAs)
- ***winter oats*** *20082624*

Approval information
- Diflufenican and flupyrsulfuron-methyl included in Annex I under EC Directive 91/414

Efficacy guidance
- Best results achieved from applications made in good growing conditions
- Good spray cover of weeds must be obtained
- Best control of blackgrass obtained from application from the 2-leaf stage but tank mixing or sequencing with a partner product is recommended as a resistance management strategy (see below)
- Growth of weeds is inhibited within hours of treatment but visible symptoms may not be apparent for up to 4 wk
- Product has moderate residual life in soil. Under normal moisture conditions susceptible weeds germinating soon after treatment will be controlled
- Product may be used on all soil types except those with more than 10% organic matter, but residual activity and weed control is reduced by high soil temperatures and when soil is very dry and on highly alkaline soils
- Flupyrsulfuron-methyl is a member of the ALS-inhibitor group of herbicides. To avoid the build up of resistance do not use any product containing an ALS-inhibitor herbicide with claims for control of grass weeds more than once on any crop
- Use these products as part of a resistance management strategy that includes cultural methods of control and does not use ALS inhibitors as the sole chemical method of grass weed control

Restrictions
- Maximum number of treatments 1 per crop
- Do not use on wheat undersown with grasses, clover or other legumes, or any other broad-leaved crop
- Do not apply within 7 d of rolling
- Do not treat broadcast crops. Seed should be covered by at least 25 mm of settled soil
- Do not treat any crop suffering from drought, waterlogging, pest or disease attack, nutrient deficiency, or any other stress factors
- Do not use on soils with more than 10% organic matter
- Specific restrictions apply to use in sequence or tank mixture with other sulfonylurea or ALS-inhibiting herbicides. See label for details

Crop-specific information
- Latest use: before 1st node detectable stage (GS 31) for winter wheat
- Under certain conditions slight chlorosis and stunting may occur from which recovery is rapid

Following crops guidance
- Only cereals, oilseed rape or field beans may be sown in the yr of harvest of a treated crop. Ploughing must be carried out before drilling any non-cereal crop
- An interval of at least 20 wk must elapse after treatment before planting wheat, barley, oilseed rape, peas, field beans, sugar beet, potatoes, carrots, brassicas or onions in the following spring, and only after ploughing
- In the event of failure of a treated crop only winter wheat or spring wheat may be sown within 3 mth of treatment, and only after ploughing and cultivating to at least 15 cm. After 3 mth the normal guidance for following crops may be observed
- Successive treatments with any products containing diflufenican can lead to soil build-up and inversion ploughing to 150 mm must precede sowing any following non-cereal crop. Even where ploughing occurs some crops may be damaged

Environmental safety
- Dangerous for the environment
- Very toxic to aquatic organisms
- LERAP Category B
- Take extreme care to avoid damage by drift onto broad-leaved plants outside the target area

- Do not drain or flush spraying equipment onto land planted, or intended for planting, with trees or crops other than cereals
- Poor sprayer cleanout practices may result in inadequate removal of chemical deposits which may lead to damage of subsequently treated non-cereal crops. See label for guidance on cleaning

Hazard classification and safety precautions

Hazard Dangerous for the environment
Transport code 9
Packaging group III
UN number 3077
Risk phrases R50, R53a
Operator protection U08, U19a, U20b
Environmental protection E15a, E16a, E38
Storage and disposal D09a, D12a

180 diflufenican + flurtamone

A contact and residual herbicide mixture for cereals
HRAC mode of action code: F1 + F1

See also flurtamone

Products

1	Bacara	Bayer CropScience	100:250 g/l	SC	10744
2	Graduate	Bayer CropScience	80:400 g/l	SC	10776

Uses

- Annual dicotyledons in ***grass seed crops*** *(off-label)*, ***spring barley***, ***spring rye*** *(off-label)*, ***triticale*** *(off-label)*, ***winter barley***, ***winter rye*** *(off-label)*, ***winter wheat*** [1, 2]; ***winter oats*** *(off-label)* [2]
- Annual grasses in ***grass seed crops*** *(off-label)*, ***spring rye*** *(off-label)*, ***triticale*** *(off-label)*, ***winter rye*** *(off-label)* [1, 2]; ***winter oats*** *(off-label)* [2]
- Annual meadow grass in ***spring barley***, ***winter barley***, ***winter wheat*** [1, 2]; ***winter oats*** *(off-label)* [2]
- Blackgrass in ***spring barley***, ***winter barley***, ***winter wheat*** [1]
- Loose silky bent in ***spring barley***, ***winter barley***, ***winter wheat*** [1]
- Volunteer oilseed rape in ***spring barley***, ***winter barley***, ***winter wheat*** [1, 2]; ***winter oats*** *(off-label)* [2]

Specific Off-Label Approvals (SOLAs)

- ***grass seed crops*** *20060813* [1], *20060812* [2]
- ***spring rye*** *20060813* [1], *20060812* [2]
- ***triticale*** *20060813* [1], *20060812* [2]
- ***winter oats*** *20082060* [2]
- ***winter rye*** *20060813* [1], *20060812* [2]

Approval information

- Diflufenican and flurtamone included in Annex I under EC Directive 91/414
- Accepted by BBPA for use on malting barley

Efficacy guidance

- Apply pre-emergence or from when crop has first leaf unfolded before susceptible weeds pass recommended size
- Best results obtained on firm, fine seedbeds with adequate soil moisture present at and after application. Increase water volume where crop or weed foliage is dense
- Good weed control requires ash, trash and burnt straw to be buried during seed bed preparation
- Loose fluffy seedbeds should be rolled before application and the final seed bed should be fine and firm without large clods
- Speed of control depends on weather conditions and activity can be slow under cool conditions
- Always follow WRAG guidelines for preventing and managing herbicide resistant weeds. See Section 5 for more information

Restrictions

- Maximum number of treatments 1 per crop

- Crops should be drilled to a normal depth of 25 mm and the seed well covered. Do not treat broadcast crops as uncovered seed may be damaged
- Do not treat spring sown cereals, durum wheat, oats, undersown cereals or those to be undersown
- Do not treat frosted crops or when frost is imminent. Severe frost after application, or any other stress, may lead to transient discoloration or scorch
- Do not use on Sands or Very Light soils or those that are very stony or gravelly
- Do not use on waterlogged soils, or on crops subject to temporary waterlogging by heavy rainfall, as there is risk of persistent crop damage which may result in yield loss
- Do not use on soils with more than 10% organic matter
- Do not harrow at any time after application and do not roll autumn treated crops until spring

Crop-specific information

- Latest use: before 2nd node detectable (GS32)
- Take particular care to match spray swaths otherwise crop discoloration and biomass reduction may occur which may lead to yield reduction

Following crops guidance

- In the event of crop failure winter wheat may be redrilled immediately after normal cultivation, and winter barley may be sown after ploughing. Fields must be ploughed to a depth of 15 cm and 20 wk must elapse before sowing spring crops of wheat, barley, oilseed rape, peas, field beans or potatoes
- After normal harvest autumn cereals can be drilled after ploughing. Thorough mixing of the soil must take place before drilling field beans, leaf brassicae or winter oilseed rape. For sugar beet seed crops and winter onions complete inversion of the furrow slice is essential
- Do not broadcast or direct drill oilseed rape or other brassica crops as a following crop on treated land. See label for detailed advice on preparing land for subsequent autumn cropping in the normal rotation
- Successive treatments with any products containing diflufenican can lead to soil build-up and inversion ploughing to 150 mm must precede sowing any following non-cereal crop. Even where ploughing occurs some crops may be damaged

Environmental safety

- Dangerous for the environment
- Very toxic to aquatic organisms
- Extremely dangerous to fish or other aquatic life. Do not contaminate surface waters or ditches with chemical or used container
- LERAP Category B

Hazard classification and safety precautions

Hazard Dangerous for the environment
Transport code 9
Packaging group III
UN number 3082
Risk phrases R50, R53a
Operator protection A, H; U20c
Environmental protection E13a, E16a, E16b, E38
Storage and disposal D05, D09a, D10b, D12a

181 diflufenican + glyphosate

A foliar non-selective herbicide mixture
HRAC mode of action code: F1 + G

See also *glyphosate*

Products

Pistol	Bayer Environ.	40:250 g/l	SC	12173

Uses

- Annual and perennial weeds in ***natural surfaces not intended to bear vegetation***, ***permeable surfaces overlying soil***

SEE SECTION 3 FOR PRODUCTS ALSO REGISTERED

Approval information

- Diflufenican and glyphosate included in Annex I under EC Directive 91/414

Efficacy guidance

- Treat when weeds actively growing from Mar to end Sep and before they begin to senesce
- Performance may be reduced if application is made to plants growing under stress, such as drought or water-logging
- Pre-emergence activity is reduced on soils containing more than 10% organic matter or where organic debris has collected
- Perennial weeds such as docks, perennial sowthistle and willowherb are best treated before flowering or setting seed
- Perennial weeds emerging from established rootstocks after treatment will not be controlled
- A rainfree period of at least 6 h (preferably 24 h) should follow spraying for optimum control
- For optimum control do not cultivate or rake after treatment

Restrictions

- Maximum number of treatments 1 per yr
- May only be used on porous surfaces overlying soil. Must not be used if an impermeable membrane lies between the porous surface and the soil, and must not be used on any non-porous man-made surfaces
- Do not add any wetting agent or adjuvant oil
- Do not spray in windy weather

Following crops guidance

- A period of at least 6 mth must be allowed after treatment of sites that are to be cleared and grubbed before sowing and planting. Soil should be ploughed or dug first to ensure thorough mixing and dilution of any herbicide residues

Environmental safety

- Dangerous for the environment
- Very toxic to aquatic organisms
- LERAP Category B
- Avoid drift onto non-target plants
- Heavy rain after application may wash product onto sensitive areas such as newly sown grass or areas about to be planted

Hazard classification and safety precautions

Hazard Dangerous for the environment
Transport code 9
Packaging group III
UN number 3082
Risk phrases R50, R53a
Operator protection A, C; U20b
Environmental protection E15a, E16a, E16b, E38
Storage and disposal D01, D02, D09a, D11a, D12a

182 diflufenican + iodosulfuron-methyl-sodium + mesosulfuron-methyl

A contact and residual herbicide mixture containing sulfonyl ureas for winter wheat
HRAC mode of action code: F1 + B + B

- *See also iodosulfuron-methyl-sodium*
  mesosulfuron-methyl

Products

Othello	Bayer CropScience	50:2.5:7.5 g/l	OD	12695

Uses

- Annual dicotyledons in ***winter wheat***
- Annual meadow grass in ***winter wheat***
- Cleavers in ***winter wheat***
- Mayweeds in ***winter wheat***

- Rough meadow grass in ***winter wheat***
- Volunteer oilseed rape in ***winter wheat***

Approval information

- Diflufenican, iodosulfuron-methyl-sodium and mesosulfuron-methyl included in Annex I under EC Directive 91/414

Efficacy guidance

- Optimum control obtained when all weeds are emerged at spraying. Activity is primarily via foliar uptake and good spray coverage of the target weeds is essential
- Translocation occurs readily within the target weeds and growth is inhibited within hours of treatment but symptoms may not be apparent for up to 4 wk, depending on weed species, timing of treatment and weather conditions
- Iodosulfuron-methyl and mesosulfuron-methyl are both members of the ALS-inhibitor group of herbicides. To avoid the build up of resistance do not use any product containing an ALS-inhibitor herbicide with claims for control of grass weeds more than once on any crop
- Use this product as part of a Resistance Management Strategy that includes cultural methods of control and does not use ALS inhibitors as the sole chemical method of weed control in successive crops. See Section 5 for more information

Restrictions

- Maximum number of treatments 1 per crop
- Do not use on crops undersown with grasses, clover or other legumes or any other broad-leaved crop
- Do not use where annual grass weeds other than annual meadow grass and rough meadow grass are present
- Do not use as a stand-alone treatment for control of common chickweed or common poppy. Only use mixtures with non ALS-inhibitor herbicides for these weeds
- Do not use as the sole means of weed control in successive crops
- Specific restrictions apply to use in sequence or tank mixture with other sulfonylurea or ALS-inhibiting herbicides. See label for details
- Do not apply to crops under stress from any cause
- Do not apply when rain is imminent or during periods of frosty weather
- Specified adjuvant must be used. See label

Crop-specific information

- Latest use: before 2nd node detectable for winter wheat
- Transitory crop effects may occur, particularly on overlaps and after late season/spring applications. Recovery is normally complete and yield not affected

Following crops guidance

- In the event of crop failure winter or spring wheat may be drilled after normal cultivation and an interval of 6 wk
- Winter wheat, winter barley or winter oilseed rape may be drilled in the autumn following normal harvest of a treated crop. Spring wheat, spring barley, spring oilseed rape or sugar beet may be drilled in the following spring
- Where the product has been applied in sequence with a permitted ALS-inhibitor herbicide (see label) only winter or spring wheat or barley, or sugar beet may be sown as following crops
- Successive treatments with any products containing diflufenican can lead to soil build-up and inversion ploughing to 150 mm must precede sowing any following non-cereal crop. Even where ploughing occurs some crops may be damaged

Environmental safety

- Dangerous for the environment
- Very toxic to aquatic organisms
- Take extreme care to avoid drift outside the target area
- LERAP Category B

Hazard classification and safety precautions

Hazard Irritant, Dangerous for the environment
Transport code 9
Packaging group III
UN number 3082
Risk phrases R36, R50, R53a

Operator protection A, C; U05a, U11, U20b
Environmental protection E15b, E16a, E38
Storage and disposal D01, D02, D05, D09a, D10a, D12a

183 diflufenican + mecoprop-P

A mixture of shoot absorbed pyridinecarboxamide herbicide and a translocated phenoxycarboxylic acid herbicide for winter cereals
HRAC mode of action code: F1 + O

See also mecoprop-P

Products

Pixie	Nufarm UK	33.3:500 g/l	SC	13884

Uses

- Annual dicotyledons in ***winter barley***, ***winter wheat***

Approval information

- Diflufenican and mecoprop-P included in Annex I under EC Directive 91/414
- Accepted by BBPA for use on malting barley

Hazard classification and safety precautions

Hazard Harmful, Dangerous for the environment
Transport code 9
Packaging group III
UN number 3082
Risk phrases R22a, R41, R43, R50, R53a
Operator protection A, C, H; U05a, U08, U11, U19a, U20b
Environmental protection E15a, E16a, E34, E38
Storage and disposal D01, D02, D09a, D10b, D12a

184 diflufenican + pendimethalin

A mixture of pyridinecarboxamide and dinitroanaline herbicides, see Section 3
HRAC mode of action code: F1 + K1

See also pendimethalin

Products

1	Churchill	Makhteshim	300:18.75	EC	14139
2	Omaha	Makhteshim	30:300 g/l	EC	14234

Uses

- Annual dicotyledons in ***triticale***, ***winter barley***, ***winter rye***, ***winter wheat*** [1, 2]
- Annual grasses in ***triticale***, ***winter barley***, ***winter rye***, ***winter wheat*** [1]
- Annual meadow grass in ***triticale***, ***winter barley***, ***winter rye***, ***winter wheat*** [2]

Approval information

- Diflufenican and pendimethalin included in Annex I under EC Directive 91/414

Efficacy guidance

- Do not apply pre-emergence to winter crops drilled after 20th November
- Do not treat broadcast crops

Following crops guidance

- Do not drill autumn-sown broad leaved crops following application to the previous crop [1]
- Plough to 150 mm and thoroughly mix the soil before planting any following crop

Hazard classification and safety precautions

Hazard Toxic, Flammable, Dangerous for the environment
Transport code 3
Packaging group III
UN number 1993
Risk phrases R20 [1]; R36, R37, R38 [2]; R43, R50, R53a, R61 [1, 2]
Operator protection A, C, H; U08, U13, U19a, U20b [1, 2]; U14, U15, U19c [2]; U19b [1]

Environmental protection E15a, E16a, E34, E38 [1, 2]; E23 [1]
Storage and disposal D09a, D10b, D12a
Medical advice M04a

185 dimethenamid-p

A chloroacetamide herbicide available only in mixtures
HRAC mode of action code: K3

186 dimethenamid-p + metazachlor

A soil acting herbicide mixture for oilseed rape
HRAC mode of action code: K3 + K3

See also metazachlor

Products

1	Muntjac	BASF	200:200 g/l	EC	14540
2	Springbok	BASF	200:200 g/l	EC	12983

Uses
- Annual dicotyledons in ***winter oilseed rape***
- Chickweed in ***winter oilseed rape***
- Cleavers in ***winter oilseed rape***
- Common storksbill in ***winter oilseed rape***
- Field speedwell in ***winter oilseed rape***
- Poppies in ***winter oilseed rape***
- Scented mayweed in ***winter oilseed rape***
- Shepherd's purse in ***winter oilseed rape***

Approval information
- Dimethenamid-p and metazachlor included in Annex I under EC Directive 91/414

Efficacy guidance
- Best results obtained from treatments to fine, firm and moist seedbeds
- Apply pre- or post-emergence of the crop and ideally before weed emergence
- Residual weed control may be reduced under prolonged dry conditions
- Weeds germinating from depth may not be controlled

Restrictions
- Maximum total dose on winter oilseed rape equivalent to one full dose treatment
- Do not disturb soil after application
- Do not treat broadcast crops until they have attained two fully expanded cotyledons
- Do not use on Sands, Very Light soils, or soils containing >10% organic matter
- Do not apply when heavy rain forecast or on soils waterlogged or prone to waterlogging
- Do not treat crops suffering from stress from any cause
- Do not use pre-emergence when crop seed has started to germinate or if not covered with 15 mm of soil

Crop-specific information
- Latest use: before 7th true leaf for winter oilseed rape
- All varieties of winter oilseed rape may be treated

Following crops guidance
- Any crop may follow a normally harvested treated winter oilseed rape crop. Ploughing is not essential before a following cereal crop but is required for all other crops
- In the event of failure of a treated crop winter wheat (excluding durum) or winter barley may be drilled in the same autumn, and any cereal (excluding durum wheat), spring oilseed rape, peas or field beans may be sown in the following spring. Ploughing to at least 150 mm should precede in all cases

Environmental safety
- Dangerous for the environment
- Very toxic to aquatic organisms

- LERAP Category B
- Take extreme care to avoid spray drift onto non-crop plants outside the target area
- Some pesticides pose a greater threat of contamination of water than others and metazachlor is one of these pesticides. Take special care when applying metazachlor near water and do not apply if heavy rain is forecast

Hazard classification and safety precautions

Hazard Harmful, Dangerous for the environment
Transport code 9
Packaging group III
UN number 3082
Risk phrases R22a, R36, R38, R43, R50, R53a
Operator protection A, C, H; U05a, U08, U20c
Environmental protection E15a, E16a, E34, E38
Storage and disposal D01, D02, D09a, D10c, D12a
Medical advice M03, M05a

187 dimethenamid-p + metazachlor + quinmerac

A soil-acting herbicide mixture for use in oilseed rape
HRAC mode of action code: K3 + K3 + O

See also *metazachlor*
quinmerac

Products

1	Elk	BASF	200:200:100	SE	13974
2	Katamaran Turbo	BASF	200:200:100	SE	13999
3	Shadow	BASF	200:200:100	SE	14000

Uses

- Annual dicotyledons in ***winter oilseed rape***
- Annual grasses in ***winter oilseed rape***
- Cleavers in ***winter oilseed rape***

Approval information

- Dimethenamid-p and metazachlor included in Annex I under EC Directive 91/414

Environmental safety

- Some pesticides pose a greater threat of contamination of water than others and metazachlor is one of these pesticides. Take special care when applying metazachlor near water and do not apply if heavy rain is forecast

Hazard classification and safety precautions

Hazard Irritant, Dangerous for the environment
Transport code 9
Packaging group III
UN number 3082
Risk phrases R43, R50, R53a
Operator protection A, H; U05a, U08, U14, U20c
Environmental protection E07a, E15b, E16a, E34, E38
Storage and disposal D01, D02, D09a, D10c, D12a
Medical advice M03, M05a

188 dimethoate

A contact and systemic organophosphorus insecticide and acaricide
IRAC mode of action code: 1B

Products

1	BASF Dimethoate 40	BASF	400 g/l	EC	00199
2	Danadim Progress	Headland	400 g/l	EC	12208

Uses

- Aphids in ***fodder beet*** *(excluding Myzus persicae)*, ***lettuce***, ***mangels*** *(excluding Myzus persicae)*, ***ornamental plant production***, ***red beet*** *(excluding Myzus persicae)*, ***spring rye***, ***spring wheat***, ***sugar beet*** *(excluding Myzus persicae)*, ***triticale***, ***winter rye***, ***winter wheat*** [1, 2]; ***grass seed crops***, ***mangel seed crops***, ***sugar beet seed crops*** *(excluding Myzus persicae)* [1]
- Cabbage aphid in ***broccoli***, ***brussels sprouts***, ***calabrese***, ***cauliflowers*** [1, 2]
- General insect control in ***calabrese*** *(off-label)*, ***celeriac*** *(off-label)*, ***chinese cabbage*** *(off-label)*, ***collards*** *(off-label)*, ***compost*** *(off-label - for watercress propagation)*, ***kale*** *(off-label)*, ***kohlrabi*** *(off-label)*, ***protected broccoli*** *(off-label)*, ***protected cabbages*** *(off-label)*, ***protected calabrese*** *(off-label)*, ***protected cauliflowers*** *(off-label)*, ***protected celeriac*** *(off-label)*, ***protected chinese cabbage*** *(off-label)*, ***protected collards*** *(off-label)*, ***protected kale*** *(off-label)*, ***protected kohlrabi*** *(off-label)*, ***protected roscoff cauliflowers*** *(off-label)*, ***protected savoy cabbage*** *(off-label)*, ***protected spinach*** *(off-label)*, ***roscoff cauliflowers*** *(off-label)*, ***savoy cabbage*** *(off-label)* [2]; ***compost*** *(off-label - for watercress progagation)* [1]
- Insect pests in ***calabrese*** *(off-label)*, ***celeriac*** *(off-label)*, ***chinese cabbage*** *(off-label)*, ***collards*** *(off-label)*, ***durum wheat*** *(off-label)*, ***kale*** *(off-label)*, ***kohlrabi*** *(off-label)*, ***protected broccoli*** *(off-label)*, ***protected brussels sprouts*** *(off-label)*, ***protected cabbages*** *(off-label)*, ***protected calabrese*** *(off-label)*, ***protected cauliflowers*** *(off-label)*, ***protected celeriac*** *(off-label)*, ***protected chinese cabbage*** *(off-label)*, ***protected collards*** *(off-label)*, ***protected kale*** *(off-label)*, ***protected kohlrabi*** *(off-label)*, ***protected roscoff cauliflowers*** *(off-label)*, ***protected savoy cabbage*** *(off-label)*, ***protected spinach*** *(off-label)*, ***roscoff cauliflowers*** *(off-label)*, ***savoy cabbage*** *(off-label)* [1]
- Leaf miner in ***fodder beet***, ***mangels***, ***ornamental plant production***, ***red beet***, ***sugar beet*** [1, 2]
- Red spider mites in ***ornamental plant production*** [1, 2]
- Wheat bulb fly in ***spring wheat***, ***winter wheat*** [1, 2]

Specific Off-Label Approvals (SOLAs)

- ***calabrese*** *940389* [1], *20050682* [2]
- ***celeriac*** *940389* [1], *20050682* [2]
- ***chinese cabbage*** *940389* [1], *20050682* [2]
- ***collards*** *940389* [1], *20050682* [2]
- ***compost*** *(for watercress progagation) 940394* [1], *(for watercress propagation) 20050683* [2]
- ***durum wheat*** *20060806* [1]
- ***kale*** *940389* [1], *20050682* [2]
- ***kohlrabi*** *940389* [1], *20050682* [2]
- ***protected broccoli*** *940389* [1], *20050682* [2]
- ***protected brussels sprouts*** *940389* [1]
- ***protected cabbages*** *940389* [1], *20050682* [2]
- ***protected calabrese*** *940389* [1], *20050682* [2]
- ***protected cauliflowers*** *940389* [1], *20050682* [2]
- ***protected celeriac*** *940389* [1], *20050682* [2]
- ***protected chinese cabbage*** *940389* [1], *20050682* [2]
- ***protected collards*** *940389* [1], *20050682* [2]
- ***protected kale*** *940389* [1], *20050682* [2]
- ***protected kohlrabi*** *940389* [1], *20050682* [2]
- ***protected roscoff cauliflowers*** *940389* [1], *20050682* [2]
- ***protected savoy cabbage*** *940389* [1], *20050682* [2]
- ***protected spinach*** *940389* [1], *20050682* [2]
- ***roscoff cauliflowers*** *940389* [1], *20050682* [2]
- ***savoy cabbage*** *940389* [1], *20050682* [2]

Approval information

- Dimethoate included in Annex I under EC Directive 91/414
- Approved for aerial application on cereals, sugar beet [1]. See Section 5 for more information
- In 2006 PSD required that all products containing this active ingredient should carry the following warning in the main area of the container label: "Dimethoate is an anticholinesterase organophosphate. Handle with care"

Efficacy guidance

- Chemical has quick knock-down effect and systemic activity lasts for up to 14 d

- With some crops, products differ in range of pests listed as controlled. Uses section above provides summary. See labels for details
- For most pests apply when pest first seen and repeat 2-3 wk later or as necessary. Timing and number of sprays varies with crop and pest. See labels for details
- Best results achieved when crop growing vigorously. Systemic activity reduced when crops suffering from drought or other stress
- In hot weather apply in early morning or late evening
- Where aphids or spider mites resistant to organophosphorus compounds occur control is unlikely to be satisfactory and repeat treatments may result in lower levels of control

Restrictions

- Contains an anticholinesterase organophosphorus compound. Do not use if under medical advice not to work with such compounds
- Maximum number of treatments 6 per crop for Brussels sprouts, broccoli, cauliflower, calabrese, lettuce; 4 per crop for grass seed crops, cereals; 2 per crop for beet crops, triticale, rye; 1 per crop for mangel and sugar beet seed crops
- On beet crops only one treatment per crop may be made for control of leaf miners (max 84 g a.i./ha) and black bean aphid (max 400 g a.i./ha)
- In beet crops resistant strains of peach-potato aphid (*Myzus persicae*) are common and dimethoate products must not be used to control this pest
- Test for varietal susceptibility on all unusual plants or new cultivars
- Do not tank mix with alkaline materials. See label for recommended tank-mixes
- Consult processor before spraying crops grown for processing
- Must not be applied to cereals if any product containing a pyrethroid insecticide or dimethoate has been sprayed after the start of ear emergence (GS 51)

Crop-specific information

- Latest use: before 30 Jun in yr of harvest for beet crops; flowering just complete stage (GS 30) for wheat, rye, triticale [1]; before 31 Mar in yr of harvest for aerial application to cereals
- HI brassicas 21 d; grass seed crops, lettuce, cereals 14 d; ornamental plant production 7 d

Environmental safety

- Dangerous for the environment
- Harmful to aquatic organisms
- Flammable
- Harmful to game, wild birds and animals
- Harmful to livestock. Keep all livestock out of treated areas for at least 7 d
- Likely to cause adverse effects on beneficial arthropods
- High risk to non-target insects or other arthropods. Do not treat cereals after 1 Apr within 6 m of edge of crop
- High risk to bees. Do not apply to crops in flower or to those in which bees are actively foraging. Do not apply when flowering weeds are present
- Surface residues may also cause bee mortality following spraying
- Dangerous to fish or other aquatic life. Do not contaminate surface waters or ditches with chemical or used container
- LERAP Category A
- Keep in original container, tightly closed, in a safe place, under lock and key

Hazard classification and safety precautions

Hazard Harmful, Flammable
Transport code 3 [2]
Packaging group III [2]
UN number 1993 [2]
Risk phrases R20, R21, R48 [1]; R22a, R43 [1, 2]; R52 [2]
Operator protection A, H, M [1, 2]; C, J [1]; U02a, U04a, U19a [1, 2]; U05a, U10, U13, U20a [1]; U08, U14, U15, U20b [2]
Environmental protection E06c [1, 2] (7 d); E10b, E12a, E12e, E13b, E16c, E16d, E22a, E34 [1, 2]; E18 [1]
Storage and disposal D01, D02, D05, D09b [1, 2]; D08, D10c [1]; D12b [2]
Medical advice M01, M03 [1, 2]; M05a [1]

189 dimethomorph

A cinnamic acid fungicide with translaminar activity available only in mixtures
FRAC mode of action code: 40

190 dimethomorph + mancozeb

A systemic and protectant fungicide for potato blight control
FRAC mode of action code: 40 + M3

See also *mancozeb*

Products

1	Dunkirke	AgChem Access	7.5:66.7% w/w	WG	13796
2	Invader	BASF	7.5:66.7% w/w	WG	11978

Uses

- Blight in ***potatoes*** [1, 2]
- Downy mildew in ***bulb onions*** *(off-label)*, ***chinese cabbage*** *(off-label)*, ***chives*** *(off-label)*, ***choi sum*** *(off-label)*, ***cress*** *(off-label)*, ***frise*** *(off-label)*, ***herbs (see appendix 6)*** *(off-label)*, ***lamb's lettuce*** *(off-label)*, ***lettuce*** *(off-label)*, ***pak choi*** *(off-label)*, ***poppies for morphine production*** *(off-label)*, ***radicchio*** *(off-label)*, ***salad brassicas*** *(off-label - for baby leaf production)*, ***salad onions*** *(off-label)*, ***scarole*** *(off-label)*, ***shallots*** *(off-label)*, ***tatsoi*** *(off-label)* [2]
- White tip in ***leeks*** *(off-label)* [2]

Specific Off-Label Approvals (SOLAs)

- ***bulb onions*** *20090214* [2]
- ***chinese cabbage*** *20063044* [2]
- ***chives*** *20063044* [2]
- ***choi sum*** *20063044* [2]
- ***cress*** *20063044* [2]
- ***frise*** *20063044* [2]
- ***herbs (see appendix 6)*** *20063044* [2]
- ***lamb's lettuce*** *20063044* [2]
- ***leeks*** *20073434* [2]
- ***lettuce*** *20063044* [2]
- ***pak choi*** *20063044* [2]
- ***poppies for morphine production*** *20073433* [2]
- ***radicchio*** *20063044* [2]
- ***salad brassicas*** *(for baby leaf production) 20063044* [2]
- ***salad onions*** *20062953* [2]
- ***scarole*** *20063044* [2]
- ***shallots*** *20090214* [2]
- ***tatsoi*** *20063044* [2]

Approval information

- Dimethomorph and mancozeb included in Annex I under EC Directive 91/414

Efficacy guidance

- Commence treatment as soon as there is a risk of blight infection
- In the absence of a warning treatment should start before the crop meets along the rows
- Repeat treatments every 7-14 d depending in the degree of infection risk
- Irrigated crops should be regarded as at high risk and treated every 7 d
- For best results good spray coverage of the foliage is essential
- To minimise the likelihood of development of resistance these products should be used in a planned Resistance Management strategy. See Section 5 for more information

Restrictions

- Maximum total dose equivalent to eight full dose treatments on potatoes

Crop-specific information

- HI 7 d for all crops

Environmental safety

- Dangerous for the environment
- Very toxic to aquatic organisms
- LERAP Category B

Hazard classification and safety precautions

Hazard Irritant, Dangerous for the environment
Transport code 9
Packaging group III
UN number 3077
Risk phrases R36, R37, R38, R43, R50, R53a
Operator protection A; U02a, U04a, U05a, U08, U13, U19a, U20b
Environmental protection E16a, E16b, E38
Storage and disposal D01, D02, D05, D09a, D12a
Medical advice M05a

191 dimoxystrobin

A protectant strobilurin fungicide for cereals available only in mixtures
FRAC mode of action code: 11

192 dimoxystrobin + epoxiconazole

A protectant and curative fungicide mixture for cereals
FRAC mode of action code: 11 + 3

See also epoxiconazole

Products

Swing Gold	BASF	133:50 g/l	SC	11658

Uses

- Brown rust in ***winter wheat***
- Foliar disease control in ***durum wheat*** *(off-label)*, ***grass seed crops*** *(off label)*, ***spring rye*** *(off-label)*, ***triticale*** *(off-label)*, ***winter rye*** *(off-label)*
- Fusarium ear blight in ***winter wheat***
- Glume blotch in ***winter wheat***
- Septoria leaf blotch in ***winter wheat***
- Tan spot in ***winter wheat***

Specific Off-Label Approvals (SOLAs)

- ***durum wheat*** *20062642*
- ***grass seed crops*** *20062642*
- ***spring rye*** *20062642*
- ***triticale*** *20062642*
- ***winter rye*** *20062642*

Approval information

- Dimoxystrobin and epoxiconazole included in Annex I under EC Directive 91/414

Efficacy guidance

- For best results apply from the start of ear emergence
- Dimoxystrobin is a member of the QoI cross resistance group. Product should be used preventatively and not relied on for its curative potential
- Use product as part of an Integrated Crop Management strategy incorporating other methods of control, including where appropriate other fungicides with a different mode of action. Do not apply more than two foliar applications of QoI containing products to any cereal crop
- There is a significant risk of widespread resistance occurring in *Septoria tritici* populations in UK. Failure to follow resistance management action may result in reduced levels of disease control
- Epoxiconazole is a DMI fungicide. Resistance to some DMI fungicides has been identified in Septoria leaf blotch which may seriously affect performance of some products. For further advice contact a specialist advisor and visit the Fungicide Resistance Action Group (FRAG)-UK website

Restrictions

- Maximum number of treatments 1 per crop
- Do not apply before the start of ear emergence

Crop-specific information

- Latest use: up to and including flowering (GS 69)

Environmental safety

- Dangerous for the environment
- Very toxic to aquatic organisms
- LERAP Category B
- Avoid drift onto neighbouring crops

Hazard classification and safety precautions

Hazard Harmful, Dangerous for the environment
Transport code 9
Packaging group III
UN number 3082
Risk phrases R20, R22a, R40, R50, R53a, R63
Operator protection A; U05a, U20b
Environmental protection E15a, E16a, E34, E38
Storage and disposal D01, D02, D05, D09a, D10c, D12a
Medical advice M05a

193 dinocap

A protectant dinitrophenyl fungicide for powdery mildew control, products withdrawn 2008
FRAC mode of action code: 29

194 diquat

A non-residual bipyridyl contact herbicide and crop desiccant
HRAC mode of action code: D

Products

1	Agriguard Diquat	AgriGuard	200 g/l	SL	13763
2	Brogue	AgriGuard	200 g/l	SL	13328
3	Brogue	AgriGuard	200 g/l	SL	14465
4	ChemSource Diquat	ChemSource	200 g/l	SL	14533
5	Clayton Diquat	Clayton	200 g/l	SL	12739
6	Clayton Diquat 200	Clayton	200 g/l	SL	13942
7	Clayton IQ	Clayton	200 g/l	SL	14519
8	Dragoon	Hermoo	200 g/l	SL	13927
9	Dragoon Gold	Hermoo	200 g/l	SL	14053
10	Knoxdoon	AgChem Access	200 g/l	SL	13987
11	Mission	AgriChem BV	200 g/l	SL	13411
12	Quad	Q-Chem	200 g/l	SL	13396
13	Reglone	Syngenta	200 g/l	SL	10534
14	Retro	Syngenta	200 g/L	SL	13841
15	Roquat 20	Q-Chem	200 g/l	SL	13474
16	Standon Googly	Standon	200 g/l	SL	13281
17	Woomera	Hermoo	200 g/l	SL	14009

Uses

- Annual and perennial weeds in ***all edible crops (outdoor)*** *(around only)*, ***all non-edible crops (outdoor)*** *(around only)*, ***combining peas*** *(dry harvested only)*, ***hops***, ***linseed***, ***ornamental plant production***, ***potatoes*** *(weed control or dessication)*, ***red clover*** *(seed crop)*, ***spring barley*** *(animal feed only)*, ***spring field beans***, ***spring oats*** *(animal feed only)*, ***spring oilseed rape***, ***sugar beet***, ***white clover*** *(seed crop)*, ***winter barley*** *(animal feed only)*, ***winter field beans***, ***winter oats*** *(animal feed only)*, ***winter oilseed rape*** [11]; ***forest nurseries*** *(off-label)* [12]
- Annual dicotyledons in ***all edible crops (outdoor)*** *(around)*, ***all non-edible crops (outdoor)*** *(around)*, ***ornamental plant production***, ***sugar beet*** [1, 2, 4-10, 12, 13, 15-17]; ***fodder beet*** [9]; ***lucerne*** *(off-label)* [13]; ***poppies for morphine production*** *(off-label)* [12]
- Basal defoliation in ***hops*** [7]

SEE SECTION 3 FOR PRODUCTS ALSO REGISTERED

- Chemical stripping in ***hops*** [12, 15]
- Chickweed in ***all edible crops (outdoor)***, ***all non-edible crops (outdoor)***, ***ornamental plant production***, ***potatoes***, ***sugar beet*** [14]
- Desiccation in ***borage for oilseed production*** *(off-label)*, ***french beans*** *(off-label)*, ***hemp for oilseed production*** *(off-label)*, ***herbs (see appendix 6)*** *(off-label)*, ***mustard*** *(off-label)*, ***poppies for morphine production*** *(off-label)*, ***poppies for oilseed production*** *(off-label)*, ***sesame*** *(off-label)*, ***sunflowers*** *(off-label)* [12]; ***combining peas***, ***linseed***, ***potatoes***, ***red clover***, ***spring barley***, ***spring field beans***, ***spring oats***, ***spring oilseed rape***, ***white clover***, ***winter barley***, ***winter field beans***, ***winter oats***, ***winter oilseed rape*** [3]; ***crambe*** *(off-label)* [13]; ***lupins*** *(off-label)*, ***sweet potato*** *(off-label)* [12, 13]
- General weed control in ***chives*** *(off-label)*, ***herbs (see appendix 6)*** *(off-label)*, ***parsley*** *(off-label)*, ***poppies for morphine production*** *(off-label)* [13]; ***potatoes*** [1, 2, 4-10, 12, 13, 15-17]
- Grass weeds in ***all edible crops (outdoor)*** *(around only)*, ***all non-edible crops (outdoor)*** *(around only)*, ***combining peas*** *(dry harvested)*, ***hops***, ***linseed***, ***ornamental plant production***, ***potatoes*** *(weed control and dessication)*, ***red clover*** *(seed crop)*, ***spring barley*** *(animal feed)*, ***spring field beans***, ***spring oats*** *(animal feed)*, ***spring oilseed rape***, ***sugar beet***, ***white clover*** *(seed crop)*, ***winter barley*** *(animal feed)*, ***winter field beans***, ***winter oats*** *(animal feed)*, ***winter oilseed rape*** [11]
- Growth regulation in ***hops*** [6]
- Harvest management/desiccation in ***combining peas***, ***linseed***, ***red clover***, ***spring barley***, ***spring field beans***, ***spring oats***, ***spring oilseed rape***, ***white clover***, ***winter barley***, ***winter field beans***, ***winter oats***, ***winter oilseed rape*** [3]
- Haulm destruction in ***potatoes*** [1, 2, 4-10, 12, 13, 15-17]
- Perennial dicotyledons in ***lucerne*** *(off-label)* [13]
- Pre-harvest desiccation in ***borage for oilseed production*** *(off-label)*, ***brown mustard*** *(off-label)*, ***french beans*** *(off-label)*, ***hemp for oilseed production*** *(off-label)*, ***poppies for morphine production*** *(off-label)*, ***poppies for oilseed production*** *(off-label)*, ***sesame*** *(off-label)*, ***sunflowers*** *(off-label)*, ***white mustard*** *(off-label)* [13]; ***broad beans*** [4-6, 10, 13, 16]; ***clover seed crops***, ***combining peas***, ***laid barley*** *(stockfeed only)*, ***laid oats*** *(stockfeed only)*, ***linseed***, ***spring field beans***, ***spring oilseed rape***, ***winter field beans***, ***winter oilseed rape*** [1, 2, 4-10, 12, 13, 15-17]

Specific Off-Label Approvals (SOLAs)

- ***borage for oilseed production*** *20073545* [12], *20062390* [13]
- ***brown mustard*** *20062387* [13]
- ***chives*** *20062388* [13]
- ***crambe*** *20082092* [13]
- ***forest nurseries*** *20082921* [12]
- ***french beans*** *20073544* [12], *20062389* [13]
- ***hemp for oilseed production*** *20073541* [12], *20062386* [13]
- ***herbs (see appendix 6)*** *20073543* [12], *20062388* [13]
- ***lucerne*** *20082842* [13]
- ***lupins*** *20073547* [12], *20073232* [13]
- ***mustard*** *20073542* [12]
- ***parsley*** *20062388* [13]
- ***poppies for morphine production*** *20073540* [12], *20062385* [13]
- ***poppies for oilseed production*** *20073542* [12], *20062387* [13]
- ***sesame*** *20073542* [12], *20062387* [13]
- ***sunflowers*** *20073542* [12], *20062387* [13]
- ***sweet potato*** *20073546* [12], *20070734* [13]
- ***white mustard*** *20062387* [13]

Approval information

- Diquat included in Annex I under EC Directive 91/414
- All approvals for aquatic weed control expired in 2004

Efficacy guidance

- Acts rapidly on green parts of plants and rainfast in 15 min
- Best results for potato desiccation achieved by spraying in bright light and low humidity conditions
- For weed control in row crops apply as overall spray before crop emergence or before transplanting or as an inter-row treatment using a sprayer designed to prevent contamination of crop foliage by spray

- Use of authorised non-ionic wetter for improved weed control essential for most uses except potato desiccation

Restrictions
- Maximum number of treatments 1 per crop in most situations - see labels for details
- Do not apply potato haulm destruction treatment when soil dry. Tubers may be damaged if spray applied during or shortly after dry periods. See label for details of maximum allowable soil-moisture deficit and varietal drought resistance scores
- Do not add wetters to desiccant sprays for potatoes
- Consult processor before adding non-ionic or other wetter on peas. Staining may be increased
- Do not use treated straw or haulm as animal feed or bedding within 4 d of spraying
- Do not apply through ULV or mistblower equipment

Crop-specific information
- Latest use: before crop emergence for sugar beet and before transplanting ornamentals. Not specified for other crops where use varies according to situation
- HI zero when used as a crop desiccant but see label for advisory intervals
- For pre-harvest desiccation, apply to potatoes when tubers the desired size and to other crops when mature or approaching maturity. See label for details of timing and of period to be left before harvesting potatoes
- Apply as a chemical stripping treatment in hops when shoots have reached top wire [12, 15]
- Treatment of hops under stress or when leaves are wet with a heavy dew may cause damage [12, 15]
- If potato tubers are to be stored leave 14 d after treatment before lifting

Environmental safety
- Dangerous for the environment
- Very toxic to aquatic organisms
- Keep all livestock out of treated areas and away from treated water for at least 24 h
- Do not feed treated straw or haulm to livestock within 4 days of spraying
- Do not use on crops if the straw is to be used as animal feed/bedding
- Do not dump surplus herbicide in water or ditch bottoms

Hazard classification and safety precautions

Hazard Very toxic [5]; Toxic [1-4, 6-10, 12-17]; Dangerous for the environment [1-17]
Transport code 8
Packaging group III
UN number 1760
Risk phrases R22a [1-10, 12-17]; R23 [1-4, 6-17]; R26 [5]; R36 [11, 12, 15]; R37, R38, R50, R53a [1-17]; R43 [6, 11, 12, 15]; R48 [1-10, 12, 13, 15-17]; R69 [11, 14]
Operator protection A [1-17]; C [1-3, 11, 12, 15]; D [16]; H [3-11, 13, 14, 16, 17]; M [1-15, 17]; P [14]; U02a, U04a, U05a, U08, U19a, U20b [1-17]; U04c [11]; U10 [1-4, 7-10, 12-17]; U15 [12, 15]
Environmental protection E06c [1-17] (24 h); E08a [1-3, 12, 15] (4 d); E09 [4]; E09 [5, 6] (4 d); E09 [7-11, 13, 14, 16, 17]; E15a [11]; E15b [1-10, 12, 13, 15-17]; E34 [1-17]; E38 [1-4, 7-17]
Storage and disposal D01, D02, D09a, D12a [1-17]; D05 [1-10, 12, 13, 15-17]; D10a [11]; D10b [5, 6]; D10c [1-4, 7-10, 12-17]
Medical advice M04a

195 dithianon

A protectant and eradicant dicarbonitrile fungicide for scab control
FRAC mode of action code: M9

Products

1	Dithianon Flowable	BASF	750 g/l	SC	10219
2	Dithianon WG	BASF	70% w/w	WG	12538

Uses
- Scab in ***apples***, ***pears*** [1, 2]; ***quinces*** *(off-label)* [1]

Specific Off-Label Approvals (SOLAs)
- ***quinces*** *20061038* [1]

SEE SECTION 3 FOR PRODUCTS ALSO REGISTERED

Approval information

- Dithianon has not been included in Annex 1 but the date of withdrawal has yet to be set.

Efficacy guidance

- Apply at bud-burst and repeat every 7-14 d until danger of scab infection ceases
- Application at high rate within 48 h of a Mills period prevents new infection
- Spray programme also reduces summer infection with apple canker

Restrictions

- Maximum number of treatments on apples and pears 8 per crop [1]; 12 per crop [2]
- Do not use on Golden Delicious apples after green cluster
- Do not mix with lime sulphur or highly alkaline products

Crop-specific information

- HI 4 wk for apples, pears

Environmental safety

- Dangerous for the environment
- Very toxic to aquatic organisms

Hazard classification and safety precautions

Hazard Harmful, Dangerous for the environment
Transport code 9
Packaging group III
UN number 3077 [2]; 3082 [1]
Risk phrases R20 [1]; R22a, R50, R53a [1, 2]; R41 [2]
Operator protection A [1, 2]; D [2]; H [1]; U05a [1, 2]; U08, U19a, U20a [1]; U11 [2]
Environmental protection E15a, E34, E38
Storage and disposal D01, D02, D05 [1, 2]; D09a, D10b [1]; D12a [2]
Medical advice M03 [1]; M05a [1, 2]

196 dithianon + pyraclostrobin

A protectant fungicide mixture for apples and pears
FRAC mode of action code: M9 + 11

See also *pyraclostrobin*

Products

Maccani	BASF	12:4% w/w	WG	13227

Uses

- Powdery mildew in ***apples***
- Scab in ***apples, pears***

Approval information

- Pyraclostrobin included in Annex I under EC Directive 91/414

Efficacy guidance

- Best resullts obtained from treatments applied from bud burst before disease development and repeated at 10 d intervals
- Water volume should be adjusted according to size of tree and leaf area to ensure good coverage
- Pyraclostrobin is a member of the QoI cross resistance group. Product should be used preventatively at the full recommended dose and not relied on for its curative potential
- Where field performance may be adversely affected apply QoI containing fungicides in mixtures in strict alternation with fungicides from a different cross-resistance group

Restrictions

- Maximum number of treatments 4 per crop. However where the total number of fungicide treatments is less than 12, the maximum number of applications of any QoI containing products should be three
- Consult processors before use on crops destined for fruit preservation

Crop-specific information

- HI 35 d for apples, pears
- Some apple varieties may be considered to be susceptible to russetting caused by the dithianon component during fruit development

Environmental safety

- Dangerous for the environment
- Very toxic to aquatic organisms
- Broadcast air-assisted LERAP (40 m)

Hazard classification and safety precautions

Hazard Harmful, Dangerous for the environment
Transport code 9
Packaging group III
UN number 3077
Risk phrases R20, R22a, R36, R50, R53a
Operator protection A, C, D; U05a, U11, U20b
Environmental protection E15b, E34, E38; E17b (40 m)
Storage and disposal D01, D02, D05, D08, D11a, D12a
Medical advice M03, M05a

197 diuron

A residual urea herbicide for non-crop areas and woody crops; approvals expired 2008
HRAC mode of action code: C2

198 diuron + glyphosate

A non-selective residual herbicide mixture for non-crop and amenity use; approvals expired 2008
HRAC mode of action code: C2 + G

See also *glyphosate*

199 dodine

A protectant and eradicant guanidine fungicide
FRAC mode of action code: M7

Products

1	Clayton Scabius	Clayton	450 g/l	SC	13339
2	Dodifun 400 SC	Hermoo	400 g/l	SC	11657
3	Greencrop Budburst	Greencrop	450 g/l	SC	11042
4	Radspor FL	Truchem	450 g/l	SC	01685

Uses

- Currant leaf spot in ***bilberries*** *(off-label)*, ***blueberries*** *(off-label)*, ***cranberries*** *(off-label)*, ***gooseberries*** *(off-label)*, ***redcurrants*** *(off-label)*, ***whitecurrants*** *(off-label)* [4]; ***blackcurrants*** [1-4]
- Scab in ***apples***, ***pears*** [1-4]; ***quinces*** *(off-label)* [4]

Specific Off-Label Approvals (SOLAs)

- ***bilberries*** *20060617* [4]
- ***blueberries*** *20060617* [4]
- ***cranberries*** *20060617* [4]
- ***gooseberries*** *20060617* [4]
- ***quinces*** *20060616* [4]
- ***redcurrants*** *20060617* [4]
- ***whitecurrants*** *20060617* [4]

Efficacy guidance

- Apply protective spray on apples and pears at bud-burst and at 10-14 d intervals until late Jun to early Jul
- Apply post-infection spray within 36 h of rain responsible for initiating infection. Where scab already present spray prevents production of spores
- On blackcurrants commence spraying at early grape stage and repeat at 2-3 wk intervals, and at least once after picking

SEE SECTION 3 FOR PRODUCTS ALSO REGISTERED

Restrictions

- Do not apply in very cold weather (under 5 °C) or under slow drying conditions to pears or dessert apples during bloom or immediately after petal fall
- Do not mix with lime sulphur or tetradifon
- Consult processors before use on crops grown for processing

Crop-specific information

- Latest use: early Jul for culinary apples; pre-blossom for dessert apples and pears

Environmental safety

- Dangerous for the environment
- Very toxic to aquatic organisms

Hazard classification and safety precautions

Hazard Harmful, Dangerous for the environment
Transport code 9
Packaging group III
UN number 3082
Risk phrases R21 [2, 3]; R22a, R36, R38, R43, R50, R53a [1-4]; R37, R41 [1, 4]
Operator protection A, C; U05a, U08, U19a, U20b [1-4]; U11 [1, 4]
Environmental protection E15a, E34 [1-4]; E38 [4]
Storage and disposal D01, D02, D09a, D10b [1-4]; D05 [1-3]; D12a [1]
Medical advice M03 [1, 4]; M04a [2, 3]; M05a [1]

200 epoxiconazole

A systemic, protectant and curative triazole fungicide for use in cereals
FRAC mode of action code: 3

See also *boscalid + epoxiconazole*
carbendazim + epoxiconazole
dimoxystrobin + epoxiconazole

Products

1	Agriguard Epoxiconazole	AgriGuard	125 g/l	SC	09407
2	Clayton Oust	Clayton	125 g/l	SC	11588
3	Cortez	Makhteshim	125 g/l	SC	13932
4	Epic	BASF	125 g/l	SC	12136
5	Epoxi	Goldengrass	125 g/l	SC	14292
6	Epoxi 125 SC	Goldengrass	125 g/l	SC	14442
7	Landgold Epoxiconazole	Goldengrass	125 g/l	SC	14025
8	Opus	BASF	125 g/l	SC	12057
9	Oracle	ChemSource	125 g/l	SC	14470
10	Oropa	AgChem Access	125 g/l	SC	13435
11	Overture	AgriGuard	125 g/l	SC	12587
12	Rubric	Headland	125 g/l	SC	14118
13	Standon Epoxiconazole	Standon	125 g/l	SC	09517
14	Torrent	AgriGuard	125 g/l	SC	14374

Uses

- Brown rust in ***spring barley***, ***winter barley***, ***winter wheat*** [1-14]; ***spring rye***, ***spring wheat***, ***triticale***, ***winter rye*** [2-12, 14]
- Eyespot in ***winter barley*** *(reduction)*, ***winter wheat*** *(reduction)* [1, 2, 4-14]
- Foliar disease control in ***durum wheat*** *(off-label)* [3, 4, 8]; ***grass seed crops*** *(off-label)* [1-3, 8, 13]
- Fusarium ear blight in ***spring wheat***, ***winter wheat*** [2-11, 14]; ***winter wheat*** *(reduction)* [1, 13]
- Glume blotch in ***spring wheat***, ***triticale*** [2-12, 14]; ***winter wheat*** [1-14]
- Net blotch in ***spring barley***, ***winter barley*** [1-14]
- Powdery mildew in ***durum wheat*** *(off-label)*, ***grass seed crops*** *(off-label)* [12]; ***spring barley***, ***winter barley***, ***winter wheat*** [1-14]; ***spring oats***, ***spring rye***, ***spring wheat***, ***triticale***, ***winter oats***, ***winter rye*** [2-12, 14]
- Rhynchosporium in ***spring barley***, ***winter barley*** [1-14]; ***spring rye***, ***winter rye*** [2-12, 14]
- Septoria leaf blotch in ***spring wheat***, ***triticale*** [2-12, 14]; ***winter wheat*** [1-14]
- Sooty moulds in ***spring wheat*** *(reduction)* [2-12, 14]; ***winter wheat*** *(reduction)* [1-14]

- Yellow rust in ***spring barley***, ***winter barley***, ***winter wheat*** [1-14]; ***spring rye***, ***spring wheat***, ***triticale***, ***winter rye*** [2-12, 14]

Specific Off-Label Approvals (SOLAs)

- ***durum wheat*** *20080981* [3], *20061053* [4], *20061051* [8], *20090236* [12]
- ***grass seed crops*** *20061056* [1], *20061055* [2], *20080981* [3], *20061051* [8], *20090236* [12], *20061050* [13]

Approval information

- Accepted by BBPA for use on malting barley (before ear emergence only)
- Epoxiconazole included in Annex I under EC Directive 91/414

Efficacy guidance

- Apply at the start of foliar disease attack
- Optimum effect against eyespot achieved by spraying between leaf-sheath erect and second node detectable stages (GS 30-32)
- Best control of ear diseases of wheat obtained by treatment during ear emergence
- Mildew control improved by use of tank mixtures. See label for details
- For Septoria spray after third node detectable stage (GS 33) when weather favouring disease development has occurred
- Epoxiconazole is a DMI fungicide. Resistance to some DMI fungicides has been identified in Septoria leaf blotch which may seriously affect performance of some products. For further advice contact a specialist advisor and visit the Fungicide Resistance Action Group (FRAG)-UK website

Restrictions

- Maximum total dose equivalent to two full dose treatments
- Product may cause damage to broad-leaved plant species
- Avoid spray drift onto neighbouring crops

Crop-specific information

- Latest use: up to and including flowering just complete (GS 69) in wheat, rye, triticale; up to and including emergence of ear just complete (GS 59) in barley, oats

Environmental safety

- Dangerous for the environment
- Very toxic to aquatic organisms
- LERAP Category B

Hazard classification and safety precautions

Hazard Harmful [2-14]; Dangerous for the environment [1-14]
Transport code 9
Packaging group III
UN number 3082
Risk phrases R38 [2, 4-14]; R40, R50, R62, R63 [2-14]; R51 [1]; R53a [1-14]
Operator protection A; U05a [1-14]; U20a [1]; U20b [2-14]
Environmental protection E15a, E16a [1-14]; E16b [1, 11, 13]; E38 [3-6, 8-10, 12, 14]
Storage and disposal D01, D02, D05, D09a [1-14]; D08 [7]; D10b [1, 7, 13]; D10c [2-6, 8-12, 14]; D12a [2-6, 8-10, 12, 14]
Medical advice M05a [2-10, 12, 14]; M05b [13]

201 epoxiconazole + fenpropimorph

A systemic, protectant and curative fungicide mixture for cereals
FRAC mode of action code: 3 + 5

See also *fenpropimorph*

Products

1	Eclipse	BASF	84:250 g/l	SE	11731
2	Greencrop Galore	Greencrop	84:250 g/l	SE	09561
3	Opus Team	BASF	84:250 g/l	SE	11759
4	Standon Epoxifen	Standon	84:250 g/l	SE	08972

Uses

- Brown rust in ***spring barley***, ***winter barley***, ***winter wheat*** [1-4]; ***spring rye***, ***spring wheat***, ***triticale***, ***winter rye*** [1, 3]

SEE SECTION 3 FOR PRODUCTS ALSO REGISTERED

- Eyespot in ***winter barley*** *(reduction)*, ***winter wheat*** *(reduction)* [1-4]
- Foliar disease control in ***durum wheat*** *(off-label)* [1, 3]; ***grass seed crops*** *(off-label)* [1-4]
- Fusarium ear blight in ***spring wheat*, *winter wheat*** [1, 3]; ***winter wheat*** *(reduction)* [2, 4]
- Glume blotch in ***spring wheat*, *triticale*** [1, 3]; ***winter wheat*** [1-4]
- Net blotch in ***spring barley*, *winter barley*** [1-4]
- Powdery mildew in ***spring barley*, *winter barley*, *winter wheat*** [1-4]; ***spring oats*, *spring rye*, *spring wheat*, *triticale*, *winter oats*, *winter rye*** [1, 3]
- Rhynchosporium in ***spring barley*, *winter barley*** [1-4]; ***spring rye*, *winter rye*** [1, 3]
- Septoria leaf blotch in ***spring wheat*, *triticale*** [1, 3]; ***winter wheat*** [1-4]
- Sooty moulds in ***spring wheat*** *(reduction)* [1, 3]; ***winter wheat*** *(reduction)* [1-3]
- Yellow rust in ***spring barley*, *winter barley*, *winter wheat*** [1-4]; ***spring rye*, *spring wheat*, *triticale*, *winter rye*** [1, 3]

Specific Off-Label Approvals (SOLAs)

- ***durum wheat*** *20062644* [1], *20062645* [3]
- ***grass seed crops*** *20062644* [1], *20061077* [2], *20062645* [3], *20061081* [4]

Approval information

- Epoxiconazole and fenpropimorph included in Annex I under EC Directive 91/414
- Accepted by BBPA for use on malting barley

Efficacy guidance

- Apply at the start of foliar disease attack
- Optimum effect against eyespot achieved by spraying between leaf-sheath erect and second node detectable stages (GS 30-32)
- Best control of ear diseases obtained by treatment during ear emergence
- For Septoria spray after third node detectable stage (GS 33) when weather favouring disease development has occurred
- Epoxiconazole is a DMI fungicide. Resistance to some DMI fungicides has been identified in Septoria leaf blotch which may seriously affect performance of some products. For further advice contact a specialist advisor and visit the Fungicide Resistance Action Group (FRAG)-UK website

Restrictions

- Maximum total dose equivalent to two full dose treatments
- Product may cause damage to broad-leaved plant species

Crop-specific information

- Latest use: up to and including flowering just complete (GS 69) in wheat, rye, triticale; up to and including emergence of ear just complete (GS 59) in barley, oats

Environmental safety

- Dangerous for the environment
- Toxic to aquatic organisms
- LERAP Category B
- Avoid spray drift onto neighbouring crops

Hazard classification and safety precautions

Hazard Harmful, Dangerous for the environment
Transport code 9
Packaging group III
UN number 3082
Risk phrases R20, R43 [1-3]; R38 [4]; R40, R62, R63 [1, 3, 4]; R51, R53a [1-4]
Operator protection A [1-4]; H [1, 3]; U05a, U20b [1-4]; U14 [2]; U19a [2, 4]
Environmental protection E15a, E16a [1-4]; E16b [2, 4]; E38 [1, 3]
Storage and disposal D01, D02, D05, D09a [1-4]; D10a [2]; D10b [4]; D10c, D12a [1, 3]
Medical advice M05a [1, 3]; M05b [4]

202 epoxiconazole + fenpropimorph + kresoxim-methyl

A protectant, systemic and curative fungicide mixture for cereals
FRAC mode of action code: 3 + 5 + 11

See also *fenpropimorph*
kresoxim-methyl

Products

1	Bullseye	AgriGuard	125:150:125 g/l	SE	12278
2	Mantra	BASF	125:150:125 g/l	SE	11728
3	Standon Kresoxim Super	Standon	125:150:125 g/l	SE	09794

Uses

- Brown rust in ***spring barley***, ***winter barley***, ***winter wheat*** [1-3]; ***spring rye***, ***spring wheat***, ***triticale***, ***winter rye*** [1, 2]
- Eyespot in ***spring oats*** *(reduction)*, ***spring rye*** *(reduction)*, ***triticale*** *(reduction)*, ***winter oats*** *(reduction)*, ***winter rye*** *(reduction)* [1, 2]; ***winter barley*** *(reduction)*, ***winter wheat*** *(reduction)* [1-3]
- Foliar disease control in ***durum wheat*** *(off-label)*, ***grass seed crops*** *(off-label)* [2]
- Fusarium ear blight in ***spring wheat*** *(reduction)* [1, 2]; ***winter wheat*** *(reduction)* [1-3]
- Glume blotch in ***spring wheat***, ***triticale*** [1, 2]; ***winter wheat*** [1-3]
- Net blotch in ***spring barley***, ***winter barley*** [1-3]
- Powdery mildew in ***spring barley***, ***winter barley*** [1-3]; ***spring oats***, ***spring rye***, ***triticale***, ***winter oats***, ***winter rye*** [1, 2]
- Rhynchosporium in ***spring barley***, ***winter barley*** [1-3]; ***spring rye***, ***winter rye*** [1, 2]
- Septoria leaf blotch in ***spring wheat***, ***triticale*** [1, 2]; ***winter wheat*** [1-3]
- Sooty moulds in ***spring wheat*** *(reduction)* [1, 2]; ***winter wheat*** *(reduction)* [1-3]
- Yellow rust in ***spring barley***, ***winter barley***, ***winter wheat*** [1-3]; ***spring rye***, ***spring wheat***, ***triticale***, ***winter rye*** [1, 2]

Specific Off-Label Approvals (SOLAs)

- ***durum wheat*** *20061101* [2]
- ***grass seed crops*** *20061101* [2]

Approval information

- Kresoxim-methyl, epoxiconazole and fenpropimorph included in Annex I under EC Directive 91/414
- Accepted by BBPA for use on malting barley (before ear emergence only)

Efficacy guidance

- For best results spray at the start of foliar disease attack and repeat if infection conditions persist
- Optimum effect against eyespot obtained by treatment between leaf sheaths erect and first node detectable stages (GS 30-32)
- For protection against ear diseases apply during ear emergence
- Epoxiconazole is a DMI fungicide. Resistance to some DMI fungicides has been identified in Septoria leaf blotch which may seriously affect performance of some products. For further advice contact a specialist advisor and visit the Fungicide Resistance Action Group (FRAG)-UK website
- Kresoxim-methyl is a member of the QoI cross resistance group. Product should be used preventatively and not relied on for its curative potential
- Use product as part of an Integrated Crop Management strategy incorporating other methods of control, including where appropriate other fungicides with a different mode of action. Do not apply more than two foliar applications of QoI containing products to any cereal crop
- There is a significant risk of widespread resistance occurring in *Septoria tritici* populations in UK. Failure to follow resistance management action may result in reduced levels of disease control
- Strains of barley powdery mildew resistant to QoIs are common in the UK

Restrictions

- Maximum total dose equivalent to two full dose treatments

Crop-specific information

- Latest use: mid flowering (GS 65) for wheat, rye, triticale; completion of ear emergence (GS 59) for barley, oats

Environmental safety

- Dangerous for the environment
- Very toxic to aquatic organisms
- LERAP Category B
- Avoid spray drift onto neighbouring crops. Product may damage broad-leaved species

Hazard classification and safety precautions

Hazard Harmful, Dangerous for the environment
Transport code 9
Packaging group III
UN number 3082
Risk phrases R40, R50, R53a, R62, R63 [1-3]; R43 [3]
Operator protection A [1-3]; H [1, 2]; U05a, U14, U20b
Environmental protection E15a, E16a [1-3]; E16b [3]; E38 [1, 2]
Storage and disposal D01, D02, D05, D09a [1-3]; D08, D10b [3]; D10c, D12a [1, 2]
Medical advice M05a [1, 2]; M05b [3]

203 epoxiconazole + fenpropimorph + metrafenone

A broad spectrum fungicide mixture for cereals
FRAC mode of action code: 3 + 5 + U8

See also *fenpropimorph*
metrafenone

Products

Capalo	BASF	62.5:200:75 g/l	SE	13170

Uses

- Brown rust in ***spring barley***, ***spring rye***, ***spring wheat***, ***triticale***, ***winter barley***, ***winter rye***, ***winter wheat***
- Crown rust in ***spring oats***, ***winter oats***
- Fusarium ear blight in ***spring wheat*** *(reduction)*, ***winter wheat*** *(reduction)*
- Glume blotch in ***spring wheat***, ***triticale***, ***winter wheat***
- Net blotch in ***spring barley***, ***winter barley***
- Powdery mildew in ***spring barley***, ***spring oats***, ***spring rye***, ***spring wheat***, ***triticale***, ***winter barley***, ***winter oats***, ***winter rye***, ***winter wheat***
- Rhynchosporium in ***spring barley***, ***spring rye***, ***winter barley***, ***winter rye***
- Septoria leaf blotch in ***spring wheat***, ***triticale***, ***winter wheat***
- Sooty moulds in ***spring wheat*** *(reduction)*, ***winter wheat*** *(reduction)*
- Yellow rust in ***spring barley***, ***spring rye***, ***spring wheat***, ***triticale***, ***winter barley***, ***winter rye***, ***winter wheat***

Approval information

- Epoxiconazole, fenpropimorph and metrafenone included in Annex I under EC Directive 91/414

Efficacy guidance

- For best results spray at the start of foliar disease attack and repeat if infection conditions persist
- Optimum effect against eyespot obtained by treatment between leaf sheaths erect and first node detectable stages (GS 30-32)
- For protection against ear diseases apply during ear emergence
- Treatment for control of Septoria diseases should normally be made after third node detectable stage if the disease is present and favourable weather for disease development has occurred
- Epoxiconazole is a DMI fungicide. Resistance to some DMI fungicides has been identified in Septoria leaf blotch which may seriously affect performance of some products. For further advice contact a specialist advisor and visit the Fungicide Resistance Action Group (FRAG)-UK website
- Should be used as part of a resistance management strategy that includes mixtures or sequences effective against mildew and non-chemical methods
- Where mildew well established at time of treatment product should be mixed with fenpropimorph. See label

Restrictions

- Maximum number of treatments 2 per crop
- Do not apply more than two applications of any products containing metrafenone to any one crop

Crop-specific information

- Latest use: up to and including beginning of flowering (GS 61) for rye, wheat, triticale; up to and including ear emergence complete (GS 59) for barley, oats

Following crops guidance

- Cereals. oilseed rape, sugar beet, linseed, maize, clover, field beans, peas, turnips, carrots, cauliflowers, onions, lettuce or potatoes may be sown following normal harvest of a treated crop

Environmental safety

- Dangerous for the environment
- Very toxic to aquatic organisms
- LERAP Category B
- Avoid spray drift onto neighbouring crops. Broad-leaved species may be damaged

Hazard classification and safety precautions

Hazard Harmful, Dangerous for the environment
Transport code 9
Packaging group III
UN number 3082
Risk phrases R38, R40, R43, R50, R53a, R62, R63
Operator protection A, H; U05a, U08, U14, U15, U19a, U20b
Environmental protection E16a, E34, E38
Storage and disposal D01, D02, D05, D08, D09a, D10c, D12a
Medical advice M03

204 epoxiconazole + fenpropimorph + pyraclostrobin

A protectant and systemic fungicide mixture for cereals
FRAC mode of action code: 3 + 5 + 11

See also *fenpropimorph*
pyraclostrobin

Products

Diamant	BASF	43:214:114 g/l	SE	14149

Uses

- Brown rust in ***spring barley***, ***spring wheat***, ***winter barley***, ***winter wheat***
- Crown rust in ***spring oats***, ***winter oats***
- Fusarium ear blight in ***spring wheat*** *(reduction)*, ***winter wheat*** *(reduction)*
- Glume blotch in ***spring wheat***, ***winter wheat***
- Net blotch in ***spring barley***, ***winter barley***
- Powdery mildew in ***spring barley***, ***spring oats***, ***winter barley***, ***winter oats***
- Rhynchosporium in ***spring barley***, ***winter barley***
- Septoria leaf blotch in ***spring wheat***, ***winter wheat***
- Yellow rust in ***spring barley***, ***spring wheat***, ***winter barley***, ***winter wheat***

Approval information

- Epoxiconazole, fenpropimorph and pyraclostrobin included in Annex I under EC Directive 91/414
- Accepted by BBPA for use on malting barley (before ear emergence only)

Efficacy guidance

- For best results spray at the start of foliar disease attack and repeat if infection conditions persist
- For protection against ear diseases apply during ear emergence
- Epoxiconazole is a DMI fungicide. Resistance to some DMI fungicides has been identified in Septoria leaf blotch which may seriously affect performance of some products. For further advice contact a specialist advisor and visit the Fungicide Resistance Action Group (FRAG)-UK website
- Pyraclostrobin is a member of the QoI cross resistance group. Product should be used preventatively and not relied on for its curative potential

- Use product as part of an Integrated Crop Management strategy incorporating other methods of control, including where appropriate other fungicides with a different mode of action. Do not apply more than two foliar applications of QoI containing products to any cereal crop
- There is a significant risk of widespread resistance occurring in *Septoria tritici* populations in UK. Failure to follow resistance management action may result in reduced levels of disease control
- Strains of barley powdery mildew resistant to QoIs are common in the UK

Restrictions

- Maximum number of treatments 2 per crop

Crop-specific information

- Latest use: up to and including ear emergence just complete (GS 59) for barley, oats; up to and including flowering just complete (GS 69) for wheat

Environmental safety

- Dangerous for the environment
- Very toxic to aquatic organisms
- Dangerous to fish or other aquatic life. Do not contaminate surface waters or ditches with chemical or used container
- LERAP Category B
- Avoid spray drift onto neighbouring crops. Product may damage broad-leaved species

Hazard classification and safety precautions

Hazard Harmful, Dangerous for the environment
Transport code 9
Packaging group III
UN number 3082
Risk phrases R20, R22a, R38, R40, R50, R53a, R63
Operator protection A; U05a, U14, U20b
Environmental protection E16a, E34, E38
Storage and disposal D01, D02, D05, D09a, D10c, D12a
Medical advice M03, M05a

205 epoxiconazole + kresoxim-methyl

A protectant, systemic and curative fungicide mixture for cereals
FRAC mode of action code: 3 + 11

See also *kresoxim-methyl*

Products

1	Clayton Gantry	Clayton	125:125 g/l	SC	09482
2	Landmark	BASF	125:125 g/l	SC	11730
3	Serial Duo	AgriGuard	125:125	SC	12252
4	Standon Kresoxim-Epoxiconazole	Standon	125:125 g/l	SC	09281

Uses

- Brown rust in ***spring barley, winter barley, winter wheat*** [1-4]; ***spring rye, spring wheat, triticale, winter rye*** [2, 3]
- Eyespot in ***spring oats*** *(reduction)*, ***spring rye*** *(reduction)*, ***triticale*** *(reduction)*, ***winter oats*** *(reduction)*, ***winter rye*** *(reduction)* [2, 3]; ***winter barley*** *(reduction)*, ***winter wheat*** *(reduction)* [1-4]
- Foliar disease control in ***durum wheat*** *(off-label)* [2]; ***grass seed crops*** *(off-label)* [1, 2, 4]
- Fusarium ear blight in ***spring wheat*** *(reduction)* [2, 3]; ***winter wheat*** *(reduction)* [1-4]
- Glume blotch in ***spring wheat, triticale*** [2, 3]; ***winter wheat*** [1-4]
- Net blotch in ***spring barley, winter barley*** [1-4]
- Powdery mildew in ***spring barley, winter barley*** [1-4]; ***spring oats, spring rye, spring wheat, triticale, winter oats, winter rye, winter wheat*** [2, 3]
- Rhynchosporium in ***spring barley, winter barley*** [1-4]; ***spring rye, winter rye*** [2, 3]
- Septoria leaf blotch in ***spring wheat, triticale*** [2, 3]; ***winter wheat*** [1-4]
- Sooty moulds in ***spring wheat*** *(reduction)* [2, 3]; ***winter wheat*** *(reduction)* [1-4]
- Yellow rust in ***spring barley, winter barley, winter wheat*** [1-4]; ***spring rye, spring wheat, triticale, winter rye*** [2, 3]

Specific Off-Label Approvals (SOLAs)
- ***durum wheat*** *20061084* [2]
- ***grass seed crops*** *20061047* [1], *20061084* [2], *20061085* [4]

Approval information
- Epoxiconazole and kresoxim-methyl included in Annex I under EC Directive 91/414
- Accepted by BBPA for use on malting barley (before ear emergence only)

Efficacy guidance
- For best results spray at the start of foliar disease attack and repeat if infection conditions persist
- Optimum effect against eyespot obtained by treatment between leaf sheaths erect and first node detectable stages (GS 30-32)
- For protection against ear diseases apply during ear emergence
- Epoxiconazole is a DMI fungicide. Resistance to some DMI fungicides has been identified in Septoria leaf blotch which may seriously affect performance of some products. For further advice contact a specialist advisor and visit the Fungicide Resistance Action Group (FRAG)-UK website
- Kresoxim-methyl is a member of the QoI cross resistance group. Product should be used preventatively and not relied on for its curative potential
- Use product as part of an Integrated Crop Management strategy incorporating other methods of control, including where appropriate other fungicides with a different mode of action. Do not apply more than two foliar applications of QoI containing products to any cereal crop
- There is a significant risk of widespread resistance occurring in *Septoria tritici* populations in UK. Failure to follow resistance management action may result in reduced levels of disease control
- Strains of barley powdery mildew resistant to QoIs are common in the UK

Restrictions
- Maximum total dose equivalent to two full dose treatments

Crop-specific information
- Latest use: mid flowering (GS 65) for wheat, rye, triticale; completion of ear emergence (GS 59) for barley, oats

Environmental safety
- Dangerous for the environment
- Very toxic to aquatic organisms
- LERAP Category B
- Avoid spray drift onto neighbouring crops. Product may damage broad-leaved species

Hazard classification and safety precautions

Hazard Harmful, Dangerous for the environment
Transport code 9
Packaging group III
UN number 3082
Risk phrases R40, R50, R53a, R62, R63 [1-4]; R43 [1, 4]
Operator protection A; U05a, U14, U20b
Environmental protection E15a, E16a [1-4]; E16b [2-4]; E38 [2, 3]
Storage and disposal D01, D02, D09a [1-4]; D05, D10b [1, 4]; D08, D10c [2, 3]; D12a [1-3]
Medical advice M05a [2, 3]; M05b [1, 4]

206 epoxiconazole + kresoxim-methyl + pyraclostrobin

A protectant, systemic and curative fungicide mixture for cereals
FRAC mode of action code: 3 + 11 + 11

See also *kresoxim-methyl*
pyraclostrobin

Products

Opponent	BASF	50:67:133 g/l	SE	10877

Uses
- Brown rust in ***spring barley***, ***spring wheat***, ***winter barley***, ***winter wheat***
- Crown rust in ***spring oats***, ***winter oats***

SEE SECTION 3 FOR PRODUCTS ALSO REGISTERED

- Foliar disease control in ***durum wheat*** *(off-label)*, ***grass seed crops*** *(off-label)*, ***spring rye*** *(off-label)*, ***triticale*** *(off-label)*, ***winter rye*** *(off-label)*
- Glume blotch in ***spring wheat***, ***winter wheat***
- Net blotch in ***spring barley***, ***winter barley***
- Rhynchosporium in ***spring barley***, ***winter barley***
- Septoria leaf blotch in ***spring wheat***, ***winter wheat***
- Yellow rust in ***spring barley***, ***spring wheat***, ***winter barley***, ***winter wheat***

Specific Off-Label Approvals (SOLAs)

- ***durum wheat*** *20060539*
- ***grass seed crops*** *20060539*
- ***spring rye*** *20060539*
- ***triticale*** *20060539*
- ***winter rye*** *20060539*

Approval information

- Epoxiconazole, kresoxim-methyl and pyraclostrobin included in Annex I under EC Directive 91/414
- Accepted by BBPA for use on malting barley (before ear emergence only)

Efficacy guidance

- For best results apply at the start of foliar disease attack
- Good reduction of Fusarium ear blight can be obtained from treatment during flowering
- Yield response may be obtained in the absence of visual disease symptoms
- Best results on Septoria leaf blotch when treated in the latent phase
- Epoxiconazole is a DMI fungicide. Resistance to some DMI fungicides has been identified in Septoria leaf blotch which may seriously affect performance of some products. For further advice contact a specialist advisor and visit the Fungicide Resistance Action Group (FRAG)-UK website
- Kresoxim-methyl and pyraclostrobin are members of the QoI cross resistance group. Product should be used preventatively and not relied on for its curative potential
- Use product as part of an Integrated Crop Management strategy incorporating other methods of control, including where appropriate other fungicides with a different mode of action. Do not apply more than two foliar applications of QoI containing products to any cereal crop
- There is a significant risk of widespread resistance occurring in *Septoria tritici* populations in UK. Failure to follow resistance management action may result in reduced levels of disease control

Restrictions

- Maximum total dose equivalent to two full dose treatments

Crop-specific information

- Latest use: before grain watery ripe (GS 71) for wheat; up to and including emergence of ear just complete (GS 59) for barley and oats

Environmental safety

- Dangerous for the environment
- Very toxic to aquatic organisms
- Dangerous to fish or other aquatic life. Do not contaminate surface waters or ditches with chemical or used container
- LERAP Category B

Hazard classification and safety precautions

Hazard Harmful, Dangerous for the environment
Transport code 9
Packaging group III
UN number 3082
Risk phrases R20, R22a, R40, R50, R53a
Operator protection A; U05a, U14, U20b
Environmental protection E13b, E16a, E16b, E34, E38
Storage and disposal D01, D02, D05, D09a, D10c, D12a
Medical advice M05a

207 epoxiconazole + metconazole

A triazole mixture for use in cereals
FRAC mode of action code: 3 + 3

See also metconazole

Products

1	Brutus	BASF	37.5:27.5 g/l	EC	14353
2	Icarus	BASF	37.5:25.0 g/l	EC	14471

Uses

- Brown rust in ***durum wheat***, ***spring barley***, ***spring rye***, ***spring wheat***, ***triticale***, ***winter barley***, ***winter rye***, ***winter wheat*** [1, 2]
- Eyespot in ***durum wheat*** *(reduction)*, ***spring rye*** *(reduction)*, ***triticale*** *(reduction)*, ***winter barley*** *(reduction)*, ***winter rye*** *(reduction)*, ***winter wheat*** *(reduction)* [1, 2]
- Fusarium ear blight in ***durum wheat*** *(good reduction)*, ***spring wheat*** *(good reduction)*, ***triticale*** *(good reduction)*, ***winter wheat*** *(good reduction)* [1, 2]
- Glume blotch in ***durum wheat***, ***spring wheat***, ***triticale***, ***winter wheat*** [1, 2]
- Net blotch in ***spring barley***, ***winter barley*** [1, 2]
- Powdery mildew in ***durum wheat*** *(moderate control)*, ***spring barley*** *(moderate control)*, ***spring wheat*** *(moderate control)*, ***triticale*** *(moderate control)*, ***winter barley*** *(moderate control)*, ***winter wheat*** *(moderate control)* [1, 2]; ***spring rye*** *(moderate control)*, ***winter rye*** *(moderate control)* [1]; ***spring rye*** *(moderate reduction)*, ***winter rye*** *(moderate reduction)* [2]
- Ramularia leaf spots in ***spring barley*** [2]; ***spring barley*** *(moderate control)* [1]; ***winter barley*** *(moderate control)* [1, 2]
- Rhynchosporium in ***spring barley***, ***winter barley*** [1, 2]
- Septoria leaf blotch in ***durum wheat***, ***spring wheat***, ***triticale***, ***winter wheat*** [1, 2]
- Sooty moulds in ***durum wheat*** *(good reduction)*, ***spring wheat*** *(good reduction)*, ***winter wheat*** *(good reduction)* [1, 2]
- Tan spot in ***durum wheat*** *(moderate control)*, ***spring wheat*** *(moderate control)*, ***winter wheat*** *(moderate control)* [1, 2]
- Yellow rust in ***durum wheat***, ***spring barley***, ***spring rye***, ***spring wheat***, ***triticale***, ***winter barley***, ***winter rye***, ***winter wheat*** [1, 2]

Approval information

- Epoxiconazole and Metconazole included in Annex I under EC Directive 91/414

Following crops guidance

- After trreating a cereal crop, oilseed rape, cereals, sugar beet, linseed, maize, clover, beans, peas, carrots, potatoes, lettuce, cabbage, sunflower, ryegrass or onions may be sown as a following crop. The effect on other crops has not been evaluated.

Hazard classification and safety precautions

Hazard Harmful, Dangerous for the environment
Transport code 9
Packaging group III
UN number 3082
Risk phrases R40, R43, R51, R53a, R63
Operator protection A, H; U05a, U14, U20b
Environmental protection E15b, E16a, E34
Storage and disposal D01, D02, D09a, D10c
Medical advice M05a

208 epoxiconazole + metrafenone

A broad spectrum fungicide mixture for cereals
FRAC mode of action code: 3 + U8

See also *metrafenone*

Products

1	Ceando	BASF	83:100 g/l	SC	13271
2	Cloister	BASF	83:100 g/l	SC	14358

Uses

- Brown rust in ***spring barley***, ***spring rye***, ***spring wheat***, ***triticale***, ***winter barley***, ***winter rye***, ***winter wheat***
- Crown rust in ***spring oats***, ***winter oats***
- Fusarium ear blight in ***spring wheat*** *(reduction)*, ***winter wheat*** *(reduction)*
- Glume blotch in ***spring wheat***, ***triticale***, ***winter wheat***
- Net blotch in ***spring barley***, ***winter barley***
- Powdery mildew in ***spring barley***, ***spring oats***, ***spring rye***, ***spring wheat***, ***triticale***, ***winter barley***, ***winter oats***, ***winter rye***, ***winter wheat***
- Rhynchosporium in ***spring barley***, ***spring rye***, ***winter barley***, ***winter rye***
- Septoria leaf blotch in ***spring wheat***, ***triticale***, ***winter wheat***
- Sooty moulds in ***spring wheat*** *(reduction)*, ***winter wheat*** *(reduction)*
- Yellow rust in ***spring barley***, ***spring rye***, ***spring wheat***, ***triticale***, ***winter barley***, ***winter rye***, ***winter wheat***

Approval information

- Epoxiconazole and metrafenone included in Annex I under EC Directive 91/414

Efficacy guidance

- For best results spray at the start of foliar disease attack and repeat if infection conditions persist
- Optimum effect against eyespot obtained by treatment between leaf sheaths erect and first node detectable stages (GS 30-32)
- For protection against ear diseases apply during ear emergence
- Treatment for control of Septoria diseases should normally be made after third node detectable stage if the disease is present and favourable weather for disease development has occurred
- Epoxiconazole is a DMI fungicide. Resistance to some DMI fungicides has been identified in Septoria leaf blotch which may seriously affect performance of some products. For further advice contact a specialist advisor and visit the Fungicide Resistance Action Group (FRAG)-UK website
- Should be used as part of a resistance management strategy that includes mixtures or sequences effective against mildew and non-chemical methods
- Where mildew well established at time of treatment product should be mixed with fenpropimorph. See label

Restrictions

- Maximum number of treatments 2 per crop
- Do not apply more than two applications of any products containing metrafenone to any one crop

Crop-specific information

- Latest use: up to and including beginning of flowering (GS 61) for rye, wheat, triticale; up to and including ear emergence complete (GS 59) for barley, oats

Following crops guidance

- Cereals. oilseed rape, sugar beet, linseed, maize, clover, field beans, peas, turnips, carrots, cauliflowers, onions, lettuce or potatoes may be sown following normal harvest of a treated crop

Environmental safety

- Dangerous for the environment
- Very toxic to aquatic organisms
- LERAP Category B
- Avoid spray drift onto neighbouring crops. Broad-leaved species may be damaged

Hazard classification and safety precautions

Hazard Harmful, Dangerous for the environment
UN number N/C

Risk phrases R40, R50, R53a, R62, R63
Operator protection A, H; U05a, U20b
Environmental protection E15a, E16a, E34, E38
Storage and disposal D01, D02, D05, D08, D09a, D10c
Medical advice M03, M05a

209 epoxiconazole + prochloraz

A systemic, protectant and curative triazole fungicide mixture for use in cereals
FRAC mode of action code: 3

See also *prochloraz*

Products

Ennobe	BASF	62.5 + 225 g/l	DC	14340

Uses

- Brown rust in ***spring barley***, ***spring rye***, ***spring wheat***, ***triticale***, ***winter barley***, ***winter rye***, ***winter wheat***
- Eyespot in ***spring rye***, ***triticale***, ***winter barley***, ***winter rye***, ***winter wheat***
- Fusarium ear blight in ***spring wheat*** *(good reduction)*, ***triticale*** *(good reduction)*, ***winter wheat*** *(good reduction)*
- Glume blotch in ***spring wheat***, ***triticale***, ***winter wheat***
- Net blotch in ***spring barley***, ***winter barley***
- Powdery mildew in ***spring barley*** *(moderate control)*, ***spring rye*** *(moderate control)*, ***spring wheat*** *(moderate control)*, ***triticale*** *(moderate control)*, ***winter barley***, ***winter rye*** *(moderate control)*, ***winter wheat*** *(moderate control)*
- Rhynchosporium in ***spring barley***, ***spring rye***, ***winter barley***, ***winter rye***
- Septoria leaf blotch in ***spring wheat***, ***triticale***, ***winter wheat***
- Sooty moulds in ***spring wheat*** *(good reduction)*, ***winter wheat*** *(good reduction)*
- Yellow rust in ***spring barley***, ***spring rye***, ***spring wheat***, ***triticale***, ***winter barley***, ***winter rye***, ***winter wheat***

Approval information

- Epoxiconazole included in Annex I under EC Directive 91/414
- Prochloraz has not been included in Annex 1 but the withdrawal or inclusion date has yet to be set. Most current approvals set to expire 2013.

Efficacy guidance

- Use in mixture with Comaptibility Agent CA 259 when tank-mixing with other products. Order of mixing is not important.

Hazard classification and safety precautions

Hazard Harmful, Dangerous for the environment
Transport code 9
Packaging group III
UN number 3082
Risk phrases R20, R22a, R36, R38, R40, R50, R53a, R62, R63
Operator protection A, H; U05a
Environmental protection E15b, E16a, E38
Storage and disposal D01, D02, D08, D10c, D12a
Medical advice M05a

210 epoxiconazole + pyraclostrobin

A protectant, systemic and curative fungicide mixture for cereals
FRAC mode of action code: 3 + 11

See also *pyraclostrobin*

Products

1	Envoy	BASF	62.5:85 g/l	SE	12750
2	Gemstone	BASF	62.5:80	SE	13918
3	Opera	BASF	50:133 g/l	SE	12167

Uses

- Brown rust in ***spring barley***, ***spring wheat***, ***winter barley***, ***winter wheat*** [1-3]
- Cercospora leaf spot in ***sugar beet*** [3]
- Crown rust in ***spring oats***, ***winter oats*** [2, 3]; ***spring oats*** *(qualified minor use)*, ***winter oats*** *(qualified minor use)* [1]
- Foliar disease control in ***durum wheat*** *(off-label)*, ***fodder beet*** *(off-label)*, ***grass seed crops*** *(off-label)*, ***spring rye*** *(off-label)*, ***triticale*** *(off-label)*, ***winter rye*** *(off-label)* [3]
- Fusarium ear blight in ***spring wheat*** *(apply during flowering)*, ***winter wheat*** *(apply during flowering)* [2]; ***spring wheat*** *(reduction)*, ***winter wheat*** *(reduction)* [1]
- Glume blotch in ***spring wheat***, ***winter wheat*** [1-3]
- Net blotch in ***spring barley***, ***winter barley*** [1-3]
- Powdery mildew in ***sugar beet*** [3]
- Ramularia leaf spots in ***sugar beet*** [3]
- Rhynchosporium in ***spring barley***, ***winter barley*** [1-3]
- Rust in ***sugar beet*** [3]
- Septoria leaf blotch in ***spring wheat***, ***winter wheat*** [1-3]
- Tan spot in ***spring wheat***, ***winter wheat*** [1, 2]
- Yellow rust in ***spring barley***, ***spring wheat***, ***winter barley***, ***winter wheat*** [1-3]

Specific Off-Label Approvals (SOLAs)

- ***durum wheat*** *20062653* [3]
- ***fodder beet*** *20062654* [3]
- ***grass seed crops*** *20062653* [3]
- ***spring rye*** *20062653* [3]
- ***triticale*** *20062653* [3]
- ***winter rye*** *20062653* [3]

Approval information

- Epoxiconazole and pyraclostrobin included in Annex I under EC Directive 91/414
- Accepted by BBPA for use on malting barley (before ear emergence only)

Efficacy guidance

- For best results apply at the start of foliar disease attack
- Good reduction of Fusarium ear blight can be obtained from treatment during flowering
- Yield response may be obtained in the absence of visual disease symptoms
- Epoxiconazole is a DMI fungicide. Resistance to some DMI fungicides has been identified in Septoria leaf blotch which may seriously affect performance of some products. For further advice contact a specialist advisor and visit the Fungicide Resistance Action Group (FRAG)-UK website
- Pyraclostrobin is a member of the QoI cross resistance group. Product should be used preventatively and not relied on for its curative potential
- Use product as part of an Integrated Crop Management strategy incorporating other methods of control, including where appropriate other fungicides with a different mode of action. Do not apply more than two foliar applications of QoI containing products to any cereal crop
- There is a significant risk of widespread resistance occurring in *Septoria tritici* populations in UK. Failure to follow resistance management action may result in reduced levels of disease control

Restrictions

- Maximum number of treatments 2 per crop
- Do not use on sugar beet crops grown for seed [3]

Crop-specific information

- Latest use: before grain watery ripe (GS 71) for wheat; up to and including emergence of ear just complete (GS 59) for barley and oats
- HI 6 wk for sugar beet [3]

Environmental safety

- Dangerous for the environment
- Very toxic to aquatic organisms
- LERAP Category B

Hazard classification and safety precautions

Hazard Harmful, Dangerous for the environment
Transport code 9
Packaging group III
UN number 3082
Risk phrases R20, R22a, R38, R40, R43, R50, R53a [1-3]; R62, R63 [1, 2]
Operator protection A [1-3]; H [2]; U05a, U14 [1-3]; U20b [1, 3]; U20c [2]
Environmental protection E15a, E16a, E34, E38 [1-3]; E16b [3]
Storage and disposal D01, D02, D09a, D10c, D12a [1-3]; D05 [3]; D08 [2]
Medical advice M03, M05a

211 esfenvalerate

A contact and ingested pyrethroid insecticide
IRAC mode of action code: 3

Products

1	Greencrop Cajole Ultra	Greencrop	25 g/l	EC	12967
2	Standon Hounddog	Standon	25 g/l	EC	13201
3	Sumi-Alpha	Interfarm	25 g/l	EC	14023
4	Sven	Interfarm	25 g/l	EC	14035

Uses

- Aphids in ***grass seed crops*** *(off-label)* [3]; ***spring barley***, ***spring wheat***, ***winter barley***, ***winter wheat*** [1-4]

Specific Off-Label Approvals (SOLAs)

- ***grass seed crops*** *20081266* [3]

Approval information

- Esfenvalerate included in Annex I under EC Directive 91/414
- Accepted by BBPA for use on malting barley

Efficacy guidance

- For best reduction of spread of barley yellow dwarf virus winter sown crops at high risk (e.g. after grass or in areas with history of BYDV) should be treated when aphids first seen or by mid-Oct. Otherwise treat in late Oct-early Nov
- High risk winter sown crops will need a second treatment
- Spring sown crops should be treated from the 2-3 leaf stage if aphids are found colonising in the crop and a further application may be needed before the first node stage if they reinfest
- Product also recommended between onset of flowering and milky ripe stages (GS 61-73) for control of summer cereal aphids

Restrictions

- Maximum number of treatments 3 per crop of which 2 may be in autumn
- Do not use if another pyrethroid or dimethoate has been applied to crop after start of ear emergence (GS 51)
- Do not leave spray solution standing in spray tank

Crop-specific information

- Latest use: 31 Mar in yr of harvest (winter use dose); early milk stage for barley (summer use); late milk stage for wheat (summer use)

Environmental safety

- Dangerous for the environment
- Very toxic to aquatic organisms

- High risk to non-target insects or other arthropods. Do not spray within 6 m of the field boundary
- Flammable
- LERAP Category A
- Store product in dark away from direct sunlight

Hazard classification and safety precautions

Hazard Harmful, Flammable, Dangerous for the environment
Transport code 3 [3, 4]; 6.1 [1, 2]
Packaging group II [3, 4]; III [1, 2]
UN number 1993 [3, 4]; 3351 [1, 2]
Risk phrases R20, R22a, R41, R43, R50, R53a
Operator protection A, C, H; U04a, U05a, U08, U11, U14, U19a, U20b
Environmental protection E15a, E22a [1, 3, 4]; E15b, E22b [2]; E16c, E16d, E34 [1-4]; E38 [3, 4]
Storage and disposal D01, D02, D09a [1-4]; D05 [1, 2]; D07, D10b [1]; D10c, D12a [2-4]
Medical advice M03 [1-4]; M05b [2]

212 ethofumesate

A benzofuran herbicide for grass weed control in various crops
HRAC mode of action code: N

See also *bromoxynil + ethofumesate + ioxynil*
chloridazon + ethofumesate
desmedipham + ethofumesate + metamitron + phenmedipham
desmedipham + ethofumesate + phenmedipham

Products

1	Alpha Ethofumesate	Makhteshim	500 g/l	SC	13055
2	Barclay Keeper 500 SC	Barclay	500 g/l	SC	14016
3	Catalyst	AgriGuard	500 g/l	SC	13266
4	Ethosat 500	Makhteshim	500 g/l	SC	13050
5	Fumesate 500 SC	Hermoo	500 g/l	SC	14321
6	Kubist Flo	Nufarm UK	500 g/l	SC	12987
7	Nortron Flo	Bayer CropScience	500 g/l	SC	12986
8	Oblix 500	AgriChem BV	500 g/l	SC	12349

Uses

- Annual dicotyledons in ***amenity grassland***, ***grass seed crops*** [1, 2, 4]; ***fodder beet***, ***red beet***, ***sugar beet*** [1-8]; ***forest nurseries*** *(off-label)*, ***game cover*** *(off-label)*, ***ornamental specimens*** *(off-label)* [7]; ***managed amenity turf*** *(off-label)* [4]; ***mangels*** [1-4, 6-8]
- Annual grasses in ***amenity grassland***, ***grass seed crops*** [2]; ***fodder beet***, ***red beet***, ***sugar beet*** [2, 5, 8]; ***forest nurseries*** *(off-label)* [4, 7]; ***game cover*** *(off-label)*, ***ornamental specimens*** *(off-label)* [7]; ***mangels*** [2, 8]
- Annual meadow grass in ***amenity grassland***, ***grass seed crops*** [1, 4]; ***fodder beet***, ***mangels***, ***red beet***, ***sugar beet*** [1, 3, 4, 6, 7]; ***managed amenity turf*** *(off-label)* [4]
- Blackgrass in ***amenity grassland***, ***grass seed crops*** [1, 4]; ***fodder beet***, ***mangels***, ***red beet***, ***sugar beet*** [1, 3, 4, 6, 7]; ***managed amenity turf*** *(off-label)* [4]
- Fat hen in ***poppies for morphine production*** *(off-label)* [7]
- Grass weeds in ***poppies for morphine production*** *(off-label)* [7]
- Mayweeds in ***poppies for morphine production*** *(off-label)* [7]

Specific Off-Label Approvals (SOLAs)

- ***forest nurseries*** *20082846* [4], *20082919* [7]
- ***game cover*** *20082919* [7]
- ***managed amenity turf*** *20081026* [4]
- ***ornamental specimens*** *20082919* [7]
- ***poppies for morphine production*** *20070077* [7]

Approval information

- Ethofumesate included in Annex I under EC Directive 91/414

Efficacy guidance

- Most products may be applied pre- or post-emergence of crop or weeds but some restricted to pre-emergence or post-emergence use only. Check label

- Some products recommended for use only in mixtures. Check label
- Volunteer cereals not well controlled pre-emergence, weed grasses should be sprayed before fully tillered
- Grass crops may be sprayed during rain or when wet. Not recommended in very dry conditions or prolonged frost

Restrictions

- Maximum number of treatments 1 pre- plus 1 or 2 post-emergence per crop or yr for beet crops and grass seed crops; 2 per yr for amenity turf and established grassland. The maximum total dose must not exceed 1.0 kg ethofumesate per hectare in any three year period
- Do not use on Sands or Heavy soils, Very Light soils containing a high percentage of stones, or soils with more than 5-10% organic matter (percentage varies according to label)
- Do not use on swards reseeded without ploughing
- Clovers will be killed or severely checked
- Do not graze or cut grass for 14 d after, or roll less than 7 d before or after spraying

Crop-specific information

- Latest use: before crops meet across rows for beet crops and mangels; not specified for other crops
- Apply in beet crops in tank mixes with other pre- or post-emergence herbicides. Recommendations vary for different mixtures. See label for details
- Safe timing on beet crops varies with other ingredient of tank mix. See label for details
- In grass crops apply to moist soil as soon as possible after sowing or post-emergence when crop in active growth, normally mid-Oct to mid-Dec. See label for details
- May be used in Italian, hybrid and perennial ryegrass, timothy, cocksfoot, meadow fescue and tall fescue. Apply pre-emergence to autumn-sown leys, post-emergence after 2-3 leaf stage. See label for details

Following crops guidance

- In the event of failure of a treated crop only sugar beet, red beet, fodder beet and mangels may be redrilled within 3 mth of application
- Any crop may be sown 3 mth after application of mixtures in beet crops following ploughing, 5 mth after application in grass crops

Environmental safety

- Dangerous for the environment
- Toxic to aquatic organisms
- Do not empty into drains

Hazard classification and safety precautions

Hazard Dangerous for the environment [1-4, 6, 7]

Transport code 9 [1-4, 6, 7]

Packaging group III [1-4, 6, 7]

UN number 3082 [1-4, 6, 7]; N/C [5, 8]

Risk phrases R51 [1-4, 6, 7]; R52 [5, 8]; R53a [1-8]

Operator protection H [8]; U05a [2, 8]; U08 [1-4]; U09a [5-7]; U19a [1-7]; U20a [1, 3, 4, 6, 7]; U20b [2, 5]

Environmental protection E13c [3]; E15a [1, 3, 4, 6]; E15b [2, 5, 8]; E19b [1, 4]; E23 [8]; E34 [1, 2, 4, 8]; E38 [5-7]

Storage and disposal D01, D02 [8]; D05 [1, 3-7]; D09a [1-8]; D10a [2, 5, 8]; D10b [1, 3, 4, 6]; D12a [1, 2, 4-8]

213 ethofumesate + metamitron

A contact and residual herbicide mixture for beet crops
HRAC mode of action code: N + C1

See also *metamitron*

Products

1	Goltix Super	Makhteshim	150:350 g/l	SC	12216
2	Goltix Uno	Makhteshim	150:350 g/l	SC	12602
3	Oblix MT	AgriChem BV	150:350 g/l	SC	14510
4	Torero	Makhteshim	150:350 g/l	SC	11158

Uses

- Annual dicotyledons in ***fodder beet, mangels, sugar beet***
- Annual meadow grass in ***fodder beet, mangels, sugar beet***

Approval information

- Ethofumesate included in Annex I under EC Directive 91/414

Efficacy guidance

- Best results obtained from a series of treatments applied as an overall fine spray commencing when earliest germinating weeds are no larger than fully expanded cotyledon and the majority of the crop at fully expanded cotyledon
- Apply subsequent sprays as each new flush of weeds reaches early cotyledon and continue until weed emergence ceases (maximum 3 sprays)
- Product may be used on all soil types but residual activity may be reduced on those with more than 5% organic matter

Restrictions

- Maximum total dose equivalent to three full dose treatments

Crop-specific information

- Latest use: before crop leaves meet between rows
- Crop tolerance may be reduced by stress caused by growing conditions, effects of pests, disease or other pesticides, nutrient deficiency etc

Following crops guidance

- Beet crops may be sown at any time after treatment. Any other crop may be sown after mouldboard ploughing to 15 cm and a minimum interval of 3 mth after treatment

Environmental safety

- Dangerous for the environment
- Very toxic to aquatic organisms
- Do not empty into drains

Hazard classification and safety precautions

Hazard Harmful [3]; Dangerous for the environment [1-4]
Transport code 9
Packaging group III
UN number 3082
Risk phrases R22a [3]; R51, R53a [1-4]
Operator protection U05a, U19a, U20a [1-4]; U08, U14, U15 [1, 2, 4]
Environmental protection E13c, E34 [1, 2, 4]; E15a [3]; E19b [1-4]
Storage and disposal D01, D02, D09a, D10b, D12a [1-4]; D05 [1, 2, 4]
Medical advice M05a [3]

214 ethofumesate + metamitron + phenmedipham

A contact and residual herbicide mixture for sugar beet
HRAC mode of action code: N + C1 + C1

See also *metamitron*
phenmedipham

Products

Phemo	AgriChem BV	51:153:51 g/l	SE	14193

Uses

- Annual dicotyledons in ***fodder beet, mangels, red beet, sugar beet***
- Annual meadow grass in ***fodder beet, mangels, red beet, sugar beet***

Approval information

- Ethofumesate and phenmedipham included in Annex I under EC Directive 91/414

Efficacy guidance

- Best results obtained from a series of treatments applied as an overall fine spray commencing when earliest germinating weeds are no larger than fully expanded cotyledon
- Apply subsequent sprays as each new flush of weeds reaches early cotyledon and continue until weed emergence ceases (maximum 3 sprays)

- Product must be applied with Actipron
- Product may follow certain pre-emergence treatments and be used in conjunction with other post-emergence sprays - see label for details
- Where a pre-emergence band spray has been applied, the first treatment should be timed according to the size of the weeds in the untreated area between the rows

Restrictions
- Maximum number of treatments 3 per crop; maximum total dose 6 kg product per ha
- Product may be used on all soil types but residual activity may be reduced on those with more than 5% organic matter
- Crop tolerance may be reduced by stress caused by growing conditions, effects of pests, disease or other pesticides, nutrient deficiency etc
- Only beet crops should be sown within 4 mth of last treatment; winter cereals may be sown after this interval
- Any spring crop may be sown in the year following use
- Mould-board ploughing to 150 mm followed by thorough cultivation recommended before planting any crop

Crop-specific information
- Latest use: before crop leaves meet between rows

Environmental safety
- Irritating to eyes
- Harmful to fish or other aquatic life. Do not contaminate surface waters or ditches with chemical or used container

Hazard classification and safety precautions

Hazard Dangerous for the environment
Transport code 9
Packaging group III
UN number 3082
Risk phrases R51, R53a
Operator protection A; U05a, U20c
Environmental protection E15a, E34, E38
Storage and disposal D01, D02, D09a, D11a, D11b, D12a

215 ethofumesate + phenmedipham

A contact and residual herbicide for use in beet crops
HRAC mode of action code: N + C1

See also *phenmedipham*

Products

1	Betosip Combi	Sipcam	190:200 g/l	SC	13254
2	Duo 400 SC	AgriGuard	200:200 g/l	SC	12561
3	Duo 400 SC	AgriGuard	200:200 g/l	SC	14232
4	Fenlander 2	Makhteshim	94:97 g/l	EC	14030
5	Powertwin	Makhteshim	200:200 g/l	SC	14004
6	Twin	Makhteshim	94:97 g/l	EC	13998

Uses
- Annual dicotyledons in ***fodder beet*, *mangels*, *sugar beet*** [1-6]; ***red beet*** [2]
- Annual meadow grass in ***fodder beet*, *mangels*, *sugar beet*** [1-6]; ***red beet*** [2]
- Blackgrass in ***fodder beet*, *mangels*, *sugar beet*** [1-6]; ***red beet*** [2]

Approval information
- Ethofumesate and phenmedipham included in Annex I under EC Directive 91/414
- Approval expiry 31 Aug 2010 [2]

Efficacy guidance
- Best results achieved by repeat applications to cotyledon stage weeds. Larger susceptible weeds not killed by first treatment usually checked and controlled by second application
- Apply on all soil types at 5-10 d intervals
- On soils with more than 5-10% organic matter residual activity may be reduced

Restrictions

- Maximum number of treatments normally 3 per crop or maximum total dose equivalent to three full dose treatments, or less - see labels for details
- Must be used in conjunction with specified adjuvant. See label [1, 2, 5]
- Do not spray wet foliage or if rain imminent
- Spray must be applied low volume to avoid crystallisation and nozzle blockage. See label for details
- Spray in evening if daytime temperatures above 21 °C expected
- Avoid or delay treatment if frost expected within 7 d
- Avoid or delay treating crops under stress from wind damage, manganese or lime deficiency, pest or disease attack etc

Crop-specific information

- Latest use: before crop foliage meets in the rows
- Check from which recovery may not be complete may occur if treatment made during conditions of sharp diurnal temperature fluctuation

Following crops guidance

- Beet crops may be sown at any time after treatment. Any other crop may be sown after mouldboard ploughing to 15 cm and a minimum interval of 3 mth after treatment

Environmental safety

- Dangerous for the environment [2, 4-6]
- Toxic (harmful [1]) to aquatic organisms
- Do not empty into drains
- Extra care necessary to avoid drift because product is recommended for use as a fine spray

Hazard classification and safety precautions

Hazard Harmful [2-4, 6]; Irritant [5]; Dangerous for the environment [2-6]
Transport code 9
Packaging group III
UN number 3082
Risk phrases R20 [4]; R37, R40 [6]; R43, R51 [2-6]; R52 [1]; R53a [1-6]
Operator protection A [2-6]; H [2, 3, 5]; U05a, U08, U19a, U20a [1-6]; U14 [2, 3, 5]
Environmental protection E15a, E34 [1-6]; E19b [2-6]; E38 [4]
Storage and disposal D01, D09a, D10b, D12a [1-6]; D02 [1, 4-6]; D05 [1-3, 5]
Medical advice M05a [2-6]

216 ethoprophos

An organophosphorus nematicide and insecticide
IRAC mode of action code: 1B

Products

Mocap 10G	Bayer CropScience	10% w/w	GR	10003

Uses

- Potato cyst nematode in ***potatoes***
- Wireworm in ***potatoes***

Approval information

- Ethoprophos included in Annex I under EC Directive 91/414
- In 2006 PSD required that all products containing this active ingredient should carry the following warning in the main area of the container label: "Ethoprofos is an anticholinesterase organophosphate. Handle with care"

Efficacy guidance

- Broadcast shortly before or during final soil preparation with suitable fertilizer spreader and incorporate immediately to 10-15 cm. Deeper incorporation may reduce efficacy. See label for details
- Treatment can be applied on all soil types. Control of pests reduced on organic soils
- Effectiveness dependent on soil moisture. Drought after application may reduce control
- Where pre-planting nematode populations are high, loss of crop vigour and subsequent loss of yield may occur in spite of treatment

Restrictions
- Contains an anticholinesterase organophosphorous compound. Do not use if under medical advice not to work with such compounds
- Maximum number of treatments 1 per crop
- Do not harvest crops for human or animal consumption for at least 8 wk after application
- Vehicles with a closed cab must be used when applying and incorporating
- Do not apply by hand or hand held equipment

Crop-specific information
- Latest use: pre-planting of crop
- Field scale application must only be through positive displacement type specialist granule applicators or dual purpose fertilizer applicators with microgranule setting

Environmental safety
- Dangerous for the environment
- Toxic to aquatic organisms
- Dangerous to livestock. Keep all livestock out of treated areas/away from treated water for at least 13 wk. Bury or remove spillages
- Dangerous to game, wild birds and animals
- Product supplied in returnable/refillable containers. Follow instructions on label

Hazard classification and safety precautions

Hazard Toxic, Dangerous for the environment
Transport code 6.1
Packaging group II
UN number 2783
Risk phrases R22a, R23, R43, R51, R53a
Operator protection A, D, H; U02a, U04a, U05a, U07, U08, U13, U14, U19a, U20a
Environmental protection E06b (13 wk); E10a, E13b, E34, E36a, E38
Consumer protection C02a (8 wk)
Storage and disposal D01, D02, D09a, D12a, D14
Medical advice M01, M03, M04a

217 ethylene

A gas used for fruit ripening and potato storage

Products

1	Ethylene	Biofresh	99.9% w/v	LI	14177
2	ethylene	various	-	GA	-

Uses
- Fruit ripening in ***fruit crops*** *(in store)* [2]
- Sprout suppression in ***potatoes*** [1]; ***potatoes*** *(in store)* [2]

Approval information
- Ethylene is being supported for review in the fourth stage of the EC Review Programme under Directive 91/414. Whether or not it is included in Annex I, the existing commodity chemical approval will have to be revoked because substances listed in Annex I must be approved for marketing and use under the EC regime. If ethylene is included in Annex I, products containing it will need to gain approval in the normal way if they carry label claims for pesticidal activity

Efficacy guidance
- For use in stored fruit or potatoes after harvest

Restrictions
- Handling and release of ethylene must only be undertaken by operators suitably trained and competent to carry out the work
- Operators must vacate treated areas immediately after ethylene introduction
- Unprotected persons must be excluded from the treated areas until atmospheres have been thoroughly ventilated for 15 minutes minimum before re-entry
- Ambient atmospheric ethylene concentration must not exceed 1000 ppm. Suitable self-contained breathing apparatus must be worn in atmospheres containing ethylene in excess of 1000 ppm
- A minimum 3 d post treatment period is required before removal of treated crop from storage

- Ethylene treatment must only be undertaken in fully enclosed storage areas that are air tight with appropriate air circulation and venting facilities

218 etoxazole

A mite growth inhibitor
IRAC mode of action code: 10B

Products

Borneo	Interfarm	110 g/l	SC	13919

Uses

- Mites in ***protected aubergines***, ***protected tomatoes***
- Spider mites in ***protected ornamentals*** *(off-label)*, ***protected strawberries*** *(off-label)*
- Two-spotted spider mite in ***protected ornamentals*** *(off-label)*, ***protected strawberries*** *(off-label)*

Specific Off-Label Approvals (SOLAs)

- ***protected ornamentals*** *20081216*
- ***protected strawberries*** *20092400*

Approval information

- Etoxazole included in Annex I under EC Directive 91/414

Hazard classification and safety precautions

Hazard Dangerous for the environment
Transport code 9
Packaging group III
UN number 3082
Risk phrases R50, R53a
Operator protection A; U05a, U20b
Environmental protection E15a, E34
Storage and disposal D01, D02, D09b, D11a, D12b

219 famoxadone

A strobilurin fungicide available only in mixtures
FRAC mode of action code: 11

See also *cymoxanil + famoxadone*

220 famoxadone + flusilazole

A contact, preventative and curative fungicide mixture for cereals and oilseed rape
FRAC mode of action code: 11 + 3

See also *flusilazole*

Products

1	Charisma	DuPont	100:106.7 g/l	EC	10415
2	Jenga	DuPont	100:106.7 g/l	EC	14357
3	Medley	DuPont	100:106.7 g/l	EC	10933

Uses

- Brown rust in ***spring barley***, ***winter barley***, ***winter wheat*** [1, 3]
- Foliar disease control in ***durum wheat*** *(off-label)*, ***grass seed crops*** *(off-label)*, ***triticale*** *(off-label)* [1, 3]; ***spring rye*** *(off-label)*, ***winter rye*** *(off-label)* [1]
- Glume blotch in ***winter wheat*** [1, 3]
- Light leaf spot in ***spring oilseed rape***, ***winter oilseed rape*** [1-3]
- Net blotch in ***spring barley***, ***winter barley*** [1, 3]
- Phoma in ***spring oilseed rape*** *(suppression)*, ***winter oilseed rape*** *(suppression)* [1-3]
- Rhynchosporium in ***spring barley***, ***winter barley*** [1, 3]
- Septoria leaf blotch in ***winter wheat*** [1, 3]
- Yellow rust in ***spring barley***, ***winter barley***, ***winter wheat*** [1, 3]

Specific Off-Label Approvals (SOLAs)

- ***durum wheat*** *20060505* [1], *20060506* [3]
- ***grass seed crops*** *20060505* [1], *20060506* [3]
- ***spring rye*** *20060505* [1]
- ***triticale*** *20060505* [1], *20060506* [3]
- ***winter rye*** *20060505* [1]

Approval information

- Famoxadone included in Annex I under EC Directive 91/414
- Flusilazole has been reinstated in Annex 1 under Directive 91/414 following an appeal
- Accepted by BBPA for use on malting barley

Efficacy guidance

- Best results obtained from application at early stage of disease development before infection spreads to new growth
- Famoxadone is a member of the QoI cross resistance group. Product should be used preventatively and not relied on for its curative potential
- Use product as part of an Integrated Crop Management strategy incorporating other methods of control, including where appropriate other fungicides with a different mode of action. Do not apply more than two foliar applications of QoI containing products to any cereal crop
- There is a significant risk of widespread resistance occurring in *Septoria tritici* populations in UK. Failure to follow resistance management action may result in reduced levels of disease control
- Flusilazole is a DMI fungicide. Resistance to some DMI fungicides has been identified in Septoria leaf blotch which may seriously affect performance of some products. For further advice contact a specialist advisor and visit the Fungicide Resistance Action Group (FRAG)-UK website

Restrictions

- Maximum number of treatments 2 per crop for wheat and barley; 1 per crop for oilseed rape
- Do not apply to crops under stress
- Do not apply during frosty weather

Crop-specific information

- Latest use: before flowering (GS 60) for winter wheat; before quarter ear emergence (GS 53) for barley; up to decline of flowering (GS 67) for oilseed rape

Environmental safety

- Dangerous for the environment
- Very toxic to aquatic organisms
- Dangerous to fish or other aquatic life. Do not contaminate surface waters or ditches with chemical or used container
- LERAP Category B

Hazard classification and safety precautions

Hazard Toxic, Dangerous for the environment
Transport code 9
Packaging group III
UN number 3082
Risk phrases R36, R40, R53a [1-3]; R50, R63 [2]; R51, R61 [1, 3]
Operator protection A, C [1-3]; D, M, P [2]; U05a, U19a, U20b [1-3]; U16a, U23a [2]
Environmental protection E13b, E16a, E16b, E38
Storage and disposal D01, D02, D05, D09a, D10b, D12a
Medical advice M03 [1, 3]; M04a [1-3]

221 fatty acids

A soap concentrate insecticide and acaricide

Products

1	Finalsan	Certis	198.6 g/l	EC	13102
2	Savona	Koppert	49% w/w	SL	06057

Uses

- Algae in ***amenity grassland, amenity vegetation, managed amenity turf, natural surfaces not intended to bear vegetation, ornamental plant production, permeable surfaces overlying soil*** [1]
- Annual and perennial weeds in ***amenity grassland, amenity vegetation, managed amenity turf, natural surfaces not intended to bear vegetation, ornamental plant production, permeable surfaces overlying soil*** [1]
- Aphids in ***broad beans, brussels sprouts, cabbages, cucumbers, fruit trees, lettuce, peas, peppers, protected aubergines*** *(off-label)*, ***protected brassica seedlings*** *(off-label - baby leaf production)*, ***protected cayenne peppers*** *(off-label)*, ***protected frise*** *(off-label)*, ***protected herbs (see appendix 6)*** *(off-label)*, ***protected radicchio*** *(off-label)*, ***protected scarole*** *(off-label)*, ***protected spinach*** *(off-label)*, ***protected tomatoes, pumpkins, runner beans, tomatoes (outdoor), woody ornamentals*** [2]
- Insect pests in ***protected aubergines*** *(off-label)*, ***protected brassica seedlings*** *(off-label)*, ***protected cayenne peppers*** *(off-label)*, ***protected frise*** *(off-label)*, ***protected herbs (see appendix 6)*** *(off-label)*, ***protected radicchio*** *(off-label)*, ***protected scarole*** *(off-label)*, ***protected spinach*** *(off-label)* [2]
- Mealybugs in ***broad beans, brussels sprouts, cabbages, cucumbers, fruit trees, lettuce, peas, peppers, protected tomatoes, pumpkins, runner beans, tomatoes (outdoor), woody ornamentals*** [2]
- Moss in ***amenity grassland, amenity vegetation, managed amenity turf, natural surfaces not intended to bear vegetation, ornamental plant production, permeable surfaces overlying soil*** [1]
- Scale insects in ***broad beans, brussels sprouts, cabbages, cucumbers, fruit trees, lettuce, peas, peppers, protected tomatoes, pumpkins, runner beans, tomatoes (outdoor), woody ornamentals*** [2]
- Spider mites in ***broad beans, brussels sprouts, cabbages, cucumbers, fruit trees, lettuce, peas, peppers, protected tomatoes, pumpkins, runner beans, tomatoes (outdoor), woody ornamentals*** [2]
- Whitefly in ***broad beans, brussels sprouts, cabbages, cucumbers, fruit trees, lettuce, peas, peppers, protected tomatoes, pumpkins, runner beans, tomatoes (outdoor), woody ornamentals*** [2]

Specific Off-Label Approvals (SOLAs)

- ***protected aubergines*** *20060653* [2]
- ***protected brassica seedlings*** *(baby leaf production) 20060655* [2], *20060655* [2]
- ***protected cayenne peppers*** *20060654* [2]
- ***protected frise*** *20060655* [2]
- ***protected herbs (see appendix 6)*** *20060655* [2]
- ***protected radicchio*** *20060655* [2]
- ***protected scarole*** *20060655* [2]
- ***protected spinach*** *20060655* [2]

Efficacy guidance

- Use only soft or rain water for diluting spray
- Pests must be sprayed directly to achieve any control. Spray all plant parts thoroughly to run off
- For glasshouse use apply when insects first seen and repeat as necessary. For scale insects apply several applications at weekly intervals after egg hatch
- To control whitefly spray when required and use biological control after 12 h

Restrictions

- Do not use on new transplants, newly rooted cuttings or plants under stress
- Do not use on specified susceptible shrubs. See label for details

Crop-specific information

- HI zero

Environmental safety

- Harmful to fish or other aquatic life. Do not contaminate surface waters or ditches with chemical or used container

Hazard classification and safety precautions

Hazard Irritant [1]

Risk phrases R36 [1]
Operator protection A, C, P [1]; U05a, U11, U15, U16b, U19a [1]; U20c [2]
Environmental protection E13c [2]; E15a, E16b, E19b [1]
Storage and disposal D01, D02 [1]; D05 [1, 2]; D09a, D10a [2]
Medical advice M05a [1]

222 fenamidone

A strobilurin fungicide available only in mixtures
FRAC mode of action code: 11

223 fenamidone + propamocarb hydrochloride

A systemic fungicide mixture for potatoes
FRAC mode of action code: 11 + 28

See also *propamocarb hydrochloride*

Products

1	Consento	Bayer CropScience	75:375 g/l	SC	11889
2	Prompto	Bayer CropScience	75:375 g/l	SC	14391

Uses
- Blight in ***potatoes*** [1]
- Late blight in ***potatoes*** [2]

Approval information
- Fenamidone and propamocarb hydrochloride included in Annex I under EC Directive 91/414

Efficacy guidance
- Apply as soon as there is risk of blight infection or immediately after a blight warning
- Fenamidone is a member of the QoI cross resistance group. To minimise the likelihood of development of resistance these products should be used in a planned Resistance Management strategy. In addition before application consult and adhere to the latest FRAG-UK resistance guidance on application of QoI fungicides to potatoes. See Section 5 for more information

Restrictions
- Maximum number of treatments (including any other QoI containing fungicide) 6 per yr but no more than three should be applied consecutively. Use in alternation with fungicides from a different cross-resistance group
- Observe a 7 d interval between treatments
- Use only as a protective treatment. Do not use when blight has become readily visible (1% leaf area destroyed)

Crop-specific information
- HI 7 d for potatoes
- All varieties of early, main and seed crop potatoes may be sprayed [2]

Environmental safety
- Dangerous for the environment
- Very toxic to aquatic organisms
- Dangerous to fish or other aquatic life. Do not contaminate surface waters or ditches with chemical or used container
- LERAP Category B

Hazard classification and safety precautions
Hazard Irritant, Dangerous for the environment
Transport code 9
Packaging group III
UN number 3082
Risk phrases R36, R50, R53a
Operator protection A, C; U05a, U08, U11 [1, 2]; U19a, U20b, U20c [2]; U20a [1]
Environmental protection E13b [1]; E15a [2]; E16a, E38 [1, 2]
Storage and disposal D01, D02, D09a, D10b, D12a [1, 2]; D05 [1]

224 fenarimol

A systemic curative and protective pyrimidine fungicide
FRAC mode of action code: 3

225 fenbuconazole

A systemic protectant and curative triazole fungicide for top fruit and grapevines
FRAC mode of action code: 3

Products

1	Agrovista Radni	Agrovista	50 g/l	EW	14404
2	Indar 5 EW	Landseer	50 g/l	EW	09518

Uses

- Blossom wilt in ***cherries*** *(off-label)*, ***mirabelles*** *(off-label)*, ***plums*** *(off-label)* [2]
- Brown rot in ***cherries*** *(off-label)*, ***mirabelles*** *(off-label)*, ***plums*** *(off-label)* [2]
- Powdery mildew in ***apples*** *(reduction)*, ***pears*** *(reduction)* [1, 2]; ***wine grapes*** *(off-label)* [2]
- Scab in ***apples***, ***pears*** [1, 2]

Specific Off-Label Approvals (SOLAs)

- ***cherries*** *20031372* [2]
- ***mirabelles*** *20031372* [2]
- ***plums*** *20031372* [2]
- ***wine grapes*** *20032081* [2]

Approval information

- Fenbuconazole was not supported for inclusion in Annex 1 and approvals will be withdrawn but final use dates have yet to be set

Efficacy guidance

- Most effective when used as part of a routine preventative programme from bud burst to onset of petal fall
- After petal fall, tank mix with other protectant fungicides to enhance scab control
- See label for recommended spray intervals. In periods of rapid growth or high disease pressure, a 7 d interval should be used

Restrictions

- Maximum total dose equivalent to ten full doses per yr on apples, pears
- Consult processors before using on pears for processing
- Do not harvest for human or animal consumption for at least 4 wk after last application

Crop-specific information

- HI 28 d for apples, pears; 21 d for grapevines; 3 d for cherries, mirabelles, plums
- Safe to use on all main commercial varieties of apples and pears in UK

Environmental safety

- Dangerous for the environment
- Toxic to aquatic organisms

Hazard classification and safety precautions

Hazard Irritant, Dangerous for the environment
Transport code 3
Packaging group III
UN number 1993
Risk phrases R36, R51, R53a [1, 2]; R41 [2]
Operator protection A, C; U05a, U08, U11 [1, 2]; U20a [2]; U20b [1]
Environmental protection E15a, E38
Consumer protection C02a [1] (4 w); C02a [2] (4 wk)
Storage and disposal D01, D02, D11a, D12a [1, 2]; D05 [2]

226 fenhexamid

A protectant hydroxyanilide fungicide for soft fruit and a range of horticultural crops
FRAC mode of action code: 17

Products

1	Agrovista Fenamid	Agrovista	50% w/w	WG	13733
2	Teldor	Bayer CropScience	50% w/w	WG	11229

SECTION 2

Uses

- Botrytis in ***bilberries*** *(off-label)*, ***blackberries***, ***blackcurrants***, ***blueberries*** *(off-label)*, ***cherries*** *(off-label)*, ***gooseberries***, ***herbs (see appendix 6)*** *(off-label)*, ***loganberries***, ***plums*** *(off-label)*, ***protected aubergines*** *(off-label)*, ***protected courgettes*** *(off-label)*, ***protected cucumbers*** *(off-label)*, ***protected gherkins*** *(off-label)*, ***protected peppers*** *(off-label)*, ***protected squashes*** *(off-label)*, ***protected tomatoes*** *(off-label)*, ***raspberries***, ***redcurrants***, ***rubus hybrids***, ***strawberries***, ***tomatoes (outdoor)*** *(off-label)*, ***whitecurrants*** [1, 2]; ***frise*** *(off-label)*, ***hops*** *(off-label)*, ***lamb's lettuce*** *(off-label)*, ***leaf brassicas*** *(off-label - baby leaf production)*, ***protected forest nurseries*** *(off-label)*, ***rocket*** *(off-label)*, ***scarole*** *(off-label)*, ***soft fruit*** *(off-label)*, ***wine grapes*** [2]; ***outdoor grapes***, ***protected chilli peppers*** *(off-label)* [1]

Specific Off-Label Approvals (SOLAs)

- ***bilberries*** *20080485* [1], *20061290* [2]
- ***blueberries*** *20080485* [1], *20061290* [2]
- ***cherries*** *20080479* [1], *20031866* [2]
- ***frise*** *20082062* [2]
- ***herbs (see appendix 6)*** *20082056* [1], *20082062* [2]
- ***hops*** *20082926* [2]
- ***lamb's lettuce*** *20082062* [2]
- ***leaf brassicas*** *(baby leaf production) 20082062* [2]
- ***plums*** *20080479* [1], *20031866* [2]
- ***protected aubergines*** *20080482* [1], *20042087 expires 30 Nov 2010* [2]
- ***protected chilli peppers*** *20080481* [1]
- ***protected courgettes*** *20080480* [1], *20042085 expires 30 Nov 2010* [2]
- ***protected cucumbers*** *20080480* [1], *20042085 expires 30 Nov 2010* [2]
- ***protected forest nurseries*** *20082926* [2]
- ***protected gherkins*** *20080480* [1], *20042085 expires 30 Nov 2010* [2]
- ***protected peppers*** *20080481* [1], *20042086 expires 30 Nov 2010* [2]
- ***protected squashes*** *20080480* [1], *20042085 expires 30 Nov 2010* [2]
- ***protected tomatoes*** *20080482* [1], *20042087 expires 30 Nov 2010* [2]
- ***rocket*** *20082062* [2]
- ***scarole*** *20082062* [2]
- ***soft fruit*** *20082926* [2]
- ***tomatoes (outdoor)*** *20080483* [1], *20042399* [2]

Approval information

- Fenhexamid included in Annex I under EC Directive 91/414

Efficacy guidance

- Use as part of a programme of sprays throughout the flowering period to achieve effective control of Botrytis
- To minimise possibility of development of resistance, no more than two sprays of the product may be applied consecutively. Other fungicides from a different chemical group should then be used for at least two consecutive sprays. If only two applications are made on grapevines, only one may include fenhexamid
- Complete spray cover of all flowers and fruitlets throughout the blossom period is essential for successful control of Botrytis
- Spray programmes should normally start at the start of flowering

Restrictions

- Maximum number of treatments 2 per yr on grapevines; 4 per yr on other listed crops but no more than 2 sprays may be applied consecutively

Crop-specific information

- HI 1 d for strawberries, raspberries, loganberries, blackberries, Rubus hybrids; 3 d for cherries; 7 d for blackcurrants, redcurrants, whitecurrants, gooseberries; 21d for outdoor grapes

Environmental safety

- Dangerous for the environment
- Harmful to fish or other aquatic life. Do not contaminate surface waters or ditches with chemical or used container

Hazard classification and safety precautions

Hazard Dangerous for the environment
Transport code 9
Packaging group III
UN number 3077
Risk phrases R53a [2]
Operator protection A, H [1]; U08, U19a, U20b
Environmental protection E13c, E38 [2]; E15a [1]
Storage and disposal D05, D09a, D11a [1, 2]; D12a [2]

227 fenoxaprop-P-ethyl

An aryloxyphenoxypropionate herbicide for use in wheat
HRAC mode of action code: A

Products

1	Cheetah Super	Bayer CropScience	55 g/l	EW	08723
2	Foxtrot EW	Headland	69 g/l	EW	13243
3	Oskar	Headland	69 g/l	EW	13344
4	Triumph	Bayer CropScience	120 g/l	EC	10902
5	Warrant	Bayer CropScience	83 g/l	EC	13806

Uses

- Awned canary grass in ***spring wheat, winter wheat*** [5]
- Blackgrass in ***spring barley, winter barley*** [2, 3]; ***spring wheat, winter wheat*** [1-5]
- Canary grass in ***spring barley, winter barley*** [2, 3]; ***spring wheat, winter wheat*** [1-4]
- Grass weeds in ***grass seed crops*** *(off-label)*, ***spring rye*** *(off-label)*, ***triticale*** *(off-label)*, ***winter rye*** *(off-label)* [1, 4]
- Rough meadow grass in ***spring barley, winter barley*** [2, 3]; ***spring wheat, winter wheat*** [1-5]
- Wild oats in ***spring barley, winter barley*** [2, 3]; ***spring wheat, winter wheat*** [1-5]

Specific Off-Label Approvals (SOLAs)

- ***grass seed crops*** *20080838* [1], *20080839* [4]
- ***spring rye*** *20080838* [1], *20080839* [4]
- ***triticale*** *20080838* [1], *20080839* [4]
- ***winter rye*** *20080838* [1], *20080839* [4]

Approval information

- Fenoxaprop-P-ethyl included in Annex I under EC Directive 91/414

Efficacy guidance

- Treat weeds from 2 fully expanded leaves up to flag leaf ligule just visible; for awned canary-grass and rough meadow-grass from 2 leaves to the end of tillering
- A second application may be made in spring where susceptible weeds emerge after an autumn application
- Spray is rainfast 1 h after application
- Dry conditions resulting in moisture stress may reduce effectiveness
- Fenoxaprop-P-ethyl is an ACCase inhibitor herbicide. To avoid the build up of resistance do not apply products containing an ACCase inhibitor herbicide more than twice to any crop. In addition do not use any product containing fenoxaprop-P-ethyl in mixture or sequence with any other product containing the same ingredient
- Use these products as part of a resistance management strategy that includes cultural methods of control and does not use ACCase inhibitors as the sole chemical method of grass weed control

- Applying a second product containing an ACCase inhibitor to a crop will increase the risk of resistance development; only use a second ACCase inhibitor to control different weeds at a different timing
- Always follow WRAG guidelines for preventing and managing herbicide resistant weeds. See Section 5 for more information

Restrictions

- Maximum total dose equivalent to one or two full dose treatments depending on product used
- Do not apply to barley, durum wheat, undersown crops or crops to be undersown
- Do not roll or harrow within 1 wk of spraying
- Do not spray crops under stress, suffering from drought, waterlogging or nutrient deficiency or those grazed or if soil compacted
- Avoid spraying immediately before or after a sudden drop in temperature or a period of warm days/cold nights
- Do not mix with hormone weedkillers

Crop-specific information

- Latest use: before flag leaf sheath extending (GS 41)
- Treat from crop emergence to flag leaf fully emerged (GS 41).
- Product may be sprayed in frosty weather provided crop hardened off but do not spray wet foliage or leaves covered with ice
- Broadcast crops should be sprayed post-emergence after plants have developed well-established root system

Environmental safety

- Dangerous for the environment
- Very toxic (toxic [1]) to aquatic organisms

Hazard classification and safety precautions

Hazard Irritant [2-5]; Dangerous for the environment [1-5]
Transport code 9
Packaging group III
UN number 3082
Risk phrases R36, R50 [4]; R38 [2-5]; R43 [2, 3]; R51 [1-3, 5]; R53a [1-5]
Operator protection A, C, H; U05a, U20b [1-5]; U08 [2-5]; U09a [1]; U11 [5]; U14 [2, 3, 5]; U19a [2, 3]
Environmental protection E13c [1, 4]; E15a [2, 3, 5]; E34 [2, 3]; E38 [1-5]
Storage and disposal D01, D02, D09a, D10b [1-5]; D05 [2, 3, 5]; D12a [1, 4, 5]

228 fenoxycarb

An insect specific growth regulator for top fruit
IRAC mode of action code: 7B

Products

Insegar WG	Syngenta	25% w/w	WG	09789

Uses

- Summer-fruit tortrix moth in ***apples, pears***

Efficacy guidance

- Best results from application at 5th instar stage before pupation. Product prevents transformation from larva to pupa
- Correct timing best identified from pest warnings
- Because of mode of action rapid knock-down of pest is not achieved and larvae continue to feed for a period after treatment
- Adequate water volume necessary to ensure complete coverage of leaves

Restrictions

- Maximum number of treatments 2 per crop
- Consult processors before use

Crop-specific information

- HI 42 d for apples, pears
- Use on all varieties of apples and pears

SEE SECTION 3 FOR PRODUCTS ALSO REGISTERED

Environmental safety

- Dangerous for the environment
- Toxic to aquatic organisms
- High risk to bees. Do not apply to crops in flower or to those in which bees are actively foraging. Do not apply when flowering weeds are present
- Risk to certain non-target insects or other arthropods. See directions for use
- Broadcast air-assisted LERAP (8 m)
- Apply to minimise off-target drift to reduce effects on non-target organisms. Some margin of safety to beneficial arthropods is indicated.

Hazard classification and safety precautions

Hazard Dangerous for the environment
Transport code 9
Packaging group III
UN number 3077
Risk phrases R51, R53a
Operator protection A, C, H; U05a, U20a, U23a
Environmental protection E12a, E12e, E15a, E22b, E38; E17b (8 m)
Storage and disposal D01, D02, D07, D09a, D11a, D12a

229 fenpropidin

A systemic, curative and protective piperidine (morpholine) fungicide
FRAC mode of action code: 5

See also difenoconazole + fenpropidin

Products

1	Alpha Fenpropidin 750 EC	Makhteshim	750 g/l	EC	12843
2	Instinct	Headland	750 g/l	EC	12317
3	Tern	Syngenta	750 g/l	EC	08660

Uses

- Brown rust in ***spring barley***, ***winter barley*** [1-3]
- Foliar disease control in ***durum wheat*** *(off-label)* [1, 3]; ***spring rye*** *(off-label)*, ***triticale*** *(off-label)*, ***winter rye*** *(off-label)* [3]
- Powdery mildew in ***spring barley***, ***spring wheat***, ***winter barley***, ***winter wheat*** [1-3]
- Rhynchosporium in ***spring barley*** *(moderate control)*, ***winter barley*** *(moderate control)* [1-3]
- Yellow rust in ***spring barley*** *(moderate control)*, ***spring wheat*** *(moderate control)*, ***winter barley*** *(moderate control)*, ***winter wheat*** *(moderate control)* [1-3]

Specific Off-Label Approvals (SOLAs)

- ***durum wheat*** *20070459* [1], *20060533* [3]
- ***spring rye*** *20060533* [3]
- ***triticale*** *20060533* [3]
- ***winter rye*** *20060533* [3]

Approval information

- Accepted by BBPA for use on malting barley
- Fenpropidin included in Annex I under EC Directive 91/414

Efficacy guidance

- Best results obtained when applied at early stage of disease development. See label for details of recommended timing alone and in mixtures
- Disease control enhanced by vapour-phase activity. Control can persist for 4-6 wk
- Alternate with triazole fungicides to discourage build-up of resistance

Restrictions

- Maximum number of treatments 3 per crop (up to 2 in yr of harvest) for winter crops; 2 per crop for spring crops
- Treated crops must not be harvested for human or animal consumption for at least 5 wk after the last application

Crop-specific information

- Latest use: up to and including ear emergence complete (GS 59).
- HI 5 wk

Environmental safety

- Dangerous for the environment
- Very toxic to aquatic organisms

Hazard classification and safety precautions

Hazard Harmful, Dangerous for the environment
Transport code 6.1 [1]; 9 [2, 3]
Packaging group III
UN number 2902 [1]; 3082 [2, 3]
Risk phrases R20, R21 [1]; R22a [1, 3]; R37 [3]; R38 [2]; R41 [1, 2]; R50, R53a [1-3]
Operator protection A, C, H; U02a, U05a [1-3]; U04a, U10 [1, 3]; U09a [2]; U11, U19a, U20b [1, 2]; U20a [3]
Environmental protection E15a, E34 [2, 3]; E19b [1]; E38 [1-3]
Consumer protection C02a [2, 3] (5 wk)
Storage and disposal D01, D02, D05, D09a, D12a [1-3]; D10c [2]
Medical advice M03

230 fenpropidin + tebuconazole

A broad spectrum fungicide mixture for cereals
FRAC mode of action code: 5 + 3

See also *tebuconazole*

Products

Eros	Makhteshim	150:100 g/l	EC	14112

Uses

- Ear diseases in ***spring wheat***, ***winter wheat***
- Glume blotch in ***spring wheat***, ***winter wheat***
- Septoria leaf blotch in ***spring wheat***, ***winter wheat***

Approval information

- Fenpropidin included in Annex I under EC Directive 91/414
- Accepted by BBPA for use on malting barley

Efficacy guidance

- Best disease control and yield benefit obtained when applied at early stage of disease development before infection spreads to new growth
- To protect the flag leaf and ear from Septoria diseases apply from flag leaf emergence to ear fully emerged (GS 37-59)
- Applications once foliar symptoms of *Septoria tritici* are already present on upper leaves will be less effective
- Resistance to some DMI fungicides has been identified in Septoria leaf blotch which may seriously affect the performance of some products.

Restrictions

- Maximum total dose 2.5 l/ha per crop
- Occasional slight temporary leaf speckling may occur on wheat but this has not been shown to reduce yield response or disease control
- Do not treat durum wheat
- Do not apply when temperatures are high [1]

Crop-specific information

- Latest use: Before grain milky ripe stage (GS 73) for winter wheat; up to and including ear emergence just complete (GS 59) for barley and spring wheat
- HI 5 wk for crops for human or animal consumption

Environmental safety

- Harmful if swallowed and in contact with skin
- Irritating to skin

- Risk of serious damage to eyes
- Extremely dangerous to fish or other aquatic life. Do not contaminate surface waters or ditches with chemical or used container

Hazard classification and safety precautions

Hazard Harmful, Dangerous for the environment
Transport code 6.1
Packaging group III
UN number 2902
Risk phrases R20, R21, R22a, R51, R53a, R63, R69
Operator protection A, C, H; U02a, U04a, U05a, U11, U20a
Environmental protection E15a, E34, E38
Storage and disposal D02, D09a, D12a
Medical advice M03

231 fenpropimorph

A contact and systemic morpholine fungicide
FRAC mode of action code: 5

See also *azoxystrobin + fenpropimorph*
epoxiconazole + fenpropimorph
epoxiconazole + fenpropimorph + kresoxim-methyl
epoxiconazole + fenpropimorph + metrafenone
epoxiconazole + fenpropimorph + pyraclostrobin

Products

1	Clayton Spigot	Clayton	750 g/l	EC	11560
2	Corbel	BASF	750 g/l	EC	00578
3	Crebol	AgChem Access	750 g/l	EC	13922
4	Raven	AgriGuard	750 g/l	EC	13188
5	Standon Fenpropimorph 750	Standon	750 g/l	EC	08965

Uses

- Alternaria in ***carrots*** *(off-label)*, ***horseradish*** *(off-label)*, ***parsley root*** *(off-label)*, ***parsnips*** *(off-label)*, ***salsify*** *(off-label)* [2]
- Brown rust in ***spring barley***, ***spring wheat***, ***triticale***, ***winter barley***, ***winter wheat*** [1-5]
- Crown rot in ***carrots*** *(off-label)*, ***horseradish*** *(off-label)*, ***parsley root*** *(off-label)*, ***parsnips*** *(off-label)*, ***salsify*** *(off-label)* [2]
- Foliar disease control in ***durum wheat*** *(off-label)* [2]; ***grass seed crops*** *(off-label)* [1, 5]
- Powdery mildew in ***bilberries*** *(off-label)*, ***blackcurrants*** *(off-label)*, ***blueberries*** *(off-label)*, ***carrots*** *(off-label)*, ***cranberries*** *(off-label)*, ***dewberries*** *(off-label)*, ***gooseberries*** *(off-label)*, ***hops*** *(off-label)*, ***horseradish*** *(off-label)*, ***loganberries*** *(off-label)*, ***parsley root*** *(off-label)*, ***parsnips*** *(off-label)*, ***raspberries*** *(off-label)*, ***redcurrants*** *(off-label)*, ***salsify*** *(off-label)*, ***strawberries*** *(off-label)*, ***tayberries*** *(off-label)*, ***whitecurrants*** *(off-label)* [2]; ***spring barley***, ***spring oats***, ***spring wheat***, ***winter barley***, ***winter oats***, ***winter rye***, ***winter wheat*** [1-5]
- Rhynchosporium in ***spring barley***, ***winter barley*** [1-5]
- Rust in ***fodder beet*** *(off-label)*, ***red beet*** *(off-label)*, ***sugar beet seed crops*** *(off-label)* [2]; ***leeks*** [1]
- Yellow rust in ***spring barley***, ***spring wheat***, ***triticale***, ***winter barley***, ***winter wheat*** [1-5]

Specific Off-Label Approvals (SOLAs)

- ***bilberries*** *20040804* [2]
- ***blackcurrants*** *20040804* [2]
- ***blueberries*** *20040804* [2]
- ***carrots*** *20023753* [2]
- ***cranberries*** *20040804* [2]
- ***dewberries*** *20040804* [2]
- ***durum wheat*** *20061460* [2]
- ***fodder beet*** *20023757* [2]
- ***gooseberries*** *20040804* [2]
- ***grass seed crops*** *20061060* [1], *20061057* [5]
- ***hops*** *20023759* [2]

- ***horseradish*** *20023753* [2]
- ***loganberries*** *20040804* [2]
- ***parsley root*** *20023753* [2]
- ***parsnips*** *20023753* [2]
- ***raspberries*** *20040804* [2]
- ***red beet*** *20023751* [2]
- ***redcurrants*** *20040804* [2]
- ***salsify*** *20023753* [2]
- ***strawberries*** *20040804* [2]
- ***sugar beet seed crops*** *20023757* [2]
- ***tayberries*** *20040804* [2]
- ***whitecurrants*** *20040804* [2]

Approval information

- Accepted by BBPA for use on malting barley and hops
- Fenpropimorph included in Annex I under EC Directive 91/414

Efficacy guidance

- On all crops spray at start of disease attack. See labels for recommended tank mixes. Follow-up treatments may be needed if disease pressure remains high
- Product rainfast after 2 h

Restrictions

- Maximum number of treatments 2 per crop for spring cereals; 3 or 4 per crop for winter cereals; 6 per crop for leeks
- Consult processors before using on crops for processing

Crop-specific information

- HI cereals 5 wk
- Scorch may occur on cereals if applied during frosty weather or in high temperatures
- Some leaf spotting may occur on undersown clovers
- Leeks should be treated every 2-3 wk as required, up to 6 applications

Environmental safety

- Dangerous for the environment
- Very toxic to aquatic organisms

Hazard classification and safety precautions

Hazard Harmful [1-4]; Irritant [5]; Dangerous for the environment [1-5]
Transport code 3 [4]; 9 [1-3, 5]
Packaging group III
UN number 1993 [4]; 3082 [1-3, 5]
Risk phrases R22a [1]; R38, R53a [1-5]; R50 [2-4]; R51 [1, 5]; R63 [1-4]
Operator protection A [1-5]; C [1, 5]; U05a, U08, U14, U15, U19a, U20b
Environmental protection E15a [1-5]; E38 [2-4]
Storage and disposal D01, D02, D09a [1-5]; D05 [1, 5]; D08 [2-4]; D10b [5]; D10c, D12a [1-4]
Medical advice M03 [4]; M05a [1-3, 5]

232 fenpropimorph + flusilazole

A broad-spectrum eradicant and protectant fungicide mixture for cereals
FRAC mode of action code: 5 + 3

See also flusilazole

Products

1	Colstar	DuPont	375:160 g/l	EC	12175
2	Pluton	DuPont	375:160 g/l	EC	12200

Uses

- Brown rust in ***spring barley***, ***winter barley***, ***winter wheat*** [1, 2]
- Foliar disease control in ***durum wheat*** *(off-label)*, ***grass seed crops*** *(off-label)* [1, 2]; ***spring rye*** *(off-label)*, ***triticale*** *(off-label)*, ***winter rye*** *(off-label)* [1]
- Glume blotch in ***winter wheat*** [1, 2]
- Net blotch in ***spring barley***, ***winter barley*** [1, 2]

- Powdery mildew in ***spring barley***, ***winter barley***, ***winter wheat*** [1, 2]
- Rhynchosporium in ***spring barley***, ***winter barley*** [1, 2]
- Septoria leaf blotch in ***winter wheat*** [1, 2]
- Yellow rust in ***spring barley***, ***winter barley***, ***winter wheat*** [1, 2]

Specific Off-Label Approvals (SOLAs)

- ***durum wheat*** *20060977* [1], *20060985* [2]
- ***grass seed crops*** *20060977* [1], *20060985* [2]
- ***spring rye*** *20060977* [1]
- ***triticale*** *20060977* [1]
- ***winter rye*** *20060977* [1]

Approval information

- Fenpropimorph and flusilazole included in Annex I under EC Directive 91/414
- Flusilazole has been reinstated in Annex 1 under Directive 91/414 following an appeal
- Accepted by BBPA for use on malting barley

Efficacy guidance

- Disease control is more effective if treatment made at an early stage of disease development
- Treat winter cereals in spring or early summer before diseases spread to new growth
- Spring barley should be treated when diseases are first evident
- Treatment may be repeated after 3-4 wk if necessary
- Flusilazole is a DMI fungicide. Resistance to some DMI fungicides has been identified in Septoria leaf blotch which may seriously affect performance of some products. For further advice contact a specialist advisor and visit the Fungicide Resistance Action Group (FRAG)-UK website

Restrictions

- Maximum number of treatments 2 per crop of barley or wheat
- Do not apply to crops under stress
- Do not apply during frosty weather

Crop-specific information

- Latest use: before beginning of anthesis (GS 60) for winter wheat; up to and including completion of ear emergence (GS 59) for barley

Environmental safety

- Dangerous for the environment
- Toxic to aquatic organisms
- Dangerous to fish or other aquatic life. Do not contaminate surface waters or ditches with chemical or used container

Hazard classification and safety precautions

Hazard Toxic, Dangerous for the environment
Transport code 9
Packaging group III
UN number 3082
Risk phrases R36, R40, R51, R53a, R61
Operator protection A, C; U05a, U11, U19a, U20b
Environmental protection E13b, E15a, E38
Storage and disposal D01, D02, D05, D09a, D10b, D12a
Medical advice M04a

233 fenpropimorph + kresoxim-methyl

A protectant and systemic fungicide mixture for cereals
FRAC mode of action code: 5 + 11

See also kresoxim-methyl

Products

1	Ensign	BASF	300:150 g/l	SE	11729
2	Greencrop Monsoon	Greencrop	300:150 g/l	SE	09573
3	Standon Kresoxim FM	Standon	300:150 g/l	SE	08922

Uses

- Foliar disease control in ***durum wheat*** *(off-label)* [1]; ***grass seed crops*** *(off-label)* [2, 3]

- Glume blotch in ***spring wheat*** *(reduction)*, ***triticale*** *(reduction)* [1]; ***winter wheat*** *(reduction)* [1-3]
- Powdery mildew in ***spring barley***, ***winter barley*** [1-3]; ***spring oats***, ***spring rye***, ***triticale***, ***winter oats***, ***winter rye*** [1]
- Rhynchosporium in ***spring barley***, ***winter barley*** [1-3]; ***spring rye***, ***triticale***, ***winter rye*** [1]
- Septoria leaf blotch in ***spring wheat*** *(reduction)*, ***triticale*** *(reduction)* [1]; ***winter wheat*** *(reduction)* [1-3]

Specific Off-Label Approvals (SOLAs)

- ***durum wheat*** *20062643* [1]
- ***grass seed crops*** *20060978* [2], *20060984* [3]

Approval information

- Fenpropimorph and kresoxim-methyl included in Annex I under EC Directive 91/414
- Accepted by BBPA for use on malting barley

Efficacy guidance

- For best results spray at the start of foliar disease attack and repeat if infection conditions persist
- Kresoxim-methyl is a member of the QoI cross resistance group. Product should be used preventatively and not relied on for its curative potential
- Use product as part of an Integrated Crop Management strategy incorporating other methods of control, including where appropriate other fungicides with a different mode of action. Do not apply more than two foliar applications of QoI containing products to any cereal crop
- There is a significant risk of widespread resistance occurring in *Septoria tritici* populations in UK. Strains of barley powdery mildew resistant to QoIs are common in UK. Failure to follow resistance management action may result in reduced levels of disease control
- Strains of barley powdery mildew resistant to QoIs are common in the UK

Restrictions

- Maximum total dose equivalent to two full dose treatments for all crops

Crop-specific information

- Latest use: completion of ear emergence (GS 59) for barley and oats; completion of flowering (GS 69) for wheat, rye and triticale

Environmental safety

- Dangerous for the environment
- Very toxic to aquatic organisms

Hazard classification and safety precautions

Hazard Harmful, Dangerous for the environment
Transport code 9
Packaging group III
UN number 3082
Risk phrases R40, R50, R53a [1-3]; R43 [2, 3]; R63 [1, 2]
Operator protection A [1-3]; H [1]; U05a, U14, U20b [1-3]; U19a [2]
Environmental protection E15a [1-3]; E38 [1]
Storage and disposal D01, D02, D05, D09a [1-3]; D10b [2, 3]; D10c, D12a [1]
Medical advice M05a [1, 3]

234 fenpropimorph + pyraclostrobin

A protectant and curative fungicide mixture for cereals
FRAC mode of action code: 5 +11

See also *pyraclostrobin*

Products

Jenton	BASF	375:100 g/l	EC	11898

Uses

- Brown rust in ***spring barley***, ***spring wheat***, ***winter barley***, ***winter wheat***
- Crown rust in ***spring oats***, ***winter oats***
- Drechslera leaf spot in ***spring wheat***, ***winter wheat***
- Foliar disease control in ***grass seed crops*** *(off-label)*, ***triticale*** *(off-label)*
- Glume blotch in ***spring wheat***, ***winter wheat***

- Net blotch in ***spring barley*, *winter barley***
- Powdery mildew in ***spring barley*, *spring oats*, *winter barley*, *winter oats***
- Rhynchosporium in ***spring barley*, *winter barley***
- Septoria leaf blotch in ***spring wheat*, *winter wheat***
- Yellow rust in ***spring barley*, *spring wheat*, *winter barley*, *winter wheat***

Specific Off-Label Approvals (SOLAs)

- ***grass seed crops*** *20062650*
- ***triticale*** *20062650*

Approval information

- Fenpropimorph and pyraclostrobin included in Annex I under EC Directive 91/414
- Accepted by BBPA for use on malting barley

Efficacy guidance

- Best results obtained from treatment at the start of foliar disease attack
- Yield response may be obtained in the absence of visual disease
- Pyraclostrobin is a member of the QoI cross resistance group. Product should be used preventatively and not relied on for its curative potential
- Use product as part of an Integrated Crop Management strategy incorporating other methods of control, including where appropriate other fungicides with a different mode of action. Do not apply more than two foliar applications of QoI containing products to any cereal crop
- There is a significant risk of widespread resistance occurring in *Septoria tritici* populations in UK. Strains of barley powdery mildew resistant to QoIs are common in UK. Failure to follow resistance management action may result in reduced levels of disease control

Restrictions

- Maximum number of treatments 2 per crop

Crop-specific information

- Latest use: up to and including emergence of ear just complete (GS 59) for barley

Environmental safety

- Dangerous for the environment
- Very toxic to aquatic organisms
- LERAP Category B

Hazard classification and safety precautions

Hazard Harmful, Dangerous for the environment
Transport code 9
Packaging group III
UN number 3082
Risk phrases R20, R22a, R36, R38, R50, R53a, R63
Operator protection A, C; U05a, U14, U20b
Environmental protection E15a, E16a, E34, E38
Storage and disposal D01, D02, D05, D09a, D10c, D12a
Medical advice M03, M05a

235 fenpropimorph + quinoxyfen

A systemic fungicide mixture for cereals
FRAC mode of action code: 5 + 13

See also *quinoxyfen*

Products

Orka	Dow	250:66.7 g/l	EW	08879

Uses

- Powdery mildew in ***durum wheat*, *grass seed crops*** *(off-label)*, ***spring barley*, *spring oats*, *spring rye*, *spring wheat*, *triticale*, *winter barley*, *winter oats*, *winter rye*, *winter wheat***

Specific Off-Label Approvals (SOLAs)

- ***grass seed crops*** *20060979*

Approval information

- Fenpropimorph and quinoxyfen included in Annex I under Directive 91/414
- Accepted by BBPA for use on malting barley

Efficacy guidance

- For best results treat at early stage of disease development before infection spreads to new crop growth. Further treatment may be necessary if disease pressure remains high
- For control of established infections and broad spectrum disease control use in tank mixtures. See label
- Product rainfast after 1 h
- Systemic activity may be reduced in severe drought

Restrictions

- Maximum total dose 3.0 l product per ha
- Apply only in the spring from mid-tillering stage (GS 25)

Crop-specific information

- Latest use: when first awns visible (GS 49)
- Crop scorch may occur when treatment made in high temperatures

Environmental safety

- Dangerous for the environment
- Very toxic to aquatic organisms
- LERAP Category B

Hazard classification and safety precautions

Hazard Harmful, Dangerous for the environment
Transport code 9
Packaging group III
UN number 3082
Risk phrases R43, R50, R53a, R61
Operator protection A, H; U05a, U14
Environmental protection E15a, E16a, E34, E38
Storage and disposal D01, D02, D12a

236 fenpyroximate

A mitochondrial electron transport inhibitor (METI) acaricide for apples
IRAC mode of action code: 21

Products

Sequel	Certis	51.3 g/l	SC	12657

Uses

- Fruit tree red spider mite in ***apples, plums*** *(off-label)*
- Red spider mites in ***plums*** *(off-label)*
- Tarsonemid mites in ***protected strawberries*** *(off-label)*, ***strawberries*** *(off-label)*

Specific Off-Label Approvals (SOLAs)

- ***plums*** *20060444*
- ***protected strawberries*** *20070902*
- ***strawberries*** *20070902*

Approval information

- Fenpyroximate included in Annex I under EC Directive 91/414

Efficacy guidance

- Kills motile stages of fruit tree red spider mite. Best results achieved if applied in warm weather
- Total spray cover of trees essential. Use higher volumes for large trees
- Apply when majority of winter eggs have hatched

Restrictions

- Maximum number of treatments 1 per yr or total dose equivalent to one full dose treatment
- Other mitochondrial electron transport inhibitor (METI) acaricides should not be applied to the same crop in the same calendar yr either separately or in mixture

SEE SECTION 3 FOR PRODUCTS ALSO REGISTERED

- Do not apply when apple crops or pollinator are in flower
- Consult processor before use on crops for processing

Crop-specific information

- HI 2 wk

Environmental safety

- Dangerous for the environment
- Very toxic to aquatic organisms
- Risk to non-target insects or other arthropods
- Broadcast air-assisted LERAP (40 m)

Hazard classification and safety precautions

Hazard Harmful, Dangerous for the environment
Transport code 9
Packaging group III
UN number 3082
Risk phrases R20, R36, R43, R50, R53a
Operator protection A, C, H; U05a, U08, U14, U15, U20b
Environmental protection E13b, E22c; E17b (40 m)
Storage and disposal D01, D02, D05, D09a, D10c, D12a

237 ferric phosphate

A molluscicide bait for controlling slugs and snails

Products

1	Ferramol	Certis	1% w/w	GB	12274
2	Sluggo	Omex	1% w/w	GB	12529
3	Sluxx	Certis	2.97% w/w	GB	14462

Uses

- Slugs in ***all edible crops (outdoor)***, ***all non-edible crops (outdoor)***, ***protected crops*** [1, 2]
- Slugs and snails in ***all edible crops (outdoor and protected)***, ***all non-edible crops (outdoor)***, ***amenity vegetation*** [3]
- Snails in ***all edible crops (outdoor)***, ***all non-edible crops (outdoor)***, ***protected crops*** [1, 2]

Efficacy guidance

- Treat as soon as damage first seen preferably in early evening. Repeat as necessary to maintain control
- Best results obtained from moist soaked granules. This will occur naturally on moist soils or in humid conditions
- Ferric phosphate does not cause excessive slime secretion and has no requirement to collect moribund slugs from the soil surface
- Active ingredient is degraded by micro-organisms to beneficial plant nutrients

Restrictions

- Maximum total dose equivalent to four full dose treatments

Crop-specific information

- Latest use not specified for any crop

Hazard classification and safety precautions

UN number N/C
Operator protection U05a [1-3]; U20a [3]; U20b [1, 2]
Environmental protection E15a [1, 2]; E34 [1-3]
Storage and disposal D01, D09a

238 ferrous sulphate

A herbicide/fertilizer combination for moss control in turf

See also dichlorprop-P + ferrous sulphate + MCPA

Products

1	Elliott's Lawn Sand	Elliott	9.3% w/w	SA	04860
2	Elliott's Mosskiller	Elliott	24.6% w/w	GR	04909
3	Ferromex Mosskiller Concentrate	Omex	16.4% w/w	SL	13180
4	Greenmaster Autumn	Scotts	18.2% w/w	GR	12196
5	Greentec Mosskiller	Headland Amenity	9% w/w	GR	12518
6	SHL Lawn Sand	Sinclair	5.4% w/w	SA	05254
7	Taylors Lawn Sand	Rigby Taylor	4.1% w/w	SA	04451
8	Vitax Microgran 2	Vitax	10.9% w/w	MG	04541
9	Vitax Turf Tonic	Vitax	15% w/w	SA	04354

Uses

- Moss in ***amenity grassland*** [3, 5]; ***managed amenity turf*** [1-9]

Efficacy guidance

- For best results apply when turf is actively growing and the soil is moist
- Fertilizer component of most products encourages strong root growth and tillering
- Mow 3 d before treatment and do not mow for 3-4 d afterwards
- Water after 2 d if no rain
- Rake out dead moss thoroughly 7-14 d after treatment. Re-treatment may be necessary for heavy infestations

Restrictions

- Maximum number of treatments - see labels
- Do not apply during drought or when heavy rain expected
- Do not apply in frosty weather or when the ground is frozen
- Do not walk on treated areas until well watered
- Do not apply product undiluted [3]

Crop-specific information

- If spilt on paving, concrete, clothes etc brush off immediately to avoid discolouration
- Observe label restrictions for interval before cutting after treatment

Environmental safety

- Harmful to fish or other aquatic life. Do not contaminate surface waters or ditches with chemical or used container [4]

Hazard classification and safety precautions

UN number N/C
Operator protection A [3]; U20a [1, 2]; U20b [3, 5, 8, 9]; U20c [4, 6, 7]
Environmental protection E13c [4]; E15a [1, 2, 5, 7-9]; E15b [3]
Storage and disposal D01 [3, 5, 6, 9]; D09a [1-9]; D11a [1, 2, 4-9]; D12a [3]

239 ferrous sulphate + MCPA + mecoprop-P

A translocated herbicide and moss killer mixture
HRAC mode of action code: O + O

See also MCPA
mecoprop-P

Products

Renovator Pro	Scotts	0.49:0.29:16.3% w/w	GR	12204

Uses

- Annual dicotyledons in ***managed amenity turf***
- Moss in ***managed amenity turf***
- Perennial dicotyledons in ***managed amenity turf***

Approval information

- MCPA and mecoprop-P included in Annex I under EC Directive 91/414

Efficacy guidance

- Apply from Apr to Sep when weeds are growing
- For best results apply when light showers or heavy dews are expected
- Apply with a suitable calibrated fertilizer distributor
- Retreatment may be necessary after 6 wk if weeds or moss persist
- For best control of moss scarify vigorously after 2 wk to remove dead moss
- Where regrowth of moss or weeds occurs a repeat treatment may be made after 6 wk
- Avoid treatment of wet grass or during drought. If no rain falls within 48 h water in thoroughly

Restrictions

- Maximum number of treatments 3 per yr
- Do not treat new turf until established for 6 mth
- The first 4 mowings after treatment should not be used to mulch cultivated plants unless composted at least 6 mth
- Avoid walking on treated areas until it has rained or they have been watered
- Do not re-seed or turf within 8 wk of last treatment
- Do not cut grass for at least 3 d before and at least 4 d after treatment
- Do not apply during freezing conditions or when rain imminent

Environmental safety

- Keep livestock out of treated areas for up to two weeks following treatment and until poisonous weeds, such as ragwort, have died down and become unpalatable
- Harmful to fish or other aquatic life. Do not contaminate surface waters or ditches with chemical or used container
- Do not empty into drains
- Some pesticides pose a greater threat of contamination of water than others and mecoprop-P is one of these pesticides. Take special care when applying mecoprop-P near water and do not apply if heavy rain is forecast

Hazard classification and safety precautions

UN number N/C
Operator protection U20b
Environmental protection E07a, E13c, E19b
Storage and disposal D01, D09a, D12a
Medical advice M05a

240 fipronil

A phenylpyrazole insecticide for horticulture; products withdrawn since approval expires 2009
IRAC mode of action code: 2B

241 flazasulfuron

A sulfonylurea herbicide for non-crop use

Products

1	Chikara Weed Control	Belchim	25% w/w	WG	14189
2	Paradise	Pan Agriculture	25% w/w	WG	14504

Uses

- Annual and perennial weeds in ***natural surfaces not intended to bear vegetation***, ***permeable surfaces overlying soil***, ***railway tracks***

Hazard classification and safety precautions

Hazard Dangerous for the environment
Transport code 9
Packaging group III
UN number 3077
Risk phrases R50, R53a
Operator protection A, H, M; U02a, U05a, U20b [1, 2]; U08 [1]; U09a [2]

Environmental protection E15a, E16a, E16b, E38
Storage and disposal D01, D02, D09a, D11a, D12a

242 flonicamid

A selective feeding blocker aphicide
IRAC mode of action code: 9C

Products

1	Mainman	Belchim	50% w/w	WG	13123
2	Teppeki	Belchim	50% w/w	WG	12402

Uses
- Aphids in ***apples***, ***hops*** *(off-label)*, ***pears*** [1]; ***ornamental plant production*** *(off-label)* [1, 2]; ***potatoes***, ***winter wheat*** [2]

Specific Off-Label Approvals (SOLAs)
- ***hops*** *20090927* [1]
- ***ornamental plant production*** *20082898* [1], *20082906* [2]

Efficacy guidance
- Apply when warning systems forecast significant aphid infestations
- Persistence of action is 21 d
- Do not apply more than two consecutive treatments of flonicamid. If further treatment is needed use an insecticide with a different mode of action

Restrictions
- Maximum number of treatments 2 per crop for potatoes, winter wheat
- Must not be applied to winter wheat before 50% ear emerged stage (GS 53)
- Use a maximum of two consecutive applications of [1] in apples and pears. If further treatments are required, use an insecticide with a different mode of action before applying the final application of [1]

Crop-specific information
- HI: potatoes 14 d; winter wheat 28 d

Environmental safety
- Dangerous for the environment
- Harmful to aquatic organisms

Hazard classification and safety precautions

Hazard Dangerous for the environment [2]
UN number N/C
Risk phrases R52, R53a
Operator protection A, H [1, 2]; E, M [1]; U05a [1, 2]; U13, U14, U15 [1]
Environmental protection E15a
Storage and disposal D01, D02, D12b [1, 2]; D03, D08 [1]; D07 [2]

243 florasulam

A triazolopyrimidine herbicide for cereals
HRAC mode of action code: B

See also *2,4-D + florasulam*
clopyralid + florasulam + fluroxypyr

Products

1	Barton WG	Dow	25% w/w	WG	13284
2	Boxer	Dow	50 g/l	SC	09819

Uses
- Annual dicotyledons in ***farm forestry*** *(off-label)*, ***forest nurseries*** *(off-label)*, ***game cover*** *(off-label)*, ***grass seed crops*** *(off-label)*, ***ornamental plant production*** *(off-label)*, ***spring rye*** *(off-label)*, ***triticale*** *(off-label)*, ***winter rye*** *(off-label)* [2]; ***spring barley***, ***spring oats***, ***spring wheat***, ***winter barley***, ***winter oats***, ***winter wheat*** [1, 2]

- Chickweed in ***spring barley, spring oats, spring wheat, winter barley, winter oats, winter wheat*** [1, 2]
- Cleavers in ***spring barley, spring oats, spring wheat, winter barley, winter oats, winter wheat*** [1, 2]
- Mayweeds in ***spring barley, spring oats, spring wheat, winter barley, winter oats, winter wheat*** [1, 2]
- Volunteer oilseed rape in ***spring barley, spring oats, spring wheat, winter barley, winter oats, winter wheat*** [1, 2]

Specific Off-Label Approvals (SOLAs)

- ***farm forestry*** *20082826* [2]
- ***forest nurseries*** *20082826* [2]
- ***game cover*** *20082826* [2]
- ***grass seed crops*** *20060997* [2]
- ***ornamental plant production*** *20082826* [2]
- ***spring rye*** *20060997* [2]
- ***triticale*** *20060997* [2]
- ***winter rye*** *20060997* [2]

Approval information

- Florasulam included in Annex I under EC Directive 91/414
- Accepted by BBPA for use on malting barley

Efficacy guidance

- Best results obtained from treatment of small actively growing weeds in good conditions
- Apply in autumn or spring once crop has 3 leaves
- Product is mainly absorbed by leaves of weeds and is effective on all soil types
- Florasulam is a member of the ALS-inhibitor group of herbicides

Restrictions

- Maximum total dose on any crop equivalent to one full dose treatment
- Do not roll or harrow within 7 d before or after application
- Do not spray when crops under stress from cold, drought, pest damage, nutrient deficiency or any other cause
- Specific restrictions apply to use in sequence or tank mixture with other sulfonylurea or ALS-inhibiting herbicides. See label for details

Crop-specific information

- Latest use: up to and including flag leaf just visible stage for all crops

Following crops guidance

- Where the product has been used in mixture with certain named products (see label) only cereals or grass may be sown in the autumn following harvest [2]
- Unless otherwise restricted cereals, oilseed rape, field beans, grass or vegetable brassicas as transplants may be sown as a following crop in the same calendar yr as treatment. Oilseed rape may show some temporary reduction of vigour after a dry summer, but yields are not affected
- In addition to the above, linseed, peas, sugar beet, potatoes, maize, clover (for use in grass/clover mixtures) or carrots may be sown in the calendar yr following treatment
- In the event of failure of a treated crop in spring only spring wheat, spring barley, spring oats, maize or ryegrass may be sown

Environmental safety

- Dangerous for the environment
- Very toxic to aquatic organisms
- See label for detailed instructions on tank cleaning

Hazard classification and safety precautions

Hazard Dangerous for the environment
Transport code 9
Packaging group III
UN number 3077 [1]; 3082 [2]
Risk phrases R50, R53a
Operator protection U05a
Environmental protection E15a [2]; E15b [1]; E34, E38 [1, 2]
Storage and disposal D01, D02, D05, D09a, D10b, D12a

244 florasulam + fluroxypyr

A post-emergence herbicide mixture for cereals
HRAC mode of action code: B + O

See also fluroxypyr

Products

1	Cabadex	Headland Amenity	2.5:100 g/l	SE	13948
2	GF 184	Dow	2.5:100 g/l	SE	10878
3	Hiker	Dow	1.0:100 g/l	SE	11451
4	Hunter	Dow	2.5:100 g/l	SE	12836
5	Starane Gold	Dow	1.0:100 g/l	SE	10879
6	Starane Vantage	Dow	1.0:100 g/l	SE	10922
7	Starane XL	Dow	2.5:100 g/l	SE	10921

Uses

- Annual and perennial weeds in ***amenity grassland, lawns, managed amenity turf*** [1]
- Annual dicotyledons in ***durum wheat*** *(off-label)*, ***spring rye*** *(off-label)*, ***triticale*** *(off-label)*, ***winter rye*** *(off-label)* [3, 5-7]; ***game cover*** *(off-label)* [2, 5, 7]; ***grass seed crops*** *(off-label)* [2, 3, 5-7]; ***ornamental plant production*** *(off-label)* [7]; ***spring barley, spring oats, spring wheat, winter barley, winter oats, winter wheat*** [2, 4, 7]
- Buttercups in ***amenity grassland, lawns, managed amenity turf*** [1]
- Chickweed in ***spring barley, spring oats, spring wheat, winter barley, winter oats, winter wheat*** [2-7]
- Cleavers in ***spring barley, spring oats, spring wheat, winter barley, winter oats, winter wheat*** [2-7]
- Clover in ***amenity grassland, lawns, managed amenity turf*** [1]
- Daisies in ***amenity grassland, lawns, managed amenity turf*** [1]
- Dandelions in ***amenity grassland, lawns, managed amenity turf*** [1]
- Mayweeds in ***spring barley, spring oats, spring wheat, winter barley, winter oats, winter wheat*** [2, 4, 7]
- Plantains in ***amenity grassland, lawns, managed amenity turf*** [1]

Specific Off-Label Approvals (SOLAs)

- ***durum wheat*** *20072810* [3], *20072811* [5], *20072812* [6], *20072815* [7]
- ***game cover*** *20082858* [2], *20082878* [5], *20082904* [7]
- ***grass seed crops*** *20072809* [2], *20072810* [3], *20072811* [5], *20072812* [6], *20072815* [7]
- ***ornamental plant production*** *20082904* [7]
- ***spring rye*** *20072810* [3], *20072811* [5], *20072812* [6], *20072815* [7]
- ***triticale*** *20072810* [3], *20072811* [5], *20072812* [6], *20072815* [7]
- ***winter rye*** *20072810* [3], *20072811* [5], *20072812* [6], *20072815* [7]

Approval information

- Florasulam and fluroxypyr included in Annex I under EC Directive 91/414
- Accepted by BBPA for use on malting barley

Efficacy guidance

- Best results obtained when weeds are small and growing actively
- Products are mainly absorbed through weed foliage. Cleavers emerging after application will not be controlled
- Florasulam is a member of the ALS-inhibitor group of herbicides

Restrictions

- Maximum total dose equivalent to one full dose treatment
- Do not roll or harrow 7 d before or after application
- Do not spray when crops are under stress from cold, drought, pest damage or nutrient deficiency
- Do not apply through CDA applicators
- Specific restrictions apply to use in sequence or tank mixture with other sulfonylurea or ALS-inhibiting herbicides. See label for details

Crop-specific information

- Latest use: before flag leaf sheath extended (before GS 41) for spring barley and spring wheat; before flag leaf sheath opening (before GS 47) for winter barley and winter wheat; before second node detectable (before GS 32) for winter oats

Following crops guidance

- Cereals, oilseed rape, field beans or grass may follow treated crops in the same yr. Oilseed rape may suffer temporary vigour reduction after a dry summer
- In addition to the above, linseed, peas, sugar beet, potatoes, maize or clover may be sown in the calendar yr following treatment
- In the event of failure of a treated crop in the spring, only spring cereals, maize or ryegrass may be planted

Environmental safety

- Dangerous for the environment
- Toxic to aquatic organisms
- Take extreme care to avoid drift onto non-target crops or plants

Hazard classification and safety precautions

Hazard Irritant, Dangerous for the environment
Transport code 9
Packaging group III
UN number 3082
Risk phrases R36, R38, R53a, R67 [1-7]; R50 [1]; R51 [2-7]
Operator protection A, C; U05a, U11, U19a [1-7]; U08, U14, U20b [1]
Environmental protection E15a [2-7]; E15b [1]; E34, E38 [1-7]
Storage and disposal D01, D02, D09a, D10c, D12a

245 florasulam + pyroxsulam

A mixture of two triazolopirimidine sulfonamides for winter wheat
HRAC mode of action code: B

Products

Broadway Star	Dow	1.42:7.08% w/w	WG	14319

Uses

- Charlock in ***winter wheat***
- Chickweed in ***winter wheat***
- Cleavers in ***winter wheat***
- Field pansy in ***winter wheat***
- Field speedwell in ***winter wheat***
- Geranium species in ***winter wheat***
- Ivy-leaved speedwell in ***winter wheat***
- Mayweeds in ***winter wheat***
- Poppies in ***winter wheat***
- Ryegrass in ***winter wheat***
- Sterile brome in ***winter wheat***
- Volunteer oilseed rape in ***winter wheat***
- Wild oats in ***winter wheat***

Approval information

- Florasulam and pyroxsulam included in Annex I under EC Directive 91/414

Efficacy guidance

- Rainfast within 1 hour of application
- Requires an authorised adjuvant at application and recommended for use in a programme with herbicides employing a different mode of action.

Crop-specific information

- Crop injury may occur if applied in tank mixture with plant growth regulators - allow a minimum interval of 7 days.
- Crop injury may occur if applied in tank mixture with OP insecticides - allow a minimum interval of 14 days

Following crops guidance

- Crop failure before 1st Feb - allow 6 weeks to elapse and then drill spring wheat, spring barley, grass or maize.
- Crop failure after 1st Feb - Plough and allow 6 weeks to elapse before drilling grass or maize.

Environmental safety

- Take extreme care to avoid drift on to susceptible crops, non-target plants or waterways

Hazard classification and safety precautions

Hazard Dangerous for the environment
Transport code 9
Packaging group III
UN number 3077
Risk phrases R50, R53a
Operator protection U05a, U14, U15, U20a
Environmental protection E15b, E16a, E34, E38, E39
Storage and disposal D01, D02, D09a

246 fluazifop-P-butyl

A phenoxypropionic acid grass herbicide for broadleaved crops
HRAC mode of action code: A

Products

1	Clayton Maximus	Clayton	125 g/l	EC	12543
2	Fusilade Max	Syngenta	125 g/l	EC	11519
3	Greencrop Bantry	Greencrop	125 g/l	EC	12737

Uses

- Annual grasses in ***almonds*** *(off-label)*, ***apples*** *(off-label)*, ***apricots*** *(off-label)*, ***artichokes*** *(off-label)*, ***asparagus*** *(off-label)*, ***bilberries*** *(off-label)*, ***blackberries*** *(off-label)*, ***blueberries*** *(off-label)*, ***borage for oilseed production*** *(off-label)*, ***broad beans*** *(off-label)*, ***cabbages*** *(off-label)*, ***canary flower (echium spp.)*** *(off-label)*, ***cherries*** *(off-label)*, ***chicory*** *(off-label)*, ***choi sum*** *(off-label)*, ***cob nuts*** *(off-label)*, ***collards*** *(off-label)*, ***cranberries*** *(off-label)*, ***evening primrose*** *(off-label)*, ***field margins***, ***figs*** *(off-label)*, ***flax***, ***flax for industrial use***, ***garlic*** *(off-label)*, ***grapevines*** *(off-label)*, ***haricot beans*** *(off-label)*, ***hazel nuts*** *(off-label)*, ***herbs (see appendix 6)*** *(off-label)*, ***honesty*** *(off-label)*, ***linseed for industrial use***, ***lucerne*** *(off-label)*, ***lupins*** *(off-label)*, ***mallow (althaea spp.)*** *(off-label)*, ***mustard*** *(off-label)*, ***navy beans*** *(off-label)*, ***non-edible flowers*** *(off-label)*, ***ornamental plant production*** *(off-label)*, ***pak choi*** *(off-label)*, ***parsley root*** *(off-label)*, ***parsnips*** *(off-label)*, ***peaches*** *(off-label)*, ***pears*** *(off-label)*, ***plums*** *(off-label)*, ***protected non-edible flowers*** *(off-label)*, ***protected ornamentals*** *(off-label)*, ***quinces*** *(off-label)*, ***red beet*** *(off-label)*, ***redcurrants*** *(off-label)*, ***rubus hybrids*** *(off-label)*, ***shallots*** *(off-label)*, ***spring greens*** *(off-label)*, ***tic beans*** *(off-label)*, ***walnuts*** *(off-label)*, ***whitecurrants*** *(off-label)* [2]; ***blackcurrants***, ***bulb onions***, ***carrots***, ***combining peas***, ***farm forestry***, ***fodder beet***, ***gooseberries***, ***hops***, ***linseed***, ***raspberries***, ***spring field beans***, ***spring oilseed rape***, ***spring oilseed rape for industrial use***, ***strawberries***, ***sugar beet***, ***swedes*** *(stockfeed only)*, ***turnips*** *(stockfeed only)*, ***vining peas***, ***winter field beans***, ***winter oilseed rape***, ***winter oilseed rape for industrial use*** [1-3]; ***kale*** *(stockfeed only)* [1, 3]
- Barren brome in ***field margins*** [2]
- Blackgrass in ***hops***, ***spring field beans***, ***winter field beans*** [1-3]
- Green cover in ***land temporarily removed from production*** [1-3]
- Perennial grasses in ***almonds*** *(off-label)*, ***apples*** *(off-label)*, ***apricots*** *(off-label)*, ***artichokes*** *(off-label)*, ***asparagus*** *(off-label)*, ***bilberries*** *(off-label)*, ***blackberries*** *(off-label)*, ***blueberries*** *(off-label)*, ***borage for oilseed production*** *(off-label)*, ***broad beans*** *(off-label)*, ***cabbages*** *(off-label)*, ***canary flower (echium spp.)*** *(off-label)*, ***cherries*** *(off-label)*, ***chicory*** *(off-label)*, ***choi sum*** *(off-label)*, ***cob nuts*** *(off-label)*, ***collards*** *(off-label)*, ***cranberries*** *(off-label)*, ***evening primrose*** *(off-label)*, ***figs*** *(off-label)*, ***flax***, ***flax for industrial use***, ***garlic*** *(off-label)*, ***grapevines*** *(off-label)*, ***haricot beans*** *(off-label)*, ***hazel nuts*** *(off-label)*, ***herbs (see appendix 6)*** *(off-label)*, ***honesty*** *(off-label)*, ***linseed for industrial use***, ***lucerne*** *(off-label)*, ***lupins*** *(off-label)*, ***mallow (althaea spp.)*** *(off-label)*, ***mustard*** *(off-label)*, ***navy beans*** *(off-label)*, ***non-edible flowers*** *(off-label)*, ***ornamental plant production*** *(off-label)*, ***pak choi*** *(off-label)*, ***parsley root*** *(off-label)*, ***parsnips*** *(off-label)*, ***peaches*** *(off-label)*, ***pears*** *(off-label)*, ***plums*** *(off-label)*, ***protected non-edible flowers*** *(off-label)*, ***protected ornamentals*** *(off-label)*, ***quinces*** *(off-label)*, ***red beet*** *(off-label)*, ***redcurrants*** *(off-label)*, ***rubus hybrids*** *(off-label)*, ***shallots*** *(off-label)*, ***spring greens*** *(off-label)*, ***tic beans*** *(off-label)*, ***walnuts*** *(off-label)*, ***whitecurrants*** *(off-label)* [2]; ***blackcurrants***, ***bulb onions***, ***carrots***, ***combining peas***, ***farm forestry***, ***fodder beet***, ***gooseberries***, ***hops***, ***linseed***, ***raspberries***, ***spring***

field beans, spring oilseed rape, spring oilseed rape for industrial use, strawberries, sugar beet, swedes *(stockfeed only)*, ***turnips*** *(stockfeed only)*, ***vining peas, winter field beans, winter oilseed rape, winter oilseed rape for industrial use*** [1-3]; ***kale*** *(stockfeed only)* [1, 3]
- Volunteer cereals in ***blackcurrants, bulb onions, carrots, combining peas, farm forestry, fodder beet, gooseberries, hops, linseed, raspberries, spring field beans, spring oilseed rape, spring oilseed rape for industrial use, strawberries, sugar beet, swedes*** *(stockfeed only)*, ***turnips*** *(stockfeed only)*, ***vining peas, winter field beans, winter oilseed rape, winter oilseed rape for industrial use*** [1-3]; ***field margins, flax, flax for industrial use, linseed for industrial use*** [2]; ***kale*** *(stockfeed only)* [1, 3]
- Wild oats in ***blackcurrants, bulb onions, carrots, combining peas, farm forestry, fodder beet, gooseberries, hops, linseed, raspberries, spring field beans, spring oilseed rape, spring oilseed rape for industrial use, strawberries, sugar beet, swedes*** *(stockfeed only)*, ***turnips*** *(stockfeed only)*, ***vining peas, winter field beans, winter oilseed rape, winter oilseed rape for industrial use*** [1-3]; ***field margins, flax, flax for industrial use, linseed for industrial use*** [2]; ***kale*** *(stockfeed only)* [1, 3]

Specific Off-Label Approvals (SOLAs)

- ***almonds*** *20080323* [2]
- ***apples*** *20080323* [2]
- ***apricots*** *20080323* [2]
- ***artichokes*** *20080324* [2]
- ***asparagus*** *20080325* [2]
- ***bilberries*** *20063272* [2]
- ***blackberries*** *20063274* [2]
- ***blueberries*** *20063272* [2]
- ***borage for oilseed production*** *20081832* [2]
- ***broad beans*** *20081777* [2]
- ***cabbages*** *20080328* [2]
- ***canary flower (echium spp.)*** *20081832* [2]
- ***cherries*** *20080323* [2]
- ***chicory*** *20080326* [2]
- ***choi sum*** *20081777* [2]
- ***cob nuts*** *20080323* [2]
- ***collards*** *20081777* [2]
- ***cranberries*** *20063272* [2]
- ***evening primrose*** *20081832* [2]
- ***figs*** *20080323* [2]
- ***garlic*** *20081777* [2]
- ***grapevines*** *20080327* [2]
- ***haricot beans*** *20081777* [2]
- ***hazel nuts*** *20080323* [2]
- ***herbs (see appendix 6)*** *20091176* [2]
- ***honesty*** *20081832* [2]
- ***lucerne*** *20080845* [2]
- ***lupins*** *20063268* [2]
- ***mallow (althaea spp.)*** *20063273* [2]
- ***mustard*** *20081832* [2]
- ***navy beans*** *20081777* [2]
- ***non-edible flowers*** *20081777* [2]
- ***ornamental plant production*** *20081777* [2]
- ***pak choi*** *20081777* [2]
- ***parsley root*** *20063273* [2]
- ***parsnips*** *20081777* [2]
- ***peaches*** *20080323* [2]
- ***pears*** *20080323* [2]
- ***plums*** *20080323* [2]
- ***protected non-edible flowers*** *20081777* [2]
- ***protected ornamentals*** *20081777* [2]
- ***quinces*** *20080323* [2]
- ***red beet*** *20081777* [2]

- ***redcurrants*** *20063272* [2]
- ***rubus hybrids*** *20063274* [2]
- ***shallots*** *20063271* [2]
- ***spring greens*** *20081777* [2]
- ***tic beans*** *20081777* [2]
- ***walnuts*** *20080323* [2]
- ***whitecurrants*** *20063272* [2]

Approval information

- Accepted by BBPA for use on hops

Efficacy guidance

- Best results achieved by application when weed growth active under warm conditions with adequate soil moisture.
- Spray weeds from 2-expanded leaf stage to fully tillered, couch from 4 leaves when majority of shoots have emerged, with a second application if necessary
- Control may be reduced under dry conditions. Do not cultivate for 2 wk after spraying couch
- Annual meadow grass is not controlled
- May also be used to remove grass cover crops
- Fluazifop-P-butyl is an ACCase inhibitor herbicide. To avoid the build up of resistance do not apply products containing an ACCase inhibitor herbicide more than twice to any crop. In addition do not use any product containing fluazifop-P-butyl in mixture or sequence with any other product containing the same ingredient
- Use these products as part of a resistance management strategy that includes cultural methods of control and does not use ACCase inhibitors as the sole chemical method of grass weed control

Restrictions

- Maximum number of treatments 1 per crop or yr for all crops
- Do not sow cereals or grass crops for at least 8 wk after application of high rate or 2 wk after low rate
- Do not apply through CDA sprayer, with hand-held equipment or from air
- Avoid treatment before spring growth has hardened or when buds opening
- Do not treat bush and cane fruit or hops between flowering and harvest
- Consult processors before treating crops intended for processing
- Oilseed rape, linseed and flax for industrial use must not be harvested for human or animal consumption nor grazed
- Do not use for forestry establishment on land not previously under arable cultivation or improved grassland
- Treated vegetation in field margins, land temporarily removed from production etc, must not be grazed or harvested for human or animal consumption and unprotected persons must be kept out of treated areas for at least 24 h

Crop-specific information

- Latest use: before 50% ground cover for swedes, turnips; before 5 leaf stage for spring oilseed rape; before flowering for blackcurrants, gooseberries, hops, raspberries, strawberries; before flower buds visible for field beans, peas, linseed, flax, winter oilseed rape; 2 wk before sowing cereals or grass for field margins, land temporarily removed from production
- HI beet crops, kale, carrots 8 wk; onions 4 wk; oilseed rape for industrial use 2 wk
- Apply to sugar and fodder beet from 1-true leaf to 50% ground cover
- Apply to winter oilseed rape from 1-true leaf to established plant stage
- Apply to spring oilseed rape from 1-true leaf but before 5-true leaves
- Apply in fruit crops after harvest. See label for timing details on other crops
- Before using on onions or peas use crystal violet test to check that leaf wax is sufficient

Environmental safety

- Dangerous for the environment
- Very toxic to aquatic organisms

Hazard classification and safety precautions

Hazard Harmful, Dangerous for the environment
Transport code 9
Packaging group III
UN number 3082
Risk phrases R38, R50, R53a, R63

SEE SECTION 3 FOR PRODUCTS ALSO REGISTERED

Operator protection A, C, H, M; U05a, U08, U20b
Environmental protection E15a [1-3]; E38 [2]
Storage and disposal D01, D02, D05, D09a, D12a [1-3]; D10b [1]; D10c [2, 3]
Medical advice M05b [1]

247 fluazinam

A dinitroaniline fungicide for use in potatoes
FRAC mode of action code: 29

Products

1	Blizzard	AgriGuard	500 g/l	SC	13831
2	Blizzard	AgriGuard	500 g/l	SC	14363
3	Clayton Solstice	Clayton	500 g/l	SC	13943
4	Floozee	AgChem Access	500 g/l	SC	13856
5	Greencrop Solanum	Greencrop	500 g/l	SC	11052
6	Nando 500 SC	Nufarm UK	500 g/l	SC	14372
7	Ohayo	Syngenta	500 g/l	SC	13734
8	Shirlan	Syngenta	500 g/l	SC	10573
9	Tizca	Headland	500 g/l	SC	13877
10	Volley	Makhteshim	500 g/l	SC	13591

Uses

- Blight in ***potatoes*** [1-10]; ***seed potatoes*** *(off-label)* [1]
- Phytophthora root rot in ***blackberries*** *(off-label)*, ***container-grown blackberry*** *(off-label)*, ***container-grown raspberry*** *(off-label)*, ***container-grown rubus hybrids*** *(off-label)*, ***protected blackberries*** *(off-label)*, ***protected raspberries*** *(off-label)*, ***protected rubus hybrids*** *(off-label)*, ***raspberries*** *(off-label)*, ***rubus hybrids*** *(off-label)* [10]
- Powdery scab in ***seed potatoes*** *(off-label)* [8]
- Root rot in ***blackberries*** *(off-label)*, ***protected blackberries*** *(off-label)*, ***protected raspberries*** *(off-label)*, ***protected rubus hybrids*** *(off-label)*, ***raspberries*** *(off-label)*, ***rubus hybrids*** *(off-label)* [8, 9]
- Scab in ***seed potatoes*** *(off-label)* [9, 10]

Specific Off-Label Approvals (SOLAs)

- ***blackberries*** *20032168* [8], *20080597* [9], *20073759* [10]
- ***container-grown blackberry*** *20073759* [10]
- ***container-grown raspberry*** *20073759* [10]
- ***container-grown rubus hybrids*** *20073759* [10]
- ***protected blackberries*** *20032168* [8], *20080597* [9], *20073759* [10]
- ***protected raspberries*** *20032168* [8], *20080597* [9], *20073759* [10]
- ***protected rubus hybrids*** *20032168* [8], *20080597* [9], *20073759* [10]
- ***raspberries*** *20032168* [8], *20080597* [9], *20073759* [10]
- ***rubus hybrids*** *20032168* [8], *20080597* [9], *20073759* [10]
- ***seed potatoes*** *20081060* [1], *20060893* [8], *20080598* [9], *20073757* [10]

Approval information

- Fluazinam included in Annex I under EC Directive 91/414

Efficacy guidance

- Commence treatment at the first blight risk warning (before blight enters the crop). Products are rainfast within 1 h
- In the absence of a warning, treatment should start before foliage of adjacent plants meets in the rows
- Spray at 7-14 d intervals (5-10 d intervals [8]) depending on severity of risk (see label)
- Ensure complete coverage of the foliage and stems, increasing volume as haulm growth progresses, in dense crops and if blight risk increases

Restrictions

- Maximum number of treatments 10 per crop on potatoes [1, 5]
- Maximum total dose equivalent to 7.5 full dose treatments [8, 10]
- Do not use with hand-held sprayers

Crop-specific information

- HI 0 - 10 d for potatoes. Check label

- Ensure complete kill of potato haulm before lifting and do not lift crops for storage while there is any green tissue left on the leaves or stem bases

Environmental safety

- Dangerous for the environment
- Very toxic to aquatic organisms
- LERAP Category B

Hazard classification and safety precautions

Hazard Harmful [6]; Irritant [1-5, 7-9]; Dangerous for the environment [1-10]
Transport code 9
Packaging group III
UN number 3082
Risk phrases R36 [1-8]; R41 [9]; R43 [1-9]; R50, R53a [1-10]
Operator protection A, C, H; U02a, U04a, U08 [3-8, 10]; U05a [1-9]; U11 [9]; U14 [3, 4, 6-9]; U15, U20a [3, 4, 6-8]; U20b [5, 9, 10]
Environmental protection E15a, E16a [3-10]; E16b, E34 [3, 4, 6-8, 10]; E38 [1-4, 6-10]
Storage and disposal D01, D02 [1-9]; D05 [3-9]; D09a [3-10]; D10a [10]; D10b [5]; D10c [3, 4, 6-9]; D12a [1-4, 6-10]
Medical advice M03 [3-8, 10]; M05a [1-4, 6-8]

248 fluazinam + metalaxyl-M

A mixture of contact and systemic fungicides for potatoes
FRAC mode of action code: 29 + 4

See also metalaxyl-M

Products

Epok	Belchim	400:200 g/l	EC	11997

Uses

- Blight in ***potatoes***

Approval information

- Fluazinam and metalaxyl-M included in Annex I under EC Directive 91/414

Efficacy guidance

- Commence treatment at the first blight risk warning (before blight enters the crop). Products are rainfast within 1 h
- In the absence of a warning, treatment should start before foliage of adjacent plants meets in the rows
- Spray at 7-14 d intervals depending on severity of risk (see label)
- Ensure complete coverage of the foliage and stems, increasing volume as haulm growth progresses, in dense crops and if blight risk increases
- Metalaxyl-M works best on young actively growing foliage and efficacy declines with the onset of senescence. Therefore it is recommended that the product is only used for the first part of the blight control program, which should be completed with a reliable protectant fungicide

Restrictions

- Maximum number of treatments 3 per crop

Crop-specific information

- HI 7d for potatoes

Environmental safety

- Dangerous for the environment
- Very toxic to aquatic organisms
- LERAP Category B

Hazard classification and safety precautions

Hazard Harmful, Dangerous for the environment
Transport code 9
Packaging group III
UN number 3082
Risk phrases R20, R36, R38, R40, R43, R50, R53a

Operator protection A, C, H; U05a, U07, U11, U14, U15, U19a, U20a
Environmental protection E15a, E16a, E16b, E34, E38
Storage and disposal D01, D02, D09a, D10b, D12a
Medical advice M05a

249 fludioxonil

A phenylpyrrole fungicide seed treatment for wheat and barley
FRAC mode of action code: 12

See also *chlorothalonil + fludioxonil + propiconazole*
cymoxanil + fludioxonil + metalaxyl-M
cyprodinil + fludioxonil
difenoconazole + fludioxonil

Products

Beret Gold	Syngenta	25 g/l	FS	12625

Uses

- Bunt in ***spring wheat*** *(seed treatment)*, ***winter wheat*** *(seed treatment)*
- Covered smut in ***spring barley*** *(seed treatment)*, ***winter barley*** *(seed treatment)*
- Fusarium foot rot and seedling blight in ***spring barley*** *(seed treatment)*, ***spring oats*** *(seed treatment)*, ***spring wheat*** *(seed treatment)*, ***winter barley*** *(seed treatment)*, ***winter oats*** *(seed treatment)*, ***winter wheat*** *(seed treatment)*
- Leaf stripe in ***spring barley*** *(seed treatment - reduction)*, ***winter barley*** *(seed treatment - reduction)*
- Pyrenophora leaf spot in ***spring oats*** *(seed treatment)*, ***winter oats*** *(seed treatment)*
- Seed-borne diseases in ***durum wheat*** *(off-label)*, ***spring rye*** *(off-label)*, ***triticale*** *(off-label)*, ***winter rye*** *(off-label)*
- Septoria seedling blight in ***spring wheat*** *(seed treatment)*, ***winter wheat*** *(seed treatment)*
- Snow mould in ***spring barley*** *(seed treatment)*, ***spring oats*** *(seed treatment)*, ***spring wheat*** *(seed treatment)*, ***winter barley*** *(seed treatment)*, ***winter oats*** *(seed treatment)*, ***winter wheat*** *(seed treatment)*

Specific Off-Label Approvals (SOLAs)

- ***durum wheat*** *20071445*
- ***spring rye*** *20071445*
- ***triticale*** *20071445*
- ***winter rye*** *20071445*

Approval information

- Fludioxonil included in Annex I under EC Directive 91/414
- Accepted by BBPA for use on malting barley

Efficacy guidance

- Apply direct to seed using conventional seed treatment equipment. Continuous flow treaters should be calibrated using product before use
- Effective against benzimidazole-resistant strains of *Microdochium nivale*

Restrictions

- Maximum number of treatments 1 per seed batch
- Do not apply to cracked, split or sprouted seed
- Sow treated seed within 6 mth

Crop-specific information

- Latest use: before drilling
- Product may reduce flow rate of seed through drill. Recalibrate with treated seed before drilling

Environmental safety

- Dangerous for the environment
- Toxic to aquatic organisms
- Do not use treated seed as food or feed
- Treated seed harmful to game and wildlife

Hazard classification and safety precautions

Hazard Dangerous for the environment
Transport code 9
Packaging group III
UN number 3082
Risk phrases R51, R53a
Operator protection A, H; U05a, U20b
Environmental protection E03, E15a, E38
Storage and disposal D01, D02, D05, D09a, D11a, D12a
Treated seed S02, S05, S07

250 fludioxonil + flutriafol

A fungicide seed treatment mixture for cereals
FRAC mode of action code: 12 + 3

See also *flutriafol*

Products

Beret Multi	Syngenta	25:25 g/l	FS	13017

Uses

- Bunt in ***spring wheat*** *(seed treatment)*, ***winter wheat*** *(seed treatment)*
- Covered smut in ***spring barley*** *(seed treatment)*, ***winter barley*** *(seed treatment)*
- Fusarium foot rot and seedling blight in ***spring barley*** *(seed treatment)*, ***spring oats*** *(seed treatment)*, ***spring wheat*** *(seed treatment)*, ***winter barley*** *(seed treatment)*, ***winter oats*** *(seed treatment)*, ***winter wheat*** *(seed treatment)*
- Leaf stripe in ***spring barley*** *(seed treatment)*, ***winter barley*** *(seed treatment)*
- Loose smut in ***spring barley*** *(seed treatment)*, ***spring wheat*** *(seed treatment)*, ***winter barley*** *(seed treatment)*, ***winter wheat*** *(seed treatment)*
- Pyrenophora leaf spot in ***spring oats*** *(seed treatment)*, ***winter oats*** *(seed treatment)*
- Septoria seedling blight in ***spring wheat*** *(seed treatment)*, ***winter wheat*** *(seed treatment)*
- Snow mould in ***spring barley*** *(seed treatment)*, ***spring wheat*** *(seed treatment)*, ***winter barley*** *(seed treatment)*, ***winter wheat*** *(seed treatment)*

Approval information

- Fludioxonil included in Annex I under EC Directive 91/414

Efficacy guidance

- Apply directly to the seed using conventional seed treatment equipment
- Calibrate continuous flow treaters before use
- Effective against benzimidazole-resistant strains of *Microdochium nivale*

Restrictions

- Maximum number of treatments 1 per seed batch
- Do not apply to cracked, split or sprouted seed

Crop-specific information

- Latest use: before drilling for all cereals
- Treated seed may affect the flow rate through drills. Recalibrate with treated seed before drilling

Environmental safety

- Dangerous for the environment
- Toxic to aquatic organisms

Hazard classification and safety precautions

Hazard Dangerous for the environment
Transport code 9
Packaging group III
UN number 3082
Risk phrases R51, R53a
Operator protection A, H; U05a, U20c
Environmental protection E03, E15a, E38

Storage and disposal D01, D02, D05, D09a, D10c, D11a, D12a
Treated seed S02, S05, S07

251 fludioxonil + metalaxyl M

A seed dressing for use in forage maize
FRAC mode of action code: 12 + 4

Products

Maxim XL	Syngenta	25:9.69 g/l	FS	14136

Uses

- Damping off in ***forage maize***
- Fusarium in ***forage maize***
- Pythium in ***forage maize***

Approval information

- Fludioxonil and metalaxyl-M included in Annex I under EC Directive 91/414

Restrictions

- For advice on resistance management, refer to the latest Fungicide Resistance Action Group (FRAG) guidelines

Hazard classification and safety precautions

UN number N/C
Risk phrases R52, R53a
Operator protection A, H; U05a
Environmental protection E34, E38
Storage and disposal D01, D02, D05, D09a, D10c, D11a, D12a
Treated seed S01, S02, S03, S04a, S05, S07

252 fludioxonil + metalaxyl-M + thiamethoxam

A seed dressing for use in oilseed rape, fodder rape and mustard
FRAC mode of action code: 12 + 4 + 4A

See also *metalaxyl-M*
thiamethoxam

Products

Cruiser OSR	Syngenta	8:32.3:280 g/l	FS	14496

Uses

- Alternaria in ***fodder rape*** *(moderate control)*, ***mustard*** *(moderate control)*, ***spring oilseed rape*** *(moderate control)*, ***winter oilseed rape*** *(moderate control)*
- Cabbage stem flea beetle in ***fodder rape*** *(reduction of damage)*, ***mustard*** *(reduction of damage)*, ***spring oilseed rape*** *(reduction of damage)*, ***winter oilseed rape*** *(reduction of damage)*
- Downy mildew in ***fodder rape***, ***mustard***, ***spring oilseed rape***, ***winter oilseed rape***
- Flea beetle in ***fodder rape*** *(reduction of damage)*, ***mustard*** *(reduction of damage)*, ***spring oilseed rape*** *(reduction of damage)*, ***winter oilseed rape*** *(reduction of damage)*
- Phoma lingam in ***fodder rape***, ***mustard***, ***spring oilseed rape***, ***winter oilseed rape***
- Pythium in ***fodder rape*** *(reduction)*, ***mustard*** *(reduction)*, ***spring oilseed rape*** *(reduction)*, ***winter oilseed rape*** *(reduction)*

Approval information

- Fludioxonil, metalaxyl-M and thiamethoxam included in Annex I under EC Directive 91/414

Efficacy guidance

- Contains a neonicotinoid insecticide. The first subsequent foliar insecticide spray should contain an insecticide with a different mode of action to combat resistance - consult the IRAC website for details of modes of action.
- Contains metalaxyl-M, a phenylamide fungicide. Do not use fungicide sprays containing metalaxyl-M on crops grown from treated seed

Crop-specific information

- For use on all varieties of oilseed rape, fodder rape and mustard [1]

Hazard classification and safety precautions

Hazard Dangerous for the environment
UN number N/C
Risk phrases R50, R53a
Operator protection A, D, H; U05a, U07, U20b, U20e
Environmental protection E15b, E34, E38
Storage and disposal D01, D02, D09a, D10a, D10d, D11a, D12a, D14
Treated seed S01, S02, S03, S04a, S04b, S05, S06b, S07, S08, S09

253 fludioxonil + tefluthrin

A fungicide and insecticide seed treatment mixture for cereals
FRAC mode of action code: 12

See also *tefluthrin*

Products

Austral Plus	Syngenta	10:40 g/l	FS	13314

Uses

- Bunt in ***spring wheat, winter wheat***
- Covered smut in ***spring barley, winter barley***
- Fusarium foot rot and seedling blight in ***spring barley, spring oats, spring wheat, winter barley, winter oats, winter wheat***
- Leaf stripe in ***spring barley*** *(partial control)*, ***winter barley*** *(partial control)*
- Pyrenophora leaf spot in ***spring oats, winter oats***
- Seed-borne diseases in ***triticale seed crop*** *(off-label)*
- Septoria seedling blight in ***spring wheat, winter wheat***
- Snow mould in ***spring barley, spring wheat, winter barley, winter wheat***
- Wheat bulb fly in ***spring barley, spring wheat, winter barley, winter wheat***
- Wireworm in ***spring barley, spring oats, spring wheat, winter barley, winter oats, winter wheat***

Specific Off-Label Approvals (SOLAs)

- ***triticale seed crop*** *20080982*

Approval information

- Fludioxonil included in Annex I under EC Directive 91/414

Efficacy guidance

- Apply direct to seed using conventional seed treatment equipment. Continuous flow treaters should be calibrated using product before use
- Best results obtained from seed drilled into a firm even seedbed
- Tefluthrin is released into soil after drilling and repels or kills larvae of wheat bulb fly and wireworm attacking below ground. Pest control may be reduced by deep or shallow drilling
- Where egg counts of wheat bulb fly or population counts of wireworm indicate a high risk of severe attack follow up spray treatments may be needed
- Control of leaf stripe in barley may not be sufficient in crops grown for seed certification
- Effective against benzimidazole-resistant strains of *Microdochium nivale*
- Under adverse soil or environmental conditions seed rates should be increased to compensate for possible reduced germination capacity

Restrictions

- Maximum number of treatments 1 per batch
- Store treated seed in cool dry conditions and drill within 3 mth
- Do not use on seed above 16% moisture content or on sprouted, cracked or damaged seed

Crop-specific information

- Latest use: before drilling

Environmental safety

- Dangerous for the environment
- Very toxic to aquatic organisms

Hazard classification and safety precautions

Hazard Dangerous for the environment
Transport code 9
Packaging group III
UN number 3082
Risk phrases R50, R53a
Operator protection A, C, D, H; U02a, U04a, U05a, U07, U08, U14, U20b
Environmental protection E03, E15b, E34, E38
Storage and disposal D01, D02, D05, D09b, D10c, D11a, D12a
Treated seed S02, S04b, S05, S07, S08

254 flufenacet

A broad spectrum oxyacetamide herbicide available only in mixtures
HRAC mode of action code: K3

See also diflufenican + flufenacet

255 flufenacet + isoxaflutole

A residual herbicide mixture for maize
HRAC mode of action code: K3 + F2

See also isoxaflutole

Products

1	Amethyst	AgChem Access	48:10% w/w	WG	14079
2	Cadou Star	Bayer CropScience	48:10% w/w	WG	13242

Uses

- Annual dicotyledons in ***forage maize, grain maize*** [1, 2]; ***forage maize (under plastic mulches)*** *(off-label)*, ***game cover*** *(off-label)*, ***hops*** *(off-label)*, ***miscanthus*** *(off-label)*, ***ornamental plant production*** *(off-label)*, ***sweetcorn*** *(off-label)*, ***sweetcorn under plastic mulches*** *(off-label)*, ***top fruit*** *(off-label)* [2]
- Annual grasses in ***forage maize (under plastic mulches)*** *(off-label)*, ***game cover*** *(off-label)*, ***hops*** *(off-label)*, ***miscanthus*** *(off-label)*, ***ornamental plant production*** *(off-label)*, ***sweetcorn*** *(off-label)*, ***sweetcorn under plastic mulches*** *(off-label)*, ***top fruit*** *(off-label)* [2]
- Grass weeds in ***forage maize, grain maize*** [1, 2]

Specific Off-Label Approvals (SOLAs)

- ***forage maize (under plastic mulches)*** *20080960* [2]
- ***game cover*** *20082829* [2]
- ***hops*** *20082829* [2]
- ***miscanthus*** *20082829* [2]
- ***ornamental plant production*** *20082829* [2]
- ***sweetcorn*** *20080798* [2]
- ***sweetcorn under plastic mulches*** *20080960* [2]
- ***top fruit*** *20082829* [2]

Approval information

- Flufenacet and isoxaflutole included in Annex I under EC Directive 91/414

Efficacy guidance

- Best results obtained from applications to a fine, firm seedbed in the presence of some soil moisture
- Efficacy may be reduced on cloddy seedbeds or under prolonged dry conditions
- Established perennial grasses and broad-leaved weeds growing from rootstocks will not be controlled

- Always follow WRAG guidelines for preventing and managing herbicide resistant weeds. See Section 5 for more information

Restrictions

- Maximum number of treatments 1 per crop
- Do not use on soils both above 70% sand and less than 2% organic matter
- Do not use on maize crops intended for seed production

Crop-specific information

- Latest use: before crop emergence
- Ideally treatment should be made within 4 d of sowing, before tha maize seeds have germinated

Following crops guidance

- In the event of failure of a treated crop maize, sweet corn or potatoes may be grown after surface cultivation or ploughing
- After normal harvest of a treated crop oats or barley may be sown after ploughing and wheat, rye, triticale, sugar beet, field beans, peas, soya, sorghum or sunflowers may be sown after surface cultivation or ploughing.

Environmental safety

- Dangerous for the environment
- Very toxic to aquatic organisms
- Take care to avoid drift over other crops. Beet crops and sunflowers are particularly sensitive
- LERAP Category B

Hazard classification and safety precautions

Hazard Harmful, Dangerous for the environment
Transport code 9
Packaging group III
UN number 3077
Risk phrases R22a, R36, R43, R48, R50, R53a, R63
Operator protection A, C, H; U05a, U11, U20b
Environmental protection E15b, E16a, E34, E38
Storage and disposal D01, D02, D05, D09a, D10a, D12a
Medical advice M03

256 flufenacet + metribuzin

A herbicide mixture for potatoes
HRAC mode of action code: K3 + C1

See also metribuzin

Products

Artist	Bayer CropScience	24:17.5% w/w	WP	11239

Uses

- Annual dicotyledons in ***bilberries*** *(off-label)*, ***blackcurrants*** *(off-label)*, ***blueberries*** *(off-label)*, ***cranberries*** *(off-label)*, ***early potatoes***, ***gooseberries*** *(off-label)*, ***maincrop potatoes***, ***redcurrants*** *(off-label)*, ***ribes hybrids*** *(off-label)*
- Annual grasses in ***bilberries*** *(off-label)*, ***blackcurrants*** *(off-label)*, ***blueberries*** *(off-label)*, ***cranberries*** *(off-label)*, ***gooseberries*** *(off-label)*, ***redcurrants*** *(off-label)*, ***ribes hybrids*** *(off-label)*
- Annual meadow grass in ***early potatoes***, ***maincrop potatoes***

Specific Off-Label Approvals (SOLAs)

- ***bilberries*** *20092641*
- ***blackcurrants*** *20092641*
- ***blueberries*** *20092641*
- ***cranberries*** *20092641*
- ***gooseberries*** *20092641*
- ***redcurrants*** *20092641*
- ***ribes hybrids*** *20092641*

Approval information

- Flufenacet and metribuzin included in Annex I under EC Directive 91/414

SEE SECTION 3 FOR PRODUCTS ALSO REGISTERED

Efficacy guidance

- Product acts through root uptake and needs sufficient soil moisture at and shortly after application
- Effectiveness is reduced under dry soil conditions
- Residual activity is reduced on mineral soils with a high organic matter content and on peaty or organic soils
- Ensure application is made evenly to both sides of potato ridges
- Perennial weeds are not controlled

Restrictions

- Maximum total dose equivalent to one full dose treatment
- Potatoes must be sprayed before emergence of crop and weeds
- See label for list of tolerant varieties. Do not treat Maris Piper grown on Sands or Very Light soils
- Do not use on Sands
- On stony or gravelly soils there is risk of crop damage especially if heavy rain falls soon after application

Crop-specific information

- Latest use: before potato crop emergence
- Consult processor before use on crops for processing

Following crops guidance

- Before drilling or planting any succeeding crop soil must be mouldboard ploughed to at least 15 cm as soon as possible after lifting and no later than end Dec
- In W Cornwall on soils with more than 5% organic matter treated early potatoes may be followed by summer planted brassica crops 14 wk after treatment and after mouldboard ploughing. Elsewhere cereals or winter beans may be grown in the same year if at least 16 wk have elapsed since treatment
- In the yr following treatment any crop may be grown except lettuce or radish, or vegetable brassica crops on silt soils in Lincs

Environmental safety

- Dangerous for the environment
- Very toxic to aquatic organisms
- Take care to avoid spray drift onto neighbouring crops, especially lettuce or brassicas

Hazard classification and safety precautions

Hazard Harmful, Dangerous for the environment
Transport code 9
Packaging group III
UN number 3077
Risk phrases R22a, R43, R48, R50, R53a
Operator protection A, D, H; U05a, U08, U13, U14, U19a, U20b
Environmental protection E15a, E38
Storage and disposal D01, D02, D09a, D11a, D12a
Medical advice M03

257 flufenacet + pendimethalin

A broad spectrum residual and contact herbicide mixture for winter cereals
HRAC mode of action code: K3 + K1

See also *pendimethalin*

Products

1	Crystal	BASF	60:300 g/l	EC	13914
2	Ice	BASF	60:300 g/l	EC	13930
3	Shooter	BASF	60:300	EC	14106
4	Trooper	BASF	60:300 g/l	EC	13924

Uses

- Annual dicotyledons in ***game cover*** *(off-label)* [1]; ***winter barley***, ***winter wheat*** [1-4]
- Annual grasses in ***winter barley***, ***winter wheat*** [1, 2, 4]
- Annual meadow grass in ***game cover*** *(off-label)* [1]; ***winter barley***, ***winter wheat*** [1-4]
- Blackgrass in ***game cover*** *(off-label)* [1]; ***winter barley***, ***winter wheat*** [1, 2, 4]

- Chickweed in ***winter barley*, *winter wheat*** [1, 2, 4]
- Corn marigold in ***winter barley*, *winter wheat*** [1, 2, 4]
- Field speedwell in ***winter barley*, *winter wheat*** [1, 2, 4]
- Ivy-leaved speedwell in ***winter barley*, *winter wheat*** [1, 2, 4]

Specific Off-Label Approvals (SOLAs)

- ***game cover*** *20090450* [1]

Approval information

- Flufenacet and pendimethalin included in Annex I under EC Directive 91/414
- Accepted by BBPA for use on malting barley

Efficacy guidance

- Best results achieved when applied from pre-emergence of weeds to 2-leaf stage but post emergence treatment is not recommended on clay soils
- Product requires some soil moisture to be activated ideally from rain within 7 d of application. Prolonged dry conditions may reduce residual control
- Product is slow acting and final level of weed control may take some time to appear
- For effective weed control seed bed preparations should ensure even incorporation of any trash, straw and ash to 15 cm
- Efficacy may be reduced on soils with more than 6% organic matter
- Always follow WRAG guidelines for preventing and managing herbicide resistant weeds. See Section 5 for more information

Restrictions

- Maximum total dose equivalent to one full dose treatment
- For pre-emergence treatments seed should be covered with at least 32 mm settled soil. Shallow drilled crops should be treated post-emergence only
- Do not treat undersown crops
- Avoid spraying during periods of prolonged or severe frosts
- Do not use on stony or gravelly soils or those with more than 10% organic matter
- Pre-emergence treatment may only be used on crops drilled before 30 Nov. All crops must be treated before 31 Dec in yr of planting.
- Concentrated or diluted product may stain clothing or skin

Crop-specific information

- Latest use: before third tiller stage (GS 23) and before 31 Dec in yr of planting
- Very wet weather before and after treatment may result in loss of crop vigour and reduced yield, particularly where soils become waterlogged

Following crops guidance

- Any crop may follow a failed or normally harvested treated crop provided ploughing to at least 15 cm is carried out beforehand

Environmental safety

- Dangerous for the environment
- Very toxic to aquatic organisms
- Risk to certain non-target insects or other arthropods
- LERAP Category B
- Some products supplied in small volume returnable packs. Follow instructions for use

Hazard classification and safety precautions

Hazard Harmful, Dangerous for the environment
Transport code 3
Packaging group III
UN number 1993
Risk phrases R22a, R38, R50, R53a [1-4]; R22b, R40 [1, 2, 4]
Operator protection A, C, H; U02a, U05a [1-4]; U14, U20b [3]; U20c [1, 2, 4]
Environmental protection E15a [1, 2, 4]; E15b [3]; E16a, E16b, E22b, E34, E38 [1-4]
Storage and disposal D01, D02, D09a, D12a [1-4]; D08, D10c [3]; D10a [1, 2, 4]
Medical advice M03 [1-4]; M05b [1, 2, 4]

258 flumioxazine

A phenyphthalimide herbicide for winter wheat
HRAC mode of action code: E

Products

1	Digital	Interfarm	300 g/l	SC	13561
2	Guillotine	Interfarm	300 g/l	SC	13562
3	Sumimax	Interfarm	300 g/l	SC	13548

Uses

- Annual dicotyledons in ***bulb onions*** *(off-label)*, ***carrots*** *(off-label)*, ***ornamental plant production*** *(off-label)*, ***parsnips*** *(off-label)*, ***vining peas*** *(off-label)*, ***winter wheat*** [1-3]; ***farm forestry*** *(off-label)*, ***forest nurseries*** *(off-label)* [1]; ***hops*** *(off-label)*, ***soft fruit*** *(off-label)*, ***top fruit*** *(off-label)* [2, 3]
- Annual grasses in ***winter oats*** *(off-label)* [2]
- Annual meadow grass in ***winter wheat*** [1-3]
- Blackgrass in ***winter oats*** *(off-label)* [2]
- Chickweed in ***winter wheat*** [1-3]
- Cleavers in ***winter wheat*** [1-3]
- Groundsel in ***bulb onions*** *(off-label)*, ***carrots*** *(off-label)*, ***ornamental plant production*** *(off-label)*, ***parsnips*** *(off-label)*, ***vining peas*** *(off-label)* [1-3]; ***farm forestry*** *(off-label)*, ***forest nurseries*** *(off-label)* [1]; ***hops*** *(off-label)*, ***soft fruit*** *(off-label)*, ***top fruit*** *(off-label)* [2, 3]
- Loose silky bent in ***winter wheat*** [1-3]
- Mayweeds in ***winter wheat*** [1-3]
- Ryegrass in ***winter oats*** *(off-label)* [2]
- Speedwells in ***winter wheat*** [1-3]
- Volunteer oilseed rape in ***bulb onions*** *(off-label)*, ***carrots*** *(off-label)*, ***ornamental plant production*** *(off-label)*, ***parsnips*** *(off-label)*, ***vining peas*** *(off-label)*, ***winter wheat*** [1-3]; ***farm forestry*** *(off-label)*, ***forest nurseries*** *(off-label)* [1]; ***hops*** *(off-label)*, ***soft fruit*** *(off-label)*, ***top fruit*** *(off-label)* [2, 3]
- Volunteer potatoes in ***bulb onions*** *(off-label)*, ***carrots*** *(off-label)*, ***ornamental plant production*** *(off-label)*, ***parsnips*** *(off-label)*, ***vining peas*** *(off-label)* [1-3]; ***farm forestry*** *(off-label)*, ***forest nurseries*** *(off-label)* [1]; ***hops*** *(off-label)*, ***soft fruit*** *(off-label)*, ***top fruit*** *(off-label)* [2, 3]

Specific Off-Label Approvals (SOLAs)

- ***bulb onions*** *20091114* [1], *20091116* [2], *20091108* [3]
- ***carrots*** *20091112* [1], *20091117* [2], *20091111* [3]
- ***farm forestry*** *20082844* [1]
- ***forest nurseries*** *20082844* [1]
- ***hops*** *20082897* [2], *20082881* [3]
- ***ornamental plant production*** *20082844* [1], *20082897* [2], *20082881* [3]
- ***parsnips*** *20091112* [1], *20091117* [2], *20091111* [3]
- ***soft fruit*** *20082897* [2], *20082881* [3]
- ***top fruit*** *20082897* [2], *20082881* [3]
- ***vining peas*** *20091113* [1], *20091115* [2], *20091109* [3]
- ***winter oats*** *20092506* [2]

Efficacy guidance

- Best results obtained from applications to moist soil early post-emergence of the crop when weeds are germinating up to 1 leaf stage
- Flumioxazin is contact acting and relies on weeds germinating in moist soil and coming into sufficient contact with the herbicide
- Flumioxazin remains in the surface layer of the soil and does not migrate to lower layers
- Contact activity is long lasting during the autumn
- Seed beds should be firm, fine and free from clods. Crops should be drilled to 25 mm and well covered by soil
- Efficacy is reduced on soils with more than 10% organic matter
- Flumioxazin is a member of the protoporphyrin oxidase (PPO) inhibitor herbicides. To avoid the build up of resistance do not use PPO-inhibitor herbicides more than once on any crop
- Use as part of a resistance management strategy that includes cultural methods of control and does not use PPO-inhibitors as the sole chemical method of grass weed control

Restrictions

- Maximum number of treatments 1 per crop for winter wheat
- Only apply to crops that have been hardened by cool weather. Do not treat crops with lush or soft growth
- Do not apply to waterlogged soils or to soils with more than 10% organic matter
- Do not follow an application with another pesticide for 14 d
- Do not treat undersown crops
- Do not treat broadcast crops until they are past the 3 leaf stage
- Do not treat crops under stress for any reason, or during prolonged frosty weather
- Do not roll or harrow for 2 wk before treatment or at any time afterwards

Crop-specific information

- Latest use: before 5th true leaf stage (GS 15) for winter wheat
- Light discolouration of leaf margins can occur after treatment and treatment of soft crops will cause transient leaf bleaching

Following crops guidance

- Any crop may be sown following normal harvest of a treated crop
- In the event of failure of a treated crop spring cereals, spring oilseed rape, sugar beet, maize or potatoes may be redrilled after ploughing

Environmental safety

- Dangerous for the environment
- Very toxic to aquatic organisms
- LERAP Category B
- Take extreme care to avoid drift onto plants outside the target area

Hazard classification and safety precautions

Hazard Toxic, Dangerous for the environment
Transport code 9
Packaging group III
UN number 3082
Risk phrases R50, R53a, R61
Operator protection A, H; U05a, U19a
Environmental protection E15b, E16a, E34, E38
Storage and disposal D01, D02, D05, D07, D09a, D12b
Medical advice M04a

259 fluopicolide

An benzamide fungicide available only in mixtures
FRAC mode of action code: 43

260 fluopicolide + propamocarb hydrochloride

A protectant and systemic fungicide mixture for potato blight
FRAC mode of action code: 43 + 28

See also propamocarb hydrochloride

Products

Infinito	Bayer CropScience	62.5:625 g/l	SC	12644

Uses

- Blight in ***potatoes***

Approval information

- Propamocarb hydrochloride included in Annex I under EC Directive 91/414

Efficacy guidance

- Commence spray programme before infection appears as soon as weather conditions favourable for disease development occur. At latest the first treatment should be made as the foliage meets along the rows
- Repeat treatments at 7-10 day intervals according to disease incidence and weather conditions

- Reduce spray interval if conditions are conducive to the spread of blight
- Spray as soon as possible after irrigation
- Increase water volume in dense crops
- When used from full canopy development to haulm desiccation as part of a full blight protection programme tubers will be protected from late blight after harvest and tuber blight incidence will be reduced
- To reduce the development of resistance product should be used in single or block applications with fungicides from a different cross-resistance group

Restrictions

- Maximum total dose on potatoes equivalent to four full dose treatments
- Do not apply if rainfall or irrigation is imminent. Product is rainfast in 1 h provided spray has dried on leaf
- Do not apply as a curative treatment when blight is present in the crop
- Do not apply more than 3 consecutive treatments of the product

Crop-specific information

- HI 7 d for potatoes
- All varieties of potatoes, including seed crops, may be treated

Environmental safety

- Dangerous for the environment
- Toxic to aquatic organisms

Hazard classification and safety precautions

Hazard Dangerous for the environment
Transport code 9
Packaging group III
UN number 3082
Risk phrases R51, R53a
Operator protection A, H; U05a, U08, U11, U19a, U20a
Environmental protection E15a, E38
Storage and disposal D01, D02, D09a, D10b, D12a

261 fluoxastrobin

A protectant stobilurin fungicide available in mixtures
FRAC mode of action code: 11

262 fluoxastrobin + prothioconazole

A strobilurin and triazole fungicide mixture for cereals
FRAC mode of action code: 11 + 3

See also prothioconazole

Products

1	Clayton Edge	Clayton	100:100 g/l	EC	12911
2	Fandango	Bayer CropScience	100:100 g/l	EC	12276
3	Firefly	Bayer CropScience	100:50 g/l	EC	13692
4	Kurdi	AgChem Access	100+100 g/l	EC	13531
5	Maestro	Bayer CropScience	100:100 g/l	EC	12307
6	Redigo Twin	Bayer CropScience	37.5:37.5 g/l	FS	12314
7	Redigo Twin TXC	Bayer CropScience	112.5;112.5 g/l	FS	14259

Uses

- Brown rust in ***spring barley***, ***winter barley*** [1, 2, 4, 5]; ***spring wheat*** [2-5]; ***winter rye***, ***winter wheat*** [1-5]
- Bunt in ***durum wheat***, ***spring oats***, ***spring rye***, ***spring wheat***, ***triticale***, ***winter oats***, ***winter rye***, ***winter wheat*** [7]; ***durum wheat*** *(seed treatment)*, ***spring rye*** *(seed treatment)*, ***triticale*** *(seed treatment)*, ***winter rye*** *(seed treatment)*, ***winter wheat*** *(seed treatment)* [6]
- Covered smut in ***spring oats***, ***winter oats*** [7]; ***winter oats*** *(seed treatment)* [6]
- Crown rust in ***spring oats***, ***winter oats*** [2, 4, 5]

- Eyespot in ***spring barley*** *(reduction)*, ***winter barley*** *(reduction)* [1, 2, 4, 5]; ***spring oats***, ***winter oats*** [2, 4, 5]; ***spring wheat*** [5]; ***spring wheat*** *(reduction)* [2-4]; ***winter rye*** *(reduction)*, ***winter wheat*** *(reduction)* [1-5]
- Foliar disease control in ***forest nurseries*** *(off-label)* [2]
- Fusarium foot rot and seedling blight in ***durum wheat*** *(seed treatment)*, ***spring rye*** *(seed treatment)*, ***triticale*** *(seed treatment)*, ***winter oats*** *(seed treatment)*, ***winter rye*** *(seed treatment)*, ***winter wheat*** *(seed treatment)* [6]
- Fusarium root rot in ***spring wheat*** *(reduction)*, ***winter wheat*** *(reduction)* [2, 4]
- Glume blotch in ***spring wheat*** [2-4]; ***spring wheat*** *(reduction)* [5]; ***winter wheat*** [1-5]
- Late ear diseases in ***spring barley***, ***winter barley***, ***winter wheat*** [1, 2, 4, 5]; ***spring wheat*** [2, 4, 5]
- Loose smut in ***durum wheat***, ***spring oats***, ***spring rye***, ***spring wheat***, ***triticale***, ***winter oats***, ***winter rye***, ***winter wheat*** [7]; ***durum wheat*** *(seed treatment)*, ***winter oats*** *(seed treatment)*, ***winter wheat*** *(seed treatment)* [6]
- Net blotch in ***spring barley***, ***winter barley*** [1, 2, 4, 5]
- Powdery mildew in ***spring barley***, ***winter barley*** [1, 2, 4, 5]; ***spring oats***, ***winter oats*** [2, 4, 5]; ***spring wheat*** [2-5]; ***winter rye***, ***winter wheat*** [1-5]
- Rhynchosporium in ***spring barley***, ***winter barley*** [1, 2, 4, 5]; ***winter rye*** [1-5]
- Seedling blight and foot rot in ***durum wheat***, ***spring oats***, ***spring rye***, ***spring wheat***, ***triticale***, ***winter oats***, ***winter rye***, ***winter wheat*** [7]
- Septoria leaf blotch in ***spring wheat*** [2-5]; ***winter wheat*** [1-5]
- Sharp eyespot in ***spring wheat*** *(reduction)*, ***winter wheat*** [2, 4]
- Sooty moulds in ***spring wheat*** *(reduction)*, ***winter wheat*** *(reduction)* [2, 4]
- Take-all in ***spring wheat*** *(reduction)*, ***winter barley*** *(reduction)*, ***winter wheat*** *(reduction)* [2, 4, 5]
- Tan spot in ***spring wheat*** [2, 4]; ***winter wheat*** [1, 2, 4, 5]
- Yellow rust in ***spring wheat*** [2-5]; ***winter wheat*** [1-5]

Specific Off-Label Approvals (SOLAs)

- ***forest nurseries*** *20090226* [2]

Approval information

- Fluoxastrobin and prothioconazole included in Annex I under EC Directive 91/414
- Accepted by BBPA for use on malting barley

Efficacy guidance

- Seed treatments must be applied by manufacturer's recommended treatment application equipment [6]
- Treated cereal seed should preferably be drilled in the same season [6]
- Follow-up treatments will be needed later in the season to give protection against air-borne and splash-borne diseases [6]
- Best results on foliar diseases obtained from treatment at early stages of disease development. Further treatment may be needed if disease attack is prolonged [1, 2, 5]
- Foliar applications to established infections of any disease are likely to be less effective [1, 2, 5]
- Best control of cereal ear diseases obtained by treatment during ear emergence [1, 2, 5]
- Fluoxastrobin is a member of the QoI cross resistance group. Foliar product should be used preventatively and not relied on for its curative potential [1, 2, 5]
- Use product as part of an Integrated Crop Management strategy incorporating other methods of control, including where appropriate other fungicides with a different mode of action. Do not apply more than two foliar applications of QoI containing products to any cereal crop
- There is a significant risk of widespread resistance occurring in *Septoria tritici* populations in UK. Failure to follow resistance management action may result in reduced levels of disease control
- Strains of wheat and barley powdery mildew resistant to QoIs are common in the UK. Control of wheat mildew can only be relied on from the triazole component
- Prothioconazole is a DMI fungicide. Resistance to some DMI fungicides has been identified in Septoria leaf blotch which may seriously affect performance of some products. For further advice contact a specialist advisor and visit the Fungicide Resistance Action Group (FRAG)-UK website
- Where specific control of wheat mildew is required this should be achieved through a programme of measures including products recommended for the control of mildew that contain a fungicide from a different cross-resistance group and applied at a dose that will give robust control

Restrictions

- Maximum number of seed treatments one per batch [6]

- Maximum total dose of foliar sprays equivalent to two full dose treatments for the crop [1-3, 5]
- Seed treatment must be fully re-dispersed and homogeneous before use [6]
- Do not use on seed with more than 16% moisture content, or on sprouted, cracked or skinned seed [6]
- All seed batches should be tested to ensure they are suitable for treatment [6]

Crop-specific information

- Latest use: pre-drilling for seed treatment [6]; before grain milky ripe for spray treatments on wheat and rye; beginning of flowering for barley, oats [1, 2, 5]
- Some transient leaf chlorosis may occur after treatment of wheat or barley but this has not been found to affect yield [1, 2, 5]

Environmental safety

- Dangerous for the environment
- Toxic to aquatic organisms
- Risk to non-target insects or other arthropods. Avoid spraying within 6 m of the field boundary to reduce the effects on non-target insects or other arthropods [1, 2, 5]
- LERAP Category B [1, 2, 5]

Hazard classification and safety precautions

Hazard Irritant [6]; Dangerous for the environment [1-7]
Transport code 9
Packaging group III
UN number 3082
Risk phrases R43 [6]; R51, R53a [1-7]
Operator protection A [1-6]; H [6, 7]; U05a [1-7]; U09a, U20b [3]; U09b, U20a [1, 2, 4, 5]; U14 [6]; U20c, U25 [7]
Environmental protection E03 [6]; E15a, E34 [1-7]; E16a [1-5]; E22c [1, 2, 4, 5]; E38 [2-7]
Storage and disposal D01, D12a [1-7]; D02 [1, 6, 7]; D05, D09a [1-6]; D09b, D10d, D14 [7]; D10b [1-5]; D11a [6]
Treated seed S01, S02, S03, S04b, S05, S07, S08 [6, 7]; S04a, S06a [6]; S06b [7]
Medical advice M03 [1-5]

263 fluoxastrobin + prothioconazole + trifloxystrobin

A triazole and strobilurin fungicide mixture for cereals
FRAC mode of action code: 11 + 3 + 11

See also *prothioconazole*
trifloxystrobin

Products

Jaunt	Bayer CropScience	75:150:75 g/l	EC	12350

Uses

- Brown rust in ***spring barley***, ***winter barley***, ***winter wheat***
- Eyespot in ***spring barley*** *(reduction)*, ***winter barley*** *(reduction)*, ***winter wheat*** *(reduction)*
- Glume blotch in ***winter wheat***
- Late ear diseases in ***spring barley***, ***winter barley***, ***winter wheat***
- Net blotch in ***spring barley***, ***winter barley***
- Powdery mildew in ***spring barley***, ***winter barley***, ***winter wheat***
- Rhynchosporium in ***spring barley***, ***winter barley***
- Septoria leaf blotch in ***winter wheat***
- Tan spot in ***winter wheat***
- Yellow rust in ***winter wheat***

Approval information

- Fluoxastrobin, prothioconazole and trifloxystrobin included in Annex I under EC Directive 91/414
- Accepted by BBPA for use on malting barley

Efficacy guidance

- Best results obtained from treatment at early stages of disease development. Further treatment may be needed if disease attack is prolonged
- Applications to established infections of any disease are likely to be less effective

- Best control of cereal ear diseases obtained by treatment during ear emergence
- Fluoxastrobin and trifloxystrobin are members of the QoI cross resistance group. Product should be used preventatively and not relied on for its curative potential
- Use product as part of an Integrated Crop Management strategy incorporating other methods of control, including where appropriate other fungicides with a different mode of action. Do not apply more than two foliar applications of QoI containing products to any cereal crop
- There is a significant risk of widespread resistance occurring in *Septoria tritici* populations in UK. Failure to follow resistance management action may result in reduced levels of disease control
- Strains of wheat and barley powdery mildew resistant to QoIs are common in the UK. Control of wheat mildew can only be relied on from the triazole component
- Where specific control of wheat mildew is required this should be achieved through a programme of measures including products recommended for the control of mildew that contain a fungicide from a different cross-resistance group and applied at a dose that will give robust control
- Prothioconazole is a DMI fungicide. Resistance to some DMI fungicides has been identified in Septoria leaf blotch which may seriously affect performance of some products. For further advice contact a specialist advisor and visit the Fungicide Resistance Action Group (FRAG)-UK website

Restrictions
- Maximum total dose equivalent to two full dose treatments

Crop-specific information
- Latest use: before grain milky ripe for winter wheat; up to beginning of anthesis (GS 61) for barley

Environmental safety
- Dangerous for the environment
- Very toxic to aquatic organisms
- Risk to non-target insects or other arthropods. Avoid spraying within 6 m of the field boundary to reduce the effects on non-target insects or other arthropods
- LERAP Category B

Hazard classification and safety precautions

Hazard Irritant, Dangerous for the environment
Transport code 9
Packaging group III
UN number 3082
Risk phrases R37, R50, R53a
Operator protection A, C, H; U05a, U09b, U19a, U20b
Environmental protection E15a, E16a, E22c, E34, E38
Storage and disposal D01, D02, D05, D09a, D10b, D12a
Medical advice M03

264 flupyrsulfuron-methyl

A sulfonylurea herbicide for winter wheat and winter barley
HRAC mode of action code: B

See also *carfentrazone-ethyl + flupyrsulfuron-methyl*
diflufenican + flupyrsulfuron-methyl

Products

Lexus SX	DuPont	50% w/w	WG	12979

Uses
- Annual dicotyledons in ***winter barley***, ***winter linseed*** *(off-label)*, ***winter wheat***
- Blackgrass in ***winter barley***, ***winter linseed*** *(off-label)*, ***winter wheat***

Specific Off-Label Approvals (SOLAs)
- ***winter linseed*** *20063714*

Approval information
- Flupyrsulfuron-methyl included in Annex I under EC Directive 91/414

Efficacy guidance
- Best results achieved from applications made in good growing conditions
- Good spray cover of weeds must be obtained

SEE SECTION 3 FOR PRODUCTS ALSO REGISTERED

- For pre-emergence control of blackgrass application must be made in tank mixture with specified products (see label). Best control of blackgrass post-emergence in wheat obtained from application from the 1-leaf stage
- Winter barley must be treated pre-emergence and used in mixture. See label
- Growth of weeds is inhibited within hours of treatment but visible symptoms may not be apparent for up to 4 wk
- Flupyrsulfuron-methyl has moderate residual life in soil. Under normal moisture conditions susceptible weeds germinating soon after treatment will be controlled
- May be used on all soil types but residual activity and weed control is reduced on highly alkaline soils
- Flupyrsulfuron-methyl is a member of the ALS-inhibitor group of herbicides. To avoid the build up of resistance do not use any product containing an ALS-inhibitor herbicide with claims for control of grass weeds more than once on any crop
- Use these products as part of a resistance management strategy that includes cultural methods of control and does not use ALS inhibitors as the sole chemical method of grass weed control

Restrictions

- Maximum number of treatments 1 per crop
- Do not use on wheat undersown with grasses, clover or other legumes, or any other broad-leaved crop
- Do not apply within 7 d of rolling
- Do not treat any crop suffering from drought, waterlogging, pest or disease attack, nutrient deficiency, or any other stress factors
- Specific restrictions apply to use in sequence or tank mixture with other sulfonylurea or ALS-inhibiting herbicides. See label for details

Crop-specific information

- Latest use: pre-emergence or before 1st node detectable (GS 31) for wheat; pre-emergence for barley
- Under certain conditions chlorosis and stunting may occur, from which recovery is rapid

Following crops guidance

- Only cereals, oilseed rape, field beans, clover or grass may be sown in the yr of harvest of a treated crop
- In the event of crop failure only winter or spring wheat may be sown within 3 mth of treatment. Land should be ploughed and cultivated to 15 cm minimum before resowing. After 3 mth crops may be sown as shown above

Environmental safety

- Dangerous for the environment
- Very toxic to aquatic organisms
- Take extreme care to avoid drift onto broad-leaved plants outside the target area or onto ponds, waterways or ditches, or onto land intended for cropping
- Spraying equipment should not be drained or flushed onto land planted, or to be planted, with trees or crops other than cereals and should be thoroughly cleansed after use - see label for instructions

Hazard classification and safety precautions

Hazard Dangerous for the environment
Transport code 9
Packaging group III
UN number 3077
Risk phrases R50, R53a
Operator protection U08, U19a, U20b
Environmental protection E15a, E38
Storage and disposal D09a, D11a, D12a

265 flupyrsulfuron-methyl + thifensulfuron-methyl

A sulfonylurea herbicide mixture for winter wheat and winter oats
HRAC mode of action code: B + B

See also *thifensulfuron-methyl*

Products

1	Lancer	Headland	10:40% w/w	WG	13031
2	Lexus Millenium	DuPont	10:40% w/w	WG	09206

Uses

- Annual dicotyledons in ***winter oats, winter wheat***
- Blackgrass in ***winter oats, winter wheat***

Approval information

- Flupyrsulfuron-methyl and thifensulfuron-methyl included in Annex I under EC Directive 91/414

Efficacy guidance

- Best results obtained when applied to small actively growing weeds
- Good spray cover of weeds must be obtained
- Increased degradation of active ingredient in high soil temperatures reduces residual activity
- Product has moderate residual life in soil. Under normal moisture conditions susceptible weeds germinating soon after treatment will be controlled
- Product may be used on all soil types but residual activity and weed control is reduced on highly alkaline soils
- Blackgrass should be treated from 1 leaf up to first node stage
- Flupyrsulfuron-methyl and thifensulfuron-methyl are members of the ALS-inhibitor group of herbicides. To avoid the build up of resistance do not use any product containing an ALS-inhibitor herbicide with claims for control of grass weeds more than once on any crop
- Use these products as part of a resistance management strategy that includes cultural methods of control and does not use ALS inhibitors as the sole chemical method of grass weed control

Restrictions

- Maximum number of treatments 1 per crop
- Do not use on wheat undersown with grasses or legumes, or any other broad-leaved crop
- Do not apply within 7 d of rolling
- Do not treat any crop suffering from drought, waterlogging, pest or disease attack, nutrient deficiency, or any other stress factors
- Contact contract agents before use on crops grown for seed
- Specific restrictions apply to use in sequence or tank mixture with other sulfonylurea or ALS-inhibiting herbicides. See label for details

Crop-specific information

- Latest use: before 1st node detectable (GS 31) on wheat; before 31 Dec in yr of sowing on oats [2]
- Slight chlorosis, speckling and stunting may occur in certain conditions. Recovery is rapid and yield not affected

Following crops guidance

- Only cereals, oilseed rape, field beans or grass may be sown in the yr of harvest of a treated crop. Any crop may be sown the following spring
- In the event of crop failure only winter wheat may be sown before normal harvest date. Land should be ploughed and cultivated to 15 cm minimum before resowing

Environmental safety

- Dangerous for the environment
- Very toxic to aquatic organisms
- LERAP Category B

Hazard classification and safety precautions

Hazard Irritant, Dangerous for the environment
Transport code 9
Packaging group III
UN number 3077
Risk phrases R43, R50, R53a

SEE SECTION 3 FOR PRODUCTS ALSO REGISTERED

Operator protection A, H; U05a, U08, U14, U19a, U20b
Environmental protection E15a, E16a, E38
Storage and disposal D01, D02, D09a, D11a, D12a

266 flupyrsulfuron-methyl + tribenuron-methyl

A foliar applied sulfonylurea herbicide mixture; product withdrawn 2008
HRAC mode of action code: B + B

See also tribenuron-methyl

267 fluquinconazole

A protectant, eradicant and systematic triazole fungicide for winter cereals
FRAC mode of action code: 3

Products

1	Flamenco	BASF	100 g/l	SC	11699
2	Galmano	Bayer CropScience	167 g/l	FS	11650
3	Jockey Solo	BASF	167 g/l	FS	14281

Uses

- Brown rust in ***winter wheat*** [1]
- Bunt in ***winter wheat*** *(seed treatment)* [2]
- Foliar disease control in ***grass seed crops*** *(off-label)*, ***triticale*** *(off-label)* [1]
- Glume blotch in ***winter wheat*** [1]
- Powdery mildew in ***winter wheat*** [1]
- Seed-borne diseases in ***durum wheat*** *(off-label)*, ***triticale*** *(off-label)* [2]; ***winter barley***, ***winter wheat*** [3]
- Septoria leaf blotch in ***winter wheat*** [1]; ***winter wheat*** *(seed treatment)* [2]
- Take-all in ***winter wheat*** *(seed treatment - reduction)* [2]
- Yellow rust in ***winter wheat*** [1]; ***winter wheat*** *(seed treatment)* [2]

Specific Off-Label Approvals (SOLAs)

- ***durum wheat*** *20063065* [2]
- ***grass seed crops*** *20063472* [1]
- ***triticale*** *20063472* [1], *20063065* [2]

Approval information

- Fluquinconazole not included in Annex 1 but the date of withdrawal has yet to be set.
- Accepted by BBPA on malting barley before ear emergence only

Efficacy guidance

- Best results obtained from spray treatments when disease first becomes active in crop but before infection spreads to younger leaves [1]
- Adequate disease protection throughout the season will usually require a programme of at least two fungicide treatments [1]
- May also be applied as a protectant at end of ear emergence (but no later) if crop still disease free [1]
- With seed treatments ensure good even coverage of seed to obtain reliable disease control [2]
- Seed treatment provides early control of Septoria leaf blotch and yellow rust but may require foliar treatment for later infection [2]
- Fluquinconazole is a DMI fungicide. Resistance to some DMI fungicides has been identified in Septoria leaf blotch which may seriously affect performance of some products. For further advice contact a specialist advisor and visit the Fungicide Resistance Action Group (FRAG)-UK website

Restrictions

- Maximum total spray dose equivalent to two full doses [1]
- Maximum number of seed treatments 1 per batch [2]
- Do not apply spray treatments after the start of anthesis (GS 59) [1]
- Do not mix with any other products [2]
- Do not treat cracked, split or sprouted seed [2]
- Do not use treated seed as food or feed [2]

Crop-specific information

- Latest use: before beginning of anthesis (GS 59) for spray treatment; before drilling for seed treatment
- Ensure good spray coverage and increase volume in dense crops [1]
- Delayed emergence may result when treated seed is drilled into heavy or poorly drained soils which then become wet or waterlogged [2]

Environmental safety

- Dangerous for the environment
- Toxic to aquatic organisms
- Dangerous to fish or other aquatic life. Do not contaminate surface waters or ditches with chemical or used container
- Special PPE requirements and precautions apply where product supplied in returnable packs. Check label

Hazard classification and safety precautions

Hazard Toxic, Dangerous for the environment

Transport code 9

Packaging group III

UN number 3082

Risk phrases R22a, R51, R53a [1-3]; R25b [3]; R36, R43 [1]; R48 [1, 2]

Operator protection A, H [1-3]; C, M [1]; D [2, 3]; U05a [1-3]; U07 [2]; U11, U14, U15 [1]; U20b [2, 3]; U25 [3]

Environmental protection E13b, E36a [2]; E15a [1, 3]; E34, E38 [1-3]

Storage and disposal D01, D02, D12a [1-3]; D05 [1, 2]; D08 [2]; D09a, D14 [2, 3]; D10c [1]; D10d [3]

Treated seed S01, S03, S04b, S06a [2]; S02, S05, S07, S08 [2, 3]

Medical advice M03 [1, 2]; M04a [1, 3]

268 fluquinconazole + prochloraz

A broad-spectrum triazole mixture for use as a spray and seed treatment in winter cereals

FRAC mode of action code: 3 + 3

See also *prochloraz*

Products

1	Epona	BASF	167:31 g/l	FS	12444
2	Foil	BASF	54:174 g/l	SE	11700
3	Galmano Plus	Bayer CropScience	167:31 g/l	FS	11645
4	Jockey	BASF	167:31 g/l	FS	11689

Uses

- Brown rust in ***winter barley***, ***winter wheat*** [2]; ***winter wheat*** *(seed treatment - moderate control)* [1]
- Bunt in ***winter wheat*** *(seed treatment)* [1, 3, 4]
- Covered smut in ***winter barley*** *(seed treatment)* [4]
- Foliar disease control in ***durum wheat*** *(off-label)*, ***grass seed crops*** *(off-label)*, ***spring rye*** *(off-label)*, ***triticale*** *(off-label)*, ***winter rye*** *(off-label)* [2]
- Fusarium foot rot and seedling blight in ***winter barley*** *(seed treatment)* [4]
- Fusarium root rot in ***winter wheat*** *(seed treatment)* [1, 3, 4]
- Loose smut in ***winter barley*** *(seed treatment)* [4]
- Powdery mildew in ***winter barley***, ***winter wheat*** *(moderate control)* [2]
- Rhynchosporium in ***winter barley*** [2]
- Seed-borne diseases in ***durum wheat*** *(off-label)*, ***triticale*** *(off-label)* [3]
- Septoria leaf blotch in ***winter wheat*** [2]; ***winter wheat*** *(seed treatment)* [3, 4]; ***winter wheat*** *(seed treatment - moderate control)* [1]
- Take-all in ***winter barley*** *(seed treatment - reduction)* [4]; ***winter wheat*** *(seed treatment - reduction)* [1, 3, 4]
- Yellow rust in ***winter barley*** *(seed treatment)* [4]; ***winter wheat*** [2]; ***winter wheat*** *(seed treatment)* [3, 4]; ***winter wheat*** *(seed treatment - moderate control)* [1]

Specific Off-Label Approvals (SOLAs)

- ***durum wheat*** *20060982* [2], *20061093* [3]
- ***grass seed crops*** *20060982* [2]
- ***spring rye*** *20060982* [2]
- ***triticale*** *20060982* [2], *20061093* [3]
- ***winter rye*** *20060982* [2]

Approval information

- Accepted by BBPA for use on malting barley (before ear emergence only)

Efficacy guidance

- Best results obtained from spray treatment when disease first becomes active in crop but before infection spreads to younger leaves [2]
- When treating seed ensure good even coverage of seed to obtain reliable disease control [1, 3, 4]
- Seed treatment provides early control of *Septoria* leaf spot and rust but may require foliar treatment for later infection [1, 3, 4]
- Fluquinconazole and prochloraz are DMI fungicides. Resistance to some DMI fungicides has been identified in Septoria leaf blotch which may seriously affect performance of some products. For further advice contact a specialist advisor and visit the Fungicide Resistance Action Group (FRAG)-UK website

Restrictions

- Maximum number of treatments 1 per seed batch (seed dressing)
- Maximum total dose equivalent to two full doses (spray)
- Do not treat cracked, split or sprouted seed [1, 3, 4]
- Do not use treated seed as food or feed [1, 3, 4]

Crop-specific information

- Latest use: before 1st awns visible (GS 47) for winter barley; before beginning of anthesis (GS 59) for winter wheat; before drilling (seed treatment)
- Ensure good spray coverage and increase volume in dense crops [2]
- Delayed emergence may result when treated seed is drilled into heavy or poorly drained soils which then become wet or waterlogged [1, 3, 4]

Environmental safety

- Dangerous for the environment
- Toxic to aquatic organisms
- Dangerous to fish or other aquatic life. Do not contaminate surface waters or ditches with chemical or used container
- Product supplied in returnable packs for which special PPE requirements and precautions apply. Check label

Hazard classification and safety precautions

Hazard Toxic [1, 3, 4]; Harmful, Irritant [2]; Dangerous for the environment [1-4]

Transport code 9

Packaging group III

UN number 3082

Risk phrases R22a, R48, R51, R53a [1-4]; R36 [2, 3]; R43, R66 [2]

Operator protection A, H [1-4]; C [2]; D [3, 4]; G [1]; U05a [1-4]; U07 [3, 4]; U20b [1, 3, 4]

Environmental protection E13b [3]; E15a [1, 2, 4]; E34, E38 [1-4]; E36a [3, 4]

Storage and disposal D01, D02, D09a, D12a [1-4]; D05 [2, 3]; D08 [1, 3, 4]; D10c [2]; D11a [1, 4]; D14 [3, 4]

Treated seed S01, S03, S04b, S06a [3, 4]; S02, S05, S07, S08 [1, 3, 4]

Medical advice M03 [2, 3]; M04a [1, 4]; M05b [2]

269 fluroxypyr

A post-emergence pyridinecarboxylic acid herbicide
HRAC mode of action code: O

See also *2,4-D + dicamba + fluroxypyr*
aminopyralid + fluroxypyr
clopyralid + florasulam + fluroxypyr
clopyralid + fluroxypyr + MCPA
clopyralid + fluroxypyr + triclopyr
florasulam + fluroxypyr

SECTION 2

Products

1	Agriguard Fluroxypyr	AgriGuard	200 g/l	EC	12525
2	Agriguard Fluroxypyr	AgriGuard	200 g/l	EC	13335
3	Awac	Globachem	200 g/l	EC	13296
4	Barclay Hudson	Barclay	180 g/l	EC	13606
5	Barclay Hurler	Barclay	200 g/l	EC	13458
6	Casino	AgriGuard	25% w/w	WG	14352
7	Crescent	AgriGuard	200 g/l	EC	12559
8	Crescent	AgriGuard	200 g/l	EC	13330
9	Flurostar 200	Globachem	200 g/l	EC	12903
10	Fluxyr 200 EC	AgriChem BV	200 g/l	EC	14086
11	Greencrop Reaper	Greencrop	200 g/l	EC	12261
12	Hatchet	AgriGuard	200 g/l	EC	12524
13	Hatchet Xtra	AgriGuard	200 g/l	EC	13213
14	Klever	Globachem	200 g/l	EC	13295
15	Minstrel	United Phosphorus	200 g/l	EC	13745
16	Skora	AgChem Access	200 g/l	EC	13422
17	Standon Homerun 2	Standon	200 g/l	EC	13890
18	Starane 2	Dow	200 g/l	EC	12018
19	Tandus	Nufarm UK	200 g/l	EC	13432
20	Tomahawk	Makhteshim	200 g/l	EC	09249

Uses

- Annual and perennial weeds in ***durum wheat, forage maize, permanent grassland, rotational grass, spring barley, spring oats, spring rye, spring wheat, triticale, winter barley, winter oats, winter rye, winter wheat*** [17]
- Annual dicotyledons in ***bulb onions*** *(off-label)*, ***farm forestry*** *(off-label)*, ***forest nurseries*** *(off-label)*, ***garlic*** *(off-label)*, ***leeks*** *(off-label)*, ***ornamental plant production*** *(off-label)*, ***shallots*** *(off-label)* [18]; ***durum wheat, forage maize, spring barley, spring oats, spring wheat, triticale, winter barley, winter oats, winter rye, winter wheat*** [1-16, 18-20]; ***game cover*** *(off-label)* [18, 20]; ***grassland*** [10, 15]; ***miscanthus*** *(off-label)* [1, 18]; ***newly sown grass leys*** [17, 19]; ***permanent grassland, rotational grass*** [1-9, 11-14, 16, 18-20]; ***spring rye*** [1-9, 11-16, 18, 20]
- Black bindweed in ***bulb onions*** *(off-label)*, ***garlic*** *(off-label)*, ***shallots*** *(off-label)* [18]; ***durum wheat, spring barley, spring oats, spring wheat, triticale, winter barley, winter oats, winter rye, winter wheat*** [1-9, 11-20]; ***forage maize, spring rye*** [1-9, 11-18, 20]; ***grassland*** [15]; ***newly sown grass leys*** [17]; ***permanent grassland, rotational grass*** [1-9, 11-14, 16-18, 20]
- Chickweed in ***bulb onions*** *(off-label)*, ***garlic*** *(off-label)*, ***shallots*** *(off-label)* [18]; ***durum wheat, forage maize, spring barley, spring oats, spring wheat, triticale, winter barley, winter oats, winter rye, winter wheat*** [1-9, 11-20]; ***grassland*** [15]; ***newly sown grass leys*** [17, 19]; ***permanent grassland, rotational grass*** [1-9, 11-14, 16-20]; ***spring rye*** [1-9, 11-18, 20]
- Cleavers in ***apple orchards*** *(off-label)*, ***bulb onions*** *(off-label)*, ***garlic*** *(off-label)*, ***pear orchards*** *(off-label)*, ***poppies for morphine production*** *(off-label)*, ***shallots*** *(off-label)* [18]; ***durum wheat, spring barley, spring oats, spring wheat, triticale, winter barley, winter oats, winter rye, winter wheat*** [1-9, 11-20]; ***forage maize, spring rye*** [1-9, 11-18, 20]; ***grassland*** [15]; ***millet*** *(off-label - grown for wild bird seed production)* [20]; ***newly sown grass leys*** [17]; ***permanent grassland, rotational grass*** [1-9, 11-14, 16-18, 20]
- Docks in ***apple orchards*** *(off-label)*, ***pear orchards*** *(off-label)* [18]; ***durum wheat, spring barley, spring oats, spring wheat, triticale, winter barley, winter oats, winter rye, winter wheat*** [1-9, 11-20]; ***forage maize, spring rye*** [1-9, 11-18, 20]; ***permanent grassland, rotational grass*** [1-9, 11-14, 16-18, 20]

- Forget-me-not in ***bulb onions*** *(off-label)*, ***garlic*** *(off-label)*, ***shallots*** *(off-label)* [18]; ***durum wheat, spring barley, spring oats, spring wheat, triticale, winter barley, winter oats, winter rye, winter wheat*** [1-9, 11-20]; ***forage maize, spring rye*** [1-9, 11-18, 20]; ***grassland*** [15]; ***newly sown grass leys*** [17]; ***permanent grassland, rotational grass*** [1-9, 11-14, 16-18, 20]
- Hemp-nettle in ***durum wheat, spring barley, spring oats, spring wheat, triticale, winter barley, winter oats, winter rye, winter wheat*** [1-9, 11-20]; ***forage maize, spring rye*** [1-9, 11-18, 20]; ***grassland*** [15]; ***newly sown grass leys*** [17]; ***permanent grassland, rotational grass*** [1-9, 11-14, 16-18, 20]
- Poppies in ***poppies for morphine production*** *(off-label)* [18]
- Stinging nettle in ***apple orchards*** *(off-label)*, ***pear orchards*** *(off-label)* [18]
- Volunteer potatoes in ***bulb onions*** *(off-label)*, ***leeks*** *(off-label)*, ***poppies for morphine production*** *(off-label)*, ***sweetcorn*** *(off-label)* [18]; ***durum wheat, forage maize, permanent grassland, rotational grass, spring rye, triticale, winter oats, winter rye*** [17]; ***spring barley, spring oats, spring wheat*** [17, 19]; ***winter barley, winter wheat*** [1-9, 11-20]

Specific Off-Label Approvals (SOLAs)

- ***apple orchards*** *20040988 expires 30 Nov 2010* [18]
- ***bulb onions*** *20040986 expires 30 Nov 2010* [18], *20081404* [18]
- ***farm forestry*** *20082925* [18]
- ***forest nurseries*** *20082925* [18]
- ***game cover*** *20082925* [18], *20082907* [20]
- ***garlic*** *20081404* [18]
- ***leeks*** *20040987 expires 30 Nov 2010* [18]
- ***millet*** *(grown for wild bird seed production) 20080936* [20]
- ***miscanthus*** *20082803* [1], *20082925* [18]
- ***ornamental plant production*** *20082925* [18]
- ***pear orchards*** *20040988 expires 30 Nov 2010* [18]
- ***poppies for morphine production*** *20052059 expires 30 Nov 2010* [18]
- ***shallots*** *20081404* [18]
- ***sweetcorn*** *20051696 expires 30 Nov 2010* [18]

Approval information

- Fluroxypyr included in Annex I under EC Directive 91/414
- Accepted by BBPA for use on malting barley

Efficacy guidance

- Best results achieved under good growing conditions in a strongly competing crop
- A number of tank mixtures with HBN and other herbicides are recommended for use in autumn and spring to extend range of species controlled. See label for details
- Spray is rainfast in 1 h

Restrictions

- Maximum number of treatments 1 per crop or yr or maximum total dose equivalent to one full dose treatment
- Do not apply in any tank-mix on triticale or forage maize
- Do not use on crops undersown with clovers or other legumes
- Do not treat crops suffering stress caused by any factor
- Do not roll or harrow for 7 d before or after treatment
- Do not spray if frost imminent
- Do not apply before 1 Mar in yr of harvest [19]
- Do not use on crops grown for seed production [19]
- Straw from treated crops must not be returned directly to the soil but must be removed and used only for livestock bedding

Crop-specific information

- Latest use: before flag leaf sheath opening (GS 47) for winter wheat and barley; before flag leaf sheath extending (GS 41) for spring wheat and barley; before second node detectable (GS 32) for oats, rye, triticale and durum wheat; before 7 leaves unfolded and before buttress roots appear for maize
- Apply to new leys from 3 expanded leaf stage
- Timing varies in tank mixtures. See label for details
- Crops undersown with grass may be sprayed provided grasses are tillering

Following crops guidance

- Clovers, peas, beans and other legumes must not be sown for 12 mth following treatment at the highest dose

Environmental safety

- Dangerous for the environment
- Vey toxic to aquatic organisms
- Flammable
- Keep livestock out of treated areas for at least 3 d following treatment and until poisonous weeds, such as ragwort, have died down and become unpalatable
- Wash spray equipment thoroughly with water and detergent immediately after use. Traces of product can damage susceptible plants sprayed later

Hazard classification and safety precautions

Hazard Harmful [1-5, 7-20]; Irritant [6]; Flammable [1-5, 7-18, 20]; Dangerous for the environment [1-20]
Transport code 3 [1-5, 7-9, 11-20]; 9 [6, 10]
Packaging group III
UN number 1933 [18]; 1993 [1-5, 7-9, 11-17, 19, 20]; 3077 [6]; 3082 [10]
Risk phrases R22b, R67 [1-5, 7-20]; R36 [1-3, 6-9, 12-14]; R37 [1-5, 7-10, 12-18, 20]; R38 [1-5, 7-9, 12-14]; R43 [1-5, 7-9, 12-14, 19]; R50 [2, 3, 6, 8, 9, 13, 14]; R51 [1, 4, 5, 7, 10-12, 15-20]; R53a [1-20]
Operator protection A [1-9, 12-15, 17, 19]; C [1-3, 6-9, 12-15, 17]; H [1-5, 7-9, 12-15, 17, 19]; U05a [1-9, 12-14]; U08, U19a [1-20]; U11 [1-3, 6-9, 12-14]; U14 [1-5, 7-9, 12-14, 19]; U20b [1-18, 20]
Environmental protection E06a [6, 8] (3 d); E07a [1-3, 7, 9-20]; E07c [4, 5] (14 d); E15a [10, 11, 15-18, 20]; E15b [1-9, 12-14, 19]; E34 [1-9, 12-14, 20]; E38 [2-5, 8-10, 13-20]
Storage and disposal D01, D02 [1-9, 12-14]; D05 [1-5, 8, 9, 11, 13, 14, 20]; D09a [1-20]; D10a [4, 5]; D10b [1-3, 6-20]; D12a [1-5, 7-10, 12-20]
Treated seed S06a [19]
Medical advice M03 [1-9, 12-14]; M05b [1-5, 7-20]

270 fluroxypyr + mecoprop-P

A post-emergence herbicide for broadleaved weeds in amenity turf
HRAC mode of action code: O + O

See also *mecoprop-P*

271 fluroxypyr + thifensulfuron-methyl + tribenuron-methyl

A foliar applied herbicide mixture for cereals with some residual activity
HRAC mode of action code: O + B + B

See also *thifensulfuron-methyl*
tribenuron-methyl

Products

GEX 353	DuPont	200 g/l + 50.25% w/w	KK	12755

Uses

- Annual dicotyledons in ***spring barley***, ***spring wheat***, ***winter barley***, ***winter wheat***

Approval information

- Fluroxypyr, thifensulfuron-methyl and tribenuron-methyl included in Annex I under EC Directive 91/414
- Accepted by BBPA for use on malting barley

Efficacy guidance

- Best results obtained when weeds small and actively growing. This is particularly important for cleavers
- Ensure good spray cover
- Susceptible species cease growth immediately after application but may take 2 wk to show symptoms
- Weed control may be reduced in very dry conditions

- Rain within 4 hr may reduce effectiveness
- Thifensulfuron-methyl and tribenuron-methyl are members of the ALS-inhibitor group of herbicides and products should be used in a planned Resistance Management strategy. See Section 5 for more information

Restrictions

- Maximum number of treatments 1 per crop for all cereals
- Must only be applied after 1 Feb in yr of harvest and should not be applied more than once to any cereal crop
- Do not use on any crop suffering stress from drought, waterlogging, frost, deficiency, pest or disease attack or any other cause
- Do not use on crops undersown with grasses, clover or legumes, or any other broad leaved crop
- Specific restrictions apply to use in sequence or tank mixture with other sulfonylurea or ALS-inhibiting herbicides. See label for details
- Do not apply within 7 d of rolling

Crop-specific information

- Latest use: before flag leaf extending (GS 39) for all crops

Following crops guidance

- Only cereals, field beans or oilseed rape may be sown in the same calendar yr as harvest of a treated crop
- In the event of failure of a treated crop only a cereal crop may be sown within 3 mth

Environmental safety

- Dangerous for the environment
- Very toxic to aquatic organisms
- Extremely dangerous to fish or other aquatic life. Do not contaminate surface waters or ditches with chemical or used container
- Keep livestock out of treated areas for at least 3 days and until foliage of any poisonous weeds such as ragwort has died and become unpalatable
- Take extreme care to avoid damage by drift onto broad-leaved plants outside the target area, or onto ponds, waterways or ditches
- Spraying equipment should not be drained or flushed onto land planted, or to be planted, with trees or crops other than cereals and should be thoroughly cleansed after use - see label for instructions

Hazard classification and safety precautions

Hazard Harmful, Flammable, Dangerous for the environment
Transport code 3
Packaging group II
UN number 1993
Risk phrases R22b, R37, R43, R50, R53a, R67
Operator protection A, C, H; U08, U19a, U20b
Environmental protection E07c (3 d); E13a, E16a, E16b, E38
Storage and disposal D05, D09a, D10b, D12a
Medical advice M05b

272 fluroxypyr + triclopyr

A foliar acting herbicide for docks in grassland
HRAC mode of action code: O + O

See also *triclopyr*

Products

1	Dovekie	AgChem Access	100:100 g/l	EC	14121
2	Doxstar	Dow	100:100 g/l	EC	11063

Uses

- Chickweed in ***rotational grass***
- Docks in ***permanent grassland***, ***rotational grass***

Approval information

- Fluroxypyr and triclopyr included in Annex I under EC Directive 91/414

Efficacy guidance

- Seedling docks in new leys are controlled up to 50 mm diameter. In established grass apply in spring or autumn or, at lower dose, in spring and autumn on docks up to 200 mm. A second application in the subsequent yr may be needed
- Allow 2-3 wk after cutting or grazing to allow sufficient regrowth of docks to occur before spraying
- Control may be reduced if rain falls within 2 h of application
- To allow maximum translocation to the roots of docks do not cut grass for 28 d after spraying

Restrictions

- Maximum total dose equivalent to one full dose treatment
- Do not roll or harrow for 10 d before or 7 d after spraying
- Do not spray in drought, very hot or very cold weather

Crop-specific information

- Latest use: 7 d before grazing or harvest of grass
- Grass less than one yr old may be treated at half dose from the third leaf visible stage
- Clover will be killed or severely checked by treatment

Following crops guidance

- Do not sow kale, turnips, swedes or grass mixtures containing clover by direct drilling or minimum cultivation techniques within 6 wk of application

Environmental safety

- Dangerous for the environment
- Toxic to aquatic organisms
- Keep livestock out of treated areas for at least 7 d following treatment and until poisonous weeds, such as ragwort, have died down and become unpalatable
- Do not allow drift to come into contact with crops, amenity plantings, gardens, ponds, lakes or watercourses
- Wash spray equipment thoroughly with water and detergent immediately after use. Traces of product can damage susceptible plants sprayed later

Hazard classification and safety precautions

Hazard Harmful, Flammable, Dangerous for the environment
Transport code 3
Packaging group III
UN number 1993
Risk phrases R22a, R22b, R37, R38, R43, R51, R53a, R67
Operator protection A, C; U02a, U05a, U08, U14, U19a, U20b
Environmental protection E07a, E15a, E34, E38
Consumer protection C01
Storage and disposal D01, D02, D05, D09a, D10b, D12a
Medical advice M05b

273 flurtamone

A carotenoid synthesis inhibitor available only in mixtures
HRAC mode of action code: F1

See also diflufenican + flurtamone

274 flusilazole

A systemic, protective and curative triazole fungicide for cereals, oilseed rape and sugar beet
FRAC mode of action code: 3

See also *carbendazim + flusilazole*
chlorothalonil + flusilazole
famoxadone + flusilazole
fenpropimorph + flusilazole

Products

1	Capitan 25	DuPont	250 g/l	EW	13067
2	Genie 25	DuPont	250 g/l	EW	13065
3	Lyric	DuPont	250 g/l	EW	13022
4	Nustar 25	DuPont	250 g/l	EW	13921
5	Punch 25	DuPont	250 g/l	EW	14497
6	Sanction 25	DuPont	250 g/l	EW	13066

Uses

- Brown rust in ***spring barley***, ***winter barley***, ***winter wheat*** [1-6]
- Canker in ***brussels sprouts*** *(off-label)*, ***swedes*** *(off-label)*, ***turnips*** *(off-label)* [1-3, 6]
- Eyespot in ***forage maize*** *(off-label)* [1-3, 6]; ***spring barley***, ***winter barley***, ***winter wheat*** [1-6]
- Foliar disease control in ***brussels sprouts*** *(off-label)*, ***fodder beet*** *(off-label)*, ***forage maize*** *(off-label)*, ***swedes*** *(off-label)*, ***turnips*** *(off-label)* [4]; ***grain maize*** *(off-label)* [3, 4]
- Glume blotch in ***spring barley***, ***winter barley*** [1-3, 6]; ***winter wheat*** [1-4, 6]
- Light leaf spot in ***spring oilseed rape***, ***winter oilseed rape*** [1-6]
- Mildew in ***spring barley*** *(moderate control)*, ***winter barley*** *(moderate control)*, ***winter wheat*** [5]
- Net blotch in ***spring barley***, ***winter barley*** [1-6]
- Powdery mildew in ***spring barley***, ***winter barley*** [1-3, 6]; ***spring barley*** *(moderate control)*, ***winter barley*** *(moderate control)* [4]; ***sugar beet*** [1-6]; ***winter wheat*** [1-4, 6]
- Rhynchosporium in ***spring barley***, ***winter barley*** [1-3, 6]; ***spring barley*** *(moderate control)*, ***winter barley*** *(moderate control)* [4, 5]
- Rust in ***sugar beet*** [1-6]
- Septoria leaf blotch in ***spring barley***, ***winter barley*** [1-3, 6]; ***winter wheat*** [1-6]
- Yellow rust in ***spring barley***, ***winter barley***, ***winter wheat*** [1-6]

Specific Off-Label Approvals (SOLAs)

- ***brussels sprouts*** *20070102* [1], *20070101* [2], *20063998* [3], *20081104* [4], *20070100* [6]
- ***fodder beet*** *20091618* [4]
- ***forage maize*** *20080197* [1], *20080198* [2], *20080178* [3], *20081102* [4], *20080196* [6]
- ***grain maize*** *20081044* [3], *20081102* [4]
- ***swedes*** *20073284* [1], *20073283* [2], *20073270* [3], *20081103* [4], *20073285* [6]
- ***turnips*** *20073284* [1], *20073283* [2], *20073270* [3], *20081103* [4], *20073285* [6]

Approval information

- Flusilazole has been reinstated in Annex 1 under Directive 91/414 following an appeal
- Accepted by BBPA for use on malting barley

Efficacy guidance

- Best results obtained from treatment at onset of disease development
- Best control of eyespot in cereals achieved by spraying between leaf-sheath erect and second node detectable stages (GS 30-32)
- Product active against both MBC-sensitive and MBC-resistant strains of eyespot
- Rain occurring within 2 h after application may reduce effectiveness
- See label for recommended tank-mixes to give broader spectrum control
- Flusilazole is a DMI fungicide. Resistance to some DMI fungicides has been identified in Septoria leaf blotch which may seriously affect performance of some products. For further advice contact a specialist advisor and visit the Fungicide Resistance Action Group (FRAG)-UK website

Restrictions

- Maximum number of treatments (including other products containing flusilazole) on cereals and oilseed rape depends on dose and timing - see labels for details. High rate must not be used more than once in any crop

- Maximum total dose on sugar beet equivalent to one full dose
- Do not apply to crops under stress or during frosty weather

Crop-specific information

- Latest use: high dose before 3rd node detectable (GS 33) plus reduced dose before early milk stage (GS 72) on winter wheat and barley; before first flowers open (GS 4,0) on oilseed rape
- HI 7 wk for sugar beet
- Treat oilseed rape in autumn and/or spring at stem extension stage
- On sugar beet apply at an early stage of disease development, usually in early Aug

Environmental safety

- Dangerous for the environment
- Toxic to aquatic organisms
- Harmful to fish or other aquatic life. Do not contaminate surface waters or ditches with chemical or used container

Hazard classification and safety precautions

Hazard Toxic, Dangerous for the environment
Transport code 9
Packaging group III
UN number 3082
Risk phrases R22a, R38, R40, R51, R53a, R61
Operator protection A, C [1-6]; H [4]; M [4, 5]; P [5]; U05a, U11, U19a [1-6]; U16a, U23a [4]; U19c [5]; U20a [1-3, 6]; U20b [4, 5]
Environmental protection E13c, E34, E38
Storage and disposal D01, D02, D10b, D12a [1-6]; D05 [1-3, 5, 6]; D09a [4, 5]
Medical advice M04a

275 flutolanil

An carboxamide fungicide for treatment of potato seed tubers
FRAC mode of action code: 7

Products

1	Rhino	Certis	460 g/l	FS	13101
2	Rhino	Certis	460 g/l	SC	14311
3	Rhino DS	Certis	6% w/w	DS	12763

Uses

- Black scurf in ***potatoes*** *(tuber treatment)* [1-3]
- Rhizoctonia in ***seed potatoes*** *(off-label)* [1]
- Stem canker in ***potatoes*** *(tuber treatment)* [1-3]

Specific Off-Label Approvals (SOLAs)

- ***seed potatoes*** *20081149* [1]

Approval information

- Flutolanil included in Annex I under EC Directive 91/414

Efficacy guidance

- Apply to clean tubers before chitting, prior to planting, or at planting
- Apply flowable concentrate through canopied, hydraulic or spinning disc equipment (with or without electrostatics) mounted on a rolling conveyor or table [1]
- Flowable concentrate may be diluted with water up to 2.0 l per tonne to improve tuber coverage. Disease in areas not covered by spray will not be controlled [1]
- Dry powder may be applied at hopper filling, at box filling, to the box surface, to chitting trays or via an on-planter applicator [3]

Restrictions

- Maximum number of treatments 1 per batch of seed tubers
- Check with processor before use on crops for processing
- Treated tubers must be planted within 3 mth [3]

Crop-specific information

- Latest use: at planting
- HI 12 wk for potatoes [1]

SEE SECTION 3 FOR PRODUCTS ALSO REGISTERED

- Seed tubers should be of good quality and free from bacterial rots, physical damage or virus infection, and should not be sprouted to such an extent that mechanical damage to the shoots will occur during treatment or planting

Environmental safety

- Harmful to aquatic organisms

Hazard classification and safety precautions

Hazard Irritant
Transport code 9 [1, 2]
Packaging group III [1, 2]
UN number 3082 [1, 2]; N/C [3]
Risk phrases R36, R52, R53a [3]; R43, R51 [1, 2]
Operator protection A, H [1-3]; C [3]; U04a, U05a, U19a, U20a [1-3]; U14 [1, 2]
Environmental protection E15a, E34, E38
Storage and disposal D01, D02, D09a, D12a [1-3]; D05, D10a [1, 2]; D11a [3]
Treated seed S01, S03, S04a, S05 [1-3]; S02 [3]
Medical advice M03

276 flutriafol

A broad-spectrum triazole fungicide for cereals
FRAC mode of action code: 3

See also *chlorothalonil + flutriafol*
fludioxonil + flutriafol

Products

1	Consul	Headland	125 g/l	SC	12976
2	Pointer	Headland	125 g/l	SC	12975

Uses

- Brown rust in ***spring barley***, ***winter barley***, ***winter wheat***
- Powdery mildew in ***spring barley***, ***winter barley***, ***winter wheat***
- Rhynchosporium in ***spring barley***, ***winter barley***
- Septoria leaf blotch in ***winter wheat***
- Yellow rust in ***spring barley***, ***winter barley***, ***winter wheat***

Approval information

- Accepted by BBPA for use on malting barley
- Accepted by BBPA on malting barley

Efficacy guidance

- Best results obtained from treatment in early stages of disease development. See label for detailed guidance on spray timing for specific diseases and the need for repeat treatments
- Tank mix options available on the label to broaden activity spectrum
- Good spray coverage essential for optimum performance
- Flutriafol is a DMI fungicide. Resistance to some DMI fungicides has been identified in Septoria leaf blotch which may seriously affect performance of some products. For further advice contact a specialist advisor and visit the Fungicide Resistance Action Group (FRAG)-UK website

Restrictions

- Maximum number of treatments 2 per crop (including other products containing flutriafol)

Crop-specific information

- Latest use: before early grain milky ripe stage (GS 73)
- Flag leaf tip scorch on wheat caused by stress may be increased by fungicide treatment

Environmental safety

- Harmful to aquatic organisms
- Harmful to fish or other aquatic life. Do not contaminate surface waters or ditches with chemical or used container

Hazard classification and safety precautions

Hazard Harmful
UN number N/C

Risk phrases R43 [1]; R48, R52 [1, 2]
Operator protection A, H; U05a, U09a, U19a, U20b
Environmental protection E13c, E38
Storage and disposal D01, D02, D09a, D10c

277 fosetyl-aluminium

A systemic phosphonic acid fungicide for various horticultural crops
FRAC mode of action code: 33

Products

1	Aliette 80 WG	Certis	80% w/w	WG	13130
2	Standon Fullstop	Standon	80% w/w	WG	13571

Uses

- Collar rot in ***apples*** [1, 2]
- Crown rot in ***apples*** [1, 2]; ***protected strawberries*** *(off-label)*, ***strawberries*** *(off-label)* [1]
- Downy mildew in ***beet leaves*** *(off-label)*, ***chard*** *(off-label)*, ***chives*** *(off-label)*, ***combining peas*** *(off-label - seed treatment)*, ***endives*** *(off-label)*, ***herbs (see appendix 6)*** *(off-label)*, ***lamb's lettuce*** *(off-label)*, ***lupins*** *(off-label)*, ***parsley*** *(off-label)*, ***protected brassica seedlings*** *(off-label)*, ***protected chives*** *(off-label)*, ***protected endives*** *(off-label)*, ***protected herbs (see appendix 6)*** *(off-label)*, ***protected lamb's lettuce*** *(off-label)*, ***protected lettuce*** *(off-label)*, ***protected parsley*** *(off-label)*, ***protected salad brassicas*** *(off-label - for baby leaf production)*, ***protected scarole*** *(off-label)*, ***salad brassicas*** *(off-label - for baby leaf production)*, ***salad onions*** *(off-label)*, ***scarole*** *(off-label)*, ***spinach*** *(off-label)*, ***spinach beet*** *(off-label)*, ***tatsoi*** *(off-label)*, ***vining peas*** *(off-label - seed treatment)*, ***watercress*** *(off-label - during propagation)*, ***wine grapes*** *(off-label)* [1]; ***broad beans***, ***broccoli*** *(off-label)*, ***brussels sprouts*** *(off-label)*, ***cabbages*** *(off-label)*, ***calabrese*** *(off-label)*, ***cauliflowers*** *(off-label)*, ***chinese cabbage*** *(off-label)*, ***choi sum*** *(off-label)*, ***collards*** *(off-label)*, ***hops***, ***kale*** *(off-label)*, ***pak choi*** *(off-label)* [1, 2]; ***protected broccoli*** *(off-label)*, ***protected brussels sprouts*** *(off-label)*, ***protected cabbages*** *(off-label)*, ***protected calabrese*** *(off-label)*, ***protected cauliflowers*** *(off-label)*, ***protected chinese cabbage*** *(off-label)*, ***protected collards*** *(off-label)*, ***protected kale*** *(off-label)*, ***protected lettuce***, ***protected pak choi*** *(off-label)*, ***watercress*** *(off-label)* [2]
- Foliar disease control in ***amenity grassland*** *(off-label)*, ***beet leaves*** *(off-label)*, ***chard*** *(off-label)*, ***chicory*** *(off-label)*, ***endives*** *(off-label)*, ***frise*** *(off-label)*, ***herbs (see appendix 6)*** *(off-label)*, ***lamb's lettuce*** *(off-label)*, ***managed amenity turf*** *(off-label)*, ***outdoor grapes*** *(off-label)*, ***protected endives*** *(off-label)*, ***protected frise*** *(off-label)*, ***protected lamb's lettuce*** *(off-label)*, ***protected radicchio*** *(off-label)*, ***protected salad brassicas*** *(off-label)*, ***protected scarole*** *(off-label)*, ***protected spinach*** *(off-label)*, ***protected strawberries*** *(off-label)*, ***radicchio*** *(off-label)*, ***salad brassicas*** *(off-label)*, ***salad onions*** *(off-label)*, ***scarole*** *(off-label)*, ***spinach*** *(off-label)*, ***spinach beet*** *(off-label)*, ***strawberries*** *(off-label)* [2]
- Phytophthora in ***chicory*** *(off-label - for forcing)*, ***watercress*** *(off-label - during propagation)* [1]
- Phytophthora root rot in ***capillary benches*** [1]; ***protected pot plants*** [1, 2]
- Phytophthora stem rot in ***capillary benches*** [1]; ***protected pot plants*** [1, 2]
- Phytophthora wilt in ***ornamental plant production*** [1, 2]
- Pythium in ***managed amenity turf*** *(off-label)*, ***watercress*** *(off-label - during propagation)* [1]
- Red core in ***protected strawberries*** *(off-label)*, ***strawberries*** *(off-label)* [1]; ***strawberries*** [1, 2]
- Root rot in ***capillary benches*** [1]; ***protected pot plants*** [1, 2]
- Seed-borne diseases in ***combining peas*** *(off-label)*, ***lupins*** *(off-label)*, ***protected herbs (see appendix 6)*** *(off-label)*, ***vining peas*** *(off-label)* [2]

Specific Off-Label Approvals (SOLAs)

- ***amenity grassland*** *20080270* [2]
- ***beet leaves*** *20063520* [1], *20080267* [2]
- ***broccoli*** *20082031* [1], *20080271* [2]
- ***brussels sprouts*** *20082031* [1], *20080271* [2]
- ***cabbages*** *20082031* [1], *20080271* [2]
- ***calabrese*** *20082031* [1], *20080271* [2]
- ***cauliflowers*** *20082031* [1], *20080271* [2]
- ***chard*** *20063520* [1], *20080267* [2]
- ***chicory*** *(for forcing) 20063521* [1], *20080268* [2]

SEE SECTION 3 FOR PRODUCTS ALSO REGISTERED

- ***chinese cabbage*** *20082031* [1], *20080271* [2]
- ***chives*** *20063522* [1]
- ***choi sum*** *20082031* [1], *20080271* [2]
- ***collards*** *20082031* [1], *20080271* [2]
- ***combining peas*** *(seed treatment) 20063518* [1], *20080265* [2]
- ***endives*** *20063522* [1], *20080269* [2]
- ***frise*** *20080269* [2]
- ***herbs (see appendix 6)*** *20063520* [1], *20080267* [2]
- ***kale*** *20082031* [1], *20080271* [2]
- ***lamb's lettuce*** *20063522* [1], *20080269* [2]
- ***lupins*** *20063518* [1], *20080265* [2]
- ***managed amenity turf*** *20063523* [1], *20080270* [2]
- ***outdoor grapes*** *20080266* [2]
- ***pak choi*** *20082031* [1], *20080271* [2]
- ***parsley*** *20063522* [1]
- ***protected brassica seedlings*** *20082031* [1]
- ***protected broccoli*** *20080271* [2]
- ***protected brussels sprouts*** *20080271* [2]
- ***protected cabbages*** *20080271* [2]
- ***protected calabrese*** *20080271* [2]
- ***protected cauliflowers*** *20080271* [2]
- ***protected chinese cabbage*** *20080271* [2]
- ***protected chives*** *20063522* [1]
- ***protected collards*** *20080271* [2]
- ***protected endives*** *20063522* [1], *20080269* [2]
- ***protected frise*** *20080269* [2]
- ***protected herbs (see appendix 6)*** *20063522* [1], *20080269* [2]
- ***protected kale*** *20080271* [2]
- ***protected lamb's lettuce*** *20063522* [1], *20080269* [2]
- ***protected lettuce*** *20063522* [1]
- ***protected pak choi*** *20080271* [2]
- ***protected parsley*** *20063522* [1]
- ***protected radicchio*** *20080269* [2]
- ***protected salad brassicas*** *(for baby leaf production) 20063522* [1], *20080269* [2]
- ***protected scarole*** *20063522* [1], *20080269* [2]
- ***protected spinach*** *20080269* [2]
- ***protected strawberries*** *20063517* [1], *20080264* [2]
- ***radicchio*** *20080269* [2]
- ***salad brassicas*** *(for baby leaf production) 20063522* [1], *20080269* [2]
- ***salad onions*** *20063515* [1], *20080263* [2]
- ***scarole*** *20063522* [1], *20080269* [2]
- ***spinach*** *20063520* [1], *20080267* [2]
- ***spinach beet*** *20063520* [1], *20080267* [2]
- ***strawberries*** *20063518* [1], *20080264* [2]
- ***tatsoi*** *20082031* [1]
- ***vining peas*** *(seed treatment) 20063518* [1], *20080265* [2]
- ***watercress*** *(during propagation) 20071260* [1], *20080272* [2]
- ***wine grapes*** *20063519* [1]

Approval information

- Fosetyl-aluminium included in Annex I under EC Directive 91/414
- Accepted by BBPA for use on hops

Restrictions

- Maximum number of treatments 1 per seed batch for peas; 1 per batch of compost for lettuce; 1 per yr for strawberries (root dip or foliar spray); 1 per yr for apples (bark paste), 2 per yr for apples (foliar spray); 2 per crop for broad beans, peas; 2 per yr for hops (basal spray); 6 per yr for hops (foliar spray); 1 per crop during propagation, 2 per crop after planting out for lettuce
- Check tolerance of ornamental species before large-scale treatment

Crop-specific information

- Latest use: pre-planting for strawberry runners; pre-sowing for protected lettuce
- HI apples 5 mth (bark paste) or 4 wk (spray); broad beans 17 d; hops 14 d
- Spray young orchards for crown rot protection after blossom when first leaves fully open and repeat after 4-6 wk. Apply as paste to bark of apples to control collar rot
- Apply to broad beans when infection appears (usually at flowering) and 14 d later. Consult before treating crops to be processed
- Use on strawberries only between harvest and 31 Dec
- Spray autumn-planted strawberry runners 2-3 wk after planting or use dip treatment at planting. Spray established crops in late summer/early autumn after picking and repeat annually
- Apply to hops as early season basal spray or as foliar spray every 10-14 d from when training is completed
- Use by wet incorporation in blocking compost for protected lettuce only from Sep to Apr and follow all directions carefully to avoid severe crop injury. Crop maturity may be delayed by a few days
- Apply as drench to rooted cuttings of hardy nursery stock after first potting and repeat mthly. Up to 6 applications may be needed
- See label for details of application to capillary benches and trays of young plants.

Environmental safety

- Harmful to aquatic organisms

Hazard classification and safety precautions

Hazard Irritant [2]
UN number N/C
Risk phrases R36, R52 [2]
Operator protection A, C, H; U11 [2]; U20c [1, 2]
Environmental protection E15a
Storage and disposal D01, D02, D12a [2]; D09a, D11a [1, 2]

278 fosetyl-aluminium + propamocarb hydrochloride

A systemic and protectant fungicide mixture for use in horticulture
FRAC mode of action code: 33 + 28

See also *propamocarb hydrochloride*

Products

Previcur Energy	Bayer CropScience	310:530 g/l	SL	13342

Uses

- Damping off in ***chard*** *(off-label)*, ***herbs (see appendix 6)*** *(off-label)*, ***leaf brassicas*** *(off-label - baby leaf production)*, ***ornamental plant production*** *(off-label)*, ***protected aubergines*** *(off-label)*, ***protected cabbages*** *(off-label)*, ***protected cayenne peppers*** *(off-label)*, ***protected courgettes*** *(off-label)*, ***protected gherkins*** *(off-label)*, ***protected herbs (see appendix 6)*** *(off-label)*, ***protected marrows*** *(off-label)*, ***protected melons*** *(off-label)*, ***protected peppers*** *(off-label)*, ***protected pumpkins*** *(off-label)*, ***protected salad brassicas*** *(off-label - baby leaf production)*, ***protected squashes*** *(off-label)*, ***protected watermelon*** *(off-label)*, ***spinach beet*** *(off-label)*
- Downy mildew in ***chard*** *(off-label)*, ***frise*** *(off-label)*, ***herbs (see appendix 6)*** *(off-label)*, ***leaf brassicas*** *(off-label - baby leaf production)*, ***protected chicory*** *(off-label)*, ***protected herbs (see appendix 6)*** *(off-label)*, ***protected lettuce***, ***protected salad brassicas*** *(off-label - baby leaf production)*, ***scarole*** *(off-label)*, ***spinach***, ***spinach beet*** *(off-label)*
- Pythium in ***protected cucumbers***, ***protected lettuce***, ***protected tomatoes***

Specific Off-Label Approvals (SOLAs)

- ***chard*** *20082648*
- ***frise*** *20082654*
- ***herbs (see appendix 6)*** *20082668*
- ***leaf brassicas*** *(baby leaf production) 20082668*
- ***ornamental plant production*** *20082667*
- ***protected aubergines*** *20082653*
- ***protected cabbages*** *20082650*

- ***protected cayenne peppers*** *20082652*
- ***protected chicory*** *20082649*
- ***protected courgettes*** *20082653*
- ***protected gherkins*** *20082653*
- ***protected herbs (see appendix 6)*** *20082655*
- ***protected marrows*** *20082651*
- ***protected melons*** *20082651*
- ***protected peppers*** *20082652*
- ***protected pumpkins*** *20082651*
- ***protected salad brassicas*** *(baby leaf production) 20082655*
- ***protected squashes*** *20082651*
- ***protected watermelon*** *20082651*
- ***scarole*** *20082654*
- ***spinach beet*** *20082648*

Approval information

- Fosetyl-aluminium and propamocarb hydrochloride included in Annex I under EC Directive 91/414

Hazard classification and safety precautions

Hazard Irritant
UN number N/C
Risk phrases R43
Operator protection A, H, M; U04a, U05a, U08, U14, U19a, U20b
Environmental protection E15a
Storage and disposal D09a, D11a
Medical advice M03

279 fosthiazate

An organophosphorus contact nematicide for potatoes
IRAC mode of action code: 1B

Products

Nemathorin 10G	Syngenta	10% w/w	FG	11003

Uses

- Nematodes in ***hops*** *(off-label)*, ***ornamental plant production*** *(off-label)*, ***soft fruit*** *(off-label)*
- Potato cyst nematode in ***potatoes***
- Spraing vectors in ***potatoes*** *(reduction)*
- Wireworm in ***potatoes*** *(reduction)*

Specific Off-Label Approvals (SOLAs)

- ***hops*** *20082912*
- ***ornamental plant production*** *20082912*
- ***soft fruit*** *20082912*

Approval information

- Fosthiazate included in Annex I under EC Directive 91/414
- In 2006 PSD required that all products containing this active ingredient should carry the following warning in the main area of the container label: "Fosthiazate is an anticholinesterase organophosphate. Handle with care"

Efficacy guidance

- Treatment may be in furrow or applied and incorporated in one operation to a uniform depth of 10-15 cm. Deeper incorporation will reduce control
- Application best achieved using equipment such as Horstine Farmery Microband Applicator, Matco or Stocks Micrometer applicators together with a rear mounted powered rotary cultivator
- Granules must not become wet or damp before use

Restrictions

- Contains an anticholinesterase organophosphorus compound. Do not use if under medical advice not to work with such compounds
- Maximum number of treatments 1 per crop

- Product must only be applied using tractor-mounted/drawn direct placement machinery. Do not use air assisted broadcast machinery other than that referenced on the product label
- Do not allow granules to stand overnight in the application hopper
- Do not apply more than once every four years on the same area of land
- Do not use on crops to be harvested less than 17 wk after treatment
- Consult before using on crops intended for processing

Crop-specific information
- Latest use: at planting
- HI 17 wk for potatoes

Environmental safety
- Dangerous for the environment
- Toxic to aquatic organisms
- Dangerous to game, wild birds and animals
- Dangerous to livestock. Keep all livestock out of treated areas for at least 13 wk
- Incorporation to 10-15 cm and ridging up of treated soil must be carried out immediately after application. Powered rotary cultivators are preferred implements for incorporation but discs, power, spring tine or Dutch harrows may be used provided two passes are made at right angles
- To protect birds and wild mammals remove spillages
- Failure completely to bury granules immediately after application is hazardous to wildlife
- To protect groundwater do not apply any product containing fosthiazate more than once every four yr

Hazard classification and safety precautions

Hazard Harmful, Dangerous for the environment
Transport code 9
Packaging group III
UN number 3077
Risk phrases R22a, R43, R51, R53a
Operator protection A, E, G, H, K, M; U02a, U04a, U05a, U09a, U13, U14, U19a, U20a
Environmental protection E06b (13 wk); E15b, E34, E38
Storage and disposal D01, D02, D05, D09a, D11a, D14
Medical advice M01, M03, M05a

280 fuberidazole

A benzimidazole (MBC) fungicide available only in mixtures
FRAC mode of action code: 1

281 fuberidazole + imidacloprid + triadimenol

A broad spectrum systemic fungicide and insecticide seed treatment for winter cereals
FRAC mode of action code: 1 + 3

See also imidacloprid

Products

Tripod Plus	Makhteshim	15:117:125 g/l	LS	13168

Uses
- Blue mould in ***winter wheat*** *(seed treatment)*
- Brown foot rot in ***winter barley*** *(seed treatment)*
- Brown rust in ***winter barley*** *(seed treatment)*, ***winter wheat*** *(seed treatment)*
- Bunt in ***winter wheat*** *(seed treatment)*
- Covered smut in ***winter barley*** *(seed treatment)*
- Fusarium foot rot and seedling blight in ***winter barley*** *(seed treatment - reduction)*, ***winter oats*** *(seed treatment - reduction)*, ***winter wheat*** *(seed treatment - reduction)*
- Loose smut in ***winter barley*** *(seed treatment)*, ***winter oats*** *(seed treatment)*, ***winter wheat*** *(seed treatment)*
- Net blotch in ***winter barley*** *(seed treatment - seed-borne only)*

- Powdery mildew in ***winter barley*** *(seed treatment)*, ***winter oats*** *(seed treatment)*, ***winter wheat*** *(seed treatment)*
- Pyrenophora leaf spot in ***winter oats*** *(seed treatment)*
- Septoria leaf blotch in ***winter wheat*** *(seed treatment)*
- Septoria seedling blight in ***winter wheat*** *(seed treatment - reduction)*
- Snow rot in ***winter barley*** *(seed treatment - reduction)*
- Virus vectors in ***winter barley*** *(seed treatment)*, ***winter oats*** *(seed treatment)*, ***winter wheat*** *(seed treatment)*
- Wireworm in ***winter barley*** *(reduction of damage)*, ***winter oats*** *(reduction of damage)*, ***winter wheat*** *(reduction of damage)*
- Yellow rust in ***winter barley*** *(seed treatment)*, ***winter wheat*** *(seed treatment)*

Approval information

- Fuberidazole and imidacloprid included in Annex I under EC Directive 91/414
- Accepted by BBPA for use on malting barley

Efficacy guidance

- Apply through recommended seed treatment machinery
- Evenness of seed cover improved by simultaneous application of equal volumes of product and water
- Calibrate drill for treated seed and drill at 4 cm into firm, well prepared seedbed
- Use minimum 125 kg treated seed per ha and increase seed rate as drilling season progresses. Lower drilling rates and/or early drilling affect duration of BYDV protection needed and may require follow-up aphicide treatment
- When aphid activity unusually late, or is heavy and prolonged in areas of high risk, and mild weather predominates, follow-up treatment may be required
- In addition to seed-borne diseases early attacks of various foliar, air-borne diseases are controlled or suppressed. See label for details

Restrictions

- Maximum number of treatments 1 per batch of seed
- Do not use on naked oats
- Do not drill treated winter wheat seed after end of Nov
- Do not handle seed unnecessarily
- Do not use treated seed as food or feed
- Treated seed should not be broadcast, but drilled to a depth of 4 cm in a well prepared seedbed
- Germination tests should be done on all batches of seed to be treated to ensure seed viability and suitability for treatment
- Do not use on seed with moisture content above 16%, on sprouted, cracked or skinned seed or on seed already treated with another seed treatment

Crop-specific information

- Latest use: before drilling
- Store treated seed in cool, dry, well-ventilated store and drill as soon as possible, preferably in season of purchase
- Treatment may accentuate effects of adverse seedbed conditions on crop emergence

Environmental safety

- Dangerous to fish or other aquatic life. Do not contaminate surface waters or ditches with chemical or used container
- Dangerous to birds, game and other wildlife. Treated seed should not be left on the soil surface. Bury spillages
- Seed should be drilled to a depth of 4 cm into a well-prepared seed bed
- If seed is present on the soil surface, or if spills have occurred, the field should be harrowed and rolled if conditions are appropriate to ensure good incorporation

Hazard classification and safety precautions

Transport code 9
Packaging group III
UN number 3082
Risk phrases R52, R53a
Operator protection A, H; U04a, U05a, U07, U13, U14, U20b
Environmental protection E03, E36a, E38

Storage and disposal D01, D02, D05, D09a, D12a
Treated seed S01, S02, S03, S04c, S05, S06a, S07, S08

282 fuberidazole + triadimenol

A broad spectrum systemic fungicide seed treatment for cereals
FRAC mode of action code: 1 + 3

Products

Tripod	Makhteshim	22.5:187.5 g/l	FS	13014

Uses

- Blue mould in ***spring wheat*** *(seed treatment)*, ***triticale*** *(seed treatment)*, ***winter wheat*** *(seed treatment)*
- Brown foot rot in ***spring barley*** *(seed treatment)*, ***spring oats*** *(seed treatment)*, ***spring rye*** *(seed treatment)*, ***spring wheat*** *(seed treatment)*, ***triticale*** *(seed treatment)*, ***winter barley*** *(seed treatment)*, ***winter oats*** *(seed treatment)*, ***winter rye*** *(seed treatment)*, ***winter wheat*** *(seed treatment)*
- Bunt in ***spring wheat*** *(seed treatment)*, ***winter wheat*** *(seed treatment)*
- Covered smut in ***spring barley*** *(seed treatment)*, ***winter barley*** *(seed treatment)*
- Leaf stripe in ***spring barley*** *(seed treatment)*
- Loose smut in ***spring barley*** *(seed treatment)*, ***spring oats*** *(seed treatment)*, ***spring wheat*** *(seed treatment)*, ***winter barley*** *(seed treatment)*, ***winter oats*** *(seed treatment)*, ***winter wheat*** *(seed treatment)*
- Pyrenophora leaf spot in ***spring oats*** *(seed treatment)*, ***winter oats*** *(seed treatment)*

Approval information

- Fuberidazole included in Annex I under EC Directive 91/414
- Accepted by BBPA for use on malting barley

Efficacy guidance

- Apply through recommended seed treatment machinery
- Calibrate drill for treated seed and drill at 2.5-4 cm into firm, well prepared seedbed
- Product will give control of early attacks of foliar air-borne diseases such as powdery mildew, yellow rust, brown rust, crown rust, Septoria. On autumn-sown crops control of rust is prolonged and usually extends over winter, leading to a delayed build-up of the disease in the following spring, often resulting in fewer fungicide sprays being required to maintain control during spring/summer
- Product will often delay and reduce the likelihood and extent of lodging in autumn sown wheat.

Restrictions

- Maximum number of treatments 1 per batch of seed
- Do not use on naked oats
- Do not drill treated winter wheat or rye seed after end of Nov. Seed rate should be increased as drilling season progresses
- Germination tests should be done on all batches of seed to be treated to ensure seed viability and suitability for treatment
- Do not use on seed with moisture content above 16%, on sprouted, cracked or skinned seed or on seed already treated with another seed treatment
- Do not use treated seed as food or feed

Crop-specific information

- Latest use: before drilling
- Treated spring wheat may be drilled in autumn up to end of Nov or from Feb onwards
- Store treated seed in cool, dry, well-ventilated store and drill as soon as possible, preferably in season of purchase
- Treatment may accentuate effects of adverse seedbed conditions on crop emergence

Environmental safety

- Dangerous for the environment
- Harmful to aquatic organisms

Hazard classification and safety precautions

Hazard Dangerous for the environment

Transport code 9
Packaging group III
UN number 3082
Risk phrases R52, R53a
Operator protection A; U20b
Environmental protection E03, E15a, E34
Storage and disposal D05, D09a, D11a
Treated seed S01, S02, S03, S04a, S05, S06a, S07

283 gibberellins

A plant growth regulator for use in top fruit

Products

1	Berelex	Interfarm	1 g per tablet	TB	14123
2	GIBB 3	Globachem	10% w/w	TB	12290
3	GIBB Plus	Globachem	10 g/l	SL	11875
4	Gistar	Hermoo	10 g/l	SL	11397
5	Novagib	Fine	10 g/l	SL	08954
6	Regulex 10 SG	Interfarm	10% w/w	SG	14116

Uses

- Growth regulation in ***crab apples*** *(off-label)*, ***ornamental crop production (seed)*** *(off-label)*, ***quinces*** *(off-label)* [6]; ***farm forestry*** *(off-label)*, ***forest*** *(off-label)* [3, 6]; ***pears*** *(off-label)* [5, 6]
- Improved germination in ***nothofagus*** *(do not exceed 2.5 g per 5 litres when used as a seed soak)* [6]
- Increasing fruit set in ***pears*** [1, 2]
- Increasing germination in ***nothofagus*** *(qualified minor use)* [3, 4]
- Increasing yield in ***celery (outdoor)*** [1, 2]; ***cherries*** *(off-label)*, ***protected cherries*** *(off-label)*, ***rhubarb*** [1]; ***protected rhubarb*** [2]
- Reducing fruit russeting in ***apples*** [3-5]; ***pears*** *(off-label)*, ***quinces*** *(off-label)* [3]
- Russet reduction in ***apples*** [6]

Specific Off-Label Approvals (SOLAs)

- ***cherries*** *20091072* [1]
- ***crab apples*** *20091078* [6]
- ***farm forestry*** *20071675* [3], *20091076* [6]
- ***forest*** *20071675* [3], *20091076* [6]
- ***ornamental crop production (seed)*** *20091077* [6]
- ***pears*** *20061627* [3], *20012756* [5], *20091078* [6]
- ***protected cherries*** *20091072* [1]
- ***quinces*** *20061627* [3], *20091078* [6]

Efficacy guidance

- For optimum results on apples spray under humid, slow drying conditions and ensure good spray cover [3-5]
- Treat apples and pears immediately after completion of petal fall and repeat as directed on the label [3-5]
- Nothofagus seeds should be stirred into the solution, soaked for 24 hr and then sown immediately [3, 4]
- Fruit set in pears can be improved when blossom is spare, setting is poor or where frost has killed many flowers [2]

Restrictions

- Maximum total dose varies with crop and product. Check label
- Prepared spray solutions are unstable. Do not leave in the sprayer during meal breaks or overnight
- Return bloom may be reduced in yr following treatment
- Consult before treating apple crops grown for processing [3-5]
- Avoid storage at temperatures above 32 °C [2]

Crop-specific information

- HI zero for apples

- Apply to apples at completion of petal fall and repeat 3 or 4 times at 7-10 d intervals. Number of sprays and spray interval depend on weather conditions and dose (see labels) [3, 5]
- Good results achieved on apple varieties Cox's Orange Pippin, Discovery, Golden Delicious and Karmijn. For other cultivars test on a small number of trees [3, 5]
- Split treatment on pears allows second spray to be omitted if pollinating conditions become very favourable [2]
- Pear variety Conference usually responds well to treatment; Doyenne du Comice can be variable [2]
- On celery the interval between application and harvest should never exceed 4 wk to avoid risk of premature seeding and root splitting [2]
- Treat washed rhubarb crowns on transfer to forcing shed [2]

Hazard classification and safety precautions

UN number N/C

Operator protection U05a [6]; U08 [2]; U09a [1]; U20b [5]; U20c [1-4, 6]

Environmental protection E15a

Storage and disposal D01 [5, 6]; D02 [6]; D03, D05, D07 [5]; D09a [1-6]; D10b [3, 5, 6]; D10c [4]

284 glufosinate-ammonium

A non-selective, non-residual phosphinic acid contact herbicide

HRAC mode of action code: H

Products

1	Basta	Bayer CropScience	150 g/l	SL	13820
2	Finale	Bayer Environ.	120 g/l	SL	10092
3	Harvest	Bayer CropScience	150 g/l	SL	07321
4	Kaspar	Certis	150 g/l	SL	11214
5	Kibosh	AgChem Access	150 g/l	SL	13976

Uses

- Annual and perennial weeds in ***all tree nuts***, ***apple orchards***, ***bilberries***, ***blueberries***, ***cane fruit***, ***cherries***, ***cranberries***, ***cultivated land/soil***, ***currants***, ***damsons***, ***fallows***, ***forest***, ***non-crop farm areas***, ***pear orchards***, ***plums***, ***potatoes*** *(not for seed)*, ***strawberries***, ***stubbles***, ***sugar beet***, ***vegetables***, ***wine grapes*** [3]
- Annual dicotyledons in ***all tree nuts***, ***apple orchards***, ***cane fruit***, ***cherries***, ***cultivated land/soil***, ***currants***, ***damsons***, ***fallows***, ***forest***, ***headlands*** *(uncropped)*, ***non-crop farm areas***, ***pear orchards***, ***plums***, ***strawberries***, ***sugar beet***, ***vegetables***, ***ware potatoes***, ***wine grapes*** [4, 5]; ***nursery stock***, ***woody ornamentals*** [2]; ***potatoes*** [1]
- Annual grasses in ***all tree nuts***, ***apple orchards***, ***cane fruit***, ***cherries***, ***cultivated land/soil***, ***currants***, ***damsons***, ***fallows***, ***forest***, ***headlands*** *(uncropped)*, ***non-crop farm areas***, ***pear orchards***, ***plums***, ***strawberries***, ***sugar beet***, ***vegetables***, ***ware potatoes***, ***wine grapes*** [4, 5]; ***nursery stock***, ***woody ornamentals*** [2]; ***potatoes*** [1]
- Green cover in ***land temporarily removed from production*** [3-5]
- Harvest management/desiccation in ***combining peas***, ***ware potatoes*** [4, 5]; ***combining peas*** *(not for seed)*, ***potatoes*** [3]; ***linseed***, ***spring field beans***, ***spring oilseed rape***, ***winter field beans***, ***winter oilseed rape*** [3-5]
- Line marking preparation in ***managed amenity turf*** [2]
- Perennial dicotyledons in ***all tree nuts***, ***apple orchards***, ***cane fruit***, ***cherries***, ***currants***, ***damsons***, ***fallows***, ***forest***, ***headlands*** *(uncropped)*, ***land temporarily removed from production***, ***non-crop farm areas***, ***pear orchards***, ***plums***, ***strawberries***, ***wine grapes*** [4, 5]; ***nursery stock***, ***woody ornamentals*** [2]
- Perennial grasses in ***all tree nuts***, ***apple orchards***, ***cane fruit***, ***cherries***, ***currants***, ***damsons***, ***fallows***, ***forest***, ***headlands*** *(uncropped)*, ***land temporarily removed from production***, ***non-crop farm areas***, ***pear orchards***, ***plums***, ***strawberries***, ***wine grapes*** [4, 5]; ***nursery stock***, ***woody ornamentals*** [2]
- Sward destruction in ***permanent grassland*** [3-5]

Approval information

- Glufosinate-ammonium included in Annex I under EC Directive 91/414
- Accepted by BBPA for use on malting barley and hops

Efficacy guidance

- Activity quickest under warm, moist conditions. Light rainfall 3-4 h after application will not affect activity. Do not spray wet foliage or if rain likely within 6 h
- For weed control uses treat when weeds growing actively. Deep rooted weeds may require second treatment
- On uncropped headlands apply in May/Jun to prevent weeds invading field
- Glufosinate kills all green tissue but does not harm mature bark

Restrictions

- Maximum number of treatments 4 per crop for potatoes (including 2 desiccant uses); 2 per crop (including 1 desiccant use) for oilseed rape, dried peas, field beans, linseed, wheat and barley; 3 per yr for fruit and forestry; 2 per yr for strawberries, non-crop land and land temporarily removed from production, round fruit trees, canes or bushes; 1 per crop for sugar beet, vegetables and other crops; 1 per yr for grassland destruction, on cultivated land prior to planting edible crops
- Pre-harvest desiccation sprays should not be used on seed crops of wheat, barley, peas or potatoes but may be used on seed crops of oilseed rape, field beans and linseed [3, 4]
- Do not desiccate potatoes in exceptionally wet weather or in saturated soil. See label for details
- Do not spray potatoes after emergence if grown from small or diseased seed or under very dry conditions
- Application for weed control and line-marking uses in sports turf must be between 1 Mar and 30 Sep [2]
- Do not spray hedge bottoms or on broadcast crops [3]

Crop-specific information

- Latest use: 30 Sep for use on non-crop land, top fruit, soft fruit, cane fruit, bush fruit, forestry; pre-drilling, pre-planting or pre-emergence in sugar beet, vegetables and other crops; before winter dormancy for grassland destruction.
- HI potatoes, oilseed rape, combining peas, field beans, linseed 7 d; wheat, barley 14 d
- Ploughing or other cultivations can follow 4 h after spraying
- Crops can normally be sown/planted immediately after spraying or sprayed post-drilling. On sand, very light or immature peat soils allow at least 3 d before sowing/planting or expected emergence
- For weed control in potatoes apply pre-emergence or up to 10% emergence on earlies and seed crops, up to 40% on maincrop, on plants up to 15 cm high
- In sugar beet and vegetables apply just before crop emergence, using stale seedbed technique
- In top and soft fruit, grapevines and forestry apply between 1 Mar and 30 Sep as directed sprays
- For grass destruction apply before winter dormancy occurs. Heavily grazed fields should show active regrowth. Plough from the day after spraying
- Apply pre-harvest desiccation treatments 10-21 d before harvest (14-21 d for oilseed rape). See label for timing details on individual crops
- For potato haulm desiccation apply to listed varieties (not seed crops) at onset of senescence, 14-21 d before harvest

Environmental safety

- Harmful to fish or other aquatic life. Do not contaminate surface waters or ditches with chemical or used container
- Keep livestock out of treated areas until foliage of any poisonous weeds such as ragwort has died and become unpalatable
- Risk to certain non-target insects or other arthropods [3, 4]
- Treated pea haulm may be fed to livestock from 7 d after spraying, treated grain from 14 d [3]
- Product supplied in small volume returnable container - see label for filling and mixing instructions [3]

Hazard classification and safety precautions

Hazard Harmful [1, 3-5]; Irritant [2]
Transport code 6.1
Packaging group III
UN number 2902
Risk phrases R21, R22a, R41 [1, 3-5]; R36 [2]
Operator protection A [1, 3-5]; B [2]; C, H, M; U04a, U11, U13, U14, U15, U19a [1, 3-5]; U05a, U08 [1-5]; U20a [3-5]; U20b [1, 2]
Environmental protection E07a [1-5]; E13c [2, 4, 5]; E15a, E22c [1, 3]; E22b [4, 5]; E34 [1, 3-5]

Storage and disposal D01, D02, D09a, D10b [1-5]; D05 [2]; D12a [3-5]
Medical advice M03 [1, 3-5]

285 glyphosate

A translocated non-residual glycine derivative herbicide
HRAC mode of action code: G

See also *diflufenican + glyphosate*
diuron + glyphosate

Products

1	Amega Duo	Nufarm UK	540 g/l	SL	13358
2	Amega Duo	Nufarm UK	540 g/l	SL	14131
3	Asteroid	Headland	360 g/l	SL	11118
4	Barclay Gallup 360	Barclay	360 g/l	SL	12659
5	Barclay Gallup Amenity	Barclay	360 g/l	SL	13250
6	Barclay Gallup Biograde 360	Barclay	360 g/l	SL	12660
7	Barclay Gallup Biograde Amenity	Barclay	360 g/l	SL	12716
8	Barclay Gallup Hi-Aktiv	Barclay	490 g/l	SL	12661
9	Buggy XTG	Sipcam	36% w/w	SG	12951
10	CDA Vanquish Biactive	Bayer Environ.	120 g/l	UL	12586
11	Charger	AgChem Access	360 g/l	SL	14107
12	Charger B	AgChem Access	360 g/l	SL	14269
13	Charger C	AgChem Access	360 g/l	SL	14216
14	Clinic Ace	Nufarm UK	360 g/l	SL	12980
15	Clinic Ace	Nufarm UK	360 g/l	SL	14040
16	Clipper	Makhteshim	360 g/l	SL	13217
17	Credit DST	Nufarm UK	540 g/l	SL	13822
18	Credit DST	Nufarm UK	540 g/l	SL	14066
19	Discman Biograde	Barclay	216 g/l	SL	12856
20	Dow Agrosciences Glyphosate 360	Dow	360 g/l	SL	12720
21	Envision	Headland	450 g/l	SL	10569
22	Glyfo TDI	Q-Chem	360 g/l	SL	13940
23	Glyfos	Headland	360 g/l	SL	10995
24	Glyfos Dakar	Headland	68% w/w	SG	13054
25	Glyfos Dakar Pro	Headland Amenity	68% w/w	SG	13147
26	Glyfos Proactive	Nomix Enviro	360 g/l	SL	11976
27	Glyfos Supreme	Headland	450 g/l	SL	12371
28	Glyfosat 36	Goldengrass	360 g/l	SL	14297
29	Glyphogan	Makhteshim	360 g/l	SL	12668
30	Glypho-Rapid 450	Barclay	450 g/l	SL	13882
31	Greenaway Gly-490	Greenaway	490 g/l	SL	12718
32	Hilite	Nomix Enviro	144 g/l	RH	13871
33	Kernel	Headland	480 g/l	SL	10993
34	Landgold Glyphosate 360	Goldengrass	360 g/l	SL	13977
35	Manifest	Headland	360 g/l	SL	11041
36	Nomix Conqueror	Nomix Enviro	144 g/l	UL	12370
37	Nufosate Ace	Nufarm UK	360 g/l	SL	13794
38	Proliance Quattro	Syngenta	360 g/l	SL	13670
39	Roundup Ace	Monsanto	450 g/l	SL	12772
40	Roundup Biactive	Monsanto	360 g/l	SL	10320
41	Roundup Energy	Monsanto	450 g/l	SL	12945
42	Roundup Klik	Monsanto	450 g/l	SL	12866
43	Roundup Max	Monsanto	68% w/w	SG	12952
44	Roundup Metro	Monsanto	360 g/l	SL	13989
45	Roundup Pro Biactive	Monsanto	360 g/l	SL	10330
46	Roundup ProBiactive 450	Monsanto	450 g/l	SL	12778
47	Roundup Ultimate	Monsanto	540 g/l	SL	13081
48	Samurai	Monsanto	360 g/l	SL	12674
49	Slingshot	ChemSource	360 g/l	SL	14505
50	Stacato	Sipcam	360 g/l	SL	12675
51	Stirrup	Nomix Enviro	144 g/l	RH	13875
52	Tangent	Headland Amenity	450 g/l	SL	11872
53	Touchdown Quattro	Syngenta	360 g/l	SL	10608
54	Typhoon 360	Makhteshim	360 g/l	SL	12932

Uses

- Annual and perennial weeds in ***all edible crops (outdoor and protected)*** [1, 2, 14, 15, 17, 18, 30, 37, 50]; ***all edible crops (outdoor)*** [22, 44]; ***all edible crops (outdoor)*** *(before planting)*, ***all non-edible crops (outdoor)*** *(before planting)* [4, 6, 11, 12, 16, 24, 27, 28, 49, 53]; ***all non-edible crops (outdoor)***, ***grassland***, ***leeks***, ***sugar beet***, ***swedes***, ***turnips*** [1, 2, 14, 15, 17, 18, 22, 30, 37, 44, 50]; ***almonds*** *(off-label)*, ***apples*** *(off-label)*, ***apricots*** *(off-label)*, ***cherries*** *(off-label)*, ***crab apples*** *(off-label)*, ***peaches*** *(off-label)*, ***pears*** *(off-label)*, ***plums*** *(off-label)*, ***quinces*** *(off-label)*, ***wine grapes*** *(off-label)* [40]; ***amenity grassland***, ***amenity grassland*** *(wiper application)*, ***amenity vegetation*** *(wiper application)*, ***broad-leaved trees***, ***conifers***, ***farm buildings/yards***, ***fencelines***, ***land clearance***, ***non-crop areas***, ***paths and drives***, ***walls***, ***woody ornamentals*** [45]; ***amenity vegetation*** [1, 2, 4, 6, 10-12, 14, 15, 17, 18, 28, 31, 32, 36, 37, 44-46, 49-51]; ***apple orchards***, ***pear orchards*** [1-4, 6, 8, 9, 11-18, 20, 21, 23, 24, 27-31, 33-35, 37, 39, 41, 42, 47-50, 54]; ***apples***, ***pears*** [22, 36, 44]; ***asparagus*** [4, 6, 11, 12, 28, 44]; ***asparagus*** *(off-label)*, ***bilberries*** *(off-label)*, ***blackcurrants*** *(off-label)*, ***blueberries*** *(off-label)*, ***chestnuts*** *(off-label)*, ***cob nuts*** *(off-label)*, ***cranberries*** *(off-label)*, ***gooseberries*** *(off-label)*, ***hazel nuts*** *(off-label)*, ***redcurrants*** *(off-label)*, ***rhubarb*** *(off-label)*, ***vaccinium spp.*** *(off-label)*, ***walnuts*** *(off-label)*, ***whitecurrants*** *(off-label)* [3, 40]; ***buckwheat*** *(off-label - grown for wild bird seed production)*, ***canary seed*** *(off-label - grown for wild bird seed production)*, ***hemp*** *(off-label - grown for wild bird seed production)*, ***millet*** *(off-label - grown for wild bird seed production)*, ***quinoa*** *(off-label - grown for wild bird seed production)*, ***sorghum*** *(off-label - grown for wild bird seed production)* [39, 42]; ***bulb onions***, ***vining peas*** [1, 2, 14, 15, 17, 18, 37, 44, 50]; ***cherries*** [1-4, 6, 8, 9, 11-18, 20-24, 27-29, 31, 33, 35-37, 39, 41, 42, 44, 47-50, 54]; ***combining peas***, ***linseed***, ***spring barley***, ***spring field beans***, ***spring oilseed rape***, ***spring wheat***, ***winter barley***, ***winter field beans***, ***winter oilseed rape***, ***winter wheat*** [1-3, 14, 15, 17, 18, 22, 30, 37, 44, 50, 53]; ***cultivated land/soil*** [3, 20, 48]; ***damsons*** [1-4, 6, 8, 9, 11-18, 20-24, 27-29, 31, 33-37, 39-42, 44, 47-50, 54]; ***durum wheat*** [1-3, 14, 15, 17, 18, 30, 37, 44, 50]; ***enclosed waters***, ***open waters*** [5, 7, 22, 30, 36]; ***farm forestry*** [22]; ***farm forestry*** *(off-label)*, ***game cover*** *(off-label)*, ***hops*** *(off-label)*, ***miscanthus*** *(off-label)*, ***soft fruit*** *(off-label)*, ***top fruit*** *(off-label)*, ***woad*** *(off-label)* [39-43, 47]; ***field crops*** *(wiper application)*, ***non-crop farm areas*** [20, 48]; ***forest*** [1, 2, 4-9, 11-23, 26-33, 35-37, 44-46, 49-52, 54]; ***forestry plantations*** [1, 2, 14, 15, 17, 18, 37, 50]; ***grapevines*** *(off-label)* [3]; ***green cover on land temporarily removed from production*** [1, 2, 14, 15, 17, 18, 30, 32, 37, 44, 50]; ***hard surfaces*** [4-8, 11, 12, 16, 19, 22, 27, 28, 30-32, 36, 38, 44-46, 49, 51, 52]; ***hemp*** *(off-label)* [40, 41, 43, 47]; ***industrial sites***, ***road verges*** [26, 45]; ***land immediately adjacent to aquatic areas*** [5, 7, 8, 22, 30, 31, 36, 46]; ***land not intended to bear vegetation*** [8-10, 13, 20, 21, 23, 26, 31, 33-35, 40, 43, 45, 48, 54]; ***land prior to cultivation*** [27, 52]; ***managed amenity turf*** *(pre-establishment)* [32, 51]; ***managed amenity turf*** *(pre-establishment only)* [36]; ***mustard*** [1, 2, 14, 15, 17, 18, 22, 30, 37, 44, 50, 53]; ***natural surfaces not intended to bear vegetation*** [1, 2, 4-8, 11, 12, 14-19, 22, 27, 28, 30-32, 36-38, 44, 46, 49-52]; ***onions*** [22, 30]; ***ornamental plant production*** [36, 51]; ***ornamental plant production*** *(off-label)* [14]; ***ornamental specimens*** *(off-label)* [45, 46]; ***permanent grassland*** [9, 20, 29, 34, 39-42, 48]; ***permeable surfaces overlying soil*** [1, 2, 4-8, 11, 12, 14-19, 22, 27, 28, 30-32, 37, 38, 44, 46, 49-52]; ***plums*** [1-4, 6, 8, 9, 11-18, 20-24, 27-29, 31, 33-37, 39, 41, 42, 44, 47-50, 54]; ***spring oats***, ***winter oats*** [1-3, 14, 15, 17, 18, 22, 30, 37, 44, 50]; ***stubbles*** [3, 4, 6, 8, 9, 11, 12, 20, 28, 29, 31, 34, 39-43, 47-49, 53]; ***water or waterside areas*** [19]
- Annual dicotyledons in ***bulb onions***, ***sugar beet***, ***swedes*** [4, 6, 11, 12, 24, 34, 39-43, 47]; ***combining peas*** [4, 6, 11, 12, 16, 24, 43]; ***cultivated land/soil*** [8, 31]; ***durum wheat***, ***linseed***, ***mustard***, ***spring field beans***, ***spring oilseed rape***, ***winter field beans*** [4, 6, 11, 12, 16, 24, 34, 39-43, 47]; ***leeks*** [4, 6, 11, 12, 24, 39-43, 47]; ***peas*** [34, 40]; ***permanent grassland*** [34, 39-42, 47]; ***spring barley***, ***spring oats***, ***spring wheat***, ***winter barley***, ***winter oats***, ***winter wheat*** [4, 6, 11-13, 16, 21, 23, 24, 27, 33-35, 39-43, 47, 54]; ***turnips*** [24, 34, 39-43, 47]; ***vining peas*** [4, 6, 11, 12, 39, 41-43, 47]; ***winter oilseed rape*** [4, 6, 11, 12, 16, 24, 28, 34, 39-43, 47, 49]
- Annual grasses in ***bulb onions***, ***durum wheat***, ***linseed***, ***mustard***, ***spring barley***, ***spring field beans***, ***spring oats***, ***spring oilseed rape***, ***spring wheat***, ***sugar beet***, ***swedes***, ***turnips***, ***winter barley***, ***winter field beans***, ***winter oats***, ***winter oilseed rape***, ***winter wheat*** [24, 34, 39-43, 47]; ***combining peas*** [24, 43]; ***cultivated land/soil*** [8, 13, 21, 23, 31, 33, 35, 54]; ***leeks*** [24, 39-43, 47]; ***peas*** [34, 40]; ***permanent grassland*** [34, 39-42, 47]; ***vining peas*** [39, 41-43, 47]
- Aquatic weeds in ***open waters*** [8, 31, 46]
- Black bent in ***combining peas***, ***spring field beans***, ***spring oilseed rape***, ***stubbles***, ***winter field beans***, ***winter oilseed rape*** [8, 13, 21, 23, 27, 31, 33, 35, 54]; ***durum wheat***, ***spring barley***,

spring oats*, *spring wheat*, *winter barley*, *winter oats*, *winter wheat [8, 31]; ***linseed*** [13, 21, 23, 27, 33, 35, 54]

- Bolters in ***sugar beet*** *(wiper application)* [8, 20, 24, 31, 34, 39-42, 47, 48]
- Bracken in ***amenity vegetation*, *fencelines*, *land clearance*, *non-crop areas*, *paths and drives*** [45]; ***forest*** [4, 6, 8, 9, 11-13, 16, 20, 21, 23, 25-29, 31, 33, 35, 45, 46, 49, 52, 54]; ***tolerant conifers*** [26, 45]
- Brambles in ***tolerant conifers*** [26, 45]
- Canary grass in ***aquatic areas*** [24]; ***enclosed waters*, *land immediately adjacent to aquatic areas*** [25]
- Chemical thinning in ***forest*** [4, 6, 8, 11-13, 16, 21, 23, 25, 27-29, 31, 33, 35, 45, 46, 49, 52, 54]; ***forest*** *(stump treatment)* [9]
- Couch in ***combining peas*, *spring field beans*, *spring oilseed rape*, *stubbles*, *winter field beans*, *winter oilseed rape*** [3, 8, 9, 13, 20, 21, 23, 24, 27, 29, 31, 33-35, 39-42, 47, 48, 53, 54]; ***durum wheat*** [3, 8, 24, 31, 34, 39-42, 47]; ***forest*** [8, 25, 29, 31, 45, 46]; ***linseed*** [3, 9, 13, 21, 23, 24, 27, 29, 33-35, 39-42, 47, 53, 54]; ***mustard*** [9, 29, 34, 39-42, 47, 53]; ***permanent grassland*** [9]; ***spring barley*, *spring wheat*, *winter barley*, *winter wheat*** [3, 8, 9, 20, 24, 29, 31, 34, 39-42, 47, 48, 53]; ***spring oats*, *winter oats*** [3, 8, 9, 20, 24, 29, 31, 34, 39-42, 47, 48]
- Creeping bent in ***aquatic areas*** [24]; ***combining peas*, *spring field beans*, *spring oilseed rape*, *stubbles*, *winter field beans*, *winter oilseed rape*** [8, 13, 21, 23, 27, 31, 33, 35, 54]; ***durum wheat*, *spring barley*, *spring oats*, *spring wheat*, *winter barley*, *winter oats*, *winter wheat*** [8, 31]; ***enclosed waters*, *land immediately adjacent to aquatic areas*** [25]; ***linseed*** [13, 21, 23, 27, 33, 35, 54]
- Desiccation in ***buckwheat*** *(off-label - for wild bird seed production)*, ***canary seed*** *(off-label - for wild bird seed production)*, ***hemp*** *(off-label - for wild bird seed production)*, ***millet*** *(off-label - for wild bird seed production)*, ***quinoa*** *(off-label - for wild bird seed production)*, ***sorghum*** *(off-label - for wild bird seed production)* [41, 43]; ***lupins*** *(off-label)*, ***poppies for morphine production*** *(off-label)* [41]
- Destruction of short term leys in ***grassland*** [32]
- Grass weeds in ***all edible crops (outdoor and protected)*, *all non-edible crops (outdoor)*, *apple orchards*, *combining peas*, *durum wheat*, *enclosed waters*, *grassland*, *green cover on land temporarily removed from production*, *land immediately adjacent to aquatic areas*, *leeks*, *linseed*, *mustard*, *onions*, *open waters*, *pear orchards*, *spring barley*, *spring field beans*, *spring oats*, *spring oilseed rape*, *spring wheat*, *sugar beet*, *swedes*, *turnips*, *winter barley*, *winter field beans*, *winter oats*, *winter oilseed rape*, *winter wheat*** [30]; ***forest*, *hard surfaces*, *natural surfaces not intended to bear vegetation*, *permeable surfaces overlying soil*** [19, 30]; ***water or waterside areas*** [19]
- Green cover in ***land temporarily removed from production*** [3, 4, 6, 8, 9, 11-13, 16, 21, 23, 24, 27, 28, 31, 33-35, 39-43, 47, 49, 53, 54]
- Harvest management/desiccation in ***combining peas*, *spring field beans*, *winter field beans*** [4, 6, 9, 11, 12, 16, 20, 28, 29, 34, 39-42, 47-49]; ***durum wheat*** [4, 6, 11, 12, 16, 20, 24, 28, 34, 39-42, 47-49]; ***linseed*** [4, 6, 8, 9, 11, 12, 16, 20, 28, 29, 31, 34, 39-42, 47-49]; ***mustard*** [4, 6, 9, 11, 12, 16, 20, 28, 29, 34, 39-42, 47-49, 53]; ***spring barley*, *winter barley*** [4, 6, 9, 11-13, 16, 21, 23, 24, 27-29, 33-35, 39-42, 47, 49, 54]; ***spring oats*, *spring wheat*, *winter oats*, *winter wheat*** [4, 6, 9, 11-13, 16, 20, 21, 23, 24, 27-29, 33-35, 39-42, 47-49, 54]; ***spring oilseed rape*, *winter oilseed rape*** [3, 4, 6, 8, 9, 11-13, 16, 20, 21, 23, 24, 27-29, 31, 33-35, 39-42, 47-49, 53, 54]
- Heather in ***forest*** [4, 6, 8, 9, 11-13, 16, 20, 21, 23, 25-28, 31, 33, 35, 46, 49, 52, 54]
- Perennial dicotyledons in ***combining peas*, *linseed*, *spring field beans*, *spring oilseed rape*, *winter field beans*, *winter oilseed rape*** [13, 21, 23, 24, 27, 33, 35, 54]; ***forest*** [52]; ***forest*** *(wiper application)* [45]; ***permanent grassland*** *(wiper application)* [34, 39-42, 47]
- Perennial grasses in ***aquatic areas*** [3, 8, 9, 20, 21, 23, 26, 31, 33, 35, 45, 52]; ***combining peas*, *linseed*, *spring field beans*, *stubbles*, *winter field beans*** [24]; ***enclosed waters*, *land immediately adjacent to aquatic areas*** [52]; ***tolerant conifers*** [26, 45]
- Pre-harvest desiccation in ***combining peas*, *linseed*, *spring barley*, *spring field beans*, *spring wheat*, *winter barley*, *winter field beans*, *winter wheat*** [53]; ***poppies for morphine production*** *(off-label)* [40]
- Reeds in ***aquatic areas*** [3, 8, 9, 20, 21, 23, 24, 26, 31, 33, 35, 40, 45, 52]; ***enclosed waters*, *land immediately adjacent to aquatic areas*** [8, 25, 31, 46, 52]
- Rhododendrons in ***forest*** [4, 6, 8, 9, 11-13, 16, 20, 21, 23, 25-28, 31, 33, 35, 45, 46, 49, 52, 54]

- Rushes in ***aquatic areas*** [3, 8, 9, 20, 21, 23, 24, 26, 31, 33, 35, 40, 45, 52]; ***enclosed waters, land immediately adjacent to aquatic areas*** [8, 25, 31, 46, 52]; ***forest*** [29]
- Sedges in ***aquatic areas*** [3, 8, 9, 20, 21, 23, 24, 26, 31, 33, 35, 40, 45, 52]; ***enclosed waters, land immediately adjacent to aquatic areas*** [8, 25, 31, 46, 52]
- Sucker control in ***apple orchards, damsons, pear orchards, plums*** [34, 39-42, 48]; ***cherries*** [39-42, 48]
- Sward destruction in ***amenity grassland*** [45]; ***permanent grassland*** [3, 4, 6, 8, 9, 11-13, 16, 20, 21, 23, 24, 27-29, 31, 33-35, 39-42, 47-49, 53, 54]; ***rotational grass*** [4, 6, 11, 12, 16, 28, 49, 53]
- Total vegetation control in ***all edible crops (outdoor)*** *(pre-sowing/planting)*, ***all non-edible crops (outdoor)*** *(pre-sowing/planting)* [39, 41, 42, 46]; ***amenity vegetation*** [10, 26, 45]; ***fencelines, industrial sites, road verges*** [26, 45]; ***hard surfaces*** [25, 26, 45]; ***land not intended to bear vegetation*** [10, 26, 45, 53]; ***natural surfaces not intended to bear vegetation, permeable surfaces overlying soil*** [25]
- Volunteer cereals in ***bulb onions, linseed, mustard, spring field beans, sugar beet, swedes, turnips, winter field beans*** [24, 34, 39-43, 47]; ***combining peas*** [24, 43]; ***cultivated land/soil*** [8, 13, 21, 23, 31, 33, 35, 54]; ***durum wheat, spring barley, spring oats, spring oilseed rape, spring wheat, winter barley, winter oats, winter oilseed rape, winter wheat*** [34, 39-43, 47]; ***leeks*** [24, 39-43, 47]; ***peas*** [34, 40]; ***stubbles*** [3, 8, 9, 13, 20, 21, 23, 24, 27, 29, 31, 33-35, 39-42, 47, 48, 53, 54]; ***vining peas*** [39, 41-43, 47]
- Volunteer potatoes in ***stubbles*** [3, 8, 9, 20, 24, 29, 31, 34, 39-42, 47, 48, 53]
- Waterlilies in ***aquatic areas*** [3, 8, 9, 20, 21, 23, 24, 26, 31, 33, 35, 45, 52]; ***enclosed waters*** [8, 25, 31, 46]
- Weed beet in ***sugar beet*** *(wiper application)* [24, 48]
- Wild oats in ***durum wheat, spring barley, spring oats, spring wheat, winter barley, winter oats, winter wheat*** [20, 48]
- Woody weeds in ***forest*** [4, 6, 8, 9, 11-13, 16, 20, 21, 23, 25-29, 31, 33, 35, 45, 46, 49, 52, 54]; ***tolerant conifers*** [26, 45]

Specific Off-Label Approvals (SOLAs)

- ***almonds*** *20072045* [40]
- ***apples*** *20072045* [40]
- ***apricots*** *20072045* [40]
- ***asparagus*** *20071476* [3], *20062036* [40]
- ***bilberries*** *20071475* [3], *20062035* [40]
- ***blackcurrants*** *20071475* [3], *20062035* [40]
- ***blueberries*** *20071475* [3], *20062035* [40]
- ***buckwheat*** *(grown for wild bird seed production) 20072936* [39], *(grown for wild bird seed production) 20072934* [42], *(for wild bird seed production) 20070582* [43]
- ***canary seed*** *(grown for wild bird seed production) 20072936* [39], *(grown for wild bird seed production) 20072934* [42], *(for wild bird seed production) 20070582* [43]
- ***cherries*** *20072045* [40]
- ***chestnuts*** *20071477* [3], *20072045* [40]
- ***cob nuts*** *20071477* [3], *20072045* [40]
- ***crab apples*** *20072045* [40]
- ***cranberries*** *20071475* [3], *20062035* [40]
- ***farm forestry*** *20082887* [39], *20082869* [40], *20082888* [41], *20082889* [42], *20082891* [43], *20082892* [47]
- ***game cover*** *20082887* [39], *20082869* [40], *20082888* [41], *20082889* [42], *20082891* [43], *20082892* [47]
- ***gooseberries*** *20071475* [3], *20062035* [40]
- ***grapevines*** *20071479* [3]
- ***hazel nuts*** *20071477* [3], *20072045* [40]
- ***hemp*** *(grown for wild bird seed production) 20072937* [39], *20082869* [40], *20082888* [41], *(grown for wild bird seed production) 20072935* [42], *(for wild bird seed production) 20070581* [43], *20082891* [43], *20082892* [47]
- ***hops*** *20082887* [39], *20082869* [40], *20082888* [41], *20082889* [42], *20082891* [43], *20082892* [47]
- ***millet*** *(grown for wild bird seed production) 20072936* [39], *(grown for wild bird seed production) 20072934* [42], *(for wild bird seed production) 20070582* [43]

- ***miscanthus*** *20082887* [39], *20082869* [40], *20082888* [41], *20082889* [42], *20082891* [43], *20082892* [47]
- ***ornamental plant production*** *20082840* [14]
- ***ornamental specimens*** *20082877* [45], *20082890* [46]
- ***peaches*** *20072045* [40]
- ***pears*** *20072045* [40]
- ***plums*** *20072045* [40]
- ***poppies for morphine production*** *20062030* [40], *20081058* [41]
- ***quinces*** *20072045* [40]
- ***quinoa*** *(grown for wild bird seed production) 20072937* [39], *(grown for wild bird seed production) 20072935* [42], *(for wild bird seed production) 20070581* [43]
- ***redcurrants*** *20071475* [3], *20062035* [40]
- ***rhubarb*** *20071478* [3], *20062029* [40]
- ***soft fruit*** *20082887* [39], *20082869* [40], *20082888* [41], *20082889* [42], *20082891* [43], *20082892* [47]
- ***sorghum*** *(grown for wild bird seed production) 20072936* [39], *(grown for wild bird seed production) 20072934* [42], *(for wild bird seed production) 20070582* [43]
- ***top fruit*** *20082887* [39], *20082869* [40], *20082888* [41], *20082889* [42], *20082891* [43], *20082892* [47]
- ***vaccinium spp.*** *20071475* [3], *20062035* [40]
- ***walnuts*** *20071477* [3], *20072045* [40]
- ***whitecurrants*** *20071475* [3], *20062035* [40]
- ***wine grapes*** *20062031* [40]
- ***woad*** *20082887* [39], *20082869* [40], *20082888* [41], *20082889* [42], *20082891* [43], *20082892* [47]

Approval information

- Glyphosate included in Annex I under EC Directive 91/414
- Accepted by BBPA for use on malting barley and hops

Efficacy guidance

- For best results apply to actively growing weeds with enough leaf to absorb chemical
- For most products a rainfree period of at least 6 h (preferably 24 h) should follow spraying
- Adjuvants are obligatory for some products and recommended for some uses with others. See labels
- Mixtures with other pesticides or fertilizers may lead to reduced control.
- Products are formulated as isopropylamine, ammonium, potassium, or trimesium salts of glyphosate and may vary in the details of efficacy claims. See individual product labels
- Some products in ready-to-use formulations for use through hand-held applicators. See label for instructions [10, 32, 51]
- When applying products through rotary atomisers the spray droplet spectra must have a minimum Volume Median Diameter (VMD) of 200 microns
- With wiper application weeds should be at least 10 cm taller than crop
- Annual weed grasses should have at least 5 cm of leaf and annual broad-leaved weeds at least 2 expanded true leaves
- Perennial grass weeds should have 4-5 new leaves and be at least 10 cm long when treated. Perennial broad-leaved weeds should be treated at or near flowering but before onset of senescence
- Volunteer potatoes and polygonums are not controlled by harvest-aid rates
- Bracken must be treated at full frond expansion
- Fruit tree suckers best treated in late spring
- Chemical thinning treatment can be applied as stump spray or stem injection
- In order to allow translocation, do not cultivate before spraying and do not apply other pesticides, lime, fertilizer or farmyard manure within 5 d of treatment
- Recommended intervals after treatment and before cultivation vary. See labels

Restrictions

- Maximum total dose per crop or season normally equivalent to one full dose treatment on field and edible crops and no restriction for non-crop uses. However some older labels indicate a maximum number of treatments. Check for details
- Do not treat cereals grown for seed or undersown crops

- Consult grain merchant before treating crops grown on contract or intended for malting
- Do not use treated straw as a mulch or growing medium for horticultural crops
- For use in nursery stock, shrubberies, orchards, grapevines and tree nuts care must be taken to avoid contact with the trees. Do not use in orchards established less than 2 yr and keep off low-lying branches
- Certain conifers may be sprayed overall in dormant season. See label for details
- Use a tree guard when spraying in established forestry plantations
- Do not spray root suckers in orchards in late summer or autumn
- Do not use under glass or polythene as damage to crops may result
- Do not mix, store or apply in galvanised or unlined mild steel containers or spray tanks
- Do not leave diluted chemical in spray tanks for long periods and make sure that tanks are well vented

Crop-specific information

- Harvest intervals: 4 wk for blackcurrants, blueberries; 14 d for linseed, oilseed rape; 8 d for mustard; 5-7 d for all other edible crops. Check label for exact details
- Latest use: for most products 2-14 d before cultivating, drilling or planting a crop in treated land; after harvest (post-leaf fall) but before bud formation in the following season for nuts and most fruit and vegetable crops; before fruit set for grapevines. See labels for details

Following crops guidance

- Decaying remains of plants killed by spraying must be dispersed before direct drilling
- Crops may be drilled 48 h after application. Trees and shrubs may be planted 7 d after application.

Environmental safety

- Products differ in their hazard and environmental safety classification. See labels
- Do not dump surplus herbicide in water or ditch bottoms or empty into drains
- Check label for maximum permitted concentration in treated water
- The Environment Agency or Local River Purification Authority must be consulted before use in or near water
- LERAP Category B [47]
- Take extreme care to avoid drift and possible damage to neighbouring crops or plants
- Treated poisonous plants must be removed before grazing or conserving
- Do not use in covered areas such as greenhouses or under polythene
- For field edge treatment direct spray away from hedge bottoms
- Do not dump surplus herbicide in water or ditch bottoms
- Some products require livestock to be excluded from treated areas and do not permit treated forage to be used for hay, silage or bedding. Check label for details

Hazard classification and safety precautions

Hazard Harmful [12, 20, 28, 48-50]; Irritant [4, 5, 9, 11, 13-16, 22, 29, 37, 43, 44, 47, 51, 54]; Dangerous for the environment [4, 5, 11-16, 20, 22, 28, 29, 32, 33, 35-37, 43, 44, 47-51, 53, 54]

Transport code 9 [1, 2, 4, 5, 14-16, 20, 23, 26-29, 32-35, 37, 43, 44, 47, 48, 50, 51, 54]

Packaging group III [1, 2, 4, 5, 14-16, 20, 23, 26-29, 32-35, 37, 43, 44, 47, 48, 50, 51, 54]

UN number 3077 [43]; 3082 [1, 2, 4, 5, 14-16, 20, 23, 26-29, 32-35, 37, 44, 47, 48, 50, 51, 54]; N/C [3, 6-13, 17-19, 21, 24, 25, 30, 36, 38-42, 45, 46, 49, 52, 53]

Risk phrases R20 [4, 11, 12, 20, 48, 50]; R36 [20, 29, 51]; R38 [44, 47]; R41 [4, 9, 11-16, 22, 28, 37, 43, 48-50, 54]; R50 [32, 36, 44, 47]; R51 [4, 11-16, 20, 22, 28, 29, 33, 35, 37, 43, 48-51, 54]; R52 [24, 25, 27, 39, 41, 42, 46, 53]; R53a [1, 2, 4, 11-18, 20, 22, 24, 25, 27-29, 32, 33, 35-37, 39, 41-44, 46-51, 53, 54]

Operator protection A [1-54]; C [3-9, 11-16, 20-23, 26-31, 33-35, 37, 39-43, 45, 46, 48-50, 52, 54]; D [3, 21, 23, 26, 33, 35, 52]; F [29, 48]; H [1-11, 13, 16-19, 21-27, 29-36, 40, 43-48, 51-54]; M [1-13, 16-27, 29-36, 39-48, 51, 52, 54]; N [13, 20, 22, 29, 48, 54]; U02a [1, 2, 5-8, 14, 15, 17-20, 22-25, 27-33, 35-37, 43, 49-51]; U05a [1, 2, 4, 11-13, 16-18, 20, 22, 29, 32, 36, 43, 44, 47, 48, 51, 53, 54]; U08 [5-8, 16, 20, 22-25, 27, 29, 31, 33, 35, 43]; U09a [13, 19, 30, 32, 36, 51, 54]; U11 [4, 9, 11-16, 20, 22, 24, 25, 28, 29, 37, 43, 48-51, 54]; U14 [22]; U15 [29]; U16b [51]; U19a [5-8, 10, 13, 16, 19, 20, 22-25, 27, 29-33, 35, 36, 43, 51, 54]; U20a [4, 11, 12, 19, 30, 34, 40, 47, 48, 53]; U20b [1-3, 5-8, 10, 13-18, 20-29, 31-33, 35-39, 41-46, 49-52, 54]

Environmental protection E06a [38] (2 weeks); E06d, E08b [53]; E13c [8, 23, 29, 31, 32, 36]; E15a [3, 10, 20, 21, 26, 27, 29, 33-35, 40, 43, 45, 47, 52]; E15b [1, 2, 4-7, 9, 11, 12, 14, 15, 17-19, 22, 24, 25, 28, 30, 32, 36-39, 41, 42, 44, 46, 48-51, 53]; E16a, E16b [44, 47]; E19a [3, 8, 20, 21, 23,

31, 33, 35, 52]; E19b [13, 54]; E34 [10, 17, 18, 27, 29]; E36a [17, 18]; E38 [4, 9, 11, 12, 14, 15, 20, 22, 24, 25, 27-29, 37, 43, 44, 47-51, 53]
Storage and disposal D01 [1-7, 9-13, 16-20, 22-27, 29, 30, 32-36, 39-48, 51-54]; D02 [1, 2, 4-7, 9-13, 16-20, 22, 24-26, 29, 30, 32, 34, 36, 39-48, 51, 53, 54]; D05 [1, 2, 4-8, 10-12, 14, 15, 17-19, 26-32, 36, 37, 39, 41-43, 45-51, 53]; D07 [43]; D09a [3-16, 19-54]; D10a [5-7, 10]; D10b [3, 8, 13, 16, 20-22, 29-31, 33, 52, 54]; D10c [1, 2, 4, 11, 12, 14, 15, 17, 18, 23, 26-28, 34, 35, 37-42, 44-50, 53]; D11a [9, 19, 24, 25, 32, 36, 43, 51]; D12a [4, 11-15, 20, 22, 24, 25, 27-29, 32, 36, 37, 43, 44, 47-51, 53, 54]; D12b [33, 35]; D14 [17, 18]
Medical advice M03 [53]; M05a [13-16, 28, 37, 49, 50, 54]

286 glyphosate + sulfosulfuron

A translocated non-residual glycine derivative herbicide + a sulfonyl urea herbicide
HRAC mode of action code: B + G

See also *sulfosulfuron*

Products

Nomix Dual	Nomix Enviro	120:2.22 g/l	RH	13420

Uses
- Annual and perennial weeds in ***amenity vegetation***, ***natural surfaces not intended to bear vegetation***, ***permeable surfaces overlying soil***

Approval information
- Glyphosate and sulfosulfuron included in Annex I under EC Directive 91/414

Hazard classification and safety precautions
Hazard Dangerous for the environment
Transport code 9
Packaging group III
UN number 3082
Risk phrases R50, R53a
Operator protection A, H, M; U02a, U09a, U19a, U20b
Environmental protection E15b
Storage and disposal D01, D05, D09a, D11a

287 guazatine

A guanidine fungicide seed dressing for cereals
FRAC mode of action code: M7

Products

1	Panoctine	Makhteshim	300 g/l	LS	11404
2	Ravine	Makhteshim	300 g/l	LS	12211

Uses
- Brown foot rot in ***spring barley*** *(seed treatment - reduction)*, ***spring oats*** *(seed treatment - reduction)*, ***winter barley*** *(seed treatment - reduction)*, ***winter oats*** *(seed treatment - reduction)* [1]
- Foot rot in ***spring barley*** *(seed treatment)*, ***winter barley*** *(seed treatment)* [1]
- Fusarium foot rot and seedling blight in ***spring barley*** *(seed treatment - reduction)*, ***spring oats*** *(seed treatment - reduction)*, ***spring wheat*** *(seed treatment - reduction)*, ***winter barley*** *(seed treatment - reduction)*, ***winter oats*** *(seed treatment - reduction)*, ***winter wheat*** *(seed treatment - reduction)* [1, 2]
- Leaf stripe in ***spring barley*** *(seed treatment)*, ***winter barley*** *(seed treatment)* [1]
- Net blotch in ***spring barley*** *(seed treatment)*, ***winter barley*** *(seed treatment)* [1]
- Pyrenophora leaf spot in ***spring oats*** *(seed treatment)*, ***winter oats*** *(seed treatment)* [1]
- Septoria seedling blight in ***spring wheat*** *(seed treatment)*, ***winter wheat*** *(seed treatment)* [2]; ***spring wheat*** *(seed treatment - partial control)*, ***winter wheat*** *(seed treatment - partial control)* [1]

Approval information

- Accepted by BBPA for use on malting barley
- Guazatine not included in Annex 1 and current approvals set to expire 2013.

Efficacy guidance

- Apply with conventional seed treatment machinery

Restrictions

- Maximum number of treatments 1 per batch
- Do not treat grain with moisture content above 16% and do not allow moisture content of treated seed to exceed 16%
- Do not apply to cracked, split or sprouted seed
- Do not use treated seed as food or feed

Crop-specific information

- Latest use: pre-drilling
- After treating, bag seed immediately and keep in dry, draught-free store
- Treatment may lower germination capacity, particularly if seed grown, harvested or stored under adverse conditions

Environmental safety

- Dangerous for the environment
- Very toxic to aquatic organisms
- Harmful to fish or other aquatic life. Do not contaminate surface waters or ditches with chemical or used container

Hazard classification and safety precautions

Hazard Harmful, Dangerous for the environment
Transport code 9
Packaging group III
UN number 3082
Risk phrases R20, R22a, R37, R41, R43, R50, R53a
Operator protection A, C, D, H; U05a, U20b [1, 2]; U09a, U14, U15 [1]
Environmental protection E13c [1]; E15a [2]; E34, E38 [1, 2]
Storage and disposal D01, D02, D05, D09a, D10a [1, 2]; D14 [1]
Treated seed S01, S02, S04b, S05, S06a, S07
Medical advice M03 [1]; M04a [1, 2]

288 guazatine + imazalil

A fungicide seed treatment for barley and oats
FRAC mode of action code: M7 + 3

See also imazalil

Products

1	Panoctine Plus	Makhteshim	300:25 g/l	LS	11757
2	Ravine Plus	Makhteshim	300:25 g/l	LS	11832

Uses

- Brown foot rot in ***spring barley*** *(seed treatment)*, ***winter barley*** *(seed treatment)*
- Covered smut in ***spring barley*** *(seed treatment)*, ***winter barley*** *(seed treatment)*
- Leaf stripe in ***spring barley*** *(seed treatment)*, ***winter barley*** *(seed treatment)*
- Net blotch in ***spring barley*** *(seed treatment - moderate control)*, ***winter barley*** *(seed treatment - moderate control)*
- Pyrenophora leaf spot in ***spring oats*** *(seed treatment)*, ***winter oats*** *(seed treatment)*
- Seedling blight and foot rot in ***spring barley*** *(seed treatment - moderate control)*, ***winter barley*** *(seed treatment - moderate control)*
- Snow mould in ***spring barley*** *(seed treatment)*, ***winter barley*** *(seed treatment)*

Approval information

- Imazalil included in Annex I under EC Directive 91/414
- Accepted by BBPA for use on malting barley

Efficacy guidance

- Apply with conventional seed treatment machinery

Restrictions

- Maximum number of treatments 1 per batch
- Do not mix with other formulations
- Do not treat grain with moisture content above 16% and do not allow moisture content of treated seed to exceed 16%
- Do not apply to cracked, split or sprouted seed
- Do not store treated seed for more than 6 mth
- Treatment may lower germination capacity, particularly if seed grown, harvested or stored under adverse conditions

Crop-specific information

- Latest use: pre-drilling

Environmental safety

- Dangerous for the environment
- Very toxic to aquatic organisms
- Do not use treated seed as food or feed

Hazard classification and safety precautions

Hazard Toxic, Dangerous for the environment
Transport code 6.1
Packaging group III
UN number 2902
Risk phrases R22a, R23, R37, R41, R43, R50, R53a
Operator protection A, C, D, H; U05a, U10, U11, U14, U19a, U20b
Environmental protection E15a, E34, E38
Storage and disposal D01, D02, D09a, D10b
Treated seed S02, S04b, S05, S07
Medical advice M04a

289 hymexazol

A systemic heteroaromatic fungicide for pelleting sugar beet seed
FRAC mode of action code: 32

Products

Tachigaren 70 WP	Summit Agro	70% w/w	WP	12568

Uses

- Aphanomyces cochlioides in ***red beet*** *(off-label - seed treatment)*
- Black leg in ***sugar beet*** *(seed treatment)*

Specific Off-Label Approvals (SOLAs)

- ***red beet*** *(seed treatment) 20062835*

Efficacy guidance

- Incorporate into pelleted seed using suitable seed pelleting machinery

Restrictions

- Maximum number of treatments 1 per batch of seed
- Do not use treated seed as food or feed

Crop-specific information

- Latest use: before planting sugar beet seed

Environmental safety

- Harmful to aquatic organisms
- Harmful to fish or other aquatic life. Do not contaminate surface waters or ditches with chemical or used container
- Treated seed harmful to game and wildlife

Hazard classification and safety precautions

Hazard Irritant, Highly flammable
Transport code 3

SEE SECTION 3 FOR PRODUCTS ALSO REGISTERED

Packaging group III
UN number 1325
Risk phrases R41, R52, R53a
Operator protection A, C, F; U05a, U11, U20b
Environmental protection E03, E13c
Storage and disposal D01, D02, D09a, D11a
Treated seed S01, S02, S03, S04a, S05

290 imazalil

A systemic and protectant imidazole fungicide
FRAC mode of action code: 3

See also guazatine + imazalil

Products

1	Fungaflor 100EC	Certis	100 g/l	EC	12978
2	Fungazil 100 SL	BASF	100 g/l	SL	11762
3	Fungazil 50LS	Certis	50 g/l	LS	14069
4	IMA 100 SL	Q-Chem	100 g/l	LS	13559
5	Imaz 100 SL	AgriChem BV	100 g/l	SL	13322
6	Imaz 200 EC	AgriChem BV	200 g/l	EC	12991
7	Magnate 100 SL	Makhteshim	100 g/l	SL	11705

Uses

- Dry rot in ***seed potatoes, ware potatoes*** [2, 4, 7]
- Fusarium in ***potatoes, seed potatoes*** [5]
- Gangrene in ***potatoes*** [5]; ***seed potatoes*** [2, 4, 5, 7]; ***ware potatoes*** [2, 4, 7]
- Leaf stripe in ***spring barley, winter barley*** [3]
- Net blotch in ***spring barley, winter barley*** [3]
- Powdery mildew in ***protected cucumbers*** [1, 6]; ***protected hops*** *(off-label)*, ***protected ornamentals*** *(off-label - only ornamentals grown on synthetic/inert material)*, ***protected soft fruit*** *(off-label)*, ***protected tomatoes*** *(off-label - only tomatoes grown on synthetic/inert material)* [1]
- Silver scurf in ***potatoes*** [5]; ***seed potatoes*** [2, 4, 5, 7]; ***ware potatoes*** [2, 4, 7]
- Skin spot in ***seed potatoes, ware potatoes*** [2, 4, 7]

Specific Off-Label Approvals (SOLAs)

- ***protected hops*** *20082854 expires 30 Jun 2010* [1]
- ***protected ornamentals*** *(only ornamentals grown on synthetic/inert material) 20082854 expires 30 Jun 2010* [1]
- ***protected soft fruit*** *20082854 expires 30 Jun 2010* [1]
- ***protected tomatoes*** *(only tomatoes grown on synthetic/inert material) 20072799 expires 30 Jun 2010* [1]

Approval information

- Imazalil included in Annex I under EC Directive 91/414
- Approval expiry 30 Jun 2010 [1]

Efficacy guidance

- For best control of skin and wound diseases of ware potatoes treat as soon as possible after harvest, preferably within 7-10 d, before any wounds have healed [2, 4, 7]
- Treat protected cucumbers as soon as mildew is present and repeat at 7-10 d intervals depending on disease pressure [1]
- Ensure spray has dried on treated cucumbers before nightfall [1]
- Treatment of protected cucumbers will also have activity against Botrytis and black stem rot [1]

Restrictions

- Maximum number of treatments 1 per batch of ware tubers [2, 4, 7]; 2 per batch of seed tubers [2, 4, 7]; 3 per crop of protected cucumbers [1]
- Consult processor before treating potatoes for processing [2, 4, 7]
- Do not treat protected cucumbers in bright sunny conditions [2, 4, 7]

Crop-specific information

- Latest use: during storage and before chitting for seed potatoes [2, 4, 7]
- HI 1 d for potatoes [4, 7]; 72 h for protected cucumbers [1]

- Apply to clean soil-free potatoes post-harvest before putting into store, or at first grading. A further treatment may be applied in early spring before planting [2, 4, 7]
- Apply through canopied hydraulic or spinning disc equipment preferably diluted with up to two litres water per tonne of potatoes to obtain maximum skin cover and penetration [2, 4, 7]
- Use on ware potatoes subject to discharges of imazalil from potato washing plants being within emission limits set by the UK monitoring authority [2, 4, 7]

Environmental safety

- Dangerous for the environment
- Toxic to aquatic organisms
- Do not empty into drains [2, 4, 7]
- Personal protective equipment requirements may vary for each pack size. Check label

Hazard classification and safety precautions

Hazard Harmful [1, 6]; Irritant [2-5, 7]; Flammable [3]; Dangerous for the environment [1, 4-7]
Transport code 9 [1, 3-7]
Packaging group III [1, 3-7]
UN number 3082 [1, 3-7]; N/C [2]
Risk phrases R22a [1, 6]; R36 [3, 7]; R41 [2, 4, 5]; R43 [1]; R50 [7]; R51 [1, 4-6]; R52 [2]; R53a [1, 2, 4-7]; R66, R67 [6]
Operator protection A [1-7]; C [2, 4, 5, 7]; D [7]; H [1-6]; U02a [1]; U04a [1, 2, 4, 5, 7]; U05a [1-7]; U09a [3]; U11 [2, 4, 5]; U14 [1, 2, 4, 6]; U15 [6]; U19a [1, 2, 4, 6, 7]; U20a [1, 2, 4, 7]; U20b [3, 5, 6]
Environmental protection E13c [3, 7]; E15a, E19b [2, 4]; E15b [1, 5, 6]; E34 [2, 4-7]; E38 [1, 5]
Consumer protection C01 [5]
Storage and disposal D01, D02, D09a [1-7]; D05 [2-4, 7]; D10a [3, 5, 7]; D10b [1, 6]; D10c [2, 4]; D12a [1, 2, 4, 5]; D12b [6]
Treated seed S01, S03, S05 [2-5, 7]; S02 [2-5]; S04a [2-4, 7]; S04b [5]; S06a [2, 4, 7]; S07 [3]
Medical advice M03 [3, 5-7]

291 imazalil + pencycuron

A fungicide mixture for treatment of seed potatoes
FRAC mode of action code: 3 + 20

See also *pencycuron*

Products

Monceren IM	Bayer CropScience	0.6:12.5% w/w	DS	11426

Uses

- Black scurf in ***potatoes*** *(tuber treatment)*
- Silver scurf in ***potatoes*** *(tuber treatment - reduction)*
- Stem canker in ***potatoes*** *(tuber treatment - reduction)*

Approval information

- Imazalil included in Annex I under EC Directive 91/414

Efficacy guidance

- Apply to clean seed tubers during planting (see label for suitable method) or sprinkle over tubers in chitting trays before loading into planter. It is essential to obtain an even distribution over tubers
- If seed tubers become damp from light rain distribution of product should not be affected. Tubers in the hopper should be covered if a shower interrupts planting

Restrictions

- Maximum number of treatments 1 per batch of tubers
- Operators must wear suitable respiratory equipment and gloves when handling product and when riding on planter. Wear gloves when handling treated tubers
- Do not use on tubers which have previously been treated with a dry powder seed treatment or hot water
- Do not use treated tubers for human or animal consumption

Crop-specific information

- Latest use: immediately before planting
- May be used on seed tubers previously treated with a liquid fungicide but not before 8 wk have elapsed if this contained imazalil

Environmental safety

- Harmful to aquatic organisms
- Harmful to fish or other aquatic life. Do not contaminate surface waters or ditches with chemical or used container
- Treated seed harmful to game and wildlife

Hazard classification and safety precautions

UN number N/C
Risk phrases R52, R53a
Operator protection A, D; U20b
Environmental protection E03, E13c
Storage and disposal D09a, D11a
Treated seed S01, S02, S03, S04a, S05, S06a

292 imazalil + thiabendazole

A fungicide mixture for treatment of seed potatoes; product approval expired 2008
FRAC mode of action code: 3 + 1

See also thiabendazole

293 imazamox

An imidazolinone contact and residual herbicide available only in mixtures
HRAC mode of action code: B

294 imazamox + pendimethalin

A pre-emergence broad-spectrum herbicide mixture for legumes
HRAC mode of action code: B + K1

See also pendimethalin

Products

Nirvana	BASF	16.7:250 g/l	EC	14256

Uses

- Annual dicotyledons in ***combining peas***, ***spring field beans***, ***vining peas***, ***winter field beans***

Approval information

- Imazamox and pendimethalin included in Annex I under EC Directive 91/414

Efficacy guidance

- Best results obtained from applications to fine firm seedbeds in the presence of adequate moisture
- Weed control may be reduced on cloddy seedbeds and on soils with over 6% organic matter
- Residual control may be reduced under prolonged dry conditions

Restrictions

- Maximum number of treatments 1 per crop
- Seed must be drilled to at least 2.5 cm of settled soil
- Do not use on soils containing more than 10% organic matter
- Do apply to soils that are waterlogged or are prone to waterlogging
- Do not apply if heavy rain is forecast
- Do not soil incorporate the product or disturb the soil after application
- Consult processors before use on crops destined for processing

Crop-specific information

- Latest use: pre-crop emergence for all crops
- Inadequately covered seed may result in cupping of the leaves after application from which recovery is normally complete
- Crop damage may occur on stony or gravelly soils especially if heavy rain follows treatment

Following crops guidance

- Winter wheat or winter barley may be drilled as a following crop provided 3 mth have elapsed since treatment and the land has been at least cultivated by a non-inversion technique such as discing
- Winter oilseed rape or other brassica crops should not be drilled as following crops
- A minimum of 12 mth must elapse between treatment and sowing red beet, sugar beet or spinach

Environmental safety

- Dangerous for the environment
- Very toxic to aquatic organisms
- LERAP Category B

Hazard classification and safety precautions

Hazard Irritant, Dangerous for the environment
Transport code 9
Packaging group III
UN number 3082
Risk phrases R38, R43, R50, R53a
Operator protection A, H; U05a, U14, U20b
Environmental protection E15b, E16a, E38
Storage and disposal D01, D02, D09a, D10c, D12a
Medical advice M03

295 imazaquin

An imidazolinone herbicide and plant growth regulator available only in mixtures
HRAC mode of action code: B

See also chlormequat + imazaquin

296 imidacloprid

A neonicotinoid insecticide for seed, soil, peat or foliar treatment
IRAC mode of action code: 4A

See also beta-cyfluthrin + imidacloprid
fuberidazole + imidacloprid + triadimenol

Products

1	Admire	Bayer CropScience	70% w/w	WG	11234
2	Couraze	Solufeed	5% w/w	GR	13862
3	Imidasect 5 GR	Fargro	5% w/w	GR	13581
4	Imidasect 5 GR	Fargro	5% w/w	GR	14142
5	Intercept 70WG	Scotts	70% w/w	WG	08585
6	Merit Turf	Bayer Environ.	0.5% w/w	GR	12415
7	Nuprid 600FS	Nufarm UK	600 g/l	FS	13855

Uses

- Aphids in ***amenity vegetation*** *(off-label)* [1]; ***bedding plants***, ***herbaceous perennials***, ***ornamental plant production***, ***pot plants*** [5]; ***fodder beet***, ***sugar beet*** [7]; ***ornamental plant production*** *(container grown)* [3, 4]; ***protected ornamentals*** [2-4]
- Chafer grubs in ***managed amenity turf*** [6]
- Damson-hop aphid in ***hops*** [1]
- Flea beetle in ***fodder beet***, ***sugar beet*** [7]
- Glasshouse whitefly in ***bedding plants***, ***herbaceous perennials***, ***ornamental plant production***, ***pot plants*** [5]; ***ornamental plant production*** *(container grown)*, ***protected ornamentals*** [3, 4]
- Leaf miner in ***fodder beet***, ***sugar beet*** [7]
- Leatherjackets in ***managed amenity turf*** [6]
- Millipedes in ***fodder beet***, ***sugar beet*** [7]
- Pygmy beetle in ***fodder beet***, ***sugar beet*** [7]

- Sciarid flies in ***bedding plants, herbaceous perennials, ornamental plant production, pot plants*** [5]; ***ornamental plant production*** *(container grown)*, ***protected ornamentals*** [3, 4]
- Springtails in ***fodder beet, sugar beet*** [7]
- Symphylids in ***fodder beet, sugar beet*** [7]
- Tobacco whitefly in ***bedding plants, herbaceous perennials, ornamental plant production, pot plants*** [5]; ***ornamental plant production*** *(container grown)*, ***protected ornamentals*** [3, 4]
- Vine weevil in ***bedding plants, herbaceous perennials, ornamental plant production, pot plants*** [5]; ***ornamental plant production*** *(container grown)*, ***protected ornamentals*** [3, 4]
- Whitefly in ***protected ornamentals*** [2]

Specific Off-Label Approvals (SOLAs)

- ***amenity vegetation*** *20080298* [1]

Approval information

- Imidacloprid is included in Annex I under EC Directive 91/414
- Accepted by BBPA for use on hops

Efficacy guidance

- Treated seed should be drilled within the season of purchase [7]
- Base of hop plants should be free of weeds and debris at application [1]
- Hop bines emerging away from the main stock or adjacent to poles may require a separate application [1]
- Uptake and movement within hops requires soil moisture and good growing conditions [1]
- Control may be impaired in plantations greater than 3640 plants/ha [1]
- To minimise likelihood of resistance in hops do not treat all the crop in any one yr [1]
- Best results on managed amenity turf achieved when treatment is followed by sufficient irrigation to move chemical through the thatch to wet the top 2.5 cm of soil [6]
- To avoid disturbing uniformity of treatment on turf do not mow until irrigation or rainfall has occurred [6]
- To minimise likelihood of resistance when using in compost adopt a planned programme to alternate with pesticides of different types or use other measures [3, 5]
- When applied as drench or incorporated as granules in moist compost imidacloprid is readily absorbed and translocated to aerial parts of plant [3, 5]

Restrictions

- Maximum number of treatments 1 per batch of seed [7] or growing medium [3, 5]; 1 per yr for hops and managed amenity turf [1, 6]
- Must not be used in compost that has already been treated with an imidacloprid-containing product, nor should compost be subsequently re-treated with imidacloprid [3, 5]
- Product formulated for use only as a compost incorporation treatment into peat-based growing media using suitable automated equipment [3]
- For use only on container grown ornamentals [3, 5]
- Users should test treat a small number of all new or unaccustomed species or varieties to satisfy themselves of the safety of the treatment before applying to the bulk of the crop [3]
- Product must not be used on crops for human or animal consumption and treated compost must not be re-used for this purpose [3, 5]
- Do not use treated seed as food or feed
- Do not treat saturated or waterlogged turf, or when raining or in windy conditions [6]
- Avoid puddling or run-off of irrigation water after treatment [6]
- Do not use grass clippings from treated turf in situ, or use them for mulch [6]

Crop-specific information

- Latest use: before bines reach 2 m or before end 1st wk Jun for hops; before drilling for sugar beet; before sowing or planting for brassicas, bedding plants, hardy ornamental nursery stock, pot plants
- Apply to sugar beet seed as part of the normal commercial pelleting process using special treatment machinery [7]
- Apply to hops as a directed stem base spray before most bines reach a height of 2 m. If necessary treat both sides of the crown at half the normal concentration [1]
- The rate used per hop plant must be adjusted in accordance with the plant population if greater than 3640 per ha. See label for details [1]
- On managed amenity turf application should be made prior to egg-laying of the target pests [6]

- Low levels of leaf scorch on some *Agrostis* turf species if application is made in mixture with fertilisers [6]

Environmental safety

- Harmful to aquatic organisms [1]
- High risk to bees. Do not apply to crops in flower or to those in which bees are actively foraging. Do not apply when flowering weeds are present [1, 5]
- Treated seed harmful to game and wildlife [7]

Hazard classification and safety precautions

Hazard Harmful [1, 5, 7]; Irritant [6]
Transport code 9 [7]
Packaging group III [7]
UN number 3082 [7]; N/C [1-6]
Risk phrases R22a [1, 5, 7]; R36 [6]; R43 [7]; R52 [1]
Operator protection A [1-4, 6, 7]; C, H [7]; U04a, U13, U14 [7]; U05a [1, 3-7]; U20b [1-3, 5]; U20c [4, 6, 7]
Environmental protection E12a, E12e [1, 5]; E15a [1-7]; E34 [1, 2, 5, 7]
Storage and disposal D01 [1-3, 5-7]; D02 [1, 3, 5-7]; D09a, D11a [1-7]; D12a [4]
Medical advice M03 [7]

297 indol-3-ylacetic acid

A plant growth regulator for promoting rooting of cuttings

Products

1	Rhizopon A Powder	Rhizopon	1% w/w	DP	09087
2	Rhizopon A Tablets	Rhizopon	50 mg a.i.	WT	09088

Uses

- Rooting of cuttings in ***ornamental plant production***

Efficacy guidance

- Apply by dipping end of prepared cuttings into powder or dissolved tablet solution
- Shake off excess powder and make planting holes to prevent powder stripping off [1]
- Consult manufacturer for details of application by spray or total immersion

Restrictions

- Maximum number of treatments 1 per cutting
- Store product in a cool, dark and dry place
- Use solutions once only. Discard after use [2]
- Use plastic, not metallic, container for solutions [2]

Crop-specific information

- Latest use: before cutting insertion

Hazard classification and safety precautions

UN number N/C
Operator protection U19a [1]; U20a [1, 2]
Environmental protection E15a
Storage and disposal D09a, D11a

298 4-indol-3-ylbutyric acid

A plant growth regulator promoting the rooting of cuttings

Products

1	Chryzoplus Grey 0.8%	Rhizopon	0.8% w/w	DP	09094
2	Chryzopon Rose 0.1%	Rhizopon	0.1% w/w	DP	09092
3	Chryzosan White 0.6%	Rhizopon	0.6% w/w	DP	09093
4	Chryzotek Beige	Rhizopon	0.4% w/w	DP	09081
5	Chryzotop Green	Rhizopon	0.25% w/w	DP	09085
6	Rhizopon AA Powder (0.5%)	Rhizopon	0.5% w/w	DP	09082
7	Rhizopon AA Powder (1%)	Rhizopon	1% w/w	DP	09084
8	Rhizopon AA Powder (2%)	Rhizopon	2% w/w	DP	09083

Products – continued

9	Rhizopon AA Tablets	Rhizopon	50 mg a.i.	WT	09086
10	Seradix 1	Certis	0.1% w/w	DP	11330
11	Seradix 2	Certis	0.3% w/w	DP	11331
12	Seradix 3	Certis	0.8% w/w	DP	11332

Uses

- Rooting of cuttings in ***ornamental plant production***

Efficacy guidance

- Dip base of cuttings into powder immediately before planting
- Powders or solutions of different concentration are required for different types of cutting. Lowest concentration for softwood, intermediate for semi-ripe, highest for hardwood
- See label for details of concentration and timing recommended for different species
- Use of planting holes recommended for powder formulations to ensure product is not removed on insertion of cutting. Cuttings should be watered in if necessary

Restrictions

- Maximum number of treatments 1 per situation
- Use of too strong a powder or solution may cause injury to cuttings
- No unused moistened powder should be returned to container

Crop-specific information

- Latest use: before cutting insertion for ornamental specimens

Hazard classification and safety precautions

UN number N/C
Operator protection A, C [10-12]; U14, U15 [10-12]; U19a, U20a [1-12]
Environmental protection E15a
Storage and disposal D01 [10-12]; D09a, D11a [1-12]

299 4-indol-3-ylbutyric acid + 1-naphthylacetic acid

A plant growth regulator for promoting rooting of cuttings

See also 1-naphthylacetic acid

Products

Synergol	Certis	5.0:5.0 g/l	SL	07386

Uses

- Rooting of cuttings in ***ornamental plant production***

Efficacy guidance

- Dip base of cuttings into diluted concentrate immediately before planting
- Suitable for hardwood and softwood cuttings
- See label for details of concentration and timing for different species

Hazard classification and safety precautions

Hazard Harmful
UN number N/C
Risk phrases R22a
Operator protection A, C; U05a, U13, U20b
Environmental protection E15a, E34
Storage and disposal D01, D02, D09a, D11a, D12a
Medical advice M03

300 Indoxacarb

An oxadiazine insecticide for caterpillar control in a range of crops
IRAC mode of action code: 22A

Products

Steward	DuPont	30% w/w	WG	13149

SECTION 2

Uses

- Caterpillars in ***all edible seed crops grown outdoors*** *(off-label)*, ***all non-edible seed crops grown outdoors*** *(off-label)*, ***apples***, ***broccoli***, ***brussels sprouts*** *(off-label)*, ***cabbages***, ***cauliflowers***, ***cherries*** *(off-label)*, ***forest nurseries*** *(off-label)*, ***grapevines*** *(off-label)*, ***hops*** *(off-label)*, ***ornamental plant production*** *(off-label)*, ***pears***, ***protected aubergines***, ***protected courgettes***, ***protected cucumbers***, ***protected forest nurseries*** *(off-label)*, ***protected hops*** *(off-label)*, ***protected marrows***, ***protected melons***, ***protected ornamentals***, ***protected peppers***, ***protected pumpkins***, ***protected soft fruit*** *(off-label)*, ***protected squashes***, ***protected tomatoes***, ***protected top fruit*** *(off-label)*, ***soft fruit*** *(off-label)*, ***sweetcorn*** *(off-label)*, ***top fruit*** *(off-label)*
- Diamond-back moth in ***sweetcorn*** *(off-label)*

Specific Off-Label Approvals (SOLAs)

- ***all edible seed crops grown outdoors*** *20082905*
- ***all non-edible seed crops grown outdoors*** *20082905*
- ***brussels sprouts*** *20081485*
- ***cherries*** *20092300*
- ***forest nurseries*** *20082905*
- ***grapevines*** *20072289*
- ***hops*** *20082905*
- ***ornamental plant production*** *20082905*
- ***protected forest nurseries*** *20082905*
- ***protected hops*** *20082905*
- ***protected soft fruit*** *20082905*
- ***protected top fruit*** *20082905*
- ***soft fruit*** *20082905*
- ***sweetcorn*** *20072288*
- ***top fruit*** *20082905*

Approval information

- Indoxacarb included in Annex I under EC Directive 91/414

Efficacy guidance

- Best results in brassica and protected crops obtained from treatment when first caterpillars are detected, or when damage first seen, or 7-10 d after trapping first adults in pheromone traps
- In apples and pears apply at egg-hatch
- Subsequent treatments in all crops may be applied at 8-14 d intervals
- Indoxacarb acts by ingestion and contact. Only larval stages are controlled but there is some ovicidal action against some species

Restrictions

- Maximum number of treatments 3 per crop or yr for apples, pears, brassica crops; 6 per yr for protected crops
- Do not apply to any crop suffering from stress from any cause

Crop-specific information

- HI: brassica crops, protected crops 1 d; apples, pears 7 d

Environmental safety

- Dangerous for the environment
- Toxic to aquatic organisms
- Broadcast air-assisted LERAP
- In accordance with good agricultural practice apply in early morning or late evening when bees are less active

Hazard classification and safety precautions

Hazard Harmful, Dangerous for the environment
Transport code 9
Packaging group III
UN number 3077
Risk phrases R22a, R51, R53a
Operator protection A, D; U04a, U05a
Environmental protection E15b, E34, E38; E17b (18 m)

SEE SECTION 3 FOR PRODUCTS ALSO REGISTERED

Storage and disposal D01, D02, D09a, D12a
Medical advice M03

301 iodosulfuron-methyl-sodium

A post-emergence sulfonylurea herbicide for winter sown cereals
HRAC mode of action code: B

See also *amidosulfuron + iodosulfuron-methyl-sodium*
diflufenican + iodosulfuron-methyl-sodium + mesosulfuron-methyl

Products

Hussar	Bayer CropScience	5% w/w	WG	12364

Uses

- Annual dicotyledons in ***durum wheat*** *(off-label)*, ***grass seed crops*** *(off-label)*
- Chickweed in ***spring barley***, ***triticale***, ***winter rye***, ***winter wheat***
- Cleavers in ***triticale***, ***winter rye***, ***winter wheat***
- Fat hen in ***spring barley***
- Field pansy in ***spring barley***
- Field speedwell in ***spring barley***, ***triticale***, ***winter rye***, ***winter wheat***
- Hemp-nettle in ***spring barley***
- Italian ryegrass in ***triticale*** *(from seed)*, ***winter rye*** *(from seed)*, ***winter wheat*** *(from seed)*
- Ivy-leaved speedwell in ***triticale***, ***winter rye***, ***winter wheat***
- Mayweeds in ***spring barley***, ***triticale***, ***winter rye***, ***winter wheat***
- Perennial ryegrass in ***triticale*** *(from seed)*, ***winter rye*** *(from seed)*, ***winter wheat*** *(from seed)*
- Red dead-nettle in ***spring barley***, ***triticale***, ***winter rye***, ***winter wheat***
- Ryegrass in ***durum wheat*** *(off-label - from seed)*, ***grass seed crops*** *(off-label - from seed)*
- Volunteer oilseed rape in ***spring barley***

Specific Off-Label Approvals (SOLAs)

- ***durum wheat*** *(from seed) 20080912, 20080912*
- ***grass seed crops*** *(from seed) 20080912, 20080912*

Approval information

- Iodosulfuron-methyl-sodium included in Annex I under EC Directive 91/414
- Accepted by BBPA for use on malting barley

Efficacy guidance

- Best results obtained from treatment in warm weather when soil is moist and the weeds are growing actively
- Weeds must be present at application to be controlled
- Weed control is slow especially under cool dry conditions
- Dry conditions resulting in moisture stress may reduce effectiveness
- Occasionally weeds may only be stunted but they will normally have little or no competitive effect on the crop
- Iodosulfuron is a member of the ALS-inhibitor group of herbicides. To avoid the build up of resistance do not use any product containing an ALS-inhibitor herbicide with claims for control of grass weeds more than once on any crop
- Use these products as part of a resistance management strategy that includes cultural methods of control and does not use ALS inhibitors as the sole chemical method of grass weed control

Restrictions

- Maximum number of treatments 1 per crop
- Must only be applied between 1 Feb in yr of harvest and specified latest time of application
- Do not apply to undersown crops, or crops to be undersown
- Do not roll or harrow within 1 wk of spraying
- Do not spray crops under stress from any cause or if the soil is compacted
- Treat broadcast crops after the plants have a well-established root system
- Do not spray if rain imminent or frost expected
- Do not apply in mixture or in sequence with any other ALS inhibitor

Crop-specific information

- Latest use: before third node detectable (GS 33)

Following crops guidance

- No restrictions apply to the sowing of cereal crops or sugar beet in the spring of the yr following treatment

Environmental safety

- Dangerous for the environment
- Very toxic to aquatic organisms
- Dangerous to fish or other aquatic life. Do not contaminate surface waters or ditches with chemical or used container
- LERAP Category B
- Take extreme care to avoid damage by drift onto broad-leaved plants outside the target area or onto ponds, waterways and ditches
- Observe carefully label instructions for sprayer cleaning

Hazard classification and safety precautions

Hazard Irritant, Dangerous for the environment
Transport code 9
Packaging group III
UN number 3077
Risk phrases R41, R50, R53a
Operator protection A, C, H; U05a, U08, U11, U14, U15, U20b
Environmental protection E15b, E16a, E16b, E38
Storage and disposal D01, D02, D10a, D12a

302 iodosulfuron-methyl-sodium + mesosulfuron-methyl

A sulfonyl urea herbicide mixture for winter wheat
HRAC mode of action code: B + B

See also *mesosulfuron-methyl*

Products

1	Atlantis WG	Bayer CropScience	0.6:3.0% w/w	WG	12478
2	Greencrop Doonbeg 2	Greencrop	0.6:3.0 % w/w	WG	13025
3	Nemo	AgChem Access	0.6:3.0% w/w	WG	14140
4	Pacifica	Bayer CropScience	1.0:3.0% w/w	WG	12049

Uses

- Annual dicotyledons in ***durum wheat*** *(off-label)*, ***spring rye*** *(off-label)*, ***triticale*** *(off-label)*, ***winter rye*** *(off-label)* [1, 4]; ***spring wheat*** *(off-label)* [1]
- Annual grasses in ***durum wheat*** *(off-label)*, ***spring rye*** *(off-label)*, ***triticale*** *(off-label)*, ***winter rye*** *(off-label)* [1, 4]; ***spring wheat*** *(off-label)* [1]
- Annual meadow grass in ***winter wheat*** [1-4]
- Blackgrass in ***winter wheat*** [1-4]
- Chickweed in ***winter wheat*** [1-4]
- Italian ryegrass in ***winter wheat*** [1-4]
- Mayweeds in ***winter wheat*** [1-4]
- Perennial ryegrass in ***winter wheat*** [1-4]
- Rough meadow grass in ***winter wheat*** [1-4]
- Wild oats in ***winter wheat*** [1-4]

Specific Off-Label Approvals (SOLAs)

- ***durum wheat*** *20063529* [1], *20081264* [4]
- ***spring rye*** *20063529* [1], *20081264* [4]
- ***spring wheat*** *20082946* [1]
- ***triticale*** *20063529* [1], *20081264* [4]
- ***winter rye*** *20063529* [1], *20081264* [4]

Approval information

- Iodosulfuron-methyl-sodium and mesosulfuron-methyl included in Annex I under EC Directive 91/414

Efficacy guidance

- Optimum grass weed control obtained when all grass weeds are emerged at spraying. Activity is primarily via foliar uptake and good spray coverage of the target weeds is essential
- Translocation occurs readily within the target weeds and growth is inhibited within hours of treatment but symptoms may not be apparent for up to 4 wk, depending on weed species, timing of treatment and weather conditions
- Residual activity is important for best results and is optimised by treatment on fine moist seedbeds. Avoid application under very dry conditions
- Residual efficacy may be reduced by high soil temperatures and cloddy seedbeds
- Iodosulfuron-methyl and mesosulfuron-methyl are both members of the ALS-inhibitor group of herbicides. To avoid the build up of resistance do not use any product containing an ALS-inhibitor herbicide with claims for control of grass weeds more than once on any crop
- Use these products as part of a Resistance Management Strategy that includes cultural methods of control and does not use ALS inhibitors as the sole chemical method of grass weed control. See Section 5 for more information

Restrictions

- Maximum number of treatments 1 per crop with a maximum total dose equivalent to one full dose treatment
- Do not use on crops undersown with grasses, clover or other legumes or any other broad-leaved crop
- Do not use as a stand-alone treatment for blackgrass, ryegrass or chickweed control
- Do not use as the sole means of weed control in successive crops
- Do not use in mixture or in sequence with any other ALS-inhibitor herbicide except those (if any) specified on the label
- Do not mix with products containing isoproturon, chlorotoluron or linuron and allow at least 28 d after use of a residual urea herbicide before application, and at least 10 d after application before use of a residual urea herbicide [2]
- Do not apply earlier than 1 Feb in the yr of harvest [4]
- Do not apply to crops under stress from any cause
- Do not apply when rain is imminent or during periods of frosty weather
- Specified adjuvant must be used. See label

Crop-specific information

- Latest use: flag leaf just visible (GS 39)
- Winter wheat may be treated from the two-leaf stage of the crop
- Safety to crops grown for seed not established

Following crops guidance

- In the event of crop failure sow only winter wheat in the same cropping season
- Only winter wheat or winter barley (or winter oilseed rape [1]) may be sown in the year of harvest of a treated crop
- Spring wheat, spring barley, sugar beet or spring oilseed rape may be drilled in the following spring. Plough before drilling oilseed rape

Environmental safety

- Dangerous for the environment
- Very toxic to aquatic organisms
- Dangerous to fish or other aquatic life. Do not contaminate surface waters or ditches with chemical or used container
- LERAP Category B
- Take extreme care to avoid drift onto plants outside the target area or on to ponds, waterways or ditches

Hazard classification and safety precautions

Hazard Irritant, Dangerous for the environment
Transport code 9
Packaging group III
UN number 3077
Risk phrases R40 [4]; R41, R50, R53a [1-4]
Operator protection A, C, H; U05a, U11, U20b
Environmental protection E15a [1, 3, 4]; E15b [2]; E16a, E38 [1-4]
Storage and disposal D01, D02, D05, D09a, D10a, D12a

303 ioxynil

A contact acting HBN herbicide for use in turf and onions
HRAC mode of action code: C3

See also *bromoxynil + diflufenican + ioxynil*
bromoxynil + ethofumesate + ioxynil
bromoxynil + ioxynil
dichlorprop-P + ioxynil

Products

1	Totril	Bayer CropScience	225 g/l	EC	13100
2	Totril	Bayer CropScience	225 g/l	EC	14633

Uses
- Annual dicotyledons in ***bulb onions, garlic, leeks, salad onions, shallots*** [1, 2]; ***chives*** *(off-label)* [1]

Specific Off-Label Approvals (SOLAs)
- ***chives*** *20063162* [1]

Approval information
- Ioxynil included in Annex I under EC Directive 91/414
- Ioxynil was reviewed in 1995 and approvals for home garden use, and most hand held applications revoked
- Approval expiry 30 Nov 2010 [1]

Efficacy guidance
- Best results on seedling to 4-leaf stage weeds in active growth during mild weather

Restrictions
- Maximum number of treatments up to 4 per crop at split doses on onions; 1 per crop on carrots, garlic, leeks, shallots
- Do not apply by hand-held equipment or at concentrations higher than those recommended

Crop-specific information
- Latest use: pre-emergence for carrots, parsnips
- HI bulb onions, salad onions, shallots, garlic, leeks 14 d
- Apply to sown onion crops as soon as possible after plants have 3 true leaves or to transplanted crops when established

Environmental safety
- Dangerous for the environment
- Very toxic to aquatic organisms
- Flammable
- Harmful to bees. Do not apply to crops in flower or to those in which bees are actively foraging. Do not apply when flowering weeds are present
- Keep livestock out of treated areas for at least 6 wk after treatment and until foliage of any poisonous weeds such as ragwort has died and become unpalatable

Hazard classification and safety precautions

Hazard Harmful, Flammable, Dangerous for the environment
Transport code 3
Packaging group III
UN number 1993
Risk phrases R22a, R22b, R36, R37, R43, R50, R53a, R63, R66, R67
Operator protection A, C; U05a, U08, U19a, U20b, U23a
Environmental protection E06a (6 wk); E12d, E12e, E15a, E34, E38
Storage and disposal D01, D02, D05, D09a, D10b, D12a
Medical advice M03, M05b

304 ipconazole

A triazole fungicide for disease control in cereals
FRAC mode of action code: 3

Products

Rancona 15 ME	Chemtura	15 g/l	MS	14407

Uses

- Seed-borne diseases in ***spring barley***, ***spring wheat***, ***winter barley***, ***winter wheat***
- Soil-borne diseases in ***spring barley***, ***spring wheat***, ***winter barley***, ***winter wheat***

Approval information

- Annex 1 approval is pending
- Accepted by BBPA for use on malting barley

Hazard classification and safety precautions

UN number N/C
Operator protection A, H; U05a, U09a, U20b
Environmental protection E03, E15b, E34
Storage and disposal D01, D02, D09a, D10a
Treated seed S01, S02, S03, S04d, S05, S07
Medical advice M03

305 iprodione

A protectant dicarboximide fungicide with some eradicant activity
FRAC mode of action code: 2

Products

1	Chipco Green	Bayer Environ.	255 g/l	SC	13843
2	Green Turf King	Standon	255 g/l	SC	14348
3	Mascot Rayzor	Rigby Taylor	255 g/l	SC	14329
4	Panto	Pan Amenity	255 g/l	SC	14307
5	Prime-Turf	ChemSource	255 g/l	SC	14617
6	Rovral AquaFlo	BASF	500 g/l	SC	14206
7	Rovral WG	BASF	75% w/w	WG	13811

Uses

- Alternaria in ***brassica seed crops*** *(pre-storage only)*, ***brussels sprouts***, ***cauliflowers***, ***chicory*** *(off-label)*, ***dwarf beans*** *(off-label)*, ***french beans*** *(off-label)*, ***garlic*** *(off-label)*, ***grapevines*** *(off-label)*, ***mange-tout peas*** *(off-label)*, ***protected herbs (see appendix 6)*** *(off-label)*, ***runner beans*** *(off-label)*, ***shallots*** *(off-label)*, ***spring oilseed rape***, ***winter oilseed rape***, ***witloof*** *(off-label)* [7]; ***broccoli*** *(off-label)*, ***brussels sprouts*** *(off-label)*, ***cabbages*** *(off-label)*, ***calabrese*** *(off-label)*, ***cauliflowers*** *(off-label)*, ***chinese cabbage*** *(off-label)*, ***collards*** *(off-label)*, ***kale*** *(off-label)* [6]
- Botrytis in ***bulb onions***, ***cabbages*** *(off-label)*, ***chicory*** *(off-label)*, ***dwarf beans*** *(off-label)*, ***forest nurseries*** *(off-label)*, ***french beans*** *(off-label)*, ***garlic*** *(off-label)*, ***grapevines*** *(off-label)*, ***hops*** *(off-label)*, ***lettuce***, ***mange-tout peas*** *(off-label)*, ***ornamental plant production***, ***pears*** *(off-label)*, ***protected aubergines*** *(off-label)*, ***protected forest nurseries*** *(off-label)*, ***protected herbs (see appendix 6)*** *(off-label)*, ***protected hops*** *(off-label)*, ***protected lettuce***, ***protected rhubarb*** *(off-label)*, ***protected soft fruit*** *(off-label)*, ***protected strawberries***, ***protected tomatoes***, ***protected top fruit*** *(off-label)*, ***raspberries***, ***red beet*** *(off-label)*, ***runner beans*** *(off-label)*, ***salad onions***, ***shallots*** *(off-label)*, ***soft fruit*** *(off-label)*, ***spring oilseed rape***, ***strawberries***, ***top fruit*** *(off-label)*, ***winter oilseed rape***, ***witloof*** *(off-label)* [7]
- Brown patch in ***amenity grassland***, ***managed amenity turf*** [1, 5]
- Collar rot in ***bulb onions***, ***garlic*** *(off-label)*, ***salad onions***, ***shallots*** *(off-label)* [7]
- Disease control in ***managed amenity turf*** [3, 4]
- Dollar spot in ***amenity grassland***, ***managed amenity turf*** [1, 5]
- Fusarium patch in ***amenity grassland*** [1, 5]; ***managed amenity turf*** [1, 2, 5]
- Melting out in ***amenity grassland***, ***managed amenity turf*** [1]
- Red thread in ***amenity grassland*** [1, 5]; ***managed amenity turf*** [1, 2, 5]
- Seed-borne diseases in ***ornamental crop production (seed)***, ***seed potatoes*** [6]
- Snow mould in ***amenity grassland***, ***managed amenity turf*** [1]

Specific Off-Label Approvals (SOLAs)

- ***broccoli*** *20090708* [6]
- ***brussels sprouts*** *20090708* [6]
- ***cabbages*** *20090708* [6], *20091845* [7]
- ***calabrese*** *20090708* [6]
- ***cauliflowers*** *20090708* [6]
- ***chicory*** *20081759* [7]
- ***chinese cabbage*** *20090708* [6]
- ***collards*** *20090708* [6]
- ***dwarf beans*** *20081760* [7]
- ***forest nurseries*** *20082902* [7]
- ***french beans*** *20081760* [7]
- ***garlic*** *20081768* [7]
- ***grapevines*** *20081758* [7]
- ***hops*** *20082902* [7]
- ***kale*** *20090708* [6]
- ***mange-tout peas*** *20081760* [7]
- ***pears*** *20081761* [7]
- ***protected aubergines*** *20081767* [7]
- ***protected forest nurseries*** *20082902* [7]
- ***protected herbs (see appendix 6)*** *20081757* [7]
- ***protected hops*** *20082902* [7]
- ***protected rhubarb*** *20081763* [7]
- ***protected soft fruit*** *20082902* [7]
- ***protected top fruit*** *20082902* [7]
- ***red beet*** *20081762* [7]
- ***runner beans*** *20081760* [7]
- ***shallots*** *20081768* [7]
- ***soft fruit*** *20082902* [7]
- ***top fruit*** *20082902* [7]
- ***witloof*** *20081759* [7]

Approval information

- Iprodione included in Annex I under EC Directive 91/414
- Accepted by BBPA for use on malting barley

Efficacy guidance

- Many diseases require a programme of 2 or more sprays at intervals of 2-4 wk. Recommendations vary with disease and crop - see label for details
- Use as a drench to control cabbage storage diseases. Spray ornamental pot plants and cucumbers to run-off
- Apply turf and amenity grass treatments after mowing to dry grass, free of dew. Do not mow again for at least 24 h [1]

Restrictions

- Maximum number of treatments or maximum total dose equivalent to 1 per batch for seed treatments; 1 per crop on cereals, cabbage (as drench), 2 per crop on field beans and stubble turnips; 3 per crop on brassicas (including seed crops), oilseed rape, protected winter lettuce (Oct-Feb); 4 per crop on strawberries, grapevines, salad onions, cucumbers; 5 per crop on raspberries; 6 per crop on bulb onions, tomatoes, turf; 7 per crop on lettuce (Mar-Sep)
- A minimum of 3 wk must elapse between treatments on leaf brassicas
- Treated brassica seed crops not to be used for human or animal consumption
- See label for pot plants showing good tolerance. Check other species before applying on a large scale
- Do not treat oilseed rape seed that is cracked or broken or of low viability
- Do not treat oats

Crop-specific information

- Latest use: pre-planting for seed treatments; at planting for potatoes; before grain watery ripe (GS 69) for wheat and barley; 21 d before consumption by livestock for stubble turnips

- HI strawberries, protected tomatoes 1 d; outdoor tomatoes 2 d; bulb onions, salad onions, raspberries, lettuce 7 d; brassicas, brassica seed crops, oilseed rape, mustard, field beans, stubble turnips 21 d
- Turf and amenity grass may become temporarily yellowed if frost or hot weather follows treatment
- Personal protective equipment requirements may vary for each pack size. Check label

Environmental safety

- Dangerous for the environment
- Very toxic to aquatic organisms
- See label for guidance on disposal of spent drench liquor
- Treatment harmless to *Encarsia* or *Phytoseiulus* being used for integrated pest control

Hazard classification and safety precautions

Hazard Harmful, Dangerous for the environment
Transport code 9
Packaging group III
UN number 3077 [7]; 3082 [1-6]
Risk phrases R36 [1, 3-5, 7]; R38 [1, 3-5]; R40 [1, 3-7]; R50, R53a [1-7]
Operator protection A [1-7]; C [1, 3-5, 7]; H [6, 7]; P [2]; U04a, U08 [1, 3-5]; U05a [1-7]; U09a [2]; U11 [6, 7]; U16b, U19a, U20b [7]; U20c [1-6]
Environmental protection E15a [1, 3-6]; E15b, E16a [2, 7]; E16b [2]; E17b [7] (18 m); E34 [1-7]; E38 [1, 3-7]
Storage and disposal D01, D02, D09a [1-7]; D05, D10b [1, 3-5]; D10a [2]; D11a [7]; D12a [1, 3-7]
Treated seed S01, S02, S03, S04a, S05 [6]
Medical advice M03 [1, 3-6]; M04a [2]; M05a [2, 6, 7]

306 iprodione + thiophanate-methyl

A protectant and systemic fungicide for oilseed rape
FRAC mode of action code: 2 + 1

See also thiophanate-methyl

Products

Compass	BASF	167:167 g/l	SC	11740

Uses

- Alternaria in ***carrots*** *(off-label)*, ***horseradish*** *(off-label)*, ***parsnips*** *(off-label)*, ***spring oilseed rape***, ***winter oilseed rape***
- Botrytis in ***meadowfoam (limnanthes alba)*** *(off-label)*
- Chocolate spot in ***spring field beans***, ***winter field beans***
- Crown rot in ***carrots*** *(off-label)*, ***horseradish*** *(off-label)*, ***parsnips*** *(off-label)*
- Grey mould in ***spring oilseed rape***, ***winter oilseed rape***
- Light leaf spot in ***spring oilseed rape***, ***winter oilseed rape***
- Sclerotinia stem rot in ***spring oilseed rape***, ***winter oilseed rape***
- Stem canker in ***spring oilseed rape***, ***winter oilseed rape***

Specific Off-Label Approvals (SOLAs)

- ***carrots*** *20040525*
- ***horseradish*** *20040525*
- ***meadowfoam (limnanthes alba)*** *20040507*
- ***parsnips*** *20040525*

Approval information

- Iprodione and thiophanate-methyl included in Annex I under EC Directive 91/414

Efficacy guidance

- Timing of sprays on oilseed rape varies with disease, see label for details
- For season-long control product should be applied as part of a disease control programme

Restrictions

- Maximum number of treatments (refers to total sprays containing benomyl, carbendazim or thiophanate methyl) 2 per crop for oilseed rape, field beans, meadowfoam; 3 per crop for carrots, horseradish, parsnips

Crop-specific information

- Latest use: before end of flowering for oilseed rape, field beans
- HI meadowfoam, field beans, oilseed rape 3 wk; carrots, horseradish, parsnips 28 d
- Treatment may extend duration of green leaf in winter oilseed rape

Environmental safety

- Dangerous for the environment
- Very toxic to aquatic organisms

Hazard classification and safety precautions

Hazard Harmful, Dangerous for the environment
Transport code 9
Packaging group III
UN number 3082
Risk phrases R40, R43, R50, R53a, R68
Operator protection A, C, H, M; U05a, U20c
Environmental protection E15a, E38
Storage and disposal D01, D02, D08, D09a, D10b, D12a
Medical advice M05a

307 isoproturon

A residual urea herbicide for use in cereals; All approvals expired 30 June 2009.
HRAC mode of action code: C2

308 isoxaben

A soil-acting benzamide herbicide for use in grass and fruit
HRAC mode of action code: L

Products

1	Agriguard Isoxaben	AgriGuard	125 g/l	SC	11652
2	Flexidor 125	Landseer	125 g/l	SC	10946

Uses

- Annual dicotyledons in ***almonds*** *(off-label)*, ***amenity vegetation***, ***apple orchards***, ***blackberries***, ***blackcurrants***, ***cherries***, ***chestnuts*** *(off-label)*, ***forestry transplants***, ***gooseberries***, ***hazel nuts*** *(off-label)*, ***hops***, ***ornamental plant production***, ***pear orchards***, ***plums***, ***raspberries***, ***rubus hybrids*** *(off-label)*, ***strawberries***, ***walnuts*** *(off-label)*, ***wine grapes*** [1, 2]; ***bilberries*** *(off-label)*, ***blueberries*** *(off-label)*, ***canary grass*** *(off-label - grown for game cover)*, ***carrots*** *(off-label)*, ***clovers*** *(off-label - grown for game cover)*, ***courgettes*** *(off-label)*, ***cranberries*** *(off-label)*, ***forage maize*** *(off-label - grown for game cover)*, ***marrows*** *(off-label)*, ***millet*** *(off-label - grown for game cover)*, ***protected carrots*** *(off-label - temporary protection)*, ***protected courgettes*** *(off-label)*, ***protected marrows*** *(off-label)*, ***protected ornamentals*** *(off-label)*, ***protected parsnips*** *(off-label - temporary protection)*, ***protected pumpkins*** *(off-label)*, ***protected squashes*** *(off-label)*, ***pumpkins*** *(off-label)*, ***quinoa*** *(off-label - grown for game cover)*, ***redcurrants*** *(off-label)*, ***squashes*** *(off-label)*, ***sunflowers*** *(off-label - grown for game cover)*, ***whitecurrants*** *(off-label)* [2]; ***table grapes*** [1]
- Cleavers in ***asparagus*** *(off-label)* [2]
- Fat hen in ***asparagus*** *(off-label)* [2]
- Groundsel in ***asparagus*** *(off-label)* [2]
- Knotgrass in ***asparagus*** *(off-label)* [2]

Specific Off-Label Approvals (SOLAs)

- ***almonds*** *20061109* [1], *20061111* [2]
- ***asparagus*** *20050893* [2]
- ***bilberries*** *20061113* [2]
- ***blueberries*** *20061113* [2]
- ***canary grass*** *(grown for game cover) 20050896* [2]
- ***carrots*** *20050895* [2]
- ***chestnuts*** *20061109* [1], *20061111* [2]

- ***clovers*** *(grown for game cover) 20050896* [2]
- ***courgettes*** *20050894* [2]
- ***cranberries*** *20061113* [2]
- ***forage maize*** *(grown for game cover) 20050896* [2]
- ***hazel nuts*** *20061109* [1], *20061111* [2]
- ***marrows*** *20050894* [2]
- ***millet*** *(grown for game cover) 20050896* [2]
- ***protected carrots*** *(temporary protection) 20050892* [2]
- ***protected courgettes*** *20050894* [2]
- ***protected marrows*** *20050894* [2]
- ***protected ornamentals*** *20050891* [2]
- ***protected parsnips*** *(temporary protection) 20050892* [2]
- ***protected pumpkins*** *20050894* [2]
- ***protected squashes*** *20050894* [2]
- ***pumpkins*** *20050894* [2]
- ***quinoa*** *(grown for game cover) 20050896* [2]
- ***redcurrants*** *20061113* [2]
- ***rubus hybrids*** *20061110* [1], *20061112* [2]
- ***squashes*** *20050894* [2]
- ***sunflowers*** *(grown for game cover) 20050896* [2]
- ***walnuts*** *20061109* [1], *20061111* [2]
- ***whitecurrants*** *20061113* [2]

Approval information

- Accepted by BBPA for use on hops
- Isoxaben not included in Annex 1.

Efficacy guidance

- When used alone apply pre-weed emergence
- Effectiveness is reduced in dry conditions. Weed seeds germinating at depth are not controlled
- Activity reduced on soils with more than 10% organic matter. Do not use on peaty soils
- Various tank mixtures are recommended for early post-weed emergence treatment (especially for grass weeds). See label for details

Restrictions

- Maximum number of treatments 1 per crop for all edible crops; 2 per yr on amenity vegetation and non-edible crops

Crop-specific information

- Latest use: before 1 Apr in yr of harvest for other edible crops

Following crops guidance

- See label for details of crops which may be sown in the event of failure of a treated crop

Environmental safety

- Keep all livestock out of treated areas for at least 50 d

Hazard classification and safety precautions

UN number N/C
Operator protection A, C; U05a, U20b
Environmental protection E06a (50 d); E15a
Storage and disposal D01, D09a, D11a [1, 2]; D05 [2]

309 isoxaben + terbuthylazine

A contact and residual herbicide for use in peas and spring field beans
HRAC mode of action code: L + C1

See also *terbuthylazine*

Products

Skirmish	Syngenta	75:420 g/l	SC	08444

Uses

- Annual dicotyledons in ***broad beans*** *(off-label)*, ***combining peas***, ***lupins*** *(off-label)*, ***spring field beans***, ***vining peas***
- Annual grasses in ***broad beans*** *(off-label)*

Specific Off-Label Approvals (SOLAs)

- ***broad beans*** *20070076*
- ***lupins*** *20060540*

Efficacy guidance

- May be applied pre- or post-emergence of crop but before second node stage (GS 102)
- Product will give residual control of germinating weeds on mineral soils for up to 8 wk
- Product is slow acting and control may not be evident for 7-10 d or more after spraying
- Best results achieved when soil surface damp and with a fine, firm tilth. Do not use on very cloddy or stony soil
- Rain after spraying will normally improve weed control but excessive rainfall, very dry conditions or unusually low soil temperatures may lead to unsatisfactory control

Restrictions

- Maximum total dose equivalent to one full dose treatment on peas and spring field beans
- Do not use on forage peas
- Do not use on soils lighter than Coarse Sandy Loam, on very stony soils or soils with more than 10% organic matter
- Pea seed should be covered by at least 25 mm of soil
- Heavy rain after application may cause some crop damage, especially on light soils. Do not use on soils where surface water is likely to accumulate
- For post-emergence treatment a crystal violet test for cuticle wax is advised

Crop-specific information

- Latest use: before second node stage (GS 102) for peas; pre-emergence for spring field beans; beans that are emerging at application will be severely damaged or killed
- Product may be used on all varieties of spring sown vining and combining peas and spring field beans but pea varieties Vedette and Printana may be damaged from which recovery may not be complete

Environmental safety

- Dangerous for the environment
- Very toxic to aquatic organisms

Hazard classification and safety precautions

Hazard Harmful, Dangerous for the environment
Transport code 9
Packaging group III
UN number 3082
Risk phrases R22a, R50, R53a
Operator protection A; U05a, U20a
Environmental protection E15a, E38
Storage and disposal D01, D02, D05, D07, D09a, D10c, D12a

310 isoxaflutole

An isoxazole herbicide available only in mixtures
HRAC mode of action code: F2

See also flufenacet + isoxaflutole

311 kresoxim-methyl

A protectant strobilurin fungicide for apples
FRAC mode of action code: 11

See also *epoxiconazole + fenpropimorph + kresoxim-methyl*
epoxiconazole + kresoxim-methyl
epoxiconazole + kresoxim-methyl + pyraclostrobin
fenpropimorph + kresoxim-methyl

Products

1	Beem WG	AgChem Access	50% w/w	WG	13807
2	Kresoxy 50 WG	AgriGuard	50% w/w	WG	12275
3	Stroby WG	BASF	50% w/w	WG	08653

Uses

- Black spot in ***protected roses, roses*** [1-3]
- Powdery mildew in ***all edible seed crops grown outdoors*** *(off-label)*, ***all non-edible seed crops grown outdoors*** *(off-label)*, ***gooseberries*** *(off-label)*, ***hops*** *(off-label)*, ***protected soft fruit*** *(off-label)*, ***quinces*** *(off-label)*, ***redcurrants*** *(off-label)*, ***soft fruit*** *(off-label)*, ***whitecurrants*** *(off-label)* [3]; ***apples*** *(reduction)*, ***blackcurrants, protected roses, protected strawberries, roses, strawberries*** [1-3]
- Scab in ***apples*** [1-3]

Specific Off-Label Approvals (SOLAs)

- ***all edible seed crops grown outdoors*** *20082917* [3]
- ***all non-edible seed crops grown outdoors*** *20082917* [3]
- ***gooseberries*** *20060656* [3]
- ***hops*** *20082917* [3]
- ***protected soft fruit*** *20082917* [3]
- ***quinces*** *20060657* [3]
- ***redcurrants*** *20060656* [3]
- ***soft fruit*** *20082917* [3]
- ***whitecurrants*** *20060656* [3]

Approval information

- Kresoxim-methyl included in Annex I under EC Directive 91/414
- Accepted by BBPA on malting barley
- Approved for use in ULV systems

Efficacy guidance

- Activity is protectant. Best results achieved from treatments prior to disease development. See label for timing details on each crop. Treatments should be repeated at 10-14 d intervals but note limitations below
- To minimise the likelihood of development of resistance to strobilurin fungicides these products should be used in a planned Resistance Management strategy. See Section 5 for more information
- Product may be applied in ultra low volumes (ULV) but disease control may be reduced

Restrictions

- Maximum number of treatments 4 per yr on apples; 3 per yr on other crops. See notes in Efficacy about limitations on consecutive treatments
- Consult before using on crops intended for processing

Crop-specific information

- HI 14 d for blackcurrants, protected strawberries, strawberries; 35 d for apples
- On apples do not spray product more than twice consecutively and separate each block of two consecutive treatments with at least two applications from a different cross-resistance group. For all other crops do not apply consecutively and use a maximum of once in every three fungicide sprays
- Product should not be used as final spray of the season on apples

Environmental safety

- Dangerous for the environment
- Very toxic to aquatic organisms

- Broadcast air-assisted LERAP (5 m)
- Harmless to ladybirds and predatory mites
- Harmless to honey bees and may be applied during flowering. Nevertheless local beekeepers should be notified when treatment of orchards in flower is to occur

Hazard classification and safety precautions

Hazard Harmful, Dangerous for the environment
Transport code 9
Packaging group III
UN number 3077
Risk phrases R40, R50, R53a
Operator protection A, H, J [2]; U05a, U20b
Environmental protection E15a, E38 [1-3]; E16b [2]; E17b [1-3] (5 m)
Storage and disposal D01, D02, D09a, D10c, D12a [1-3]; D05 [2]; D08 [1, 3]
Medical advice M05a

312 lambda-cyhalothrin

A quick-acting contact and ingested pyrethroid insecticide
IRAC mode of action code: 3

Products

1	Clayton Lanark	Clayton	100 g/l	CS	12942
2	Clayton Sparta	Clayton	50 g/l	EC	13457
3	Hallmark with Zeon Technology	Syngenta	100 g/l	CS	12629
4	Jackpot 2	AgriGuard	100 g/l	CS	12888
5	Karate 2.5WG	Syngenta	2.5% w/w	GR	14060
6	Markate 50	Agrovista	50 g/l	EC	13529
7	Seal Z	AgChem Access	100 g/l	CS	14201
8	Warrior	Syngenta	100 g/l	CS	13857

Uses

- Aphids in ***all edible seed crops grown outdoors*** *(off-label)*, ***all non-edible seed crops grown outdoors*** *(off-label)*, ***borage for oilseed production*** *(off-label)*, ***crambe*** *(off-label)*, ***farm forestry*** *(off-label)*, ***forest nurseries*** *(off-label)*, ***hops*** *(off-label)*, ***lupins*** *(off-label)*, ***miscanthus*** *(off-label)*, ***ornamental plant production*** *(off-label)*, ***protected forest nurseries*** *(off-label)*, ***protected ornamentals*** *(off-label)*, ***protected pak choi*** *(off-label)*, ***protected tat soi*** *(off-label)*, ***soft fruit*** *(off-label)*, ***top fruit*** *(off-label)* [3]; ***borage*** *(off-label)* [1]; ***canary flower (echium spp.)*** *(off-label)*, ***chives*** *(off-label)*, ***evening primrose*** *(off-label)*, ***frise*** *(off-label)*, ***grass seed crops*** *(off-label)*, ***herbs (see appendix 6)*** *(off-label)*, ***honesty*** *(off-label)*, ***lamb's lettuce*** *(off-label)*, ***mustard*** *(off-label)*, ***parsley*** *(off-label)*, ***salad brassicas*** *(off-label - for baby leaf production)*, ***scarole*** *(off-label)*, ***spring linseed*** *(off-label)*, ***spring rye*** *(off-label)*, ***triticale*** *(off-label)*, ***winter linseed*** *(off-label)*, ***winter rye*** *(off-label)* [1, 3]; ***chestnuts*** *(off-label)*, ***cob nuts*** *(off-label)*, ***hazel nuts*** *(off-label)*, ***walnuts*** *(off-label)* [3, 6]; ***combining peas***, ***vining peas*** [2, 6]; ***durum wheat*** [1, 3-8]; ***edible podded peas***, ***spring field beans***, ***spring oilseed rape***, ***sugar beet***, ***winter field beans***, ***winter oilseed rape*** [6]; ***potatoes***, ***spring wheat*** [1-4, 6-8]; ***spring barley***, ***spring oats*** [1, 3, 4, 7, 8]; ***winter barley***, ***winter oats*** [1, 3-5, 7, 8]; ***winter wheat*** [1-8]
- Barley yellow dwarf virus vectors in ***durum wheat***, ***winter barley***, ***winter oats***, ***winter wheat*** [1, 3-5, 7, 8]; ***spring barley***, ***spring wheat*** [4, 8]; ***spring oats*** [8]
- Bean beetle in ***broad beans*** *(off-label)* [1, 3]
- Beet leaf miner in ***fodder beet*** *(off-label)* [1, 3]; ***sugar beet*** [1, 3, 4, 7, 8]
- Beet virus yellows vectors in ***spring oilseed rape*** [1, 3, 4, 7, 8]; ***winter oilseed rape*** [1, 3-5, 7, 8]
- Beetles in ***combining peas***, ***durum wheat***, ***edible podded peas***, ***potatoes***, ***spring field beans***, ***spring oilseed rape***, ***spring wheat***, ***sugar beet***, ***vining peas***, ***winter field beans***, ***winter oilseed rape***, ***winter wheat*** [6]
- Brassica pod midge in ***spring oilseed rape***, ***winter oilseed rape*** [2]
- Cabbage seed weevil in ***spring oilseed rape***, ***winter oilseed rape*** [1-4, 7]
- Cabbage stem flea beetle in ***spring oilseed rape*** [1, 3, 4, 7, 8]; ***winter oilseed rape*** [1, 3-5, 7, 8]
- Cabbage stem weevil in ***celeriac*** *(off-label)* [6]; ***radishes*** *(off-label)* [1, 3, 6]
- Capsids in ***blackberries*** *(off-label)*, ***dewberries*** *(off-label)*, ***raspberries*** *(off-label)*, ***rubus hybrids*** *(off-label)* [1, 3, 6]; ***grapevines*** *(off-label)* [3, 6]

- Carrot fly in ***carrots, parsnips*** [8]; ***carrots*** *(off-label)*, ***parsnips*** *(off-label)*, ***radishes*** *(off-label)* [6]; ***celeriac*** *(off-label)*, ***celery (outdoor)*** *(off-label)*, ***horseradish*** *(off-label)*, ***mallow (althaea spp.)*** *(off-label)*, ***parsley root*** *(off-label)* [1, 3, 6]; ***fennel*** *(off-label)* [3]
- Caterpillars in ***broccoli, brussels sprouts, cabbages, calabrese, cauliflowers*** [1, 3, 4, 7, 8]; ***celery (outdoor)*** *(off-label)*, ***combining peas, durum wheat, dwarf beans*** *(off-label)*, ***edible podded peas, french beans*** *(off-label)*, ***navy beans*** *(off-label)*, ***potatoes, red beet*** *(off-label)*, ***runner beans*** *(off-label)*, ***spring field beans, spring oilseed rape, spring wheat, sugar beet, vining peas, winter field beans, winter oilseed rape, winter wheat*** [6]; ***chestnuts*** *(off-label)*, ***cob nuts*** *(off-label)*, ***hazel nuts*** *(off-label)*, ***walnuts*** *(off-label)* [3]; ***wine grapes*** *(off-label)* [1]
- Clay-coloured weevil in ***blackberries*** *(off-label)*, ***dewberries*** *(off-label)*, ***raspberries*** *(off-label)*, ***rubus hybrids*** *(off-label)* [1, 3, 6]
- Cutworms in ***carrots, lettuce*** [1, 3, 4, 7]; ***celeriac*** *(off-label)*, ***fodder beet*** *(off-label)*, ***red beet*** *(off-label)* [1, 3, 6]; ***chicory*** *(off-label - for forcing)* [3]; ***fennel*** *(off-label)* [1, 3]; ***outdoor lettuce*** [8]; ***parsnips*** [3, 7]; ***radishes*** *(off-label)* [6]; ***sugar beet*** [1-4, 7, 8]
- Flea beetle in ***fodder beet*** *(off-label)* [1, 3, 6]; ***poppies for morphine production*** *(off-label)* [1, 3]; ***spring oilseed rape, sugar beet*** [1-4, 7, 8]; ***winter oilseed rape*** [1-5, 7, 8]
- Frit fly in ***sweetcorn*** *(off-label)* [1, 3, 6]
- Grain aphid in ***durum wheat, spring barley, spring oats, spring wheat, winter barley, winter oats, winter wheat*** [8]
- Insect pests in ***all edible seed crops grown outdoors*** *(off-label)*, ***all non-edible seed crops grown outdoors*** *(off-label)*, ***crambe*** *(off-label)*, ***farm forestry*** *(off-label)*, ***forest nurseries*** *(off-label)*, ***hops*** *(off-label)*, ***miscanthus*** *(off-label)*, ***ornamental plant production*** *(off-label)*, ***protected forest nurseries*** *(off-label)*, ***protected ornamentals*** *(off-label)*, ***protected pak choi*** *(off-label)*, ***protected tat soi*** *(off-label)*, ***soft fruit*** *(off-label)*, ***top fruit*** *(off-label)* [3]; ***blackberries*** *(off-label)*, ***blackcurrants*** *(off-label)*, ***borage for oilseed production*** *(off-label)*, ***canary flower (echium spp.)*** *(off-label)*, ***dewberries*** *(off-label)*, ***evening primrose*** *(off-label)*, ***gooseberries*** *(off-label)*, ***honesty*** *(off-label)*, ***horseradish*** *(off-label)*, ***mallow (althaea spp.)*** *(off-label)*, ***mustard*** *(off-label)*, ***parsley root*** *(off-label)*, ***raspberries*** *(off-label)*, ***redcurrants*** *(off-label)*, ***rubus hybrids*** *(off-label)*, ***spring linseed*** *(off-label)*, ***whitecurrants*** *(off-label)*, ***winter linseed*** *(off-label)* [3, 6]; ***broad beans*** *(off-label)*, ***bulb onions*** *(off-label)*, ***carrots*** *(off-label)*, ***celeriac*** *(off-label)*, ***celery (outdoor)*** *(off-label)*, ***chestnuts*** *(off-label)*, ***cob nuts*** *(off-label)*, ***dwarf beans*** *(off-label)*, ***fodder beet*** *(off-label)*, ***french beans*** *(off-label)*, ***garlic*** *(off-label)*, ***grass seed crops*** *(off-label)*, ***hazel nuts*** *(off-label)*, ***herbs (see appendix 6)*** *(off-label)*, ***leeks*** *(off-label)*, ***navy beans*** *(off-label)*, ***parsnips*** *(off-label)*, ***radishes*** *(off-label)*, ***red beet*** *(off-label)*, ***runner beans*** *(off-label)*, ***salad onions*** *(off-label)*, ***shallots*** *(off-label)*, ***short rotation coppice willow*** *(off-label)*, ***spring rye*** *(off-label)*, ***sweetcorn*** *(off-label)*, ***triticale*** *(off-label)*, ***walnuts*** *(off-label)*, ***winter rye*** *(off-label)* [6]
- Leaf midge in ***blackcurrants*** *(off-label)*, ***gooseberries*** *(off-label)*, ***redcurrants*** *(off-label)*, ***whitecurrants*** *(off-label)* [6]
- Leaf miner in ***fodder beet*** *(off-label)* [6]; ***sugar beet*** [2]
- Mangold fly in ***sugar beet*** [8]
- Midges in ***blackcurrants*** *(off-label)*, ***gooseberries*** *(off-label)*, ***redcurrants*** *(off-label)*, ***whitecurrants*** *(off-label)* [1, 3]
- Orange blossom midge in ***spring wheat, winter wheat*** [8]
- Pea and bean weevil in ***combining peas, vining peas*** [1-3, 7, 8]; ***edible podded peas*** [1, 3, 7, 8]; ***peas*** [4]; ***spring field beans, winter field beans*** [1-4, 7, 8]
- Pea aphid in ***combining peas, edible podded peas, vining peas*** [1, 3, 7, 8]; ***peas*** [4]
- Pea midge in ***combining peas, edible podded peas, vining peas*** [1, 3, 7]; ***peas*** [4]
- Pea moth in ***combining peas, vining peas*** [1-3, 7, 8]; ***edible podded peas*** [1, 3, 7, 8]; ***peas*** [4]
- Pear midge in ***combining peas, edible podded peas, vining peas*** [8]
- Pear sucker in ***pears*** [1-4, 7, 8]; ***winter wheat*** [2]
- Pod midge in ***spring oilseed rape, winter oilseed rape*** [1, 3, 4, 7, 8]
- Pollen beetle in ***poppies for morphine production*** *(off-label)* [1, 3]; ***spring oilseed rape, winter oilseed rape*** [1-4, 7, 8]
- Sawflies in ***blackcurrants*** *(off-label)*, ***gooseberries*** *(off-label)*, ***redcurrants*** *(off-label)*, ***whitecurrants*** *(off-label)* [1, 3, 6]; ***short rotation coppice willow*** *(off-label)* [6]
- Seed beetle in ***broad beans*** *(off-label)* [6]
- Seed weevil in ***spring oilseed rape, winter oilseed rape*** [8]
- Silver Y moth in ***celery (outdoor)*** *(off-label)*, ***dwarf beans*** *(off-label)*, ***french beans*** *(off-label)*, ***navy beans*** *(off-label)*, ***red beet*** *(off-label)*, ***runner beans*** *(off-label)* [1, 3, 6]

- Thrips in ***broad beans*** *(off-label)*, ***bulb onions*** *(off-label)*, ***garlic*** *(off-label)*, ***leeks*** *(off-label)*, ***shallots*** *(off-label)* [1, 3]; ***salad onions*** *(off-label)* [3]
- Wasps in ***grapevines*** *(off-label)* [3, 6]
- Weevils in ***combining peas, durum wheat, edible podded peas, potatoes, spring field beans, spring oilseed rape, spring wheat, sugar beet, vining peas, winter field beans, winter oilseed rape, winter wheat*** [6]
- Wheat-blossom midge in ***winter wheat*** [1, 3, 7]
- Whitefly in ***broccoli, brussels sprouts, cabbages, calabrese, cauliflowers*** [1, 3, 4, 7, 8]
- Willow aphid in ***short rotation coppice willow*** *(off-label)* [3, 6]
- Willow beetle in ***short rotation coppice willow*** *(off-label)* [3, 6]
- Willow sawfly in ***short rotation coppice willow*** *(off-label)* [3, 6]
- Yellow cereal fly in ***winter wheat*** [1-4, 7, 8]

Specific Off-Label Approvals (SOLAs)

- ***all edible seed crops grown outdoors*** *20082944* [3]
- ***all non-edible seed crops grown outdoors*** *20082944* [3]
- ***blackberries*** *20073266* [6]
- ***blackcurrants*** *20073269* [6]
- ***borage for oilseed production*** *20073258* [6]
- ***broad beans*** *20073234* [6]
- ***bulb onions*** *20073256* [6]
- ***canary flower (echium spp.)*** *20073258* [6]
- ***carrots*** *20080201* [6]
- ***celeriac*** *20080204* [6]
- ***celery (outdoor)*** *20073257* [6]
- ***chestnuts*** *20080206* [6]
- ***cob nuts*** *20080206* [6]
- ***crambe*** *20081046* [3]
- ***dewberries*** *20073266* [6]
- ***dwarf beans*** *20080202* [6]
- ***evening primrose*** *20073258* [6]
- ***farm forestry*** *20082944* [3]
- ***fodder beet*** *20073233* [6]
- ***forest nurseries*** *20082944* [3]
- ***french beans*** *20080202* [6]
- ***garlic*** *20073256* [6]
- ***gooseberries*** *20073269* [6]
- ***grapevines*** *20080205* [6]
- ***grass seed crops*** *20073260* [6]
- ***hazel nuts*** *20080206* [6]
- ***herbs (see appendix 6)*** *20073259* [6]
- ***honesty*** *20073258* [6]
- ***hops*** *20082944* [3]
- ***horseradish*** *20080201* [6]
- ***leeks*** *20073256* [6]
- ***mallow (althaea spp.)*** *20080201* [6]
- ***miscanthus*** *20082944* [3]
- ***mustard*** *20073258* [6]
- ***navy beans*** *20080202* [6]
- ***ornamental plant production*** *20082944* [3]
- ***parsley root*** *20080201* [6]
- ***parsnips*** *20080201* [6]
- ***protected forest nurseries*** *20082944* [3]
- ***protected ornamentals*** *20082944* [3]
- ***protected pak choi*** *20081263* [3]
- ***protected tat soi*** *20081263* [3]
- ***radishes*** *20080204* [6]
- ***raspberries*** *20073266* [6]
- ***red beet*** *20073254* [6]
- ***redcurrants*** *20073269* [6]

- ***rubus hybrids*** *20073266* [6]
- ***runner beans*** *20080202* [6]
- ***salad onions*** *20073256* [6]
- ***shallots*** *20073256* [6]
- ***short rotation coppice willow*** *20073268* [6]
- ***soft fruit*** *20082944* [3]
- ***spring linseed*** *20073258* [6]
- ***spring rye*** *20073260* [6]
- ***sweetcorn*** *20080203* [6]
- ***top fruit*** *20082944* [3]
- ***triticale*** *20073260* [6]
- ***walnuts*** *20080206* [6]
- ***whitecurrants*** *20073269* [6]
- ***winter linseed*** *20073258* [6]
- ***winter rye*** *20073260* [6]

Approval information

- Lambda-cyhalothrin included in Annex I under EC Directive 91/414
- Accepted by BBPA for use on malting barley

Efficacy guidance

- Best results normally obtained from treatment when pest attack first seen. See label for detailed recommendations on each crop
- Timing for control of barley yellow dwarf virus vectors depends on specialist assessment of the level of risk in the area
- Repeat applications recommended in some crops where prolonged attack occurs, up to maximum total dose. See label for details
- Where strains of aphids resistant to lambda-cyhalothrin occur control is unlikely to be satisfactory
- Addition of wetter recommended for control of certain pests in brassicas and oilseed rape
- Use of sufficient water volume to ensure thorough crop penetration recommended for optimum results
- Use of drop-legged sprayer gives improved results in crops such as Brussels sprouts

Restrictions

- Maximum number of applications or maximum total dose per crop varies see labels
- Do not apply to a cereal crop if any product containing a pyrethroid insecticide or dimethoate has been applied to the crop after the start of ear emergence (GS 51)
- Do not spray cereals in the spring/summer (ie after 1 Apr) within 6 m of edge of crop

Crop-specific information

- Latest use before late milk stage on cereals; before end of flowering for winter oilseed rape
- HI 3 d for radishes, red beet; 7 d for lettuce [1, 3]; 14 d for carrots and parsnips; 25 d for peas, field beans; 6 wk for spring oilseed rape; 8 wk for sugar beet

Environmental safety

- Dangerous for the environment
- Very toxic to aquatic organisms
- Flammable [2]
- Risk to certain non-target insects or other arthropods [3]
- To protect non-target arthropods respect an untreated buffer zone of 5 m to non-crop land
- LERAP Category B
- Broadcast air-assisted LERAP (25 m) [1, 3]; 38 m [2]

Hazard classification and safety precautions

Hazard Harmful, Dangerous for the environment [1-8]; Corrosive [6]; Flammable [2]
Transport code 3 [2]; 9 [1, 3-5, 7, 8]
Packaging group III [1-5, 7, 8]
UN number 1993 [2]; 3077 [5]; 3082 [1, 3, 4, 7, 8]
Risk phrases R20, R22a, R53a [1-8]; R21, R37, R67 [2]; R22b [2, 6]; R34, R51 [6]; R36, R38 [2, 5]; R43 [1, 3-5, 7, 8]; R50 [1-5, 7, 8]
Operator protection A, H [1-8]; C [2, 4-6]; J, K, M [4, 5]; L [4]; U02a [1, 3-8]; U04a [6]; U05a [1-8]; U08 [1-4, 6-8]; U09a [5]; U11 [5, 6]; U14, U20b [1, 3-5, 7, 8]; U19a, U20a [2, 6]

Environmental protection E12a [6]; E15b [2, 5, 6, 8]; E16a [1, 3-5, 7, 8]; E16b [1, 3, 4, 7]; E16c [2, 6]; E16d [6, 8]; E17b [1] (25 m); E17b [2] (38 m); E17b [3, 4, 7] (25 m); E17b [8]; E22a [5, 6]; E22b [3, 6-8]; E34 [1-4, 6-8]; E38 [1, 3-8]

Storage and disposal D01, D02 [1-8]; D05, D07 [2]; D09a [1-4, 6-8]; D10b [2, 6]; D10c [1, 3, 4, 7, 8]; D12a [1-5, 7, 8]

Medical advice M03 [1, 3-5, 7, 8]; M04a [6]; M05b [2, 6]

313 lambda-cyhalothrin + pirimicarb

An insecticide mixture combining translaminar, contact, fumigant and stomach activity

IRAC mode of action code: 3 + 1A

See also pirimicarb

Products

1	Clayton Groove	Clayton	5:100 g/l	EC	13246
2	Dovetail	Syngenta	5:100 g/l	EC	12550
3	Mortice	AgriGuard	5:100 g/l	EC	12961

Uses

- Aphids in ***borage*** *(off-label)*, ***canary flower (echium spp.)*** *(off-label)*, ***canary seed*** *(off-label)*, ***evening primrose*** *(off-label)*, ***fodder beet*** *(off-label)*, ***grass seed crops*** *(off-label)*, ***honesty*** *(off-label)*, ***horseradish*** *(off-label)*, ***lupins*** *(off-label)*, ***mallow (althaea spp.)*** *(off-label)*, ***mustard*** *(off-label)*, ***parsnips*** *(off-label)*, ***protected chives*** *(off-label)*, ***protected frise*** *(off-label)*, ***protected herbs (see appendix 6)*** *(off-label)*, ***protected parsley*** *(off-label)*, ***protected radicchio*** *(off-label)*, ***protected salad brassicas*** *(off-label - for baby leaf production)*, ***protected scarole*** *(off-label)*, ***spring linseed*** *(off-label)*, ***winter linseed*** *(off-label)* [2]; ***broccoli***, ***brussels sprouts***, ***cabbages***, ***calabrese***, ***carrots***, ***cauliflowers***, ***lettuce***, ***peas***, ***potatoes***, ***spring barley***, ***spring oats***, ***spring rye***, ***spring wheat***, ***sugar beet***, ***triticale***, ***winter barley***, ***winter oats***, ***winter rye***, ***winter wheat*** [1-3]; ***durum wheat*** [2, 3]
- Bean aphid in ***spring field beans***, ***winter field beans*** [2]
- Beet virus yellows vectors in ***spring oilseed rape***, ***winter oilseed rape*** [1-3]
- Cabbage seed weevil in ***spring oilseed rape***, ***winter oilseed rape*** [1-3]
- Cabbage stem flea beetle in ***spring oilseed rape***, ***winter oilseed rape*** [1-3]
- Caterpillars in ***broccoli***, ***brussels sprouts***, ***cabbages***, ***calabrese***, ***cauliflowers*** [1-3]
- Cutworms in ***carrots***, ***lettuce***, ***potatoes***, ***sugar beet*** [1-3]; ***fodder beet*** *(off-label)*, ***horseradish*** *(off-label)*, ***mallow (althaea spp.)*** *(off-label)*, ***parsnips*** *(off-label)* [2]
- Flea beetle in ***fodder beet*** *(off-label)* [2]; ***spring oilseed rape***, ***sugar beet***, ***winter oilseed rape*** [1-3]
- Insect pests in ***canary seed*** *(off-label)* [2]
- Leaf miner in ***fodder beet*** *(off-label)* [2]; ***sugar beet*** [1-3]
- Mealy aphid in ***broccoli***, ***brussels sprouts***, ***cabbages***, ***calabrese***, ***cauliflowers***, ***spring oilseed rape***, ***winter oilseed rape*** [1-3]
- Pea and bean weevil in ***peas***, ***spring field beans***, ***winter field beans*** [1-3]
- Pea midge in ***peas*** [1-3]
- Pea moth in ***peas*** [1-3]
- Pod midge in ***spring oilseed rape***, ***winter oilseed rape*** [1-3]
- Pollen beetle in ***broccoli***, ***brussels sprouts***, ***cabbages***, ***calabrese***, ***cauliflowers***, ***spring oilseed rape***, ***winter oilseed rape*** [1-3]
- Whitefly in ***broccoli***, ***brussels sprouts***, ***cabbages***, ***calabrese***, ***cauliflowers*** [1-3]

Specific Off-Label Approvals (SOLAs)

- ***borage*** *20060641* [2]
- ***canary flower (echium spp.)*** *20060641* [2]
- ***canary seed*** *20073139* [2]
- ***evening primrose*** *20060641* [2]
- ***fodder beet*** *20060642* [2]
- ***grass seed crops*** *20060640* [2]
- ***honesty*** *20060641* [2]
- ***horseradish*** *20060639* [2]
- ***lupins*** *20060638* [2]

- ***mallow (althaea spp.)*** *20060639* [2]
- ***mustard*** *20060641* [2]
- ***parsnips*** *20060639* [2]
- ***protected chives*** *20060643* [2]
- ***protected frise*** *20060643* [2]
- ***protected herbs (see appendix 6)*** *20060643* [2]
- ***protected parsley*** *20060643* [2]
- ***protected radicchio*** *20060643* [2]
- ***protected salad brassicas*** *(for baby leaf production) 20060643* [2]
- ***protected scarole*** *20060643* [2]
- ***spring linseed*** *20060641* [2]
- ***winter linseed*** *20060641* [2]

Approval information

- Lambda-cyhalothrin and pirimicarb included in Annex I under EC Directive 91/414
- Accepted by BBPA for use on malting barley
- In 2006 PSD required that all products containing pirimicarb should carry the following warning in the main area of the container label: "Pirimicarb is an anticholinesterase carbamate. Handle with care"

Efficacy guidance

- Best results obtained from treatment when pest attack first seen or after warning issued
- Repeat applications recommended in some crops where prolonged attacks occur, up to maximum total dose
- Addition of a non-ionic surfactant that is not an organosilicon recommended for certain uses in brassicas and oilseed rape. See label
- Use of drop-leg sprayers improves efficacy in crops such as Brussels sprouts
- Control unlikely to be satisfactory if aphids resistant to lambda-cyhalothrin or pirimicarb present

Restrictions

- Maximum total dose equivalent to two full dose treatments on carrots, lettuce, peas, field beans; three full dose treatments on brassicas, oilseed rape, cereals; four full dose treatments on sugar beet; eight full dose treatments on potatoes
- Must not be applied to cereals if any product containing a pyrethroid insecticide or dimethoate has been sprayed after the start of ear emergence (GS 51)

Crop-specific information

- Latest use: before late milky ripe (GS 77) for cereals; before end of flowering for oilseed rape
- HI sugar beet 8 wk; carrots 14 d; lettuce 7 d; brassicas, potatoes, peas, field beans 3 d

Environmental safety

- Dangerous for the environment
- Very toxic to aquatic organisms
- Do not spray cereals after 1 Apr within 6 m of the edge of the crop
- Harmful to livestock. Keep all livestock out of treated areas for at least 7 d following treatment
- Keep in original container, tightly closed, in a safe place, under lock and key
- LERAP Category A

Hazard classification and safety precautions

Hazard Harmful, Dangerous for the environment [1-3]; Flammable [3]
Transport code 9
Packaging group III
UN number 3082
Risk phrases R20, R22a, R38, R50, R53a [1-3]; R22b, R37, R41, R43 [3]
Operator protection A, C, D, E, H, J, K, M; U02a, U14 [1, 2]; U05a, U08, U19a, U20b [1-3]
Environmental protection E06a [1, 3] (7 d); E06c [2] (7 d); E15a, E16c, E16d, E34 [1-3]; E22c [1]; E38 [2]
Storage and disposal D01, D02, D09b [1-3]; D05, D10c, D12a [1, 2]; D11a [3]
Medical advice M02, M03 [1-3]; M05b [1, 2]

314 lenacil

A residual, soil-acting uracil herbicide for beet crops
HRAC mode of action code: C1

See also chloridazon + lenacil

Products

1	Agriguard Lenacil	AgriGuard	80% w/w	WP	10488
2	Clayton Lenacil 80 W	Clayton	80% w/w	WP	09488
3	Clayton Lenaflo	Clayton	440 g/l	SC	11031
4	Fernpath Lenzo Flo	AgriGuard	440 g/l	SC	11919
5	Lenazar Flo	Hermoo	440 g/l	SC	12286
6	Venzar Flowable	DuPont	440 g/l	SC	06907

SECTION 2

Uses

- Annual dicotyledons in ***blackberries*** *(off-label)*, ***blackcurrants***, ***established woody ornamentals***, ***gooseberries***, ***raspberries***, ***roses***, ***rubus hybrids*** *(off-label)*, ***strawberries*** [2]; ***blackcurrants*** *(off-label)*, ***blueberries*** *(off-label)*, ***cranberries*** *(off-label)*, ***farm woodland*** *(off-label)*, ***gooseberries*** *(off-label)*, ***herbs (see appendix 6)*** *(off-label)*, ***redcurrants*** *(off-label)*, ***ribes hybrids*** *(off-label)*, ***spinach*** *(off-label)*, ***spinach beet*** *(off-label)*, ***whitecurrants*** *(off-label)* [6]; ***fodder beet***, ***mangels***, ***red beet***, ***sugar beet*** [1-6]
- Annual grasses in ***blackberries*** *(off-label)*, ***rubus hybrids*** *(off-label)* [2]; ***blackcurrants*** *(off-label)*, ***blueberries*** *(off-label)*, ***cranberries*** *(off-label)*, ***gooseberries*** *(off-label)*, ***herbs (see appendix 6)*** *(off-label)*, ***redcurrants*** *(off-label)*, ***ribes hybrids*** *(off-label)*, ***spinach*** *(off-label)*, ***spinach beet*** *(off-label)*, ***whitecurrants*** *(off-label)* [6]
- Annual meadow grass in ***blackberries*** *(off-label)*, ***blackcurrants***, ***established woody ornamentals***, ***gooseberries***, ***raspberries***, ***roses***, ***rubus hybrids*** *(off-label)*, ***strawberries*** [2]; ***farm woodland*** *(off-label)* [6]; ***fodder beet***, ***mangels***, ***red beet***, ***sugar beet*** [1-6]

Specific Off-Label Approvals (SOLAs)

- ***blackberries*** *20091657 expires 31 Dec 2010* [2]
- ***blackcurrants*** *970704* [6]
- ***blueberries*** *970704* [6]
- ***cranberries*** *970704* [6]
- ***farm woodland*** *971282* [6]
- ***gooseberries*** *970704* [6]
- ***herbs (see appendix 6)*** *970703* [6]
- ***redcurrants*** *970704* [6]
- ***ribes hybrids*** *970704* [6]
- ***rubus hybrids*** *20091657 expires 31 Dec 2010* [2]
- ***spinach*** *970703* [6]
- ***spinach beet*** *970703* [6]
- ***whitecurrants*** *970704* [6]

Approval information

- Lenacil included in Annex I under EC Directive 91/414
- Approval expiry 31 Dec 2010 [5]

Efficacy guidance

- Best results, especially from pre-emergence treatments, achieved on fine, even, firm and moist soils free from clods. Continuing presence of moisture from rain or irrigation gives improved residual control of later germinating weeds. Effectiveness may be reduced by dry conditions
- On beet crops may be used pre- or post-emergence, alone or in mixture to broaden weed spectrum
- Apply overall or as band spray to beet crops pre-drilling incorporated, pre- or post-emergence
- All labels have limitations on soil types that may be treated. Residual activity reduced on soils with high OM content

Restrictions

- Maximum number of treatments in beet crops 1 pre-emergence + 3 post-emergence per crop; 2 per yr on established woody ornamentals and roses
- See label for soil type restrictions

- Do not use any other residual herbicide within 3 mth of the initial application to fruit or ornamental crops
- Do not treat crops under stress from drought, low temperatures, nutrient deficiency, pest or disease attack, or waterlogging

Crop-specific information

- Latest use: pre-emergence for red beet, fodder beet, spinach, spinach beet, mangels; before leaves meet over rows when used on these crops post-emergence; 24 h after planting new strawberry runners or before flowering for established strawberry crops, blackcurrants, gooseberries, raspberries [2]
- Heavy rain after application to beet crops may cause damage especially if followed by very hot weather
- Reduction in beet stand may occur where crop emergence or vigour is impaired by soil capping or pest attack
- Strawberry runner beds to be treated should be level without depressions around the roots
- New soft fruit cuttings should be planted at least 15 cm deep and firmed before treatment
- Check varietal tolerance of ornamentals before large scale treatment

Following crops guidance

- Succeeding crops should not be planted or sown for at least 4 mth (6 mth on organic soils) after treatment following ploughing to at least 150 mm.
- Only beet crops, mangels or strawberries may be sown within 4 months of treatment and no further applications of lenacil should be made for at least 4 months [6]

Environmental safety

- Dangerous for the environment
- Very toxic to aquatic organisms

Hazard classification and safety precautions

Hazard Irritant [1-4]; Dangerous for the environment [1-6]
Transport code 9
Packaging group III
UN number 3077 [1, 2]; 3082 [3-6]
Risk phrases R36, R38 [1-4]; R37 [1, 2, 4]; R50, R53a [2-6]
Operator protection A [1-6]; C [1-5]; U05a, U08, U19a [1-6]; U14, U15 [1, 4]; U20a [1, 3-6]; U20b [2]
Environmental protection E13c, E38 [4-6]; E15a [1-3]; E34 [1]
Storage and disposal D01, D02, D09a [1-6]; D05, D12a [3-6]; D10a [4-6]; D10b [1, 3]; D11a [2]

315 lenacil + triflusulfuron-methyl

A foliar and residual herbicide mixture for sugar beet
HRAC mode of action code: C1 + B

See also triflusulfuron-methyl

Products

Safari Lite WSB	DuPont	71.4:5.4% w/w	WG	12169

Uses

- Annual dicotyledons in ***fodder beet*** *(off-label)*, ***sugar beet***
- Annual meadow grass in ***fodder beet*** *(off-label)*
- Volunteer oilseed rape in ***sugar beet***

Specific Off-Label Approvals (SOLAs)

- ***fodder beet*** *20060936*

Approval information

- Lenacil and triflusulfuron-methyl included in Annex I under EC Directive 91/414

Efficacy guidance

- Best results obtained when weeds are small and growing actively
- Product recommended for use in a programme of treatments in tank mixture with a suitable herbicide partner to broaden the weed spectrum

- Ensure good spray cover of weeds. Apply when first weeds have emerged provided crop has reached cotyledon stage
- Susceptible plants cease to grow almost immediately after treatment and symptoms can be seen 5-10 d later
- Weed control may be reduced in very dry soil conditions
- Triflusulfuron-methyl is a member of the ALS-inhibitor group of herbicides

Restrictions

- Maximum number of treatments on sugar beet 3 per crop and do not apply more than 4 applications of any product containing triflusulfuron-methyl
- Do not apply to crops suffering from stress caused by drought, water-logging, low temperatures, pest or disease attack, nutrient deficiency or any other factors affecting crop growth
- Do not use on Sands, stony or gravelly soils or on soils with more than 10% organic matter
- Do not apply when temperature above or likely to exceed 21 °C on day of spraying or under conditions of high light intensity

Crop-specific information

- Latest use: before crop leaves meet between rows for sugar beet

Following crops guidance

- Only cereals may be sown in the same calendar yr as a treated sugar beet crop. Any crop may be sown in the following spring
- In the event of crop failure sow only sugar beet within 4 mth of treatment

Environmental safety

- Dangerous for the environment
- Very toxic to aquatic organisms
- LERAP Category B
- Take extreme care to avoid drift onto broad-leaved plants outside the target area or onto surface waters or ditches, or land intended for cropping
- Spraying equipment should not be drained or flushed onto land planted, or to be planted, with trees or crops other than cereals and should be thoroughly cleansed after use - see label for instructions

Hazard classification and safety precautions

Hazard Irritant, Dangerous for the environment
Transport code 9
Packaging group III
UN number 3077
Risk phrases R36, R37, R38, R43, R50, R53a
Operator protection A, C; U05a, U08, U11, U19a, U20b, U22a
Environmental protection E15a, E16a, E38
Storage and disposal D01, D02, D09a, D12a

316 linuron

A contact and residual urea herbicide for various field crops
HRAC mode of action code: C2

See also *2,4-DB + linuron + MCPA*
clomazone + linuron

Products

1	Afalon	Makhteshim	450 g/l	SC	14187
2	Alpha Linuron 50SC	Makhteshim	500 g/l	SC	14215
3	Luas	AgriGuard	500 g/l	SC	14359

Uses

- Annual dicotyledons in ***bulb onions*** *(off-label)*, ***celeriac*** *(off-label)*, ***garlic*** *(off-label)*, ***herbs (see appendix 6)*** *(off-label)*, ***leeks*** *(off-label)*, ***mallow (althaea spp.)*** *(off-label)*, ***ornamental plant production*** *(off-label)*, ***parsley root*** *(off-label)*, ***shallots*** *(off-label)* [1, 2]; ***carrots***, ***parsley***, ***parsnips***, ***potatoes*** [1-3]; ***celery (outdoor)*** *(off-label)*, ***dwarf beans***, ***forest nurseries*** *(off-label)*, ***french beans***, ***herbs for medicinal uses (see appendix 6)*** *(off-label)*, ***runner beans*** [2]; ***combining peas***, ***dandelions*** *(off-label)*, ***dwarf beans*** *(off-label)*, ***french beans*** *(off-label)*, ***game***

cover *(off-label)*, ***ginseng root*** *(off-label)*, ***liquorice*** *(off-label)*, ***nettle*** *(off-label)*, ***runner beans*** *(off-label)*, ***spring field beans***, ***valerian root*** *(off-label)* [1]
- Annual grasses in ***bulb onions*** *(off-label)*, ***celeriac*** *(off-label)*, ***dandelions*** *(off-label)*, ***dwarf beans*** *(off-label)*, ***french beans*** *(off-label)*, ***game cover*** *(off-label)*, ***garlic*** *(off-label)*, ***ginseng root*** *(off-label)*, ***herbs (see appendix 6)*** *(off-label)*, ***leeks*** *(off-label)*, ***liquorice*** *(off-label)*, ***mallow (althaea spp.)*** *(off-label)*, ***nettle*** *(off-label)*, ***ornamental plant production*** *(off-label)*, ***parsley root*** *(off-label)*, ***runner beans*** *(off-label)*, ***shallots*** *(off-label)*, ***valerian root*** *(off-label)* [1]
- Annual meadow grass in ***bulb onions*** *(off-label)*, ***celeriac*** *(off-label)*, ***celery (outdoor)*** *(off-label)*, ***dwarf beans***, ***forest nurseries*** *(off-label)*, ***french beans***, ***garlic*** *(off-label)*, ***herbs (see appendix 6)*** *(off-label)*, ***herbs for medicinal uses (see appendix 6)*** *(off-label)*, ***leeks*** *(off-label)*, ***mallow (althaea spp.)*** *(off-label)*, ***ornamental plant production*** *(off-label)*, ***parsley root*** *(off-label)*, ***runner beans***, ***shallots*** *(off-label)* [2]; ***carrots***, ***parsley***, ***parsnips***, ***potatoes*** [1-3]; ***combining peas***, ***spring field beans*** [1]

Specific Off-Label Approvals (SOLAs)
- ***bulb onions*** *20090875* [1], *20090865* [2]
- ***celeriac*** *20090885* [1], *20090884* [2]
- ***celery (outdoor)*** *20091918* [2]
- ***dandelions*** *20090861* [1]
- ***dwarf beans*** *20090859* [1]
- ***forest nurseries*** *20082805* [2]
- ***french beans*** *20090859* [1]
- ***game cover*** *20082802* [1]
- ***garlic*** *20090875* [1], *20090865* [2]
- ***ginseng root*** *20090861* [1]
- ***herbs (see appendix 6)*** *20090860* [1], *20090847* [2]
- ***herbs for medicinal uses (see appendix 6)*** *20090848* [2]
- ***leeks*** *20090864* [1], *20090850* [2]
- ***liquorice*** *20090861* [1]
- ***mallow (althaea spp.)*** *20090876* [1], *20090866* [2]
- ***nettle*** *20090861* [1]
- ***ornamental plant production*** *20090877* [1], *20090867* [2]
- ***parsley root*** *20090876* [1], *20090866* [2]
- ***runner beans*** *20090859* [1]
- ***shallots*** *20090875* [1], *20090865* [2]
- ***valerian root*** *20090861* [1]

Approval information
- Linuron included in Annex I under EC Directive 91/414
- All approvals for use of linuron on parsley were revoked in June 2005
- Accepted by BBPA for use on malting barley

Efficacy guidance
- Many weeds controlled pre-emergence or post-emergence to 2-3 leaf stage, some (annual meadow grass, mayweed) only susceptible pre-emergence. See label for details
- Best results achieved by application to firm, moist soil of fine tilth
- Little residual effect on soil with more than 10% organic matter

Restrictions
- Maximum number of treatments 1 per crop on most crops or 1 pre-em + 1 post-em for carrots, parsley, parsnips
- Maximum total dose equivalent to one full dose treatment
- Do not use on undersown cereals or crops grown on Sands or Very Light soils or soils heavier than Sandy Clay Loam or with more than 10% organic matter
- Do not apply to emerged crops of carrots, parsnips or parsley under stress
- Do not apply by hand-held sprayers

Crop-specific information
- Latest use: pre-emergence for most crops; pre- or up to 40% emergence for potatoes (products differ, see label for details).
- HI onions, garlic 8 wk; celeriac 12 wk; leeks 16 wk
- Drill spring cereals at least 3 cm deep and apply pre-emergence of crop or weeds

- Apply to potatoes well earthed up to a rounded ridge pre-crop emergence and do not cultivate after spraying
- Apply to carrots at any time after drilling on organic soils, and within 4 d of drilling on other soils. Apply post-emergence as soon as weeds appear but after first rough leaf stage.
- Recommendations for parsnips, parsley and celery vary. See label for details

Following crops guidance

- Potatoes, carrots and parsnips may be planted at any time after application. Lettuce should not be grown within 12 mth of treatment. Transplanted brassicas may be grown from 3 mth after treatment

Environmental safety

- Dangerous for the environment
- Very toxic to aquatic organisms
- Keep livestock out of treated areas for at least 5 mth [11590]
- LERAP Category B

Hazard classification and safety precautions

Hazard Toxic [1, 2]; Harmful [3]; Dangerous for the environment [1-3]
Transport code 9
Packaging group III
UN number 3082
Risk phrases R22a, R40, R48, R50, R53a, R61, R62 [1, 3]
Operator protection A, H; U05a, U13, U14, U15, U19a [1-3]; U08, U19c, U20b [2]
Environmental protection E15b, E16a, E38
Storage and disposal D02, D05, D09a, D11a, D12a [1-3]; D08 [1, 3]
Medical advice M04a

317 magnesium phosphide

A phosphine generating compound used to control insect pests in stored commodities

Products

Degesch Plates	Rentokil	56% w/w	GE	07603

Uses

- Insect pests in ***stored grain***

Approval information

- Accepted by BBPA for use in stores for malting barley

Efficacy guidance

- Product acts as fumigant by releasing poisonous hydrogen phosphide gas on contact with moisture in the air
- Place plates on the floor or wall of the building or on the surface of the commodity. Exposure time varies depending on temperature and pest. See label

Restrictions

- Magnesium phosphide is subject to the Poisons Rules 1982 and the Poisons Act 1972. See Section 5 for more information
- Only to be used by professional operators trained in the use of magnesium phosphide and familiar with the precautionary measures to be observed. See label for full precautions

Environmental safety

- Highly flammable
- Prevent access to buildings under fumigation by livestock, pets and other non-target mammals and birds
- Dangerous to fish or other aquatic life. Do not contaminate surface waters or ditches with chemical or used container
- Keep in original container, tightly closed, in a safe place, under lock and key
- Do not allow plates or their spent residues to come into contact with food other than raw cereal grains
- Remove used plates after treatment. Do not bulk spent plates and residues: spontaneous ignition could result
- Keep livestock out of treated areas

SEE SECTION 3 FOR PRODUCTS ALSO REGISTERED

Hazard classification and safety precautions

Hazard Very toxic, Highly flammable
Transport code 4.3
Packaging group I
UN number 2011
Risk phrases R21, R26, R28
Operator protection A, D, H; U01, U05b, U07, U13, U19a, U20a
Environmental protection E02a (4 h min); E02b, E13b, E34
Storage and disposal D01, D02, D05, D07, D09b, D11b
Medical advice M04a

318 maleic hydrazide

A pyridazine plant growth regulator suppressing sprout and bud growth

Products

Fazor	Dow	60% w/w	GR	13679

Uses

- Annual grasses in ***amenity grassland*** *(off-label)*, ***hops*** *(off-label)*, ***ornamental specimens*** *(off-label)*, ***soft fruit*** *(off-label)*
- Growth regulation in ***bulb onions***, ***carrots*** *(off-label)*, ***garlic*** *(off-label)*, ***parsnips*** *(off-label)*, ***potatoes***, ***shallots*** *(off-label)*

Specific Off-Label Approvals (SOLAs)

- ***amenity grassland*** *20082790*
- ***carrots*** *20081326*
- ***garlic*** *20072795*
- ***hops*** *20082848*
- ***ornamental specimens*** *20082848*
- ***parsnips*** *20081326*
- ***shallots*** *20072795*
- ***soft fruit*** *20082848*

Approval information

- Maleic hydrazide included in Annex I under EC Directive 91/414

Efficacy guidance

- Apply to grass at any time of yr when growth active, best when growth starting in Apr-May and repeated when growth recommences
- Uniform coverage and dry weather necessary for effective results
- Accurate timing essential for good results on potatoes but rain or irrigation within 24 h may reduce effectiveness on onions and potatoes
- Mow 2-3 d before and 5-10 d after spraying for best results. Need for mowing reduced for up to 6 wk
- When used for suppression of volunteer potatoes treatment may also give some suppression of sprouting in store but separate treatment will be necessary if sprouting occurs

Restrictions

- Maximum number of treatments 2 per yr on amenity grass, land not intended for cropping and land adjacent to aquatic areas; 1 per crop on onions, potatoes and on or around tree trunks
- Do not apply in drought or when crops are suffering from pest, disease or herbicide damage. Do not treat fine turf or grass seeded less than 8 mth previously
- Do not treat potatoes within 3 wk of applying a haulm desiccant or if temperatures above 26 °C
- Consult processor before use on potato crops for processing

Crop-specific information

- Latest use: 3 wk before haulm destruction for potatoes; before 50% necking for onions
- HI onions 1 wk; potatoes 3 wk
- Apply to onions at 10% necking and not later than 50% necking stage when the tops are still green
- Only treat onions in good condition and properly cured, and do not treat more than 2 wk before maturing. Treated onions may be stored until Mar but must then be removed to avoid browning
- Apply to second early or maincrop potatoes at least 3 wk before haulm destruction

- Only treat potatoes of good keeping quality; not on seed, first earlies or crops grown under polythene

Environmental safety

- Only apply to grass not to be used for grazing
- Do not use treated water for irrigation purposes within 3 wk of treatment or until concentration in water falls below 0.02 ppm
- Maximum permitted concentration in water 2 ppm
- Do not dump surplus product in water or ditch bottoms
- Avoid drift onto nearby vegetables, flowers or other garden plants

Hazard classification and safety precautions

Transport code 9
Packaging group III
UN number 3077
Operator protection A, H; U08, U20b
Environmental protection E15a
Storage and disposal D09a, D11a

319 mancozeb

A protective dithiocarbamate fungicide for potatoes and other crops
FRAC mode of action code: M3

See also *benalaxyl + mancozeb*
benthiavalicarb-isopropyl + mancozeb
chlorothalonil + mancozeb
cymoxanil + mancozeb
dimethomorph + mancozeb

Products

1	Dithane 945	Dow	80% w/w	WP	14547
2	Dithane 945	Interfarm	80% w/w	WP	14621
3	Dithane NT Dry Flowable	Interfarm	75% w/w	WG	12565
4	Hotspur	AgriGuard	75% w/w	WG	13449
5	Karamate Dry Flo Newtec	Landseer	75% w/w	WG	12691
6	Laminator FL	Interfarm	455 g/l	SC	11072
7	Malvi	AgChem Access	75% w/w	WG	12862
8	Mewati	AgChem Access	80% w/w	WP	13029
9	Micene 80	Sipcam	80% w/w	WP	09112
10	Micene DF	Sipcam	77% w/w	WG	09957
11	Penncozeb 80 WP	United Phosphorus	80% w/w	WP	11622
12	Penncozeb 80 WP	United Phosphorus	80% w/w	WP	14580
13	Penncozeb WDG	Nufarm UK	77% w/w	WG	12156
14	Quell Flo	Interfarm	455 g/l	SC	09894
15	SpudGun	AgriGuard	80% w/w	WP	13447
16	Yankee	AgriGuard	75% w/w	WG	13905

Uses

- Black spot in ***roses*** [5]
- Blight in ***potatoes*** [1-4, 6-16]
- Brown rust in ***durum wheat*** *(useful control)*, ***spring wheat*** *(useful control)*, ***winter wheat*** *(useful control)* [1, 2]; ***spring barley***, ***winter barley*** [9-11, 13]; ***spring barley*** *(reduction)*, ***winter barley*** *(reduction)* [12]; ***spring wheat***, ***winter wheat*** [3, 6-8, 10, 11, 13-15]
- Currant leaf spot in ***blackcurrants***, ***gooseberries*** [5]
- Disease control in ***grass seed crops*** *(off-label)* [13]
- Downy mildew in ***bulb onions*** *(off-label)*, ***poppies for morphine production*** *(off-label)*, ***shallots*** *(off-label)* [3, 13]; ***lettuce***, ***protected lettuce*** [5, 15]; ***winter oilseed rape*** [3, 6-8, 14, 15]
- Fire in ***tulips*** [5]
- Foliar disease control in ***bulb onions*** *(off-label)*, ***poppies for morphine production*** *(off-label)*, ***shallots*** *(off-label)* [6, 14]; ***chives*** *(off-label)*, ***frise*** *(off-label)*, ***herbs (see appendix 6)*** *(off-label)*, ***lamb's lettuce*** *(off-label)*, ***parsley*** *(off-label)*, ***protected chives*** *(off-label)*, ***protected herbs (see appendix 6)*** *(off-label)*, ***protected parsley*** *(off-label)*, ***radicchio*** *(off-label)*, ***redcurrants*** *(off-label)*, ***whitecurrants*** *(off-label)*, ***wine grapes*** *(off-label)* [5]; ***durum wheat*** *(off-label)* [3]; ***grapevines***

(off-label) [13]; ***grass seed crops*** *(off-label)* [3, 9-12, 14]; ***spring rye*** *(off-label)*, ***winter rye*** *(off-label)* [14]

- Glume blotch in ***spring wheat***, ***winter wheat*** [10, 11, 13]
- Net blotch in ***spring barley*** [9-11, 13]; ***spring barley*** *(moderate control)*, ***winter barley*** *(moderate control)* [12]; ***winter barley*** [6, 9-11, 13]
- Ray blight in ***chrysanthemums*** [5]
- Rhynchosporium in ***spring barley*** [10, 11, 13]; ***winter barley*** [6, 10, 11, 13, 15]
- Rust in ***carnations***, ***geraniums***, ***roses*** [5]
- Scab in ***apples*** [5, 9-13]; ***pears*** [5]
- Septoria leaf blotch in ***durum wheat*** *(reduction)* [1, 2]; ***spring wheat***, ***winter wheat*** [3, 6-11, 13-15]; ***spring wheat*** *(reduction)*, ***winter wheat*** *(reduction)* [1, 2, 12]
- Sooty moulds in ***spring barley*** [10, 11, 13]; ***spring wheat***, ***winter wheat*** [3, 6-8, 10, 11, 13-15]; ***winter barley*** [6, 10, 11, 13]
- White mould in ***daffodils grown for galanthamine production*** *(off-label)* [13]
- White tip in ***leeks*** *(off-label)* [13]
- Yellow rust in ***spring barley***, ***spring wheat***, ***winter barley*** [10, 11, 13]; ***winter wheat*** [3, 6-8, 10, 11, 13-15]

Specific Off-Label Approvals (SOLAs)

- ***bulb onions*** *20081946* [3], *20081908* [6], *20091841* [13], *20081907* [14]
- ***chives*** *20061978* [5]
- ***daffodils grown for galanthamine production*** *20091831* [13]
- ***durum wheat*** *20052532* [3]
- ***frise*** *20061978* [5]
- ***grapevines*** *20091835* [13]
- ***grass seed crops*** *20052532* [3], *20052492* [9], *20052487* [10], *20052530* [11], *20092074* [12], *20091839* [13], *20052496* [14]
- ***herbs (see appendix 6)*** *20061978* [5]
- ***lamb's lettuce*** *20061978* [5]
- ***leeks*** *20091837* [13]
- ***parsley*** *20061978* [5]
- ***poppies for morphine production*** *20060458* [3], *20081909* [6], *20091828* [13], *20081910* [14]
- ***protected chives*** *20061978* [5]
- ***protected herbs (see appendix 6)*** *20061978* [5]
- ***protected parsley*** *20061978* [5]
- ***radicchio*** *20061978* [5]
- ***redcurrants*** *20061979* [5]
- ***shallots*** *20081946* [3], *20081908* [6], *20091841* [13], *20081907* [14]
- ***spring rye*** *20052496* [14]
- ***whitecurrants*** *20061979* [5]
- ***wine grapes*** *20063434* [5]
- ***winter rye*** *20052496* [14]

Approval information

- Mancozeb included in Annex I under EC Directive 91/414
- Approved for aerial application on potatoes [3, 9, 10, 13, 14]. See Section 5 for more information
- Accepted by BBPA for use on malting barley
- Approval expiry 30 Apr 2010 [7]

Efficacy guidance

- Mancozeb is a protectant fungicide and will give moderate control, suppression or reduction of the cereal diseases listed if treated before they are established but in many cases mixture with carbendazim is essential to achieve satisfactory results. See labels for details
- May be recommended for suppression or control of mildew in cereals depending on product and tank mix. See label for details

Restrictions

- Maximum number of treatments varies with crop and product used - check labels for details
- Check labels for minimum interval that must elapse between treatments
- On protected lettuce only 2 post-planting applications of mancozeb or of any combination of products containing EBDC fungicide (mancozeb, maneb, thiram, zineb) either as a spray or a dust are permitted within 2 wk of planting out and none thereafter.

- Avoid treating wet cereal crops or those suffering from drought or other stress
- Keep dry formulations away from fire and sparks
- Use dry formulations immediately. Do not store

Crop-specific information

- Latest use: before early milk stage (GS 73) for cereals; before 6 true leaf stage and before 31 Dec for winter oilseed rape.
- HI potatoes 7 d; outdoor lettuce 14 d; protected lettuce 21 d; apples, blackcurrants 28 d
- Apply to potatoes before haulm meets across rows (usually mid-Jun) or at earlier blight warning, and repeat every 7-14 d depending on conditions and product used (see label)
- May be used on potatoes up to desiccation of haulm
- On oilseed rape apply as soon as disease develops between cotyledon and 5-leaf stage (GS 1,0-1,5)
- Apply to cereals from 4-leaf stage to before early milk stage (GS 71). Recommendations vary, see labels for details
- Treat winter oilseed rape before 6 true leaf stage (GS 1,6) and before 31 Dec

Environmental safety

- Dangerous for the environment
- Very toxic to aquatic organisms
- Harmful to fish or other aquatic life. Do not contaminate surface waters or ditches with chemical or used container
- Do not empty into drains

Hazard classification and safety precautions

Hazard Harmful [11, 13]; Irritant [1-10, 14-16]; Dangerous for the environment [1-5, 7, 8, 11, 13-16]

Transport code 9

Packaging group III

UN number 3077 [1-3, 5, 7-13, 15, 16]; 3082 [4, 6, 14]

Risk phrases R36, R38 [9, 10]; R37 [1-3, 5, 7-11, 13-15]; R42 [11, 13]; R43 [1-8, 11, 13-16]; R50, R53a [1-5, 7, 8, 11, 13-16]

Operator protection A [1-11, 13-16]; C [4, 9, 10]; D [1-3, 5, 7-10, 15, 16]; H [1, 2]; U05a [1-11, 13-16]; U08 [6, 9, 10, 14]; U13 [9, 10]; U14 [1-3, 5, 7, 8, 14]; U19a [9-11, 13]; U20a [9]; U20b [1-8, 10, 14-16]

Environmental protection E13c [6, 9, 10]; E15a [1-5, 7, 8, 11, 13-16]; E19b [11, 13]; E34 [6, 14]; E38 [1-5, 7, 8, 14-16]

Storage and disposal D01 [1-3, 5, 7-11, 13, 14]; D02, D09a, D11a [1-11, 13-16]; D05 [1-7, 9, 10, 14, 15]; D07 [6]; D12a [1-5, 7, 8, 14-16]

320 mancozeb + metalaxyl-M

A systemic and protectant fungicide mixture
FRAC mode of action code: M3 + 4

See also metalaxyl-M

Products

1	Clayton Erase	Clayton	64:4% w/w	WG	13836
2	Fubol Gold WG	Syngenta	64:4% w/w	WG	10184

Uses

- Blight in ***potatoes*** [1, 2]
- Downy mildew in ***red beet*** *(off-label)*, ***rhubarb*** *(off-label)*, ***salad onions*** *(off-label)* [2]
- Fungus diseases in ***chives*** *(off-label)*, ***herbs (see appendix 6)*** *(off-label)*, ***lettuce*** *(off-label)*, ***parsley*** *(off-label)*, ***protected chives*** *(off-label)*, ***protected herbs (see appendix 6)*** *(off-label)*, ***protected lettuce*** *(off-label)*, ***protected parsley*** *(off-label)* [2]
- Phytophthora fruit rot in ***apples*** *(off-label - applied to orchard floor)* [2]
- White blister in ***cabbages*** *(off-label)*, ***chinese cabbage*** *(off-label)*, ***choi sum*** *(off-label)*, ***collards*** *(off-label)*, ***kale*** *(off-label)*, ***pak choi*** *(off-label)*, ***protected radishes*** *(off-label)*, ***radishes*** *(off-label)* [2]

Specific Off-Label Approvals (SOLAs)

- ***apples*** *(applied to orchard floor) 20011610* [2]

SEE SECTION 3 FOR PRODUCTS ALSO REGISTERED

- ***cabbages*** *20011610* [2]
- ***chinese cabbage*** *20063643* [2]
- ***chives*** *20032142* [2]
- ***choi sum*** *20063643* [2]
- ***collards*** *20063643* [2]
- ***herbs (see appendix 6)*** *20032142* [2]
- ***kale*** *20063643* [2]
- ***lettuce*** *20032142* [2]
- ***pak choi*** *20063643* [2]
- ***parsley*** *20032142* [2]
- ***protected chives*** *20032142* [2]
- ***protected herbs (see appendix 6)*** *20032142* [2]
- ***protected lettuce*** *20032142* [2]
- ***protected parsley*** *20032142* [2]
- ***protected radishes*** *20011610* [2]
- ***radishes*** *20011610* [2]
- ***red beet*** *20071256* [2]
- ***rhubarb*** *20080653* [2]
- ***salad onions*** *20032324* [2]

Approval information

- Mancozeb and metalaxyl-M included in Annex I under EC Directive 91/414

Efficacy guidance

- Commence potato blight programme before risk of infection occurs as crops begin to meet along the rows and repeat every 7-14 d according to blight risk. Do not exceed a 14 d interval between sprays
- If infection risk conditions occur earlier than the above growth stage commence spraying potatoes immediately
- Complete the potato blight programme using a protectant fungicide starting no later than 10 d after the last phenylamide spray. At least 2 such sprays should be applied
- To minimise the likelihood of development of resistance these products should be used in a planned Resistance Management strategy. See Section 5 for more information

Crop-specific information

- Latest use: before end of active potato haulm growth or before end Aug, whichever is earlier
- HI 7 d for potatoes
- After treating early potatoes destroy and remove any remaining haulm after harvest to minimise blight pressure on neighbouring maincrop potatoes

Environmental safety

- Dangerous for the environment
- Very toxic to aquatic organisms
- Do not harvest crops for human consumption for at least 7 d after final application

Hazard classification and safety precautions

Hazard Harmful, Dangerous for the environment
Transport code 9
Packaging group III
UN number 3077
Risk phrases R37, R43, R50, R53a, R63
Operator protection A; U05a, U08, U20b
Environmental protection E15a, E34, E38
Consumer protection C02a (7 d)
Storage and disposal D01, D02, D05, D09a, D11a, D12a

321 mancozeb + zoxamide

A protectant fungicide mixture for potatoes
FRAC mode of action code: M3 + 22

See also zoxamide

Products

1	Electis 75WG	Gowan	66.7:8.3% w/w	WG	14195
2	Roxam 75WG	Gowan	66.7:8.3% w/w	WG	14191

Uses
- Blight in ***potatoes***

Approval information
- Mancozeb and zoxamide included in Annex I under EC Directive 91/414

Efficacy guidance
- Apply as protectant spray on potatoes immediately risk of blight in district or as crops begin to meet along the rows and repeat every 7-14 d according to blight risk
- Do not use if potato blight present in crop. Products are not curative
- Spray irrigated potato crops as soon as possible after irrigation once the crop leaves are dry

Restrictions
- Maximum number of treatments 10 per crop for potatoes

Crop-specific information
- HI 7 d for potatoes

Environmental safety
- Dangerous for the environment
- Very toxic to aquatic organisms
- LERAP Category B
- Keep away from fire and sparks

Hazard classification and safety precautions

Hazard Irritant, Dangerous for the environment
Transport code 9
Packaging group III
UN number 3077
Risk phrases R37, R43, R50, R53a
Operator protection A, H; U05a, U14, U20b
Environmental protection E15a, E16a, E16b, E38
Storage and disposal D01, D02, D05, D09a, D11a, D12a

322 mandipropamid

A mandelamide fungicide for the control of potato blight
FRAC mode of action code: 40

Products

Revus	Syngenta	250 g/l	SC	13484

Uses
- Blight in ***potatoes***
- Disease control in ***ornamental plant production*** *(off-label)*

Specific Off-Label Approvals (SOLAs)
- ***ornamental plant production*** *20082867*

Efficacy guidance
- Mandipropamid acts preventatively by preventing spore germination and inhibiting mycelial growth during incubation. Apply immediately after blight warning or as soon as local conditions favour disease development but before blight enters the crop
- Spray at 7-10 d intervals reducing the interval as blight risk increases
- Spray programme should include a complete haulm desiccant to prevent tuber infection at harvest
- Eliminate other potential infection sources

- To minimise the likelihood of development of resistance this product should be used in a planned Resistance Management strategy. See Section 5 for more information
- See label for details of tank mixtures that may be used as part of a resistance management strategy

Restrictions

- Maximum number of treatments 4 per crop. Do not apply more than 3 treatments of this, or any other fungicide in the same resistance category, consecutively

Crop-specific information

- HI 3 d for potatoes

Hazard classification and safety precautions

UN number N/C
Risk phrases R52, R53a
Operator protection A, H; U05a, U20b
Environmental protection E15b, E34, E38
Storage and disposal D01, D02, D05, D09a, D10c, D12a
Medical advice M03, M05a

323 MCPA

A translocated phenoxycarboxylic acid herbicide for cereals and grassland
HRAC mode of action code: O

See also *2,4-D + dichlorprop-P + MCPA + mecoprop-P*
2,4-D + MCPA
2,4-DB + linuron + MCPA
2,4-DB + MCPA
bentazone + MCPA + MCPB
clopyralid + 2,4-D + MCPA
clopyralid + diflufenican + MCPA
clopyralid + fluroxypyr + MCPA
dicamba + dichlorprop-P + MCPA
dicamba + MCPA + mecoprop-P
dichlorprop-P + ferrous sulphate + MCPA
dichlorprop-P + MCPA
dichlorprop-P + MCPA + mecoprop-P
ferrous sulphate + MCPA + mecoprop-P

Products

1	Agritox 50	Nufarm UK	500 g/l	SL	07400
2	Agroxone	Headland	500 g/l	SL	09947
3	Campbell's MCPA 50	United Phosphorus	500 g/l	SL	00381
4	Headland Spear	Headland	500 g/l	SL	07115
5	HY-MCPA	Agrichem	500 g/l	SL	06293
6	Tasker 75	Headland	750 g/l	SL	10544

Uses

- Annual dicotyledons in ***amenity grassland, managed amenity turf*** [1, 2, 4]; ***durum wheat*** *(off-label)* [4]; ***grass seed crops*** [1, 2, 6]; ***land not intended to bear vegetation*** [1, 2]; ***linseed*** [1, 2, 5, 6]; ***permanent grassland, spring barley, spring oats, spring wheat, winter barley, winter oats, winter wheat*** [1-6]; ***rotational grass*** [2, 5, 6]; ***spring rye, undersown barley, undersown oats, undersown rye, undersown wheat*** [1]; ***spring rye*** *(off-label)*, ***winter rye*** *(off-label)* [2]; ***triticale*** *(off-label)* [1-3]; ***undersown barley*** *(red clover or grass)*, ***undersown wheat*** *(red clover or grass)* [2, 6]; ***undersown spring cereals, undersown winter cereals*** [5]; ***winter rye*** [1, 3, 4]
- Buttercups in ***amenity grassland, managed amenity turf*** [4, 5]; ***permanent grassland*** [3-5]
- Charlock in ***grass seed crops*** [6]; ***linseed*** [5, 6]; ***permanent grassland*** [1, 2, 6]; ***rotational grass*** [2, 5, 6]; ***spring barley, spring oats, spring wheat, winter barley, winter oats, winter wheat*** [1-6]; ***spring rye*** [1]; ***undersown barley*** *(red clover or grass)*, ***undersown wheat*** *(red clover or grass)* [2, 6]; ***winter rye*** [1, 3, 4]
- Dandelions in ***permanent grassland*** [3, 5]

- Docks in ***permanent grassland*** [3-5]
- Fat hen in ***amenity grassland, hard surfaces, managed amenity turf, natural surfaces not intended to bear vegetation, permeable surfaces overlying soil*** [5]; ***grass seed crops*** [6]; ***linseed*** [5, 6]; ***permanent grassland*** [1, 2, 6]; ***rotational grass*** [2, 5, 6]; ***spring barley, spring oats, spring wheat, winter barley, winter oats, winter wheat*** [1-6]; ***spring rye*** [1]; ***undersown barley*** *(red clover or grass)*, ***undersown wheat*** *(red clover or grass)* [2, 6]; ***winter rye*** [1, 3, 4]
- Hemp-nettle in ***grass seed crops*** [6]; ***linseed*** [5, 6]; ***permanent grassland*** [1, 2, 6]; ***rotational grass, undersown barley*** *(red clover or grass)*, ***undersown wheat*** *(red clover or grass)* [2, 6]; ***spring barley, spring oats, spring wheat, winter barley, winter oats, winter wheat*** [1-6]; ***spring rye*** [1]; ***winter rye*** [1, 3, 4]
- Perennial dicotyledons in ***amenity grassland, managed amenity turf*** [1, 2, 4]; ***grass seed crops, linseed*** [6]; ***land not intended to bear vegetation*** [1, 2]; ***permanent grassland, spring barley, spring oats, spring wheat, winter barley, winter oats, winter wheat*** [1-6]; ***rotational grass*** [5, 6]; ***spring rye*** [1]; ***undersown barley*** *(red clover or grass)*, ***undersown wheat*** *(red clover or grass)* [2, 6]; ***winter rye*** [1, 3, 4]
- Plantains in ***amenity grassland, grass seed crops, managed amenity turf, permanent grassland*** [5]
- Ragwort in ***permanent grassland*** [4]
- Rushes in ***permanent grassland*** [4]
- Stinging nettle in ***hard surfaces, natural surfaces not intended to bear vegetation, permeable surfaces overlying soil*** [5]; ***industrial sites, land not intended to bear vegetation*** [4]
- Thistles in ***hard surfaces, natural surfaces not intended to bear vegetation, permeable surfaces overlying soil*** [5]; ***industrial sites, land not intended to bear vegetation*** [4]; ***permanent grassland*** [3-5]
- Wild radish in ***grass seed crops*** [6]; ***linseed*** [5, 6]; ***permanent grassland*** [1, 2, 6]; ***rotational grass, undersown barley*** *(red clover or grass)*, ***undersown wheat*** *(red clover or grass)* [2, 6]; ***spring barley, spring oats, spring wheat, winter barley, winter oats, winter wheat*** [1-6]; ***spring rye*** [1]; ***winter rye*** [1, 3, 4]

Specific Off-Label Approvals (SOLAs)

- ***durum wheat*** *20052272* [4]
- ***spring rye*** *20052264* [2]
- ***triticale*** *20052270* [1], *20052264* [2], *20052266* [3]
- ***winter rye*** *20052264* [2]

Approval information

- MCPA included in Annex I under EC Directive 91/414
- Accepted by BBPA for use on malting barley

Efficacy guidance

- Best results achieved by application to weeds in seedling to young plant stage under good growing conditions when crop growing actively
- Spray perennial weeds in grassland before flowering. Most susceptible growth stage varies between species. See label for details
- Do not spray during cold weather, drought, if rain or frost expected or if crop wet

Restrictions

- Maximum number of treatments normally 1 per crop or yr except grass (2 per yr) for some products. See label
- Do not treat grass within 3 mth of germination and preferably not in the first yr of a direct sown ley or after reseeding
- Do not use on cereals before undersowing
- Do not roll, harrow or graze for a few days before or after spraying; see label
- Do not use on grassland where clovers are an important part of the sward
- Do not use on any crop suffering from stress or herbicide damage
- Avoid spray drift onto nearby susceptible crops

Crop-specific information

- Latest use: before 1st node detectable (GS 31) for cereals; 4-6 wk before heading for grass seed crops; before crop 15-25 cm high for linseed
- Apply to winter cereals in spring from fully tillered, leaf sheath erect stage to before first node detectable (GS 31)

- Apply to spring barley and wheat from 5-leaves unfolded (GS 15), to oats from 1-leaf unfolded (GS 11) to before first node detectable (GS 31)
- Apply to cereals undersown with grass after grass has 2-3 leaves unfolded
- Recommendations for crops undersown with legumes vary. Red clover may withstand low doses after 2-trifoliate leaf stage, especially if shielded by taller weeds but white clover is more sensitive. See label for details
- Apply to grass seed crops from 2-3 leaf stage to 5 wk before head emergence
- Temporary wilting may occur on linseed but without long term effects

Following crops guidance

- Do not direct drill brassicas or legumes within 6 wk of spraying grassland

Environmental safety

- Harmful to aquatic organisms
- MCPA is active at low concentrations. Take extreme care to avoid drift onto neighbouring crops, especially beet crops, brassicas, most market garden crops including lettuce and tomatoes under glass, pears and vines
- Keep livestock out of treated areas for at least 2 wk and until foliage of poisonous weeds such as ragwort has died and become unpalatable

Hazard classification and safety precautions

Hazard Harmful
Transport code 9 [3]
Packaging group III [3]
UN number 3082 [3]; N/C [1, 2, 4-6]
Risk phrases R20, R21 [2, 3]; R22a [2-6]; R41 [1, 3-6]; R52 [4, 5]; R53a [5]
Operator protection A, C [1-5]; U05a, U08, U11 [1-6]; U15 [4]; U20a [3]; U20b [1, 2, 4-6]
Environmental protection E06a [5] (2 wk); E07a [1-4, 6]; E15a, E34 [1-6]; E38 [4]
Storage and disposal D01, D02, D05, D09a [1-6]; D10a [1, 2, 5, 6]; D10b [4]; D10c [3]
Medical advice M03 [1-6]; M05a [3, 5]

324 MCPA + MCPB

A translocated herbicide for undersown cereals, grassland and various legumes
HRAC mode of action code: O + O

See also *MCPB*

Products

1	Bellmac Plus	United Phosphorus	38:262 g/l	SL	07521
2	Impetus	Headland	25:275 g/l	SL	11021
3	Tropotox Plus	Nufarm UK	37.5:262.5 g/l	SL	11142

Uses

- Annual dicotyledons in ***all cereals, clover seed crops, direct-sown seedling clovers, dredge corn containing peas*** [3]; ***combining peas, red clover, spring barley, spring oats, spring wheat, vining peas, white clover, winter barley, winter oats, winter wheat*** [2]; ***permanent grassland*** [2, 3]; ***rotational grass, undersown spring cereals, undersown winter cereals*** [1-3]; ***sainfoin, triticale*** *(off-label)* [1, 3]
- Perennial dicotyledons in ***all cereals, clover seed crops, direct-sown seedling clovers, dredge corn containing peas*** [3]; ***combining peas, red clover, spring barley, spring oats, spring wheat, vining peas, white clover, winter barley, winter oats, winter wheat*** [2]; ***permanent grassland*** [2, 3]; ***rotational grass, undersown spring cereals, undersown winter cereals*** [1-3]; ***sainfoin*** [1, 3]; ***triticale*** *(off-label)* [1]

Specific Off-Label Approvals (SOLAs)

- ***triticale*** *20052379* [1], *20052385* [3]

Approval information

- MCPA and MCPB included in Annex I under EC Directive 91/414
- Accepted by BBPA for use on malting barley

Efficacy guidance

- Best results achieved by application to weeds in seedling to young plant stage under good growing conditions when crop growing actively. Spray perennials when adequate leaf surface before flowering. Retreatment often needed in following year
- Spray leys and sainfoin before crop provides cover for weeds
- Rain, cold or drought may reduce effectiveness

Restrictions

- Maximum number of treatments 1 per crop or yr
- Do not spray clovers for seed
- Do not spray peas [1]
- Do not roll or harrow for a few days before or after spraying

Crop-specific information

- Latest use: before first node detectable (GS 31) for cereals; before flower buds visible for peas
- Apply to cereals from 2-expanded leaf stage to before jointing (GS 12-30) and, where undersown, after 1-trifoliate leaf stage of clover
- Apply to direct-sown seedling clover after 1-trifoliate leaf stage
- Apply to mature white clover for fodder at any stage. Do not spray red clover after flower stalk has begun to form
- Apply to sainfoin after first trifoliate leaf stage

Environmental safety

- Harmful to aquatic organisms
- Harmful to fish or other aquatic life. Do not contaminate surface waters or ditches with chemical or used container
- Keep livestock out of treated areas until foliage of any poisonous weeds such as ragwort has died and become unpalatable
- Take extreme care to avoid drift onto neighbouring crops, especially beet crops, brassicas, most market garden crops including lettuce and tomatoes under glass, pears and vines

Hazard classification and safety precautions

Hazard Harmful
Transport code 9 [3]
Packaging group III [3]
UN number 3082 [3]; N/C [1, 2]
Risk phrases R20, R21 [2]; R22a [1-3]; R38, R53a [1]; R41, R52 [1, 2]
Operator protection A, C [2]; U05a [2, 3]; U08, U20b [1-3]; U11, U14, U15 [1, 2]
Environmental protection E07a, E34 [1-3]; E13c [1]; E15a [2, 3]
Storage and disposal D01, D10b [2, 3]; D02 [2]; D05, D09a [1-3]; D10c [1]
Medical advice M03 [2]; M05a [1-3]

325 MCPA + mecoprop-P

A translocated selective herbicide for amenity grass
HRAC mode of action code: O + O

See also *mecoprop-P*

Products

1	Cleanrun Pro	Scotts	0.49:0.29% w/w	GR	12083
2	Greenmaster Extra	Scotts	0.49:0.29 % w/w	GR	11563

Uses

- Annual dicotyledons in ***managed amenity turf***
- Perennial dicotyledons in ***managed amenity turf***

Approval information

- MCPA and mecoprop-P included in Annex I under EC Directive 91/414

Efficacy guidance

- Apply from Apr to Sep, when weeds growing actively and have large leaf area available for chemical absorption

Restrictions

- Maximum total dose 105 g/sq m per yr. The total amount of mecoprop-P applied in a single yr must not exceed the maximum total dose approved for any single product for use on turf
- Avoid contact with cultivated plants
- Do not use first 4 mowings as compost or mulch unless composted for 6 mth
- Do not treat newly sown or turfed areas for at least 6 mth
- Do not reseed bare patches for 8 wk after treatment
- Do not apply when heavy rain expected or during prolonged drought. Irrigate after 1-2 d unless rain has fallen
- Do not mow within 2-3 d of treatment
- Treat areas planted with bulbs only after the foliage has died down
- Avoid walking on treated areas until it has rained or irrigation has been applied

Crop-specific information

- Granules contain NPK fertilizer to encourage grass growth

Environmental safety

- Take extreme care to avoid drift onto neighbouring crops, especially beet crops, brassicas, most market garden crops including lettuce and tomatoes under glass, pears and vines
- Harmful to fish or other aquatic life. Do not contaminate surface waters or ditches with chemical or used container
- Keep livestock out of treated areas for at least 2 wk and until foliage of any poisonous weeds such as ragwort has died and become unpalatable
- Some pesticides pose a greater threat of contamination of water than others and mecoprop-P is one of these pesticides. Take special care when applying mecoprop-P near water and do not apply if heavy rain is forecast

Hazard classification and safety precautions

UN number N/C
Operator protection A, C, H, M; U20b
Environmental protection E07a, E13c, E19b
Storage and disposal D01, D09a, D12a
Medical advice M05a

326 MCPB

A translocated phenoxycarboxylic acid herbicide
HRAC mode of action code: O

See also *bentazone + MCPA + MCPB*
MCPA + MCPB

Products

1	Bellmac Straight	United Phosphorus	400 g/l	SL	07522
2	Butoxone	Headland	400 g/l	SL	14406
3	Tropotox	Nufarm UK	400 g/l	SL	14450

Uses

- Annual dicotyledons in ***bilberries*** *(off-label)*, ***blackberries*** *(off-label)*, ***clover seed crops***, ***cranberries*** *(off-label)*, ***durum wheat*** *(off-label)*, ***gooseberries*** *(off-label)*, ***peas***, ***redcurrants*** *(off-label)*, ***spring rye*** *(off-label)*, ***triticale*** *(off-label)*, ***whitecurrants*** *(off-label)*, ***winter rye*** *(off-label)* [1]; ***blackberries***, ***gooseberries***, ***loganberries***, ***permanent grassland***, ***raspberries***, ***spring barley***, ***spring oats***, ***spring wheat***, ***white clover*** *(seed crops)*, ***winter barley***, ***winter oats***, ***winter wheat*** [2]; ***blackcurrants***, ***rotational grass***, ***undersown spring cereals***, ***undersown winter cereals*** [1, 2]; ***combining peas***, ***vining peas*** [2, 3]
- Docks in ***combining peas***, ***vining peas*** [3]
- Perennial dicotyledons in ***bilberries*** *(off-label)*, ***blackberries*** *(off-label)*, ***clover seed crops***, ***cranberries*** *(off-label)*, ***durum wheat*** *(off-label)*, ***gooseberries*** *(off-label)*, ***peas***, ***redcurrants*** *(off-label)*, ***spring rye*** *(off-label)*, ***triticale*** *(off-label)*, ***whitecurrants*** *(off-label)*, ***winter rye*** *(off-label)* [1]; ***blackberries***, ***combining peas***, ***gooseberries***, ***loganberries***, ***permanent grassland***, ***raspberries***, ***spring barley***, ***spring oats***, ***spring wheat***, ***vining peas***, ***white clover***

(seed crops), ***winter barley***, ***winter oats***, ***winter wheat*** [2]; ***blackcurrants***, ***rotational grass***, ***undersown spring cereals***, ***undersown winter cereals*** [1, 2]
- Thistles in ***combining peas***, ***vining peas*** [3]

Specific Off-Label Approvals (SOLAs)
- ***bilberries*** *20052381* [1]
- ***blackberries*** *20052381* [1]
- ***cranberries*** *20052381* [1]
- ***durum wheat*** *20052382* [1]
- ***gooseberries*** *20052381* [1]
- ***redcurrants*** *20052381* [1]
- ***spring rye*** *20052382* [1]
- ***triticale*** *20052382* [1]
- ***whitecurrants*** *20052381* [1]
- ***winter rye*** *20052382* [1]

Approval information
- MCPB included in Annex I under EC Directive 91/414
- Accepted by BBPA for use on malting barley

Efficacy guidance
- Best results achieved by spraying young seedling weeds in good growing conditions
- Best results on perennials by spraying before flowering
- Effectiveness may be reduced by rain within 12 h, by very cold or dry conditions

Restrictions
- Maximum number of treatments 1 per crop or yr.
- Do not roll or harrow for 7-10 d before or after treatment (check label)

Crop-specific information
- Latest use: first node detectable stage (GS 31) for cereals; before flower buds appear in terminal leaf (GS 201) for peas; before flower buds form for clover; before weeds damaged by frost for cane and bush fruit
- Apply to undersown cereals from 2-leaves unfolded to first node detectable (GS 12-31), and after first trifoliate leaf stage of clover
- Red clover seedlings may be temporarily damaged but later growth is normal
- Apply to white clover seed crops in Mar to early Apr, not after mid-May, and allow 3 wk before cutting and closing up for seed
- Apply to peas from 3-6 leaf stage but before flower bud detectable (GS 103-201). Consult PGRO (see Appendix 2) or label for information on susceptibility of cultivars.
- Do not use on leguminous crops not mentioned on the label
- Apply to cane and bush fruit after harvest and after shoot growth ceased but before weeds are damaged by frost, usually in late Aug or Sep; direct spray onto weeds as far as possible

Environmental safety
- Harmful to aquatic organisms
- Harmful to fish or other aquatic life. Do not contaminate surface waters or ditches with chemical or used container
- Keep livestock out of treated areas until foliage of any poisonous weeds such as ragwort has died and become unpalatable
- Take extreme care to avoid drift onto neighbouring sensitive crops

Hazard classification and safety precautions

Hazard Harmful
Transport code 9 [3]
Packaging group III [3]
UN number 3082 [3]; N/C [1, 2]
Risk phrases R22a [1-3]; R38, R41 [1, 2]; R51 [3]; R52 [1]; R53a [1, 3]
Operator protection A, C; U05a, U08, U20b [1-3]; U11, U14, U15 [1, 2]; U19a [3]
Environmental protection E07a, E34 [1-3]; E13c [1, 2]; E15a, E38 [3]
Storage and disposal D01, D02, D09a [1-3]; D05 [1, 2]; D10b [2, 3]; D10c [1]
Medical advice M03, M05a

327 mecoprop-P

A translocated phenoxycarboxylic acid herbicide for cereals and grassland
HRAC mode of action code: O

See also *2,4-D + dichlorprop-P + MCPA + mecoprop-P*
carfentrazone-ethyl + mecoprop-P
dicamba + MCPA + mecoprop-P
dicamba + mecoprop-P
dichlorprop-P + MCPA + mecoprop-P
diflufenican + mecoprop-P
ferrous sulphate + MCPA + mecoprop-P
fluroxypyr + mecoprop-P
MCPA + mecoprop-P

Products

1	Compitox Plus	Nufarm UK	600 g/l	SL	14390
2	Duplosan KV	Nufarm UK	600 g/l	SL	13971
3	Headland Charge	Headland	600 g/l	SL	14394
4	Isomec	Nufarm UK	600 g/l	SL	14385

Uses

- Annual dicotyledons in ***amenity grassland*** [1, 2, 4]; ***grass seed crops, managed amenity turf, spring barley, spring oats, spring wheat, winter barley, winter oats, winter wheat*** [1-4]; ***permanent grassland, rotational grass*** [3]
- Chickweed in ***amenity grassland*** [1, 2, 4]; ***grass seed crops, managed amenity turf, spring barley, spring oats, spring wheat, winter barley, winter oats, winter wheat*** [1-4]; ***permanent grassland, rotational grass*** [3]
- Cleavers in ***amenity grassland*** [1, 2, 4]; ***grass seed crops, managed amenity turf, spring barley, spring oats, spring wheat, winter barley, winter oats, winter wheat*** [1-4]; ***permanent grassland, rotational grass*** [3]
- Perennial dicotyledons in ***amenity grassland, grass seed crops, managed amenity turf, spring barley, spring oats, spring wheat, winter barley, winter oats, winter wheat*** [1, 2, 4]

Approval information

- Mecoprop-P included in Annex I under EC Directive 91/414
- Accepted by BBPA for use on malting barley

Efficacy guidance

- Best results achieved by application to seedling weeds which have not been frost hardened, when soil warm and moist and expected to remain so for several days

Restrictions

- Maximum number of treatments normally 1 per crop for spring cereals and 1 per yr for newly sown grass; 2 per crop or yr for winter cereals and grass crops. Check labels for details
- The total amount of mecoprop-P applied in a single yr must not exceed the maximum total dose approved for any single product for the crop/situation
- Do not spray cereals undersown with clovers or legumes or to be undersown with legumes or grasses
- Do not spray grass seed crops within 5 wk of seed head emergence
- Do not spray crops suffering from herbicide damage or physical stress
- Do not spray during cold weather, periods of drought, if rain or frost expected or if crop wet
- Do not roll or harrow for 7 d before or after treatment

Crop-specific information

- Latest use: generally before 1st node detectable (GS 31) for spring cereals and before 3rd node detectable (GS 33) for winter cereals, but individual labels vary; 5 wk before emergence of seed head for grass seed crops
- Spray winter cereals from 1 leaf stage in autumn up to and including first node detectable in spring (GS 10-31) or up to second node detectable (GS 32) if necessary. Apply to spring cereals from first fully expanded leaf stage (GS 11) but before first node detectable (GS 31)
- Spray cereals undersown with grass after grass starts to tiller
- Spray newly sown grass leys when grasses have at least 3 fully expanded leaves and have begun to tiller. Any clovers will be damaged

Environmental safety

- Harmful to aquatic organisms
- Harmful to fish or other aquatic life. Do not contaminate surface waters or ditches with chemical or used container
- Keep livestock out of treated areas for at least 2 wk and until foliage of any poisonous weeds, such as ragwort, has died and become unpalatable
- Take extreme care to avoid drift onto neighbouring crops, especially beet crops, brassicas, most market garden crops including lettuce and tomatoes under glass, pears and vines
- Some pesticides pose a greater threat of contamination of water than others and mecoprop-P is one of these pesticides. Take special care when applying mecoprop-P near water and do not apply if heavy rain is forecast

Hazard classification and safety precautions

Hazard Harmful, Dangerous for the environment
Transport code 9
Packaging group III
UN number 3082
Risk phrases R22a, R38, R41, R51, R53a [1-4]; R52 [3]
Operator protection A, C [1-4]; H [1, 2, 4]; P [3]; U05a, U08, U11, U20b [1-4]; U15 [3]
Environmental protection E07a, E13c [3]; E15b [1, 2, 4]; E34, E38 [1-4]
Storage and disposal D01, D02 [1-4]; D05, D10b [3]; D09a, D10c [1, 2, 4]
Medical advice M03

328 mepanipyrim

An anilinopyrimidine fungicide for use in horticulture
FRAC mode of action code: 9

Products

Frupica SC	Certis	450 g/l	SC	12067

Uses

- Botrytis in ***forest nurseries*** *(off-label)*, ***protected strawberries***, ***strawberries***

Specific Off-Label Approvals (SOLAs)

- ***forest nurseries*** *20082853*

Approval information

- Mepanipyrim included in Annex I under Directive 91/414

Efficacy guidance

- Product is protectant and should be applied as a preventative spray when conditions favourable for Botrytis development occur
- To maintain Botrytis control use as part of a programme with other fungicides that control the disease
- To minimise the possibility of development of resistance adopt resistance management procedures by using products from different chemical groups as part of a mixed spray programme

Restrictions

- Maximum number of treatments 2 per crop (including other anilinopyrimidine products)
- Consult processor before use on crops for processing
- Use spray mixture immediately after preparation

Crop-specific information

- HI 3 d

Environmental safety

- Dangerous for the environment
- Very toxic to aquatic organisms
- LERAP Category B

Hazard classification and safety precautions

Hazard Dangerous for the environment
Transport code 9
Packaging group III

UN number 3082
Risk phrases R50, R53a
Operator protection A, H; U20c
Environmental protection E16a, E16b, E34
Storage and disposal D01, D02, D05, D10c, D11a, D12b

329 mepiquat chloride

A quaternary ammonium plant growth regulator available only in mixtures

See also *2-chloroethylphosphonic acid + mepiquat chloride*
chlormequat + 2-chloroethylphosphonic acid + mepiquat chloride
chlormequat + mepiquat chloride

330 mepiquat chloride + prohexadione-calcium

A growth regulator mixture for cereals

See also *prohexadione-calcium*

Products

1	Canopy	BASF	300:50 g/l	SC	13181
2	Medax Top	BASF	300:50 g/l	SC	13282
3	Standon Midget	Standon	300:50	SC	13803

Uses

- Increasing yield in ***durum wheat*** *(off-label)* [1]; ***triticale*** [1, 2]; ***winter barley***, ***winter wheat*** [1-3]
- Lodging control in ***durum wheat*** *(off-label)* [1]; ***triticale*** [1, 2]; ***winter barley***, ***winter wheat*** [1-3]

Specific Off-Label Approvals (SOLAs)

- ***durum wheat*** *20072403* [1]

Approval information

- Mepiquat chloride and prohexadione-calcium included in Annex I under EC Directive 91/414

Efficacy guidance

- Best results obtained from treatments applied to healthy crops from the beginning of stem extension

Restrictions

- Maximum total dose equivalent to one full dose treatment on all crops
- Do not apply to any crop suffering from physical stress caused by waterlogging, drought or other conditions
- Do not treat on soils with a substantial moisture deficit
- Consult grain merchant or processor before use on crops for bread making or brewing. Effects on these processes have not been tested

Crop-specific information

- Latest use: before flag leaf fully emerged on wheat and barley

Following crops guidance

- Any crop may follow a normally harvested treated crop. Ploughing is not essential.

Environmental safety

- Harmful to aquatic organisms
- Avoid spray drift onto neighbouring crops

Hazard classification and safety precautions

Hazard Harmful
UN number N/C
Risk phrases R22a, R52, R53a
Operator protection A; U05a, U20b
Environmental protection E15a, E34, E38
Storage and disposal D01, D02, D05, D09a, D10c
Medical advice M05a

331 meptyldinocap

A protectant dinitrophenyl fungicide for powdery mildew control
FRAC mode of action code: 29

Products

Kindred	Landseer	350 g/l	EC	13891

Uses
- Powdery mildew in ***apples*** *(off-label)*, ***crab apples*** *(off-label)*, ***pears*** *(off-label)*, ***quinces*** *(off-label)*, ***table grapes***, ***wine grapes***

Specific Off-Label Approvals (SOLAs)
- ***apples*** *20092664*
- ***crab apples*** *20092664*
- ***pears*** *20092664*
- ***quinces*** *20092664*

Hazard classification and safety precautions

Hazard Harmful, Flammable, Dangerous for the environment
Transport code 3
Packaging group III
UN number 1993
Risk phrases R22a, R36, R38, R43, R50, R53a, R67
Operator protection A, C, H; U04a, U05a, U11, U14, U19a, U20a
Environmental protection E15b, E16a, E34, E38
Storage and disposal D01, D02, D09a, D10a, D12a
Medical advice M05a

332 mesosulfuron-methyl

A sulfonyl urea herbicide for cereals available only in mixtures
HRAC mode of action code: B

See also *diflufenican + iodosulfuron-methyl-sodium + mesosulfuron-methyl*
iodosulfuron-methyl-sodium + mesosulfuron-methyl

333 mesotrione

A foliar applied triketone herbicide for maize
HRAC mode of action code: F2

Products

1	Callisto	Syngenta	100 g/l	SC	12323
2	Greencrop Goldcob	Greencrop	100 g/l	SC	12863
3	Kalypstowe	AgChem Access	100 g/l	SC	13844
4	Marvel	ChemSource	100 g/l	SC	14508

Uses
- Annual dicotyledons in ***forage maize*** [1-4]; ***game cover*** *(off-label)*, ***linseed*** *(off-label)*, ***sweetcorn*** *(off-label)* [1]
- Annual grasses in ***game cover*** *(off-label)*, ***sweetcorn*** *(off-label)* [1]
- Annual meadow grass in ***grain maize*** [1, 3, 4]; ***linseed*** *(off-label)* [1]
- Volunteer oilseed rape in ***forage maize*** [1-4]; ***grain maize*** [1, 3, 4]; ***linseed*** *(off-label)* [1]

Specific Off-Label Approvals (SOLAs)
- ***game cover*** *20082830* [1]
- ***linseed*** *20071706* [1]
- ***sweetcorn*** *20051893* [1]

Approval information
- Mesotrione included in Annex I under EC Directive 91/414

SEE SECTION 3 FOR PRODUCTS ALSO REGISTERED

Efficacy guidance

- Best results obtained from treatment of young actively growing weed seedlings in the presence of adequate soil moisture
- Treatment in poor growing conditions or in dry soil may give less reliable control
- Activity is mostly by foliar uptake with some soil uptake
- To minimise the possible development of resistance where continuous maize is grown the product should not be used for more than two consecutive seasons

Restrictions

- Maximum number of treatments 1 per crop of forage maize
- Do not use on seed crops or on sweetcorn varieties
- Do not spray when crop foliage wet or when excessive rainfall is expected to follow application
- Do not treat crops suffering from stress from cold or drought conditions, or when wide temperature fluctuations are anticipated

Crop-specific information

- Latest use: 8 leaves unfolded stage (GS 18) for forage maize
- Treatment under adverse conditions may cause mild to moderate chlorosis. The effect is transient and does not affect yield

Following crops guidance

- Winter wheat, durum wheat, winter barley or ryegrass may follow a normally harvested treated crop of maize. Oilseed rape may be sown provided it is preceded by deep ploughing to more than 15 cm
- In the spring following application only forage maize, ryegrass, spring wheat or spring barley may be sown
- In the event of crop failure maize may be re-seeded immediately. Some slight crop effects may be seen soon after emergence but these are normally transient

Environmental safety

- Dangerous for the environment
- Very toxic to aquatic organisms
- LERAP Category B
- Take extreme care to avoid drift onto all plants outside the target area

Hazard classification and safety precautions

Hazard Irritant, Dangerous for the environment
UN number N/C
Risk phrases R36, R50, R53a
Operator protection A, C; U05a, U20b [1-4]; U08, U19a [2]; U09a [1, 3, 4]
Environmental protection E15b, E16a, E16b [1-4]; E38 [1, 3, 4]
Storage and disposal D01, D02, D05, D09a, D12a [1-4]; D10b [2]; D10c, D11a [1, 3, 4]
Medical advice M05a [2]

334 mesotrione + terbuthylazine

A foliar and soil acting herbicide mixture for maize
HRAC mode of action code: F2 + C1

See also *terbuthylazine*

Products

1	Calaris	Syngenta	70:330 g/l	SC	12405
2	Clayton Faize	Clayton	70:330 g/l	SC	13810
3	Destiny	AgChem Access	70:330 g/l	SC	14159
4	Pan Theta	Pan Agriculture	70:330 g/l	SC	14501

Uses

- Annual dicotyledons in ***forage maize*, *grain maize*** [1, 2, 4]; ***forage maize (under plastic mulches)*** *(off-label)*, ***sweetcorn*** *(off-label)* [1]
- Annual meadow grass in ***forage maize*, *grain maize*** [1, 2, 4]; ***forage maize (under plastic mulches)*** *(off-label)*, ***sweetcorn*** *(off-label)* [1]

Specific Off-Label Approvals (SOLAs)

- ***forage maize (under plastic mulches)*** *20061439* [1]
- ***sweetcorn*** *20051892* [1]

Approval information

- Mesotrione included in Annex I under EC Directive 91/414

Efficacy guidance

- Best results obtained from treatment of young actively growing weed seedlings in the presence of adequate soil moisture
- Treatment in poor growing conditions or in dry soil may give less reliable control
- Residual weed control is reduced on soils with more than 10% organic matter
- To minimise the possible development of resistance where continuous maize is grown the product should not be used for more than two consecutive seasons

Restrictions

- Maximum number of treatments 1 per crop of forage maize
- Do not use on seed crops or on sweetcorn varieties
- Do not spray when crop foliage wet or when excessive rainfall is expected to follow application
- Do not treat crops suffering from stress from cold or drought conditions, or when wide temperature fluctuations are anticipated
- Do not apply on Sands or Very Light soils

Crop-specific information

- Latest use: 8 leaves unfolded stage (GS 18) for forage maize
- Treatment under adverse conditions may cause mild to moderate chlorosis. The effect is transient and does not affect yield

Following crops guidance

- Winter wheat, durum wheat, winter barley or ryegrass may follow a normally harvested treated crop of maize. Oilseed rape may be sown provided it is preceded by deep ploughing to more than 15 cm
- In the spring following application forage maize, ryegrass, spring wheat or spring barley may be sown
- Spinach, beet crops, peas, beans, lettuce and cabbages must not be sown in the yr following application
- In the event of crop failure maize may be re-seeded immediately. Some slight crop effects may be seen soon after emergence but these are normally transient

Environmental safety

- Dangerous for the environment
- Very toxic to aquatic organisms
- LERAP Category B
- Take extreme care to avoid drift onto all plants outside the target area

Hazard classification and safety precautions

Hazard Harmful, Dangerous for the environment [1, 2, 4]
Transport code 9
Packaging group III
UN number 3082
Risk phrases R22a, R50, R53a [1, 2, 4]
Operator protection A [1, 2, 4]; U05a [1, 2, 4]
Environmental protection E15a, E16a, E16b, E34, E38 [1, 2, 4]
Storage and disposal D01, D02, D05, D09a, D10c, D12a [1, 2, 4]

335 metalaxyl-M

A phenylamide systemic fungicide
FRAC mode of action code: 4

See also *chlorothalonil + metalaxyl-M*
cymoxanil + fludioxonil + metalaxyl-M
fluazinam + metalaxyl-M
fludioxonil + metalaxyl-M + thiamethoxam
mancozeb + metalaxyl-M

Products

1	Clayton Tine	Clayton	465 g/l	SL	14072
2	SL 567A	Syngenta	465.2 g/l	SL	12380
3	Subdue	Fargro	465 g/l	SL	12503

Uses

- Cavity spot in ***carrots***, ***parsnips*** *(off-label)* [2]; ***carrots*** *(reduction only)* [1]
- Crown rot in ***protected water lilies*** *(off-label)*, ***water lilies*** *(off-label)* [2]
- Damping off in ***watercress*** *(off-label)* [2]
- Downy mildew in ***asparagus*** *(off-label)*, ***blackberries*** *(off-label)*, ***grapevines*** *(off-label)*, ***herbs (see appendix 6)*** *(off-label)*, ***hops*** *(off-label)*, ***horseradish*** *(off-label)*, ***protected cucumbers*** *(off-label)*, ***protected herbs (see appendix 6)*** *(off-label)*, ***protected soft fruit*** *(off-label)*, ***protected spinach*** *(off-label)*, ***protected spinach beet*** *(off-label)*, ***protected water lilies*** *(off-label)*, ***raspberries*** *(off-label)*, ***rubus hybrids*** *(off-label)*, ***salad onions*** *(off-label)*, ***soft fruit*** *(off-label)*, ***spinach*** *(off-label)*, ***spinach beet*** *(off-label)*, ***water lilies*** *(off-label)*, ***watercress*** *(off-label)* [2]
- Phytophthora root rot in ***ornamental plant production***, ***protected ornamentals*** [3]
- Pythium in ***ornamental plant production***, ***protected ornamentals*** [3]
- Root malformation disorder in ***red beet*** *(off-label)* [2]
- Storage rots in ***cabbages*** *(off-label)* [2]
- White blister in ***horseradish*** *(off-label)* [2]

Specific Off-Label Approvals (SOLAs)

- ***asparagus*** *20051502* [2]
- ***blackberries*** *20072195* [2]
- ***cabbages*** *20062117* [2]
- ***grapevines*** *20051504* [2]
- ***herbs (see appendix 6)*** *20051507* [2]
- ***hops*** *20051500* [2]
- ***horseradish*** *20061040* [2]
- ***parsnips*** *20051508* [2]
- ***protected cucumbers*** *20051503* [2]
- ***protected herbs (see appendix 6)*** *20051507* [2]
- ***protected soft fruit*** *20082937* [2]
- ***protected spinach*** *20051507* [2]
- ***protected spinach beet*** *20051507* [2]
- ***protected water lilies*** *20051501* [2]
- ***raspberries*** *20072195* [2]
- ***red beet*** *20051307* [2]
- ***rubus hybrids*** *20072195* [2]
- ***salad onions*** *20072194* [2]
- ***soft fruit*** *20082937* [2]
- ***spinach*** *20051507* [2]
- ***spinach beet*** *20051507* [2]
- ***water lilies*** *20051501* [2]
- ***watercress*** *20072193* [2]

Approval information

- Metalaxyl-M included in Annex I under EC Directive 91/414
- Accepted by BBPA for use on hops

Efficacy guidance

- Best results achieved when applied to damp soil or potting media
- Treatments to ornamentals should be followed immediately by irrigation to wash any residues from the leaves and allow penetration to the rooting zone [3]
- Efficacy may be reduced in prolonged dry weather [2]
- Results may not be satisfactory on soils with high organic matter content [2]
- Control of cavity spot on carrots overwintered in the ground or lifted in winter may be lower than expected [2]
- Use in an integrated pest management strategy and, where appropriate, alternate with products from different chemical groups
- Product should ideally be used preventatively and the number of phenylamide applications should be limited to 1-2 consecutive treatments
- Always follow FRAG guidelines for preventing and managing fungicide resistance. See Section 5 for more information

Restrictions

- Maximum number of treatments on ornamentals 1 per situation for media treatment. See label for details of drench treatment of protected ornamentals [3]
- Maximum total dose on carrots equivalent to one full dose treatment [2]
- Do not use where carrots have been grown on the same site within the previous eight yrs [2]
- Consult before use on crops intended for processing [2]
- Do not re-use potting media from treated plants for subsequent crops [3]
- Disinfect pots thoroughly prior to re-use [3]

Crop-specific information

- Latest use: 6 wk after drilling for carrots [2]
- Because of the large number of species and ornamental cultivars susceptibility should be checked before large scale treatment [3]
- Treatment of *Viburnum* and *Prunus* species not recommended [3]

Environmental safety

- Harmful to aquatic organisms
- Limited evidence suggests that metalaxyl-M is not harmful to soil dwelling predatory mites

Hazard classification and safety precautions

Hazard Harmful
UN number N/C
Risk phrases R22a, R37, R52, R53a
Operator protection A [1-3]; C, H [3]; U02a, U04a, U05a, U10, U19a, U20b
Environmental protection E15b, E34, E38
Storage and disposal D01, D02, D05, D07, D09a, D10c, D12a
Medical advice M03

336 metaldehyde

A molluscicide bait for controlling slugs and snails

Products

1	Allure	Chiltern	1.5% w/w	PT	11089
2	Appeal	Chiltern	1.5% w/w	PT	12022
3	Attract	Chiltern	1.5% w/w	PT	12023
4	Carakol 3	Makhteshim	3% w/w	PT	14309
5	Carakol Plus	Makhteshim	5% w/w	PT	13980
6	CDP Minis	De Sangosse	5% w/w	PT	12955
7	Certis Deal 5	Certis	5% w/w	PT	13488
8	Certis Metaldehyde 3	Certis	3% w/w	PT	14337
9	Certis Red 3	Certis	3% w/w	PT	14061
10	Chiltern Hundreds	Chiltern	3% w/w	PT	10072
11	Cleancrop Hyde 3	Makhteshim	3% w/w	PT	14310
12	Condor 5	Doff Portland	5% w/w	PT	12484
13	Delicia Slug-lentils	Barclay	3% w/w	PT	13613
14	Desire	Chiltern	1.5% w/w	PT	14048
15	Dixie 6	Greencrop	6% w/w	PT	11790
16	Doff Horticultural Slug Killer Blue Mini Pellets	Doff Portland	3% w/w	RB	11463

SEE SECTION 3 FOR PRODUCTS ALSO REGISTERED

Products – continued

17	Duro	Makhteshim	3% w/w	PT	14347
18	Enzo	Makhteshim	3% w/w	PT	14306
19	Escar-go 3	Chiltern	3% w/w	PT	14044
20	ESP	De Sangosse	5% w/w	PT	12999
21	Gusto	Makhteshim	5% w/w	PT	13992
22	Gusto 3	Makhteshim	3% w/w	PT	14308
23	Helimax S	De Sangosse	5% w/w	RB	13109
24	Lynx H	De Sangosse	3% w/w	RB	14426
25	Lynx S	De Sangosse	5% w/w	PT	13276
26	Molotov	Chiltern	3% w/w	PT	12650
27	Monza	Makhteshim	5% w/w	PT	13600
28	Navona	Makhteshim	4% w/w	PT	13892
29	Optimol	De Sangosse	4% w/w	RB	12997
30	Oreto	Makhteshim	3% w/w	PT	14258
31	Osarex W	De Sangosse	3% w/w	RB	14428
32	Pastel M	Clayton	5% w/w	PT	13479
33	Permesso	AgChem Access	5% w/w	PT	14108
34	Pesta H	De Sangosse	3% w/w	RB	14427
35	Pesta M	De Sangosse	5% w/w	PT	13275
36	Polymet 5	Unicrop	5% w/w	RB	13480
37	Prego	Makhteshim	5% w/w	RB	13582
38	Slugdown 3	Unicrop	3% w/w	PT	14366
39	Spinner	Certis	5% w/w	PT	14127
40	Steam	Chiltern	3% w/w	PT	12473
41	Super 3	Unicrop	35 w/w	RB	14370
42	Super 5	Unicrop	5% w/w	RB	12535
43	TDS Major	De Sangosse	4% w/w	CB	13462
44	TDS Metarex Amba	De Sangosse	4% w/w	CB	13461
45	Trevi	Makhteshim	4% w/w	PT	13983
46	Trojan	Chiltern	3% w/w	PT	14224

Uses

- Slugs in ***all edible crops (outdoor and protected)*** [13]; ***all edible crops (outdoor)*** [1-3, 6, 7, 10, 12, 15, 16, 20, 21, 23, 25, 26, 29, 32, 33, 35-37, 40, 43, 44, 46]; ***all edible crops (outdoor)*** *(around)*, ***all non-edible crops (outdoor)*** *(around)* [42]; ***all non-edible crops (outdoor)*** [1-3, 6, 7, 10, 12, 13, 15, 16, 20, 21, 23, 25, 26, 29, 32, 33, 35-37, 40, 43, 44, 46]; ***amenity grassland, managed amenity turf*** [7]; ***cultivated land/soil*** [1-3, 10, 16, 26, 40, 46]; ***natural surfaces not intended to bear vegetation*** [6, 7, 12, 13, 15, 23, 25, 29, 32, 35, 36, 42-44]; ***ornamental plant production*** [21, 33, 37]; ***protected crops*** [1-3, 10, 21, 26, 33, 37, 40, 43, 44, 46]
- Slugs and snails in ***all edible crops (outdoor and protected)*** [14, 19, 27, 39]; ***all edible crops (outdoor and protected)*** *(do not apply to cauliflowers)*, ***amenity grassland, managed amenity turf*** [8, 9]; ***all edible crops (outdoor)*** [4, 11, 17, 18, 22, 24, 30, 31, 34, 38, 41]; ***all edible crops (outdoor)*** *(except potatoes and cauliflowers)* [5, 28, 45]; ***all non-edible crops (outdoor)*** [5, 14, 19, 24, 27, 28, 31, 34, 38, 39, 41, 45]; ***all non-edible crops (outdoor)*** *(Excluding potato and cauliflower)* [4, 11, 17, 18, 22, 30]; ***cauliflowers*** [24, 31, 34]; ***cultivated land/soil, protected crops*** [14, 19]; ***natural surfaces not intended to bear vegetation*** [8, 9, 24, 31, 34, 38, 41]; ***potatoes*** [4, 8, 9, 11, 17, 18, 22, 24, 30, 31, 34]
- Snails in ***all edible crops (outdoor), all non-edible crops (outdoor)*** [1-3, 6, 7, 10, 12, 15, 16, 20, 21, 23, 25, 26, 29, 32, 33, 35-37, 40, 43, 44, 46]; ***all edible crops (outdoor)*** *(around)*, ***all non-edible crops (outdoor)*** *(around)* [42]; ***amenity grassland, managed amenity turf*** [7]; ***cultivated land/soil*** [1-3, 10, 16, 26, 40, 46]; ***natural surfaces not intended to bear vegetation*** [6, 7, 12, 15, 23, 25, 29, 32, 35, 36, 42-44]; ***ornamental plant production*** [21, 33, 37]; ***protected crops*** [1-3, 10, 21, 26, 33, 37, 40, 43, 44, 46]

Approval information

- Metaldehyde has not been included in Annex 1 but the date of withdrawal has not yet been published
- Accepted by BBPA for use on malting barley and hops

Efficacy guidance

- Apply pellets by hand, fiddle drill, fertilizer distributor, by air (check label) or in admixture with seed. See labels for rates and timing.
- Best results achieved from an even spread of granules applied during mild, damp weather when slugs and snails most active. May be applied in standing crops

- To establish the need for pellet application on winter wheat or winter oilseed rape, monitor for slug activity. Where bait traps are used, use a foodstuff attractive to slugs e.g. chicken layer's mash
- Varieties of oilseed rape low in glucosinolates can be more acceptable to slugs than "single low" varieties and control may not be as good
- To prevent slug build up apply at end of season to brassicas and other leafy crops
- To reduce tuber damage in potatoes apply twice in Jul and Aug
- For information on slug trapping and damage risk assessment refer to HGCA Topic Sheets No. 84 (winter wheat) and 85 (winter oilseed rape), available from the HGCA website (www.hgca.com)

Restrictions

- Do not apply when rain imminent or water glasshouse crops within 4 d of application
- Take care to avoid lodging of pellets in the foliage when making late applications to edible crops

Crop-specific information

- Put slug traps out before cultivation, when the soil surface is visibly moist and the weather mild (5-25 °C) (see label for guidance)
- For winter wheat, a catch of 4 or more slugs/trap indicates a possible risk, where soil and weather conditions favour slug activity
- For winter oilseed rape a catch of 4 or more slugs in standing cereals, or 1 or more in cereal stubble, if other conditions were met, would indicate possible risk of damage

Environmental safety

- Dangerous to game, wild birds and animals
- Some products contain proprietary cat and dog deterrent
- Keep poultry out of treated areas for at least 7 d
- Do not use slug pellets in traps in winter wheat or winter oilseed rape since they are a potential hazard to wildlife and pets
- Some pesticides pose a greater threat of contamination of water than others and metaldehyde is one of these pesticides. Take special care when applying metaldehyde near water and do not apply if heavy rain is forecast

Hazard classification and safety precautions

UN number N/C [1-12, 14-46]

Operator protection A, H [1-15, 17-46]; J [6, 12, 15, 20, 25, 29, 32, 35, 36, 42]; U05a [1-11, 13, 14, 17-19, 22, 24-28, 30-32, 34, 35, 39, 43-46]; U15 [21, 33, 37]; U20a [4, 11, 17, 18, 21, 22, 30, 33, 37]; U20b [5, 8-10, 24, 26-28, 31, 34, 36, 45, 46]; U20c [1-3, 6, 7, 14-16, 19, 20, 23, 25, 29, 32, 35, 38-41, 43, 44]

Environmental protection E05b [1-12] (7 d); E05b [13] (Keep poultry out of treated areas for at least 7 days following application); E05b [14] (7 days); E05b [15-18] (7 d); E05b [19] (7 days); E05b [20-23] (7 d); E05b [24]; E05b [25-30] (7 d); E05b [31]; E05b [32, 33] (7 d); E05b [34]; E05b [35-46] (7 d); E06b [14, 19] (7 d); E07a, E34 [24, 31, 34]; E10a [1-7, 10-46]; E10c [8, 9]; E15a [1-46]

Storage and disposal D01 [1-11, 13, 14, 17-35, 37-46]; D02 [1-11, 13, 14, 17-19, 21-28, 30-35, 37, 39, 43-46]; D05 [6-9, 32, 39]; D07 [1-3, 8-10, 12, 14, 16, 19, 21, 23, 24, 26, 31, 33, 34, 36-44, 46]; D09a [1-12, 14, 15, 17-46]; D11a [1-12, 14-22, 24, 26-34, 36-42, 45, 46]; D11b [23, 25, 35, 43, 44]; D12a [21, 23, 33, 37]

Treated seed S04a [13, 20, 21, 23-25, 29, 31, 33-35, 37, 42-44]

Medical advice M04a [6-9, 32]; M05a [13, 21, 23, 24, 31, 33, 34, 37, 39, 43, 44]

337 metamitron

A contact and residual triazinone herbicide for use in beet crops
HRAC mode of action code: C1

See also *chloridazon + chlorpropham + metamitron*
chloridazon + metamitron
chlorpropham + metamitron
desmedipham + ethofumesate + metamitron + phenmedipham
ethofumesate + metamitron
ethofumesate + metamitron + phenmedipham

Products

1	Agriguard Metamitron	AgriGuard	70% w/w	WG	09859
2	Barclay Seismic	Barclay	70% w/w	WG	11377

SEE SECTION 3 FOR PRODUCTS ALSO REGISTERED

Products – continued

3	Bettix 70 WG	United Phosphorus	70% w/w	WG	11154
4	Bettix Flo	United Phosphorus	700 g/l	SC	11959
5	Clayton Mitrex	Clayton	70% w/w	WG	11921
6	Defiant SC	United Phosphorus	700 g/l	SC	12302
7	Fernpath Haptol	AgriGuard	70% w/w	WG	11951
8	Goldbeet	Makhteshim	90% w/w	WG	11538
9	Goltix 90	Makhteshim	90% w/w	WG	11578
10	Goltix Flowable	Makhteshim	700 g/l	SC	12851
11	Goltix WG	Makhteshim	70% w/w	WG	11539
12	Marquise	Makhteshim	70% w/w	WG	08738
13	Mitron 70 WG	Hermoo	70% w/w	WG	11516
14	Mitron 90 WG	Hermoo	90% w/w	WG	10888
15	Mitron SC	Hermoo	700 g/l	SC	11643
16	MM 70 Flo	Nufarm UK	700 g/l	SC	13318
17	Predator WG	AgriGuard	70% w/w	WG	12630
18	Skater	Makhteshim	700 g/l	SC	12857
19	Target SC	AgriChem BV	700 g/l	SC	13307

Uses

- Annual dicotyledons in ***asparagus*** *(off-label)*, ***forest*** *(off-label)* [9, 12]; ***cardoons*** *(off-label)*, ***celery (outdoor)*** *(off-label)*, ***rhubarb*** *(off-label)* [9, 18]; ***fodder beet***, ***sugar beet*** [1-19]; ***herbs (see appendix 6)*** *(off-label)* [9]; ***horseradish*** *(off-label)*, ***parsnips*** *(off-label)* [8-10]; ***mangels*** [1-12, 14-19]; ***red beet*** [1-12, 14-17, 19]; ***strawberries*** *(off-label)* [9, 10]
- Annual grasses in ***asparagus*** *(off-label)*, ***forest*** *(off-label)* [12]; ***fodder beet***, ***sugar beet*** [1-18]; ***herbs (see appendix 6)*** *(off-label)* [9]; ***horseradish*** *(off-label)*, ***parsnips*** *(off-label)* [8-10]; ***mangels*** [1-12, 14-18]; ***red beet*** [1-12, 14-17]; ***strawberries*** *(off-label)* [10]
- Annual meadow grass in ***asparagus*** *(off-label)*, ***cardoons*** *(off-label)*, ***celery (outdoor)*** *(off-label)*, ***forest*** *(off-label)*, ***herbs (see appendix 6)*** *(off-label)*, ***rhubarb*** *(off-label)*, ***strawberries*** *(off-label)* [9]; ***fodder beet***, ***sugar beet*** [1-19]; ***horseradish*** *(off-label)*, ***parsnips*** *(off-label)* [8, 9]; ***mangels*** [1-12, 14-19]; ***red beet*** [1-12, 14-17, 19]
- Fat hen in ***fodder beet***, ***sugar beet*** [1-18]; ***herbs (see appendix 6)*** *(off-label)* [9]; ***horseradish*** *(off-label)*, ***parsnips*** *(off-label)* [8, 9]; ***mangels*** [1-12, 14-18]; ***red beet*** [1-12, 14-17]
- General weed control in ***asparagus*** *(off-label)*, ***forest*** *(off-label)*, ***horseradish*** *(off-label)*, ***parsnips*** *(off-label)* [11]
- Groundsel in ***chives*** *(off-label)*, ***herbs (see appendix 6)*** *(off-label)*, ***parsley*** *(off-label)* [11, 12]
- Polygonums in ***cardoons*** *(off-label)*, ***celery (outdoor)*** *(off-label)*, ***rhubarb*** *(off-label)* [18]

Specific Off-Label Approvals (SOLAs)

- ***asparagus*** *20081387* [9], *20041758* [11], *20030737* [12]
- ***cardoons*** *20081389* [9], *20070844* [18]
- ***celery (outdoor)*** *20081389* [9], *20070844* [18]
- ***chives*** *20041756* [11], *20030737* [12]
- ***forest*** *20081386* [9], *20041757* [11], *20030737* [12]
- ***herbs (see appendix 6)*** *20081385* [9], *20041756* [11], *20030737* [12]
- ***horseradish*** *20081039* [8], *20081041* [9], *20070513* [10], *20061637* [11]
- ***parsley*** *20041756* [11], *20030737* [12]
- ***parsnips*** *20081039* [8], *20081041* [9], *20070513* [10], *20061637* [11]
- ***rhubarb*** *20081389* [9], *20070844* [18]
- ***strawberries*** *20081388* [9], *20070856* [10]

Efficacy guidance

- May be used pre-emergence alone or post-emergence in tank mixture or with an authorised adjuvant oil
- Low dose programme (LDP). Apply a series of low-dose post-weed emergence sprays, including adjuvant oil, timing each treatment according to weed emergence and size. See label for details and for recommended tank mixes and sequential treatments. On mineral soils the LDP should be preceded by pre-drilling or pre-emergence treatment
- Traditional application. Apply either pre-drilling before final cultivation with incorporation to 8-10 cm, or pre-crop emergence at or soon after drilling into firm, moist seedbed to emerged weeds from cotyledon to first true leaf stage
- On emerged weeds at or beyond 2-leaf stage addition of adjuvant oil advised
- Up to 3 post-emergence sprays may be used on soils with over 10% organic matter

- For control of wild oats and certain other weeds, tank mixes with other herbicides or sequential treatments are recommended. See label for details

Restrictions

- Maximum total dose equivalent to three full dose treatments for most products. Check label
- Using traditional method post-crop emergence on mineral soils do not apply before first true leaves have reached 1 cm long

Crop-specific information

- Latest use: before crop foliage meets across rows for beet crops
- HI herbs 6 wk
- Crop tolerance may be reduced by stress caused by growing conditions, effects of pests, disease or other pesticides, nutrient deficiency etc

Following crops guidance

- Only sugar beet, fodder beet or mangels may be drilled within 4 mth after treatment. Winter cereals may be sown in same season after ploughing, provided 16 wk passed since last treatment

Environmental safety

- Dangerous for the environment
- Very toxic to aquatic organisms
- Dangerous to fish or other aquatic life. Do not contaminate surface waters or ditches with chemical or used container
- Do not empty into drains

Hazard classification and safety precautions

Hazard Harmful [1, 3, 4, 6, 8-12, 14-17, 19]; Irritant [2, 5, 7]; Dangerous for the environment [1, 3-19]

Transport code 9

Packaging group III

UN number 3077 [1-3, 5, 7-9, 11-14, 17]; 3082 [4, 6, 10, 15, 16, 18, 19]

Risk phrases R22a [1, 3, 4, 6, 8-12, 14-17, 19]; R41 [1, 2, 5, 7]; R43 [1, 2, 4-7, 12]; R50 [3-6, 8-10, 12-15]; R51 [1, 7, 11, 16-19]; R53a [1, 3-19]

Operator protection A [1, 2, 4-7, 10-12, 15-17]; C [1, 2, 5, 7]; H [1, 2, 5, 7, 12, 17]; U05a [1-3, 5, 7, 12, 13, 16, 19]; U08 [12, 13, 16, 19]; U09a [4, 6]; U11, U15 [5]; U14 [4-6, 12, 13, 16, 19]; U19a [1, 3, 5, 11-13, 16, 19]; U20a [1, 5, 7-9, 11-14, 16]; U20b [2, 3, 10, 15, 19]; U20c [4, 6, 18]

Environmental protection E13b [3, 4, 6]; E15a [1, 2, 5, 7-19]; E19b [8-10, 14, 15, 17, 18]; E34 [10, 15, 16, 19]; E38 [1, 11, 13, 19]

Storage and disposal D01 [1-10, 12-17, 19]; D02 [1-7, 10, 12, 13, 15, 16, 19]; D05 [4-6, 16]; D09a [1-19]; D10c [4, 6, 18]; D11a [1-3, 5, 7-16, 19]; D12a [1, 3-6, 8-11, 14, 15, 17-19]; D12b [12, 16]

Medical advice M03 [3, 4, 6, 10, 15, 19]; M04a [14]; M05a [3, 4, 6, 8-10, 15, 17]

338 metam-sodium

A methyl isothiocyanate producing sterilant for glasshouse, nursery and outdoor soils
HRAC mode of action code: Z

Products

1	Discovery	United Phosphorus	510 g/l	SL	10416
2	Metam 510	Taminco	510 g/l	LI	09796
3	Sistan 51	Unicrop	510 g/l	SL	10046

Uses

- Brown rot in ***potting soils*, *soils*** [2]
- Millipedes in ***potting soils*, *soils*** [2]
- Nematodes in ***glasshouse soils*, *nursery soils*, *outdoor soils*, *potting soils*** [1, 3]
- Potato cyst nematode in ***potting soils*, *soils*** [2]
- Root rot in ***potting soils*, *soils*** [2]
- Root-knot nematodes in ***potting soils*, *soils*** [2]
- Soil pests in ***glasshouse soils*, *nursery soils*, *outdoor soils*, *potting soils*** [1, 3]
- Soil-borne diseases in ***glasshouse soils*, *nursery soils*, *outdoor soils*, *potting soils*** [1, 3]
- Symphylids in ***potting soils*, *soils*** [2]
- Weed seeds in ***glasshouse soils*, *nursery soils*, *outdoor soils*** [1, 3]; ***potting soils*** [1-3]; ***soils*** [2]
- Wireworm in ***potting soils*, *soils*** [2]

SEE SECTION 3 FOR PRODUCTS ALSO REGISTERED

Efficacy guidance

- Metam-sodium is a partial soil sterilant and acts by breaking down in contact with soil to release methyl isothiocyanate (MIT)
- Apply to glasshouse soils as a drench, or inject undiluted to 20 cm at 30 cm intervals and seal immediately, or apply to surface and rotavate
- May also be used by mixing into potting soils
- Apply when soil temperatures exceed 7 °C, preferably above 10 °C, between 1 Apr and 31 Oct. Soil must be of fine tilth, free from debris and with 'potting moisture' content. If soil is too dry postpone treatment and water soil

Restrictions

- No plants must be present during treatment
- Crops must not be planted until a cress germination test has been completed satisfactorily
- Do not treat glasshouses within 2 m of growing crops. Fumes are damaging to all plants
- Avoid using in equipment incorporating natural rubber parts

Crop-specific information

- Latest use: pre-planting of crop

Following crops guidance

- Do not plant until soil is entirely free of fumes

Environmental safety

- Dangerous for the environment
- Very toxic to aquatic organisms
- Keep unprotected persons, livestock and pets out of treated areas for at least 24 h following treatment
- When diluted breakdown commences almost immediately. Only quantities for immediate use should be made up
- Divert or block drains which could carry solution under untreated glasshouses
- After treatment allow sufficient time (several weeks) for residues to dissipate and aerate soil by forking. Time varies with soil and season. Soils with high clay or organic matter content will retain gas longer than lighter soils

Hazard classification and safety precautions

Hazard Harmful [1, 3]; Corrosive, Dangerous for the environment [1-3]
Transport code 8 [2, 3]; 9 [1]
Packaging group III
UN number 3082 [1]; 3267 [2, 3]
Risk phrases R22a, R34, R43, R50, R53a
Operator protection A, C, H, M; U02a, U08, U11 [1-3]; U04a, U05a, U16a, U19a, U20c [2]; U07, U20b [1, 3]
Environmental protection E02a, E34, E38 [2]; E15a [1-3]; E36a [1, 3]
Storage and disposal D01, D02 [1-3]; D09a, D10a, D12a [2]; D12b, D14 [1, 3]
Medical advice M03 [2]; M04a [1-3]

339 metazachlor

A residual anilide herbicide for use in brassicas, nurseries and forestry
HRAC mode of action code: K3

See also *clomazone + metazachlor*
dimethenamid-p + metazachlor
dimethenamid-p + metazachlor + quinmerac

Products

1	Agriguard Metazachlor	AgriGuard	500 g/l	SC	10417
2	Alpha Metazachlor 50 SC	Makhteshim	500 g/l	SC	10669
3	Butisan S	BASF	500 g/l	SC	11733
4	Clayton Buzz	Clayton	500 g/l	SC	12509
5	Clayton Metazachlor	Clayton	500 g/l	SC	09688
6	Clayton Metazachlor 50 SC	Clayton	500 g/l	SC	11719
7	Greencrop Monogram	Greencrop	500 g/l	SC	12048
8	Marksman	AgriGuard	500 g/l	SC	12456
9	Mashona	AgChem Access	500 g/l	SC	13006

Products – continued

10	Mezzanine	AgriGuard	500 g/l	SC	13719
11	Rapsan 500 SC	Globachem	500 g/l	SC	13916
12	Standon Metazachlor 50	Standon	500 g/l	SC	05581
13	Standon Metazachlor 500	Standon	500 g/l	SC	12012
14	Sultan 50 SC	Makhteshim	500 g/l	SC	10418

Uses

- Annual dicotyledons in ***almonds, apples, apricots, cherries, chestnuts, crab apples, damsons, forest nurseries, hazel nuts, nectarines, ornamental plant production, peaches, pears, plums, quinces, walnuts*** [9]; ***amenity vegetation*** [4]; ***borage*** *(off-label)* [2, 5-7, 12, 13]; ***borage for oilseed production*** *(off-label)* [3, 14]; ***broccoli*** [2-7, 9, 10, 12-14]; ***brussels sprouts, cabbages, cauliflowers, spring oilseed rape, swedes, turnips, winter oilseed rape*** [1-14]; ***calabrese*** [1-6, 8-14]; ***canary flower (echium spp.)*** *(off-label)* [2, 3, 5-7, 12-14]; ***chinese cabbage*** *(off-label)*, ***choi sum*** *(off-label)*, ***collards*** *(off-label)*, ***kale*** *(off-label)*, ***leeks*** *(off-label)*, ***pak choi*** *(off-label)*, ***spring linseed*** *(off-label)*, ***winter linseed*** *(off-label)* [2, 14]; ***evening primrose*** *(off-label)* [1-3, 5, 6, 12-14]; ***farm forestry*** [6, 9]; ***farm woodland*** [1-3, 8, 11, 13, 14]; ***forest*** [1-11, 13, 14]; ***honesty*** *(off-label)*, ***mustard*** *(off-label)* [1-3, 5-7, 12-14]; ***linseed*** *(off-label)* [3]; ***nursery fruit trees and bushes*** [1-8, 10, 11, 13, 14]; ***ornamental trees, shrubs*** [1-8, 11, 13, 14]; ***transplanted leeks*** *(off-label)* [2, 3, 14]
- Annual grasses in ***almonds, apples, apricots, broccoli, brussels sprouts, cabbages, calabrese, cauliflowers, cherries, chestnuts, crab apples, damsons, farm forestry, forest nurseries, hazel nuts, nectarines, ornamental plant production, peaches, pears, plums, quinces, spring oilseed rape, swedes, turnips, walnuts, winter oilseed rape*** [9]; ***amenity vegetation*** [4]; ***borage*** *(off-label)* [2, 5-7, 12, 13]; ***borage for oilseed production*** *(off-label)*, ***transplanted leeks*** *(off-label)* [3, 14]; ***canary flower (echium spp.)*** *(off-label)* [2, 3, 5-7, 12-14]; ***chinese cabbage*** *(off-label)*, ***choi sum*** *(off-label)*, ***collards*** *(off-label)*, ***kale*** *(off-label)*, ***pak choi*** *(off-label)*, ***spring linseed*** *(off-label)*, ***winter linseed*** *(off-label)* [2, 14]; ***evening primrose*** *(off-label)* [1-3, 5, 6, 12-14]; ***farm woodland*** [1-3, 8, 11, 13, 14]; ***forest*** [1-11, 13, 14]; ***honesty*** *(off-label)*, ***mustard*** *(off-label)* [1-3, 5-7, 12-14]; ***leeks*** *(off-label)* [14]; ***linseed*** *(off-label)* [3]; ***ornamental trees, shrubs*** [1-8, 11, 13, 14]
- Annual meadow grass in ***amenity vegetation*** [4]; ***borage for oilseed production*** *(off-label)*, ***canary flower (echium spp.)*** *(off-label)*, ***evening primrose*** *(off-label)*, ***honesty*** *(off-label)*, ***mustard*** *(off-label)* [3, 14]; ***broccoli, brussels sprouts, cabbages, cauliflowers, spring oilseed rape, swedes, turnips, winter oilseed rape*** [1-8, 10-14]; ***calabrese*** [1-6, 8, 10-14]; ***chinese cabbage*** *(off-label)*, ***choi sum*** *(off-label)*, ***collards*** *(off-label)*, ***kale*** *(off-label)*, ***leeks*** *(off-label)*, ***pak choi*** *(off-label)*, ***spring linseed*** *(off-label)*, ***transplanted leeks*** *(off-label)*, ***winter linseed*** *(off-label)* [14]; ***farm forestry*** [6]; ***linseed*** *(off-label)* [3]; ***nursery fruit trees and bushes*** [1-8, 10, 11, 13, 14]; ***ornamental trees, shrubs*** [1-8, 11, 13, 14]
- Blackgrass in ***amenity vegetation*** [4]; ***borage for oilseed production*** *(off-label)*, ***canary flower (echium spp.)*** *(off-label)*, ***chinese cabbage*** *(off-label)*, ***choi sum*** *(off-label)*, ***collards*** *(off-label)*, ***evening primrose*** *(off-label)*, ***honesty*** *(off-label)*, ***kale*** *(off-label)*, ***leeks*** *(off-label)*, ***mustard*** *(off-label)*, ***pak choi*** *(off-label)*, ***spring linseed*** *(off-label)*, ***transplanted leeks*** *(off-label)*, ***winter linseed*** *(off-label)* [14]; ***broccoli*** [2-7, 10, 12-14]; ***brussels sprouts, cabbages, cauliflowers, spring oilseed rape, swedes, turnips, winter oilseed rape*** [1-8, 10-14]; ***calabrese*** [1-6, 8, 10-14]; ***farm forestry*** [6]; ***nursery fruit trees and bushes*** [1-8, 10, 11, 13, 14]; ***ornamental trees, shrubs*** [1-8, 11, 13, 14]
- Groundsel in ***chinese cabbage*** *(off-label)*, ***choi sum*** *(off-label)*, ***collards*** *(off-label)*, ***kale*** *(off-label)*, ***pak choi*** *(off-label)* [3]
- Mayweeds in ***chinese cabbage*** *(off-label)*, ***choi sum*** *(off-label)*, ***collards*** *(off-label)*, ***kale*** *(off-label)*, ***pak choi*** *(off-label)* [3]
- Shepherd's purse in ***chinese cabbage*** *(off-label)*, ***choi sum*** *(off-label)*, ***collards*** *(off-label)*, ***kale*** *(off-label)*, ***pak choi*** *(off-label)* [3]

Specific Off-Label Approvals (SOLAs)

- ***borage*** *20060901* [2], *20060905* [5], *20060902* [6], *20060898* [7], *20060891* [12], *20060890* [13]
- ***borage for oilseed production*** *20060906* [3], *20081419* [14]
- ***canary flower (echium spp.)*** *20060901* [2], *20060906* [3], *20060905* [5], *20060902* [6], *20060898* [7], *20060891* [12], *20060890* [13], *20081419* [14]
- ***chinese cabbage*** *20073306* [2], *20050344* [3], *20081420* [14]
- ***choi sum*** *20073306* [2], *20050344* [3], *20081420* [14]

- ***collards*** *20073306* [2], *20050344* [3], *20081420* [14]
- ***evening primrose*** *20060911* [1], *20060901* [2], *20060906* [3], *20060905* [5], *20060902* [6], *20060891* [12], *20060890* [13], *20081419* [14]
- ***honesty*** *20060911* [1], *20060901* [2], *20060906* [3], *20060905* [5], *20060902* [6], *20060898* [7], *20060891* [12], *20060890* [13], *20081419* [14]
- ***kale*** *20073306* [2], *20050344* [3], *20081420* [14]
- ***leeks*** *20081415* [2], *20081421* [14]
- ***linseed*** *20060906* [3]
- ***mustard*** *20060911* [1], *20060901* [2], *20060906* [3], *20060905* [5], *20060902* [6], *20060898* [7], *20060891* [12], *20060890* [13], *20081419* [14]
- ***pak choi*** *20073306* [2], *20050344* [3], *20081420* [14]
- ***spring linseed*** *20060901* [2], *20081419* [14]
- ***transplanted leeks*** *20081415* [2], *20063698* [3], *20081421* [14]
- ***winter linseed*** *20060901* [2], *20081419* [14]

Approval information

- Metazachlor is included in Annex I under EC Directive 91/414

Efficacy guidance

- Activity is dependent on root uptake. For pre-emergence use apply to firm, moist, clod-free seedbed
- Some weeds (chickweed, mayweed, blackgrass etc) susceptible up to 2- or 4-leaf stage. Moderate control of cleavers achieved provided weeds not emerged and adequate soil moisture present
- Split pre- and post-emergence treatments recommended for certain weeds in winter oilseed rape on light and/or stony soils
- Effectiveness is reduced on soils with more than 10% organic matter
- Always follow WRAG guidelines for preventing and managing herbicide resistant weeds. See Section 5 for more information

Restrictions

- Maximum number of treatments 1 per crop for spring oilseed rape, swedes, turnips and brassicas; 2 per crop for winter oilseed rape (split dose treatment); 3 per yr for ornamentals, nursery stock, nursery fruit trees, forestry and farm forestry
- Do not use on sand, very light or poorly drained soils
- Do not treat protected crops or spray overall on ornamentals with soft foliage
- Do not spray crops suffering from wilting, pest or disease
- Do not spray broadcast crops or if a period of heavy rain forecast
- When used on nursery fruit trees any fruit harvested within 1 yr of treatment must be destroyed

Crop-specific information

- Latest use: pre-emergence for swedes and turnips; before 10 leaf stage for spring oilseed rape; before end of Jan for winter oilseed rape
- HI brassicas 6 wk
- On winter oilseed rape may be applied pre-emergence from drilling until seed chits, post-emergence after fully expanded cotyledon stage (GS 1,0) or by split dose technique depending on soil and weeds. See label for details
- On spring oilseed rape may also be used pre-weed-emergence from cotyledon to 10-leaf stage of crop (GS 1,0-1,10)
- With pre-emergence treatment ensure seed covered by 15 mm of well consolidated soil. Harrow across slits of direct-drilled crops
- Ensure brassica transplants have roots well covered and are well established. Direct drilled brassicas should not be treated before 3 leaf stage
- In ornamentals and hardy nursery stock apply after plants established and hardened off as a directed spray or, on some subjects, as an overall spray. See label for list of tolerant subjects. Do not treat plants in containers

Following crops guidance

- Any crop can follow normally harvested treated winter oilseed rape. See label for details of crops which may be planted after spring treatment and in event of crop failure

Environmental safety

- Dangerous for the environment
- Very toxic to aquatic organisms

- Keep livestock out of treated areas until foliage of any poisonous weeds such as ragwort has died and become unpalatable
- Keep livestock out of treated areas of swede and turnip for at least 5 wk following treatment
- Some pesticides pose a greater threat of contamination of water than others and metazachlor is one of these pesticides. Take special care when applying metazachlor near water and do not apply if heavy rain is forecast

Hazard classification and safety precautions

Hazard Harmful, Dangerous for the environment
Transport code 9
Packaging group III
UN number 3082
Risk phrases R22a, R43, R50 [1-14]; R38 [2-7, 9, 12, 14]; R41 [10, 11]; R53a [3-6, 8-12]
Operator protection A, C, H, M; U05a, U19a [1-14]; U08, U20b [1-10, 12-14]; U09a, U20a [11]; U14 [2, 3, 9, 14]; U15 [2, 14]
Environmental protection E06a [1-9] (5 wk for swedes, turnips); E06a [10] (5 weeks for swedes and turnips); E06a [11-14] (5 wk for swedes, turnips); E07a, E15a, E34 [1-14]; E38 [3, 4, 9]
Consumer protection C02a [11] (5 weeks for swedes and turnips)
Storage and disposal D01, D02, D09a [1-14]; D05 [1, 2, 4-8, 11-14]; D10b [1, 5-7, 12, 13]; D10c [2-4, 8-11, 14]; D12a [3-6, 9]; D12b [2, 14]
Medical advice M03 [1-14]; M05a [3-6, 9]

340 metazachlor + quinmerac

A residual herbicide mixture for oilseed rape
HRAC mode of action code: K3 + O

See also quinmerac

Products

1	Boomerang	BASF	400:100 g/l	SC	14043
2	Clayton Mazarac	Clayton	375:125 g/l	SC	11161
3	Naspar TDI	Q-Chem	400:100 g/l	SC	14584
4	Novall	BASF	400:100 g/l	SC	12031
5	Oryx	BASF	333:83 g/l	SC	13527
6	Standon Metazachlor-Q	Standon	375:125 g/l	SC	09676

Uses

- Annual dicotyledons in ***borage*** *(off-label)* [2, 4, 6]; ***canary flower (echium spp.)*** *(off-label)* [2, 4]; ***evening primrose*** *(off-label)*, ***honesty*** *(off-label)*, ***linseed*** *(off-label)*, ***mustard*** *(off-label)* [4]; ***winter oilseed rape*** [1-6]
- Annual grasses in ***borage*** *(off-label)* [2, 4, 6]; ***canary flower (echium spp.)*** *(off-label)* [2, 4]; ***evening primrose*** *(off-label)*, ***honesty*** *(off-label)*, ***linseed*** *(off-label)*, ***mustard*** *(off-label)* [4]; ***winter oilseed rape*** [1]
- Annual meadow grass in ***winter oilseed rape*** [2-6]
- Blackgrass in ***winter oilseed rape*** [2-6]
- Cleavers in ***winter oilseed rape*** [1-6]
- Poppies in ***winter oilseed rape*** [2-6]

Specific Off-Label Approvals (SOLAs)

- ***borage*** *20060503* [2], *20060548* [4], *20060502* [6]
- ***canary flower (echium spp.)*** *20060503* [2], *20060548* [4]
- ***evening primrose*** *20060548* [4]
- ***honesty*** *20060548* [4]
- ***linseed*** *20060548* [4]
- ***mustard*** *20060548* [4]

Approval information

- Metazachlor is included in Annex I under EC Directive 91/414

Efficacy guidance

- Activity is dependent on root uptake. Pre-emergence treatments should be applied to firm moist seedbeds. Applications to dry soil do not become effective until after rain has fallen
- Maximum activity achieved from treatment before weed emergence for some species

- Weed control may be reduced if excessive rain falls shortly after application especially on light soils
- May be used on all soil types except Sands, Very Light Soils, and soils containing more than 10% organic matter. Crop vigour and/or plant stand may be reduced on brashy and stony soils

Restrictions

- Maximum total dose equivalent to one full dose treatment
- Damage may occur in waterlogged conditions. Do not use on poorly drained soils
- Do not treat stressed crops. In frosty conditions transient scorch may occur

Crop-specific information

- Latest use: end Jan in yr of harvest
- To ensure crop safety it is essential that crop seed is well covered with soil to 15 mm. Loose or puffy seedbeds must be consolidated before treatment. Do not use on broadcast crops
- Crop vigour and possibly plant stand may be reduced if excessive rain falls shortly after treatment especially on light soils

Following crops guidance

- In the event of crop failure after use, wheat or barley may be sown in the autumn after ploughing to 15 cm. Spring cereals or brassicas may be planted after ploughing in the spring

Environmental safety

- Dangerous for the environment
- Very toxic to aquatic organisms
- Keep livestock out of treated areas until foliage of any poisonous weeds such as ragwort has died and become unpalatable
- To reduce risk of movement to water do not apply to dry soil or if heavy rain is forecast. On clay soils create a fine consolidated seedbed
- LERAP Category B
- Some pesticides pose a greater threat of contamination of water than others and metazachlor is one of these pesticides. Take special care when applying metazachlor near water and do not apply if heavy rain is forecast

Hazard classification and safety precautions

Hazard Irritant, Dangerous for the environment
Transport code 9
Packaging group III
UN number 3082
Risk phrases R43, R50, R53a
Operator protection A [1-6]; C, H [1, 3-5]; U04a [1, 3-5]; U05a, U08, U14, U19a, U20b [1-6]
Environmental protection E07a, E15a, E16a [1-6]; E16b [6]; E34 [2]; E38 [1, 3-5]
Storage and disposal D01, D02, D09a [1-6]; D05, D10b [2, 6]; D08 [1, 3-5]; D10c, D12a [1-5]
Medical advice M05a [1, 3-5]

341 metconazole

A triazole fungicide for cereals and oilseed rape
FRAC mode of action code: 3

See also epoxiconazole + metconazole

Products

1	Caramba	BASF	60 g/l	SL	10213
2	Caramba 90	BASF	90 g/l	SL	12655
3	Clayton Tunik	Clayton	60 g/l	SL	10545
4	Gringo	AgChem Access	60 g/l	SL	14467
5	Juventus	BASF	90 g/l	SL	13340
6	Pan Metconazole	Pan Agriculture	60 g/l	SL	10473
7	Redstar	AgChem Access	90 g/l	SL	14197
8	Sunorg Pro	BASF	90 g/l	SL	11112

Uses

- Alternaria in ***spring oilseed rape***, ***winter oilseed rape*** [1-8]
- Ascochyta in ***combining peas*** *(reduction)*, ***lupins*** *(qualified minor use)*, ***vining peas*** *(reduction)* [1, 2, 4, 5, 7, 8]

- Botrytis in ***combining peas*** *(reduction)*, ***lupins*** *(qualified minor use)*, ***vining peas*** *(reduction)* [1, 2, 4, 5, 7, 8]; ***spring linseed*** *(off-label)*, ***winter linseed*** *(off-label)* [1, 8]
- Brown rust in ***spring barley***, ***winter barley***, ***winter wheat*** [1-8]
- Foliar disease control in ***durum wheat*** *(off-label)*, ***spring rye*** *(off-label)*, ***triticale*** *(off-label)*, ***winter rye*** *(off-label)* [1]; ***grass seed crops*** *(off-label)* [1, 3, 6]
- Fusarium ear blight in ***winter wheat*** *(reduction)* [1-8]
- Light leaf spot in ***spring oilseed rape***, ***winter oilseed rape*** [1-8]
- Mycosphaerella in ***combining peas*** *(reduction)*, ***vining peas*** *(reduction)* [1, 2, 4, 5, 7, 8]
- Net blotch in ***spring barley*** *(reduction)*, ***winter barley*** *(reduction)* [1-8]
- Phoma in ***spring oilseed rape*** *(reduction)*, ***winter oilseed rape*** *(reduction)* [1-8]
- Powdery mildew in ***spring barley*** *(moderate control)*, ***winter barley***, ***winter wheat*** *(moderate control)* [1-8]
- Rhynchosporium in ***spring barley***, ***winter barley*** [1-8]
- Rust in ***broad beans*** *(off-label)* [1, 8]; ***combining peas***, ***lupins*** *(qualified minor use)*, ***spring field beans***, ***vining peas***, ***winter field beans*** [1, 2, 4, 5, 7, 8]
- Septoria leaf blotch in ***winter wheat*** [1-8]
- Yellow rust in ***winter wheat*** [1-8]

Specific Off-Label Approvals (SOLAs)

- ***broad beans*** *20063078* [1], *20063080* [8]
- ***durum wheat*** *20063077* [1]
- ***grass seed crops*** *20063077* [1], *20061092* [3], *20061104* [6]
- ***spring linseed*** *20063075* [1], *20063079* [8]
- ***spring rye*** *20063077* [1]
- ***triticale*** *20063077* [1]
- ***winter linseed*** *20063075* [1], *20063079* [8]
- ***winter rye*** *20063077* [1]

Approval information

- Metconazole included in Annex I under EC Directive 91/414
- Accepted by BBPA for use on malting barley

Efficacy guidance

- Best results from application to healthy, vigorous crops when disease starts to develop
- Good spray cover of the target is essential for best results. Spray volume should be increased to improve spray penetration into dense crops
- Metconazole is a DMI fungicide. Resistance to some DMI fungicides has been identified in Septoria leaf blotch which may seriously affect performance of some products. For further advice contact a specialist advisor and visit the Fungicide Resistance Action Group (FRAG)-UK website

Restrictions

- Maximum total dose equivalent to two full dose treatments on all crops
- Do not apply to oilseed rape crops that are damaged or stressed from previous treatments, adverse weather, nutrient deficiency or pest attack. Spring application may lead to reduction of crop height
- The addition of adjuvants is neither advised nor necessary and can lead to enhanced growth regulatory effects on stressed crops of oilseed rape
- Ensure sprayer is free from residues of previous treatments that may harm the crop, especially oilseed rape. Use of a detergent cleaner is advised before and after use
- Do not apply with pyrethroid insecticides on oilseed rape at flowering [2, 5, 8]

Crop-specific information

- Latest use: up to and including milky ripe stage (GS 71) for cereals; 10% pods at final size for oilseed rape
- HI: 14 d for peas, field beans, lupins
- Spring treatments on oilseed rape can reduce the height of the crop
- Treatment for Septoria leaf spot in wheat should be made before second node detectable stage and when weather favouring development of the disease has occurred. If conditions continue to favour disease development a follow-up treatment may be needed
- Treat mildew infections in cereals before 3% infection on any green leaf. A specific mildewicide will improve control of established infections
- Treat yellow rust on wheat before 1% infection on any leaf or as preventive treatment after GS 39

SEE SECTION 3 FOR PRODUCTS ALSO REGISTERED

- For brown rust in cereals spray susceptible varieties before any of top 3 leaves has more than 2% infection
- Peas should be treated at the start of flowering and repeat 3-4 wk later if required
- Field beans and lupins should be treated at petal fall and repeat 3-4 wk later if required

Following crops guidance

- Only cereals, oilseed rape, sugar beet, linseed, maize, clover, beans, peas, carrots, potatoes or onions may be sown as following crops after treatment

Environmental safety

- Dangerous for the environment
- Very toxic to aquatic organisms
- Flammable [1, 3]
- LERAP Category B but avoid treatment close to field boundary, even if permitted by LERAP assessment, to reduce effects on non-target insects or other arthropods [1, 3]

Hazard classification and safety precautions

Hazard Harmful, Flammable [1, 3, 4, 6]; Irritant [2, 5, 7, 8]; Dangerous for the environment [1-8]
Transport code 3 [1, 3]; 9 [2, 4-8]
Packaging group III
UN number 1993 [1, 3, 4, 6]; 3082 [2, 5, 7, 8]
Risk phrases R22b, R38, R41, R43, R50 [1, 3, 4, 6]; R36, R51 [2, 5, 7, 8]; R37 [3]; R53a, R63 [1-8]
Operator protection A, C [1-8]; H [1, 3, 4, 6]; U02a, U05a, U20c [1-8]; U11 [1, 3, 4, 6]; U14 [1, 4, 6]; U19a [3]
Environmental protection E15a, E34 [1-8]; E16a [1, 3, 4, 6]; E16b [1, 4, 6]; E38 [1, 2, 4-8]
Storage and disposal D01, D02, D05, D09a, D10a, D12a [1-8]; D06c [2, 5, 7, 8]
Medical advice M05b [1, 3, 4, 6]

342 methiocarb

A stomach acting carbamate molluscicide and insecticide
IRAC mode of action code: 1A

Products

1	Cobra	Interfarm	4% w/w	PT	14561
2	Decoy Wetex	Bayer CropScience	2% w/w	PT	11266
3	Draza Forte	Bayer CropScience	4% w/w	PT	13306
4	Exit Wetex	Bayer CropScience	3% w/w	PT	11276
5	Huron	Bayer CropScience	3% w/w	PT	11288
6	Karan	Bayer CropScience	3% w/w	PT	11289
7	Rivet	Bayer CropScience	3% w/w	PT	11300

Uses

- Cutworms in ***sugar beet*** *(reduction)* [1-7]
- Leatherjackets in ***all cereals*** *(reduction)*, ***potatoes*** *(reduction)*, ***sugar beet*** *(reduction)* [1-7]; ***rotational grass*** *(seed admixture)* [2, 4-7]
- Millipedes in ***sugar beet*** *(reduction)* [1-7]
- Slugs in ***all cereals***, ***all non-edible crops (outdoor)*** *(outdoor only)*, ***brussels sprouts***, ***cabbages***, ***cauliflowers***, ***forage maize***, ***leaf spinach***, ***lettuce***, ***potatoes***, ***spring oilseed rape***, ***strawberries***, ***sugar beet***, ***sunflowers***, ***winter oilseed rape*** [1-7]; ***rotational grass*** *(seed admixture)* [2, 4-7]
- Strawberry seed beetle in ***strawberries*** [1-7]

Approval information

- Methiocarb included in Annex I under EC Directive 91/414
- In 2006 PSD required that all products containing this active ingredient should carry the following warning in the main area of the container label: "Methiocarb is an anticholinesterase carbamate. Handle with care"
- Accepted by BBPA for use on malting barley

Efficacy guidance

- Use as a surface, overall application when pests active (normally mild, damp weather), pre-drilling or post-emergence. May also be used on cereals or ryegrass in admixture with seed at time of drilling

- Also reduces populations of cutworms and millipedes
- Best on potatoes in late Jul to Aug
- Apply to strawberries before strawing down to prevent seed beetles contaminating crop
- See label for details of suitable application equipment

Restrictions

- This product contains an anticholinesterase carbamate compound. Do not use if under medical advice not to work with such compounds
- Maximum number of treatments 3 per crop for potatoes; 2 per crop for cereals, cauliflowers, maize, oilseed rape, sunflowers; 1 per crop for cabbages, lettuce, leaf spinach, sugar beet; 1 per yr for strawberries. Other crops vary according to product - see labels
- Do not treat any protected crops
- Do not allow pellets to lodge in edible crops

Crop-specific information

- Latest use: before first node detectable for cereals, maize; before three visibly extended internodes for oilseed rape, sunflowers
- HI 7 d for spinach, strawberries; 14 d for Brussels sprouts, cabbages, cauliflowers, lettuce; 18 d for potatoes; 6 mth for sugar beet

Environmental safety

- Dangerous for the environment [3]
- Very toxic to aquatic organisms [3]
- Harmful to aquatic organisms [2, 4-7]
- Dangerous to game, wild birds and animals
- Risk to certain non-target insects or other arthropods
- Avoid surface broadcasting application within 6 m of field boundary to reduce effects on non-target species
- Admixed seed should be drilled and not broadcast, and may not be applied from the air

Hazard classification and safety precautions

Hazard Harmful [1-7]; Dangerous for the environment [1, 3]
Transport code 9
Packaging group III
UN number 3077
Risk phrases R22a, R53a [1-7]; R50 [1, 3]; R52 [2, 4-7]
Operator protection A, H [1-7]; J [2, 4-7]; U05a, U20b
Environmental protection E05b [1-7] (7 d); E10a, E22b, E34 [1-7]; E13b [2, 4-7]
Storage and disposal D01, D02, D09a, D11a
Medical advice M02, M03

343 methoxyfenozide

A moulting accelerating diacylhydrazine insecticide
IRAC mode of action code: 18A

Products

1	Runner	Bayer CropScience	240 g/l	SC	12940
2	Trotter	Agrovista		SC	14236

Uses

- Codling moth in ***apples***, ***pears*** [1, 2]
- Insect pests in ***hops*** *(off-label)*, ***ornamental specimens*** *(off-label)*, ***soft fruit*** *(off-label)* [1]
- Tortrix moths in ***apples***, ***pears*** [1, 2]
- Winter moth in ***apples***, ***pears*** [1, 2]

Specific Off-Label Approvals (SOLAs)

- ***hops*** *20082903* [1]
- ***ornamental specimens*** *20082903* [1]
- ***soft fruit*** *20082903* [1]

Efficacy guidance

- To achieve best results uniform coverage of the foliage and full spray penetration of the leaf canopy is important, particularly when spraying post-blossom

SEE SECTION 3 FOR PRODUCTS ALSO REGISTERED

- For maximum effectiveness on winter moth and tortrix spray pre-blossom when first signs of active larvae are seen, followed by a further spray in June if larvae of the summer generation are present
- For codling moth spray post-blossom to coincide with early to peak egg deposition. Follow-up treatments will normally be needed
- Methoxyfenozide is a moulting accelerating compound (MAC) and may be used in an anti-resistance strategy with other top fruit insecticides (including chitin biosynthesis inhibitors and juvenile hormones) which have a different mode of action
- To reduce further the likelihood of resistance development use at full recommended dose in sufficient water volume to achieve required spray penetration

Restrictions

- Maximum number of treatments 3 per yr but no more than two should be sprayed consecutively

Crop-specific information

- HI 14 d for apples, pears

Environmental safety

- Risk to non-target insects or other arthropods
- LERAP Category B
- Broadcast air-assisted LERAP (5 m)

Hazard classification and safety precautions

UN number N/C
Operator protection U20b
Environmental protection E15b, E16b, E22c; E17b (5 m)
Storage and disposal D05, D09a, D11a

344 methyl bromide

The EC Ozone Depleting Substances Regulation (2037/2000) has banned the use of methyl bromide within the UK with certain exceptions, including where a 'Critical Use Exemption' (CUE) has been granted. All other uses ceased in December 2005
IRAC mode of action code: 8A

345 1-methylcyclopropene

An inhibitor of ethylene production for use in stored apples

Products

SmartFresh	Landseer	3.3% w/w	SP	11799

Uses

- Ethylene inhibition in ***apples*** *(post-harvest use)*
- Scald in ***apples*** *(post-harvest use)*

Efficacy guidance

- Best results obtained from treatment of fruit in good condition and of proper quality for long-term storage
- Effects may be reduced in fruit that is in poor condition or ripe prior to storage or harvested late
- Product acts by releasing vapour into store when mixed with water
- Apply as soon as possible after harvest
- Treatment controls superficial scald and maintains fruit firmness and acid content for 3-6 mth in normal air, and 6-9 mth in controlled atmosphere storage
- Ethylene production recommences after removal from storage

Restrictions

- Maximum number of treatments 1 per batch of apples
- Must only be used by suitably trained and competent persons in fumigation operations
- Consult processors before treatment of fruit destined for processing or cider making
- Do not apply in mixture with other products
- Ventilate all areas thoroughly with all refrigeration fans operating at maximum power for at least 15 min before re-entry

Crop-specific information

- Latest use: 7d after harvest of apples
- Product tested on Granny Smith, Gala, Jonagold, Bramley and Cox. Consult distributor or supplier before treating other varieties

Environmental safety

- Unprotected persons must be kept out of treated stores during the 24 h treatment period
- Prior to application ensure that the store can be properly and promptly sealed

Hazard classification and safety precautions

UN number N/C
Operator protection U05a
Environmental protection E02a (24 h); E15a, E34
Consumer protection C12
Storage and disposal D01, D02, D07, D14

346 metrafenone

A benzophenone protectant and curative fungicide for cereals
FRAC mode of action code: U8

See also *epoxiconazole + fenpropimorph + metrafenone*
epoxiconazole + metrafenone

Products

1	Attenzo	BASF	300 g/l	SC	11917
2	Flexity	BASF	300 g/l	SC	11775

Uses

- Eyespot in ***spring wheat*** *(reduction)*, ***winter wheat*** *(reduction)* [1, 2]
- Powdery mildew in ***forest nurseries*** *(off-label)* [1]; ***ornamental plant production*** [2]; ***spring barley***, ***spring oats*** *(evidence of mildew control on oats is limited)*, ***spring wheat***, ***winter barley***, ***winter oats*** *(evidence of mildew control on oats is limited)*, ***winter wheat*** [1, 2]

Specific Off-Label Approvals (SOLAs)

- ***forest nurseries*** *20082814* [1]

Approval information

- Metrafenone included in Annex I under EC Directive 91/414
- Accepted by BBPA for use on malting barley

Efficacy guidance

- Best results obtained from treatment at the start of foliar disease attack
- Activity against mildew in wheat is mainly protectant with moderate curative control in the latent phase; activity is entirely protectant in barley
- Useful reduction of eyespot in wheat is obtained if treatment applied at GS 30-32
- Should be used as part of a resistance management strategy that includes mixtures or sequences effective against mildew and non-chemical methods

Restrictions

- Maximum number of treatments 2 per crop
- Avoid the use of sequential applications of Flexity unless it is used in tank mixture with other products active against powdery mildew employing a different mode of action

Crop-specific information

- Latest use: beginning of flowering (GS 31) for wheat, barley, oats

Following crops guidance

- Cereals, oilseed rape, sugar beet, linseed, maize, clover, field beans, peas, turnips, carrots, cauliflowers, onions, lettuce or potatoes may follow a treated cereal crop

Environmental safety

- Dangerous for the environment
- Toxic to aquatic organisms

Hazard classification and safety precautions

Hazard Irritant, Dangerous for the environment

UN number N/C
Risk phrases R43, R51, R53a
Operator protection A, H; U05a, U20b
Environmental protection E15a, E34
Storage and disposal D01, D02, D05, D09a, D10c
Medical advice M03

347 metribuzin

A contact and residual triazinone herbicide for use in potatoes
HRAC mode of action code: C1

See also *clomazone + metribuzin*
flufenacet + metribuzin

Products

1	Greencrop Majestic	Greencrop	70% w/w	WG	12422
2	Metribuzin WG	Goldengrass	70% w/w	WG	14315
3	Sarabi	AgChem Access	70% w/w	WG	13453
4	Sencorex WG	Bayer CropScience	70% w/w	WG	11304
5	Shotput	Makhteshim	70% w/w	WG	13788

Uses

- Annual and perennial weeds in ***asparagus*** *(off-label)*, ***sweet potato*** *(off-label)* [5]
- Annual dicotyledons in ***asparagus*** *(off-label)*, ***sweet potato*** *(off-label)* [4]; ***early potatoes***, ***maincrop potatoes*** [1-5]
- Annual grasses in ***asparagus*** *(off-label)*, ***sweet potato*** *(off-label)* [4]; ***early potatoes***, ***maincrop potatoes*** [1-5]
- Annual meadow grass in ***asparagus*** *(off-label)* [4]
- Fool's parsley in ***carrots*** *(off-label)*, ***mallow (althaea spp.)*** *(off-label)*, ***parsnips*** *(off-label)*, ***protected carrots*** *(off-label)*, ***protected parsnips*** *(off-label)* [5]
- Volunteer oilseed rape in ***early potatoes***, ***maincrop potatoes*** [1-5]
- Wild mignonette in ***carrots*** *(off-label)*, ***mallow (althaea spp.)*** *(off-label)*, ***parsnips*** *(off-label)* [4]

Specific Off-Label Approvals (SOLAs)

- ***asparagus*** *20091033* [4], *20080588* [5]
- ***carrots*** *20031887* [4], *20080589* [5]
- ***mallow (althaea spp.)*** *20031887* [4], *20080589* [5]
- ***parsnips*** *20031887* [4], *20080589* [5]
- ***protected carrots*** *20080589* [5]
- ***protected parsnips*** *20080589* [5]
- ***sweet potato*** *20070733* [4], *20080590* [5]

Approval information

- Metribuzin included in Annex I under EC Directive 91/414

Efficacy guidance

- Best results achieved on weeds at cotyledon to 1-leaf stage
- May be applied pre- or post-emergence of crop on named maincrop varieties and cv. Marfona; pre-emergence only on named early varieties
- Apply to moist soil with well-rounded ridges and few clods
- Activity reduced by dry conditions and on soils with high organic matter content
- On fen and moss soils pre-planting incorporation to 10-15 cm gives increased activity. Incorporate thoroughly and evenly
- With named maincrop and second early potato varieties on soils with more than 10% organic matter shallow pre- or post-planting incorporation may be used. See label for details
- Effective control using a programme of reduced doses is made possible by using a spray of smaller droplets, thus improving retention

Restrictions

- Maximum total dose equivalent to one full dose treatment on early potato varieties and one and a third full doses on maincrop varieties

- Only certain varieties may be treated. Apply pre-emergence only on named first earlies, pre- or post-emergence on named second earlies. On named maincrop varieties apply pre-emergence (except for certain varieties on Sands or Very Light soils) or post-emergence. See label
- All post-emergence treatments must be carried out before longest shoots reach 15 cm
- Do not cultivate after treatment
- Some recommended varieties may be sensitive to post-emergence treatment if crop under stress

Crop-specific information

- Latest use: pre-crop emergence for named early potato varieties; before most advanced shoots have reached 15 cm for post-emergence treatment of potatoes
- On stony or gravelly soils there is risk of crop damage, especially if heavy rain falls soon after application
- When days are hot and sunny delay spraying until evening

Following crops guidance

- Ryegrass, cereals or winter beans may be sown in same season provided at least 16 wk elapsed after treatment and ground ploughed to 15 cm and thoroughly cultivated as soon as possible after harvest and no later than end Dec
- In W Cornwall on soil with more than 5% organic matter early potatoes treated as recommended may be followed by summer planted brassica crops provided the soil has been ploughed, spring rainfall has been normal and at least 14 wk have elapsed since treatment
- Do not grow any vegetable brassicas, lettuces or radishes on land treated the previous yr. Other crops may be sown normally in spring of next yr

Environmental safety

- Dangerous for the environment
- Very toxic to aquatic organisms
- Do not empty into drains
- LERAP Category B

Hazard classification and safety precautions

Hazard Harmful, Dangerous for the environment
Transport code 9
Packaging group III
UN number 3077
Risk phrases R22a, R50, R53a [1-5]; R43 [5]
Operator protection A, H [5]; U05a [1, 5]; U08, U13 [1-5]; U14, U20b [5]; U19a [2-5]; U20a [1-4]
Environmental protection E15a [2-5]; E16a, E16b [1-5]; E19b [5]; E34 [1]; E38 [2-4]
Storage and disposal D01, D02 [1, 5]; D09a, D11a, D12a [1-5]
Medical advice M03 [1]; M05a [5]

348 metsulfuron-methyl

A contact and residual sulfonylurea herbicide used in cereals, linseed and set-aside
HRAC mode of action code: B

See also carfentrazone-ethyl + metsulfuron-methyl

Products

1	Asset	AgriGuard	20% w/w	SG	13902
2	Avail XT	AgriGuard	20% w/w	SG	13194
3	Cimarron	Headland	50% w/w	TB	13408
4	Finy	AgriChem BV	20% w/w	WG	12855
5	Firecrest	European Ag	20% w/w	SG	14486
6	Forge	Interfarm	20% w/w	SG	12848
7	Goldron-M	Goldengrass	20% w/w	SG	13466
8	Jubilee SX	DuPont	20% w/w	SG	12203
9	Standon Metso XT	Standon	20% w/w	SG	13880
10	Standon Mexxon xtra	Standon	20% w/w	SG	13931

Uses

- Annual dicotyledons in ***durum wheat*** [6]; ***farm forestry*** *(off-label)*, ***forest nurseries*** *(off-label)*, ***game cover*** *(off-label)*, ***miscanthus*** *(off-label)*, ***ornamental plant production*** *(off-label)* [8]; ***green cover on land temporarily removed from production*** [4, 6, 9, 10]; ***linseed, spring***

barley*, *spring oats*, *spring wheat*, *triticale*, *winter barley*, *winter oats*, *winter wheat [1, 2, 4-10]
- Chickweed in ***durum wheat*** [6]; ***linseed*, *spring barley*, *spring oats*, *spring wheat*, *triticale*, *winter barley*, *winter oats*, *winter wheat*** [1, 2, 5-8]
- Docks in ***grassland*** [3]
- Green cover in ***land temporarily removed from production*** [1, 2, 5, 7, 8]
- Mayweeds in ***durum wheat*** [6]; ***linseed*, *spring barley*, *spring oats*, *spring wheat*, *triticale*, *winter barley*, *winter oats*, *winter wheat*** [1, 2, 5-8]

Specific Off-Label Approvals (SOLAs)
- ***farm forestry*** *20082859* [8]
- ***forest nurseries*** *20082859* [8]
- ***game cover*** *20082859* [8]
- ***miscanthus*** *20082859* [8]
- ***ornamental plant production*** *20082859* [8]

Approval information
- Metsulfuron-methyl included in Annex I under EC Directive 91/414
- Accepted by BBPA for use on malting barley

Efficacy guidance
- Best results achieved on small, actively growing weeds up to 6-true leaf stage. Good spray cover is important
- Weed control may be reduced in dry soil conditions
- Commonly used in tank-mixture on wheat and barley with other cereal herbicides to improve control of resistant dicotyledons (cleavers, fumitory, ivy-leaved speedwell), larger weeds and grasses. See label for recommended mixtures
- Metsulfuron-methyl is a member of the ALS-inhibitor group of herbicides and products should be used in a planned Resistance Management strategy. See Section 5 for more information

Restrictions
- Maximum number of treatments 1 per crop or per yr (for set-aside)
- Product must only be used after 1 Feb
- Do not apply within 7 d of rolling
- Do not use on cereal crops undersown with grass or legumes
- Do not use in tank mixture on oats, triticale or linseed. On linseed allow at least 7 d before or after other treatments
- Do not tank mix with chlorpyrifos. Allow at least 14 d before or after chlorpyrifos treatments
- Consult contract agents before use on a cereal crop for seed
- Do not use on any crop suffering stress from drought, waterlogging, frost, deficiency, pest or disease attack or apply within 7 d of rolling
- Specific restrictions apply to use in sequence or tank mixture with other sulfonylurea or ALS-inhibiting herbicides. See label for details
- Spraying equipment should not be drained or flushed onto land planted, or to be planted, with trees or crops other than cereals and should be thoroughly cleansed after use - see label for instructions

Crop-specific information
- Latest use: before flag leaf sheath extending stage for cereals (GS 41); before flower buds visible or crop 30 cm tall for linseed [2, 8]; before 1 Aug in yr of treatment for land not being used for crop production [2, 8]
- Apply after 1 Feb to wheat, oats and triticale from 2-leaf (GS 12), and to barley from 3-leaf (GS 13) until flag-leaf sheath extending (GS 41)
- On linseed allow at least 7 d (10 d if crop is growing poorly or under stress) before or after other treatments
- Use in set-aside when a full green cover is established and made up predominantly of grassland, wheat, barley, oats or triticale. Do not use on seedling grasses

Following crops guidance
- Only cereals, oilseed rape, field beans or grass may be sown in same calendar year after treating cereals with the product alone. Other restrictions apply to tank mixtures. See label for details
- Only cereals should be planted within 16 mth of applying to a linseed crop or set-aside
- In the event of failure of a treated crop sow only wheat within 3 mth after treatment

Environmental safety

- Dangerous for the environment
- Very toxic to aquatic organisms
- Take extreme care to avoid damage by drift onto broad-leaved plants outside the target area, onto surface waters or ditches or onto land intended for cropping
- LERAP Category B
- A range of broad leaved species will be fully or partially controlled when used in land temporarily removed from production, hence product may not be suitable where wild flower borders or other forms of conservation headland are being developed
- Before use on land temporarily removed from production as part of grant-aided scheme, ensure compliance with the management rules
- Green cover on land temporarily removed from production must not be grazed by livestock or harvested for human or animal consumption or used for animal bedding [2, 8]

Hazard classification and safety precautions

Hazard Dangerous for the environment [1, 2, 4-10]
Transport code 9
Packaging group III
UN number 3077
Risk phrases R50, R53a [1, 2, 4-10]
Operator protection U05a [4]; U08 [2]; U09a [9, 10]; U19a [2, 9, 10]; U20b [2, 3, 9, 10]
Environmental protection E07a [3]; E15b [1-10]; E16a [1, 2, 4-10]; E16b [1]; E23 [4]; E34 [2, 9, 10]; E38 [1, 2, 5-10]
Storage and disposal D01 [2, 4, 9, 10]; D02, D12b [4]; D09a [1-4, 7, 9, 10]; D11a [1, 4, 7]; D12a [1-3, 5-10]; D17, D18 [3]
Medical advice M05a [2, 9, 10]

349 metsulfuron-methyl + thifensulfuron-methyl

A contact residual and translocated sulfonylurea herbicide mixture for use in cereals
HRAC mode of action code: B + B

See also *thifensulfuron-methyl*

Products

1	Avro SX	DuPont	2.9:42.9 % w/w	SG	14313
2	Chimera SX	DuPont	2.9:42.9% w/w	SG	13171
3	Concert SX	DuPont	4:40% w/w	SG	12288
4	Finish SX	DuPont	6.7:33.3% w/w	SG	12259
5	Harmony M SX	DuPont	4:40% w/w	SG	12258
6	Pennant	Headland	4:40% w/w	SG	13350
7	Presite SX	DuPont	6.7:33.3% w/w	SG	12291

Uses

- Annual dicotyledons in ***durum wheat*** *(off-label)* [7]; ***miscanthus*** *(off-label)* [2, 5]; ***spring barley, spring wheat, winter wheat*** [1-7]; ***winter barley*** [1, 2, 4, 7]; ***winter oats*** [4, 7]
- Black bindweed in ***spring barley, spring wheat, winter barley, winter wheat*** [1]
- Charlock in ***spring barley, spring wheat, winter barley, winter wheat*** [1]
- Chickweed in ***durum wheat*** *(off-label)* [7]; ***spring barley, spring wheat, winter wheat*** [1-7]; ***winter barley*** [1, 2, 4, 7]; ***winter oats*** [4, 7]
- Creeping thistle in ***spring barley, spring wheat, winter barley, winter wheat*** [1, 2]
- Fat hen in ***spring barley, spring wheat, winter barley, winter wheat*** [1]
- Field pansy in ***durum wheat*** *(off-label)* [7]; ***spring barley, spring wheat, winter barley, winter wheat*** [1, 4, 7]; ***winter oats*** [4, 7]
- Field speedwell in ***durum wheat*** *(off-label)* [7]; ***spring barley, spring wheat, winter barley, winter wheat*** [1, 4, 7]; ***winter oats*** [4, 7]
- Forget-me-not in ***spring barley, spring wheat, winter barley, winter wheat*** [1]
- Hemp-nettle in ***spring barley, spring wheat, winter barley, winter wheat*** [1]
- Ivy-leaved speedwell in ***durum wheat*** *(off-label)* [7]; ***spring barley, spring wheat, winter barley, winter oats, winter wheat*** [4, 7]
- Knotgrass in ***durum wheat*** *(off-label)* [7]; ***spring barley, spring wheat, winter barley, winter wheat*** [1, 4, 7]; ***winter oats*** [4, 7]

- Mayweeds in ***durum wheat*** *(off-label)* [7]; ***spring barley, spring wheat, winter wheat*** [1-7]; ***winter barley*** [1, 2, 4, 7]; ***winter oats*** [4, 7]
- Polygonums in ***spring barley, spring wheat, winter wheat*** [2, 3, 5, 6]; ***winter barley*** [2]
- Poppies in ***spring barley, spring wheat, winter barley, winter wheat*** [1]
- Red dead-nettle in ***spring barley, spring wheat, winter wheat*** [1-3, 5, 6]; ***winter barley*** [1, 2]
- Redshank in ***spring barley, spring wheat, winter barley, winter wheat*** [1]
- Shepherd's purse in ***spring barley, spring wheat, winter wheat*** [1-3, 5, 6]; ***winter barley*** [1, 2]

Specific Off-Label Approvals (SOLAs)

- ***durum wheat*** *20070452* [7]
- ***miscanthus*** *20082835* [2], *20082916* [5]

Approval information

- Metsulfuron-methyl and thifensulfuron-methyl included in Annex I under EC Directive 91/414
- Accepted by BBPA for use on malting barley

Efficacy guidance

- Best results by application to small, actively growing weeds up to 6-true leaf stage
- Ensure good spray cover
- Susceptible weeds stop growing almost immediately but symptoms may not be visible for about 2 wk
- Effectiveness may be reduced by heavy rain or if soil conditions very dry
- Metsulfuron-methyl and thifensulfuron-methyl are members of the ALS-inhibitor group of herbicides and products should be used in a planned Resistance Management strategy. See Section 5 for more information

Restrictions

- Maximum number of treatments 1 per crop
- Products may only be used after 1 Feb
- Do not use on any crop suffering stress from drought, waterlogging, frost, deficiency, pest or disease attack or any other cause
- Do not use on crops undersown with grasses, clover or legumes, or any other broad leaved crop
- Specific restrictions apply to use in sequence or tank mixture with other sulfonylurea or ALS-inhibiting herbicides. See label for details
- Do not apply within 7 d of rolling
- Consult contract agents before use on a cereal crop grown for seed

Crop-specific information

- Latest use: before flag leaf sheath extending (GS 39) for barley and wheat

Following crops guidance

- Only cereals, oilseed rape, field beans or grass may be sown in same calendar year after treatment
- Additional constraints apply after use of certain tank mixtures. See label
- In the event of crop failure sow only winter wheat within 3 mth after treatment and after ploughing and cultivating to a depth of at least 15 cm

Environmental safety

- Dangerous for the environment
- Very toxic to aquatic organisms
- LERAP Category B
- Take extreme care to avoid damage by drift onto broad-leaved plants outside the target area, or onto ponds, waterways or ditches
- Spraying equipment should not be drained or flushed onto land planted, or to be planted, with trees or crops other than cereals and should be thoroughly cleansed after use - see label for instructions

Hazard classification and safety precautions

Hazard Dangerous for the environment
Transport code 9
Packaging group III
UN number 3077
Risk phrases R50, R53a
Operator protection U05a [4, 7]; U08, U19a, U20b [1-7]
Environmental protection E15a [4, 7]; E15b [3, 5, 6]; E16a, E38 [1-7]
Storage and disposal D01, D02 [4, 7]; D09a, D12a [1-7]; D11a [3-7]

350 metsulfuron-methyl + tribenuron-methyl

A sulfonylurea herbicide mixture for cereals
HRAC mode of action code: B + B

See also *tribenuron-methyl*

Products

1	Ally Max SX	DuPont	14.3:14.3% w/w	SG	12631
2	Biplay SX	DuPont	11.1:22.2% w/w	SG	12246
3	Traton SX	DuPont	11.1:22.2% w/w	SG	12270

Uses

- Annual dicotyledons in ***durum wheat*** *(off-label)*, ***spring oats***, ***spring rye*** *(off-label)*, ***winter rye*** *(off-label)* [1]; ***spring barley***, ***spring wheat***, ***triticale***, ***winter barley***, ***winter oats***, ***winter wheat*** [1-3]
- Chickweed in ***durum wheat*** *(off-label)*, ***spring oats*** [1]; ***spring barley***, ***spring wheat***, ***triticale***, ***winter barley***, ***winter oats***, ***winter wheat*** [1-3]
- Field pansy in ***durum wheat*** *(off-label)* [1]; ***spring barley***, ***spring wheat***, ***triticale***, ***winter barley***, ***winter oats***, ***winter wheat*** [1-3]
- Field speedwell in ***spring oats*** [1]
- Hemp-nettle in ***durum wheat*** *(off-label)*, ***spring oats*** [1]; ***spring barley***, ***spring wheat***, ***triticale***, ***winter barley***, ***winter oats***, ***winter wheat*** [1-3]
- Mayweeds in ***durum wheat*** *(off-label)*, ***spring oats*** [1]; ***spring barley***, ***spring wheat***, ***triticale***, ***winter barley***, ***winter oats***, ***winter wheat*** [1-3]
- Red dead-nettle in ***durum wheat*** *(off-label)*, ***spring oats*** [1]; ***spring barley***, ***spring wheat***, ***triticale***, ***winter barley***, ***winter oats***, ***winter wheat*** [1-3]
- Volunteer oilseed rape in ***durum wheat*** *(off-label)*, ***spring barley***, ***spring oats***, ***spring wheat***, ***triticale***, ***winter barley***, ***winter oats***, ***winter wheat*** [1]
- Volunteer sugar beet in ***durum wheat*** *(off-label)*, ***spring barley***, ***spring oats***, ***spring wheat***, ***triticale***, ***winter barley***, ***winter oats***, ***winter wheat*** [1]

Specific Off-Label Approvals (SOLAs)

- ***durum wheat*** *20070453* [1]
- ***spring rye*** *20090013* [1]
- ***winter rye*** *20090013* [1]

Approval information

- Metsulfuron-methyl and tribenuron-methyl included in Annex I under EC Directive 91/414
- Accepted by BBPA for use on malting barley

Efficacy guidance

- Best results obtained when applied to small actively growing weeds
- Product acts by foliar and root uptake. Good spray cover essential but performance may be reduced when soil conditions are very dry and residual effects may be reduced by heavy rain
- Weed growth inhibited within hours of treatment and many show marked colour changes as they die back. Full effects may not be apparent for up to 4 wk
- Metsulfuron-methyl and tribenuron-methyl are members of the ALS-inhibitor group of herbicides and products should be used in a planned Resistance Management strategy. See Section 5 for more information

Restrictions

- Maximum number of treatments 1 per crop
- Product must only be used after 1 Feb and after crop has three leaves
- Do not apply to a crop suffering from drought, water-logging, low temperatures, pest or disease attack, nutrient deficiency, soil compaction or any other stress
- Do not use on crops undersown with grasses, clover or other legumes
- Do not apply within 7 d of rolling
- Specific restrictions apply to use in sequence or tank mixture with other sulfonylurea or ALS-inhibiting herbicides. See label for details

Crop-specific information

- Latest use: before flag leaf sheath extending

Following crops guidance

- Only cereals, field beans or oilseed rape may be sown in the same calendar yr as harvest of a treated crop [1]
- Only cereals, field beans, grass or oilseed rape may be sown in the same calendar yr as harvest of a treated crop [2, 3]
- In the event of failure of a treated crop only winter wheat may sown within 3 mth of treatment, and only after ploughing and cultivation to 15 cm min

Environmental safety

- Dangerous for the environment
- Very toxic to aquatic organisms
- Some non-target crops are highly sensitive. Take extreme care to avoid drift outside the target area, or onto ponds, waterways or ditches
- Spraying equipment should be thoroughly cleaned in accordance with manufacturer's instructions
- LERAP Category B [1]

Hazard classification and safety precautions

Hazard Irritant, Dangerous for the environment
Transport code 9
Packaging group III
UN number 3077
Risk phrases R43, R50, R53a
Operator protection A, H; U08, U19a, U20b
Environmental protection E15a, E38 [1-3]; E16a [1]
Storage and disposal D01, D02, D09a, D11a, D12a

351 myclobutanil

A systemic, protectant and curative triazole fungicide
FRAC mode of action code: 3

Products

1	Aristocrat	AgriGuard	200 g/l	EW	12343
2	Crassus	Landseer	200 g/l	EW	13899
3	Masalon	Rigby Taylor	45 g/l	EW	12385
4	Systhane 20EW	Landseer	200 g/l	EW	09396

Uses

- American gooseberry mildew in ***blackberries*** [2]; ***blackcurrants*** [1, 4]; ***gooseberries*** [1, 2, 4]
- Black spot in ***ornamental plant production***, ***roses*** [1, 2, 4]
- Blossom wilt in ***cherries*** *(off-label)*, ***mirabelles*** *(off-label)* [4]
- Brown rot in ***apricots*** *(off-label)*, ***peaches*** *(off-label)* [4]
- Fusarium patch in ***managed amenity turf*** [3]
- Plum rust in ***plums*** *(off-label)* [4]
- Powdery mildew in ***apples***, ***ornamental plant production***, ***pears***, ***roses***, ***strawberries*** [1, 2, 4]; ***apricots*** *(off-label)*, ***artichokes*** *(off-label)*, ***courgettes*** *(off-label)*, ***crab apples*** *(off-label)*, ***gherkins*** *(off-label)*, ***hops*** *(off-label)*, ***peaches*** *(off-label)*, ***protected aubergines*** *(off-label)*, ***protected blackberries*** *(off-label)*, ***protected courgettes*** *(off-label)*, ***protected cucumbers*** *(off-label)*, ***protected gherkins*** *(off-label)*, ***protected raspberries*** *(off-label)*, ***protected rubus hybrids*** *(off-label)*, ***protected tomatoes*** *(off-label)*, ***quinces*** *(off-label)*, ***redcurrants*** *(off-label)*, ***whitecurrants*** *(off-label)*, ***wine grapes*** *(off-label)* [4]
- Rust in ***ornamental plant production***, ***roses*** [1, 2, 4]
- Scab in ***apples***, ***pears*** [1, 2, 4]

Specific Off-Label Approvals (SOLAs)

- ***apricots*** *20070626* [4]
- ***artichokes*** *20070624* [4]
- ***cherries*** *991535* [4]
- ***courgettes*** *20070627* [4]
- ***crab apples*** *20080504* [4]
- ***gherkins*** *20070627* [4]
- ***hops*** *20021412* [4]

- ***mirabelles*** *991535* [4]
- ***peaches*** *20070626* [4]
- ***plums*** *20012459* [4]
- ***protected aubergines*** *20070625* [4]
- ***protected blackberries*** *20051189* [4]
- ***protected courgettes*** *20070627* [4]
- ***protected cucumbers*** *20070627* [4]
- ***protected gherkins*** *20070627* [4]
- ***protected raspberries*** *20051189* [4]
- ***protected rubus hybrids*** *20051189* [4]
- ***protected tomatoes*** *20070625* [4]
- ***quinces*** *20080504* [4]
- ***redcurrants*** *20080505* [4]
- ***whitecurrants*** *20080505* [4]
- ***wine grapes*** *20040116* [4]

Approval information

- Accepted by BBPA for use on hops

Efficacy guidance

- Best results achieved when used as part of routine preventive spray programme from bud burst to end of flowering in apples and pears and from just before the signs of mildew infection in blackcurrants and gooseberries [4]
- In strawberries commence spraying at, or just prior to, first flower. Post-harvest sprays may be required on mildew-susceptible varieties where mildew is present and likely to be damaging [4]
- Spray at 7-14 d intervals depending on disease pressure and dose applied [4]
- For improved scab control on apples in post-blossom period tank-mix with mancozeb or captan [4]
- Apply alone from mid-Jun for control of secondary mildew on apples and pears
- On roses spray at first signs of disease and repeat every 2 wk. In high risk areas spray when leaves emerge in spring, repeat 1 wk later and then continue normal programme [4]
- Treatment of managed amenity turf may be carried out at any time of year before, or at, the first sign of disease [3]
- Myclobutanil products should be used in conjunction with other fungicides with a different mode of action to reduce the possibility of resistance developing

Restrictions

- Maximum total dose equivalent to ten full dose treatments in apples and pears; six full dose treatments in blackcurrants, gooseberries, strawberries
- Maximum number of treatments on managed amenity turf: two per yr [3]
- Do not mow turf within 24 hr after treatment [3]

Crop-specific information

- HI 28 d (grapevines); 21 d (cherries, mirabelles); 14 d (apples, pears, blackcurrants, gooseberries, hops); 3 d (plums, protected blackberries, protected raspberries, protected Rubus hybrids, strawberries)
- Product may be applied to newly sown turf after the two-leaf stage but in view of the large number of turf grass cultivars a safety test on a small area is recommended before large scale treatment [3]

Environmental safety

- Dangerous for the environment
- Toxic (harmful [3]) to aquatic organisms
- LERAP Category B [3]

Hazard classification and safety precautions

Hazard Harmful, Dangerous for the environment [1, 2, 4]
Transport code 9 [1, 2, 4]
Packaging group III [1, 2, 4]
UN number 3082 [1, 2, 4]; N/C [3]
Risk phrases R22b, R51, R63 [1, 2, 4]; R52 [3]; R53a [1-4]
Operator protection A, H [1-4]; J [1, 2, 4]; U08, U20a [1, 3, 4]; U09a, U20b [2]; U15 [1, 2, 4]
Environmental protection E15a [1-4]; E16a [3]
Storage and disposal D01, D02 [1, 2, 4]; D05, D10c, D12a [1-4]
Medical advice M05b [1, 2, 4]

SEE SECTION 3 FOR PRODUCTS ALSO REGISTERED

352 1-naphthylacetic acid

A plant growth regulator to promote rooting of cuttings

See also 4-indol-3-ylbutyric acid + 1-naphthylacetic acid

Products

1	Rhizopon B Powder (0.1%)	Rhizopon	0.1% w/w	DP	09089
2	Rhizopon B Powder (0.2%)	Rhizopon	0.2% w/w	DP	09090
3	Rhizopon B Tablets	Rhizopon	25 mg a.i.	WT	09091

Uses

- Rooting of cuttings in ***ornamental plant production***

Efficacy guidance

- Dip moistened base of cuttings into powder immediately before planting [1, 2]
- See label for details of concentrations recommended for promotion of rooting in cuttings of different species [3]
- Dip prepared cuttings in solution for 4-24 h depending on species [3]

Restrictions

- Maximum number of treatments 1 per cutting

Crop-specific information

- Latest use: before cutting insertion for ornamental specimens

Hazard classification and safety precautions

UN number N/C
Operator protection U08, U19a, U20a
Environmental protection E15a, E34
Storage and disposal D09a, D11a

353 napropamide

A soil applied alkanamide herbicide for oilseed rape, fruit and woody ornamentals
HRAC mode of action code: K3

Products

1	AC 650	United Phosphorus	450 g/l	SC	11102
2	Devrinol	United Phosphorus	450 g/l	SC	09374
3	Jouster	AgriGuard	450 g/l	SC	12741
4	MAC-Napropamide 450 SC	AgChem Access	450 g/l	SC	13972
5	Naprop	Globachem	450 g/l	SC	13402

Uses

- Annual dicotyledons in ***bilberries*** *(off-label)*, ***blackberries*** *(off-label)*, ***blueberries*** *(off-label)*, ***chinese cabbage*** *(off-label)*, ***choi sum*** *(off-label)*, ***collards*** *(off-label)*, ***evening primrose*** *(off-label)*, ***farm forestry*** *(off-label)*, ***forest*** *(off-label)*, ***honesty*** *(off-label)*, ***mustard*** *(off-label)*, ***pak choi*** *(off-label)*, ***redcurrants*** *(off-label)*, ***rubus hybrids*** *(off-label)*, ***whitecurrants*** *(off-label)* [2]; ***blackcurrants***, ***forest nurseries***, ***gooseberries***, ***ornamental plant production***, ***raspberries***, ***strawberries*** [2-4]; ***broccoli***, ***brussels sprouts***, ***cabbages***, ***calabrese***, ***cauliflowers***, ***kale*** [2-5]; ***linseed*** *(off-label)* [1, 2]; ***winter oilseed rape*** [1, 3, 5]; ***woody ornamentals*** [2, 4]
- Annual grasses in ***bilberries*** *(off-label)*, ***blackberries*** *(off-label)*, ***blueberries*** *(off-label)*, ***chinese cabbage*** *(off-label)*, ***choi sum*** *(off-label)*, ***collards*** *(off-label)*, ***evening primrose*** *(off-label)*, ***farm forestry*** *(off-label)*, ***forest*** *(off-label)*, ***honesty*** *(off-label)*, ***mustard*** *(off-label)*, ***pak choi*** *(off-label)*, ***redcurrants*** *(off-label)*, ***rubus hybrids*** *(off-label)*, ***whitecurrants*** *(off-label)* [2]; ***blackcurrants***, ***forest nurseries***, ***gooseberries***, ***ornamental plant production***, ***raspberries***, ***strawberries*** [2-4]; ***broccoli***, ***brussels sprouts***, ***cabbages***, ***calabrese***, ***cauliflowers***, ***kale*** [2-5]; ***linseed*** *(off-label)* [1, 2]; ***winter oilseed rape*** [1, 3, 5]; ***woody ornamentals*** [2, 4]
- Cleavers in ***blackcurrants***, ***forest nurseries***, ***gooseberries***, ***ornamental plant production***, ***raspberries***, ***strawberries*** [2-4]; ***broccoli***, ***brussels sprouts***, ***cabbages***, ***calabrese***, ***cauliflowers***, ***kale*** [2-5]; ***winter oilseed rape*** [3, 5]; ***woody ornamentals*** [2, 4]
- Groundsel in ***blackcurrants***, ***forest nurseries***, ***gooseberries***, ***ornamental plant production***, ***raspberries***, ***strawberries*** [2-4]; ***broccoli***, ***brussels sprouts***, ***cabbages***, ***calabrese***, ***cauliflowers***, ***kale*** [2-5]; ***winter oilseed rape*** [3, 5]; ***woody ornamentals*** [2, 4]

Specific Off-Label Approvals (SOLAs)

- ***bilberries*** *20090403* [2]
- ***blackberries*** *20090399* [2]
- ***blueberries*** *20090403* [2]
- ***chinese cabbage*** *20090406* [2]
- ***choi sum*** *20090406* [2]
- ***collards*** *20090406* [2]
- ***evening primrose*** *20090400* [2]
- ***farm forestry*** *20090409* [2]
- ***forest*** *20090409* [2]
- ***honesty*** *20090400* [2]
- ***linseed*** *20090397* [1], *20090400* [2]
- ***mustard*** *20090400* [2]
- ***pak choi*** *20090406* [2]
- ***redcurrants*** *20090403* [2]
- ***rubus hybrids*** *20090399* [2]
- ***whitecurrants*** *20090403* [2]

Approval information

- An EU decision published on 4 Dec 2008 to revoke the approval of all products containing napropamide has been repealed and approval now lasts until 31 Dec 2013

Efficacy guidance

- Best results obtained from treatment pre-emergence of weeds but product may be used in conjunction with contact herbicides for control of emerged weeds. Otherwise remove existing weeds before application
- Weed control may be reduced where spray is mixed too deeply in the soil
- Manure, crop debris or other organic matter may reduce weed control
- Napropamide broken down by sunlight, so application during such conditions not recommended. Most crops recommended for treatment between Nov and end-Feb
- Apply to winter oilseed rape pre-emergence or as pre-drilling treatment in tank-mixture with trifluralin and incorporate within 30 min [1, 3, 5]
- Post-emergence use of specific grass weedkilller recommended where volunteer cereals are a serious problem
- Increase water volume to ensure adequate dampening of compost when treating containerised nursery stock with dense leaf canopy [2]

Restrictions

- Maximum number of treatments 1 per crop or yr
- Do not use on Sands
- Do not use on soils with more than 10% organic matter
- Strawberries, blackcurrants, gooseberries and raspberries must be treated between 1 Nov and end Feb [2, 3]
- Applications to strawberries must only be made where the mature foliage has been previously removed [2, 3]
- Applications to ornamentals and forest nurseries must be made after 1 Nov and before end Apr
- Do not treat ornamentals in containers of less than 1 litre [2]
- Some phytotoxicity seen on yellow and golden varieties of conifers and container grown alpines. On any ornamental variety treat only a small number of plants in first season [2]
- Consult processors before use on any crop for processing

Crop-specific information

- Latest use: before transplanting for brassicas; pre-emergence for winter oilseed rape; before end Feb for strawberries, bush and cane fruit; before end of Apr for field and container grown ornamental trees and shrubs
- Where minimal cultivation used to establish oilseed rape, tank-mixture may be applied directly to stubble and mixed into top 25 mm as part of surface cultivations [1]
- Apply up to 14 d prior to drilling winter oilseed rape [1, 3, 5]
- Apply to strawberries established for at least one season or to maiden crops as long as planted carefully and no roots exposed, between 1 Nov and end Feb. Do not treat runners of poor vigour or with shallow roots, or runner beds [2, 3]

- Newly planted ornamentals should have no roots exposed. Do not treat stock of poor vigour or with shallow roots. Treatments made in Mar and Apr must be followed by 25 mm irrigation within 24 h [2]
- Bush and cane fruit must be established for at least 10 mth before spraying and treated between 1 Nov and end Feb [2, 3]

Following crops guidance

- After use in fruit or ornamentals no crop can be drilled within 7 mth of treatment. Leaf, flowerhead, root and fodder brassica crops may be drilled after 7 mth; potatoes, maize, peas or dwarf beans after 9 mth; autumn sown wheat or grass after 18 mth; any crop after 2 yr
- After use in oilseed rape only oilseed rape, swedes, fodder turnips, brassicas or potatoes should be sown within 12 mth of application
- Soil should be mould-board ploughed to a depth of at least 200 mm before drilling or planting any following crop

Environmental safety

- Dangerous for the environment
- Toxic to aquatic organisms

Hazard classification and safety precautions

Hazard Irritant, Dangerous for the environment
Transport code 9
Packaging group III
UN number 3082
Risk phrases R36, R38, R51, R53a
Operator protection A [1-5]; C [2-5]; U05a, U09a, U11, U20b
Environmental protection E15a [2-5]; E34, E38 [1-5]
Storage and disposal D01, D02, D05, D09a, D10c [1-5]; D12a [1]; D12b [2-5]

354 nicosulfuron

A sulfonylurea herbicide for maize
HRAC mode of action code: B

Products

1	Agrotech-Nicosulfron 40 SC	AgChem Access	40 g/l	SC	12938
2	Bishop	AgChem Access	40 g/l	SC	14055
3	Clayton Delilah	Clayton	40 g/l	SC	12610
4	Clayton Delilah XL	Clayton	60 g/l	SC	14443
5	Clayton Myth	Clayton	40 g/l	SC	14489
6	Clayton Myth XL	Clayton	60 g/l	SC	14466
7	Greencrop Folklore	Greencrop	40 g/l	SC	12867
8	Nico 4	Goldengrass	40 g/l	SC	14293
9	Samson Extra 6%	Syngenta	60 g/l	OD	13436
10	Standon Frontrunner	Standon	40 g/l	SC	13308

Uses

- Amaranthus in ***forage maize, grain maize*** [4, 6]
- Annual dicotyledons in ***forage maize*** [1-3, 5, 7-10]; ***grain maize, sweetcorn*** *(off-label)* [9]
- Annual meadow grass in ***forage maize*** [1-8, 10]; ***grain maize*** [4, 6]
- Black bindweed in ***forage maize, grain maize*** [4, 6]
- Black nightshade in ***forage maize, grain maize*** [4, 6]
- Chickweed in ***forage maize*** [1-8, 10]; ***grain maize*** [4, 6]
- Couch in ***forage maize*** [1-3, 5, 7, 8, 10]
- Fat hen in ***forage maize, grain maize*** [4, 6]
- Grass weeds in ***forage maize, grain maize, sweetcorn*** *(off-label)* [9]
- Italian ryegrass in ***forage maize, grain maize*** [4, 6]
- Knotgrass in ***forage maize, grain maize*** [4, 6]
- Mayweeds in ***forage maize*** [1-8, 10]; ***grain maize*** [4, 6]
- Perennial ryegrass in ***forage maize, grain maize*** [4, 6]
- Redshank in ***forage maize, grain maize*** [4, 6]
- Ryegrass in ***forage maize*** [1-3, 5, 7, 8, 10]

- Shepherd's purse in ***forage maize*** [1-8, 10]; ***grain maize*** [4, 6]
- Volunteer oilseed rape in ***forage maize*** [1-3, 5, 7, 8, 10]

Specific Off-Label Approvals (SOLAs)

- ***sweetcorn*** *20080800* [9]

Approval information

- Nicosulfuron included in Annex I under EC Directive 91/414

Efficacy guidance

- Product should be applied post-emergence between 2 and 8 crop leaf stage and to emerged weeds from the 2-leaf stage
- Product acts mainly by foliar activity. Ensure good spray cover of the weeds
- Nicosulfuron is a member of the ALS-inhibitor group of herbicides. To avoid the build up of resistance do not use any product containing an ALS-inhibitor herbicide with claims for control of grass weeds more than once on any crop
- Use these products as part of a resistance management strategy that includes cultural methods of control and does not use ALS inhibitors as the sole chemical method of grass weed control

Restrictions

- Maximum number of treatments 1 per crop
- Do not use if an organophosphorus soil insecticide has been used on the same crop
- Do not mix with foliar or liquid fertilisers or specified herbicides. See label for details
- Do not apply in mixture, or sequence, with any other sulfonyl-urea containing product
- Do not treat crops under stress
- Do not apply if rainfall is forecast to occur within 6 h of application

Crop-specific information

- Latest use: up to and including 8 true leaves for maize
- Some transient yellowing may be seen from 1-2 wk after treatment
- Only healthy maize crops growing in good field conditions should be treated

Following crops guidance

- Winter wheat (not undersown) may be sown 4 mth after treatment; maize may be sown in the spring following treatment; all other crops may be sown from the next autumn

Environmental safety

- Dangerous for the environment [7, 10]
- Very toxic to aquatic organisms [7, 10]
- LERAP Category B

Hazard classification and safety precautions

Hazard Harmful [4, 6, 9]; Irritant [1-3, 5, 7, 8, 10]; Dangerous for the environment [1, 4-7, 9, 10]
Transport code 9
Packaging group III
UN number 3082
Risk phrases R20, R36, R43 [4, 6, 9]; R38 [1-3, 5, 7, 8, 10]; R50, R53a [1, 4-7, 9, 10]; R53b [5]
Operator protection A, H [1-10]; C [4, 6, 9]; U02a, U05a, U14, U20b [1-10]; U10 [1, 4-7, 9, 10]; U11, U19a [4, 6, 9]
Environmental protection E15a [2, 3, 5, 7, 8]; E15b [1, 9, 10]; E16a, E16b [1-3, 5, 7-10]; E38 [1, 7, 9, 10]
Storage and disposal D01, D02 [1-3, 5, 7-10]; D05 [1, 7, 10]; D09a [2, 3, 5, 8, 10]; D10b [2, 3, 5, 8]; D10c, D12a [1, 7, 9, 10]
Medical advice M05a

355 nicotine

A general purpose, non-persistent, contact, alkaloid insecticide
IRAC mode of action code: 4B

Products

Nicotine 40% Shreds	Dow	40% w/w	FU	05725

Uses

- Aphids in ***aubergines, protected courgettes, protected cucumbers, protected herbs (see appendix 6), protected lettuce, protected ornamentals, protected peppers, protected potatoes, protected salad onions, protected tomatoes***
- Glasshouse whitefly in ***aubergines, protected courgettes, protected cucumbers, protected herbs (see appendix 6), protected lettuce, protected ornamentals, protected peppers, protected potatoes, protected salad onions, protected tomatoes***
- Green leafhopper in ***aubergines, protected courgettes, protected cucumbers, protected herbs (see appendix 6), protected lettuce, protected ornamentals, protected peppers, protected potatoes, protected salad onions, protected tomatoes***
- Potato virus vectors in ***chitting potatoes***
- Thrips in ***aubergines, protected courgettes, protected cucumbers, protected herbs (see appendix 6), protected lettuce, protected ornamentals, protected peppers, protected potatoes, protected salad onions, protected tomatoes***

Approval information

- Following a decision by the EU published on 9 Jan 2009 to revoke the approval of all products containing nicotine, all these products now have a final use date of 8th June 2010
- Approval expiry 08 Jun 2010 [1]

Efficacy guidance

- Apply as foliar spray, taking care to cover undersides of leaves and repeat as necessary or dip young plants, cuttings or strawberry runners before planting out
- Best results achieved by spraying at air temperatures above 16 °C but do not treat in bright sunlight or windy weather [1]
- Fumigate glasshouse crops at temperatures of at least 16 °C. See label for details of recommended fumigation procedure [1]
- Potatoes may be fumigated in chitting houses to control virus-spreading aphids [1]
- May be used in integrated control systems to give partial control of organophosphorus-, organochlorine- and pyrethroid-resistant whitefly

Restrictions

- Nicotine is subject to the Poisons Rules 1982 and the Poisons Act 1972. See Section 5 for more information [1]
- Maximum number of treatments not specified
- On plants of unknown sensitivity test first on a small scale
- Keep unprotected persons out of treated glasshouses for at least 12 h
- Wear overall, hood, rubber gloves and respirator if entering glasshouse within 12 h of fumigating [1]

Crop-specific information

- HI 2 d for most crops - check labels

Environmental safety

- Dangerous to fish or other aquatic life. Do not contaminate surface waters or ditches with chemical or used container
- Dangerous to livestock. Keep all livestock out of treated areas/away from treated water for at least 12 h
- Dangerous to game, wild birds and animals
- Harmful to bees. Do not apply to crops in flower or to those in which bees are actively foraging. Do not apply when flowering weeds are present
- Keep in original container, tightly closed, in a safe place, under lock and key [1]

Hazard classification and safety precautions

Hazard Harmful, Highly flammable
Transport code 4.1 6.1
Packaging group III
UN number 2926
Risk phrases R20, R21, R22a
Operator protection A, D, H, J; U05b, U19a, U20a
Environmental protection E02a, E06b (12 h); E10a, E12d, E12e, E13b, E34
Consumer protection C02a (24 h)

Storage and disposal D01, D02, D09b, D11a
Medical advice M03

356 oxadiazon

A residual and contact oxadiazole herbicide for fruit and ornamentals
HRAC mode of action code: E

Products

1	Clayton Oxen FL	Clayton	250 g/l	EC	12861
2	Clayton Oxen G	Clayton	2% w/w	GR	12770
3	Ronstar 2G	Certis	2% w/w	GR	12965
4	Ronstar Liquid	Certis	250 g/l	EC	11215
5	Standon Roxx L	Standon	250 g/l	EC	13309

Uses

- Annual dicotyledons in ***almonds*** *(off-label)*, ***bilberries*** *(off-label)*, ***blackberries*** *(off-label)*, ***blueberries*** *(off-label)*, ***chestnuts*** *(off-label)*, ***cranberries*** *(off-label)*, ***hazel nuts*** *(off-label)*, ***redcurrants*** *(off-label)*, ***rubus hybrids*** *(off-label)*, ***walnuts*** *(off-label)*, ***whitecurrants*** *(off-label)* [4]; ***apple orchards***, ***blackcurrants***, ***gooseberries***, ***hops***, ***pear orchards***, ***raspberries***, ***wine grapes***, ***woody ornamentals*** [1, 4, 5]; ***ornamental plant production*** [2, 3]
- Annual grasses in ***almonds*** *(off-label)*, ***bilberries*** *(off-label)*, ***blackberries*** *(off-label)*, ***blueberries*** *(off-label)*, ***chestnuts*** *(off-label)*, ***cranberries*** *(off-label)*, ***hazel nuts*** *(off-label)*, ***redcurrants*** *(off-label)*, ***rubus hybrids*** *(off-label)*, ***walnuts*** *(off-label)*, ***whitecurrants*** *(off-label)* [4]; ***apple orchards***, ***blackcurrants***, ***gooseberries***, ***hops***, ***pear orchards***, ***raspberries***, ***wine grapes***, ***woody ornamentals*** [1, 4, 5]; ***ornamental plant production*** [2, 3]
- Bindweeds in ***almonds*** *(off-label)*, ***bilberries*** *(off-label)*, ***blackberries*** *(off-label)*, ***blueberries*** *(off-label)*, ***chestnuts*** *(off-label)*, ***cranberries*** *(off-label)*, ***hazel nuts*** *(off-label)*, ***redcurrants*** *(off-label)*, ***rubus hybrids*** *(off-label)*, ***walnuts*** *(off-label)*, ***whitecurrants*** *(off-label)* [4]; ***apple orchards***, ***blackcurrants***, ***gooseberries***, ***hops***, ***pear orchards***, ***raspberries***, ***wine grapes***, ***woody ornamentals*** [1, 4, 5]
- Cleavers in ***almonds*** *(off-label)*, ***bilberries*** *(off-label)*, ***blackberries*** *(off-label)*, ***blueberries*** *(off-label)*, ***chestnuts*** *(off-label)*, ***cranberries*** *(off-label)*, ***hazel nuts*** *(off-label)*, ***redcurrants*** *(off-label)*, ***rubus hybrids*** *(off-label)*, ***walnuts*** *(off-label)*, ***whitecurrants*** *(off-label)* [4]; ***apple orchards***, ***blackcurrants***, ***gooseberries***, ***hops***, ***pear orchards***, ***raspberries***, ***wine grapes***, ***woody ornamentals*** [1, 4, 5]
- Knotgrass in ***almonds*** *(off-label)*, ***bilberries*** *(off-label)*, ***blackberries*** *(off-label)*, ***blueberries*** *(off-label)*, ***chestnuts*** *(off-label)*, ***cranberries*** *(off-label)*, ***hazel nuts*** *(off-label)*, ***redcurrants*** *(off-label)*, ***rubus hybrids*** *(off-label)*, ***walnuts*** *(off-label)*, ***whitecurrants*** *(off-label)* [4]; ***apple orchards***, ***blackcurrants***, ***gooseberries***, ***hops***, ***pear orchards***, ***raspberries***, ***wine grapes***, ***woody ornamentals*** [1, 4, 5]

Specific Off-Label Approvals (SOLAs)

- ***almonds*** *20063295* [4]
- ***bilberries*** *20063296* [4]
- ***blackberries*** *20063297* [4]
- ***blueberries*** *20063296* [4]
- ***chestnuts*** *20063295* [4]
- ***cranberries*** *20063296* [4]
- ***hazel nuts*** *20063295* [4]
- ***redcurrants*** *20063296* [4]
- ***rubus hybrids*** *20063297* [4]
- ***walnuts*** *20063295* [4]
- ***whitecurrants*** *20063296* [4]

Approval information

- Accepted by BBPA for use on hops
- Oxadiazon included in Annex I under EC Directive 91/414
- Approval expiry 31 Dec 2010 [2]

Efficacy guidance

- Apply as a directed spray or as a spot treatment as directed. Avoid direct contact with the crop. See label [1, 4, 5]

SEE SECTION 3 FOR PRODUCTS ALSO REGISTERED

- Apply granules as soon after potting as possible and before crop plant is making soft growth [2, 3]
- Best results from spray treatments obtained when weeds are dry at application and for a period afterwards. Later rain or overhead watering is needed for effective results [1, 4, 5]
- Residual activity reduced on soils with more than 10% organic matter and, in these conditions, post-emergence treatment is more effective [1, 4, 5]
- Best results on bindweed when first shoots are 10-15 cm long [1, 4, 5]
- Adjust water volume to achieve good coverage when weed density is high [1, 4, 5]

Restrictions

- Maximum number of treatments 2 per yr for ornamental plant production [2, 3]
- Maximum total dose per yr equivalent to one full dose treatment on edible crops [1, 4, 5]
- See label for list of ornamental species which may be treated with granules. Treat small numbers of other species to check safety. Do not treat Hydrangea, Spiraea or Genista [2, 3]
- Do not cultivate after treatment [1, 4, 5]
- Do not treat container stock under glass or use on plants rooted in media with high sand or non-organic content. Do not apply to plants with wet foliage [2, 3]
- Consult processors before treatment of crops intended for processing [1, 4, 5]

Crop-specific information

- Latest use: Jul for apples, grapevines, hops, pears; Jun for raspberries, woody ornamentals
- Apply spray to apples and pears from Jan to Jul, avoiding young growth [1, 4, 5]
- Treat bush fruit from Jan to bud-break, avoiding bushes, grapevines in Feb/Mar before start of new growth or in Jun/Jul avoiding foliage [1, 4, 5]
- Treat hops cropped for at least 2 yr in Feb or in Jun/Jul after deleafing [1, 4, 5]
- Treat woody ornamentals from Jan to Jun, avoiding young growth. Do not spray container stock overall [1, 4, 5]

Following crops guidance

- A period of at least 6 mth must elapse between treatment at half the recommended maximum dose and planting a following crop. 12 mth must elapse after use of the maximum recommended dose. In either case soil must be ploughed to min 15 cm and cultivated before sowing or planting a following crop [1, 4, 5]

Environmental safety

- Flammable [1, 4, 5]
- Dangerous for the environment [1, 4, 5]
- Very toxic to aquatic organisms [1, 4, 5]
- Harmful to aquatic organisms [2, 3]

Hazard classification and safety precautions

Hazard Harmful, Flammable, Dangerous for the environment [1, 4, 5]
Transport code 3 [1, 4, 5]; 9 [2, 3]
Packaging group III
UN number 1993 [1, 4, 5]; 3077 [2, 3]
Risk phrases R22b, R36, R38, R50, R67 [1, 4, 5]; R41 [4]; R52 [2, 3]; R53a [1-5]
Operator protection A, C [1-5]; H [5]; U02a [1, 5]; U05a, U20b [1-5]; U08 [1, 4, 5]; U09a, U19a [2, 3]; U14, U15 [3-5]
Environmental protection E15a [1-5]; E38 [4]
Storage and disposal D01, D09a, D12a [1-5]; D02 [1, 3-5]; D05 [5]; D10a [1, 4, 5]; D11a [2, 3]
Medical advice M05b [1, 4, 5]

357 oxamyl

A soil-applied, systemic carbamate nematicide and insecticide
IRAC mode of action code: 1A

Products

1	Tevday	AgChem Access	10% w/w	GR	13783
2	Vydate 10G	DuPont	10% w/w	GR	02322

Uses

- American serpentine leaf miner in ***aubergines*** *(off-label)*, ***ornamental plant production*** *(off-label)*, ***protected broad beans*** *(off-label)*, ***protected ornamentals*** *(off-label)*, ***protected peppers*** *(off-label)*, ***protected soya beans*** *(off-label)*, ***protected tomatoes*** *(off-label)* [2]

- Aphids in ***fodder beet*** *(off-label)* [2]; ***potatoes, sugar beet*** [1, 2]
- Docking disorder vectors in ***fodder beet*** *(off-label)* [2]; ***sugar beet*** [1, 2]
- Eelworm in ***bulb onions*** *(off-label)*, ***garlic*** *(off-label)*, ***shallots*** *(off-label)* [2]
- Free-living nematodes in ***potatoes*** [1, 2]
- Insect pests in ***garlic*** *(off-label)* [2]
- Leaf miner in ***aubergines*** *(off-label)*, ***ornamental plant production*** *(off-label)*, ***protected broad beans*** *(off-label)*, ***protected ornamentals*** *(off-label)*, ***protected peppers*** *(off-label)*, ***protected soya beans*** *(off-label)*, ***protected tomatoes*** *(off-label)* [2]
- Mangold fly in ***fodder beet*** *(off-label)* [2]; ***sugar beet*** [1, 2]
- Millipedes in ***fodder beet*** *(off-label)* [2]; ***sugar beet*** [1, 2]
- Potato cyst nematode in ***potatoes*** [1, 2]
- Pygmy beetle in ***fodder beet*** *(off-label)* [2]; ***sugar beet*** [1, 2]
- South American leaf miner in ***aubergines*** *(off-label)*, ***ornamental plant production*** *(off-label)*, ***protected broad beans*** *(off-label)*, ***protected ornamentals*** *(off-label)*, ***protected peppers*** *(off-label)*, ***protected soya beans*** *(off-label)*, ***protected tomatoes*** *(off-label)* [2]
- Spraing vectors in ***potatoes*** [1, 2]
- Stem nematodes in ***bulb onions*** *(off-label)*, ***garlic*** *(off-label)*, ***protected garlic*** *(off-label)*, ***protected onions*** *(off-label)*, ***shallots*** *(off-label)* [2]; ***carrots, parsnips*** [1, 2]

Specific Off-Label Approvals (SOLAs)

- ***aubergines*** *930020* [2]
- ***bulb onions*** *20061890* [2]
- ***fodder beet*** *20061107* [2]
- ***garlic*** *20061890* [2], *920163* [2]
- ***ornamental plant production*** *930020* [2]
- ***protected broad beans*** *930020* [2]
- ***protected garlic*** *940925* [2]
- ***protected onions*** *940925* [2]
- ***protected ornamentals*** *930020* [2]
- ***protected peppers*** *930020* [2]
- ***protected soya beans*** *930020* [2]
- ***protected tomatoes*** *930020* [2]
- ***shallots*** *20061890* [2]

Approval information

- Oxamyl included in Annex I under EC Directive 91/414
- In 2006 PSD required that all products containing this active ingredient should carry the following warning in the main area of the container label: "Oxamyl is an anticholinesterase carbamate. Handle with care"

Efficacy guidance

- Apply granules with suitable applicator before drilling or planting. See label for details of recommended machines
- In potatoes incorporate thoroughly to 10 cm and plant within 3-4 d
- In sugar beet apply in seed furrow at drilling

Restrictions

- Oxamyl is subject to the Poisons Rules 1982 and the Poisons Act 1972. See Section 5 for more information
- Contains an anticholinesterase carbamate compound. Do not use if under medical advice not to work with such compounds
- Maximum number of treatments 1 per crop or yr
- Keep in original container, tightly closed, in a safe place, under lock and key
- Wear protective gloves if handling treated compost or soil within 2 wk after treatment
- Allow at least 12 h, followed by at least 1 h ventilation, before entry of unprotected persons into treated glasshouses

Crop-specific information

- Latest use: at drilling/planting for vegetables; before drilling/planting for potatoes and peas.
- HI:potatoes, sugar beet, carrots, parsnips 12 wk

Environmental safety

- Dangerous for the environment
- Toxic to aquatic organisms

SEE SECTION 3 FOR PRODUCTS ALSO REGISTERED

- Dangerous to fish or other aquatic life. Do not contaminate surface waters or ditches with chemical or used container
- Dangerous to game, wild birds and animals. Bury spillages

Hazard classification and safety precautions

Hazard Toxic, Dangerous for the environment [2]
Transport code 6.1
Packaging group II [1]; II [2]
UN number 2757
Risk phrases R23, R25, R51, R53a [2]
Operator protection A, B, H, K, M [2]; C [2] (or D); U02a, U04a, U05a, U07, U09a, U13, U19a, U20a [2]
Environmental protection E10a, E13b, E34, E36a, E38 [2]
Storage and disposal D01, D02, D09a, D11a, D14 [2]
Medical advice M04a [2]

358 paclobutrazol

A triazole plant growth regulator for ornamentals and fruit

Products

1	Agrovista Paclo	Agrovista	250 g/l	SC	13424
2	Bonzi	Syngenta Bioline	4 g/l	SC	13623
3	Cultar	Syngenta	250 g/l	SC	10523
4	Pirouette	Fine	4 g/l	SC	13073

Uses

- Control of shoot growth in ***apples, pears*** [1, 3]
- Growth regulation in ***cherries*** *(off-label)*, ***plums*** *(off-label)* [1, 3]; ***quinces*** *(off-label)* [1]
- Improving colour in ***poinsettias*** [2, 4]
- Increasing flowering in ***azaleas, bedding plants, begonias, kalanchoes, lilies, roses, tulips*** [2, 4]
- Increasing fruit set in ***apples, pears*** [1, 3]; ***quinces*** *(off-label)* [3]
- Stem shortening in ***azaleas, bedding plants, begonias, kalanchoes, lilies, poinsettias, roses, tulips*** [2, 4]

Specific Off-Label Approvals (SOLAs)

- ***cherries*** *20080461* [1], *20031235* [3]
- ***plums*** *20080461* [1], *20031235* [3]
- ***quinces*** *20080462* [1], *20062993* [3]

Efficacy guidance

- Chemical is active via both foliage and root uptake. For best results apply in dull weather when relative humidity not high

Restrictions

- Maximum number of treatments 1 per specimen for some species [4]
- Maximum total dose equivalent to 3 full dose treatments for pears and 4 full dose treatments for apples [1, 3]
- Some varietal restrictions apply in top fruit (see label for details) [1, 3]
- Do not use on trees of low vigour or under stress [1, 3]
- Do not use on trees from green cluster to 2 wk after full petal fall [1, 3]
- Do not use in underplanted orchards or those recently interplanted [1, 3]

Crop-specific information

- HI apples, pears 14 d [1, 3]
- Apply to apple and pear trees under good growing conditions as pre-blossom spray (apples only) and post-blossom at 7-14 d intervals [1, 3]
- Timing and dose of orchard treatments vary with species and cultivar. See label [1, 3]

Following crops guidance

- Chemical has residual soil activity which can affect growth of following crops. Withhold treatment from orchards due for grubbing to allow the following intervals between the last treatment and planting the next crop: apples, pears 1 yr; beans, peas, onions 2 yr; stone fruit 3 yr; cereals, grass, oilseed rape, carrots 4 yr; market brassicas 6 yr; potatoes and other crops 7 yr [1, 3]

Environmental safety

- Harmful to aquatic organisms
- Do not use on food crops [4]
- Keep livestock out of treated areas for at least 2 years after treatment [1, 3]

Hazard classification and safety precautions

Hazard Irritant [1, 3]
UN number N/C
Risk phrases R36, R52, R53a [1, 3]
Operator protection A, H, M [1, 3]; U05a, U20a [1-4]; U08 [2, 4]
Environmental protection E06a [1, 3] (2 yr); E13c [4]; E15a [2-4]; E15b [1]; E34, E38 [1, 3]
Consumer protection C01 [2, 4]; C02a [1, 3] (2 wk)
Storage and disposal D01, D02, D09a [1-4]; D05 [2, 4]; D10b [4]; D10c [1-3]; D12a [1, 3]
Medical advice M03 [1, 3]

359 paraffin oil (commodity substance)

An agent for the control of birds by egg treatment

Products

paraffin oil	various	OL	-

Uses

- Birds in ***miscellaneous pest control situations*** *(egg treatment)*

Approval information

- Approval for the use of paraffin oil as a commodity substance was granted on 25 April 1995 by Ministers under regulation 5 of the Control of Pesticides Regulations 1986
- Only to be used where a licence has been approved in accordance with Section 16(1) of the Wildlife and Countryside Act (1981)
- Paraffin oils included in Annex I under EC Directive 91/414

Efficacy guidance

- Egg treatment should be undertaken as soon as clutch is complete
- Eggs should be treated by complete immersion in liquid paraffin

Restrictions

- Use to control eggs of birds covered by licences issued by the Agriculture and Environment Departments under Section 16(1) of the Wildlife and Countryside Act (1981)
- Treat eggs once only

Hazard classification and safety precautions

Operator protection A, C

360 penconazole

A protectant triazole fungicide with antisporulant activity
FRAC mode of action code: 3

Products

1	Agrovista Penco	Agrovista	100 g/l	EC	13425
2	Topas	Syngenta	100 g/l	EC	09717
3	Topenco 100 EC	Globachem	100 g/l	EC	11972

Uses

- Powdery mildew in ***apples*** [1-3]; ***bilberries*** *(off-label)*, ***blackcurrants***, ***blueberries*** *(off-label)*, ***cranberries*** *(off-label)*, ***gooseberries*** *(off-label)*, ***hops***, ***ornamental trees***, ***protected strawberries*** *(off-label)*, ***redcurrants*** *(off-label)*, ***strawberries*** *(off-label)*, ***whitecurrants*** *(off-label)* [1, 2]
- Rust in ***roses*** [1, 2]
- Scab in ***ornamental trees*** [1, 2]

Specific Off-Label Approvals (SOLAs)

- ***bilberries*** *20080477* [1], *20061106* [2]
- ***blueberries*** *20080477* [1], *20061106* [2]

- ***cranberries*** *20080477* [1], *20061106* [2]
- ***gooseberries*** *20080477* [1], *20061106* [2]
- ***protected strawberries*** *20080478* [1], *20073651* [2]
- ***redcurrants*** *20080477* [1], *20061106* [2]
- ***strawberries*** *20080478* [1], *20073651* [2]
- ***whitecurrants*** *20080477* [1], *20061106* [2]

Approval information

- Accepted by BBPA for use on hops

Efficacy guidance

- Use as a protectant fungicide by treating at the earliest signs of disease
- Treat crops every 10-14 d (every 7-10 d in warm, humid weather) at first sign of infection or as a protective spray ensuring complete coverage. See label for details of timing
- Increase dose and volume with growth of hops but do not exceed 2000 l/ha. Little or no activity will be seen on established powdery mildew
- Antisporulant activity reduces development of secondary mildew in apples

Restrictions

- Maximum number of treatments 10 per yr for apples, 6 per yr for hops, 4 per yr for blackcurrants
- Check for varietal susceptibility in roses. Some defoliation may occur after repeat applications on Dearest

Crop-specific information

- HI apples, hops 14 d; currants, bilberries, blueberries, cranberries, gooseberries 4 wk

Environmental safety

- Dangerous for the environment
- Toxic to aquatic organisms

Hazard classification and safety precautions

Hazard Irritant, Dangerous for the environment
Transport code 3
Packaging group III
UN number 1915
Risk phrases R36, R51, R53a [1-3]; R38 [3]
Operator protection A, C; U02a, U05a, U20b [1-3]; U08 [1, 2]; U11 [3]
Environmental protection E15a [1-3]; E22c [3]; E38 [1, 2]
Storage and disposal D01, D02, D05, D09a [1-3]; D07 [1, 3]; D10b [3]; D10c, D12a [1, 2]
Medical advice M05b

361 pencycuron

A non-systemic urea fungicide for use on seed potatoes
FRAC mode of action code: 20

See also imazalil + pencycuron

Products

1	Agriguard Pencycuron	AgriGuard	12.5% w/w	DS	10445
2	Monceren DS	Bayer CropScience	12.5% w/w	DS	11292
3	Pan Pencycuron P	Pan Agriculture	12.5% w/w	DP	10494
4	Pency Flowable	Globachem	250 g/l	FS	12637
5	Standon Pencycuron DP	Standon	12.5% w/w	DS	08774

Uses

- Black scurf in ***potatoes*** *(tuber treatment)*
- Stem canker in ***potatoes*** *(tuber treatment)*

Efficacy guidance

- Provides control of tuber-borne disease and gives some reduction of stem canker
- Seed tubers should be of good quality and vigour, and free from soil deposits when treated
- Dust formulations should be applied immediately before, or during, the planting process by treating seed tubers in chitting trays, in bulk bins immediately before planting or in hopper at planting. Liquid formulations may be applied by misting equipment at any time, into or out of store but treatment over a roller table at the end of grading out, or at planting, is usually most convenient

- Whichever method is chosen, complete and even distribution on the tubers is essential for optimum efficacy
- Apply in accordance with detailed guidelines in manufacturer's literature
- If rain interrupts planting, cover dust treated tubers in hopper

Restrictions

- Maximum number of treatments 1 per batch of seed potatoes
- Treated tubers must be used only as seed and not for human or animal consumption
- Do not use on tubers previously treated with a dry powder seed treatment or hot water
- To prevent rapid absorption through damaged 'eyes' seed tubers should only be treated before the 'eyes' begin to open or immediately before or during planting [4]
- Tubers removed from cold storage must be allowed to attain a temperature of at least 8 °C before treatment [4]
- Use of suitable dust mask is mandatory when applying dust, filling the hopper or riding on planter [1, 2, 5]
- Treated tubers in bulk bins must be allowed to dry before being stored, especially when intended for cold storage [4]
- Some internal staining of boxes used to store treated tubers may occur. Ware tubers should not be stored or supplied to packers in such stained boxes [4]

Crop-specific information

- Latest use: at planting

Environmental safety

- Do not use treated seed as food or feed
- Treated seed harmful to game and wildlife

Hazard classification and safety precautions

Hazard Irritant [4]
UN number N/C
Risk phrases R43 [4]
Operator protection A [1-5]; F [1-3, 5]; U14, U20a [4]; U20b [1-3, 5]
Environmental protection E03, E15a
Storage and disposal D05 [4]; D07 [1, 3, 5]; D09a, D11a [1-5]
Treated seed S01, S02, S03, S04a, S05, S06a

362 pendimethalin

A residual dinitroaniline herbicide for cereals and other crops
HRAC mode of action code: K1

See also *diflufenican + pendimethalin*
flufenacet + pendimethalin
imazamox + pendimethalin

Products

1	Alpha Pendimethalin 330 EC	Makhteshim	330 g/l	EC	13815
2	Blazer	Headland	330 g/l	EC	14126
3	Cinder	Makhteshim	400 g/l	CS	14526
4	Fastnet	Sipcam	330 g/l	EC	14068
5	Nighthawk 330 EC	AgChem Access	330 g/l	EC	14083
6	PDM 330 EC	BASF	330 g/l	EC	13406
7	Sherman	Makhteshim	330 g/l	EC	13859
8	Shuttle	AgriGuard	330 g/l	EC	14041
9	Stamp 330 EC	AgChem Access	330 g/l	EC	13995
10	Stomp 400 SC	BASF	400 g/l	SC	13405
11	Stomp Aqua	BASF	455 g/l	CS	14664
12	Trample	AgChem Access	330 g/l	EC	14173

Uses

- Annual dicotyledons in ***almonds*** *(off-label)*, ***asparagus*** *(off-label)*, ***bilberries*** *(off-label)*, ***blueberries*** *(off-label)*, ***chestnuts*** *(off-label)*, ***cranberries*** *(off-label)*, ***cress*** *(off-label)*, ***evening primrose*** *(off-label)*, ***farm forestry*** *(off-label)*, ***fennel*** *(off-label)*, ***frise*** *(off-label)*, ***grass seed crops*** *(off-label)*, ***hazel nuts*** *(off-label)*, ***herbs (see appendix 6)*** *(off-label)*, ***lamb's lettuce*** *(off-label)*, ***lettuce*** *(off-label)*, ***lupins*** *(off-label)*, ***parsley root*** *(off-label)*, ***quinces*** *(off-label)*, ***radicchio*** *(off-label)*, **redcurrants** *(off-label)*, **runner beans** *(off-label)*, **scarole** *(off-label)*, **walnuts** *(off-label)*, **whitecurrants** *(off-label)* [1, 3, 7, 10]; **apple orchards**, **pear orchards** [1, 2, 5, 7, 10]; **apples**, **pears** [3, 11]; **blackberries**, **blackcurrants**, **broccoli**, **brussels sprouts**, **cabbages**, **calabrese**, **carrots**, **cauliflowers**, **cherries**, **gooseberries**, **leeks**, **loganberries**, **parsnips**, **plums**, **raspberries**, **strawberries** [1-3, 5, 7, 10, 11]; **broad beans** *(off-label)*, **celery (outdoor)** *(off-label)*, **dwarf beans** *(off-label)*, **french beans** *(off-label)*, **miscanthus** *(off-label)*, **navy beans** *(off-label)*, **protected forest nurseries** *(off-label)*, **rhubarb** *(off-label)*, **spring wheat** *(off-label)*, **top fruit** *(off-label)* [3, 10]; **bulb onions** [2, 3, 7, 10, 11]; **bulb onions** *(off-label)*, **leaf brassicas** *(off-label)*, **parsley** *(off-label)*, **protected cress** *(off-label)*, **protected frise** *(off-label)*, **protected herbs (see appendix 6)** *(off-label)*, **protected lamb's lettuce** *(off-label)*, **protected lettuce** *(off-label)*, **protected salad brassicas** *(off-label)*, **protected scarole** *(off-label)* [1]; **bulb onions** *(off-label - pre + post emergence treatment)*, **chives** *(off-label)*, **forest** *(off-label)*, **leeks** *(off-label)*, **salad brassicas** *(off-label - for baby leaf production)*, **sweetcorn** *(off-label - under covers)* [10]; **chervil** *(off-label)*, **coriander** *(off-label)*, **dill** *(off-label)* [1, 10]; **combining peas**, **forage maize**, **potatoes**, **spring barley**, **sunflowers**, **triticale**, **winter barley**, **winter rye**, **winter wheat** [1-12]; **daffodils grown for galanthamine production** *(off-label)*, **mallow (althaea spp.)** *(off-label)*, **onion sets** *(off-label)* [1, 3, 7]; **durum wheat** [2-12]; **edible podded peas** *(off-label)*, **vining peas** *(off-label)* [1-3, 7]; **forage maize (under plastic mulches)**, **grain maize**, **grain maize under plastic mulches** [1, 7]; **forage maize (under plastic mulches)** *(off-label)*, **grain maize under plastic mulches** *(off-label)* [6]; **forest nurseries** *(off-label)*, **ornamental plant production** *(off-label)* [3, 6, 10]; **game cover** *(off-label)* [1, 3, 10]; **garlic** *(off-label)*, **salad onions** *(off-label)*, **shallots** *(off-label)* [1-3, 7, 10]; **hops**, **parsley** [2, 10]; **hops** *(off-label)*, **sweetcorn** *(off-label)* [3]; **rubus hybrids** [1, 3, 5, 7, 10, 11]; **salad brassicas** *(off-label)* [3, 7]; **spring field beans** *(off-label)* [1-3, 6, 7, 10]; **sweetcorn under plastic mulches** *(off-label)* [1, 3]; **tayberries** [2]; **winter field beans** *(off-label)* [1, 3, 6, 7]

- Annual grasses in **almonds** *(off-label)*, **asparagus** *(off-label)*, **bilberries** *(off-label)*, **blueberries** *(off-label)*, **chestnuts** *(off-label)*, **cranberries** *(off-label)*, **cress** *(off-label)*, **daffodils grown for galanthamine production** *(off-label)*, **evening primrose** *(off-label)*, **farm forestry** *(off-label)*, **fennel** *(off-label)*, **frise** *(off-label)*, **grass seed crops** *(off-label)*, **hazel nuts** *(off-label)*, **herbs (see appendix 6)** *(off-label)*, **lamb's lettuce** *(off-label)*, **lettuce** *(off-label)*, **lupins** *(off-label)*, **mallow (althaea spp.)** *(off-label)*, **onion sets** *(off-label)*, **parsley root** *(off-label)*, **quinces** *(off-label)*, **radicchio** *(off-label)*, **redcurrants** *(off-label)*, **scarole** *(off-label)*, **walnuts** *(off-label)*, **whitecurrants** *(off-label)* [1, 3, 7]; **apples**, **blackberries**, **blackcurrants**, **broccoli**, **brussels sprouts**, **bulb onions**, **cabbages**, **calabrese**, **carrots**, **cauliflowers**, **cherries**, **gooseberries**, **leeks**, **loganberries**, **parsnips**, **pears**, **plums**, **raspberries**, **strawberries** [3, 11]; **broad beans** *(off-label)*, **dwarf beans** *(off-label)*, **french beans** *(off-label)*, **miscanthus** *(off-label)*, **navy beans** *(off-label)*, **protected forest nurseries** *(off-label)*, **spring wheat** *(off-label)*, **top fruit** *(off-label)* [3, 10]; **bulb onions** *(off-label)*, **leaf brassicas** *(off-label)*, **parsley** *(off-label)*, **protected cress** *(off-label)*, **protected frise** *(off-label)*, **protected herbs (see appendix 6)** *(off-label)*, **protected lamb's lettuce** *(off-label)*, **protected lettuce** *(off-label)*, **protected salad brassicas** *(off-label)*, **protected scarole** *(off-label)* [1]; **celery (outdoor)** *(off-label)*, **hops** *(off-label)*, **rhubarb** *(off-label)*, **sweetcorn** *(off-label)* [3]; **chervil** *(off-label)*, **coriander** *(off-label)*, **dill** *(off-label)* [1, 10]; **combining peas**, **durum wheat**, **forage maize**, **potatoes**, **spring barley**, **sunflowers**, **triticale**, **winter barley**, **winter rye**, **winter wheat** [3, 4, 6, 8, 9, 11, 12]; **edible podded peas** *(off-label)*, **garlic** *(off-label)*, **runner beans** *(off-label)*, **salad onions** *(off-label)*, **shallots** *(off-label)*, **vining peas** *(off-label)* [1-3, 7]; **forage maize (under plastic mulches)**, **grain maize**, **grain maize under plastic mulches** [7]; **forage maize (under plastic mulches)** *(off-label)*, **grain maize under plastic mulches** *(off-label)* [6]; **forest nurseries** *(off-label)*, **ornamental plant production** *(off-label)* [3, 6, 10]; **game cover** *(off-label)* [1, 3, 10]; **rubus hybrids** [1, 3, 5, 7, 11]; **salad brassicas** *(off-label)* [3, 7]; **spring field beans** *(off-label)*, **winter field beans** *(off-label)* [1-3, 6, 7]; **sweetcorn under plastic mulches** *(off-label)* [1, 3]
- Annual meadow grass in **apple orchards**, **blackberries**, **blackcurrants**, **broccoli**, **brussels sprouts**, **cabbages**, **calabrese**, **carrots**, **cauliflowers**, **cherries**, **combining peas**, **forage maize**, **gooseberries**, **leeks**, **loganberries**, **parsnips**, **pear orchards**, **plums**, **potatoes**, **raspberries**, **spring barley**, **strawberries**, **sunflowers**, **triticale**, **winter barley**, **winter rye**, **winter wheat** [1, 2, 5, 7, 10]; **bulb onions** [2, 7, 10]; **durum wheat** [2, 5, 7, 10]; **forage maize (under plastic mulches)**, **grain maize**, **grain maize under plastic mulches** [1]; **hops**, **parsley** [2, 10]; **rubus hybrids** [10]; **tayberries** [2]
- Blackgrass in **apple orchards**, **blackberries**, **blackcurrants**, **broccoli**, **brussels sprouts**, **cabbages**, **calabrese**, **carrots**, **cauliflowers**, **cherries**, **combining peas**, **gooseberries**, **leeks**,

loganberries, parsnips, pear orchards, plums, raspberries, spring barley, strawberries, sunflowers, triticale, winter barley, winter rye, winter wheat [1, 2, 5, 7, 10]; ***bulb onions*** [2, 7, 10]; ***durum wheat*** [2, 5, 7, 10]; ***forage maize (under plastic mulches), grain maize, grain maize under plastic mulches*** [1]; ***hops, parsley*** [2, 10]; ***rubus hybrids*** [10]; ***tayberries*** [2]

- Cleavers in ***winter field beans*** *(off-label)* [1, 2, 10]
- Fat hen in ***sweetcorn under plastic mulches*** *(off-label)* [1]
- Knotgrass in ***daffodils grown for galanthamine production*** *(off-label)* [1]; ***edible podded peas*** *(off-label)*, ***vining peas*** *(off-label)* [2, 10]
- Polygonums in ***edible podded peas*** *(off-label)*, ***vining peas*** *(off-label)* [2]
- Rough meadow grass in ***apple orchards, blackberries, blackcurrants, broccoli, brussels sprouts, cabbages, calabrese, carrots, cauliflowers, cherries, combining peas, forage maize, gooseberries, leeks, loganberries, parsnips, pear orchards, plums, potatoes, raspberries, spring barley, strawberries, sunflowers, triticale, winter barley, winter rye, winter wheat*** [1, 2, 5, 7, 10]; ***bulb onions*** [2, 7, 10]; ***durum wheat*** [2, 5, 7, 10]; ***forage maize (under plastic mulches), grain maize, grain maize under plastic mulches*** [1]; ***hops, parsley*** [2, 10]; ***rubus hybrids*** [10]; ***tayberries*** [2]
- Speedwells in ***daffodils grown for galanthamine production*** *(off-label)* [1]
- Volunteer oilseed rape in ***apple orchards, blackberries, blackcurrants, broccoli, brussels sprouts, cabbages, calabrese, carrots, cauliflowers, cherries, gooseberries, leeks, loganberries, parsnips, pear orchards, plums, raspberries, spring barley, strawberries, sunflowers*** [1, 2, 5, 7, 10]; ***bulb onions*** [2, 7, 10]; ***combining peas, forage maize, potatoes, triticale, winter barley, winter rye, winter wheat*** [1, 2, 5-10, 12]; ***daffodils grown for galanthamine production*** *(off-label)*, ***forage maize (under plastic mulches), grain maize, grain maize under plastic mulches*** [1]; ***durum wheat*** [2, 5-10, 12]; ***hops, parsley*** [2, 10]; ***rubus hybrids*** [1, 5, 7, 10]; ***spring field beans*** *(off-label)*, ***winter field beans*** *(off-label)* [6]; ***tayberries*** [2]
- Wild oats in ***durum wheat*** [2, 5, 7, 10]; ***forage maize (under plastic mulches), grain maize, grain maize under plastic mulches*** [1]; ***triticale, winter barley, winter rye, winter wheat*** [1, 2, 5, 7, 10]

Specific Off-Label Approvals (SOLAs)

- ***almonds*** *20090767* [1], *20092184* [3], *20090812* [7], *20071442* [10]
- ***asparagus*** *20090764* [1], *20092197* [3], *20090809* [7], *20071431* [10]
- ***bilberries*** *20090770* [1], *20092198* [3], *20090817* [7], *20071444* [10]
- ***blueberries*** *20090770* [1], *20092198* [3], *20090817* [7], *20071444* [10]
- ***broad beans*** *20092199* [3], *20081016* [10]
- ***bulb onions*** *20090765* [1], *(pre + post emergence treatment) 20071438* [10]
- ***celery (outdoor)*** *20092206* [3], *20071430* [10]
- ***chervil*** *20090772* [1], *20081808* [10]
- ***chestnuts*** *20090767* [1], *20092184* [3], *20090812* [7], *20071442* [10]
- ***chives*** *20071432* [10]
- ***coriander*** *20090772* [1], *20081808* [10]
- ***cranberries*** *20090770* [1], *20092198* [3], *20090817* [7], *20071444* [10]
- ***cress*** *20090772* [1], *20092187* [3], *20090819* [7], *20071432* [10]
- ***daffodils grown for galanthamine production*** *20090762* [1], *20092189* [3], *20090807* [7]
- ***dill*** *20090772* [1], *20081808* [10]
- ***dwarf beans*** *20092200* [3], *20080053* [10]
- ***edible podded peas*** *20090768* [1], *20090881* [2], *20092190* [3], *20090813* [7], *20071437* [10]
- ***evening primrose*** *20090759* [1], *20092191* [3], *20090804* [7], *20071434* [10]
- ***farm forestry*** *20090761* [1], *20092201* [3], *20090806* [7], *20071436* [10]
- ***fennel*** *20090757* [1], *20092186* [3], *20090802* [7], *20071430* [10]
- ***forage maize (under plastic mulches)*** *20080959* [6]
- ***forest*** *20071436* [10]
- ***forest nurseries*** *20092208* [3], *20082862* [6], *20082923* [10]
- ***french beans*** *20092200* [3], *20080053* [10]
- ***frise*** *20090772* [1], *20092187* [3], *20090819* [7], *20071432* [10]
- ***game cover*** *20082806* [1], *20092208* [3], *20082923* [10]
- ***garlic*** *20090765* [1], *20090883* [2], *20092203* [3], *20090810* [7], *20071438* [10]
- ***grain maize under plastic mulches*** *20080959* [6]
- ***grass seed crops*** *20090766* [1], *20092193* [3], *20090811* [7], *20071429* [10]
- ***hazel nuts*** *20090767* [1], *20092184* [3], *20090812* [7], *20071442* [10]

- **herbs (see appendix 6)** *20090772* [1], *20092187* [3], *20090819* [7], *20071432* [10]
- **hops** *20092208* [3]
- **lamb's lettuce** *20090772* [1], *20092187* [3], *20090819* [7], *20071432* [10]
- **leaf brassicas** *20090772* [1]
- **leeks** *20071438* [10]
- **lettuce** *20090772* [1], *20092187* [3], *20090819* [7], *20071432* [10]
- **lupins** *20090771* [1], *20092194* [3], *20090818* [7], *20071437* [10]
- **mallow (althaea spp.)** *20090772* [1], *20092187* [3], *20090819* [7]
- **miscanthus** *20092208* [3], *20082923* [10]
- **navy beans** *20092200* [3], *20080053* [10]
- **onion sets** *20090765* [1], *20092203* [3], *20090810* [7]
- **ornamental plant production** *20092208* [3], *20082862* [6], *20082923* [10]
- **parsley** *20090772* [1]
- **parsley root** *20090758* [1], *20092204* [3], *20090803* [7], *20071433* [10]
- **protected cress** *20090772* [1]
- **protected forest nurseries** *20092208* [3], *20082923* [10]
- **protected frise** *20090772* [1]
- **protected herbs (see appendix 6)** *20090772* [1]
- **protected lamb's lettuce** *20090772* [1]
- **protected lettuce** *20090772* [1]
- **protected salad brassicas** *20090772* [1]
- **protected scarole** *20090772* [1]
- **quinces** *20090769* [1], *20092205* [3], *20090814* [7], *20071443* [10]
- **radicchio** *20090772* [1], *20092187* [3], *20090819* [7], *20071432* [10]
- **redcurrants** *20090770* [1], *20092198* [3], *20090817* [7], *20071444* [10]
- **rhubarb** *20092206* [3], *20071430* [10]
- **runner beans** *20090760* [1], *20090882* [2], *20092195* [3], *20090805* [7], *20071437* [10]
- **salad brassicas** *20092187* [3], *20090819* [7], *(for baby leaf production) 20071432* [10]
- **salad onions** *20090765* [1], *20090883* [2], *20092203* [3], *20090810* [7], *20071438* [10]
- **scarole** *20090772* [1], *20092187* [3], *20090819* [7], *20071432* [10]
- **shallots** *20090765* [1], *20090883* [2], *20092203* [3], *20090810* [7], *20071438* [10]
- **spring field beans** *20090401* [1], *20090702* [2], *20092202* [3], *20070876* [6], *20090404* [7], *20071437* [10]
- **spring wheat** *20092196* [3], *20082947* [10]
- **sweetcorn** *20092207* [3], *(under covers) 20071435* [10]
- **sweetcorn under plastic mulches** *20090083* [1], *20092207* [3]
- **top fruit** *20092208* [3], *20082923* [10]
- **vining peas** *20090768* [1], *20090881* [2], *20092190* [3], *20090813* [7], *20071437* [10]
- **walnuts** *20090767* [1], *20092184* [3], *20090812* [7], *20071442* [10]
- **whitecurrants** *20090770* [1], *20092198* [3], *20090817* [7], *20071444* [10]
- **winter field beans** *20090763* [1], *20090880* [2], *20092202* [3], *20070876* [6], *20090808* [7], *20071437* [10]

Approval information

- Pendimethalin included in Annex I under EC Directive 91/414
- Accepted by BBPA for use on malting barley and hops

Efficacy guidance

- Apply as soon as possible after drilling. Weeds are controlled as they germinate and emerged weeds will not be controlled by use of the product alone
- For effective blackgrass control apply not more than 2 d after final cultivation and before weed seeds germinate
- Best results by application to fine firm, moist, clod-free seedbeds when rain follows treatment. Effectiveness reduced by prolonged dry weather after treatment
- Effectiveness reduced on soils with more than 6% organic matter. Do not use where organic matter exceeds 10%
- Any trash, ash or straw should be incorporated evenly during seedbed preparation
- Do not disturb soil after treatment
- Apply to potatoes as soon as possible after planting and ridging in tank-mix with metribuzin but note that this will restrict the varieties that may be treated

- Always follow WRAG guidelines for preventing and managing herbicide resistant weeds. See Section 5 for more information

Restrictions
- Maximum number of treatments 1 per crop or yr
- Maximum total dose equivalent to one full dose treatment on most crops; 2 full dose treatments on leeks
- May be applied pre-emergence of cereal crops sown before 30 Nov provided seed covered by at least 32 mm soil, or post-emergence to early tillering stage (GS 23)
- Do not undersow treated crops
- Do not use on crops suffering stress due to disease, drought, waterlogging, poor seedbed conditions or chemical treatment or on soils where water may accumulate

Crop-specific information
- Latest use: pre-emergence for spring barley, carrots, lettuce, fodder maize, parsnips, parsley, combining peas, potatoes, onions and leeks; before transplanting for brassicas; before leaf sheaths erect for winter cereals; before bud burst for blackcurrants, gooseberries, cane fruit, hops; before flower trusses emerge for strawberries; 14 d after transplanting for leaf herbs
- Do not use on spring barley after end Mar (mid-Apr in Scotland on some labels) because dry conditions likely. Do not apply to dry seedbeds in spring unless rain imminent
- Apply to combining peas as soon as possible after sowing. Do not spray if plumule less than 13 mm below soil surface
- Apply to potatoes up to 7 d before first shoot emerges
- Apply to drilled crops as soon as possible after drilling but before crop and weed emergence
- Apply in top fruit, bush fruit and hops from autumn to early spring when crop dormant
- In cane fruit apply to weed free soil from autumn to early spring, immediately after planting new crops and after cutting out canes in established crops
- Apply in strawberries from autumn to early spring (not before Oct on newly planted bed). Do not apply pre-planting or during flower initiation period (post-harvest to mid-Sep)
- Apply pre-emergence in drilled onions or leeks, not on Sands, Very Light, organic or peaty soils or when heavy rain forecast
- Apply to brassicas after final plant-bed cultivation but before transplanting. Avoid unnecessary soil disturbance after application and take care not to introduce treated soil into the root zone when transplanting. Follow transplanting with specified post-planting treatments - see label
- Do not use on protected crops or in greenhouses

Following crops guidance
- Before ryegrass is drilled after a very dry season plough or cultivate to at least 15 cm. If treated spring crops are to be followed by crops other than cereals, plough or cultivate to at least 15 cm
- In the event of crop failure land must be ploughed or thoroughly cultivated to at least 15 cm. See label for minimum intervals that should elapse between treatment and sowing a range of replacement crops

Environmental safety
- Dangerous for the environment
- Very toxic to aquatic organisms
- Dangerous to fish or other aquatic life. Do not contaminate surface waters or ditches with chemical or used container
- Flammable [1, 2]
- LERAP Category B [1, 2, 6, 10]

Hazard classification and safety precautions

Hazard Harmful [1, 2, 4-9, 12]; Flammable [1, 2, 5, 7]; Dangerous for the environment [1-12]

Transport code 3 [2]; 9 [1, 3-12]

Packaging group III

UN number 1993 [2]; 3082 [1, 3-12]

Risk phrases R22a [4, 6, 8, 9, 12]; R22b [1, 2, 5, 7]; R36, R38 [2, 4, 6, 8, 9, 12]; R37, R67 [2]; R50 [2-4, 6, 8-10, 12]; R51 [1, 5, 7, 11]; R53a [1-12]; R70 [11]

Operator protection A [2-6, 8-12]; C [2, 5, 6, 8, 9, 12]; H [2, 11]; U02a, U14 [10, 11]; U05a [1, 2, 4-12]; U08 [1-3, 5-7, 9-12]; U11 [2, 4, 6, 8, 9, 12]; U13 [2, 3, 6, 8-12]; U19a [1-3, 5-12]; U20a [2]; U20b [2, 6, 8-10, 12]; U20c [3]

Environmental protection E15a [1, 2, 5, 7, 8]; E15b [2, 6, 9-12]; E16a [1, 2, 5-12]; E34 [1, 2, 5-9, 12]; E38 [1-12]; E39 [11]

Storage and disposal D01, D12a [1-12]; D02 [1, 2, 4-12]; D05 [1, 2, 5, 7, 11]; D08 [6, 8-10, 12]; D09a [1, 2, 5-7, 9-12]; D10b [1, 2, 5-12]

Medical advice M03, M05b [1, 2, 5-7, 9, 12]; M05a [3, 4, 8, 10]

363 pendimethalin + picolinafen

A post-emergence broad-spectrum herbicide mixture for winter cereals
HRAC mode of action code: K1 + F1

See also *picolinafen*

Products

1	Flight	BASF	330:7.5 g/l	SC	12534
2	Orient	BASF	330:7.5 g/l	SC	12541
3	PicoMax	BASF	320:16 g/l	SC	13456
4	Picona	BASF	320:16 g/l	SC	13428
5	PicoPro	BASF	320:16 g/l	SC	13454
6	PicoStomp	BASF	320:16 g/l	SC	13455

Uses

- Annual dicotyledons in ***forest nurseries*** *(off-label)* [3, 6]; ***game cover*** *(off-label)* [4]; ***winter barley***, ***winter wheat*** [1-6]
- Annual grasses in ***forest nurseries*** *(off-label)* [3, 6]; ***game cover*** *(off-label)* [4]
- Annual meadow grass in ***winter barley***, ***winter wheat*** [1-6]
- Chickweed in ***winter barley***, ***winter wheat*** [1-6]
- Cleavers in ***winter barley***, ***winter wheat*** [3-6]
- Field speedwell in ***winter barley***, ***winter wheat*** [1-6]
- Ivy-leaved speedwell in ***winter barley***, ***winter wheat*** [1-6]
- Rough meadow grass in ***winter barley***, ***winter wheat*** [1-6]

Specific Off-Label Approvals (SOLAs)

- ***forest nurseries*** *20082863* [3], *20082865* [6]
- ***game cover*** *20082864* [4]

Approval information

- Pendimethalin and picolinafen included in Annex I under EC Directive 91/414

Efficacy guidance

- Best results obtained on crops growing in a fine, firm tilth and when rain falls within 7 d of application
- Loose or cloddy seed beds must be consolidated prior to application otherwise reduced weed control may occur
- Residual weed control may be reduced on soils with more than 6% organic matter or under prolonged dry conditions
- Always follow WRAG guidelines for preventing and managing herbicide resistant weeds. See Section 5 for more information

Restrictions

- Maximum number of treatments 1 per crop
- Do not treat undersown cereals or those to be undersown
- Do not roll emerged crops before treatment nor autumn treated crops until the following spring
- Do not use on stony or gravelly soils, on soils that are waterlogged or prone to waterlogging, or on soils with more than 10% organic matter
- Do not treat crops under stress from any cause
- Do not apply pre-emergence to crops drilled after 30 Nov [1, 2]
- Consult processor before treating crops for processing [1, 2]

Crop-specific information

- Latest use: before pseudo-stem erect stage (GS 30)
- Transient bleaching may occur after treatment but it does not lead to yield loss
- Do not use pre-emergence on crops drilled after 30th November [3]
- For pre-emergence applications, seed should be covered with a minimum of 3.2cm settled soil [4-6]

Following crops guidance

- In the event of failure of a treated crop plough to at least 15 cm to ensure any residues are evenly dispersed
- In the event of failure of a treated crop allow at least 8 wk from the time of treatment before re-drilling either spring wheat or spring barley. After a normally harvested crop there are no restrictions on following crops apart from rye-grass, as indicated below [3-6]
- In a very dry season, plough or cultivate to at least 15 cm before drilling rye-grass
- A minimum of 5 mth must follow autumn applications before sowing spring barley, spring wheat, maize, peas, potatoes, spring field beans, dwarf beans, turnips, Brussels sprouts, cabbage, calabrese, linseed, carrots, parsnips, parsley, cauliflower [1, 2]
- After spring or summer applications a minimum of 2 mth must follow before sowing spring field beans, dwarf beans, broad beans, peas, turnips, Brussels sprouts, cabbage, calabrese, cauliflower, carrots, parsnips, parsley, linseed and any crop may be sown after 5 mth except red beet, sugar beet and spinach, for which an interval of 12 mth must be allowed [1, 2]

Environmental safety

- Dangerous for the environment
- Very toxic to aquatic organisms
- Product binds strongly to soil minimising likelihood of movement into groundwater
- LERAP Category B

Hazard classification and safety precautions

Hazard Dangerous for the environment
Transport code 9
Packaging group III
UN number 3082
Risk phrases R50, R53a
Operator protection A, H [1-3]; U02a, U05a, U20c
Environmental protection E15a [1, 2]; E15b [3-6]; E16a, E34, E38 [1-6]
Storage and disposal D01, D02, D05, D09a, D10c, D12a
Medical advice M03

364 peroxyacetic acid (commodity substance)

An inorganic fungicide for use on potato tubers

Products

peroxyacetic acid	various	5% w/w	SL	-

Uses

- Fungus diseases in ***flower bulbs*** *(dip)*, ***potatoes*** *(tuber treatment)*
- Nematodes in ***flower bulbs*** *(dip)*, ***potatoes*** *(tuber treatment)*

Approval information

- Approval for the use of peroxyacetic acid as a commodity substance was granted on 18 January 2007 by Ministers under regulation 5 of the Control of Pesticides Regulations 1986
- Peroxyacetic acid will not be included in Annex I of EC Directive 91/414 but Essential Uses on flower bulbs, potato tubers and disinfection of glasshouses, warehouses and agricultural tools and equipment will be permitted in UK until December 2010.
- Approval expiry 31 Dec 2010 [1]

Efficacy guidance

- Flower bulbs and seed potato tubers must be immersed for one minute in a 50:1 dilution of proprietary 5% w/w peroxyacetic acid formulation

Environmental safety

- Used dip must be disposed of in accordance with Part Five of the Code of Practice for Using Plant Protection Products

Hazard classification and safety precautions

Operator protection A, C, D, H, M

365 petroleum oil

An insecticidal and acaricidal hydrocarbon oil

Products

Croptex Spraying Oil	Certis	710 g/l	EC	-

Uses

- Mealybugs in ***bush fruit***, ***cane fruit***, ***fruit trees***, ***hops***, ***nursery stock***, ***protected cucumbers***, ***protected pot plants***, ***protected tomatoes***, ***wine grapes***
- Red spider mites in ***bush fruit***, ***cane fruit***, ***fruit trees***, ***hops***, ***nursery stock***, ***protected cucumbers***, ***protected pot plants***, ***protected tomatoes***, ***wine grapes***
- Scale insects in ***bush fruit***, ***cane fruit***, ***fruit trees***, ***hops***, ***nursery stock***, ***protected cucumbers***, ***protected pot plants***, ***protected tomatoes***, ***wine grapes***

Approval information

- Product not controlled by Control of Pesticides Regulations because it acts by physical means only

Efficacy guidance

- Spray at 1% (0.5% on tender foliage) to wet plants thoroughly, particularly the underside of leaves, and repeat as necessary
- Petroleum oil acts by blocking pest breathing pores and making leaf surfaces inhospitable to pests seeking to attack or attach to the leaf surface
- On outdoor crops apply when dormant or on plants with a known tolerance
- On plants of unknown sensitivity test first on a small scale
- Mixtures with certain pesticides may damage crop plants. If mixing, spray a few plants to test for tolerance before treating larger areas

Restrictions

- Do not mix with sulphur or iprodione products or use such mixtures within 28 d of treatment
- Do not treat protected crops in bright sunshine unless the glass is well shaded
- Test species tolerance before large scale treatment

Crop-specific information

- HI zero
- Treat grapevines before flowering

Hazard classification and safety precautions

Hazard Harmful
Risk phrases R22a, R36, R37, R38
Operator protection A, C; U05a, U08, U11, U19a
Environmental protection E34
Storage and disposal D01, D02, D09a, D10a, D12a
Medical advice M03

366 phenmedipham

A contact phenyl carbamate herbicide for beet crops and strawberries
HRAC mode of action code: C1

See also *desmedipham + ethofumesate + metamitron + phenmedipham*
desmedipham + ethofumesate + phenmedipham
desmedipham + phenmedipham
ethofumesate + metamitron + phenmedipham
ethofumesate + phenmedipham

Products

1	Alpha Phenmedipham 320 SC	Makhteshim	320 g/l	SC	14070
2	Beetup Flo	United Phosphorus	160 g/l	SC	11916
3	Betamax	AgriGuard	160 g/l	SC	14371
4	Betanal Flow	Bayer CropScience	160 g/l	SE	13893
5	Betasana SC	United Phosphorus	160 g/l	SC	14209

Products – continued

6 Corzal	AgriChem BV	157 g/l	SE	14192
7 Dancer Flow	Sipcam	160 g/l	SE	11405
8 Herbasan Flow	Nufarm UK	160 g/l	SE	13894
9 Mandolin Flow	Nufarm UK	160 g/l	SE	13895

Uses

- Annual dicotyledons in ***beet leaves*** *(off-label)*, ***chard*** *(off-label)*, ***herbs (see appendix 6)*** *(off-label)*, ***ornamental plant production*** *(off-label)*, ***red mustard*** *(off-label)*, ***spinach*** *(off-label)*, ***spinach beet*** *(off-label)* [4]; ***fodder beet***, ***mangels***, ***red beet***, ***sugar beet*** [1-9]; ***strawberries*** [1]

Specific Off-Label Approvals (SOLAs)

- ***beet leaves*** *20082244* [4]
- ***chard*** *20082244* [4]
- ***herbs (see appendix 6)*** *20082245* [4]
- ***ornamental plant production*** *20082824* [4]
- ***red mustard*** *20082244* [4]
- ***spinach*** *20082244* [4]
- ***spinach beet*** *20082244* [4]

Approval information

- Phenmedipham included in Annex I under EC Directive 91/414
- Approval expiry 28 Feb 2010 [7]

Efficacy guidance

- Best results achieved by application to young seedling weeds, preferably cotyledon stage, under good growing conditions when low doses are effective
- If using a low-dose programme 2-3 repeat applications at 7-10 d intervals are recommended on mineral soils, 3-5 applications may be needed on organic soils
- Addition of adjuvant oil may improve effectiveness on some weeds
- Various tank-mixtures with other beet herbicides recommended. See label for details
- Use of certain pre-emergence herbicides is recommended in combination with post-emergence treatment. See label for details

Restrictions

- Maximum number of treatments and maximum total dose varies with crop and product used. See label for details
- At high temperatures (above 21 °C) reduce rate and spray after 5 pm
- Do not apply immediately after frost or if frost expected
- Do not spray wet foliage or if rain imminent
- Do not spray crops stressed by wind damage, nutrient deficiency, pest or disease attack etc. Do not roll or harrow for 7 d before or after treatment
- Do not use on strawberries under cloches or polythene tunnels
- Do not use after 31 Jul in yr of harvest [8, 9]
- Consult processor before use on crops for processing

Crop-specific information

- Latest use: before crop leaves meet between rows for beet crops; before flowering for strawberries
- Apply to beet crops at any stage as low dose/low volume spray or from fully developed cotyledon stage with full rate. Apply to red beet after fully developed cotyledon stage
- Apply to strawberries at any time when weeds in susceptible stage, except in period from start of flowering to picking

Following crops guidance

- Beet crops may follow at any time after a treated crop. 3 mth must elapse from treatment before any other crop is sown and must be preceded by mould-board ploughing to 15 cm

Environmental safety

- Dangerous for the environment
- Very toxic to aquatic organisms
- Harmful to fish or other aquatic life. Do not contaminate surface waters or ditches with chemical or used container

Hazard classification and safety precautions

Hazard Irritant [1-3, 5]; Dangerous for the environment [1-9]
Transport code 9
Packaging group III
UN number 3082
Risk phrases R36, R51 [2, 3, 5]; R38 [1]; R43 [1-3, 5]; R50 [1, 4, 6-9]; R53a [1-9]
Operator protection A [1-7]; C [2, 3, 5]; H [1-3, 5]; U05a [1-3, 6]; U08 [1-3, 5, 6]; U11, U20c [2, 5]; U14 [1, 2, 5]; U19a [1, 5]; U20a [1, 8, 9]; U20b [3, 4, 7]
Environmental protection E13c [7]; E15a [1, 3, 8, 9]; E15b [2, 4-6]; E16a [4, 6, 8, 9]; E16b [6]; E19b [1]; E23 [4]; E34 [1, 6]; E38 [1, 3, 4, 6-9]
Storage and disposal D01 [1-3, 5, 6]; D02 [2, 3, 5, 6]; D05 [1-5, 9]; D08 [7]; D09a, D12a [1-9]; D10b [1, 4, 6-9]; D10c [2, 3, 5]
Medical advice M05a [1]

367 d-phenothrin

A non-systemic pyrethroid insecticide available only in mixtures
IRAC mode of action code: 3A

368 d-phenothrin + tetramethrin

A pyrethroid insecticide mixture for control of flying insects
IRAC mode of action code: 3 + 3

See also *tetramethrin*

Products

Killgerm ULV 500	Killgerm	36.8:18.4 g/l	UL	H4647

Uses

- Flies in ***agricultural premises***
- Grain storage mite in ***grain stores***
- Mosquitoes in ***agricultural premises***
- Wasps in ***agricultural premises***

Approval information

- Product approved for ULV application. See label for details [1]

Efficacy guidance

- Close doors and windows and spray in all directions for 3-5 sec. Keep room closed for at least 10 min

Restrictions

- For use only by professional operators
- Do not use space sprays containing pyrethrins or pyrethroid more than once per week in intensive or controlled environment animal houses in order to avoid development of resistance. If necessary, use a different control method or product

Crop-specific information

- May be used in the presence of poultry and livestock

Environmental safety

- Dangerous for the environment
- Toxic to aquatic organisms
- Do not apply directly to livestock/poultry
- Remove exposed milk and collect eggs before application. Protect milk machinery and containers from contamination

Hazard classification and safety precautions

Hazard Harmful, Dangerous for the environment
Transport code 9
Packaging group III
UN number 3082
Risk phrases R22b, R51, R53a

Operator protection A, C, D, E, H; U02b, U09b, U14, U19a, U20a, U20b
Environmental protection E05a, E15a, E38
Consumer protection C06, C07, C08, C09, C11, C12
Storage and disposal D01, D05, D06a, D09a, D10a, D12a
Medical advice M05b

369 physical pest control

Products that work by physical action only

Products

1	Majestik	Certis	-	SL	-
2	SB Plant Invigorator	Fargro	-	SL	-
3	Silico-Sec	Interfarm	>90% w/w	DS	-

Uses

- Aphids in ***all edible crops (outdoor and protected)*** [2]; ***fruit crops***, ***protected fruit***, ***protected vegetables***, ***vegetables*** [1]; ***ornamental plant production***, ***protected ornamentals*** [1, 2]
- Leafhoppers in ***fruit crops***, ***ornamental plant production***, ***protected fruit***, ***protected ornamentals***, ***protected vegetables***, ***vegetables*** [1]
- Mealybugs in ***all edible crops (outdoor and protected)***, ***ornamental plant production***, ***protected ornamentals*** [2]
- Mites in ***fruit crops***, ***ornamental plant production***, ***protected fruit***, ***protected ornamentals***, ***protected vegetables***, ***vegetables*** [1]; ***stored grain*** [3]
- Powdery mildew in ***all edible crops (outdoor and protected)***, ***ornamental plant production***, ***protected ornamentals*** [2]
- Scale insects in ***all edible crops (outdoor and protected)***, ***ornamental plant production***, ***protected ornamentals*** [2]
- Spider mites in ***all edible crops (outdoor and protected)***, ***ornamental plant production***, ***protected ornamentals*** [2]
- Suckers in ***all edible crops (outdoor and protected)***, ***ornamental plant production***, ***protected ornamentals*** [2]
- Thrips in ***fruit crops***, ***ornamental plant production***, ***protected fruit***, ***protected ornamentals***, ***protected vegetables***, ***vegetables*** [1]
- Whitefly in ***all edible crops (outdoor and protected)*** [2]; ***fruit crops***, ***protected fruit***, ***protected vegetables***, ***vegetables*** [1]; ***ornamental plant production***, ***protected ornamentals*** [1, 2]

Approval information

- Products included in this profile are not subject to the Control of Pesticides Regulations/Plant Protection Products Regulations because they act by physical means only

Efficacy guidance

- Products act by physical means following direct contact with spray. They may therefore be used at any time of year on all pest growth stages
- Ensure thorough spray coverage of plant, paying special attention to growing points and the underside of leaves
- Treat as soon as target pests are seen and repeat as often as necessary

Restrictions

- No limit on the number of treatments and no minimum interval between applications
- Before large scale use on a new crop, treat a few plants to check for crop safety
- Do not treat ornamental crops when in flower

Crop-specific information

- HI zero

Environmental safety

- May be used in conjunction with biological control agents. Spray 24 h before they are introduced

Hazard classification and safety precautions

UN number N/C
Operator protection F [3]; U05a, U09a, U19a, U20c [1]; U10, U14, U15 [2]
Storage and disposal D01, D02, D05, D06d [2]

SEE SECTION 3 FOR PRODUCTS ALSO REGISTERED

370 picloram

A persistent, translocated pyridine carboxylic acid herbicide for non-crop areas
HRAC mode of action code: O

See also *2,4-D + picloram*
clopyralid + picloram

Products

1	Pantheon 2	Pan Agriculture	240 g/l	SC	14052
2	Tordon 22K	Nomix Enviro	240 g/l	SL	05790

Uses

- Annual and perennial weeds in ***land not intended for cropping*** [1]
- Annual dicotyledons in ***land not intended for cropping*** [2]
- Bracken in ***land not intended for cropping*** [1, 2]
- Brambles in ***land not intended for cropping*** [1]
- Japanese knotweed in ***land not intended for cropping*** [2]
- Perennial dicotyledons in ***land not intended for cropping*** [2]
- Woody weeds in ***land not intended for cropping*** [2]

Approval information

- Picloram included in Annex I under EC Directive 91/414
- Approval expiry 31 Dec 2010 [2]

Efficacy guidance

- May be applied at any time of year. Best results achieved by application as foliage spray in late winter to early spring
- For bracken control apply 2-4 wk before frond emergence
- Clovers are highly sensitive and eliminated at very low doses
- Persists in soil for up to 2 yr

Restrictions

- Maximum number of treatments 1 per yr on land not intended for cropping
- Do not apply around desirable trees or shrubs where roots may absorb chemical
- Do not apply on slopes where chemical may be leached onto areas of desirable plants

Environmental safety

- Harmful to fish or other aquatic life. Do not contaminate surface waters or ditches with chemical or used container
- Keep livestock out of treated areas for at least 2 wk and until foliage of any poisonous weeds such as ragwort has died and become unpalatable

Hazard classification and safety precautions

Hazard Irritant
Transport code 9
Packaging group III
UN number 3082
Risk phrases R43
Operator protection A, H; U05a, U08, U14, U20b
Environmental protection E07a, E13c
Storage and disposal D01, D02, D09a [1, 2]; D10a [1]; D10b [2]

371 picolinafen

A pyridinecarboxamide herbicide for cereals;
HRAC mode of action code: F1

See also *pendimethalin + picolinafen*

Products

Vixen	BASF	75% w/w	WG	13621

Uses
- Chickweed in ***winter barley***, ***winter wheat***
- Cleavers in ***winter barley***, ***winter wheat***
- Field pansy in ***winter barley***, ***winter wheat***
- Field speedwell in ***winter barley***, ***winter wheat***
- Ivy-leaved speedwell in ***winter barley***, ***winter wheat***
- Shepherd's purse in ***winter barley***, ***winter wheat***

Approval information
- Picolinafen included in Annex I under EC Directive 91/414

Efficacy guidance
- Picolinafen is contact acting. Ensure good spray cover of the target weeds
- Best results obtained from treatment of small actively growing weeds
- Weeds emerging after application will not be controlled
- Always follow WRAG guidelines for preventing and managing herbicide resistant weeds. See Section 5 for more information

Restrictions
- Maximum number of treatments 1 per crop
- Do not disturb soil surface after application
- Do not roll emerged crops prior to application
- Do not treat undersown crops or those to be undersown
- Do not spray during periods of prolonged or severe frost
- Do not apply to crops under stress from any cause

Crop-specific information
- Latest use: before pseudo-stem erect stage (GS 30) for wheat and barley
- Treatment may cause transient bleaching in some crops

Following crops guidance
- No restrictions on following crops after normal harvest of a treated crop
- In the event of failure of a treated crop further crops of wheat or barley may be drilled after a minimum interval of 8 wk from treatment and after ploughing to at least 15 cm

Environmental safety
- Dangerous for the environment
- Very toxic to aquatic organisms
- LERAP Category B
- Take extreme care to avoid drift onto plants outside the target area

Hazard classification and safety precautions

Hazard Dangerous for the environment
Transport code 9
Packaging group III
UN number 3077
Risk phrases R50, R53a
Operator protection A, D; U05a, U08, U13, U19a, U20b
Environmental protection E15a, E16a, E38
Storage and disposal D01, D02, D09a, D10b, D12a

372 picoxystrobin

A broad spectrum strobilurin fungicide for cereals
FRAC mode of action code: 11

See also *chlorothalonil + picoxystrobin*
cyprodinil + picoxystrobin

Products

Galileo	DuPont	250 g/l	SC	13252

Uses
- Brown rust in ***spring barley***, ***spring wheat***, ***winter barley***, ***winter wheat***
- Crown rust in ***spring oats***, ***winter oats***

SEE SECTION 3 FOR PRODUCTS ALSO REGISTERED

- Eyespot in ***spring wheat*** *(reduction)*, ***winter wheat*** *(reduction)*
- Foliar disease control in ***ornamental plant production*** *(off-label)*
- Glume blotch in ***spring wheat***, ***winter wheat***
- Late ear diseases in ***spring wheat***, ***winter wheat***
- Net blotch in ***spring barley***, ***winter barley***
- Powdery mildew in ***spring barley***, ***spring oats***, ***winter barley***, ***winter oats***
- Rhynchosporium in ***spring barley***, ***winter barley***
- Septoria leaf blotch in ***spring wheat***, ***winter wheat***
- Yellow rust in ***spring wheat***, ***winter wheat***

Specific Off-Label Approvals (SOLAs)

- ***ornamental plant production*** *20082855*

Approval information

- Picoxystrobin included in Annex I under EC Directive 91/414
- Accepted by BBPA for use on malting barley

Efficacy guidance

- Best results achieved from protectant treatments made before disease establishes in the crop
- Use of tank mixtures is recommended where disease has become established at the time of treatment
- Persistence of yellow rust control is less than for other foliar diseases but can be improved by use of an appropriate tank mixture
- Eyespot control is not reliable and an appropriate tank mix should be used where crops are at risk
- Picoxystrobin is a member of the QoI cross resistance group. Product should be used preventatively and not relied on for its curative potential
- Use product as part of an Integrated Crop Management strategy incorporating other methods of control, including where appropriate other fungicides with a different mode of action. Do not apply more than two foliar applications of QoI containing products to any cereal crop
- There is a significant risk of widespread resistance occurring in *Septoria tritici* populations in UK. Failure to follow resistance management action may result in reduced levels of disease control
- On cereal crops product must always be used in mixture with another product, recommended for control of the same target disease, that contains a fungicide from a different cross resistance group and is applied at a dose that will give robust control
- Strains of barley powdery mildew resistant to QoIs are common in the UK

Restrictions

- Maximum number of treatments 2 per crop per yr

Crop-specific information

- Latest use: grain watery ripe (GS 71)

Environmental safety

- Dangerous for the environment
- Very toxic to aquatic organisms

Hazard classification and safety precautions

Hazard Dangerous for the environment
Transport code 9
Packaging group III
UN number 3082
Risk phrases R50, R53a
Operator protection U05a, U09a, U14, U15, U20a
Environmental protection E15a, E38
Storage and disposal D01, D02, D05, D09a, D10c, D12a

373 pinoxaden

A phenylpyrazoline grass weed herbicide for cereals
HRAC mode of action code: A

See also clodinafop-propargyl + pinoxaden

Products

1	Aoraki	AgChem Access	100 g/l	EC	13528
2	Axial	Syngenta	100 g/l	EC	12521
3	Clayton Tonto	Clayton	100 g/l	EC	14436
4	Greencrop Helvick	Greencrop	100 g/l	EC	13052
5	Quail	ChemSource	100 g/l	EC	14506

Uses

- Annual grasses in ***game cover*** *(off-label)* [2]
- Blackgrass in ***spring barley***, ***winter barley*** [1-3, 5]
- Italian ryegrass in ***spring barley***, ***winter barley***, ***winter wheat*** [1-5]; ***spring wheat*** [1, 2, 5]
- Perennial grasses in ***game cover*** *(off-label)* [2]
- Perennial ryegrass in ***spring barley***, ***winter barley***, ***winter wheat*** [1-5]; ***spring wheat*** [1, 2, 5]
- Wild oats in ***spring barley***, ***winter barley***, ***winter wheat*** [1-5]; ***spring wheat*** [1, 2, 5]

Specific Off-Label Approvals (SOLAs)

- ***game cover*** *20082815* [2]

Efficacy guidance

- Best results obtained from treatment when all grass weeds have emerged. There is no residual activity
- Broad-leaved weeds are not controlled
- Treat before emerged weed competition reduces yield
- Blackgrass in winter and spring barley is also controlled when used as part of an integrated control strategy. Product is not recommended for blackgrass control in winter wheat
- Grass weed control may be reduced if rain falls within 1 hr of application
- Pinoxaden is an ACCase inhibitor herbicide. To avoid the build up of resistance do not apply products containing an ACCase inhibitor herbicide more than twice to any crop. In addition do not use any product containing pinoxaden in mixture or sequence with any other product containing the same ingredient
- Use these products as part of a resistance management strategy that includes cultural methods of control and does not use ACCase inhibitors as the sole chemical method of grass weed control
- Applying a second product containing an ACCase inhibitor to a crop will increase the risk of resistance development; only use a second ACCase inhibitor to control different weeds at a different timing
- Always follow WRAG guidelines for preventing and managing herbicide resistant weeds. See Section 5 for more information

Restrictions

- Maximum number of treatments 1 per crop
- Do not spray crops under stress or suffering from waterlogging, pest attack, disease or frost damage
- Do not spray crops undersown with grass mixtures
- Avoid the use of hormone-containing herbicides in mixture or in sequence. Allow 21 d after, or 7 d before, hormone application
- Product must be used with prescribed adjuvant. See label

Crop-specific information

- Latest use: before flag leaf extending stage (GS 41)
- Spray cereals in autumn, winter or spring from the 2-leaf stage (GS 12)

Following crops guidance

- There are no restrictions on succeeding crops in a normal rotation
- In the event of failure of a treated crop ryegrass, maize, oats or any broad-leaved crop may be planted after a minimum interval of 4 wk from application

Environmental safety

- Dangerous for the environment

- Toxic to aquatic organisms
- Do not allow spray to drift onto neighbouring crops of oats, ryegrass or maize

Hazard classification and safety precautions

Hazard Irritant, Dangerous for the environment
Transport code 9
Packaging group III
UN number 3082
Risk phrases R36, R38, R51, R53a
Operator protection A, C, H; U05a, U09a, U20b
Environmental protection E15b, E38
Storage and disposal D01, D02, D05, D09a, D10c, D11a, D12a

374 pirimicarb

A carbamate insecticide for aphid control
IRAC mode of action code: 1A

See also *lambda-cyhalothrin + pirimicarb*

Products

1	Aphox	Syngenta	50% w/w	WG	10515
2	Arena	AgriGuard	50% w/w	WG	14028
3	Clayton Pirimicarb 50	Clayton	50% w/w	WG	12910
4	Landgold Pirimicarb 50	Goldengrass	50% w/w	WG	14022
5	Phantom	Syngenta	50% w/w	WG	11954
6	Pirimate	European Ag	50% w/w	WG	14539
7	Pirimicarb 50	Goldengrass	50% w/w	WG	14295
8	Reynard	ChemSource	50% w/w	WG	14523
9	Standon Pirimicarb 50	Standon	50% w/w	WG	13290

Uses

- Aphids in ***apples, blackcurrants, cherries, chinese cabbage, collards, cucumbers, durum wheat, forest nurseries, gooseberries, ornamental plant production, pears, peppers, protected lettuce, raspberries, redcurrants, spring rye, strawberries, tomatoes (outdoor), triticale, winter rye*** [1-5, 7-9]; ***bilberries*** *(off-label)*, ***blueberries*** *(off-label)*, ***chives*** *(off-label)*, ***cranberries*** *(off-label)*, ***evening primrose*** *(off-label)*, ***mallow (althaea spp.)*** *(off-label)*, ***mustard*** *(off-label)*, ***parsley*** *(off-label)*, ***protected chives*** *(off-label)*, ***protected herbs (see appendix 6)*** *(off-label)*, ***protected lamb's lettuce*** *(off-label)*, ***protected salad brassicas*** *(off-label - for baby leaf production)*, ***protected spinach*** *(off-label)*, ***spring linseed*** *(off-label)*, ***sweetcorn*** *(off-label)*, ***winter linseed*** *(off-label)* [1]; ***blackberries*** *(off-label)*, ***celeriac*** *(off-label)*, ***celery (outdoor)*** *(off-label)*, ***chicory*** *(off-label - for forcing)*, ***choi sum*** *(off-label)*, ***courgettes*** *(off-label)*, ***endives*** *(off-label)*, ***fennel*** *(off-label)*, ***fodder beet*** *(off-label)*, ***frise*** *(off-label)*, ***gherkins*** *(off-label)*, ***herbs (see appendix 6)*** *(off-label)*, ***honesty*** *(off-label)*, ***horseradish*** *(off-label)*, ***leaf spinach*** *(off-label)*, ***marrows*** *(off-label)*, ***pak choi*** *(off-label)*, ***parsley root*** *(off-label)*, ***poppies for morphine production*** *(off-label)*, ***protected aubergines*** *(off-label)*, ***protected cayenne peppers*** *(off-label)*, ***protected courgettes*** *(off-label)*, ***protected endives*** *(off-label)*, ***protected frise*** *(off-label)*, ***protected gherkins*** *(off-label)*, ***protected parsley*** *(off-label)*, ***protected radicchio*** *(off-label)*, ***protected radishes*** *(off-label)*, ***protected scarole*** *(off-label)*, ***protected spinach beet*** *(off-label)*, ***radicchio*** *(off-label)*, ***radishes*** *(off-label)*, ***red beet*** *(off-label)*, ***rubus hybrids*** *(off-label)*, ***salad brassicas*** *(off-label - for baby leaf production)*, ***scarole*** *(off-label)*, ***spinach beet*** *(off-label)* [1, 5]; ***broad beans, broccoli, brussels sprouts, cabbages, calabrese, carrots, cauliflowers, dwarf beans, kale, parsnips, peas, potatoes, spring barley, spring field beans, spring oats, spring oilseed rape, spring wheat, sugar beet, swedes, sweetcorn, turnips, winter barley, winter field beans, winter oats, winter oilseed rape, winter wheat*** [1-9]; ***forage maize*** [1-3, 5-9]; ***lamb's lettuce*** *(off-label)*, ***leaf brassicas*** *(off-label - baby leaf production only)*, ***plums*** *(off-label)*, ***protected horseradish*** *(off-label)*, ***protected leaf spinach*** *(off-label)*, ***rhubarb*** *(off-label)*, ***salsify*** *(off-label)* [5]; ***lettuce*** [1, 2, 4, 7-9]; ***lettuce*** *(outdoor crops)* [3, 5]; ***maize*** [4]; ***protected carnations, protected chrysanthemums, protected cinerarias, protected cyclamen, protected roses, protected tomatoes*** [1, 3, 5, 7-9]; ***protected ornamentals*** [2, 4]; ***runner beans*** [1-4, 6-9]
- Leaf curling plum aphid in ***plums*** *(off-label)* [1, 5]

- Mealy plum aphid in ***plums*** *(off-label)* [1, 5]
- Thrips in ***endives*** *(off-label)*, ***protected endives*** *(off-label)* [1, 5]; ***protected parsley*** *(off-label)* [5]

Specific Off-Label Approvals (SOLAs)

- ***bilberries*** *20063151* [1]
- ***blackberries*** *20063110* [1], *20081186* [5]
- ***blueberries*** *20063151* [1]
- ***celeriac*** *20011292* [1], *20051747* [5]
- ***celery (outdoor)*** *20040680* [1], *20051744* [5]
- ***chicory*** *(for forcing) 20011283* [1], *(for forcing) 20051749* [5]
- ***chives*** *20063150* [1]
- ***choi sum*** *20063109* [1], *20081190* [5]
- ***courgettes*** *20011298* [1], *20051746* [5]
- ***cranberries*** *20063151* [1]
- ***endives*** *20011286* [1], *20051741* [5]
- ***evening primrose*** *20063155* [1]
- ***fennel*** *20011292* [1], *20051747* [5]
- ***fodder beet*** *20063152* [1], *20081187* [5]
- ***frise*** *20063150* [1], *20081189* [5]
- ***gherkins*** *20011298* [1], *20051746* [5]
- ***herbs (see appendix 6)*** *20063150* [1], *20081189* [5]
- ***honesty*** *20063155* [1], *20051742* [5]
- ***horseradish*** *20063154* [1], *20081191* [5]
- ***lamb's lettuce*** *20081189* [5]
- ***leaf brassicas*** *(baby leaf production only) 20051743* [5]
- ***leaf spinach*** *20011298* [1], *20051746* [5]
- ***mallow (althaea spp.)*** *20063154* [1]
- ***marrows*** *20011298* [1], *20051746* [5]
- ***mustard*** *20063155* [1]
- ***pak choi*** *20063109* [1], *20081190* [5]
- ***parsley*** *20063150* [1]
- ***parsley root*** *20011298* [1], *20051746* [5]
- ***plums*** *20012687* [1], *20051738* [5], *20090903* [5]
- ***poppies for morphine production*** *20030767* [1], *20051748* [5]
- ***protected aubergines*** *20063153* [1], *20081188* [5]
- ***protected cayenne peppers*** *20063153* [1], *20081188* [5]
- ***protected chives*** *20063150* [1]
- ***protected courgettes*** *20011301* [1], *20051745* [5]
- ***protected endives*** *20011286* [1], *20051741* [5]
- ***protected frise*** *20063150* [1], *20081189* [5]
- ***protected gherkins*** *20011301* [1], *20051745* [5]
- ***protected herbs (see appendix 6)*** *20063150* [1]
- ***protected horseradish*** *20051746* [5]
- ***protected lamb's lettuce*** *20063150* [1]
- ***protected leaf spinach*** *20051746* [5]
- ***protected parsley*** *20063150* [1], *20051741* [5]
- ***protected radicchio*** *20063150* [1], *20081189* [5]
- ***protected radishes*** *20011298* [1], *20051746* [5]
- ***protected salad brassicas*** *(for baby leaf production) 20063150* [1]
- ***protected scarole*** *20063150* [1], *20081189* [5]
- ***protected spinach*** *20011298* [1]
- ***protected spinach beet*** *20011298* [1], *20051746* [5]
- ***radicchio*** *20063150* [1], *20081189* [5]
- ***radishes*** *20011298* [1], *20051746* [5]
- ***red beet*** *20011292* [1], *20051747* [5]
- ***rhubarb*** *20051744* [5]
- ***rubus hybrids*** *20063110* [1], *20081186* [5]
- ***salad brassicas*** *(for baby leaf production) 20011295* [1], *(for baby leaf production) 20081189* [5]
- ***salsify*** *20051747* [5]
- ***scarole*** *20063150* [1], *20081189* [5]

- ***spinach beet*** *20011298* [1], *20051746* [5]
- ***spring linseed*** *20063155* [1]
- ***sweetcorn*** *20011298* [1]
- ***winter linseed*** *20063155* [1]

Approval information

- Pirimicarb included in Annex I under EC Directive 91/414
- Approved for aerial application on cereals [1, 5]. See Section 5 for more information
- Accepted by BBPA for use on malting barley
- In 2006 PSD required that all products containing this active ingredient should carry the following warning in the main area of the container label: "Pirimicarb is an anticholinesterase carbamate. Handle with care"

Efficacy guidance

- Chemical has contact, fumigant and translaminar activity
- Best results achieved under warm, calm conditions when plants not wilting and spray does not dry too rapidly. Little vapour activity at temperatures below 15 °C
- Apply as soon as aphids seen or warning issued and repeat as necessary
- Addition of non-ionic wetter recommended for use on brassicas
- On cucumbers and tomatoes a root drench is preferable to spraying when using predators in an integrated control programme
- Where aphids resistant to pirimicarb occur control is unlikely to be satisfactory

Restrictions

- Contains an anticholinesterase carbamate compound. Do not use if under medical advice not to work with such compounds
- Maximum number of treatments not specified in some cases but normally 2-6 depending on crop
- When treating ornamentals check safety by treating a small number of plants first
- Spray equipment must only be used where the operator's normal working position is within a closed cab on a tractor or self-propelled sprayer when making air-assisted applications to apples or other top fruit

Crop-specific information

- Latest use in accordance with harvest intervals below
- HI oilseed rape, cereals, maize, sweetcorn 14 d (sweetcorn off-label 3 d); grassland, chicory, plums, swedes and turnips 7 d; lettuce 3 d, cucumbers, tomatoes and peppers under glass 2 d; protected courgettes and gherkins 24 h; other edible crops 3 d; flowers and ornamentals zero

Environmental safety

- Dangerous for the environment
- Very toxic to aquatic organisms
- Dangerous to fish or other aquatic life. Do not contaminate surface waters or ditches with chemical or used container
- Chemical has little effect on bees, ladybirds and other insects and is suitable for use in integrated control programmes on apples and pears
- Keep all livestock out of treated areas for at least 7 d. Bury or remove spillages

Hazard classification and safety precautions

Hazard Toxic, Dangerous for the environment
Transport code 6.1
Packaging group III
UN number 2757
Risk phrases R20, R25, R36, R50, R53a
Operator protection A, C, D, H, M [1-9]; J [1, 3-9]; U05a, U08, U19a [1-9]; U20a [3, 5]; U20b [1, 2, 4, 6-9]
Environmental protection E06c [1-9] (7 d); E15a [3, 4]; E15b [1, 2, 5-9]; E34 [1-9]; E38 [1, 5-9]
Storage and disposal D01, D02, D09a, D11a [1-9]; D05 [2]; D07 [3]; D12a [1, 3, 5-9]
Medical advice M02, M04a [1, 3-9]; M03 [1, 2, 5-9]; M04b [3]

375 pirimiphos-methyl

A contact, fumigant and translaminar organophosphorus insecticide
IRAC mode of action code: 1B

Products

1	Actellic 50 EC	Syngenta	500 g/l	EC	12726
2	Actellic Smoke Generator No 20	Syngenta	22.5% w/w	FU	10540
3	Flycatcher	AgChem Access	500 g/l	EC	14135
4	Pan PMT	Pan Agriculture	500 g/l	EC	13996

Uses

- Flour beetle in ***stored grain*** [1, 3, 4]
- Flour moth in ***stored grain*** [1, 3, 4]
- Grain beetle in ***stored grain*** [1, 3, 4]
- Grain storage mite in ***stored grain*** [1, 3, 4]
- Grain storage pests in ***grain stores*** [1-4]
- Grain weevil in ***stored grain*** [1, 3, 4]
- Warehouse moth in ***stored grain*** [1, 3, 4]

Approval information

- Pirimiphos-methyl included in Annex I under EC Directive 91/414
- In 2006 PSD required that all products containing this active ingredient should carry the following warning in the main area of the container label: "Pirimiphos-methyl is an anticholinesterase organophosphate. Handle with care"
- Accepted by BBPA for use in stores for malting barley

Efficacy guidance

- Chemical acts rapidly and has short persistence in plants, but persists for long periods on inert surfaces
- Best results for protection of stored grain achieved by cleaning store thoroughly before use and employing a combination of pre-harvest and grain/seed treatments [1, 2]
- Best results for admixture treatment obtained when grain stored at 15% moisture or less. Dry and cool moist grain coming into store but then treat as soon as possible, ideally as it is loaded [1]
- Surface admixture can be highly effective on localised surface infestations but should not be relied on for long-term control unless application can be made to the full depth of the infestation [1]
- Where insect pests resistant to pirimiphos-methyl occur control is unlikely to be satisfactory

Restrictions

- Contains an organophosphorus anticholinesterase compound. Do not use if under medical advice not to work with such compounds
- Maximum number of treatments 2 per grain store [1, 2]; 1 per batch of stored grain [1]
- Do not apply surface admixture or complete admixture using hand held equipment [1]
- Surface admixture recommended only where the conditions for store preparation, treatment and storage detailed in the label can be met [1]

Crop-specific information

- Latest use: before storing grain [1, 2]
- Disinfect empty grain stores by spraying surfaces and/or fumigation and treat grain by full or surface admixture. Treat well before harvest in late spring or early summer and repeat 6 wk later or just before harvest if heavily infested. See label for details of treatment and suitable application machinery
- Treatment volumes on structural surfaces should be adjusted according to surface porosity. See label for guidelines
- Treat inaccessible areas with smoke generating product used in conjunction with spray treatment of remainder of store
- Predatory mites (*Cheyletus*) found in stored grain feeding on infestations of grain mites may survive treatment and can lead to rejection by buyers. They do not remain for long but grain should be inspected carefully before selling [1, 2]

Environmental safety

- Dangerous for the environment
- Very toxic to aquatic organisms [1]

SEE SECTION 3 FOR PRODUCTS ALSO REGISTERED

- Toxic to aquatic organisms [2]
- Highly flammable [2]
- Flammable [1]
- Ventilate fumigated or fogged spaces thoroughly before re-entry
- Unprotected persons must be kept out of fumigated areas within 3 h of ignition and for 4 h after treatment
- Wildlife must be excluded from buildings during treatment [1]
- Keep away from combustible materials [2]

Hazard classification and safety precautions

Hazard Harmful, Dangerous for the environment [1-4]; Highly flammable [2]; Flammable [1, 3, 4]
Transport code 3 [1, 3, 4]; 9 [2]
Packaging group III
UN number 1993 [1, 3, 4]; 3077 [2]
Risk phrases R20, R53a [1-4]; R22a, R22b, R36, R37, R50 [1, 3, 4]; R51 [2]
Operator protection A, C, J, M [1, 3, 4]; D, H; U04a, U09a, U16a, U24 [1, 3, 4]; U05a, U19a, U20b [1-4]; U14 [2]
Environmental protection E02a [2] (4 h); E15a, E34, E38 [1-4]
Consumer protection C09 [1, 3, 4]; C12 [1-4]
Storage and disposal D01, D02, D09a, D12a [1-4]; D05, D10c [1, 3, 4]; D07, D11a [2]
Treated seed S06a [1, 3, 4]
Medical advice M01 [1-4]; M03 [2]; M05b [1, 3, 4]

376 potassium bicarbonate (commodity substance)

An inorganic fungicide for use in horticulture

Products

potassium bicarbonate	various	>99% w/w	SL	-

Uses

- Fungus diseases in ***outdoor crops***, ***protected crops***

Approval information

- Approval for the use of potassium bicarbonate as a commodity substance was granted on 26 July 2005 by Ministers under regulation 5 of the Control of Pesticides Regulations 1986
- Potassium bicarbonate is being supported for review in the fourth stage of the EC Review Programme under Directive 91/414. Whether or not it is included in Annex I, the existing commodity chemical approval will have to be revoked because substances listed in Annex I must be approved for marketing and use under the EC regime. If potassium bicarbonate is included in Annex I, products containing it will need to gain approval in the normal way if they carry label claims for pesticidal activity
- Approval expiry 31 Dec 2010 [1]

Efficacy guidance

- Adjuvants authorised by PSD may be used in conjunction with potassium bicarbonate
- Crop phytotoxicity can occur

Restrictions

- Food grade potassium bicarbonate must only be used

Hazard classification and safety precautions

Operator protection A, C, G, H

377 prochloraz

A broad-spectrum protectant and eradicant imidazole fungicide
FRAC mode of action code: 3

See also *epoxiconazole + prochloraz*
fluquinconazole + prochloraz

Products

1	Mirage 40 EC	Makhteshim	400 g/l	EC	06770
2	Panache 40	DAPT	400 g/l	EC	10632
3	Poraz	BASF	450 g/l	EC	11701
4	Prelude 20LF	Agrichem	200 g/l	LS	04371
5	Prospero	Makhteshim	400 g/l	EC	11931
6	Scarlett	AgriGuard	400 g/l	EC	14271
7	Scotts Octave	Scotts	46% w/w	WP	09275
8	Sprint	AgriGuard	400 g/l	EC	14272
9	Warbler	Nufarm UK	400 g/l	EC	13389

Uses

- Alternaria in ***spring oilseed rape*** [1, 5, 6, 8, 9]; ***winter oilseed rape*** [1, 3, 5, 6, 8, 9]
- Disease control in ***grass seed crops*** *(off-label)* [5]; ***herbs (see appendix 6)*** *(off-label)*, ***protected herbs (see appendix 6)*** *(off-label)* [7]; ***winter linseed*** *(off-label)* [1]
- Eyespot in ***spring wheat*** [1, 2, 5, 6, 8]; ***winter barley***, ***winter wheat*** [1-3, 5, 6, 8, 9]; ***winter rye*** [1, 3, 5, 6, 8, 9]
- Foliar disease control in ***triticale*** *(off-label)* [1]
- Fungus diseases in ***hardy ornamentals***, ***ornamental plant production***, ***woody ornamentals*** [7]
- Glume blotch in ***spring wheat*** [1, 5, 6, 8]; ***winter wheat*** [1, 3, 5, 6, 8, 9]
- Grey mould in ***spring oilseed rape*** [1, 5, 6, 8, 9]; ***winter oilseed rape*** [1, 3, 5, 6, 8, 9]
- Light leaf spot in ***spring oilseed rape*** [1, 2, 5, 6, 8, 9]; ***winter oilseed rape*** [1-3, 5, 6, 8, 9]
- Net blotch in ***spring barley***, ***winter barley*** [1, 3, 5, 6, 8, 9]
- Phoma in ***spring oilseed rape*** [1, 2, 5, 6, 8, 9]; ***winter oilseed rape*** [1-3, 5, 6, 8, 9]
- Powdery mildew in ***spring barley***, ***spring wheat***, ***winter barley***, ***winter wheat*** [3, 9]; ***winter rye*** [1, 3, 5, 6, 8, 9]
- Rhynchosporium in ***spring barley***, ***winter barley***, ***winter rye*** [1, 3, 5, 6, 8, 9]
- Sclerotinia stem rot in ***spring oilseed rape*** [1, 5, 6, 8, 9]; ***winter oilseed rape*** [1, 3, 5, 6, 8, 9]
- Seed-borne diseases in ***flax*** *(seed treatment)*, ***hemp for oilseed production*** *(off-label - seed treatment)*, ***linseed*** *(seed treatment)* [4]
- Septoria leaf blotch in ***spring wheat*** [1, 5, 6, 8]; ***winter rye*** [3, 9]; ***winter wheat*** [1, 3, 5, 6, 8, 9]
- White leaf spot in ***spring oilseed rape*** [1, 5, 6, 8, 9]; ***winter oilseed rape*** [1, 3, 5, 6, 8, 9]

Specific Off-Label Approvals (SOLAs)

- ***grass seed crops*** *20062920* [5]
- ***hemp for oilseed production*** *(seed treatment) 20052562* [4]
- ***herbs (see appendix 6)*** *20010650* [7]
- ***protected herbs (see appendix 6)*** *20010650* [7]
- ***triticale*** *20062686* [1]
- ***winter linseed*** *20062687* [1]

Approval information

- Prochloraz not included in Annex 1 under EC Directive 91/414
- Accepted by BBPA for use on malting barley

Efficacy guidance

- Spray cereals at first signs of disease. Protection of winter crops through season usually requires at least 2 treatments. See label for details of rates and timing. Treatment active against strains of eyespot resistant to benzimidazole fungicides [1-3, 5, 9]
- Tank mixes with other fungicides recommended to improve control of rusts in wheat and barley. See label for details [1-3, 5, 9]
- A period of at least 3 h without rain should follow spraying [1-3, 5, 9]
- Can be used through most seed treatment machines if good even seed coverage is obtained. Check drill calibration before drilling treated seed. Use inert seed flow agent supplied by manufacturer [4]

- Use methylated spirits, rather than water, to clean residual material from seed treatment machinery, then carry out final rinse with water and detergent [4]
- Apply as drench against soil diseases, as a spray against aerial diseases, as a dip for cuttings, or as a drench at propagation. Spray applications may be repeated at 10-14 d intervals. Under mist propagation use 7 d intervals [7]
- Prochloraz is a DMI fungicide. Resistance to some DMI fungicides has been identified in Septoria leaf blotch which may seriously affect performance of some products. For further advice contact a specialist advisor and visit the Fungicide Resistance Action Group (FRAG)-UK website

Restrictions

- Maximum number of treatments 1 per batch of seed for flax, linseed [4]; 2 per crop for foliar spray uses
- Do not treat linseed varieties Linda, Bolas, Karen, Laura, Mikael, Norlin, Moonraker, Abbey, Agriace [4]
- Do not use treated seed as food or feed [4]

Crop-specific information

- Latest use: before drilling for seed treatments on flax or linseed [4]; watery ripe to early milk stage (GS 71-73) for cereals
- HI 6 wk for cereals
- Application with other fungicides may cause cereal crop scorch [1-3, 5, 9]

Environmental safety

- Dangerous for the environment
- Very toxic to aquatic organisms
- Treated seed harmful to game and wildlife. Product contains red dye for easy identification of treated seed [4]
- Flammable [2, 9]

Hazard classification and safety precautions

Hazard Harmful [3, 4, 8, 9]; Irritant [1, 2, 5-7]; Flammable [2, 8, 9]; Dangerous for the environment [1, 3-9]

Transport code 9

Packaging group III

UN number 3077 [7]; 3082 [1-6, 8, 9]

Risk phrases R22a [3, 4, 8, 9], R36 [1-8], R38 [1-3, 5-9]; R41 [9]; R43 [2]; R50 [3, 7-9]; R51 [1, 4-6]; R53a [1, 3-9]; R67 [8, 9]

Operator protection A [1-9]; C [1-4, 6-8]; U05a, U20b [1-9]; U08 [1-6, 8, 9]; U09a, U19a [7]; U11 [1, 4-6, 8, 9]; U14 [1, 3, 5, 6, 8]; U15 [1, 5, 6, 8]

Environmental protection E13b [2]; E15a [1, 3-9]; E34 [3, 9]; E38 [3, 4, 9]

Storage and disposal D01, D09a [1-9]; D02, D10b [1-3, 5-9]; D05 [1, 2, 4-6, 8, 9]; D10a [4]; D12a [3, 4, 7, 9]

Treated seed S02, S03, S04b, S05, S06a, S07 [4]

Medical advice M03, M05a [3]; M05b [1, 5, 6, 8]

378 prochloraz + propiconazole

A broad spectrum fungicide mixture for wheat, barley and oilseed rape
FRAC mode of action code: 3 + 3

See also *propiconazole*

Products

1	Bumper Excell	Makhteshim	400:90 g/l	EC	11015
2	Bumper P	Makhteshim	400:90 g/l	EC	08548
3	Greencrop Twinstar	Greencrop	400:90 g/l	EC	09516
4	MAC-Prochloraz Plus 490 EC	AgChem Access	400:90 g/l	EC	13616

Uses

- Disease control in ***grass seed crops*** *(off-label)* [1]
- Eyespot in ***spring barley***, ***spring wheat*** [1, 2, 4]; ***winter barley***, ***winter wheat*** [1-4]
- Light leaf spot in ***spring oilseed rape***, ***winter oilseed rape*** [1, 2, 4]
- Phoma in ***spring oilseed rape***, ***winter oilseed rape*** [1, 2, 4]

- Rhynchosporium in ***spring barley*** [1, 2, 4]; ***winter barley*** [1-4]
- Septoria leaf blotch in ***spring wheat*** [1, 2, 4]; ***winter wheat*** [1-4]

Specific Off-Label Approvals (SOLAs)

- ***grass seed crops*** *20052289* [1]

Approval information

- Propiconazole included in Annex I under EC Directive 91/414
- Prochloraz has not been included in Annex 1 but the withdrawal date has yet to be set. Most current approvals set to expire 2013.
- Accepted by BBPA for use on malting barley

Efficacy guidance

- Best results obtained from treatment when disease is active but not well established
- Treat wheat normally from flag leaf ligule just visible stage (GS 39) but earlier if there is a high risk of Septoria
- Treat barley from the first node detectable stage (GS 31)
- If disease pressure persists a second application may be necessary
- Prochloraz and propiconazole are DMI fungicides. Resistance to some DMI fungicides has been identified in Septoria leaf blotch which may seriously affect performance of some products. For further advice contact a specialist advisor and visit the Fungicide Resistance Action Group (FRAG)-UK website

Restrictions

- Maximum number of treatments 2 per crop

Crop-specific information

- Latest use: before grain watery ripe (GS 71) for cereals; before most seeds green for winter oilseed rape

Environmental safety

- Dangerous for the environment
- Very toxic to aquatic organisms

Hazard classification and safety precautions

Hazard Irritant, Dangerous for the environment
Transport code 9
Packaging group III
UN number 3082
Risk phrases R36, R53a [1-4]; R50 [3]; R51 [1, 2, 4]
Operator protection A, C; U05a, U08, U20b [1-4]; U11, U15 [1, 2, 4]
Environmental protection E15a
Storage and disposal D01, D02, D05, D09a, D10b
Medical advice M05a [1, 2, 4]

379 prochloraz + proquinazid + tebuconazole

A broad spectrum systemic fungicide mixture for cereals
FRAC mode of action code: 3 + U7 + 3

See also proquinazid
tebuconazole

Products

Vareon	DuPont	320:40:160 g/l	EC	14376

Uses

- Brown rust in ***spring barley***, ***spring wheat***, ***winter barley***, ***winter wheat***
- Crown rust in ***spring oats*** *(qualified minor use recommendation)*, ***winter oats*** *(qualified minor use recommendation)*
- Ear diseases in ***spring wheat*** *(reduction only)*, ***winter wheat*** *(reduction only)*
- Eyespot in ***spring barley***, ***spring wheat***, ***winter barley***, ***winter wheat***
- Glume blotch in ***spring wheat***, ***winter wheat***
- Net blotch in ***spring barley*** *(moderate control)*, ***winter barley*** *(moderate control)*

- Powdery mildew in ***spring barley, spring oats, spring wheat, winter barley, winter oats, winter wheat***
- Rhynchosporium in ***spring barley*** *(Moderate control)*, ***winter barley*** *(Moderate control)*
- Septoria leaf blotch in ***spring wheat*** *(Moderate control)*, ***winter wheat*** *(Moderate control)*
- Yellow rust in ***spring barley, spring wheat, winter barley, winter wheat***

Approval information

- Prochloraz has not been included in Annex 1 but the withdrawal date has yet to be set. Most current approvals set to expire 2013.

Efficacy guidance

- Resistance to DMI fungicides has been identified in Septoria leaf blotch and may seriously affect the performance of some products. Advice on resistance management is available from the Fungicide Action Grup (FRAG) web site

Environmental safety

- Avoid spraying within 5m of the field boundary to reduce the effect on non-target insects or other arthropods

Hazard classification and safety precautions

Hazard Harmful, Dangerous for the environment
Transport code 9
Packaging group III
UN number 3082
Risk phrases R22a, R36, R40, R43, R50, R53a, R63
Operator protection A, C, H
Environmental protection E15a, E16a, E22c, E34, E38
Storage and disposal D01, D02, D09a, D12a

380 prochloraz + tebuconazole

A broad spectrum systemic fungicide mixture for cereals, oilseed rape and amenity use
FRAC mode of action code: 3 + 3

See also tebuconazole

Products

1	Agate	Makhteshim	267:133 g/l	EC	13024
2	Agate	Makhteshim	267:133 g/l	EC	14482
3	Astute	Sherriff Amenity	267:133 g/l	EC	13825
4	Grail	Nufarm UK	300:200 g/l	EC	13385
5	Lunar	Bayer Environ.	267:132 g/l	EC	12366
6	Monkey	Makhteshim	267:133 g/l	EC	12906
7	Pan Proteb	Pan Agriculture	267:132 g/l	EC	13427
8	Throttle	Headland Amenity	267:133 g/l	EC	13900

Uses

- Brown rust in ***spring barley, spring rye, spring wheat, winter barley, winter rye, winter wheat*** [1, 2, 6]
- Eyespot in ***spring barley, spring rye, spring wheat, winter barley, winter rye, winter wheat*** [1, 2, 6]
- Fusarium patch in ***managed amenity turf*** [3, 5, 7, 8]
- Glume blotch in ***spring wheat, winter wheat*** [1, 2, 4, 6]
- Late ear diseases in ***spring wheat, winter wheat*** [1, 2, 6]
- Net blotch in ***spring barley, winter barley*** [1, 2, 6]
- Powdery mildew in ***spring barley, spring rye, spring wheat, winter barley, winter rye, winter wheat*** [1, 2, 6]
- Rhynchosporium in ***spring barley, spring rye, winter barley, winter rye*** [1, 2, 6]
- Sclerotinia stem rot in ***spring oilseed rape, winter oilseed rape*** [1, 2, 6]
- Septoria leaf blotch in ***spring wheat, winter wheat*** [1, 2, 4, 6]
- Yellow rust in ***spring barley, spring rye, spring wheat, winter barley, winter rye, winter wheat*** [1, 2, 6]

Approval information

- Accepted by BBPA for use on malting barley

- Prochloraz has not been included in Annex 1 but the withdrawal date has yet to be set. Most current approvals set to expire 2013.

Efficacy guidance

- Complete protection of winter cereals will usually require a programme of at least two treatments [1, 4, 6]
- Protection of flag leaf and ear from Septoria diseases requires treatment at flag leaf emergence and repeated when ear fully emerged [1, 4, 6]
- Two treatments separated by a 2 wk gap may be needed for difficult Fusarium patch attacks [5]
- To avoid resistance, do not apply repeated applications of the single product on the same crop against the same disease [8]
- Optimum application timing is normally when disease first seen but varies with main target disease - see labels
- Prochloraz and tebuconazole are DMI fungicides. Resistance to some DMI fungicides has been identified in Septoria leaf blotch which may seriously affect performance of some products. For further advice contact a specialist advisor and visit the Fungicide Resistance Action Group (FRAG)-UK website

Restrictions

- Maximum total dose on cereals equivalent to two full dose treatments
- Maximum number of treatments on managed amenity turf 2 per yr
- Do not apply during drought or to frozen turf [5]

Crop-specific information

- Latest use: before beginning of anthesis (GS 61) for cereals [4]; before grain milky ripe (GS 73) for cereals [1, 6]; most seeds green for oilseed rape [1, 6]
- HI 6 wk for cereals [1, 6]
- Occasionally transient leaf speckling may occur after treating wheat. Yield responses should not be affected [1]

Environmental safety

- Dangerous for the environment
- Very toxic to aquatic organisms

Hazard classification and safety precautions

Hazard Harmful, Dangerous for the environment
Transport code 6.1 [3, 7]; 9 [1, 2, 4-6, 8]
Packaging group III
UN number 2902 [3, 7]; 3082 [1, 2, 4-6, 8]
Risk phrases R22a, R36, R50 [1-3, 5-8]; R37, R51 [4]; R43, R53a [1-8]; R63 [4, 5, 7, 8]
Operator protection A, H [1-8]; C [1-3, 5-8]; M [5, 7, 8]; U05a [1-8]; U08, U20b [8]; U09b [1-3, 6]; U11 [4, 5, 7, 8]; U14 [1-7]; U19a [4]; U20a [1-4, 6]
Environmental protection E13b [1-3, 6]; E15a [4, 5, 7, 8]; E34, E38 [1-8]
Storage and disposal D01 [1-3, 5-8]; D02, D09a, D12a [1-8]; D05 [1-7]; D10b [4, 5, 7, 8]; D10c [1-3, 6]
Medical advice M03 [1-8]; M05a [5, 7]

381 prochloraz + thiram

A seed treatment fungicide mixture for oilseed rape
FRAC mode of action code: 3 + M3

See also *thiram*

Products

Agrichem Hy-Pro Duet	Agrichem	150:333 g/l	FS	14018

Uses

- Phoma in ***winter oilseed rape*** *(seed treatment)*

Approval information

- Thiram included in Annex I under EC Directive 91/414
- Prochloraz has not been included in Annex 1 but the withdrawal date has yet to be set. Most current approvals set to expire 2013.

SEE SECTION 3 FOR PRODUCTS ALSO REGISTERED

SECTION 2

Efficacy guidance

- Can be applied through most types of seed treatment machinery
- Good coverage of the seed surface essential for best results
- Product is active against seed-borne disease. It will not control leaf spotting from leaf spores on trash on the seed bed

Restrictions

- Maximum number of treatments 1 per crop
- Seed must be of satisfactory quality and moisture content
- Sow seed as soon as possible after treatment
- Do not store treated seed from one season to the next
- Do not co-apply product with water or mix with any other seed treatment product

Crop-specific information

- Latest use: pre-sowing
- Product will also control soil-borne fungi that cause damping-off of rape seedlings

Following crops guidance

- Any crop may be grown following a treated winter oilseed rape crop

Hazard classification and safety precautions

Hazard Harmful, Dangerous for the environment
Transport code 9
Packaging group III
UN number 3082
Risk phrases R20, R22a, R36, R38, R43, R50, R53a
Operator protection A, C, H; U02a, U08, U14, U15, U20a
Environmental protection E15a, E36b
Storage and disposal D01, D02, D05, D08, D09a, D12b, D13, D15
Treated seed S01, S02, S03, S04d, S05, S08
Medical advice M04a

382 prochloraz + triticonazole

A broad spectrum fungicide seed treatment mixture for winter cereals
FRAC mode of action code: 3 + 3

See also triticonazole

Products

Kinto	BASF	60:20 g/l	FS	12038

Uses

- Bunt in ***durum wheat*** *(off-label - seed treatment)*, ***winter wheat*** *(seed treatment)*
- Covered smut in ***winter barley*** *(seed treatment)*, ***winter oats*** *(seed treatment)*
- Fusarium foot rot and seedling blight in ***durum wheat*** *(off-label - seed treatment)*, ***triticale*** *(seed treatment)*, ***winter barley*** *(seed treatment)*, ***winter oats*** *(seed treatment)*, ***winter rye*** *(seed treatment)*, ***winter wheat*** *(seed treatment)*
- Leaf stripe in ***winter barley*** *(seed treatment)*
- Loose smut in ***durum wheat*** *(off-label - seed treatment)*, ***winter barley*** *(seed treatment)*, ***winter oats*** *(seed treatment)*, ***winter wheat*** *(seed treatment)*
- Septoria seedling blight in ***durum wheat*** *(off-label - seed treatment)*, ***winter wheat*** *(seed treatment)*

Specific Off-Label Approvals (SOLAs)

- ***durum wheat*** *(seed treatment) 20061094*

Approval information

- Triticonazole included in Annex I under EC Directive 91/414
- Prochloraz has not been included in Annex 1 but the withdrawal date has yet to be set. Most current approvals set to expire 2013.
- Accepted by BBPA for use on malting barley

Efficacy guidance

- Apply undiluted through conventional seed treatment machine

- Calibrate seed drill with treated seed before sowing
- Bag treated seed immediately and keep in a dry, draught free store
- Ensure good even coverage of seed to achieve best results
- Treatment may cause some slight delay in crop emergence

Restrictions

- Maximum number of treatments 1 per batch of seed
- Do not treat seed with moisture content above 16% and do not allow moisture content of treated seed to exceed 16%
- Do not treat cracked, split or sprouted seed
- Do not handle treated seed unnecessarily
- Do not use treated seed as food or feed

Crop-specific information

- Latest use: before drilling wheat, barley, oats, rye or triticale

Environmental safety

- Dangerous for the environment
- Toxic to aquatic organisms

Hazard classification and safety precautions

Hazard Dangerous for the environment
Transport code 9
Packaging group III
UN number 3082
Risk phrases R51, R53a
Operator protection A, H; U05a, U07, U20b
Environmental protection E15a, E34, E38
Storage and disposal D01, D02, D08, D09a, D11a, D12a
Treated seed S01, S02, S03, S04a, S04b, S05, S07

383 prohexadione-calcium

A cyclohexanecarboxylate growth regulator for use in apples

See also mepiquat chloride + prohexadione-calcium

Products

Regalis	BASF	10% w/w	WG	12414

Uses

- Control of shoot growth in ***apples***
- Growth regulation in ***hops*** *(off-label)*, ***ornamental plant production*** *(off-label)*, ***soft fruit*** *(off-label)*

Specific Off-Label Approvals (SOLAs)

- ***hops*** *20082866 expires 01 Oct 2010*
- ***ornamental plant production*** *20082866 expires 01 Oct 2010*
- ***soft fruit*** *20082866 expires 01 Oct 2010*

Approval information

- Prohexadione-calcium included in Annex I under EC Directive 91/414

Efficacy guidance

- For a standard orchard treat at the start of active growth and again after 3-5 wk depending on growth conditions
- Alternatively follow the first application with four reduced rate applications at two wk intervals

Restrictions

- Maximum total dose equivalent to two full dose treatments
- Do not apply in conjunction with calcium based foliar fertilisers

Crop-specific information

- HI 55 d for apples

Environmental safety

- Harmful to aquatic organisms

- Avoid spray drift onto neighbouring crops and other non-target plants
- Product does not harm natural insect predators when used as directed

Hazard classification and safety precautions

UN number N/C
Risk phrases R52, R53a
Operator protection U05a
Environmental protection E38
Storage and disposal D01, D02, D09a, D10c, D12a

384 propachlor

A pre-emergence chloroacetanilide herbicide for various horticultural crops
HRAC mode of action code: K3

See also *chlorthal-dimethyl + propachlor*

Products

Titan 480	AgriGuard	480 g/l	SC	12596

Uses

- Annual dicotyledons in ***broccoli*, *brown mustard*, *brussels sprouts*, *bulb onions*, *cabbages*, *calabrese*, *cauliflowers*, *fodder rape*, *kale*, *leeks*, *ornamental plant production*, *sage*, *spring oilseed rape*, *swedes*, *turnips*, *white mustard***
- Annual grasses in ***broccoli*, *brown mustard*, *brussels sprouts*, *bulb onions*, *cabbages*, *calabrese*, *cauliflowers*, *fodder rape*, *kale*, *leeks*, *ornamental plant production*, *sage*, *spring oilseed rape*, *swedes*, *turnips*, *white mustard***

Approval information

- Propachlor has not been included in Annex 1 and the final use date for products containing propachlor is 18 March 2010
- Approval expiry 18 Mar 2010

Efficacy guidance

- Controls germinating (not emerged) weeds for 6-8 wk
- Best results achieved by application to fine, firm, moist seedbed free of established weeds in spring, summer and early autumn
- Use higher rate on soils with more than 10% organic matter
- Recommended as tank-mix with glyphosate or chlorthal-dimethyl on onions, leeks, swedes, turnips. See label for details

Restrictions

- Maximum number of treatments 1 or 2 per crop, varies with product and crop - see label for details
- Do not use after products that de-wax the leaves of plants, especially brassica crops
- Do not use on crops under glass or polythene
- Do not use under extremely wet, dry or other adverse growth conditions

Crop-specific information

- Latest use: pre-emergence
- Apply in brassicas from drilling to time seed chits or after 3-4 true leaf stage but before weed emergence, in swedes and turnips pre-emergence only
- Apply in transplanted brassicas within 48 h of planting in warm weather. Plants must be hardened off and special care needed with block sown or modular propagated plants
- Apply in onions and leeks pre-emergence or from post-crook to young plant stage
- Apply to newly planted strawberries soon after transplanting, to weed-free soil in established crops in early spring before new weeds emerge
- Granules may be applied to onion, leek and brassica nurseries and to most flower crops after bedding out and hardening off
- Do not use pre-emergence of drilled wallflower seed

Following crops guidance

- In the event of crop failure only replant recommended crops in treated soil

Environmental safety

- Dangerous for the environment
- Very toxic to aquatic organisms

Hazard classification and safety precautions

Hazard Harmful
Transport code 9
Packaging group III
UN number 3082
Risk phrases R22a, R38, R43
Operator protection A, C; U02a, U04a, U05a, U08, U13, U19a, U20b
Environmental protection E15a, E34
Storage and disposal D01, D02, D05, D09a, D10b, D12a
Medical advice M03

385 propamocarb hydrochloride

A translocated protectant carbamate fungicide
FRAC mode of action code: 28

See also *chlorothalonil + propamocarb hydrochloride*
fenamidone + propamocarb hydrochloride
fluopicolide + propamocarb hydrochloride
fosetyl-aluminium + propamocarb hydrochloride

Products

1	Filex	Scotts	722 g/l	SL	07631
2	Flash	AgriGuard	722 g/l	SL	11989
3	Pan PCH	Pan Agriculture	722 g/l	SL	12842
4	Proplant	Fargro	722 g/l	SL	13359

Uses

- Blight in ***protected tomatoes***, ***tomatoes (outdoor)*** [2]
- Botrytis in ***cayenne pepper*** *(off-label)*, ***frise*** *(off-label)*, ***horseradish*** *(off-label)*, ***leaf brassicas*** *(off-label)*, ***protected horseradish*** *(off-label)*, ***protected radishes*** *(off-label)*, ***radishes*** *(off-label)* [4]; ***chives*** *(off-label)*, ***cress*** *(off-label)*, ***parsley*** *(off-label)*, ***protected herbs (see appendix 6)*** *(off-label)* [1]; ***herbs (see appendix 6)*** *(off-label)* [1, 4]; ***lettuce***, ***protected lettuce*** [3]
- Damping off in ***cayenne pepper*** *(off-label)*, ***frise*** *(off-label)*, ***herbs (see appendix 6)*** *(off-label)*, ***horseradish*** *(off-label)*, ***leaf brassicas*** *(off-label)*, ***protected cucumbers*** *(off-label)*, ***protected horseradish*** *(off-label)*, ***protected radishes*** *(off-label)*, ***protected watercress*** *(off-label)*, ***radishes*** *(off-label)* [4]; ***inert substrate cucumbers*** *(off-label)*, ***inert substrate peppers*** *(off-label)*, ***inert substrate tomatoes*** *(off-label)*, ***nft peppers*** *(off-label)*, ***nft tomatoes*** *(off-label)* [1]; ***protected peppers*** *(off-label)*, ***protected tomatoes*** *(off-label)* [1, 4]
- Damping off and foot rot in ***lamb's lettuce*** *(off-label)*, ***lettuce*** *(off-label)*, ***protected herbs (see appendix 6)*** *(off-label)*, ***protected lamb's lettuce*** *(off-label)*, ***protected lettuce*** *(off-label)*, ***protected scarole*** *(off-label)*, ***protected tomatoes*** *(off-label)*, ***scarole*** *(off-label)* [4]
- Downy mildew in ***broccoli***, ***brussels sprouts***, ***cabbages***, ***calabrese***, ***cauliflowers***, ***chinese cabbage*** [1-4]; ***cayenne pepper*** *(off-label)*, ***frise*** *(off-label)*, ***horseradish*** *(off-label)*, ***leaf brassicas*** *(off-label)*, ***protected horseradish*** *(off-label)* [4]; ***chives*** *(off-label)*, ***cress*** *(off-label)*, ***parsley*** *(off-label)*, ***protected herbs (see appendix 6)*** *(off-label)*, ***watercress*** *(off-label - under protection)* [1]; ***herbs (see appendix 6)*** *(off-label)*, ***protected radishes*** *(off-label)*, ***radishes*** *(off-label)* [1, 4]; ***lettuce***, ***protected lettuce*** [3, 4]
- Phytophthora in ***aubergines***, ***bedding plants***, ***broccoli***, ***brussels sprouts***, ***bulb onions***, ***cabbages***, ***calabrese***, ***cauliflowers***, ***chinese cabbage***, ***cucumbers***, ***leeks***, ***nursery stock***, ***peppers***, ***pot plants*** [1-4]; ***flower bulbs***, ***tomatoes (outdoor)*** [2-4]; ***ornamental plant production*** [1, 3, 4]; ***protected tomatoes*** [2, 4]; ***rockwool aubergines***, ***rockwool cucumbers***, ***rockwool peppers***, ***rockwool tomatoes*** [2]; ***tulips*** [1, 2]; ***watercress*** *(off-label - under protection)* [1]
- Phytophthora root rot in ***cayenne pepper*** *(off-label)*, ***frise*** *(off-label)*, ***herbs (see appendix 6)*** *(off-label)*, ***horseradish*** *(off-label)*, ***leaf brassicas*** *(off-label)*, ***protected cucumbers*** *(off-label)*,

protected horseradish *(off-label)*, ***protected peppers*** *(off-label)*, ***protected radishes*** *(off-label)*, ***protected tomatoes*** *(off-label)*, ***protected watercress*** *(off-label)*, ***radishes*** *(off-label)* [4]
- Pythium in ***aubergines, bedding plants, broccoli, brussels sprouts, bulb onions, cabbages, calabrese, cauliflowers, chinese cabbage, cucumbers, flower bulbs, leeks, nursery stock, peppers, pot plants, tomatoes (outdoor)*** [1-4]; ***ornamental plant production*** [1, 3, 4]; ***protected tomatoes*** [1, 2, 4]; ***rockwool aubergines, rockwool cucumbers, rockwool peppers, rockwool tomatoes, tulips*** [1, 2]; ***watercress*** *(off-label - under protection)* [1]
- White blister in ***horseradish*** *(off-label)*, ***protected horseradish*** *(off-label)* [1, 4]; ***protected radishes*** *(off-label)*, ***radishes*** *(off-label)* [4]

Specific Off-Label Approvals (SOLAs)
- ***cayenne pepper*** *20072948* [4]
- ***chives*** *20040626* [1]
- ***cress*** *20040626* [1]
- ***frise*** *20072949* [4]
- ***herbs (see appendix 6)*** *20040626* [1], *20072947* [4]
- ***horseradish*** *992036* [1], *20072947* [4]
- ***inert substrate cucumbers*** *992032* [1]
- ***inert substrate peppers*** *992032* [1]
- ***inert substrate tomatoes*** *992032* [1]
- ***lamb's lettuce*** *20072947* [4]
- ***leaf brassicas*** *20072949* [4]
- ***lettuce*** *20072947* [4]
- ***nft peppers*** *992032* [1]
- ***nft tomatoes*** *992032* [1]
- ***parsley*** *20040626* [1]
- ***protected cucumbers*** *20072945* [4]
- ***protected herbs (see appendix 6)*** *20040626* [1], *20072947* [4]
- ***protected horseradish*** *992036* [1], *20072947* [4]
- ***protected lamb's lettuce*** *20072947* [4]
- ***protected lettuce*** *20072947* [4]
- ***protected peppers*** *992032* [1], *20072945* [4]
- ***protected radishes*** *992030* [1], *20072947* [4]
- ***protected scarole*** *20072947* [4]
- ***protected tomatoes*** *992032* [1], *20072945* [4]
- ***protected watercress*** *20072947* [4]
- ***radishes*** *992030* [1], *20072947* [4]
- ***scarole*** *20072947* [4]
- ***watercress*** *(under protection) 20010439* [1]

Approval information
- Propamocarb hydrochloride included in Annex I under EC Directive 91/414

Efficacy guidance
- Chemical is absorbed through roots and translocated throughout plant
- Incorporate in compost before use or drench moist compost or soil before sowing, pricking out, striking cuttings or potting up
- Drench treatment can be repeated at 3-6 wk intervals
- Concentrated solution is corrosive to all metals other than stainless steel
- May also be applied in trickle irrigation systems
- To prevent root rot in tulip bulbs apply as dip for 20 min but full protection only achieved by soil drench treatment as well

Restrictions
- Maximum number of treatments 4 per crop for cucumbers, tomatoes, peppers, aubergines; 3 per crop for outdoor and protected lettuce; 1 per crop for listed brassicas; 1 compost incorporation and/or 1 drench treatment for leeks, onions; 1 dip and 1 soil drench for flower bulbs
- Do not apply to lettuce crops grown in non-soil systems
- When applied over established seedlings rinse off foliage with water (except lettuce) and do not apply under hot, dry conditions
- On plants of unknown tolerance test first on a small scale

- Do not apply in a recirculating irrigation/drip system
- Store away from seeds and fertilizers

Crop-specific information

- Latest use: before transplanting for brassicas
- HI 14 d for cucumbers, tomatoes, peppers, aubergines; 4 wk for brassicas; 19 wk for leeks, onions

Hazard classification and safety precautions

UN number N/C
Operator protection A, C, H, K, M [1-4]; B [1, 2]; U02a, U19a [3, 4]; U05a [2-4]; U08, U20b [1-4]
Environmental protection E15a [1-4]; E34 [3, 4]
Storage and disposal D01, D02, D05 [2-4]; D07 [2]; D09a [1-4]; D10b [3, 4]; D11a [1, 2]

386 propaquizafop

A phenoxy alkanoic acid foliar acting herbicide for grass weeds in a range of crops
HRAC mode of action code: A

Products

1	Bulldog	Makhteshim	100 g/l	EC	11723
2	Bulldog	Makhteshim	100 g/l	EC	14565
3	Clayton Orleans	Clayton	100 g/l	EC	14452
4	Cleancrop GYR	Makhteshim	100 g/l	EC	10646
5	Cleancrop GYR 2	Makhteshim	100 g/l	EC	14578
6	Falcon	Makhteshim	100 g/l	EC	10585
7	Falcon	Makhteshim	100 g/l	EC	14552
8	Greencrop Satchmo	Greencrop	100 g/l	EC	10748
9	Longhorn	AgChem Access	100 g/l	EC	12512
10	MAC-Propaquizafop 100 EC	AgChem Access	100 g/l	EC	13978
11	Raptor	Makhteshim	100 g/l	EC	11092
12	Raptor	Makhteshim	100 g/l	EC	14566
13	Shogun	Makhteshim	100 g/l	EC	10584
14	Shogun	Makhteshim	100 g/l	EC	14567
15	Standon Propaquizafop	Standon	100 g/l	EC	11536
16	Standon Zing PQF	Standon	100 g/l	EC	13898
17	Zealot	AgriGuard	100 g/l	EC	12548

Uses

- Annual grasses in ***beet leaves*** *(off-label)* [1, 4]; ***bog myrtle*** *(off-label)*, ***mallow (althaea spp.)*** *(off-label)* [6]; ***borage*** *(off-label)* [11]; ***borage for oilseed production*** *(off-label)* [4, 6, 13, 15]; ***bulb onions, carrots, combining peas, early potatoes, farm forestry, fodder beet, linseed, parsnips, spring oilseed rape, sugar beet, swedes, turnips, winter field beans, winter oilseed rape*** [1-17]; ***canary flower (echium spp.)*** *(off-label)* [4, 6, 11, 13, 15]; ***celery (outdoor)*** *(off-label)*, ***rhubarb*** *(off-label)* [4, 6, 11, 13]; ***cut logs, forest, forest nurseries*** [1, 2, 4-7, 10-14]; ***evening primrose*** *(off-label)*, ***garlic*** *(off-label)*, ***lupins*** *(off-label)*, ***mustard*** *(off-label)*, ***shallots*** *(off-label)* [1, 4, 6, 8, 11, 13, 15]; ***herbs (see appendix 6)*** *(off-label)*, ***leeks*** *(off-label)*, ***red beet*** *(off-label)*, ***spinach*** *(off-label)* [1, 4, 11, 13]; ***honesty*** *(off-label)* [6, 11, 13]; ***horseradish*** *(off-label)* [13]; ***maincrop potatoes*** [1-4, 6-17]; ***poppies for morphine production*** *(off-label)* [6, 11]; ***potatoes*** [5]; ***spring field beans*** [1-8, 10-17]; ***spring linseed*** *(off-label)* [4, 6, 13]
- Annual meadow grass in ***beet leaves*** *(off-label)* [1, 4, 6]; ***bog myrtle*** *(off-label)*, ***mallow (althaea spp.)*** *(off-label)* [6]; ***borage*** *(off-label)* [11]; ***borage for oilseed production*** *(off-label)* [4, 6, 13, 15]; ***bulb onions, carrots, combining peas, cut logs, early potatoes, farm forestry, fodder beet, forest, forest nurseries, linseed, parsnips, potatoes, spring field beans, spring oilseed rape, sugar beet, swedes, turnips, winter field beans, winter oilseed rape*** [5]; ***canary flower (echium spp.)*** *(off-label)* [4, 6, 11, 13, 15]; ***celery (outdoor)*** *(off-label)*, ***rhubarb*** *(off-label)* [4, 11, 13]; ***evening primrose*** *(off-label)*, ***garlic*** *(off-label)*, ***lupins*** *(off-label)*, ***mustard*** *(off-label)*, ***shallots*** *(off-label)* [1, 4, 6, 8, 11, 13, 15]; ***herbs (see appendix 6)*** *(off-label)*, ***leeks*** *(off-label)*, ***red beet*** *(off-label)*, ***spinach*** *(off-label)* [1, 4, 6, 11, 13]; ***honesty*** *(off-label)* [6, 11, 13]; ***horseradish*** *(off-label)* [13]; ***poppies for morphine production*** *(off-label)* [6, 11]; ***spring linseed*** *(off-label)* [4, 6, 13]
- Couch in ***beet leaves*** *(off-label)*, ***herbs (see appendix 6)*** *(off-label)*, ***leeks*** *(off-label)*, ***red beet*** *(off-label)*, ***spinach*** *(off-label)* [6]
- Perennial grasses in ***beet leaves*** *(off-label)* [1, 4]; ***bog myrtle*** *(off-label)*, ***mallow (althaea spp.)*** *(off-label)* [6]; ***borage*** *(off-label)* [11]; ***borage for oilseed production*** *(off-label)* [4, 6, 13, 15]; ***bulb***

onions, carrots, combining peas, early potatoes, farm forestry, fodder beet, linseed, parsnips, spring oilseed rape, sugar beet, swedes, turnips, winter field beans, winter oilseed rape [1-17]; ***canary flower (echium spp.)*** *(off-label)* [4, 6, 11, 13, 15]; ***celery (outdoor)*** *(off-label)*, ***rhubarb*** *(off-label)* [4, 6, 11, 13]; ***cut logs, forest, forest nurseries*** [1, 2, 4-7, 10-14]; ***evening primrose*** *(off-label)*, ***garlic*** *(off-label)*, ***lupins*** *(off-label)*, ***mustard*** *(off-label)*, ***shallots*** *(off-label)* [1, 4, 6, 8, 11, 13, 15]; ***herbs (see appendix 6)*** *(off-label)*, ***leeks*** *(off-label)*, ***red beet*** *(off-label)*, ***spinach*** *(off-label)* [1, 4, 11, 13]; ***honesty*** *(off-label)* [6, 11, 13]; ***horseradish*** *(off-label)* [13]; ***maincrop potatoes*** [1-4, 6-17]; ***poppies for morphine production*** *(off-label)* [6, 11]; ***potatoes*** [5]; ***spring field beans*** [1-8, 10-17]; ***spring linseed*** *(off-label)* [4, 6, 13]

- Volunteer cereals in ***beet leaves*** *(off-label)* [1, 6]; ***bog myrtle*** *(off-label)*, ***mallow (althaea spp.)*** *(off-label)* [6]; ***borage*** *(off-label)* [11]; ***borage for oilseed production*** *(off-label)* [6, 13, 15]; ***canary flower (echium spp.)*** *(off-label)* [6, 11, 13, 15]; ***celery (outdoor)*** *(off-label)*, ***rhubarb*** *(off-label)*, ***spring linseed*** *(off-label)* [6, 13]; ***evening primrose*** *(off-label)*, ***garlic*** *(off-label)*, ***lupins*** *(off-label)*, ***mustard*** *(off-label)*, ***shallots*** *(off-label)* [1, 6, 8, 11, 13, 15]; ***herbs (see appendix 6)*** *(off-label)*, ***leeks*** *(off-label)*, ***red beet*** *(off-label)*, ***spinach*** *(off-label)* [1, 6, 13]; ***honesty*** *(off-label)* [6, 11, 13]; ***horseradish*** *(off-label)* [13]; ***poppies for morphine production*** *(off-label)* [6, 11]
- Wild oats in ***celery (outdoor)*** *(off-label)*, ***rhubarb*** *(off-label)* [6]

Specific Off-Label Approvals (SOLAs)

- ***beet leaves*** *20080860* [1], *20080864* [4], *20080871* [6]
- ***bog myrtle*** *20091295* [6]
- ***borage*** *20080885* [11]
- ***borage for oilseed production*** *20080867* [4], *20080874* [6], *20080889* [13], *20080893* [15]
- ***canary flower (echium spp.)*** *20080867* [4], *20080874* [6], *20080885* [11], *20080889* [13], *20080893* [15]
- ***celery (outdoor)*** *20080864* [4], *20080872* [6], *20080881* [11], *20080887* [13]
- ***evening primrose*** *20080862* [1], *20080867* [4], *20080874* [6], *20080879* [8], *20080885* [11], *20080889* [13], *20080893* [15]
- ***garlic*** *20080861* [1], *20080866* [4], *20080875* [6], *20080880* [8], *20080886* [11], *20080891* [13], *20080894* [15]
- ***herbs (see appendix 6)*** *20080860* [1], *20080864* [4], *20080871* [6], *20080881* [11], *20080887* [13]
- ***honesty*** *20080874* [6], *20080885* [11], *20080889* [13]
- ***horseradish*** *20080890* [13]
- ***leeks*** *20080860* [1], *20080864* [4], *20080871* [6], *20080881* [11], *20080887* [13]
- ***lupins*** *20080863* [1], *20080865* [4], *20080876* [6], *20080878* [8], *20080883* [11], *20080888* [13], *20080892* [15]
- ***mallow (althaea spp.)*** *20080873* [6]
- ***mustard*** *20080862* [1], *20080867* [4], *20080874* [6], *20080879* [8], *20080885* [11], *20080889* [13], *20080893* [15]
- ***poppies for morphine production*** *20080877* [6], *20080882* [11]
- ***red beet*** *20080860* [1], *20080864* [4], *20080871* [6], *20080881* [11], *20080887* [13]
- ***rhubarb*** *20080864* [4], *20080872* [6], *20080881* [11], *20080887* [13]
- ***shallots*** *20080861* [1], *20080866* [4], *20080875* [6], *20080880* [8], *20080886* [11], *20080891* [13], *20080894* [15]
- ***spinach*** *20080860* [1], *20080864* [4], *20080871* [6], *20080881* [11], *20080887* [13]
- ***spring linseed*** *20080867* [4], *20080874* [6], *20080889* [13]

Efficacy guidance

- Apply to emerged weeds when they are growing actively with adequate soil moisture
- Activity is slower under cool conditions
- Broad-leaved weeds and any weeds germinating after treatment are not controlled
- Annual meadow grass up to 3 leaves checked at low doses and severely checked at highest dose
- Spray barley cover crops when risk of wind blow has passed and before there is serious competition with the crop
- Various tank mixtures and sequences recommended for broader spectrum weed control in oilseed rape, peas and sugar beet. See label for details
- Severe couch infestations may require a second application at reduced dose when regrowth has 3-4 leaves unfolded
- Products contain surfactants. Tank mixing with adjuvants not required or recommended
- Propaquizafop is an ACCase inhibitor herbicide. To avoid the build up of resistance do not apply products containing an ACCase inhibitor herbicide more than twice to any crop. In addition do not

use any product containing propaquizafop in mixture or sequence with any other product containing the same ingredient

- Use these products as part of a resistance management strategy that includes cultural methods of control and does not use ACCase inhibitors as the sole chemical method of grass weed control
- Applying a second product containing an ACCase inhibitor to a crop will increase the risk of resistance development; only use a second ACCase inhibitor to control different weeds at a different timing
- Always follow WRAG guidelines for preventing and managing herbicide resistant weeds. See Section 5 for more information

Restrictions

- Maximum number of treatments 1 or 2 per crop or yr. See label for details
- See label for list of tolerant tree species
- Do not treat seed potatoes

Crop-specific information

- Latest use: before crop flower buds visible for winter oilseed rape, linseed, field beans; before 8 fully expanded leaf stage for spring oilseed rape; before weeds are covered by the crop for potatoes, sugar beet, fodder beet, swedes, turnips; when flower buds visible for peas
- HI onions, carrots, early potatoes, parsnips 4 wk; combining peas, early potatoes 7 wk; sugar beet, fodder beet, maincrop potatoes, swedes, turnips 8 wk; field beans 14 wk
- Application in high temperatures and/or low soil moisture content may cause chlorotic spotting especially on combining peas and field beans
- Overlaps at the highest dose can cause damage from early applications to carrots and parsnips

Following crops guidance

- In the event of a failed treated crop an interval of 2 wk must elapse between the last application and redrilling with winter wheat or winter barley. 4 wk must elapse before sowing oilseed rape, peas or field beans, and 16 wk before sowing ryegrass or oats

Environmental safety

- Dangerous for the environment
- Toxic to aquatic organisms
- Risk to certain non-target insects or other arthropods. See directions for use

Hazard classification and safety precautions

Hazard Harmful [5]; Irritant [1-4, 6-17]; Dangerous for the environment [1-17]

Transport code 9

Packaging group III

UN number 3082

Risk phrases R20, R41 [5]; R36, R38 [1-4, 6-17]; R51, R53a [1-17]

Operator protection A [1-3, 5-12, 14-17]; C [3, 5-10, 14-17]; H [1, 2, 5, 9, 11, 12, 14]; U02a, U05a [1-4, 6-17]; U08, U19a, U20b [1-17]; U11 [6, 7, 10, 15, 16]; U12 [5]; U13 [1-7, 9-14, 17]; U14, U15 [1-7, 9-17]; U23a [8]

Environmental protection E15a, E22b [1-17]; E34 [1-4, 6, 7, 9-14, 17]; E38 [5]

Storage and disposal D01, D02, D05, D09a [1-17]; D10a, D12a [5]; D10b [1-4, 6-17]; D12b [1-4, 6, 7, 9-14]; D14 [1-3, 6, 7, 9-14]

Medical advice M05b

387 propiconazole

A systemic, curative and protectant triazole fungicide
FRAC mode of action code: 3

See also *chlorothalonil + cyproconazole + propiconazole*
chlorothalonil + fludioxonil + propiconazole
chlorothalonil + propiconazole
cyproconazole + propiconazole
prochloraz + propiconazole
propiconazole + tebuconazole

Products

1	Anode	Makhteshim	250 g/l	EC	14447
2	Banner Maxx	Syngenta	156 g/l	EC	13167
3	Bounty	AgChem Access	250 g/l	EC	14616
4	Bumper 250 EC	Makhteshim	250 g/l	EC	14399

Uses

- Anthracnose in ***amenity grassland*** *(qualified minor use)*, ***managed amenity turf*** *(qualified minor use)* [2]
- Brown patch in ***amenity grassland*** *(qualified minor use)*, ***managed amenity turf*** *(qualified minor use)* [2]
- Brown rust in ***spring barley, spring wheat, winter barley, winter rye, winter wheat*** [1, 3, 4]
- Crown rust in ***grass seed crops, permanent grassland, spring oats, winter oats*** [1, 3, 4]
- Disease control in ***honesty*** *(off-label)* [4]
- Dollar spot in ***amenity grassland, managed amenity turf*** [2]
- Downy mildew in ***garlic*** *(off-label)*, ***onions*** *(off-label)*, ***shallots*** *(off-label)* [4]
- Drechslera leaf spot in ***grass seed crops, permanent grassland*** [1, 3, 4]
- Fusarium patch in ***amenity grassland, managed amenity turf*** [2]
- Glume blotch in ***spring wheat, winter rye, winter wheat*** [1, 3, 4]
- Light leaf spot in ***spring oilseed rape*** *(reduction)*, ***winter oilseed rape*** *(reduction)* [1, 3, 4]
- Mildew in ***grass seed crops, permanent grassland*** [1, 3, 4]
- Powdery mildew in ***spring barley, spring oats, spring wheat, winter barley, winter oats, winter rye, winter wheat*** [1, 3, 4]
- Ramularia leaf spots in ***sugar beet*** *(reduction)* [1, 3, 4]
- Rhynchosporium in ***grass seed crops, permanent grassland, spring barley, winter barley, winter rye*** [1, 3, 4]
- Rust in ***ornamental plant production*** *(off-label)*, ***protected ornamentals*** *(off-label)* [4]; ***sugar beet*** [1, 3, 4]
- Septoria leaf blotch in ***spring wheat, winter rye, winter wheat*** [1, 3, 4]
- Sooty moulds in ***spring wheat, winter wheat*** [1, 3, 4]
- Yellow rust in ***spring barley, spring wheat, winter barley, winter wheat*** [1, 3, 4]

Specific Off-Label Approvals (SOLAs)

- ***garlic*** *20090705* [4]
- ***honesty*** *20090706* [4]
- ***onions*** *20090705* [4]
- ***ornamental plant production*** *20090707* [4]
- ***protected ornamentals*** *20090707* [4]
- ***shallots*** *20090705* [4]

Approval information

- Propiconazole included in Annex I under EC Directive 91/414
- Accepted by BBPA for use on malting barley

Efficacy guidance

- Best results achieved by applying at early stage of disease development. Recommended spray programmes vary with crop, disease, season, soil type and product.
- For optimum turf quality and control use in conjunction with turf management practices that promote good plant health [2]

- Propiconazole is a DMI fungicide. Resistance to some DMI fungicides has been identified in Septoria leaf blotch which may seriously affect performance of some products. For further advice contact a specialist advisor and visit the Fungicide Resistance Action Group (FRAG)-UK website

Restrictions

- Maximum number of treatments 4 per crop for wheat (up to 3 in yr of harvest), amenity grassland, managed amenity turf; 3 per crop for barley, oats, rye, triticale (up to 2 in yr of harvest); 2 per crop or yr for oilseed rape, sugar beet, grass seed crops, honesty; 1 per yr on grass for ensiling
- On oilseed rape do not apply during flowering. Apply first treatment before flowering, the second afterwards
- Grass seed crops must be treated in yr of harvest
- A minimum interval of 14 d must elapse between treatments on leeks and 21 d on sugar beet
- Avoid spraying crops or turf under stress, eg during cold weather or periods of frost

Crop-specific information

- Latest use: before 4 pairs of true leaves for honesty
- HI cereals, grass seed crops 35 d; oilseed rape, sugar beet, garlic, onions, shallots, grass for ensiling 28 d
- May be used on all common turf grass species [2]

Environmental safety

- Dangerous for the environment
- Toxic to aquatic organisms
- Risk to non-target insects or other arthropods [2]
- When treating turf with vehicle mounted or drawn hydraulic boom sprayers avoid spraying within 5 m of unmanaged land to reduce effects on non-target arthropods [2]

Hazard classification and safety precautions

Hazard Harmful [2]; Dangerous for the environment [1-4]
Transport code 9
Packaging group III
UN number 3082
Risk phrases R20, R43, R52 [2]; R51 [1, 3, 4]; R53a [1-4]
Operator protection A [1-4]; C [1, 3, 4]; H [2]; U02a, U05a, U19a, U20b [1-4]; U08 [1, 3, 4]; U09a, U14 [2]
Environmental protection E15a, E34 [1, 3, 4]; E15b, E22c, E38 [2]
Storage and disposal D01, D02, D05, D09a [1-4]; D10b [1, 3, 4]; D10c, D12a [2]
Treated seed S06a [1, 3, 4]
Medical advice M05a [1, 3, 4]

388 propiconazole + tebuconazole

A broad spectrum systemic fungicide mixture for cereals
FRAC mode of action code: 3 + 3

See also *propiconazole*
tebuconazole

Products

Cogito	Syngenta	250:250 g/l	EC	13847

Uses

- Brown rust in ***spring barley***, ***spring wheat***, ***winter barley***, ***winter wheat***
- Glume blotch in ***spring wheat***, ***winter wheat***
- Mildew in ***spring barley***, ***spring wheat***, ***winter barley***, ***winter wheat***
- Net blotch in ***spring barley***, ***winter barley***
- Rhynchosporium in ***spring barley***, ***winter barley***
- Septoria leaf blotch in ***spring wheat***, ***winter wheat***
- Yellow rust in ***spring barley***, ***spring wheat***, ***winter barley***, ***winter wheat***

Approval information

- Propiconazole included in Annex I under EC Directive 91/414
- Accepted by BBPA for use on malting barley

Efficacy guidance

- Best disease control and yield benefit obtained when applied at early stage of disease development before infection has spread to new growth
- To protect the flag leaf and ear from Septoria diseases apply from flag leaf emergence to ear fully emerged (GS 37-59)
- Applications once foliar symptoms of *Septoria tritici* are already present on upper leaves will be less effective
- Some strains of *Septoria tritici* have developed resistance to some DMI fungicides and this may lead to poor control

Restrictions

- Maximum total dose 1 l/ha per crop
- Occasional slight temporary leaf speckling may occur on wheat but this has not been shown to reduce yield response or disease control
- Do not treat durum wheat

Crop-specific information

- Before watery ripe stage (GS 71)

Environmental safety

- Irritating to eyes and skin. May cause sensitization by skin contact
- Propiconazole may produce an allergic reaction
- Dangerous to fish or other aquatic life. Do not contaminate surface waters or ditches with chemical or used container
- LERAP Category B

Hazard classification and safety precautions

Hazard Harmful, Dangerous for the environment
Transport code 9
Packaging group III
UN number 3082
Risk phrases R43, R50, R53a, R63
Operator protection A, H; U05a, U09a, U11, U14, U15, U20b
Environmental protection E15b, E16a, E16b, E38
Storage and disposal D01, D02, D09a, D10c, D12a

389 propoxycarbazone-sodium

A sulfonylaminocarbonyltriazolinone residual grass weed herbicide for winter wheat
HRAC mode of action code: B

Products

Attribut	Bayer CropScience	70% w/w	SG	12730

Uses

- Annual grasses in ***miscanthus*** *(off-label)*
- Blackgrass in ***winter wheat***
- Couch in ***winter wheat***
- Perennial grasses in ***miscanthus*** *(off-label)*

Specific Off-Label Approvals (SOLAs)

- ***miscanthus*** *20082761*

Approval information

- Propoxycarbazone-sodium included in Annex I under EC Directive 91/414

Efficacy guidance

- Activity by root and foliar absorption but depends on presence of sufficient soil moisture to ensure root uptake by weeds. Control is enhanced by use of an adjuvant to encourage foliar uptake
- Best results obtained from treatments applied when weed grasses are growing actively. Symptoms may not become apparent for 3-4 wk after application
- Ensure good even spray coverage
- To achieve best control of couch and reduction of infestation in subsequent crop treat between 2 true leaves and first node stage of the weed

- Blackgrass should be treated after using specialist blackgrass herbicides with a different mode of action
- Propoxycarbazone-sodium is a member of the ALS-inhibitor group of herbicides. To avoid the build up of resistance do not use any product containing an ALS-inhibitor herbicide with claims for control of grass weeds more than once on any crop
- Use these products as part of a resistance management strategy that includes cultural methods of control and does not use ALS inhibitors as the sole chemical method of grass weed control

Restrictions
- Maximum total dose equivalent to one full dose treatment
- Do not use in a programme with other aceto-lactase synthesis (ALS) inhibitors
- Do not apply when temperature near or below freezing
- Avoid treatment under dry soil conditions

Crop-specific information
- Latest use: third node detectable in wheat (GS 33)

Following crops guidance
- Only winter wheat, field beans or winter barley may be sown in the autumn following spring treatment
- Any crop may be grown in the spring on land treated during the previous calendar yr

Environmental safety
- Dangerous for the environment
- Very toxic to aquatic organisms
- LERAP Category B
- Take extreme care to avoid drift onto adjacent plants or land as this could result in severe damage

Hazard classification and safety precautions

Hazard Dangerous for the environment
Transport code 9
Packaging group III
UN number 3077
Risk phrases R50, R53a
Operator protection A; U05a, U20b
Environmental protection E15b, E16a, E16b, E38
Storage and disposal D01, D02, D09a, D11a, D12a

390 propyzamide

A residual benzamide herbicide for use in a wide range of crops
HRAC mode of action code: K1

Products

1	Conform	AgChem Access	400 g/l	SC	14602
2	Engage	Interfarm	50% w/w	WP	14233
3	Flomide	Interfarm	400 g/l	SC	14223
4	Kerb 50 W	Dow	50% w/w	WP	13715
5	Kerb Flo	Dow	400 g/l	SC	13716
6	Kerb Granules	Barclay	4% w/w	GR	14213
7	Menace 80 EDF	Dow	80% w/w	WG	13714
8	Pizamide	ChemSource	400 g/l	SC	14684
9	Pizza 400 SC	Goldengrass	400 g/l	SC	14430
10	Propyzamide 400	European Ag	400 g/l	SC	14685
11	Quaver Flo	Dow	400 g/l	SC	14203
12	Setanta 50 WP	AgriGuard	50% w/w	WP	14494
13	Setanta Flo	AgriGuard	400 g/l	SC	14493

Uses
- Annual dicotyledons in ***all edible seed crops grown outdoors*** *(off-label)*, ***all non-edible seed crops grown outdoors*** *(off-label)*, ***game cover*** *(off-label)*, ***hops*** *(off-label)* [4, 5]; ***almonds*** *(off-label)*, ***bilberries*** *(off-label)*, ***blueberries*** *(off-label)*, ***broccoli*** *(off-label)*, ***calabrese*** *(off-label)*, ***cauliflowers*** *(off-label)*, ***cherries*** *(off-label)*, ***chestnuts*** *(off-label)*, ***chicory root*** *(off-label)*, ***cob nuts*** *(off-label)*, ***courgettes*** *(off-label)*, ***cranberries*** *(off-label)*, ***cress*** *(off-label)*, ***endives*** *(off-label)*, ***frise*** *(off-label)*, ***hazel nuts*** *(off-label)*, ***herbs (see appendix 6)*** *(off-label)*, ***lamb's lettuce***

(*off-label*), ***leaf brassicas*** (*off-label*), ***marrows*** (*off-label*), ***mirabelles*** (*off-label*), ***protected endives*** (*off-label*), ***protected forest nurseries*** (*off-label*), ***protected herbs (see appendix 6)*** (*off-label*), ***protected lettuce*** (*off-label*), ***pumpkins*** (*off-label*), ***quinces*** (*off-label*), ***radicchio*** (*off-label*), ***salad brassicas*** (*off-label*), ***scarole*** (*off-label*), ***squashes*** (*off-label*), ***table grapes*** (*off-label*), ***walnuts*** (*off-label*), ***wine grapes*** (*off-label*) [5]; ***amenity vegetation*** [1, 4, 5, 7-11]; ***apple orchards, blackberries, blackcurrants, clover seed crops, gooseberries, loganberries, lucerne, pear orchards, plums, raspberries*** (*England only*), ***redcurrants, strawberries, sugar beet seed crops, winter field beans, winter oilseed rape*** [1-5, 7-13]; ***brassica seed crops, lettuce*** (*outdoor crops*), ***rhubarb*** (*outdoor*) [7]; ***farm forestry, hedges*** [1, 4, 5, 7-10]; ***fodder rape seed crops, kale seed crops, lettuce, rhubarb, turnip seed crops*** [1-5, 8-13]; ***forest*** [1, 2, 4-6, 8-10, 12]; ***forest nurseries*** [1, 4, 5, 7-10, 12]; ***ornamental plant production*** [12]; ***roses*** [2, 12]; ***top fruit*** (*off-label*) [4]; ***trees and shrubs*** [2, 6]; ***woody ornamentals*** [2, 4, 6, 7]
- Annual grasses in ***all edible seed crops grown outdoors*** (*off-label*), ***all non-edible seed crops grown outdoors*** (*off-label*), ***game cover*** (*off-label*), ***hops*** (*off-label*) [4, 5]; ***almonds*** (*off-label*), ***bilberries*** (*off-label*), ***blueberries*** (*off-label*), ***broccoli*** (*off-label*), ***calabrese*** (*off-label*), ***cauliflowers*** (*off-label*), ***cherries*** (*off-label*), ***chestnuts*** (*off-label*), ***chicory root*** (*off-label*), ***cob nuts*** (*off-label*), ***courgettes*** (*off-label*), ***cranberries*** (*off-label*), ***cress*** (*off-label*), ***endives*** (*off-label*), ***frise*** (*off-label*), ***hazel nuts*** (*off-label*), ***herbs (see appendix 6)*** (*off-label*), ***lamb's lettuce*** (*off-label*), ***leaf brassicas*** (*off-label*), ***marrows*** (*off-label*), ***mirabelles*** (*off-label*), ***protected endives*** (*off-label*), ***protected forest nurseries*** (*off-label*), ***protected herbs (see appendix 6)*** (*off-label*), ***protected lettuce*** (*off-label*), ***pumpkins*** (*off-label*), ***quinces*** (*off-label*), ***radicchio*** (*off-label*), ***salad brassicas*** (*off-label*), ***scarole*** (*off-label*), ***squashes*** (*off-label*), ***table grapes*** (*off-label*), ***walnuts*** (*off-label*), ***wine grapes*** (*off-label*) [5]; ***amenity vegetation*** [1, 4, 5, 7-11]; ***apple orchards, blackberries, blackcurrants, clover seed crops, gooseberries, loganberries, lucerne, pear orchards, plums, raspberries*** (*England only*), ***redcurrants, strawberries, sugar beet seed crops, winter field beans, winter oilseed rape*** [1-5, 7-13]; ***brassica seed crops, lettuce*** (*outdoor crops*), ***rhubarb*** (*outdoor*) [7]; ***farm forestry, hedges*** [1, 4, 5, 7-10]; ***fodder rape seed crops, kale seed crops, lettuce, rhubarb, turnip seed crops*** [1-5, 8-13]; ***forest*** [1, 2, 4-6, 8-10, 12]; ***forest nurseries*** [1, 4, 5, 7-10, 12]; ***ornamental plant production*** [12]; ***roses*** [2, 12]; ***top fruit*** (*off-label*) [4]; ***trees and shrubs*** [2, 6]; ***woody ornamentals*** [2, 4, 6, 7]
- Horsetails in ***forest*** [7]
- Perennial grasses in ***all edible seed crops grown outdoors*** (*off-label*), ***all non-edible seed crops grown outdoors*** (*off-label*), ***game cover*** (*off-label*), ***hops*** (*off-label*) [4, 5]; ***almonds*** (*off-label*), ***bilberries*** (*off-label*), ***blueberries*** (*off-label*), ***broccoli*** (*off-label*), ***calabrese*** (*off-label*), ***cauliflowers*** (*off-label*), ***cherries*** (*off-label*), ***chestnuts*** (*off-label*), ***chicory root*** (*off-label*), ***cob nuts*** (*off-label*), ***courgettes*** (*off-label*), ***cranberries*** (*off-label*), ***cress*** (*off-label*), ***endives*** (*off-label*), ***frise*** (*off-label*), ***hazel nuts*** (*off-label*), ***herbs (see appendix 6)*** (*off-label*), ***lamb's lettuce*** (*off-label*), ***leaf brassicas*** (*off-label*), ***marrows*** (*off-label*), ***mirabelles*** (*off-label*), ***protected endives*** (*off-label*), ***protected forest nurseries*** (*off-label*), ***protected herbs (see appendix 6)*** (*off-label*), ***protected lettuce*** (*off-label*), ***pumpkins*** (*off-label*), ***quinces*** (*off-label*), ***radicchio*** (*off-label*), ***salad brassicas*** (*off-label*), ***scarole*** (*off-label*), ***squashes*** (*off-label*), ***table grapes*** (*off-label*), ***walnuts*** (*off-label*), ***wine grapes*** (*off-label*) [5]; ***amenity vegetation*** [1, 4, 5, 7-11]; ***apple orchards, blackberries, blackcurrants, gooseberries, loganberries, pear orchards, plums, raspberries*** (*England only*), ***redcurrants*** [1-5, 7-13]; ***clover seed crops, fodder rape seed crops, kale seed crops, lettuce, lucerne, rhubarb, strawberries, sugar beet seed crops, turnip seed crops, winter field beans, winter oilseed rape*** [1-5, 8-13]; ***farm forestry, hedges*** [1, 4, 5, 7-10]; ***forest*** [1, 2, 4-10, 12]; ***forest nurseries*** [1, 4, 5, 7-10, 12]; ***ornamental plant production*** [12]; ***rhubarb*** (*outdoor*) [7]; ***roses*** [2, 12]; ***top fruit*** (*off-label*) [4]; ***trees and shrubs*** [2, 6]; ***woody ornamentals*** [2, 4, 6, 7]
- Sedges in ***forest*** [7]
- Volunteer cereals in ***sugar beet seed crops, winter field beans, winter oilseed rape*** [7]
- Wild oats in ***sugar beet seed crops, winter field beans, winter oilseed rape*** [7]

Specific Off-Label Approvals (SOLAs)

- ***all edible seed crops grown outdoors*** *20082943* [4], *20082942* [5]
- ***all non-edible seed crops grown outdoors*** *20082943* [4], *20082942* [5]
- ***almonds*** *20082420* [5]
- ***bilberries*** *20082419* [5]
- ***blueberries*** *20082419* [5]
- ***broccoli*** *20091902* [5]

- ***calabrese*** *20091902* [5]
- ***cauliflowers*** *20091902* [5]
- ***cherries*** *20082418* [5]
- ***chestnuts*** *20082420* [5]
- ***chicory root*** *20091530* [5]
- ***cob nuts*** *20082420* [5]
- ***courgettes*** *20082416* [5]
- ***cranberries*** *20082419* [5]
- ***cress*** *20082410* [5]
- ***endives*** *20082410* [5]
- ***frise*** *20082411* [5]
- ***game cover*** *20082943* [4], *20082942* [5]
- ***hazel nuts*** *20082420* [5]
- ***herbs (see appendix 6)*** *20082412* [5]
- ***hops*** *20082943* [4], *20082414* [5]
- ***lamb's lettuce*** *20082410* [5]
- ***leaf brassicas*** *20082410* [5]
- ***marrows*** *20082416* [5]
- ***mirabelles*** *20082418* [5]
- ***protected endives*** *20082415* [5]
- ***protected forest nurseries*** *20082942* [5]
- ***protected herbs (see appendix 6)*** *20082412* [5]
- ***protected lettuce*** *20082415* [5]
- ***pumpkins*** *20082416* [5]
- ***quinces*** *20082413* [5]
- ***radicchio*** *20082411* [5]
- ***salad brassicas*** *20082410* [5]
- ***scarole*** *20082411* [5]
- ***squashes*** *20082416* [5]
- ***table grapes*** *20082417* [5]
- ***top fruit*** *20082943* [4]
- ***walnuts*** *20082420* [5]
- ***wine grapes*** *20082417* [5]

Approval information

- Propyzamide included in Annex I under EC Directive 91/414
- Some products may be applied through CDA equipment. See labels for details
- Accepted by BBPA for use on hops

Efficacy guidance

- Active via root uptake. Weeds controlled from germination to young seedling stage, some species (including many grasses) also when established
- Best results achieved by winter application to fine, firm, moist soil. Rain is required after application if soil dry
- Uptake is slow and and may take up to 12 wk
- Excessive organic debris or ploughed-up turf may reduce efficacy
- For heavy couch infestations a repeat application may be needed in following winter
- Always follow WRAG guidelines for preventing and managing herbicide resistant weeds. See Section 5 for more information

Restrictions

- Maximum number of treatments 1 per crop or yr
- Maximum total dose equivalent to one full dose treatment for all crops
- Do not treat protected crops
- Apply to listed edible crops only between 1 Oct and the date specified as the latest time of application (except lettuce)
- Do not apply in windy weather and avoid drift onto non-target crops
- Do not use on soils with more than 10% organic matter except in forestry

Crop-specific information

- Latest use: labels vary but normally before 31 Dec in year before harvest for rhubarb, lucerne, strawberries and winter field beans; before 31 Jan for other crops

- HI: 6 wk for edible crops
- Apply as soon as possible after 3-true leaf stage of oilseed rape (GS 1,3) and seed brassicas, after 4-leaf stage of sugar beet for seed, within 7 d after sowing but before emergence for field beans, after perennial crops established for at least 1 season, strawberries after 1 yr
- Only apply to strawberries on heavy soils. Do not use on matted row crops
- Only apply to field beans on medium and heavy soils
- Only apply to established lucerne not less than 7 d after last cut
- In lettuce lightly incorporate in top 25 mm pre-drilling or irrigate on dry soil
- See label for lists of ornamental and forest species which may be treated

Following crops guidance

- Following an application between 1 Apr and 31 Jul at any dose the following minimum intervals must be observed before sowing the next crop: lettuce 0 wk; broad beans, chicory, clover, field beans, lucerne, radishes, peas 5 wk; brassicas, celery, leeks, oilseed rape, onions, parsley, parsnips 10 wk
- Following an application between 1 Aug and 31 Mar at any dose the following minimum intervals must be observed before sowing the next crop: lettuce 0 wk; broad beans, chicory, clover, field beans, lucerne, radishes, peas 10 wk; brassicas, celery, leeks, oilseed rape, onions, parsley, parsnips 25 wk or after 15 Jun, whichever occurs sooner
- Cereals or grasses or other crops not listed may be sown 30 wk after treatment up to 840 g ai/ha between 1 Aug and 31 Mar or 40 wk after treatment at higher doses at any time and after mouldboard ploughing to at least 15 cm
- A period of at least 9 mth must elapse between applications of propyzamide to the same land

Environmental safety

- Dangerous for the environment
- Very toxic to aquatic organisms
- Some pesticides pose a greater threat of contamination of water than others and propyzamide is one of these pesticides. Take special care when applying propyzamide near water and do not apply if heavy rain is forecast

Hazard classification and safety precautions

Hazard Harmful, Dangerous for the environment [1-10, 12, 13]
Transport code 9
Packaging group III
UN number 3077 [2, 4, 6, 7, 12]; 3082 [1, 3, 5, 8-11, 13]
Risk phrases R40, R53a [1-10, 12, 13]; R50 [1, 3-5, 7-10, 13]; R51 [2, 6, 12]
Operator protection A, H [1, 2, 4-10]; D, M [1, 4, 5, 7-10]; U05a [6]; U20c [1-5, 7-10, 12, 13]
Environmental protection E15a [1-3, 5-10, 12, 13]; E15b [4]; E34 [12, 13]; E38 [1-10, 13]
Consumer protection C02a [1-5, 7-10, 12, 13] (6 wk)
Storage and disposal D01, D02 [6]; D05, D11a [1-5, 7-10, 12, 13]; D07 [13]; D09a [1-10, 12, 13]; D12a [1-10, 13]
Medical advice M03 [6]

391 proquinazid

A quinazolinone fungicide for powdery mildew control in cereals
FRAC mode of action code: U7

See also prochloraz + proquinazid + tebuconazole

Products

1	Justice	DuPont	200 g/l	EC	12835
2	Talius	DuPont	200 g/l	EC	12752

Uses

- Powdery mildew in ***durum wheat*** *(off-label)*, ***spring barley***, ***spring oats***, ***spring rye***, ***spring wheat***, ***triticale***, ***winter barley***, ***winter oats***, ***winter rye***, ***winter wheat*** [1, 2]; ***forest nurseries*** *(off-label)*, ***soft fruit*** *(off-label)*, ***top fruit*** *(off-label)* [2]

Specific Off-Label Approvals (SOLAs)

- ***durum wheat*** *20090418* [1], *20090419* [2]
- ***forest nurseries*** *20090420* [2]

FOR FULL CONDITIONS OF USE ALWAYS READ THE PRODUCT LABEL

- ***soft fruit*** *20090420* [2]
- ***top fruit*** *20090420* [2]

Approval information

- Accepted by BBPA on malting barley

Efficacy guidance

- Best results obtained from preventive treatment before disease is established in the crop
- Where mildew has already spread to new growth a tank mix with a curative fungicide with an alternative mode of action should be used
- Use as part of an integrated crop management (ICM) strategy incorporating other methods of control or fungicides with different modes of action

Restrictions

- Maximum number of treatments 2 per crop
- Do not apply to any crop suffering from stress from any cause
- Avoid application in either frosty or hot, sunny conditions

Crop-specific information

- Latest use: before beginning of heading (GS 49) for barley, oats, rye, triticale; before full flowering (GS 65) for wheat

Environmental safety

- Dangerous for the environment
- Toxic to aquatic organisms
- Dangerous to fish or other aquatic life. Do not contaminate surface waters or ditches with chemical or used container
- LERAP Category B

Hazard classification and safety precautions

Hazard Harmful, Dangerous for the environment
Transport code 9
Packaging group III
UN number 3082
Risk phrases R38, R40, R41, R51, R53a
Operator protection A, C, H; U05a, U11
Environmental protection E13b, E16a, E16b, E34, E38
Storage and disposal D01, D02, D05, D09a, D10b, D11a, D12a
Medical advice M03, M05a

392 prosulfocarb

A thiocarbamate herbicide for grass and broad-leaved weed control in cereals
HRAC mode of action code: N

See also clodinafop-propargyl + prosulfocarb

Products

1	Defy	Syngenta	800 g/l	EC	12606
2	Dian	AgChem Access	800 g/l	EC	13567

Uses

- Annual dicotyledons in ***celeriac*** *(off-label)*, ***celery (outdoor)*** *(off-label)*, ***durum wheat*** *(off-label)*, ***poppies for morphine production*** *(off-label)*, ***spring field beans*** *(off-label)*, ***spring wheat*** *(off-label)*, ***winter field beans*** *(off-label)*, ***winter linseed*** *(off-label)* [1]; ***potatoes***, ***winter barley***, ***winter wheat*** [1, 2]
- Annual grasses in ***bulb onions*** *(off-label)*, ***carrots*** *(off-label)*, ***celeriac*** *(off-label)*, ***celery (outdoor)*** *(off-label)*, ***durum wheat*** *(off-label)*, ***garlic*** *(off-label)*, ***herbs (see appendix 6)*** *(off-label)*, ***horseradish*** *(off-label)*, ***leeks*** *(off-label)*, ***onion sets*** *(off-label)*, ***parsley root*** *(off-label)*, ***parsnips*** *(off-label)*, ***poppies for morphine production*** *(off-label)*, ***salsify*** *(off-label)*, ***shallots*** *(off-label)*, ***spring field beans*** *(off-label)*, ***spring onions*** *(off-label)*, ***spring rye*** *(off-label)*, ***spring wheat*** *(off-label)*, ***triticale*** *(off-label)*, ***winter field beans*** *(off-label)*, ***winter linseed*** *(off-label)*, ***winter rye*** *(off-label)* [1]

- Annual meadow grass in ***poppies for morphine production*** *(off-label)*, ***spring barley*** *(off-label)*, ***spring field beans*** *(off-label)*, ***spring onions*** *(off-label)*, ***spring wheat*** *(off-label)*, ***winter linseed*** *(off-label)* [1]; ***potatoes***, ***winter barley***, ***winter wheat*** [1, 2]
- Black nightshade in ***herbs (see appendix 6)*** *(off-label)* [1]
- Chickweed in ***potatoes***, ***winter barley***, ***winter wheat*** [1, 2]
- Cleavers in ***carrots*** *(off-label)*, ***herbs (see appendix 6)*** *(off-label)*, ***horseradish*** *(off-label)*, ***parsley root*** *(off-label)*, ***parsnips*** *(off-label)*, ***salsify*** *(off-label)*, ***spring field beans*** *(off-label)*, ***spring wheat*** *(off-label)*, ***winter field beans*** *(off-label)* [1]; ***potatoes***, ***winter barley***, ***winter wheat*** [1, 2]
- Fat hen in ***carrots*** *(off-label)*, ***horseradish*** *(off-label)*, ***parsley root*** *(off-label)*, ***parsnips*** *(off-label)*, ***salsify*** *(off-label)* [1]
- Field speedwell in ***potatoes***, ***winter barley***, ***winter wheat*** [1, 2]
- Fumitory in ***bulb onions*** *(off-label)*, ***garlic*** *(off-label)*, ***herbs (see appendix 6)*** *(off-label)*, ***leeks*** *(off-label)*, ***onion sets*** *(off-label)*, ***shallots*** *(off-label)*, ***spring onions*** *(off-label)*, ***spring wheat*** *(off-label)* [1]
- Ivy-leaved speedwell in ***potatoes***, ***winter barley***, ***winter wheat*** [1, 2]
- Knotgrass in ***carrots*** *(off-label)*, ***horseradish*** *(off-label)*, ***parsley root*** *(off-label)*, ***parsnips*** *(off-label)*, ***salsify*** *(off-label)* [1]
- Loose silky bent in ***potatoes***, ***winter barley***, ***winter wheat*** [1, 2]
- Mayweeds in ***carrots*** *(off-label)*, ***horseradish*** *(off-label)*, ***parsley root*** *(off-label)*, ***parsnips*** *(off-label)*, ***salsify*** *(off-label)* [1]
- Polygonums in ***bulb onions*** *(off-label)*, ***garlic*** *(off-label)*, ***leeks*** *(off-label)*, ***onion sets*** *(off-label)*, ***shallots*** *(off-label)* [1]
- Rough meadow grass in ***potatoes***, ***winter barley***, ***winter wheat*** [1, 2]
- Volunteer oilseed rape in ***spring onions*** *(off-label)* [1]

Specific Off-Label Approvals (SOLAs)

- ***bulb onions*** *20073775 expires 01 Aug 2010* [1]
- ***carrots*** *20073774* [1]
- ***celeriac*** *20073777* [1]
- ***celery (outdoor)*** *20071752* [1]
- ***durum wheat*** *20073781* [1]
- ***garlic*** *20073775 expires 01 Aug 2010* [1]
- ***herbs (see appendix 6)*** *20072073* [1]
- ***horseradish*** *20073774* [1]
- ***leeks*** *20073776 expires 01 Aug 2010* [1]
- ***onion sets*** *20073775 expires 01 Aug 2010* [1]
- ***parsley root*** *20073774* [1]
- ***parsnips*** *20073774* [1]
- ***poppies for morphine production*** *20082243* [1]
- ***salsify*** *20073774* [1]
- ***shallots*** *20073775 expires 01 Aug 2010* [1]
- ***spring barley*** *20073778* [1]
- ***spring field beans*** *20082932* [1]
- ***spring onions*** *20080085* [1]
- ***spring rye*** *20073779* [1]
- ***spring wheat*** *20090237* [1]
- ***triticale*** *20073779* [1]
- ***winter field beans*** *20073780* [1]
- ***winter linseed*** *20092012* [1]
- ***winter rye*** *20073779* [1]

Approval information

- Prosulfocarb included in Annex I under EC Directive 91/414

Efficacy guidance

- Best results obtained in cereals from treatment of crops in a firm moist seedbed, free from clods
- Pre-emergence use will reduce blackgrass populations but the product should only be used against this weed as part of a management strategy involving sequences with products of alternative modes of action
- Always follow WRAG guidelines for preventing and managing herbicide resistant weeds. See Section 5 for more information

Restrictions

- Maximum number of treatments 1 per crop for winter barley, winter wheat and potatoes
- Do not apply to crops under stress from any cause. Transient yellowing can occur from which recovery is complete
- Winter cereals must be covered by 3 cm of settled soil

Crop-specific information

- Latest use: at emergence (soil rising over emerging potato shoots) for potatoes, up to and including early tillering (GS 21) for winter barley, winter wheat
- When applied pre-emergence to cereals crop emergence may occasionally be slowed down but yield is not affected
- For potatoes, complete ridge formation before application and do not disturb treated soil afterwards
- Peas are severely damaged or killed.

Following crops guidance

- Do not sow field or broad beans within 12 mth of treatment
- In the event of failure of a treated cereal crop, winter wheat, winter barley or transplanted brassicas may be re-sown immediately. seed brassicas can be sown 14 weeks after application provided that the land is ploughed first. In the following spring sunflowers, maize, flax, spring cereals, peas, oilseed rape or soya beans may be sown without ploughing, and carrots, lettuce, onions, sugar beet or potatoes may be sown or planted after ploughing

Environmental safety

- Dangerous for the environment
- Very toxic to aquatic organisms
- LERAP Category B

Hazard classification and safety precautions

Hazard Irritant, Dangerous for the environment
Transport code 9
Packaging group III
UN number 3082
Risk phrases R38, R43, R50, R53a
Operator protection A, C, H; U02a, U05a, U08, U14, U15, U20b
Environmental protection E15b, E16a, E16b, E38
Storage and disposal D01, D02, D05, D09a, D10c, D12a
Medical advice M05a

393 prosulfuron

A contact and residual sulfonyl urea herbicide available only in mixture
HRAC mode of action code: B

See also *bromoxynil + prosulfuron*

394 prothioconazole

A systemic, protectant and curative triazole fungicide
FRAC mode of action code: 3

See also *clothianidin + prothioconazole*
clothianidin + prothioconazole + tebuconazole + triazoxide
fluoxastrobin + prothioconazole
fluoxastrobin + prothioconazole + trifloxystrobin

Products

1	Clayton Impress	Clayton	250 g/l	EC	12912
2	Proline	Bayer CropScience	250 g/l	EC	12084
3	Redigo	Bayer CropScience	100 g/l	FS	12085
4	Rudis	Bayer CropScience	480 g/l	SC	14122
5	Standon Win-Win	Standon	250 g/l	EC	12386

Uses

- Alternaria in ***cabbages*** [4]
- Brown rust in ***spring barley***, ***winter barley***, ***winter rye***, ***winter wheat*** [1, 2, 5]; ***spring wheat*** [2]
- Bunt in ***durum wheat*** *(seed treatment)*, ***spring rye*** *(seed treatment)*, ***spring wheat*** *(seed treatment)*, ***triticale*** *(seed treatment)*, ***winter rye*** *(seed treatment)*, ***winter wheat*** *(seed treatment)* [3]
- Covered smut in ***winter barley*** *(seed treatment)* [3]
- Crown rust in ***spring oats***, ***winter oats*** [2]
- Eyespot in ***spring barley***, ***winter barley***, ***winter rye***, ***winter wheat*** [1, 2, 5]; ***spring oats***, ***spring wheat***, ***winter oats*** [2]
- Foliar disease control in ***evening primrose*** *(off-label)*, ***honesty*** *(off-label)*, ***mustard*** *(off-label)* [2]
- Fusarium foot rot and seedling blight in ***durum wheat*** *(seed treatment)*, ***spring oats*** *(seed treatment)*, ***spring rye*** *(seed treatment)*, ***spring wheat*** *(seed treatment)*, ***triticale*** *(seed treatment)*, ***winter barley*** *(seed treatment)*, ***winter oats*** *(seed treatment)*, ***winter rye*** *(seed treatment)*, ***winter wheat*** *(seed treatment)* [3]
- Glume blotch in ***spring wheat*** [2]; ***winter wheat*** [1, 2, 5]
- Late ear diseases in ***spring barley***, ***winter barley***, ***winter wheat*** [1, 2, 5]; ***spring wheat*** [2]
- Leaf stripe in ***winter barley*** *(seed treatment)* [3]
- Light leaf spot in ***cabbages*** [4]
- Loose smut in ***durum wheat*** *(seed treatment)*, ***spring oats*** *(seed treatment)*, ***spring wheat*** *(seed treatment)*, ***winter barley*** *(seed treatment)*, ***winter oats*** *(seed treatment)*, ***winter wheat*** *(seed treatment)* [3]
- Net blotch in ***spring barley***, ***winter barley*** [1, 2, 5]
- Phoma in ***cabbages*** [4]
- Powdery mildew in ***cabbages*** [4]; ***spring barley***, ***winter barley***, ***winter rye***, ***winter wheat*** [1, 2, 5]; ***spring oats***, ***spring wheat***, ***winter oats*** [2]
- Purple blotch in ***leeks*** [4]
- Rhynchosporium in ***spring barley***, ***winter barley***, ***winter rye*** [1, 2, 5]
- Ring spot in ***cabbages*** [4]
- Rust in ***leeks*** [4]
- Sclerotinia stem rot in ***winter oilseed rape*** [1, 2, 5]
- Septoria leaf blotch in ***spring wheat*** [2]; ***winter wheat*** [1, 2, 5]
- Tan spot in ***spring wheat*** [2]; ***winter wheat*** [1, 2, 5]
- Yellow rust in ***spring wheat*** [2]; ***winter barley***, ***winter wheat*** [1, 2, 5]

Specific Off-Label Approvals (SOLAs)

- ***evening primrose*** *20090211* [2]
- ***honesty*** *20090211* [2]
- ***mustard*** *20090211* [2]

Approval information

- Accepted by BBPA for use on malting barley
- Prothioconazole included in Annex I under EC Directive 91/414

Efficacy guidance

- Seed treatments must be applied by manufacturer's recommended treatment application equipment [3]
- Treated cereal seed should preferably be drilled in the same season [3]
- Follow-up treatments will be needed later in the season to give protection against air-borne and splash-borne diseases [3]
- Best results on cereal foliar diseases obtained from treatment at early stages of disease development. Further treatment may be needed if disease attack is prolonged [2, 5]
- Foliar applications to established infections of any disease are likely to be less effective [2, 5]
- Best control of cereal ear diseases obtained by treatment during ear emergence [2, 5]
- Treat oilseed rape at early to full flower [2, 5]
- Prothioconazole is a DMI fungicide. Resistance to some DMI fungicides has been identified in Septoria leaf blotch which may seriously affect performance of some products. For further advice contact a specialist advisor and visit the Fungicide Resistance Action Group (FRAG)-UK website

Restrictions

- Maximum number of seed treatments one per batch [3]

- Maximum total dose equivalent to one full dose treatment on oilseed rape, two full dose treatments on barley, oats; three full dose treatments on wheat, rye [2, 5]
- Seed treatment must be fully re-dispersed and homogeneous before use [3]
- Do not use on seed with more than 16% moisture content, or on sprouted, cracked or skinned seed [3]
- All seed batches should be tested to ensure they are suitable for treatment [3]
- Treated winter barley seed must be used within the season of treatment; treated winter wheat seed should preferably be drilled in the same season [3]
- Do not make repeated treatments of the product alone to the same crop against pathogens such as powdery mildew. Use tank mixtures or alternate with fungicides having a different mode of action [3]
- Do not treat oilseed rape crops to be used for seed production [2, 5]

Crop-specific information

- Latest use: pre-drilling for seed treatments [3]; before grain milky ripe for foliar sprays on winter rye, winter wheat; beginning of flowering for barley, oats [2, 5]
- HI 56 d [2] or 65 d [5] for winter oilseed rape

Environmental safety

- Dangerous for the environment [2, 5]
- Toxic to aquatic organisms [2, 5]
- Harmful to aquatic organisms [3]
- LERAP Category B [2, 5]

Hazard classification and safety precautions

Hazard Irritant [1-3, 5]; Dangerous for the environment [1, 2, 4, 5]
Transport code 9 [1, 2, 4, 5]
Packaging group III [1, 2, 4, 5]
UN number 3082 [1, 2, 4, 5]; N/C [3]
Risk phrases R36 [1, 2, 5]; R43, R52 [3]; R51 [1, 2, 4, 5]; R53a [1-5]
Operator protection A, H [1-5]; C [1, 2, 5]; U05a, U20b [1, 2, 4, 5]; U07, U14, U20a [3]; U09b [1, 2, 5]
Environmental protection E15a [1-3, 5]; E15b [4]; E16a, E38 [1, 2, 4, 5]; E34 [1-5]; E36a [3]
Storage and disposal D01, D02, D12a [1-5]; D05, D09a [1-3, 5]; D10b [1, 2, 4, 5]; D14 [3]
Treated seed S01, S02, S03, S04a, S04b, S05, S06b, S07, S08 [3]
Medical advice M03 [1, 2, 4, 5]

395 prothioconazole + spiroxamine

A broad spectrum fungicide mixture for cereals
FRAC mode of action code: 3 + 5

See also *spiroxamine*

Products

1	Helix	Bayer CropScience	160:300 g/l	EC	12264
2	Spiral	AgChem Access	160:300 g/l	EC	13993

Uses

- Brown rust in ***spring barley***, ***winter barley***, ***winter rye***, ***winter wheat***
- Crown rust in ***spring oats***, ***winter oats***
- Eyespot in ***spring barley***, ***spring oats***, ***winter barley***, ***winter oats***, ***winter rye***, ***winter wheat***
- Glume blotch in ***winter wheat***
- Late ear diseases in ***winter rye***, ***winter wheat***
- Net blotch in ***spring barley***, ***winter barley***
- Powdery mildew in ***spring barley***, ***spring oats***, ***winter barley***, ***winter oats***, ***winter rye***, ***winter wheat***
- Rhynchosporium in ***spring barley***, ***winter barley***, ***winter rye***
- Septoria leaf blotch in ***winter wheat***
- Tan spot in ***winter wheat***
- Yellow rust in ***spring barley***, ***winter barley***, ***winter wheat***

SEE SECTION 3 FOR PRODUCTS ALSO REGISTERED

Approval information

- Prothioconazole and spiroxamine included in Annex I under EC Directive 91/414
- Accepted by BBPA for use on malting barley

Efficacy guidance

- Best results obtained from treatment at early stages of disease development. Further treatment may be needed if disease attack is prolonged
- Applications to established infections of any disease are likely to be less effective
- Best control of cereal ear diseases obtained by treatment during ear emergence
- Prothioconazole is a DMI fungicide. Resistance to some DMI fungicides has been identified in Septoria leaf blotch which may seriously affect performance of some products. For further advice contact a specialist advisor and visit the Fungicide Resistance Action Group (FRAG)-UK website

Restrictions

- Maximum total dose equivalent to two full dose treatments on barley and oats and three full dose treatments on wheat and rye

Crop-specific information

- Latest use: before grain watery ripe for rye, oats, winter wheat; up to beginning of anthesis for barley

Environmental safety

- Dangerous for the environment
- Very toxic to aquatic organisms
- LERAP Category B

Hazard classification and safety precautions

Hazard Harmful, Dangerous for the environment
Transport code 9
Packaging group III
UN number 3082
Risk phrases R20, R22a, R36, R38, R50, R53a
Operator protection A, C, H; U05a, U09b, U11, U19a, U20b
Environmental protection E15a, E16a, E34, E38
Storage and disposal D01, D02, D05, D09a, D10b, D12a
Medical advice M03

396 prothioconazole + spiroxamine + tebuconazole

A broad spectrum fungicide mixture for cereals
FRAC mode of action code: 3 + 5 + 3

See also *spiroxamine*
tebuconazole

Products

Cello	Bayer CropScience	100:250:100 g/l	EC	13178

Uses

- Brown rust in ***spring barley***, ***winter barley***, ***winter rye***, ***winter wheat***
- Crown rust in ***spring oats***, ***winter oats***
- Eyespot in ***spring barley*** *(reduction)*, ***spring oats***, ***winter barley*** *(reduction)*, ***winter oats***, ***winter rye*** *(reduction)*, ***winter wheat*** *(reduction)*
- Glume blotch in ***winter wheat***
- Late ear diseases in ***spring barley***, ***winter barley***, ***winter wheat***
- Net blotch in ***spring barley***, ***winter barley***
- Powdery mildew in ***spring barley***, ***spring oats***, ***winter barley***, ***winter oats***, ***winter rye***, ***winter wheat***
- Rhynchosporium in ***spring barley***, ***winter barley***, ***winter rye***
- Septoria leaf blotch in ***winter wheat***
- Yellow rust in ***spring barley***, ***winter barley***, ***winter wheat***

Approval information

- Prothioconazole and spiroxamine included in Annex I under EC Directive 91/414

Efficacy guidance

- Best results obtained from treatment at early stages of disease development. Further treatment may be needed if disease attack is prolonged
- Applications to established infections of any disease are likely to be less effective
- Best control of cereal ear diseases obtained by treatment during ear emergence
- Prothioconazole and tebuconazole are DMI fungicides. Resistance to some DMI fungicides has been identified in Septoria leaf blotch which may seriously affect performance of some products. For further advice contact a specialist advisor and visit the Fungicide Resistance Action Group (FRAG)-UK website

Restrictions

- Maximum total dose equivalent to two full dose treatments

Crop-specific information

- Latest use: before grain milky ripe stage for rye, wheat; up to beginning of anthesis for barley & oats

Environmental safety

- Dangerous for the environment
- Very toxic to aquatic organisms
- LERAP Category B

Hazard classification and safety precautions

Hazard Harmful, Dangerous for the environment
Transport code 9
Packaging group III
UN number 3082
Risk phrases R20, R36, R37, R38, R43, R50, R53a, R63
Operator protection A, C, H; U05a, U09b, U11, U14, U19a, U20b
Environmental protection E15a, E16a, E34, E38
Storage and disposal D01, D02, D05, D09a, D10b, D12a
Medical advice M03

397 prothioconazole + tebuconazole

A triazole fungicide mixture for cereals
FRAC mode of action code: 3 + 3

See also *tebuconazole*

Products

1	Corinth	Bayer CropScience	80;160 g/l	EC	14199
2	Kestrel	Bayer CropScience	160+80 g/L	EC	13809
3	Prosaro	Bayer CropScience	125:125 g/l	EC	12263
4	Proteb	AgChem Access	125:125 g/l	EC	13879
5	Standon Mastana	Standon	125:125	EC	13795

Uses

- Brown rust in ***spring barley***, ***winter barley***, ***winter rye***, ***winter wheat*** [2-5]; ***spring wheat*** [3]
- Crown rust in ***spring oats***, ***winter oats*** [2-4]
- Eyespot in ***spring barley***, ***winter barley***, ***winter rye***, ***winter wheat*** [2, 5]; ***spring barley*** *(reduction)*, ***winter barley*** *(reduction)*, ***winter rye*** *(reduction)*, ***winter wheat*** *(reduction)* [3, 4]; ***spring oats***, ***winter oats*** [2-4]; ***spring wheat*** [3]
- Foliar disease control in ***grass seed crops*** *(off-label)* [3]
- Fusarium ear blight in ***spring barley***, ***winter barley***, ***winter wheat*** [2]; ***winter wheat*** *(moderate control)* [5]
- Glume blotch in ***spring wheat*** [3]; ***winter wheat*** [2-5]
- Late ear diseases in ***spring barley***, ***winter barley***, ***winter wheat*** [3, 4]; ***spring wheat*** [3]
- Light leaf spot in ***oilseed rape*** [2]; ***oilseed rape*** *(moderate control only)* [1]; ***spring oilseed rape***, ***winter oilseed rape*** [3, 4]; ***winter oilseed rape*** *(moderate control)* [5]
- Net blotch in ***spring barley***, ***winter barley*** [2-5]; ***winter rye*** [5]
- Phoma in ***spring oilseed rape*** [3, 4]; ***winter oilseed rape*** [3-5]
- Phoma leaf spot in ***oilseed rape*** [1, 2]

- Powdery mildew in ***spring barley, spring oats, winter barley, winter oats, winter rye, winter wheat*** [2-5]; ***spring wheat*** [3]
- Rhynchosporium in ***spring barley, winter barley, winter rye*** [2-5]
- Sclerotinia stem rot in ***oilseed rape*** [1, 2]; ***spring oilseed rape*** [3, 4]; ***winter oilseed rape*** [3-5]
- Septoria leaf blotch in ***spring wheat*** [3]; ***winter wheat*** [2-5]
- Sooty moulds in ***spring barley, winter barley, winter wheat*** [2]; ***winter wheat*** *(reduction)* [5]
- Stem canker in ***oilseed rape*** [1]
- Tan spot in ***spring wheat*** [3]; ***winter wheat*** [2-4]
- Yellow rust in ***spring barley, winter barley, winter wheat*** [2-5]; ***spring wheat*** [3]; ***winter rye*** [5]

Specific Off-Label Approvals (SOLAs)

- ***grass seed crops*** *20081806* [3]

Approval information

- Prothioconazole included in Annex I under EC Directive 91/414
- Accepted by BBPA for use on malting barley

Efficacy guidance

- Best results on cereal foliar diseases obtained from treatment at early stages of disease development. Further treatment may be needed if disease attack is prolonged
- On oilseed rape apply a protective treatment in autumn/winter for Phoma followed by a further spray in early spring from the onset of stem elongation, if necessary. For control of Sclerotinia apply at early to full flower
- Applications to established infections of any disease are likely to be less effective
- Best control of cereal ear diseases obtained by treatment during ear emergence
- Prothioconazole and tebuconazole are DMI fungicides. Resistance to some DMI fungicides has been identified in Septoria leaf blotch which may seriously affect performance of some products. For further advice contact a specialist advisor and visit the Fungicide Resistance Action Group (FRAG)-UK website

Restrictions

- Maximum total dose equivalent to two full dose treatments on barley, oats, oilseed rape; three full dose treatments on wheat and rye

Crop-specific information

- Latest use: before grain milky ripe for rye, wheat; beginning of flowering for barley, oats
- HI 56 d for oilseed rape

Environmental safety

- Dangerous for the environment
- Toxic to aquatic organisms
- Risk to non-target insects or other arthropods. Avoid spraying within 6 m of the field boundary to reduce the effects on non-target insects or other arthropods
- LERAP Category B

Hazard classification and safety precautions

Hazard Harmful [1, 2, 5]; Irritant [3, 4]; Dangerous for the environment [1-5]
Transport code 9
Packaging group III
UN number 3082
Risk phrases R36 [1, 2]; R38, R43, R51, R53a, R63 [1-5]
Operator protection A, H [1-5]; C [1, 2]; U05a [1-5]; U09b, U20b [2-5]; U11, U20c [1]; U19a [1, 3-5]
Environmental protection E15a, E38 [1-4]; E15b [5]; E16a, E34 [1-5]; E22b [1]; E22c [2-5]
Storage and disposal D01, D02, D09a, D12a [1-5]; D05, D10b [1-4]; D10a [5]
Medical advice M03

398 prothioconazole + tebuconazole + triazoxide

A broad spectrum fungicide seed treatment for cereals
FRAC mode of action code: 3 + 3 + 35

See also *tebuconazole*
triazoxide

Products

Raxil Pro	Bayer CropScience	25:15:10 g/l	FS	12432

Uses

- Covered smut in ***spring barley*** *(seed treatment)*, ***winter barley*** *(seed treatment)*
- Fusarium foot rot and seedling blight in ***spring barley*** *(seed treatment)*, ***winter barley*** *(seed treatment)*
- Leaf stripe in ***spring barley*** *(seed treatment)*, ***winter barley*** *(seed treatment)*
- Loose smut in ***spring barley*** *(seed treatment)*, ***winter barley*** *(seed treatment)*
- Net blotch in ***spring barley*** *(seed treatment)*, ***winter barley*** *(seed treatment)*

Approval information

- Prothioconazole included in Annex I under EC Directive 91/414
- Accepted by BBPA for use on malting barley

Efficacy guidance

- Treatments must be applied by manufacturer's recommended treatment application equipment
- Treated seed should preferably be drilled in the same season
- Follow-up treatments will be needed later in the season to give protection against air-borne and splash-borne diseases

Restrictions

- Maximum number of seed treatments one per batch

Crop-specific information

- Latest use for seed treatment: pre-drilling

Environmental safety

- Harmful to aquatic organisms

Hazard classification and safety precautions

Hazard Irritant
Transport code 9
Packaging group III
UN number 3082
Risk phrases R37, R43, R52, R53a
Operator protection A, H; U07, U14, U19a, U20a
Environmental protection E15a, E34, E36a, E38
Storage and disposal D05, D09a, D14
Treated seed S01, S02, S03, S04b, S05, S06b, S07, S08

399 prothioconazole + trifloxystrobin

A triazole and strobilurin fungicide mixture for cereals
FRAC mode of action code: 3 + 11

See also *trifloxystrobin*

Products

1	Mobius	Bayer CropScience	175;150 g/l	SC	13395
2	Zephyr	Bayer CropScience	175:88 g/l	SC	13174

Uses

- Brown rust in ***spring barley***, ***winter barley***, ***winter wheat*** [1, 2]
- Ear diseases in ***winter wheat*** [1]
- Eyespot in ***spring barley*** *(reduction)*, ***winter barley*** *(reduction)* [2]; ***spring barley*** *(reduction in severity)*, ***winter barley*** *(reduction in severity)* [1]; ***winter wheat*** [1, 2]
- Glume blotch in ***winter wheat*** [1, 2]

- Net blotch in ***spring barley***, ***winter barley*** [1, 2]
- Powdery mildew in ***spring barley***, ***winter barley***, ***winter wheat*** [1, 2]
- Rhynchosporium in ***spring barley***, ***winter barley*** [1, 2]
- Septoria leaf blotch in ***winter wheat*** [1, 2]
- Yellow rust in ***spring barley***, ***winter barley***, ***winter wheat*** [1, 2]

Approval information

- Prothioconazole and trifloxystrobin included in Annex I under EC Directive 91/414
- Accepted by BBPA for use on malting barley

Efficacy guidance

- Best results obtained from treatment at early stages of disease development. Further treatment may be needed if disease attack is prolonged
- Applications to established infections of any disease are likely to be less effective
- Best control of cereal ear diseases obtained by treatment during ear emergence
- Prothioconazole is a DMI fungicide. Resistance to some DMI fungicides has been identified in Septoria leaf blotch which may seriously affect performance of some products. For further advice contact a specialist advisor and visit the Fungicide Resistance Action Group (FRAG)-UK website
- Trifloxystrobin is a member of the QoI cross resistance group. Product should be used preventatively and not relied on for its curative potential
- Use product as part of an Integrated Crop Management strategy incorporating other methods of control, including where appropriate other fungicides with a different mode of action. Do not apply more than two foliar applications of QoI containing products to any cereal crop
- There is a significant risk of widespread resistance occurring in *Septoria tritici* populations in UK. Failure to follow resistance management action may result in reduced levels of disease control
- Strains of wheat and barley powdery mildew resistant to QoIs are common in the UK. Control of wheat mildew can only be relied on from the triazole component
- Where specific control of wheat mildew is required this should be achieved through a programme of measures including products recommended for the control of mildew that contain a fungicide from a different cross-resistance group and applied at a dose that will give robust control

Restrictions

- Maximum total dose equivalent to two full dose treatments

Crop-specific information

- Latest use: before grain milky ripe for wheat; beginning of flowering for barley

Environmental safety

- Dangerous for the environment
- Very toxic to aquatic organisms
- LERAP Category B

Hazard classification and safety precautions

Hazard Irritant, Dangerous for the environment
Transport code 9
Packaging group III
UN number 3082
Risk phrases R37, R50, R53a [1, 2]; R70 [1]
Operator protection A, C, H; U05a, U09b, U19a [1, 2]; U20b [2]; U20c [1]
Environmental protection E15a, E16a, E38 [1, 2]; E34 [2]
Storage and disposal D01, D02, D09a, D10c, D12a
Medical advice M03

400 pymetrozine

A novel azomethine insecticide
IRAC mode of action code: 9B

Products

1	Chess WG	Syngenta Bioline	50% w/w	WG	13310
2	Plenum WG	Syngenta	50% w/w	WG	10652

Uses

- Aphids in ***bilberries*** *(off-label)*, ***blackberries*** *(off-label)*, ***blackcurrants*** *(off-label)*, ***blueberries*** *(off-label)*, ***celeriac*** *(off-label)*, ***celery (outdoor)*** *(off-label)*, ***chinese cabbage*** *(off-label)*, ***choi sum*** *(off-label)*, ***collards*** *(off-label)*, ***cranberries*** *(off-label)*, ***forest nurseries*** *(off-label)*, ***frise*** *(off-label)*, ***gooseberries*** *(off-label)*, ***kale*** *(off-label)*, ***lamb's lettuce*** *(off-label)*, ***lettuce*** *(off-label)*, ***loganberries*** *(off-label)*, ***pak choi*** *(off-label)*, ***radicchio*** *(off-label)*, ***raspberries*** *(off-label)*, ***redcurrants*** *(off-label)*, ***rubus hybrids*** *(off-label)*, ***salad brassicas*** *(off-label - for baby leaf production)*, ***seed potatoes***, ***soft fruit*** *(off-label)*, ***strawberries*** *(off-label)*, ***sweetcorn*** *(off-label)*, ***vaccinium spp.*** *(off-label)*, ***ware potatoes***, ***whitecurrants*** *(off-label)* [2]; ***brassica seed beds*** *(off-label)*, ***ornamental plant production***, ***protected blackberries*** *(off-label)*, ***protected blackcurrants*** *(off-label)*, ***protected celery*** *(off-label)*, ***protected chives*** *(off-label)*, ***protected courgettes*** *(off-label)*, ***protected cucumbers***, ***protected endives*** *(off-label)*, ***protected gherkins*** *(off-label)*, ***protected gooseberries*** *(off-label)*, ***protected herbs (see appendix 6)*** *(off-label)*, ***protected lettuce*** *(off-label)*, ***protected loganberries*** *(off-label)*, ***protected marrows*** *(off-label)*, ***protected melons*** *(off-label)*, ***protected ornamentals***, ***protected parsley*** *(off-label)*, ***protected pumpkins*** *(off-label)*, ***protected raspberries*** *(off-label)*, ***protected redcurrants*** *(off-label)*, ***protected ribes hybrids*** *(off-label)*, ***protected rubus hybrids*** *(off-label)*, ***protected salad brassicas*** *(off-label - for baby leaf production)*, ***protected squashes*** *(off-label)*, ***protected strawberries*** *(off-label)* [1]
- Cabbage aphid in ***chinese cabbage*** *(off-label)*, ***choi sum*** *(off-label)*, ***collards*** *(off-label)*, ***kale*** *(off-label)*, ***pak choi*** *(off-label)* [2]
- Damson-hop aphid in ***hops*** *(off-label)* [2]
- Glasshouse whitefly in ***protected tomatoes*** *(off-label)* [1]
- Mealy aphid in ***broccoli*** *(useful levels of control)*, ***brussels sprouts*** *(useful levels of control)*, ***cabbages*** *(useful levels of control)*, ***calabrese*** *(useful levels of control)*, ***cauliflowers*** *(useful levels of control)*, ***chinese cabbage*** *(off-label)*, ***choi sum*** *(off-label)*, ***collards*** *(off-label)*, ***kale*** *(off-label)*, ***pak choi*** *(off-label)* [2]
- Peach-potato aphid in ***broccoli***, ***brussels sprouts***, ***cabbages***, ***calabrese***, ***cauliflowers***, ***chinese cabbage*** *(off-label)*, ***choi sum*** *(off-label)*, ***collards*** *(off-label)*, ***kale*** *(off-label)*, ***pak choi*** *(off-label)* [2]; ***protected aubergines*** *(off-label)*, ***protected endives*** *(off-label)*, ***protected peppers*** *(off-label)* [1]
- Tobacco whitefly in ***protected tomatoes*** *(off-label)* [1]

Specific Off-Label Approvals (SOLAs)

- ***bilberries*** *20061702* [2]
- ***blackberries*** *20061633* [2]
- ***blackcurrants*** *20060946* [2]
- ***blueberries*** *20061702* [2]
- ***brassica seed beds*** *20070788* [1]
- ***celeriac*** *20051062* [2]
- ***celery (outdoor)*** *20051062* [2]
- ***chinese cabbage*** *20082246* [2]
- ***choi sum*** *20082246* [2]
- ***collards*** *20082081* [2]
- ***cranberries*** *20061702* [2]
- ***forest nurseries*** *20082920* [2]
- ***frise*** *20070060* [2]
- ***gooseberries*** *20061702* [2]
- ***hops*** *20031423* [2]
- ***kale*** *20082081* [2]
- ***lamb's lettuce*** *20070060* [2]
- ***lettuce*** *20070060* [2]
- ***loganberries*** *20061702* [2]
- ***pak choi*** *20082246* [2]
- ***protected aubergines*** *20070501* [1]
- ***protected blackberries*** *20070498* [1]
- ***protected blackcurrants*** *20070504* [1]
- ***protected celery*** *20070503* [1]
- ***protected chives*** *20070502* [1]
- ***protected courgettes*** *20081012* [1]

- ***protected endives*** *20092025* [1]
- ***protected gherkins*** *20081012* [1]
- ***protected gooseberries*** *20070504* [1]
- ***protected herbs (see appendix 6)*** *20070502* [1]
- ***protected lettuce*** *20070502* [1]
- ***protected loganberries*** *20070504* [1]
- ***protected marrows*** *20081012* [1]
- ***protected melons*** *20081012* [1]
- ***protected parsley*** *20070502* [1]
- ***protected peppers*** *20070501* [1]
- ***protected pumpkins*** *20081012* [1]
- ***protected raspberries*** *20070498* [1]
- ***protected redcurrants*** *20070504* [1]
- ***protected ribes hybrids*** *20070504* [1]
- ***protected rubus hybrids*** *20070504* [1]
- ***protected salad brassicas*** *(for baby leaf production) 20070502* [1]
- ***protected squashes*** *20081012* [1]
- ***protected strawberries*** *20070499* [1]
- ***protected tomatoes*** *20070501* [1]
- ***radicchio*** *20070060* [2]
- ***raspberries*** *20061633* [2]
- ***redcurrants*** *20061702* [2]
- ***rubus hybrids*** *20061702* [2]
- ***salad brassicas*** *(for baby leaf production) 20070060* [2]
- ***soft fruit*** *20082920* [2]
- ***strawberries*** *20060461* [2]
- ***sweetcorn*** *20041318* [2]
- ***vaccinium spp.*** *20061702* [2]
- ***whitecurrants*** *20061702* [2]

Approval information

- Pymetrozine included in Annex I under EC Directive 91/414
- Accepted by BBPA for use on hops

Efficacy guidance

- Pymetrozine moves systemically in the plant and acts by preventing feeding leading to death by starvation in 1-4 d. There is no immediate knockdown
- Aphids controlled include those resistant to organophosphorus and carbamate insecticides
- To prevent development of resistance do not use continuously or as the sole method of control
- Best results achieved by starting spraying as soon as aphids seen in crop and repeating as necessary.
- To limit spread of persistent viruses such as potato leaf roll virus, apply from 90% crop emergence

Restrictions

- Maximum total dose equivalent to two full dose treatments on ware potatoes; three full dose treatments on seed potatoes, leaf spinach [2]; four full dose treatments on cucumbers, ornamentals [1]
- Consult processors before use on potatoes for processing
- Check tolerance of ornamental species before large scale use. See label for list of species known to have been treated without damage. Visible spray deposits may be seen on leaves of some species [1]

Crop-specific information

- HI cucumbers 3 d; potatoes, leaf spinach 7 d

Environmental safety

- High risk to bees (outdoor use only). Do not apply to crops in flower or to those in which bees are actively foraging. Do not apply when flowering weeds are present
- Avoid spraying within 6 m of field boundaries to reduce effects on non-target insects or arthropods. Risk to certain non-target insects and arthropods.

Hazard classification and safety precautions

Hazard Harmful

UN number N/C

Risk phrases R40
Operator protection A [1, 2]; H [2]; U05a, U20c
Environmental protection E12a, E12e, E15a
Storage and disposal D01, D02, D09a, D11a

401 pyraclostrobin

A protectant and curative strobilurin fungicide for cereals
FRAC mode of action code: 11

See also *boscalid + pyraclostrobin*
dithianon + pyraclostrobin
epoxiconazole + fenpropimorph + pyraclostrobin
epoxiconazole + kresoxim-methyl + pyraclostrobin
epoxiconazole + pyraclostrobin
fenpropimorph + pyraclostrobin

Products

1	Comet 200	BASF	200 g/l	EC	12639
2	Flyer	BASF	250 g/l	EC	12654
3	Insignia	Vitax	20% w/w	WG	11865
4	Mascot Eland	Rigby Taylor	20% w/w	WG	14549
5	Platoon 250	BASF	250 g/l	EC	12640
6	Tucana	BASF	250 g/l	EC	10899
7	Vanguard	Sherriff Amenity	20% w/w	WG	13838
8	Vivid	BASF	250 g/l	EC	10898

Uses

- Brown rust in ***spring barley***, ***spring wheat***, ***winter barley***, ***winter wheat*** [1, 2, 5, 6, 8]
- Crown rust in ***spring oats***, ***winter oats*** [1, 2, 5, 6, 8]
- Dollar spot in ***managed amenity turf*** *(reduction)* [3]; ***managed amenity turf*** *(useful reduction)* [4, 7]
- Foliar disease control in ***durum wheat*** *(off-label)*, ***grass seed crops*** *(off-label)*, ***spring rye*** *(off-label)*, ***triticale*** *(off-label)*, ***winter rye*** *(off-label)* [6, 8]; ***forage maize*** *(off-label)* [1]; ***ornamental plant production*** *(off-label)* [8]
- Fusarium patch in ***managed amenity turf*** [3]; ***managed amenity turf*** *(moderate control)* [4]; ***managed amenity turf*** *(moderate control only)* [7]
- Glume blotch in ***spring wheat***, ***winter wheat*** [1, 2, 5, 6, 8]
- Net blotch in ***spring barley***, ***winter barley*** [1, 2, 5, 6, 8]
- Red thread in ***managed amenity turf*** [3, 4, 7]
- Rhynchosporium in ***spring barley*** *(moderate)*, ***winter barley*** *(moderate)* [1, 2, 5, 6, 8]
- Septoria leaf blotch in ***spring wheat***, ***winter wheat*** [1, 2, 5, 6, 8]
- Yellow rust in ***spring barley***, ***spring wheat***, ***winter barley***, ***winter wheat*** [1, 2, 5, 6, 8]

Specific Off-Label Approvals (SOLAs)

- ***durum wheat*** *20062652* [6], *20062649* [8]
- ***forage maize*** *20090928* [1]
- ***grass seed crops*** *20062652* [6], *20062649* [8]
- ***ornamental plant production*** *20082884* [8]
- ***spring rye*** *20062652* [6], *20062649* [8]
- ***triticale*** *20062652* [6], *20062649* [8]
- ***winter rye*** *20062652* [6], *20062649* [8]

Approval information

- Pyraclostrobin included in Annex I under EC Directive 91/414
- Accepted by BBPA for use on malting barley (before ear emergence only)

Efficacy guidance

- For best results apply at the start of disease attack on cereals [1-3, 5, 6, 8]
- For Fusarium Patch treat early as severe damage to turf can occur once the disease is established [3]
- Regular turf aeration, appropriate scarification and judicious use of nitrogenous fertiliser will assist the control of Fusarium Patch [3]

- Best results on Septoria glume blotch achieved when used as a protective treatment and against Septoria leaf blotch when treated in the latent phase [1-3, 5, 6, 8]
- Yield response may be obtained in the absence of visual disease symptoms [1-3, 5, 6, 8]
- Pyraclostrobin is a member of the QoI cross resistance group. Product should be used preventatively and not relied on for its curative potential
- Use product as part of an Integrated Crop Management strategy incorporating other methods of control, including where appropriate other fungicides with a different mode of action. Do not apply more than two foliar applications of QoI containing products to any cereal crop or to grass
- There is a significant risk of widespread resistance occurring in *Septoria tritici* populations in UK. Failure to follow resistance management action may result in reduced levels of disease control [1-3, 5, 6, 8]
- On cereal crops product must always be used in mixture with another product, recommended for control of the same target disease, that contains a fungicide from a different cross resistance group and is applied at a dose that will give robust control [1-3, 5, 6, 8]

Restrictions

- Maximum number of treatments 2 per crop on cereals [1, 2, 5, 6, 8]
- Maximum total dose on turf equivalent to two full dose treatments [3]
- Do not apply during drought conditions or to frozen turf [3]

Crop-specific information

- Latest use: before grain watery ripe (GS 71) for wheat; up to and including emergence of ear just complete (GS 59) for barley and oats [1, 2, 5, 6, 8]
- Avoid applying to turf immediately after cutting or 48 h before mowing [3]

Environmental safety

- Dangerous for the environment
- Very toxic to aquatic organisms
- LERAP Category B

Hazard classification and safety precautions

Hazard Harmful, Dangerous for the environment
Transport code 6.1 [1, 2, 5, 6, 8]; 9 [3, 4, 7]
Packaging group III
UN number 2902 [1, 2, 5, 6, 8]; 3077 [3, 4, 7]
Risk phrases R20, R50 [1-8]; R22a, R38, R53a [1, 2, 5, 6, 8]
Operator protection A [1-8]; D [3, 4, 7]; U05a, U14, U20b [1, 2, 5, 6, 8]
Environmental protection E15a [3, 7]; E15b [1, 2, 4-8]; E16a, E38 [1-8]; E16b [1-6, 8]; E16d [7]; E34 [1, 2, 5, 6, 8]
Storage and disposal D01, D02 [1, 2, 5, 6, 8]; D08, D09a, D10c, D12a [1-8]
Medical advice M03, M05a [1, 2, 5, 6, 8]

402 pyrethrins

A non-persistent, contact acting insecticide extracted from Pyrethrum
IRAC mode of action code: 3

Products

1	Dairy Fly Spray	B H & B	0.75 g/l	AL	H5579
2	Killgerm ULV 400	Killgerm	30 g/l	UL	H4838
3	Pyblast	Agropharm	30 g/l	UL	H7485
4	Pyrethrum 5 EC	Agropharm	50 g/l	EC	12685
5	Pyrethrum Spray	Killgerm	3.3 g/l	UL	H4636
6	Spruzit	Certis	4.59 g/l	EC	13438

Uses

- Aphids in ***all edible crops (outdoor), all non-edible crops (outdoor), protected crops*** [6]; ***broccoli, brussels sprouts, bush fruit, cabbages, calabrese, cane fruit, cauliflowers, lettuce, ornamental plant production, protected broccoli, protected brussels sprouts, protected bush fruit, protected cabbages, protected calabrese, protected cane fruit, protected cauliflowers, protected lettuce, protected ornamentals, protected tomatoes, tomatoes (outdoor)*** [4]
- Caterpillars in ***all edible crops (outdoor), all non-edible crops (outdoor), protected crops*** [6]; ***broccoli, brussels sprouts, bush fruit, cabbages, calabrese, cane fruit, cauliflowers, lettuce,***

ornamental plant production, protected broccoli, protected brussels sprouts, protected bush fruit, protected cabbages, protected calabrese, protected cane fruit, protected cauliflowers, protected lettuce, protected ornamentals, protected tomatoes, tomatoes (outdoor) [4]
- Flea beetle in ***broccoli, brussels sprouts, cabbages, calabrese, cauliflowers, protected broccoli, protected brussels sprouts, protected cabbages, protected calabrese, protected cauliflowers, protected tomatoes, tomatoes (outdoor)*** [4]; ***protected crops*** [6]
- Insect pests in ***dairies, farm buildings, livestock houses, poultry houses*** [1-3, 5]; ***glasshouses*** [5]
- Macrolophus caliginosus in ***protected tomatoes*** *(off-label)* [4]
- Mealybugs in ***protected crops*** [6]; ***protected tomatoes*** *(off-label)* [4]
- Scale insects in ***protected crops*** [6]
- Spider mites in ***all edible crops (outdoor), all non-edible crops (outdoor), protected crops*** [6]
- Thrips in ***all edible crops (outdoor), all non-edible crops (outdoor), protected crops*** [6]
- Whitefly in ***broccoli, brussels sprouts, cabbages, calabrese, cauliflowers, protected broccoli, protected brussels sprouts, protected cabbages, protected calabrese, protected cauliflowers, protected tomatoes, tomatoes (outdoor)*** [4]; ***protected crops*** [6]

SECTION 2

Specific Off-Label Approvals (SOLAs)
- ***protected tomatoes*** *20063026* [4]

Approval information
- Products formulated for ULV application [1-3]. May be applied through fogging machine or sprayer [1, 3, 5]. See label for details

Efficacy guidance
- For indoor fly control close doors and windows and spray or apply fog as appropriate [1-3]
- For best fly control outdoors spray during early morning or late afternoon and evening when conditions are still [1-3, 5]
- For all uses ensure good spray coverage of the target area or plants by increasing spray volume where necessary
- Best results on outdoor or protected crops achieved from treatment at first signs of pest attack in early morning or evening [6]
- To avoid possibility of development of resistance do not spray more frequently than once per week

Restrictions
- Maximum number of treatments on edible crops 3 per crop or season [4]
- For use only by professional operators [1-3, 5]
- Do not allow spray to contact open food products or food preparing equipment or utensils [1-3, 5]
- Remove exposed milk and collect eggs before application [1-3, 5]
- Do not treat plants [1-3, 5]
- Avoid direct application to open flowers [6]
- Do not mix with, or apply closely before or after, products containing dithianon or tolylfluanid [6]
- Do not use space sprays containing pyrethrins or pyrethroid more than once per week in intensive or controlled environment animal houses in order to avoid development of resistance. If necessary, use a different control method or product [1-3, 5]
- Store away from strong sunlight [4]

Crop-specific information
- HI: 24 h for edible crops [4]
- Some plant species including *Ageratum*, ferns, *Ficus, Lantana, Poinsettia, Petroselium crispum,* and some strawberries may be sensitive especially where more than one application is made. Test before large scale treatment [6]

Environmental safety
- Dangerous for the environment [2-6]
- Very toxic to aquatic organisms [1, 4, 5]; toxic to aquatic organisms [2, 3, 6]
- High risk to bees. Do not apply to crops in flower or to those in which bees are actively foraging. Do not apply when flowering weeds are present [4]
- Risk to non-target insects or other arthropods [4]
- Do not apply directly to livestock and exclude all persons and animals during treatment [1-3, 5]

- Wash spray equipment thoroughly after use to avoid traces of pyrethrum causing damage to susceptible crops sprayed later [4]
- LERAP Category A [4]

Hazard classification and safety precautions

Hazard Harmful [1, 2, 5]; Irritant [1, 3, 4]; Dangerous for the environment [2-6]
Transport code 9 [1, 3-6]
Packaging group III [1, 3-6]
UN number 3082 [1, 3-6]; N/C [2]
Risk phrases R22a, R36, R38 [1]; R22b [2, 5]; R41 [3, 4]; R50 [4]; R51 [2, 3, 5, 6]; R53a [2-6]
Operator protection A [1-6]; B [1]; C [2, 3, 5]; D [2, 3]; E [1-3]; H [2-6]; U02b, U09b, U14, U20a [2]; U05a [1, 4, 5]; U09a [1, 5]; U11 [3, 4]; U15 [3]; U19a [1-4]; U20b [1-3, 5]; U20c [4]; U20d [6]
Environmental protection E02c [3]; E02d, E05c [5]; E05a [1-3, 5]; E12a, E12e, E16c, E16d, E22c [4]; E13c [1]; E15a [2-5]; E15b [6]; E19b [3, 6]; E38 [2-4, 6]
Consumer protection C04 [1]; C05 [5]; C06 [1, 2, 5]; C07, C09, C11 [1-3, 5]; C08 [1-3]; C10 [1, 3]; C12 [2, 3, 5]
Storage and disposal D01, D09a [1-6]; D02 [1, 3-6]; D05 [2, 4]; D06a, D10a [2]; D11a [1, 4, 6]; D12a [2-6]
Medical advice M05a [3]; M05b [2, 5]

403 pyridate

A contact phenylpyridazine herbicide for cereals, maize and brassicas, see Section 3
HRAC mode of action code: C3

Products

Lentagran WP	Belchim	45% w/w	WP	14162

Uses

- Annual dicotyledons in ***asparagus*** *(off-label)*, ***broccoli*** *(off-label)*, ***calabrese*** *(off-label)*, ***cauliflowers*** *(off-label)*, ***collards*** *(off-label)*, ***fodder rape*** *(off-label)*, ***game cover*** *(off-label)*, ***garlic*** *(off-label)*, ***kale*** *(off-label)*, ***leeks*** *(off-label)*, ***lupins*** *(off-label - (yellow))*, ***oilseed rape*** *(off-label)*, ***shallots*** *(off-label)*, ***spring greens*** *(off-label)*
- Black nightshade in ***brussels sprouts***, ***bulb onions***, ***cabbages***
- Cleavers in ***brussels sprouts***, ***bulb onions***, ***cabbages***
- Fat hen in ***brussels sprouts***, ***bulb onions***, ***cabbages***

Specific Off-Label Approvals (SOLAs)

- ***asparagus*** *20091039*
- ***broccoli*** *20090786*
- ***calabrese*** *20090786*
- ***cauliflowers*** *20090786*
- ***collards*** *20090785*
- ***fodder rape*** *20093230*
- ***game cover*** *20090788*
- ***garlic*** *20092862*
- ***kale*** *20090785*
- ***leeks*** *20090784*
- ***lupins*** *((yellow)) 20090787*
- ***oilseed rape*** *20093230*
- ***shallots*** *20092862*
- ***spring greens*** *20090785*

Approval information

- Pyridate included in Annex I under EC Directive 91/414

Efficacy guidance

- Best results achieved by application to actively growing weeds at 6-8 leaf stage when temperatures are above 8 °C before crop foliage forms canopy

Restrictions

- Maximum number of treatments 1 per crop

- Do not apply in mixture with or within 14 d of any other product which may result in dewaxing of crop foliage
- Do not use on crops suffering stress from frost, drought, disease or pest attack

Crop-specific information

- Latest use: before 7 leaf stage for maize; before 5 true leaves for onions; before flower buds visible for oilseed rape
- HI 6 wk for Brussels sprouts, cabbages
- Apply to cabbages and Brussels sprouts after 4 fully expanded leaf stage. Allow 2 wk after transplanting before treating

Environmental safety

- Dangerous for the environment
- Toxic to aquatic organisms

Hazard classification and safety precautions

Hazard Irritant, Dangerous for the environment
Transport code 9
Packaging group III
UN number 3077
Risk phrases R43, R51, R53a
Operator protection A, C; U05a, U14, U22a
Environmental protection E15a, E38
Storage and disposal D01, D02, D09a, D10c, D12a

404 pyrimethanil

An anilinopyrimidine fungicide for apples and strawberries
FRAC mode of action code: 9

See also chlorothalonil + pyrimethanil

Products

Scala	BASF	400 g/l	SC	11695

Uses

- Botrytis in ***aubergines*** *(off-label)*, ***bilberries*** *(off-label)*, ***blackberries*** *(off-label)*, ***blackcurrants*** *(off-label)*, ***cranberries*** *(off-label)*, ***gooseberries*** *(off-label)*, ***protected blackberries*** *(off-label)*, ***protected herbs (see appendix 6)*** *(off-label)*, ***protected raspberries*** *(off-label)*, ***protected tomatoes*** *(off-label)*, ***quinces*** *(off-label)*, ***raspberries*** *(off-label)*, ***redcurrants*** *(off-label)*, ***strawberries***, ***tomatoes (outdoor)*** *(off-label)*, ***whitecurrants*** *(off-label)*, ***wine grapes*** *(off-label)*
- Scab in ***apples***, ***pears*** *(off-label)*

Specific Off-Label Approvals (SOLAs)

- ***aubergines*** *20040516*
- ***bilberries*** *20040519*
- ***blackberries*** *20051737*
- ***blackcurrants*** *20040519*
- ***cranberries*** *20040519*
- ***gooseberries*** *20040519*
- ***pears*** *20070200*
- ***protected blackberries*** *20073636*
- ***protected herbs (see appendix 6)*** *20073635*
- ***protected raspberries*** *20073636*
- ***protected tomatoes*** *20040516*
- ***quinces*** *20061379*
- ***raspberries*** *20051737*
- ***redcurrants*** *20040519*
- ***tomatoes (outdoor)*** *20040516*
- ***whitecurrants*** *20040519*
- ***wine grapes*** *20040517*

Approval information

- Pyrimethanil included in Annex I under EC Directive 91/414. However a Standing Committee decision in July 2007 to reduce to the limit of determination the MRL set for pyrimenthanil in cane fruit (other than blackberries and raspberries) and the MRL for brassicas grown for baby leaf production led to immediate revocation of SOLAs for these crops

Efficacy guidance

- On apples a programme of sprays will give early season control of scab. Season long control can be achieved by continuing programme with other approved fungicides
- In strawberries product should be used as part of a programme of disease control treatments which should alternate with other materials to prevent or limit development of less sensitive strains of grey mould

Restrictions

- Maximum number of treatments 5 per yr for apples; 4 per yr for protected crops; 3 per yr for grapevines; 2 per yr for cane fruit, bush fruit, strawberries, tomatoes
- Product does not taint apples. Processors should be consulted before use on strawberries

Crop-specific information

- Latest use: before end of flowering for apples
- HI protected cane fruit, strawberries 1 d; aubergines, tomatoes 3 d; protected lettuce 14 d; bush fruit, grapevines 21 d
- All varieties of apples and strawberries may be treated
- Treat apples from bud burst at 10-14 d intervals
- In strawberries start treatments at white bud to give maximum protection of flowers against grey mould and treat every 7-10 d. Product should not be used more than once in a 3 or 4 spray programme

Environmental safety

- Harmful to aquatic organisms
- Product has negligible effect on hoverflies and lacewings. Limited evidence indicates some margin of safety to *Typhlodromus pyri*
- Broadcast air-assisted LERAP (20 m)

Hazard classification and safety precautions

UN number N/C
Risk phrases R52, R53a
Operator protection U08, U20b
Environmental protection E15a; E17b (20 m)
Storage and disposal D01, D02, D05, D08, D09a, D10b, D12a

405 quinmerac

A residual herbicide available only in mixtures
HRAC mode of action code: O

See also *chloridazon + quinmerac*
dimethenamid-p + metazachlor + quinmerac
metazachlor + quinmerac

406 quinoxyfen

A systemic protectant quinoline fungicide for cereals
FRAC mode of action code: 13

See also *cyproconazole + quinoxyfen*
fenpropimorph + quinoxyfen

Products

1	Apres	Dow	500 g/l	SC	08881
2	Erysto	Dow	250 g/l	SC	08697
3	Fortress	Dow	500 g/l	SC	08279

Uses

- Powdery mildew in ***durum wheat, spring barley, spring oats, spring rye, spring wheat, sugar beet, triticale, winter barley, winter oats, winter rye, winter wheat*** [1-3]; ***fodder beet*** *(off-label)* [1]; ***forest nurseries*** *(off-label)*, ***hops*** *(off-label)*, ***ornamental plant production*** *(off-label)*, ***protected forest nurseries*** *(off-label)*, ***protected ornamentals*** *(off-label)*, ***protected strawberries*** *(off-label)*, ***strawberries*** *(off-label)* [3]; ***grass seed crops*** *(off-label)* [2]

Specific Off-Label Approvals (SOLAs)

- ***fodder beet*** *20062648* [1]
- ***forest nurseries*** *20082852* [3]
- ***grass seed crops*** *20090897* [2]
- ***hops*** *20061579* [3]
- ***ornamental plant production*** *20082852* [3]
- ***protected forest nurseries*** *20082852* [3]
- ***protected ornamentals*** *20082852* [3]
- ***protected strawberries*** *20041923* [3]
- ***strawberries*** *20041923* [3]

Approval information

- Quinoxyfen included in Annex I under Directive 91/414
- Accepted by BBPA for use on malting barley (before ear emergence only)

Efficacy guidance

- For best results treat at early stage of disease development before infection spreads to new crop growth. Further treatment may be necessary if disease pressure remains high
- Product not curative and will not control latent or established disease infections
- For broad spectrum control in cereals use in tank mixtures - see label
- Product rainfast after 1 h
- Systemic activity may be reduced in severe drought
- Use as part of an integrated crop management (ICM) strategy incorporating other methods of control or fungicides with different modes of action

Restrictions

- Maximum total dose equivalent to two full dose treatments in cereals; see labels for split dose recommendation in sugar beet
- On cereals apply only in the spring from mid-tillering stage (GS 25)

Crop-specific information

- Latest use: first awns visible stage (GS 49) for cereals
- HI sugar beet 28 d

Environmental safety

- Dangerous for the environment
- Very toxic to aquatic organisms

Hazard classification and safety precautions

Hazard Irritant, Dangerous for the environment
Transport code 9
Packaging group III
UN number 3082
Risk phrases R43, R50, R53a
Operator protection A, C, H; U05a, U14
Environmental protection E15a, E16a, E16b, E34, E38
Storage and disposal D01, D02, D12a

407 quizalofop-P-ethyl

An aryl phenoxypropionic acid post-emergence herbicide for grass weed control
HRAC mode of action code: A

Products

1	Leopard 5 EC	Makhteshim	50 g/l	EC	13720
2	Pilot Ultra	Nissan	50 g/l	SC	12929
3	Talon	Makhteshim	50 g/l	EC	13835
4	Visor	Makhteshim	50 g/l	EC	14513

SEE SECTION 3 FOR PRODUCTS ALSO REGISTERED

Uses

- Annual grasses in ***combining peas, fodder beet, mangels, red beet, spring field beans, spring oilseed rape, sugar beet, winter field beans, winter oilseed rape*** [1-4]; ***linseed, vining peas*** [1, 2, 4]; ***spring linseed, winter linseed*** [3]
- Barren brome in ***combining peas, fodder beet, mangels, red beet, spring field beans, spring linseed, spring oilseed rape, sugar beet, winter field beans, winter linseed, winter oilseed rape*** [3]
- Blackgrass in ***combining peas, fodder beet, mangels, red beet, spring field beans, spring linseed, spring oilseed rape, sugar beet, winter field beans, winter linseed, winter oilseed rape*** [3]
- Couch in ***combining peas, fodder beet, mangels, red beet, spring field beans, spring oilseed rape, sugar beet, winter field beans, winter oilseed rape*** [1-3]; ***linseed, vining peas*** [1, 2]; ***spring linseed, winter linseed*** [3]
- Perennial grasses in ***combining peas, fodder beet, linseed, mangels, red beet, spring field beans, spring oilseed rape, sugar beet, vining peas, winter field beans, winter oilseed rape*** [1, 2, 4]
- Volunteer cereals in ***combining peas, fodder beet, mangels, red beet, spring field beans, spring oilseed rape, sugar beet, winter field beans, winter oilseed rape*** [1-3]; ***linseed, vining peas*** [1, 2]; ***spring linseed, winter linseed*** [3]
- Wild oats in ***combining peas, fodder beet, mangels, red beet, spring field beans, spring linseed, spring oilseed rape, sugar beet, winter field beans, winter linseed, winter oilseed rape*** [3]

Efficacy guidance

- Best results achieved by application to emerged weeds growing actively in warm conditions with adequate soil moisture
- Weed control may be reduced under conditions such as drought that limit uptake and translocation
- Annual meadow-grass is not controlled
- For effective couch control do not hoe beet crops within 21 d after spraying
- At least 2 h without rain should follow application otherwise results may be reduced
- Quizalofop-P-ethyl is an ACCase inhibitor herbicide. To avoid the build up of resistance do not apply products containing an ACCase inhibitor herbicide more than twice to any crop. In addition do not use any product containing quizalofop-P-ethyl in mixture or sequence with any other product containing the same ingredient
- Use these products as part of a resistance management strategy that includes cultural methods of control and does not use ACCase inhibitors as the sole chemical method of grass weed control
- Applying a second product containing an ACCase inhibitor to a crop will increase the risk of resistance development; only use a second ACCase inhibitor to control different weeds at a different timing
- Always follow WRAG guidelines for preventing and managing herbicide resistant weeds. See Section 5 for more information

Restrictions

- Maximum number of treatments 1 on all recommended crops
- Do not spray crops under stress from any cause or in frosty weather
- Consult processor before use on crops recommended for processing
- An interval of at least 3 d must elapse between treatment and use of another herbicide on beet crops, 14 d on oilseed rape, 21 d on linseed and other recommended crops
- Avoid drift onto neighbouring crops

Crop-specific information

- HI 16 wk for beet crops; 11 wk for oilseed rape, linseed; 8 wk for field beans; 5 wk for peas
- In some situations treatment can cause yellow patches on foliage of peas, especially vining varieties. Symptoms usually rapidly and completely outgrown
- May cause taint in peas

Following crops guidance

- In the event of failure of a treated crop broad-leaved crops may be resown after a minimum interval of 2 wk, and cereals after 2-6 wk depending on dose applied
- Onions, leeks and maize are not recommended to follow a failed treated crop

Environmental safety

- Dangerous for the environment
- Toxic to aquatic organisms
- Flammable

Hazard classification and safety precautions

Hazard Harmful, Flammable [1, 3, 4]; Dangerous for the environment [1-4]
Transport code 3
Packaging group III
UN number 1993
Risk phrases R22b, R37, R38, R41, R67 [1, 3, 4]; R51, R53a [1-4]
Operator protection A, C [1-4]; H [2-4]; U05a, U11, U19a [1, 3, 4]; U09a [2]; U20b [1-4]
Environmental protection E13b [2]; E15a [1, 3, 4]; E34, E38 [1-4]
Storage and disposal D01, D02, D10b [1, 3, 4]; D03 [3]; D05 [1]; D09a, D12a [1-4]; D10c [2]
Medical advice M03, M05b [1, 3, 4]

408 quizalofop-P-tefuryl

An aryloxyphenoxypropionate herbicide for grass weed control
HRAC mode of action code: A

Products

Panarex	Certis	40 g/l	EC	12532

Uses

- Blackgrass in ***amenity vegetation*** *(off-label)*, ***cereal cover crops***, ***combining peas***, ***fodder beet***, ***natural surfaces not intended to bear vegetation*** *(off-label)*, ***ornamental plant production*** *(off-label)*, ***permeable surfaces overlying soil*** *(off-label)*, ***potatoes***, ***spring field beans***, ***spring linseed***, ***spring oilseed rape***, ***sugar beet***, ***winter field beans***, ***winter linseed***, ***winter oilseed rape***
- Couch in ***amenity vegetation*** *(off-label)*, ***cereal cover crops***, ***combining peas***, ***fodder beet***, ***natural surfaces not intended to bear vegetation*** *(off-label)*, ***ornamental plant production*** *(off-label)*, ***permeable surfaces overlying soil*** *(off-label)*, ***potatoes***, ***spring field beans***, ***spring linseed***, ***spring oilseed rape***, ***sugar beet***, ***winter field beans***, ***winter linseed***, ***winter oilseed rape***
- Italian ryegrass in ***cereal cover crops***, ***combining peas***, ***fodder beet***, ***potatoes***, ***spring field beans***, ***spring linseed***, ***spring oilseed rape***, ***sugar beet***, ***winter field beans***, ***winter linseed***, ***winter oilseed rape***
- Perennial ryegrass in ***amenity vegetation*** *(off-label)*, ***cereal cover crops***, ***combining peas***, ***fodder beet***, ***natural surfaces not intended to bear vegetation*** *(off-label)*, ***ornamental plant production*** *(off-label)*, ***permeable surfaces overlying soil*** *(off-label)*, ***potatoes***, ***spring field beans***, ***spring linseed***, ***spring oilseed rape***, ***sugar beet***, ***winter field beans***, ***winter linseed***, ***winter oilseed rape***
- Volunteer cereals in ***cereal cover crops***, ***combining peas***, ***fodder beet***, ***potatoes***, ***spring field beans***, ***spring linseed***, ***spring oilseed rape***, ***sugar beet***, ***winter field beans***, ***winter linseed***, ***winter oilseed rape***
- Wild oats in ***cereal cover crops***, ***combining peas***, ***fodder beet***, ***potatoes***, ***spring field beans***, ***spring linseed***, ***spring oilseed rape***, ***sugar beet***, ***winter field beans***, ***winter linseed***, ***winter oilseed rape***

Specific Off-Label Approvals (SOLAs)

- ***amenity vegetation*** *20060364*
- ***natural surfaces not intended to bear vegetation*** *20060364*
- ***ornamental plant production*** *20060364*
- ***permeable surfaces overlying soil*** *20060364*

Efficacy guidance

- Best results on annual grass weeds when growing actively and treated from 2 leaves to the start of tillering
- Treat cover crops when they have served their purpose and the threat of wind blow has passed
- Best results on couch achieved when the weed is growing actively and commencing new rhizome growth

SEE SECTION 3 FOR PRODUCTS ALSO REGISTERED

- Grass weeds germinating after treatment will not be controlled
- Treatment quickly stops growth and visible colour changes to the leaf tips appear after about 7 d. Complete kill takes 3-4 wk under good growing conditions
- Quizalofop-P-tefuryl is an ACCase inhibitor herbicide. To avoid the build up of resistance do not apply products containing an ACCase inhibitor herbicide more than twice to any crop. In addition do not use any product containing quizalofop-P-tefuryl in mixture or sequence with any other product containing the same ingredient
- Use these products as part of a resistance management strategy that includes cultural methods of control and does not use ACCase inhibitors as the sole chemical method of grass weed control
- Applying a second product containing an ACCase inhibitor to a crop will increase the risk of resistance development; only use a second ACCase inhibitor to control different weeds at a different timing
- Always follow WRAG guidelines for preventing and managing herbicide resistant weeds. See Section 5 for more information

Restrictions

- Maximum number of treatments 1 per crop
- Consult processors before use on peas or potatoes for processing
- Do not treat crops and weeds growing under stress from any cause

Crop-specific information

- HI 60 d for all crops
- Treat oilseed rape from the fully expanded cotyledon stage
- Treat linseed, peas and field beans from 2-3 unfolded leaves
- Treat sugar beet from 2 unfolded leaves

Following crops guidance

- In the event of failure of a treated crop any broad-leaved crop may be planted at any time. Cereals may be drilled from 4 wk after treatment

Environmental safety

- Dangerous for the environment
- Very toxic to aquatic organisms
- Risk to certain non-target insects or other arthropods. For advice on risk management and use in Integrated Pest Management (IPM) see directions for use
- Avoid spraying within 6 m of the field boundary to reduce effects on non-target insects and other arthropods

Hazard classification and safety precautions

Hazard Irritant, Dangerous for the environment
Transport code 9
Packaging group III
UN number 3082
Risk phrases R41, R43, R50, R53a
Operator protection A, C, H; U02a, U04a, U05a, U10, U11, U14, U15, U19a, U20b
Environmental protection E15a, E22b, E38
Storage and disposal D01, D02, D09a, D10b, D12a

409 rimsulfuron

A selective systemic sulfonylurea herbicide
HRAC mode of action code: B

Products

1	Caeser	AgChem Access	25% w/w	WG	13779
2	Rigid	AgriGuard	25% w/w	SG	12319
3	Tarot	Makhteshim	25% w/w	SG	11896
4	Titus	Makhteshim	25% w/w	SG	11895

Uses

- Annual dicotyledons in ***forage maize, potatoes***
- Volunteer oilseed rape in ***forage maize, potatoes***

Approval information

- Rimsulfuron included in Annex I under EC Directive 91/414

Efficacy guidance

- Product should be used with a suitable adjuvant or a suitable herbicide tank-mix partner. See label for details
- Product acts by foliar action. Best results obtained from good spray cover of small actively growing weeds. Effectiveness is reduced in very dry conditions
- Split application provides control over a longer period and improves control of fat hen and polygonums
- Weed spectrum can be broadened by tank mixture with other herbicides. See label for details
- Susceptible weeds cease growth immediately and symptoms can be seen 10 d later
- Rimsulfuron is a member of the ALS-inhibitor group of herbicides

Restrictions

- Maximum number of treatments 1 per crop
- Do not treat maize previously treated with organophosphorus insecticides
- Do not apply to potatoes grown for certified seed
- Consult processor before use on crops grown for processing
- Avoid high light intensity (full sunlight) and high temperatures on the day of spraying
- Do not treat during periods of substantial diurnal temperature fluctuation or when frost anticipated
- Do not apply to any crop stressed by drought, water-logging, low temperatures, pest or disease attack, nutrient or lime deficiency
- Do not apply to forage maize treated with organophosphate insecticides
- Do not apply to forage maize undersown with grass or clover

Crop-specific information

- Latest use: before most advanced potato plants are 25 cm high; before 4-collar stage of fodder maize
- All varieties of ware potatoes may be treated, but variety restrictions of any tank-mix partner must be observed
- Only certain named varieties of forage maize may be treated. See label

Following crops guidance

- Only winter wheat should follow a treated crop in the same calendar yr
- Only barley, wheat or maize should be sown in the spring of the yr following treatment
- In the second autumn after treatment any crop except brassicas or oilseed rape may be drilled

Environmental safety

- Dangerous for the environment
- Toxic to aquatic organisms
- Extremely dangerous to fish or other aquatic life. Do not contaminate surface waters or ditches with chemical or used container
- Herbicide is very active. Take particular care to avoid drift onto plants outside the target area
- Spraying equipment should not be drained or flushed onto land planted, or to be planted, with trees or crops other than potatoes or forage maize and should be thoroughly cleansed after use - see label for instructions

Hazard classification and safety precautions

Hazard Dangerous for the environment [1, 3, 4]
Transport code 9 [1, 3, 4]
Packaging group III [1, 3, 4]
UN number 3077 [1, 3, 4]; N/C [2]
Risk phrases R51, R53a [1, 3, 4]
Operator protection U08, U19a [1-4]; U20a [2]; U20b [1, 3, 4]
Environmental protection E13a, E34 [1-4]; E38 [1, 3, 4]
Storage and disposal D01, D09a, D11a [1-4]; D05 [2]; D12a [1, 3, 4]

410 rotenone

A natural, contact insecticide of low persistence. Rotenone will not be included in Annex I of EC Directive 91/414. All approvals for the storage and use of products will expire 31/10/2011.
IRAC mode of action code: 21

SEE SECTION 3 FOR PRODUCTS ALSO REGISTERED

411 silthiofam

A thiophene carboxamide fungicide seed dressing for cereals
FRAC mode of action code: 38

Products

Latitude	Monsanto	125 g/l	FS	10695

Uses

- Seed-borne diseases in ***durum wheat*** *(off-label)*, ***spring rye*** *(off-label)*, ***triticale*** *(off-label)*, ***winter rye*** *(off-label)*
- Take-all in ***spring wheat*** *(seed treatment)*, ***winter barley*** *(seed treatment)*, ***winter wheat*** *(seed treatment)*

Specific Off-Label Approvals (SOLAs)

- ***durum wheat*** *20060973*
- ***spring rye*** *20060973*
- ***triticale*** *20060973*
- ***winter rye*** *20060973*

Approval information

- Silthiofam included in Annex I under EC Directive 91/414
- Accepted by BBPA for use on malting barley

Efficacy guidance

- Apply using approved seed treatment equipment which has been accurately calibrated
- Apply simultaneously (i.e. not in mixture) with a standard seed treatment
- Drill treated seed in the season of purchase. The viability of treated seed and fungicide activity may be reduced by physical storage
- Drill treated seed at 2.5-4 cm into a well prepared firm seedbed
- As precaution against possible development of disease resistance do not treat more than three consecutive susceptible cereal crops in any one rotation

Restrictions

- Maximum number of treatments one per batch of seed
- Do not use on seed with more than 16% moisture content, or on sprouted, cracked or skinned seed
- Test germination of all seed batches before treatment

Crop-specific information

- Latest use: immediately prior to drilling

Hazard classification and safety precautions

UN number N/C
Operator protection A, H; U20b
Environmental protection E03, E15a, E34
Storage and disposal D05, D09a, D10a
Treated seed S01, S02, S04b, S05, S06a, S07

412 sodium chlorate

A non-selective inorganic herbicide for total vegetation control

Products

Doff Sodium Chlorate Weedkiller	Doff Portland	53% w/w	SP	06049

Uses

- Total vegetation control in ***non-crop areas, paths and drives***

Approval information

- Sodium chlorate has not been included in Annex I of EC Directive 91/414 and will not be granted Essential Uses.
- Approval expiry 10 May 2010

Efficacy guidance
- Active through foliar and root uptake
- Apply as overall spray at any time during growing season. Best results obtained from application to moist soil in spring or early summer

Restrictions
- Maximum number of treatments 2 per yr for non-crop areas, paths and drives
- Treated ground must not be replanted for at least 6 mth after treatment
- Do not apply before heavy rain

Environmental safety
- Dangerous for the environment
- Contact with combustible material may cause fire
- Harmful to aquatic organisms
- Keep livestock out of treated areas for at least two weeks following treatment and until poisonous weeds, such as ragwort, have died down and become unpalatable
- Clothing, paper, plant debris etc become highly inflammable when dry if contaminated with sodium chlorate
- Fire risk has been reduced by inclusion of fire depressant but product should not be used in areas of exceptionally high fire risk eg oil installations, timber yards
- Wash clothing thoroughly after use
- If clothes become contaminated do not stand near an open fire
- Must not be sold to persons under 18

Hazard classification and safety precautions

Hazard Harmful, Oxidising agent
Risk phrases R08, R22a
Operator protection A, H, M; U05a, U20b
Environmental protection E13c, E15a, E34
Storage and disposal D01, D02, D07, D09a, D11a
Medical advice M03, M05a

413 sodium hypochlorite (commodity substance)

An inorganic horticultural bactericide for use in mushrooms

Products

sodium hypochlorite	various	100% w/v	ZZ	-

Uses
- Bacterial blotch in ***mushrooms***

Approval information
- Approval for the use of sodium hypochlorite as a commodity substance was granted on 5 December 1996 by Ministers under regulation 5 of the Control of Pesticides Regulations 1986
- Sodium hypochlorite is being supported for review in the fourth stage of the EC Review Programme under Directive 91/414. Whether or not it is included in Annex I, the existing commodity chemical approval will have to be revoked because substances listed in Annex I must be approved for marketing and use under the EC regime. If sodium hypochlorite is included in Annex I, products containing it will need to gain approval in the normal way if they carry label claims for pesticidal activity

Restrictions
- Maximum concentration 315 mg/litre of water
- Mixing and loading must only take place in a ventilated area
- Must only be used by suitably trained and competent operators

Crop-specific information
- HI 1 d

Environmental safety
- Harmful to fish or other aquatic life. Do not contaminate surface waters or ditches with chemical or used container

Hazard classification and safety precautions

Operator protection A, C, H

Environmental protection E13c

414 spinosad

A selective insecticide derived from naturally occurring soil fungi (naturalyte)

IRAC mode of action code: 5

Products

1	Conserve	Fargro	120 g/l	SC	12058
2	Tracer	Landseer	480 g/l	SC	12438

Uses

- Cabbage moth in ***brussels sprouts, cabbages, cauliflowers*** [2]
- Cabbage white butterfly in ***brussels sprouts, cabbages, cauliflowers*** [2]
- Codling moth in ***apples, pears*** [2]
- Diamond-back moth in ***brussels sprouts, cabbages, cauliflowers*** [2]
- Insect pests in ***blackberries*** *(off-label)*, ***celery leaves, cress, frise, herbs (see appendix 6), hops*** *(off-label)*, ***lamb's lettuce, lettuce, ornamental specimens*** *(off-label)*, ***protected herbs (see appendix 6)*** *(off-label)*, ***radicchio, raspberries*** *(off-label)*, ***scarole, soft fruit*** *(off-label)*, ***top fruit*** *(off-label)* [2]
- Moths in ***hops*** *(off-label)*, ***ornamental specimens*** *(off-label)*, ***soft fruit*** *(off-label)*, ***top fruit*** *(off-label)* [2]
- Small white butterfly in ***brussels sprouts, cabbages, cauliflowers*** [2]
- Thrips in ***blackberries*** *(off-label)*, ***bulb onions, leeks, raspberries*** *(off-label)*, ***salad onions*** [2]
- Tortrix moths in ***apples, pears*** [2]
- Western flower thrips in ***protected aubergines, protected cucumbers, protected ornamentals, protected peppers, protected tomatoes*** [1]

Specific Off-Label Approvals (SOLAs)

- ***blackberries*** *20092118* [2]
- ***hops*** *20082908* [2]
- ***ornamental specimens*** *20082908* [2]
- ***protected herbs (see appendix 6)*** *20081290* [2]
- ***raspberries*** *20092118* [2]
- ***soft fruit*** *20082908* [2]
- ***top fruit*** *20082908* [2]

Approval information

- Spinosad included in Annex I under EC Directive 91/414

Efficacy guidance

- Product enters insects by contact from a treated surface or ingestion of treated plant material therefore good spray coverage is essential
- Some plants, for example Fuchsia flowers, can provide effective refuges from spray deposits and control of western flower thrips may be reduced [1]
- Apply to protected crops when western flower thrip nymphs or adults are first seen [1]
- Monitor western flower thrip development carefully to see whether further applications are necessary. A 2 spray programme at 5-7 d intervals may be needed when conditions favour rapid pest development [1]
- Treat top fruit and field crops when pests are first seen or at very first signs of crop damage [2]
- Ensure a rain-free period of 12 h after treatment before applying irrigation [2]
- To reduce possibility of development of resistance, adopt resistance management measures. See label and Section 5 for more information

Restrictions

- Maximum number of treatments 4 per crop for brassicas, onions, leeks [2]; 1 pre-blossom and 3 post blossom for apples, pears [2]
- Apply no more than 2 consecutive sprays. Rotate with another insecticide with a different mode of action or use no further treatment after applying the maximum number of treatments

- Establish whether any incoming plants have been treated and apply no more than 3 consecutive sprays. Maximum number of treatments 6 per structure per yr
- Avoid application in bright sunlight or into open flowers [1]

Crop-specific information
- HI 3 d for protected cucumbers, brassicas; 7 d for apples, pears, onions, leeks
- Test for tolerance on a small number of ornamentals or cucumbers before large scale treatment
- Some spotting of african violet flowers may occur

Environmental safety
- Dangerous for the environment
- Very toxic (toxic [1]) to aquatic organisms
- Whenever possible use an Integrated Pest Management system. Spinosad presents low risk to beneficial arthropods
- Product has low impact on many insect and mite predators but is harmful to adults of most parasitic wasps. Most beneficials may be introduced to treated plants when spray deposits are dry but an interval of 2 wk should elapse before introduction of parasitic wasps. See label for details
- Treatment may cause temporary reduction in abundance of insect and mite predators if present at application
- Exposure to direct spray is harmful to bees but dry spray deposits are harmless. Treatment of field crops should not be made in the heat of the day when bees are actively foraging
- LERAP Category B and 40 m broadcast air-assisted LERAP [2]

Hazard classification and safety precautions

Hazard Dangerous for the environment
Transport code 9
Packaging group III
UN number 3082
Risk phrases R50 [2]; R51 [1]; R53a [1, 2]
Operator protection A, H [2]; U05a [1]; U08, U20b [2]
Environmental protection E15a, E16a [2]; E15b [1]; E17b [2] (40 m); E34, E38 [1, 2]
Storage and disposal D01, D05 [1]; D02, D10b, D12a [1, 2]

415 spirodiclofen

A tetronic acid which inhibits lipid biosynthesis
IRAC mode of action code: 23

Products

Envidor	Bayer CropScience	240 g/l	SC	13947

Uses
- Mussel scale in ***apples***, ***pears***
- Pear sucker in ***pears***
- Red spider mites in ***apples***, ***pears***
- Rust mite in ***apples***, ***pears***
- Spider mites in ***hops*** *(off-label)*, ***ornamental crop production (seed)*** *(off-label)*, ***protected ornamentals*** *(off-label)*, ***protected strawberries*** *(off-label)*, ***soft fruit*** *(off-label)*, ***strawberries*** *(off-label)*
- Two-spotted spider mite in ***apples***, ***pears***, ***protected strawberries*** *(off-label)*, ***strawberries*** *(off-label)*

Specific Off-Label Approvals (SOLAs)
- ***hops*** *20082845*
- ***ornamental crop production (seed)*** *20092240*
- ***protected ornamentals*** *20092240*
- ***protected strawberries*** *20092241*
- ***soft fruit*** *20082845*
- ***strawberries*** *20092241*

Restrictions
- Consult processor on crops grown for processing or for cider production

Hazard classification and safety precautions

Hazard Harmful
UN number N/C
Risk phrases R40, R43, R52, R53a
Operator protection A, H; U05a, U20c
Environmental protection E12c, E12f, E15b, E22c, E34, E38; E17b (18 m)
Storage and disposal D01, D02, D09a, D10c, D12a
Medical advice M05a

416 spiromesifen

A tetronic acid insecticide for protected tomatoes
IRAC mode of action code: 23

Products

Oberon	Certis	240 g/l	SC	11819

Uses

- Red spider mites in ***protected aubergines*** *(off-label - on inert media or by NFT)*, ***protected cayenne peppers*** *(off-label - on inert media or by NFT)*, ***protected cucumbers*** *(off-label - on inert media or by NFT)*, ***protected gherkins*** *(off-label - on inert media or by NFT)*, ***protected ornamentals*** *(off-label)*, ***protected peppers*** *(off-label - on inert media or by NFT)*, ***protected strawberries*** *(off-label - on inert media or by NFT)*
- Whitefly in ***inert substrate tomatoes***, ***nft tomatoes***, ***protected aubergines*** *(off-label - on inert media or by NFT)*, ***protected cayenne peppers*** *(off-label - on inert media or by NFT)*, ***protected cucumbers*** *(off-label - on inert media or by NFT)*, ***protected gherkins*** *(off-label - on inert media or by NFT)*, ***protected ornamentals*** *(off-label)*, ***protected peppers*** *(off-label - on inert media or by NFT)*, ***protected strawberries*** *(off-label - on inert media or by NFT)*

Specific Off-Label Approvals (SOLAs)

- ***protected ornamentals*** *20041718*
- ***protected strawberries***

Efficacy guidance

- Apply to run off and ensure that all sides of the crop are covered
- Apply when infestation first observed and repeat 7-10 d later if necessary
- Always follow guidelines for preventing and managing insect resistance. See Section 5 for more information

Restrictions

- Maximum number of treatments 2 per cropping cycle
- Do not use on any outdoor crops
- Product should be used in rotation with compounds with different modes of action

Crop-specific information

- HI 3 d for tomatoes

Environmental safety

- Dangerous for the environment
- Very toxic to aquatic organisms

Hazard classification and safety precautions

Hazard Irritant, Dangerous for the environment
Transport code 9
Packaging group III
UN number 3082
Risk phrases R43, R50, R53a
Operator protection A, H; U05a, U14, U20b
Environmental protection E15a, E34, E38
Storage and disposal D01, D02, D05, D09a, D11a, D12a

417 spirotetramat

Tetramic acid derivative that acts on lipid synthesis and affects growth regulation
IRAC mode of action code: 23

Products

Movento	Bayer CropScience	150 g/l	OD	14446

Uses

- Aphids in ***broccoli***, ***brussels sprouts***, ***cabbages***, ***calabrese***, ***cauliflowers***, ***lettuce***
- Whitefly in ***broccoli***, ***brussels sprouts***, ***cabbages***, ***calabrese***, ***cauliflowers***

Approval information

- Spirotetramat has not yet been added to Annex 1 listing

Hazard classification and safety precautions

Hazard Irritant, Dangerous for the environment
Transport code 9
Packaging group III
UN number 3082
Risk phrases R43, R51, R53a
Operator protection A; U05a, U14, U20b
Environmental protection E15b, E22c, E34, E38
Storage and disposal D02, D09a, D10c, D12a
Medical advice M03

418 spiroxamine

A spiroketal amine fungicide for cereals
FRAC mode of action code: 5

See also *prothioconazole + spiroxamine*
prothioconazole + spiroxamine + tebuconazole

Products

Torch Extra	Bayer CropScience	800 g/l	EC	12140

Uses

- Brown rust in ***spring barley***, ***spring rye***, ***spring wheat***, ***winter barley***, ***winter rye***, ***winter wheat***
- Insect pests in ***forest nurseries*** *(off-label)*
- Powdery mildew in ***spring barley***, ***spring rye***, ***spring wheat***, ***winter barley***, ***winter rye***, ***winter wheat***
- Rhynchosporium in ***spring barley*** *(reduction)*, ***spring rye*** *(reduction)*, ***winter barley*** *(reduction)*, ***winter rye*** *(reduction)*
- Yellow rust in ***spring barley***, ***spring rye***, ***spring wheat***, ***winter barley***, ***winter rye***, ***winter wheat***

Specific Off-Label Approvals (SOLAs)

- ***forest nurseries*** *20082909*

Approval information

- Spiroxamine included in Annex I under EC Directive 91/414
- Accepted by BBPA for use on malting barley

Efficacy guidance

- For best results treat at an early stage of disease development before infection spreads to new growth
- To reduce the risk of development of resistance avoid repeat treatments on diseases such as powdery mildew. If necessary tank mix or alternate with other non-morpholine fungicides

Restrictions

- Maximum total dose equivalent to two full dose treatments

Crop-specific information

- Latest use: before caryopsis watery ripe (GS 71) for wheat, rye; ear emergence complete (GS 59) for barley

Environmental safety

- Dangerous for the environment
- Very toxic to aquatic organisms
- LERAP Category B

Hazard classification and safety precautions

Hazard Harmful, Dangerous for the environment
Transport code 6.1
Packaging group III
UN number 2902
Risk phrases R20, R21, R22a, R41, R43, R50, R53a
Operator protection A, C, H; U05a, U11, U13, U14, U15, U20a
Environmental protection E15a, E16a, E34, E38
Storage and disposal D01, D02, D09a, D10b, D12a
Medical advice M03

419 spiroxamine + tebuconazole

A broad spectrum systemic fungicide mixture for cereals
FRAC mode of action code: 5 + 3

See also *tebuconazole*

Products

Sage	Bayer CropScience	250:133 g/l	EW	11303

Uses

- Brown rust in ***spring barley***, ***spring rye***, ***spring wheat***, ***winter barley***, ***winter rye***, ***winter wheat***
- Foliar disease control in ***grass seed crops*** *(off-label)*
- Fusarium ear blight in ***spring wheat***, ***winter wheat***
- Glume blotch in ***spring wheat***, ***winter wheat***
- Net blotch in ***spring barley***, ***winter barley***
- Powdery mildew in ***spring barley***, ***spring rye***, ***spring wheat***, ***winter barley***, ***winter rye***, ***winter wheat***
- Rhynchosporium in ***spring barley***, ***winter barley***
- Septoria leaf blotch in ***spring wheat***, ***winter wheat***
- Sooty moulds in ***spring wheat***, ***winter wheat***
- Yellow rust in ***spring barley***, ***spring rye***, ***spring wheat***, ***winter barley***, ***winter rye***, ***winter wheat***

Specific Off-Label Approvals (SOLAs)

- ***grass seed crops*** *20060983*

Approval information

- Spiroxamine included in Annex I under EC Directive 91/414
- Accepted by BBPA for use on malting barley

Efficacy guidance

- Best results achieved from treatment at early stage of disease development before infection spreads to new growth
- To protect flag leaf and ear from Septoria apply from flag leaf emergence to ear fully emerged (GS 37-59). Earlier treatment may be necessary where there is high disease risk
- Control of rusts, powdery mildew, leaf blotch and net blotch may require second treatment 2-3 wk later
- Tebuconazole is a DMI fungicide. Resistance to some DMI fungicides has been identified in Septoria leaf blotch which may seriously affect performance of some products. For further advice contact a specialist advisor and visit the Fungicide Resistance Action Group (FRAG)-UK website

Restrictions

- Maximum total dose equivalent to two full dose treatments
- Do not use on durum wheat

Crop-specific information

- Latest use: before caryopsis watery ripe (GS 71) for rye, wheat; ear emergence complete (GS 59) for barley
- Some transient leaf speckling may occur on wheat but this has not been shown to reduce yield reponse or disease control

Environmental safety

- Dangerous for the environment
- Very toxic to aquatic organisms
- Dangerous to fish or other aquatic life. Do not contaminate surface waters or ditches with chemical or used container
- LERAP Category B

Hazard classification and safety precautions

Hazard Harmful, Dangerous for the environment
Transport code 9
Packaging group III
UN number 3082
Risk phrases R20, R22a, R38, R41, R43, R50, R53a
Operator protection A, C, H; U05a, U11, U13, U14, U15, U20b
Environmental protection E13b, E16a, E16b, E34, E38
Storage and disposal D01, D02, D05, D09a, D10b, D12a
Medical advice M03

420 strychnine hydrochloride (commodity substance)

A vertebrate control agent for destruction of moles underground

Products

strychnine	various	-	ZZ	-

Uses

- Moles in ***miscellaneous pest control situations*** *(areas of restricted public access)*, ***permanent grassland*** *(areas of restricted public access)*

Approval information

- Approval for the use of strychnine hydrochloride as a commodity substance was granted on 19 June 1997 by Ministers under regulation 5 of the Control of Pesticides Regulations 1986
- Strychnine has not been supported in the EC Review Programme (see Section 5) and use for mole control ceased to be authorised from September 2006. It has not been permitted as a plant protection product (e.g. to prevent crop damage) since 1 January 2005

Efficacy guidance

- Only for use as poison bait against moles on commercial agricultural/horticultural land where public access restricted, on grassland associated with aircraft landing strips, horse paddocks, race and golf courses and other areas specifically approved by Defra in England and the appropriate authorities in Wales and Scotland

Restrictions

- Strychnine is subject to the Poisons Rules 1982 and the Poisons Act 1972. See Section 5 for more information
- Must only be supplied to holders of a permit issued by Defra in England or the appropriate authorities in Wales and Scotland. Permits may only be issued to persons who satisfy the appropriate authority that they are trained and competent in its use
- Only to be supplied in original sealed pack in units up to 2 g. Quantities of more than 8 g may be supplied only to providers of a commercial service

Environmental safety

- Store in original container under lock and key and only on premises under control of a permit holder or of a named individual who satisfies the designated requirements of competence
- A written COSHH assessment must be made before use

- Must be prepared for application with great care so that there is no contamination of the ground surface
- Any prepared bait remaining at the end of the day must be buried
- Operators must be provided with a Chemicals (Hazard Information and Supply) (CHIP) Safety Data Sheet, obtainable from the supplier, before commencing work

Hazard classification and safety precautions

Transport code 6.1
Packaging group I
UN number 1692
Operator protection A

421 sulfosulfuron

A sulfonylurea herbicide for grass and broad-leaved weed control in winter wheat
HRAC mode of action code: B

See also glyphosate + sulfosulfuron

Products

Monitor	Monsanto	80% w/w	WG	12236

Uses

- Annual dicotyledons in ***game cover*** *(off-label)*, ***grass seed crops*** *(off-label)*, ***spring rye*** *(off-label)*, ***triticale*** *(off-label)*, ***winter rye*** *(off-label)*
- Brome grasses in ***winter wheat*** *(moderate control of barren brome)*
- Chickweed in ***winter wheat***
- Cleavers in ***winter wheat***
- Couch in ***winter wheat*** *(moderate control only)*
- Grass weeds in ***game cover*** *(off-label)*, ***grass seed crops*** *(off-label)*, ***spring rye*** *(off-label)*, ***triticale*** *(off-label)*, ***winter rye*** *(off-label)*
- Loose silky bent in ***winter wheat***
- Mayweeds in ***winter wheat***
- Onion couch in ***winter wheat*** *(moderate control only)*

Specific Off-Label Approvals (SOLAs)

- ***game cover*** *20082861*
- ***grass seed crops*** *20081279*
- ***spring rye*** *20081279*
- ***triticale*** *20081279*
- ***winter rye*** *20081279*

Approval information

- Sulfosulfuron included in Annex I under EC Directive 91/414

Efficacy guidance

- For best results treat in early spring when annual weeds are small and growing actively. Avoid treatment when weeds are dormant for any reason
- Best control of onion couch is achieved when the weed has more than two leaves. Effects on bulbils or on growth in the following yr have not been examined
- An extended period of dry weather before or after treatment may result in reduced control
- The addition of a recommended surfactant is essential for full activity
- Specific follow-up treatments may be needed for complete control of some weeds
- Use only where competitively damaging weed populations have emerged otherwise yield may be reduced
- Sulfosulfuron is a member of the ALS-inhibitor group of herbicides. To avoid the build up of resistance do not use any product containing an ALS-inhibitor herbicide with claims for control of grass weeds more than once on any crop
- Use these products as part of a resistance management strategy that includes cultural methods of control and does not use ALS inhibitors as the sole chemical method of grass weed control

Restrictions

- Maximum total dose equivalent to one treatment at full dose
- Do not treat crops under stress

- Apply only after 1 Feb and from 3 expanded leaf stage
- Do not treat durum wheat or any undersown wheat crop
- Do not use in mixture or in sequence with any other sulfonyl urea herbicide on the same crop

Crop-specific information

- Latest use: flag leaf ligule just visible for winter wheat (GS 39)

Following crops guidance

- In the autumn following treatment winter wheat, winter rye, winter oats, triticale, winter oilseed rape, winter peas or winter field beans may be sown on any soil, and winter barley on soils with less than 60% sand
- In the next spring following application crops of wheat, barley, oats, maize, peas, beans, linseed, oilseed rape, potatoes or grass may be sown
- In the second autumn following application winter linseed may be sown, and winter barley on soils with more than 60% sand
- Sugar beet or any other crop not mentioned above must not be drilled until the second spring following application
- Where winter oilseed rape is to be sown in the autumn following treatment soil cultivation to a minimum of 10 cm is recommended

Environmental safety

- Dangerous for the environment
- Very toxic to aquatic organisms
- Extremely dangerous to fish or other aquatic life. Do not contaminate surface waters or ditches with chemical or used container
- LERAP Category B
- Take extreme care to avoid drift onto broad leaved plants or other crops, or onto ponds, waterways or ditches.
- Follow detailed label instructions for cleaning the sprayer to avoid damage to sensitive crops during subsequent use

Hazard classification and safety precautions

Hazard Dangerous for the environment
UN number N/C
Risk phrases R50, R53a
Operator protection U20b
Environmental protection E15a, E16a, E16b, E34, E38
Storage and disposal D09a, D11a, D12a

422 sulphur

A broad-spectrum inorganic protectant fungicide, foliar feed and acaricide
FRAC mode of action code: M2

Products

1	Headland Sulphur	Headland	800 g/l	SC	12879
2	Kumulus DF	BASF	80% w/w	WG	04707
3	Microthiol Special	United Phosphorus	80% w/w	MG	06268
4	Solfa WG	Nufarm UK	80% w/w	WG	11602
5	Sulphur Flowable	United Phosphorus	800 g/l	SC	07526
6	Thiovit Jet	Syngenta	80% w/w	WG	10928
7	Venus	Headland	80% w/w	WG	11856

Uses

- American gooseberry mildew in ***gooseberries*** [4]
- Disease control/foliar feed in ***bilberries*** *(off-label)*, ***blueberries*** *(off-label)*, ***cranberries*** *(off-label)*, ***redcurrants*** *(off-label)*, ***whitecurrants*** *(off-label)* [2]
- Foliar feed in ***all top fruit***, ***bush fruit***, ***cane fruit***, ***edible brassicas***, ***managed amenity turf***, ***potatoes***, ***spring oats***, ***sugar beet***, ***swedes***, ***turnips***, ***vegetables***, ***winter oats*** [3]; ***permanent grassland*** [3, 4, 6]; ***spring barley***, ***spring wheat***, ***winter barley***, ***winter wheat*** [2, 3, 6, 7]; ***spring oilseed rape*** [3, 4, 6, 7]; ***winter oilseed rape*** [2-4, 6, 7]
- Gall mite in ***blackcurrants*** [2, 4, 7]
- Powdery mildew in ***apples***, ***strawberries*** [1, 2, 4, 5, 7]; ***aubergines*** *(off-label)*, ***combining peas*** *(off-label)*, ***protected chilli peppers*** *(off-label)*, ***protected chives*** *(off-label)*, ***protected parsley***

(off-label), ***vining peas*** *(off-label)* [6]; ***borage for oilseed production*** *(off-label)*, ***parsnips*** *(off-label)*, ***protected cucumbers*** *(off-label)*, ***protected herbs (see appendix 6)*** *(off-label)*, ***protected peppers*** *(off-label)*, ***protected tomatoes*** *(off-label)* [3, 4, 6]; ***fodder beet*** *(off-label)* [1, 3, 4, 6]; ***gooseberries*** [2, 7]; ***grapevines*** [3]; ***grass seed crops*** *(off-label)*, ***spring barley***, ***spring wheat***, ***winter barley***, ***winter wheat*** [1, 3-5]; ***hops*** [1, 2, 4-7]; ***pears*** [1, 5]; ***protected aubergines*** *(off-label)*, ***protected cayenne peppers*** *(off-label)*, ***turnips*** [3, 4]; ***quinces*** *(off-label)* [5]; ***spring oats***, ***winter oats*** [1, 3, 4]; ***spring oilseed rape***, ***spring rye***, ***triticale*** *(off-label)*, ***winter oilseed rape***, ***winter rye*** [1]; ***sugar beet*** [1-7]; ***swedes*** [1, 3-6]; ***turnips*** *(off-label)* [1, 5, 6]; ***wine grapes*** [2, 4, 7]

- Scab in ***apples*** [1, 2, 5, 7]; ***pears*** [1, 5]; ***quinces*** *(off-label)* [1]

Specific Off-Label Approvals (SOLAs)

- ***aubergines*** *20023652* [6]
- ***bilberries*** *20061042* [2]
- ***blueberries*** *20061042* [2]
- ***borage for oilseed production*** *20072027* [3], *20072238* [4], *20061719* [6]
- ***combining peas*** *20081361* [6]
- ***cranberries*** *20061042* [2]
- ***fodder beet*** *20062690* [1], *20072025* [3], *20072240* [4], *20061383* [6]
- ***grass seed crops*** *20062691* [1], *20061043* [3], *20061044* [4], *20061375* [5]
- ***parsnips*** *20072026* [3], *20072242* [4], *20023654* [6]
- ***protected aubergines*** *20072024* [3], *20072239* [4]
- ***protected cayenne peppers*** *20072024* [3], *20072239* [4]
- ***protected chilli peppers*** *20081360* [6]
- ***protected chives*** *20023652* [6]
- ***protected cucumbers*** *20072024* [3], *20072239* [4], *20023652* [6]
- ***protected herbs (see appendix 6)*** *20072024* [3], *20072239* [4], *20023652* [6]
- ***protected parsley*** *20023652* [6]
- ***protected peppers*** *20072024* [3], *20072239* [4], *20023652* [6]
- ***protected tomatoes*** *20072024* [3], *20072239* [4], *20023652* [6]
- ***quinces*** *20062688* [1], *20061374* [5]
- ***redcurrants*** *20061042* [2]
- ***triticale*** *20062691* [1]
- ***turnips*** *20062689* [1], *20061373* [5], *20061384* [6]
- ***vining peas*** *20081361* [6]
- ***whitecurrants*** *20061042* [2]

Approval information

- Accepted by BBPA for use on malting barley and hops (before burr)

Efficacy guidance

- Apply when disease first appears and repeat 2-3 wk later. Details of application rates and timing vary with crop, disease and product. See label for information
- Sulphur acts as foliar feed as well as fungicide and with some crops product labels vary in whether treatment recommended for disease control or growth promotion
- In grassland best results obtained at least 2 wk before cutting for hay or silage, 3 wk before grazing
- Treatment unlikely to be effective if disease already established in crop

Restrictions

- Maximum number of treatments normally 2 per crop for grassland, sugar beet, parsnips, swedes, hops, protected herbs; 3 per yr for blackcurrants, gooseberries; 4 per crop on apples, pears but labels vary
- Do not use on sulphur-shy apples (Beauty of Bath, Belle de Boskoop, Cox's Orange Pippin, Lanes Prince Albert, Lord Derby, Newton Wonder, Rival, Stirling Castle) or pears (Doyenne du Comice)
- Do not use on gooseberry cultivars Careless, Early Sulphur, Golden Drop, Leveller, Lord Derby, Roaring Lion, or Yellow Rough
- Do not use on apples or gooseberries when young, under stress or if frost imminent
- Do not use on fruit for processing, on grapevines during flowering or near harvest on grapes for wine-making
- Do not use on hops at or after burr stage
- Do not spray top or soft fruit with oil or within 30 d of an oil-containing spray

Crop-specific information

- Latest use: before burr stage for hops; before end Sep for parsnips, swedes, sugar beet; fruit swell for gooseberries; milky ripe stage for cereals
- HI cutting grass for hay or silage 2 wk; grazing grassland 3 wk

Environmental safety

- Sulphur products are attractive to livestock and must be kept out of their reach
- Do not empty into drains

Hazard classification and safety precautions

Hazard Irritant [4]
UN number N/C
Risk phrases R37 [4]
Operator protection A, H [6]; U02a, U08, U14 [6, 7]; U05a [4, 6]; U19a [4]; U20a [1]; U20b [3, 5]; U20c [2, 4, 6, 7]
Environmental protection E15a [1-7]; E19b [4]; E34 [1]
Storage and disposal D01 [3, 4, 6]; D02 [4, 6]; D05 [5, 6]; D09a [1-7]; D10b [1]; D10c [5]; D11a [2, 3, 6, 7]

423 sulphuric acid (commodity substance)

A strong acid used as an agricultural desiccant

Products

sulphuric acid	various	77% w/w	SL	-

Uses

- Haulm destruction in ***potatoes***
- Pre-harvest desiccation in ***bulbs/corms***, ***peas***

Approval information

- Approval for the use of sulphuric acid as a commodity substance was granted on 23 November 1995 by Ministers under regulation 5 of the Control of Pesticides Regulations 1986
- Sulphuric acid has not been granted listing on Annex 1 of EC Directive 91/414 and will be withdrawn for use as a plant protection product.
- Approval expiry 05 Jun 2010

Efficacy guidance

- Apply with suitable equipment between 1 Mar and 15 Nov

Restrictions

- Sulphuric acid is subject to the Poisons Rules 1982 and the Poisons Act 1972. See Section 5 for more information
- Must only be used by suitably trained operators competent in use of equipment for applying sulphuric acid
- Not to be applied using hand-held or pedestrian controlled applicators
- A written COSHH assessment must be made before use. Operators should observe OES set out in HSE guidance note EH40/90 or subsequent issues
- Operators must have liquid suitable for eye irrigation immediately available at all times throughout spraying operation
- Maximum number of treatments 3 per crop for potatoes; 1 per crop for peas; 1 per yr for bulbs and corms
- Only 'sulphur burnt' sulphuric acid to be used

Crop-specific information

- Latest use: 15 Nov for bulbs, corms, peas, potatoes

Environmental safety

- Spray must not be deposited within 1 m of public footpaths
- Written notice of any intended spraying must be given to owners of neighbouring land and readable warning notices posted beforehand and left in place for 96 h afterwards
- Unprotected persons must be kept out of treated areas for at least 96 h after treatment
- Do not apply to crops in which bees are actively foraging. Do not apply when flowering weeds are present

SEE SECTION 3 FOR PRODUCTS ALSO REGISTERED

Hazard classification and safety precautions
- **Hazard** Corrosive
- **Transport code** 8
- **Packaging group** II
- **UN number** 1830
- **Risk phrases** R34
- **Operator protection** B, C, D, H, K, M

424 tau-fluvalinate

A contact pyrethroid insecticide for cereals and oilseed rape
IRAC mode of action code: 3

Products

1	Greencrop Malin	Greencrop	240 g/l	EW	11787
2	Klartan	Makhteshim	240 g/l	EW	11074
3	Mavrik	Makhteshim	240 g/l	EW	10612
4	Revolt	Makhteshim	240 g/l	EW	13383

Uses
- Aphids in ***borage for oilseed production*** *(off-label)* [3]; ***durum wheat*** *(off-label)*, ***evening primrose*** *(off-label)*, ***grass seed crops*** *(off-label)*, ***honesty*** *(off-label)*, ***mustard*** *(off-label)*, ***spring linseed*** *(off-label)*, ***spring rye*** *(off-label)*, ***triticale*** *(off-label)*, ***winter rye*** *(off-label)* [2, 3]; ***spring barley***, ***spring oilseed rape***, ***spring wheat***, ***winter barley***, ***winter oilseed rape***, ***winter wheat*** [1-4]
- Barley yellow dwarf virus vectors in ***winter barley***, ***winter wheat*** [1-4]
- Cabbage stem flea beetle in ***winter oilseed rape*** [2-4]
- Insect pests in ***grass seed crops*** *(off-label)*, ***mustard*** *(off-label)* [1]
- Pollen beetle in ***spring oilseed rape***, ***winter oilseed rape*** [1-4]

Specific Off-Label Approvals (SOLAs)
- ***borage for oilseed production*** *20061366* [3]
- ***durum wheat*** *20061369* [2], *20061365* [3]
- ***evening primrose*** *20061370* [2], *20061366* [3]
- ***grass seed crops*** *20061368* [1], *20061369* [2], *20061365* [3]
- ***honesty*** *20061370* [2], *20061366* [3]
- ***mustard*** *20061367* [1], *20061370* [2], *20061366* [3]
- ***spring linseed*** *20061370* [2], *20061366* [3]
- ***spring rye*** *20061369* [2], *20061365* [3]
- ***triticale*** *20061369* [2], *20061365* [3]
- ***winter rye*** *20061369* [2], *20061365* [3]

Approval information
- Accepted by BBPA for use on malting barley

Efficacy guidance
- For BYDV control on winter cereals follow local warnings or spray high risk crops in mid-Oct and make repeat application in late autumn/early winter if aphid activity persists
- For summer aphid control on cereals spray once when aphids present on two thirds of ears and increasing
- On oilseed rape treat peach potato aphids in autumn in response to local warning and repeat if necessary
- Best control of pollen beetle in oilseed rape obtained from treatment at green to yellow bud stage and repeat if necessary
- Good spray cover of target essential for best results

Restrictions
- Maximum total dose equivalent to two full dose treatments on oilseed rape. See label for dose rates on cereals
- A minimum of 14 d must elapse between applications to cereals

Crop-specific information
- Latest use: before caryopsis watery ripe (GS 71) for barley; before flowering for oilseed rape; before kernel medium milk (GS 75) for wheat

Environmental safety

- Dangerous for the environment
- Very toxic to aquatic organisms
- High risk to non-target insects or other arthropods. Do not spray within 6 m of the field boundary [1-3]
- LERAP Category A
- Avoid spraying oilseed rape within 6 m of field boundary to reduce effects on certain non-target species or other arthropods
- Must not be applied to cereals if any product containing a pyrethroid insecticide or dimethoate has been sprayed after the start of ear emergence (GS 51)

Hazard classification and safety precautions

Hazard Irritant [1]; Dangerous for the environment [1-4]
Transport code 9
Packaging group III
UN number 3082
Risk phrases R36, R38, R51 [1]; R50 [2-4]; R53a [1-4]
Operator protection A, C, H; U05a, U10, U11, U19a, U20a
Environmental protection E15a, E16c, E16d, E22a
Storage and disposal D01, D02, D05, D10c [1-4]; D09a [2-4]; D09b [1]

425 tebuconazole

A systemic triazole fungicide for cereals and other field crops
FRAC mode of action code: 3

See also *chlorothalonil + tebuconazole*
clothianidin + prothioconazole + tebuconazole + triazoxide
fenpropidin + tebuconazole
prochloraz + proquinazid + tebuconazole
prochloraz + tebuconazole
propiconazole + tebuconazole
prothioconazole + spiroxamine + tebuconazole
prothioconazole + tebuconazole
prothioconazole + tebuconazole + triazoxide
spiroxamine + tebuconazole

Products

1	Alpha Tebuconazole 20 EW	Makhteshim	200 g/l	EW	12893
2	Bezel	Bayer CropScience	2% w/w	PA	12753
3	Chani	AgChem Access	250 g/l	EW	13325
4	Clayton Tebucon	Clayton	250 g/l	EW	08707
5	Deacon	Makhteshim	200 g/l	EW	14270
6	Fezan	Sipcam	250 g/l	EW	13681
7	Folicur	Bayer CropScience	250 g/l	EW	11278
8	Gizmo	Nufarm UK	60 g/l	FS	14150
9	Greencrop Tabloid	Greencrop	250 g/l	EW	11969
10	Icon	AgriGuard	250 g/l	EW	12202
11	Mitre	Makhteshim	200 g/l	EW	12901
12	Orius 20 EW	Makhteshim	200 g/l	EW	12311
13	Patriarch	AgriGuard	250 g/l	EW	13979
14	Riza	Headland	250 g/l	EW	12696
15	Standon Tebuconazole	Standon	250 g/l	EC	09056
16	Tebcon	Goldengrass	250 g/l	EW	14300

Uses

- Alternaria in ***brussels sprouts*** [12]; ***cabbages*** [1, 3, 5, 7, 9-12, 14, 16]; ***carrots***, ***horseradish*** [1, 3, 5, 7, 9, 10, 14, 16]; ***parsnips*** [14]; ***spring oilseed rape***, ***winter oilseed rape*** [1, 3-5, 7, 9-12, 14-16]
- Blight in ***blackberries*** *(off-label)*, ***raspberries*** *(off-label)*, ***rubus hybrids*** *(off-label)* [1, 7, 11, 12]
- Botrytis in ***daffodils*** *(off-label)* [12]; ***daffodils*** *(off-label - for galanthamine production)* [1, 7, 11]; ***daffodils grown for galanthamine production*** *(off-label)* [5]; ***linseed*** *(reduction)* [1, 3-5, 7, 9, 10, 15, 16]

SEE SECTION 3 FOR PRODUCTS ALSO REGISTERED

- Brown rust in ***spring barley, spring rye, spring wheat, winter barley, winter rye, winter wheat*** [1, 3-7, 9-16]; ***triticale*** *(off-label)* [12]
- Cane blight in ***blackberries*** *(off-label)*, ***raspberries*** *(off-label)*, ***rubus hybrids*** *(off-label)* [1, 5, 7, 11, 12]
- Canker in ***all tree nuts, farm forestry, forest, forest nurseries, ornamental plant production, pome fruit, stone fruit*** [2]; ***apples*** *(off-label)*, ***chestnuts*** *(off-label)*, ***cob nuts*** *(off-label)*, ***crab apples*** *(off-label)*, ***pears*** *(off-label)*, ***quinces*** *(off-label)*, ***walnuts*** *(off-label)* [1, 5, 7, 11, 12, 14]
- Chocolate spot in ***spring field beans, winter field beans*** [1, 3-5, 7, 9-12, 14-16]
- Crown rust in ***spring oats, winter oats*** [13, 14]; ***spring oats*** *(reduction)*, ***winter oats*** *(reduction)* [3, 7, 9, 10, 16]
- Disease control in ***borage for oilseed production*** *(off-label)*, ***canary flower (echium spp.)*** *(off-label)*, ***evening primrose*** *(off-label)*, ***grass seed crops*** *(off-label)*, ***honesty*** *(off-label)*, ***kohlrabi*** *(off-label)*, ***mallow (althaea spp.)*** *(off-label)*, ***mustard*** *(off-label)*, ***parsley root*** *(off-label)* [5, 12]; ***triticale*** *(off-label)* [5]
- Foliar disease control in ***borage*** *(off-label)*, ***grass seed crops*** *(off-label)* [1, 4, 7, 11, 15]; ***borage for oilseed production*** *(off-label)*, ***horseradish*** *(off-label)* [14]; ***canary flower (echium spp.)*** *(off-label)*, ***evening primrose*** *(off-label)*, ***honesty*** *(off-label)*, ***mustard*** *(off-label)* [1, 4, 7, 11, 14, 15]; ***mallow (althaea spp.)*** *(off-label)*, ***parsley root*** *(off-label)*, ***triticale*** *(off-label)* [1, 7, 11, 14]
- Fungus diseases in ***kohlrabi*** *(off-label)* [1, 7, 11]
- Fusarium ear blight in ***spring wheat, winter wheat*** [1, 3, 5, 7, 9-14, 16]; ***triticale*** *(off-label)* [12]
- Glume blotch in ***spring wheat, winter wheat*** [1, 3-7, 9-16]; ***triticale*** *(off-label)* [12]
- Leaf stripe in ***spring barley*** [8]
- Light leaf spot in ***brussels sprouts*** [12]; ***cabbages, spring oilseed rape, winter oilseed rape*** [1, 3-5, 7, 9-12, 14-16]
- Lodging control in ***spring oilseed rape, winter oilseed rape*** [1, 3, 5, 7, 9-12, 16]
- Loose smut in ***spring barley, winter barley*** [8]
- Net blotch in ***spring barley, winter barley*** [1, 3-7, 9-16]
- Phoma in ***spring oilseed rape, winter oilseed rape*** [1, 3, 5, 7, 9-12, 14, 16]
- Powdery mildew in ***brussels sprouts, triticale*** *(off-label)* [12]; ***cabbages, swedes, turnips*** [1, 3-5, 7, 9-12, 14-16]; ***carrots, linseed*** [1, 3, 5, 7, 9, 10, 14, 16]; ***chives*** *(off-label)* [1, 5, 7, 11, 12, 14]; ***horseradish*** [14]; ***parsnips*** [1, 3, 5, 7, 14, 16]; ***spring barley, spring rye, spring wheat, winter barley, winter rye, winter wheat*** [1, 3-7, 9-16]; ***spring oats, winter oats*** [3, 6, 7, 9, 10, 14-16]
- Pruning wounds in ***all tree nuts, farm forestry, forest, forest nurseries, ornamental plant production, pome fruit, stone fruit*** [2]
- Rhynchosporium in ***spring barley, spring rye, winter barley, winter rye*** [1, 3-7, 9-16]
- Ring spot in ***broccoli*** *(off-label)*, ***cauliflowers*** *(off-label)*, ***chinese cabbage*** *(off-label)*, ***choi sum*** *(off-label)*, ***pak choi*** *(off-label)* [1, 5, 7, 11, 12, 14]; ***brussels sprouts*** [12]; ***cabbages*** [1, 3-5, 7, 9-12, 14-16]; ***calabrese*** *(off-label)* [1, 7, 11, 12, 14]; ***kohlrabi*** *(off-label)* [14]; ***spring oilseed rape*** *(reduction)*, ***winter oilseed rape*** *(reduction)* [1, 3, 5, 7, 9-12, 16]
- Rust in ***broad beans*** *(off-label)*, ***chives*** *(off-label)*, ***dwarf beans*** *(off-label)*, ***french beans*** *(off-label)*, ***runner beans*** *(off-label)* [1, 5, 7, 11, 12, 14]; ***leeks*** [1, 3, 5, 7, 9-12, 14, 16]; ***spring field beans, winter field beans*** [1, 3-5, 7, 9-12, 14-16]
- Sclerotinia in ***carrots*** [1, 3, 5, 7, 9, 10, 14, 16]; ***horseradish, parsnips*** [14]
- Sclerotinia stem rot in ***spring oilseed rape, winter oilseed rape*** [1, 3-5, 7, 9-12, 14-16]
- Septoria leaf blotch in ***spring wheat, winter wheat*** [1, 3-7, 9-16]; ***triticale*** *(off-label)* [12]
- Sooty moulds in ***spring wheat, winter wheat*** [1, 3, 5, 7, 9-14, 16]
- Stem canker in ***spring oilseed rape, winter oilseed rape*** [1, 3, 5, 7, 9-12, 14, 16]
- White rot in ***bulb onions*** *(off-label)*, ***garlic*** *(off-label)*, ***salad onions*** *(off-label)*, ***shallots*** *(off-label)* [1, 5, 7, 11, 12, 14]; ***onion sets*** *(off-label)* [5, 12, 14]
- Yellow rust in ***spring barley, spring rye, spring wheat, winter barley, winter rye, winter wheat*** [1, 3-7, 9-16]; ***triticale*** *(off-label)* [12]

Specific Off-Label Approvals (SOLAs)

- ***apples*** *20071380* [1], *20090147* [5], *20082159* [7], *20071368* [11], *20071329* [12], *20070554* [14]
- ***blackberries*** *20091408* [1], *20091407* [5], *20082160* [7], *20091406* [11], *20091399* [12]
- ***borage*** *20071389* [1], *20061371* [4], *20062694* [7], *20071359* [11], *20061399* [15]
- ***borage for oilseed production*** *20090113* [5], *20071337* [12], *20070558* [14]
- ***broad beans*** *20071390* [1], *20090089* [5], *20031878* [7], *20071364* [11], *20071338* [12], *20070549* [14]
- ***broccoli*** *20071375* [1], *20090091* [5], *20031874* [7], *20071371* [11], *20071326* [12], *20070546* [14]

- **bulb onions** *20071378* [1], *20071384* [1], *20090090* [5], *20031877* [7], *20031879* [7], *20071365* [11], *20071369* [11], *20071327* [12], *20070550* [14]
- **calabrese** *20071375* [1], *20031874* [7], *20071371* [11], *20071326* [12], *20070546* [14]
- **canary flower (echium spp.)** *20071389* [1], *20061371* [4], *20090113* [5], *20062694* [7], *20071359* [11], *20071337* [12], *20070558* [14], *20061399* [15]
- **cauliflowers** *20071375* [1], *20090091* [5], *20031874* [7], *20071371* [11], *20071326* [12], *20070546* [14]
- **chestnuts** *20071393* [1], *20090116* [5], *20082158* [7], *20071372* [11], *20071321* [12], *20070555* [14]
- **chinese cabbage** *20071375* [1], *20090091* [5], *20031874* [7], *20071371* [11], *20071326* [12], *20070546* [14]
- **chives** *20082083* [1], *20090088* [5], *20082084* [7], *20082085* [11], *20082087* [12], *20082088* [14]
- **choi sum** *20082247* [1], *20090145* [5], *20082249* [7], *20082250* [11], *20082251* [12], *20082252* [14]
- **cob nuts** *20071393* [1], *20090116* [5], *20082158* [7], *20071372* [11], *20071321* [12], *20070555* [14]
- **crab apples** *20071380* [1], *20090147* [5], *20082159* [7], *20071368* [11], *20071329* [12], *20070554* [14]
- **daffodils** *(for galanthamine production) 20071379* [1], *(for galanthamine production) 20041516* [7], *(for galanthamine production) 20071370* [11], *20071328* [12]
- **daffodils grown for galanthamine production** *20090095* [5]
- **dwarf beans** *20071385* [1], *20090094* [5], *20031880* [7], *20071357* [11], *20071320* [12], *20070551* [14]
- **evening primrose** *20071389* [1], *20061371* [4], *20090113* [5], *20062694* [7], *20071359* [11], *20071337* [12], *20070558* [14], *20061399* [15]
- **french beans** *20071385* [1], *20090094* [5], *20031880* [7], *20071357* [11], *20071320* [12], *20070551* [14]
- **garlic** *20071378* [1], *20090090* [5], *20031879* [7], *20071369* [11], *20071327* [12], *20070550* [14]
- **grass seed crops** *20071381* [1], *20061372* [4], *20090144* [5], *20081764* [7], *20071362* [11], *20071340* [12], *20061400* [15]
- **honesty** *20071389* [1], *20061371* [4], *20090113* [5], *20062694* [7], *20071359* [11], *20071337* [12], *20070558* [14], *20061399* [15]
- **horseradish** *20070556* [14]
- **kohlrabi** *20071377* [1], *20090093* [5], *20031881* [7], *20071360* [11], *20071334* [12], *20070552* [14]
- **mallow (althaea spp.)** *20071391* [1], *20090143* [5], *20062692* [7], *20071363* [11], *20071339* [12], *20070556* [14]
- **mustard** *20071389* [1], *20061371* [4], *20090113* [5], *20062694* [7], *20071359* [11], *20071337* [12], *20070558* [14], *20061399* [15]
- **onion sets** *20090096* [5], *20071336* [12], *20070548* [14]
- **pak choi** *20082247* [1], *20090145* [5], *20082249* [7], *20082250* [11], *20082251* [12], *20082252* [14]
- **parsley root** *20071391* [1], *20090143* [5], *20062692* [7], *20071363* [11], *20071339* [12], *20070556* [14]
- **pears** *20071380* [1], *20090147* [5], *20082159* [7], *20071368* [11], *20071329* [12], *20070554* [14]
- **quinces** *20071380* [1], *20090147* [5], *20082159* [7], *20071368* [11], *20071329* [12], *20070554* [14]
- **raspberries** *20091408* [1], *20091407* [5], *20082160* [7], *20091406* [11], *20091399* [12]
- **rubus hybrids** *20091408* [1], *20091407* [5], *20082160* [7], *20091406* [11], *20091399* [12]
- **runner beans** *20071385* [1], *20090094* [5], *20031880* [7], *20071357* [11], *20071320* [12], *20070551* [14]
- **salad onions** *20071387* [1], *20090092* [5], *20042408* [7], *20071361* [11], *20071322* [12], *20070553* [14]
- **shallots** *20071378* [1], *20090090* [5], *20031879* [7], *20071369* [11], *20071327* [12], *20070550* [14]
- **triticale** *20071388* [1], *20090148* [5], *20062693* [7], *20071366* [11], *20071335* [12], *20070557* [14]
- **walnuts** *20071393* [1], *20090116* [5], *20082158* [7], *20071372* [11], *20071321* [12], *20070555* [14]

Approval information

- Accepted by BBPA for use on malting barley

Efficacy guidance

- For best results apply at an early stage of disease development before infection spreads to new crop growth

- To protect flag leaf and ear from Septoria diseases apply from flag leaf emergence to ear fully emerged (GS 37-59). Earlier application may be necessary where there is a high risk of infection
- Improved control of established cereal mildew can be obtained by tank mixture with fenpropimorph
- For light leaf spot control in oilseed rape apply in autumn/winter with a follow-up spray in spring/ summer if required
- For control of most other diseases spray at first signs of infection with a follow-up spray 2-4 wk later if necessary. See label for details
- For disease control in cabbages a 3-spray programme at 21-28 d intervals will give good control
- Tebuconazole is a DMI fungicide. Resistance to some DMI fungicides has been identified in Septoria leaf blotch which may seriously affect performance of some products. For further advice contact a specialist advisor and visit the Fungicide Resistance Action Group (FRAG)-UK website

Restrictions

- Maximum number of treatments 1 per wound for all pome fruit, all stone fruit, all tree nuts, forestry, ornamental plant production [2]
- Maximum total dose equivalent to 1 full dose treatment on linseed; 2 full dose treatments on wheat, barley, rye, oats, field beans, swedes, turnips, onions; 2.25 full dose treatments on cabbages; 2.5 full dose treatments on oilseed rape; 3 full dose treatments on leeks, parsnips, carrots
- Do not treat durum wheat
- Apply only to listed oat varieties (see label) and do not apply to oats in tank mixture
- Do not apply before swedes and turnips have a root diameter of 2.5 cm, or before heart formation in cabbages, or before button formation in Brussels sprouts
- Consult processor before use on crops for processing

Crop-specific information

- Latest use: before grain milky-ripe for cereals (GS 71); when most seed green-brown mottled for oilseed rape (GS 6,3); before brown capsule for linseed
- HI field beans, swedes, turnips, linseed, winter oats 35 d; market brassicas, carrots, horseradish, parsnips 21 d; leeks 14 d
- Some transient leaf speckling on wheat or leaf reddening/scorch on oats may occur but this has not been shown to reduce yield response to disease control

Environmental safety

- Dangerous for the environment
- Toxic to aquatic organisms

Hazard classification and safety precautions

Hazard Harmful [3, 4, 7-11, 13-16]; Irritant [1, 5, 6, 12]; Dangerous for the environment [3, 4, 7, 9-11, 13-16]

Transport code 9

Packaging group III

UN number 3082

Risk phrases R20 [3, 4, 7, 11, 13, 16]; R22a, R41, R51 [3, 4, 7, 9-11, 13-16]; R36 [1, 5, 6, 12]; R38 [1, 5, 6, 9, 10, 12, 14, 15]; R52 [1, 5, 6, 8, 12]; R53a [1, 3-16]; R63 [4, 8, 13]

Operator protection A, H [1-16]; C [1, 3-7, 9-16]; U05a, U20a [1, 3-7, 9-12, 14-16]; U11 [1, 3-7, 9-16]; U14, U15 [1, 5, 6, 12]; U19a [4, 9, 10, 14, 15]; U20b [2, 8]

Environmental protection E03, E15b [8]; E15a [1, 3-7, 9-16]; E34 [1, 3, 5-7, 9-12, 14, 16]; E38 [3, 7, 8, 11, 13, 16]

Storage and disposal D01, D02 [1, 3-7, 9-16]; D05 [2-4, 7, 9-11, 14-16]; D09a [1-12, 14-16]; D10b [3, 4, 7, 9-11, 14-16]; D11a [2, 8]; D12a [3, 7, 11, 13, 16]

Treated seed S01, S02, S03, S04a, S05, S07 [8]

Medical advice M03 [3, 4, 7, 9-11, 14-16]; M05a [13]

426 tebuconazole + triadimenol

A broad spectrum systemic fungicide for cereals
FRAC mode of action code: 3 + 3

Products

1	Silvacur	Bayer CropScience	250:125 g/l	EC	11309
2	Veto F	Bayer CropScience	225:75 g/l	EC	14274

Uses

- Brown rust in ***spring barley***, ***spring rye***, ***spring wheat***, ***winter barley***, ***winter rye***, ***winter wheat*** [1, 2]
- Crown rust in ***spring oats***, ***winter oats*** [1, 2]
- Foliar disease control in ***grass seed crops*** *(off-label)* [1]
- Fusarium ear blight in ***spring wheat***, ***winter wheat*** [1, 2]
- Glume blotch in ***spring wheat***, ***winter wheat*** [1, 2]
- Net blotch in ***spring barley***, ***winter barley*** [1, 2]
- Powdery mildew in ***spring barley***, ***spring oats***, ***spring rye***, ***spring wheat***, ***winter barley***, ***winter oats***, ***winter rye***, ***winter wheat*** [1, 2]
- Rhynchosporium in ***spring barley***, ***winter barley*** [1, 2]
- Septoria leaf blotch in ***spring wheat***, ***winter wheat*** [1, 2]
- Sooty moulds in ***spring wheat***, ***winter wheat*** [1, 2]
- Yellow rust in ***spring barley***, ***spring rye***, ***spring wheat***, ***winter barley***, ***winter rye***, ***winter wheat*** [1, 2]

Specific Off-Label Approvals (SOLAs)

- ***grass seed crops*** *20061023* [1]

Approval information

- Accepted by BBPA for use on malting barley

Efficacy guidance

- For best results apply at an early stage of disease development before infection spreads to new crop growth
- To protect flag leaf and ear from Septoria diseases apply from flag leaf emergence to ear fully emerged (GS 37-59). Earlier application may be necessary where there is a high risk of infection
- For control of rust, powdery mildew, leaf and net blotch apply at first signs of disease with a second application 2-3 wk later if necessary
- Tebuconazole is a DMI fungicide. Resistance to some DMI fungicides has been identified in Septoria leaf blotch which may seriously affect performance of some products. For further advice contact a specialist advisor and visit the Fungicide Resistance Action Group (FRAG)-UK website

Restrictions

- Maximum total dose equivalent to two full dose treatments
- Do not use on durum wheat
- Use only on listed varieties of oats. See label [1]

Crop-specific information

- Latest use: before grain milky-ripe (GS 71)
- Some transient leaf speckling may occur on wheat or leaf reddening/scorch on oats but this has not been shown to reduce yield response or disease control

Environmental safety

- Harmful to aquatic organisms
- Harmful to fish or other aquatic life. Do not contaminate surface waters or ditches with chemical or used container

Hazard classification and safety precautions

Hazard Harmful [1]
Transport code 9 [1]
Packaging group III [1]
UN number 3082 [1]; N/C [2]
Risk phrases R20, R36 [1]; R52, R53a [1, 2]
Operator protection A, C, H; U05a, U09b, U20a
Environmental protection E13c [1, 2]; E34 [1]

Storage and disposal D01, D02, D05, D09a [1, 2]; D10b [1]; D10c [2]

Medical advice M03 [2]

427 tebuconazole + trifloxystrobin

A conazole and stobilurin fungicide mixture for wheat and vegetable crops

FRAC mode of action code: 3 + 11

See also *trifloxystrobin*

Products

1	Dedicate	Bayer Environ.	200:100 g/l	SC	13612
2	Mascot Fusion	Rigby Taylor	200:100 g/l	EC	14113
3	Nativo 75 WG	Bayer CropScience	50:25% w/w	WG	13057

Uses

- Alternaria in ***broccoli, brussels sprouts, cabbages, calabrese, carrots, cauliflowers*** [3]
- Anthracnose in ***amenity grassland, managed amenity turf*** [2]
- Dollar spot in ***amenity grassland, managed amenity turf*** [1, 2]
- Foliar disease control in ***horseradish*** *(off-label)*, ***parsley root*** *(off-label)*, ***parsnips*** *(off-label)*, ***salsify*** *(off-label)* [3]
- Fusarium in ***amenity grassland, managed amenity turf*** [1]
- Fusarium patch in ***amenity grassland, managed amenity turf*** [2]
- Light leaf spot in ***broccoli, brussels sprouts, cabbages, calabrese, cauliflowers*** [3]
- Phoma leaf spot in ***broccoli, brussels sprouts, cabbages, calabrese, cauliflowers*** [3]
- Powdery mildew in ***broccoli, brussels sprouts, cabbages, calabrese, carrots, cauliflowers, wine grapes*** *(off-label)* [3]
- Purple blotch in ***leeks*** [3]
- Red thread in ***amenity grassland, managed amenity turf*** [1, 2]
- Ring spot in ***broccoli, brussels sprouts, cabbages, calabrese, cauliflowers*** [3]
- Rust in ***leeks*** [3]
- Sclerotinia rot in ***carrots*** [3]
- White blister in ***broccoli, brussels sprouts, cabbages, calabrese, cauliflowers*** [3]
- White tip in ***leeks*** [3]

Specific Off-Label Approvals (SOLAs)

- ***horseradish*** *20064163* [3]
- ***parsley root*** *20064163* [3]
- ***parsnips*** *20064163* [3]
- ***salsify*** *20064163* [3]
- ***wine grapes*** *20090730* [3]

Approval information

- Trifloxystrobin included in Annex I under EC Directive 91/414

Efficacy guidance

- Best results obtained from treatment at early stages of disease development. Further treatment may be needed if disease attack is prolonged
- Applications to established infections of any disease are likely to be less effective
- Treatment may give some control of *Stemphyllium botryosum* and leaf blotch on leeks [3]
- Tebuconazole is a DMI fungicide. Resistance to some DMI fungicides has been identified in Septoria leaf blotch which may seriously affect performance of some products. For further advice contact a specialist advisor and visit the Fungicide Resistance Action Group (FRAG)-UK website
- Trifloxystrobin is a member of the QoI cross resistance group. Product should be used preventatively and not relied on for its curative potential
- Use product as part of an Integrated Crop Management strategy incorporating other methods of control, including where appropriate other fungicides with a different mode of action. Do not apply more than two foliar applications of QoI containing products to any cereal crop, broccoli, calabrese or cauliflower. Do not apply more than three applications to Brussels sprouts, cabbage, carrots or leeks
- There is a significant risk of widespread resistance occurring in *Septoria tritici* populations in UK. Failure to follow resistance management action may result in reduced levels of disease control

- Strains of wheat powdery mildew resistant to QoIs are common in the UK. Control of wheat mildew can only be relied on from the triazole component
- Where specific control of wheat mildew is required this should be achieved through a programme of measures including products recommended for the control of mildew that contain a fungicide from a different cross-resistance group and applied at a dose that will give robust control
- Do not apply to grass during drought conditions nor to frozen turf [1]

Restrictions
- Maximum number of treatments 2 per crop for broccoli, calabrese, cauliflowers; 3 per crop for Brussels sprouts, cabbages, carrots, leeks [3]
- In addition to the maximum number of treatments per crop a maximum of 3 applications may be applied to the same ground in one calendar year [3]
- Consult processor before use on vegetable crops for processing [3]

Crop-specific information
- Latest use: before milky ripe stage on winter wheat
- HI 35 d for wheat; 21 d for all other crops
- Performance against leaf spot diseases of brassicas, *Alternaria* leaf blight of carrots, and rust in leeks may be improved by mixing with an approved sticker/wetter [3]

Environmental safety
- Dangerous for the environment
- Very toxic to aquatic organisms

Hazard classification and safety precautions

Hazard Harmful, Dangerous for the environment
Transport code 9
Packaging group III
UN number 3077 [3]; 3082 [1, 2]
Risk phrases R37 [1, 2]; R50, R53a, R63 [1-3]; R70 [1]
Operator protection A [1-3]; H [2, 3]; P [1]; U05a [1, 3]; U09a [1]; U09b [2, 3]; U19a [1, 2]; U20b [1-3]
Environmental protection E15a, E34, E38
Storage and disposal D01, D02, D09a, D10b, D12a [1-3]; D05 [2]
Medical advice M03

428 tebufenpyrad

A pyrazole mitochondrial electron transport inhibitor (METI) aphicide and acaricide
IRAC mode of action code: 21

Products

Masai	BASF	20% w/w	WB	13082

Uses
- Bud mite in ***almonds*** *(off-label)*, ***chestnuts*** *(off-label)*, ***cob nuts*** *(off-label)*, ***hazel nuts*** *(off-label)*, ***walnuts*** *(off-label)*
- Damson-hop aphid in ***hops***, ***plums*** *(off-label)*
- Gall mite in ***bilberries*** *(off-label)*, ***blackcurrants*** *(off-label)*, ***blueberries*** *(off-label)*, ***cranberries*** *(off-label)*, ***gooseberries*** *(off-label)*, ***redcurrants*** *(off-label)*, ***vaccinium spp.*** *(off-label)*, ***whitecurrants*** *(off-label)*
- Red spider mites in ***apples***, ***pears***
- Two-spotted spider mite in ***hops***, ***protected roses***, ***strawberries***

Specific Off-Label Approvals (SOLAs)
- ***almonds*** *20080132*
- ***bilberries*** *20080131*
- ***blackcurrants*** *20080131*
- ***blueberries*** *20080131*
- ***chestnuts*** *20080132*
- ***cob nuts*** *20080132*
- ***cranberries*** *20080131*
- ***gooseberries*** *20080131*

- ***hazel nuts*** *20080132*
- ***plums*** *20080132*
- ***redcurrants*** *20080131*
- ***vaccinium spp.*** *20080131*
- ***walnuts*** *20080132*
- ***whitecurrants*** *20080131*

Approval information

- Tebufenpyrad included in Annex 1 under EC Directive 91/414.
- Accepted by BBPA for use on hops

Efficacy guidance

- Acts on eggs (except winter eggs) and all motile stages of spider mites up to adults
- Treat spider mites from 80% egg hatch but before mites become established
- For effective control total spray cover of the crop is required
- Product can be used in a programme to give season-long control of damson-hop aphids coupled with mite control
- Where aphids resistant to tebufenpyrad occur in hops control is unlikely to be satisfactory and repeat treatments may result in lower levels of control. Where possible use different active ingredients in a programme

Restrictions

- Maximum total dose equivalent to one full dose treatment on apples, pears, strawberries; 3 full dose treatments on hops
- Other mitochondrial electron transport inhibitor (METI) acaricides should not be applied to the same crop in the same calendar yr either separately or in mixture
- Do not treat apples before 90% petal fall
- Small-scale testing of rose varieties to establish tolerance recommended before use
- Inner liner of container must not be removed

Crop-specific information

- Latest use: end of burr stage for hops
- HI strawberries 3 d; apples, bilberries, blackcurrants, blueberries, cranberries, gooseberries, pears, redcurrants, whitecurrants 7d; blackberries, plums, raspberries 21 d
- Product has no effect on fruit quality or finish

Environmental safety

- Dangerous for the environment
- Very toxic to aquatic organisms
- High risk to bees. Do not apply to crops in flower or to those in which bees are actively foraging. Do not apply when flowering weeds are present
- LERAP Category B
- Broadcast air-assisted LERAP (18 m)

Hazard classification and safety precautions

Hazard Harmful, Dangerous for the environment
Transport code 9
Packaging group III
UN number 3077
Risk phrases R20, R22a, R50, R53a
Operator protection A; U02a, U05a, U09a, U13, U14, U20b
Environmental protection E12a, E12e, E15a, E16a, E16b, E38; E17b (18 m)
Storage and disposal D01, D02, D09a, D11a, D12a
Medical advice M05a

429 teflubenzuron

A benzoylurea insecticide for use on ornamentals
IRAC mode of action code: 15

Products

Nemolt	Fargro	150 g/l	SC	10226

Uses

- Alder flea beetle in ***amenity vegetation*** *(off-label)*, ***forest nurseries*** *(off-label)*, ***protected ornamentals*** *(off-label)*
- Browntail moth in ***ornamental plant production***
- Cabbage moth in ***broccoli*** *(off-label)*, ***brussels sprouts*** *(off-label)*, ***cauliflowers*** *(off-label)*, ***chinese cabbage*** *(off-label)*
- Cabbage root fly in ***broccoli*** *(off-label)*, ***brussels sprouts*** *(off-label)*, ***cauliflowers*** *(off-label)*, ***chinese cabbage*** *(off-label)*
- Cabbage white butterfly in ***broccoli*** *(off-label)*, ***brussels sprouts*** *(off-label)*, ***cauliflowers*** *(off-label)*, ***chinese cabbage*** *(off-label)*
- Caterpillars in ***amenity vegetation*** *(off-label)*, ***forest nurseries*** *(off-label)*, ***ornamental plant production***, ***protected aubergines*** *(off-label)*, ***protected courgettes*** *(off-label)*, ***protected cucumbers*** *(off-label)*, ***protected melons*** *(off-label)*, ***protected ornamentals*** *(off-label)*, ***protected peppers*** *(off-label)*, ***protected tomatoes*** *(off-label)*
- Diamond-back moth in ***broccoli*** *(off-label)*, ***brussels sprouts*** *(off-label)*, ***cauliflowers*** *(off-label)*, ***chinese cabbage*** *(off-label)*
- Insect pests in ***amenity vegetation*** *(off-label)*, ***broccoli*** *(off-label)*, ***brussels sprouts*** *(off-label)*, ***cauliflowers*** *(off-label)*, ***chinese cabbage*** *(off-label)*, ***forest nurseries*** *(off-label)*, ***protected aubergines*** *(off-label)*, ***protected courgettes*** *(off-label)*, ***protected cucumbers*** *(off-label)*, ***protected melons*** *(off-label)*, ***protected ornamentals*** *(off-label)*, ***protected peppers*** *(off-label)*, ***protected tomatoes*** *(off-label)*
- Moths in ***amenity vegetation*** *(off-label)*, ***forest nurseries*** *(off-label)*, ***protected ornamentals*** *(off-label)*
- Thrips in ***protected aubergines*** *(off-label)*, ***protected courgettes*** *(off-label)*, ***protected cucumbers*** *(off-label)*, ***protected melons*** *(off-label)*, ***protected peppers*** *(off-label)*, ***protected tomatoes*** *(off-label)*
- Western flower thrips in ***amenity vegetation*** *(off-label)*, ***forest nurseries*** *(off-label)*, ***protected ornamentals*** *(off-label)*
- Whitefly in ***ornamental plant production***, ***protected aubergines*** *(off-label)*, ***protected courgettes*** *(off-label)*, ***protected cucumbers*** *(off-label)*, ***protected melons*** *(off-label)*, ***protected peppers*** *(off-label)*, ***protected tomatoes*** *(off-label)*

Specific Off-Label Approvals (SOLAs)

- ***amenity vegetation*** *20072120*
- ***broccoli*** *20072121*
- ***brussels sprouts*** *20072121*
- ***cauliflowers*** *20072121*
- ***chinese cabbage*** *20072121*
- ***forest nurseries*** *20072120*
- ***protected aubergines*** *20072119*
- ***protected courgettes*** *20072119*
- ***protected cucumbers*** *20072119*
- ***protected melons*** *20072119*
- ***protected ornamentals*** *20072120*
- ***protected peppers*** *20072119*
- ***protected tomatoes*** *20072119*

Efficacy guidance

- Product acts as larval stomach poison interfering with moulting process leading to cessation of feeding and larval death
- Apply as soon as first stage larvae seen. This will often coincide with the peak of moth flight

Restrictions

- Maximum number of treatments 3 per yr
- Test specific varieties before carrying out extensive treatments

Environmental safety

- Dangerous for the environment
- Very toxic to aquatic organisms
- Limited evidence suggests some margin of safety to *Encarsia formosa*. Effects on other parasites and predators not fully tested

SEE SECTION 3 FOR PRODUCTS ALSO REGISTERED

Hazard classification and safety precautions

Hazard Dangerous for the environment
Transport code 9
Packaging group III
UN number 3082
Risk phrases R50, R53a
Operator protection A, C; U05a, U20c
Environmental protection E15a, E38
Storage and disposal D01, D02, D05, D09a, D11a, D12a

430 tefluthrin

A soil acting pyrethroid insecticide seed treatment
IRAC mode of action code: 3

See also fludioxonil + tefluthrin

Products

Force ST	Syngenta	200 g/l	CF	11752

Uses

- Bean seed fly in ***bulb onions*** *(off-label - seed treatment)*, ***chard*** *(off-label - seed treatment)*, ***chives*** *(off-label - seed treatment)*, ***green mustard*** *(off-label - seed treatment)*, ***herbs (see appendix 6)*** *(off-label - seed treatment)*, ***leaf spinach*** *(off-label - seed treatment)*, ***leeks*** *(off-label - seed treatment)*, ***miduna*** *(off-label - seed treatment)*, ***mizuna*** *(off-label - seed treatment)*, ***pak choi*** *(off-label - seed treatment)*, ***parsley*** *(off-label - seed treatment)*, ***red mustard*** *(off-label - seed treatment)*, ***salad onions*** *(off-label - seed treatment)*, ***spinach beet*** *(off-label - seed treatment)*, ***tatsoi*** *(off-label - seed treatment)*
- Carrot fly in ***carrots*** *(off-label - seed treatment)*, ***parsnips*** *(off-label - seed treatment)*
- Millipedes in ***fodder beet*** *(seed treatment)*, ***sugar beet*** *(seed treatment)*
- Onion fly in ***bulb onions*** *(off-label - seed treatment)*, ***leeks*** *(off-label - seed treatment)*, ***salad onions*** *(off-label - seed treatment)*
- Pygmy beetle in ***fodder beet*** *(seed treatment)*, ***sugar beet*** *(seed treatment)*
- Springtails in ***fodder beet*** *(seed treatment)*, ***sugar beet*** *(seed treatment)*
- Symphylids in ***fodder beet*** *(seed treatment)*, ***sugar beet*** *(seed treatment)*

Specific Off-Label Approvals (SOLAs)

- ***bulb onions*** *(seed treatment) 20050546*
- ***carrots*** *(seed treatment) 20050547*
- ***chard*** *(seed treatment) 20050545*
- ***chives*** *(seed treatment) 20050545*
- ***green mustard*** *(seed treatment) 20050545*
- ***herbs (see appendix 6)*** *(seed treatment) 20050545*
- ***leaf spinach*** *(seed treatment) 20050545*
- ***leeks*** *(seed treatment) 20050546*
- ***miduna*** *(seed treatment) 20050545*
- ***mizuna*** *(seed treatment) 20050545*
- ***pak choi*** *(seed treatment) 20050545*
- ***parsley*** *(seed treatment) 20050545*
- ***parsnips*** *(seed treatment) 20050547*
- ***red mustard*** *(seed treatment) 20050545*
- ***salad onions*** *(seed treatment) 20050546*
- ***spinach beet*** *(seed treatment) 20050545*
- ***tatsoi*** *(seed treatment) 20050545*

Approval information

- Accepted by BBPA for use on malting barley

Efficacy guidance

- Apply during process of pelleting beet seed. Consult manufacturer for details of specialist equipment required

- Micro-capsule formulation allows slow release to provide a protection zone around treated seed during establishment

Restrictions
- Maximum number of treatments 1 per batch of seed
- Sow treated seed as soon as possible. Do not store treated seed from one drilling season to next
- If used in areas where soil erosion by wind or water likely, measures must be taken to prevent this happening
- Can cause a transient tingling or numbing sensation to exposed skin. Avoid skin contact with product, treated seed and dust throughout all operations in the seed treatment plant and at drilling

Crop-specific information
- Latest use: before drilling seed
- Treated seed must be drilled within the season of treatment

Environmental safety
- Dangerous for the environment
- Very toxic to aquatic organisms
- Keep treated seed secure from people, domestic stock/pets and wildlife at all times during storage and use
- Treated seed harmful to game and wild life. Bury spillages
- In the event of seed spillage clean up as much as possible into the related seed sack and bury the remainder completely
- Do not apply treated seed from the air
- Keep livestock out of areas drilled with treated seed for at least 80 d

Hazard classification and safety precautions

Hazard Harmful, Dangerous for the environment
Transport code 9
Packaging group III
UN number 3082
Risk phrases R20, R22b, R43, R50, R53a
Operator protection A, D, E, H; U02a, U04a, U05a, U08, U20b
Environmental protection E06a (80 d); E15a, E34, E38
Storage and disposal D01, D02, D05, D09a, D11a, D12a
Treated seed S02, S03, S04b, S05, S07
Medical advice M05b

431 tepraloxydim

A systemic post-emergence herbicide for control of annual grass weeds
HRAC mode of action code: A

Products

1	Aramo	BASF	50 g/l	EC	10280
2	Clayton Rally	Clayton	50 g/l	EC	13415
3	Landgold Tepraloxydim	Goldengrass	50 g/l	EC	14026
4	Omarra	AgChem Access	50 g/l	EC	13693
5	Standon Tonga	Standon	50 g/l	EC	13394
6	Tepra	Goldengrass	50 g/l	EC	14160

Uses
- Annual grasses in ***all edible seed crops grown outdoors*** *(off-label)*, ***all non-edible seed crops grown outdoors*** *(off-label)*, ***borage*** *(off-label)*, ***canary flower (echium spp.)*** *(off-label)*, ***evening primrose*** *(off-label)*, ***forest nurseries*** *(off-label)*, ***game cover*** *(off-label)*, ***golf courses*** *(off-label)*, ***honesty*** *(off-label)*, ***hops*** *(off-label)*, ***lupins*** *(off-label)*, ***mustard*** *(off-label)*, ***ornamental plant production*** *(off-label)*, ***swedes*** *(off-label)*, ***top fruit*** *(off-label)*, ***turnips*** *(off-label)* [1]; ***bulb onions***, ***cabbages***, ***carrots***, ***cauliflowers***, ***combining peas***, ***fodder beet***, ***land temporarily removed from production***, ***leeks***, ***linseed***, ***spring field beans***, ***sugar beet***, ***vining peas***, ***winter field beans***, ***winter oilseed rape*** [1, 2, 4, 6]; ***flax***, ***linseed for industrial use***, ***winter oilseed rape for industrial use*** [1, 4]; ***garlic*** *(off-label)*, ***mallow (althaea spp.)*** *(off-label)*, ***parsnips*** *(off-label)*, ***red beet*** *(off-label)*, ***shallots*** *(off-label)* [1, 6]
- Annual meadow grass in ***all edible seed crops grown outdoors*** *(off-label)*, ***all non-edible seed crops grown outdoors*** *(off-label)*, ***forest nurseries*** *(off-label)*, ***game cover*** *(off-label)*, ***golf***

courses *(off-label)*, ***hops*** *(off-label)*, ***ornamental plant production*** *(off-label)*, ***poppies for morphine production*** *(off-label)*, ***salad onions*** *(off-label)*, ***swedes*** *(off-label)*, ***top fruit*** *(off-label)*, ***turnips*** *(off-label)* [1]; ***bulb onions, cabbages, carrots, cauliflowers, combining peas, fodder beet, leeks, linseed, spring field beans, sugar beet, vining peas, winter field beans, winter oilseed rape*** [3, 5]; ***garlic*** *(off-label)*, ***mallow (althaea spp.)*** *(off-label)*, ***parsnips*** *(off-label)*, ***red beet*** *(off-label)*, ***shallots*** *(off-label)* [6]

- Blackgrass in ***bulb onions, cabbages, carrots, cauliflowers, combining peas, fodder beet, leeks, linseed, spring field beans, sugar beet, vining peas, winter field beans, winter oilseed rape*** [3, 5]
- Green cover in ***land temporarily removed from production*** [3, 5]
- Perennial grasses in ***all edible seed crops grown outdoors*** *(off-label)*, ***all non-edible seed crops grown outdoors*** *(off-label)*, ***borage*** *(off-label)*, ***canary flower (echium spp.)*** *(off-label)*, ***evening primrose*** *(off-label)*, ***forest nurseries*** *(off-label)*, ***game cover*** *(off-label)*, ***golf courses*** *(off-label)*, ***honesty*** *(off-label)*, ***hops*** *(off-label)*, ***lupins*** *(off-label)*, ***mustard*** *(off-label)*, ***ornamental plant production*** *(off-label)*, ***swedes*** *(off-label)*, ***top fruit*** *(off-label)*, ***turnips*** *(off-label)* [1]; ***bulb onions, cabbages, carrots, cauliflowers, combining peas, fodder beet, land temporarily removed from production, leeks, linseed, spring field beans, sugar beet, vining peas, winter field beans, winter oilseed rape*** [1, 2, 4, 6]; ***flax, linseed for industrial use, winter oilseed rape for industrial use*** [1, 4]; ***garlic*** *(off-label)*, ***mallow (althaea spp.)*** *(off-label)*, ***parsnips*** *(off-label)*, ***red beet*** *(off-label)*, ***shallots*** *(off-label)* [1, 6]
- Volunteer barley in ***bulb onions, cabbages, carrots, cauliflowers, combining peas, fodder beet, leeks, linseed, spring field beans, sugar beet, vining peas, winter field beans, winter oilseed rape*** [3, 5]
- Volunteer cereals in ***bulb onions, cabbages, carrots, cauliflowers, combining peas, fodder beet, land temporarily removed from production, leeks, linseed, spring field beans, sugar beet, vining peas, winter field beans, winter oilseed rape*** [1, 2, 4, 6]; ***flax, linseed for industrial use, winter oilseed rape for industrial use*** [1, 4]
- Volunteer wheat in ***bulb onions, cabbages, carrots, cauliflowers, combining peas, fodder beet, leeks, linseed, spring field beans, sugar beet, vining peas, winter field beans, winter oilseed rape*** [3, 5]
- Wild oats in ***red beet*** *(off-label)* [1]

Specific Off-Label Approvals (SOLAs)

- ***all edible seed crops grown outdoors*** *20082813* [1]
- ***all non-edible seed crops grown outdoors*** *20082813* [1]
- ***borage*** *20064019* [1]
- ***canary flower (echium spp.)*** *20064019* [1]
- ***evening primrose*** *20064019* [1]
- ***forest nurseries*** *20082813* [1]
- ***game cover*** *20082813* [1]
- ***garlic*** *20064021* [1], *20092128* [6]
- ***golf courses*** *20091527* [1]
- ***honesty*** *20064019* [1]
- ***hops*** *20082813* [1]
- ***lupins*** *20064022* [1]
- ***mallow (althaea spp.)*** *20064023* [1], *20092129* [6]
- ***mustard*** *20064019* [1]
- ***ornamental plant production*** *20082813* [1]
- ***parsnips*** *20064023* [1], *20092129* [6]
- ***poppies for morphine production*** *20064026* [1]
- ***red beet*** *20071844* [1], *20092127* [6]
- ***shallots*** *20064021* [1], *20092128* [6]
- ***swedes*** *20090082* [1]
- ***top fruit*** *20082813* [1]
- ***turnips*** *20090082* [1]

Approval information

- Tepraloxydim included in Annex I under EC Directive 91/414

Efficacy guidance

- Best results obtained from applications when weeds small and have not begun to compete with crop
- Only emerged weeds are controlled
- Cool conditions slow down activity and very dry conditions reduce activity by interfering with uptake and translocation
- Foliar death of susceptible weeds evident after 3-4 wks in warm conditions
- Reduced doses must not be used on resistant grass weed populations
- Tepraloxydim is an ACCase inhibitor herbicide. To avoid the build up of resistance do not apply products containing an ACCase inhibitor herbicide more than twice to any crop. In addition do not use any product containing tepraloxydim in mixture or sequence with any other product containing the same ingredient
- Use these products as part of a resistance management strategy that includes cultural methods of control and does not use ACCase inhibitors as the sole chemical method of grass weed control
- Applying a second product containing an ACCase inhibitor to a crop will increase the risk of resistance development; only use a second ACCase inhibitor to control different weeds at a different timing
- Always follow WRAG guidelines for preventing and managing herbicide resistant weeds. See Section 5 for more information

Restrictions

- Maximum number of treatments 1 per crop or yr
- Consult processors or contract agents before treatment of crops intended for processing or for seed
- Do not apply to crops damaged or stressed by factors such as previous herbicide treatments, pest or disease attack
- Do not spray if rain or frost expected, or if foliage is wet
- Do not treat oilseed rape with very low vigour and poor yield potential. Overlapping on oilseed rape may cause damage and reduce yields
- On peas a satisfactory crystal violet leaf wax test must be carried out before treatment. Winter varieties may be treated only in the spring
- For sugar beet, fodder beet, linseed and green cover on land temporarily removed from production, applications are prohibited between 1 Nov and 31 Mar
- For field beans, peas, leeks, bulb onions, carrots, cabbages and cauliflowers applications are prohibited between 1 Nov and 1 Mar

Crop-specific information

- Latest use : before end Nov or before crop has 9 true leaves, whichever is first for winter oilseed rape; before flower buds visible for flax [1], linseed; before head formation for cabbages, cauliflowers
- HI 3 wk for carrots; 4 wk for bulb onions, leeks; 5 wk for peas; 8 wk for fodder beet, field beans, sugar beet
- Treatment prohibited between 1 Nov and 31 Mar on golf courses (off-label use), sugar beet, fodder beet, linseed and green cover on land temporarily removed from production
- Green cover must be fully established on land temporarily removed from production before treatment, and treated plants must not be grazed by livestock or harvested for human or animal consumption

Following crops guidance

- In the event of failure of a treated crop, wheat or barley may be drilled after 2 wk following normal seedbed cultivations, maize or Italian rye-grass may be planted after 8 wk following cultivation to 20 cm
- Graminaceous crops other than those mentioned above should not follow a treated crop in the rotation.
- Broad leaved crops may be planted at any time following a failed or normally harvested treated crop

Environmental safety

- Dangerous for the environment
- Toxic to aquatic organisms

Hazard classification and safety precautions

Hazard Harmful, Dangerous for the environment

SEE SECTION 3 FOR PRODUCTS ALSO REGISTERED

Transport code 9
Packaging group III
UN number 3082
Risk phrases R22b, R38, R51, R53a, R63, R66 [1-6]; R40, R62 [2, 3, 5]
Operator protection A; U05a, U08, U20b, U23a [1-6]; U19a [3, 5]
Environmental protection E15a [1, 2, 4, 6]; E15b [3, 5]; E38 [1, 4, 6]
Storage and disposal D01, D02, D09a, D10c, D12a [1-6]; D05 [2, 3, 5]; D07 [2]; D08 [1, 3-6]
Medical advice M04a [3, 5]; M05b [1-6]

432 terbuthylazine

A triazine herbicide available only in mixtures
HRAC mode of action code: C1

See also *bromoxynil + terbuthylazine*
isoxaben + terbuthylazine
mesotrione + terbuthylazine

433 tetraconazole

A systemic, protectant and curative triazole fungicide for cereals; approval expired 2008
FRAC mode of action code: 3

434 tetramethrin

A contact acting pyrethroid insecticide
IRAC mode of action code: 3

See also *d-phenothrin + tetramethrin*

Products

Killgerm Py-Kill W	Killgerm	10.2% w/v	EC	H4632

Uses
- Flies in ***agricultural premises***, ***livestock houses***

Efficacy guidance
- Dilute in accordance with directions and apply as space or surface spray

Restrictions
- For use only by professional operators
- Maximum number of treatments 1 per wk in agricultural premises and livestock houses
- Do not apply directly on food or livestock
- Remove exposed milk before application. Protect milk machinery and containers from contamination
- Do not use space sprays containing pyrethrins or pyrethroid more than once per wk in intensive or controlled environment animal houses in order to avoid development of resistance. If necessary, use a different control method or product

Environmental safety
- Dangerous for the environment
- Toxic to aquatic organisms

Hazard classification and safety precautions
Hazard Harmful, Flammable, Dangerous for the environment
UN number N/C
Risk phrases R22b, R41, R51, R53a
Operator protection A, H; U02b, U09a, U19a, U20b
Environmental protection E15a
Consumer protection C05, C06, C07, C08, C09, C10, C11
Storage and disposal D01, D09a, D12a
Medical advice M05b

435 thiabendazole

A systemic, curative and protectant benzimidazole (MBC) fungicide
FRAC mode of action code: 1

See also imazalil + thiabendazole

Products

1	Hykeep	Agrichem	2% w/w	DP	12704
2	Storite Clear Liquid	Frontier	220 g/l	SL	12706
3	Storite Excel	Frontier	500 g/l	SC	12705
4	Tezate 220 SL	AgriChem BV	220 g/l	SL	14007

Uses

- Dry rot in ***potatoes***, ***seed potatoes*** [4]; ***seed potatoes*** *(tuber treatment - post-harvest)* [3]; ***ware potatoes*** *(tuber treatment - post-harvest)* [1, 3]
- Fusarium basal neck rot in ***narcissi*** [4]; ***narcissi*** *(off-label - post-lifting or pre-planting dip)*, ***narcissi*** *(post-lifting spray)* [2]
- Gangrene in ***potatoes***, ***seed potatoes*** [4]; ***seed potatoes*** *(tuber treatment - post-harvest)* [3]; ***ware potatoes*** *(tuber treatment - post-harvest)* [1, 3]
- Silver scurf in ***potatoes***, ***seed potatoes*** [4]; ***seed potatoes*** *(tuber treatment - post-harvest)* [3]; ***ware potatoes*** *(tuber treatment - post-harvest)* [1, 3]
- Skin spot in ***potatoes***, ***seed potatoes*** [4]; ***seed potatoes*** *(tuber treatment - post-harvest)* [3]; ***ware potatoes*** *(tuber treatment - post-harvest)* [1, 3]

Specific Off-Label Approvals (SOLAs)

- ***narcissi*** *(post-lifting or pre-planting dip) 20070924* [2]

Approval information

- Thiabendazole included in Annex I under EC Directive 91/414
- Use of thiabendazole products on ware potatoes requires that the discharge of thiabendazole to receiving water from washing plants is kept within emission limits set by the UK monitoring authority

Efficacy guidance

- For best results tuber treatments should be applied as soon as possible after lifting and always within 24 hr
- Dust treatments should be applied evenly over the whole tuber surface
- Thiabendazole should only be used on ware potatoes where there is a likely risk of disease during the storage period and in combination with good storage hygiene and maintenance [1, 3]
- Benzimidazole tolerant strains of silver scurf and skin spot are common in UK and tolerant strains of dry rot have been reported. To reduce the chance of such strains increasing benzimidazole based products should not be used more than once in the cropping cycle [1, 3]
- On narcissi treat as the crop goes into store and only if fungicidal treatment is essential. Adopt a resistance management strategy [2]

Restrictions

- Maximum number of treatments 1 per batch for ware or seed potato tuber treatments [1, 3]; 1 per yr for narcissi bulbs [2]
- Treated seed potatoes must not be used for food or feed [3]
- Do not remove treated potatoes from store for sale, processing or consumption for at least 21 d after application [1, 3]
- Do not mix with any other product
- Off-label use as post-lifting cold water dip or pre-planting hot water dip not to be used on narcissus bulbs grown in the Isles of Scilly [2]

Crop-specific information

- Latest use: 21 d before removal from store for sale, processing or consumption for ware potatoes [1, 3]
- Apply to potatoes as soon as possible after harvest using suitable equipment and always within 2 wk of lifting provided the skins are set. See label for details [1, 3]
- Potatoes should only be treated by systems that provide an accurate dose to tubers not carrying excessive quantities of soil [1, 3]

- Use as a post-lifting treatment to reduce basal and neck rot in narcissus bulbs. Ensure bulbs are clean [2]

Environmental safety

- Dangerous for the environment
- Toxic to aquatic organisms

Hazard classification and safety precautions

Hazard Irritant [3]; Dangerous for the environment [1-4]
Transport code 9
Packaging group III
UN number 3077 [1]; 3082 [2-4]
Risk phrases R43 [3]; R51 [2-4]; R52 [1]; R53a [1-4]
Operator protection A, H [1-4]; C, D [2-4]; U05a [2-4]; U14 [3]; U19a [1-4]; U20b [4]; U20c [1-3]
Environmental protection E15b [1-4]; E38 [2-4]
Storage and disposal D01, D09a, D12a [1-4]; D02, D10c [2-4]; D05 [2, 3]; D07, D11a [1]
Treated seed S03, S05 [4]; S04a [2, 4]

436 thiabendazole + thiram

A fungicide seed dressing mixture for field and vegetable crops
FRAC mode of action code: 1 + M3

See also *thiram*

437 thiacloprid

A chloronicotinyl insecticide for use in agriculture and horticulture
IRAC mode of action code: 4A

Products

1	Agrovista Reggae	Agrovista	480 g/l	SC	13706
2	Biscaya	Bayer CropScience	240 g/l	OD	12471
3	Calypso	Bayer CropScience	480 g/l	SC	11257
4	Exemptor	Scotts	10% w/w	GR	13122
5	Standon Zero Tolerance	Standon	240 g/l	OD	13546

Uses

- Aphids in ***all edible crops (outdoor)*** *(off-label)*, ***all non-edible crops (outdoor)*** *(off-label)*, ***chinese cabbage*** *(off-label)*, ***choi sum*** *(off-label)*, ***collards*** *(off-label)*, ***kale*** *(off-label)*, ***ornamental plant production*** *(off-label)*, ***pak choi*** *(off-label)*, ***peas***, ***potatoes***, ***spring greens*** *(off-label)*, ***tatsoi*** *(off-label)* [2]; ***bedding plants***, ***hardy ornamental nursery stock***, ***ornamental plant production***, ***pot plants***, ***protected ornamentals*** [4]; ***cherries*** *(off-label - under temporary rain covers)*, ***mirabelles*** *(off-label - under temporary rain covers)*, ***protected chives*** *(off-label)*, ***protected parsley*** *(off-label)*, ***protected salad brassicas*** *(off-label - for baby leaf production)* [3]; ***leaf brassicas*** *(off-label - baby leaf production)*, ***protected lamb's lettuce*** *(off-label)* [1]; ***protected herbs (see appendix 6)*** *(off-label)*, ***protected lettuce*** *(off-label)* [1, 3]; ***seed potatoes***, ***ware potatoes*** [2, 5]
- Beetles in ***bedding plants***, ***hardy ornamental nursery stock***, ***ornamental plant production***, ***pot plants***, ***protected ornamentals*** [4]
- Bud mite in ***almonds*** *(off-label)*, ***chestnuts*** *(off-label)*, ***cob nuts*** *(off-label)*, ***hazel nuts*** *(off-label)*, ***walnuts*** *(off-label)* [1, 3]
- Cabbage aphid in ***broccoli*** *(off-label)*, ***brussels sprouts*** *(off-label)*, ***cabbages*** *(off-label)*, ***calabrese*** *(off-label)*, ***cauliflowers*** *(off-label)*, ***chinese cabbage*** *(off-label)*, ***choi sum*** *(off-label)*, ***collards*** *(off-label)*, ***kale*** *(off-label)*, ***pak choi*** *(off-label)*, ***spring greens*** *(off-label)*, ***swedes*** *(off-label)*, ***tatsoi*** *(off-label)*, ***turnips*** *(off-label)* [2]
- Capsids in ***strawberries*** *(off-label)* [1, 3]
- Common green capsid in ***blackberries*** *(off-label)*, ***protected blackberries*** *(off-label)*, ***protected raspberries*** *(off-label)*, ***raspberries*** *(off-label)*, ***rubus hybrids*** *(off-label)* [3]
- Damson-hop aphid in ***plums*** *(off-label)* [1, 3]
- Insect pests in ***bilberries*** *(off-label)*, ***blackberries*** *(off-label)*, ***blackcurrants*** *(off-label)*, ***blueberries*** *(off-label)*, ***cherries*** *(off-label - under temporary protective rain covers)*, ***cranberries***

(off-label), ***gooseberries*** *(off-label)*, ***mirabelles*** *(off-label - under temporary protective rain covers)*, ***protected blackberries*** *(off-label)*, ***protected raspberries*** *(off-label)*, ***raspberries*** *(off-label)*, ***redcurrants*** *(off-label)*, ***rubus hybrids*** *(off-label)*, ***whitecurrants*** *(off-label)* [1]; ***courgettes*** *(off-label)*, ***gherkins*** *(off-label)*, ***hops*** *(off-label)*, ***marrows*** *(off-label)*, ***ornamental plant production*** *(off-label)*, ***protected blueberry*** *(off-label)*, ***protected forest nurseries*** *(off-label)*, ***soft fruit*** *(off-label)*, ***top fruit*** *(off-label)* [3]; ***pears*** *(off-label)* [1, 3]

- Leaf miner in ***protected aubergines*** *(off-label)*, ***protected courgettes*** *(off-label)*, ***protected cucumbers*** *(off-label)*, ***protected ornamentals*** *(off-label)*, ***protected peppers*** *(off-label)*, ***protected tomatoes*** *(off-label)* [1]
- Mealy aphid in ***broccoli*** *(off-label)*, ***cabbages*** *(off-label)*, ***calabrese*** *(off-label)*, ***cauliflowers*** *(off-label)*, ***chinese cabbage*** *(off-label)*, ***choi sum*** *(off-label)*, ***collards*** *(off-label)*, ***kale*** *(off-label)*, ***pak choi*** *(off-label)*, ***spring greens*** *(off-label)*, ***swedes*** *(off-label)*, ***tatsoi*** *(off-label)*, ***turnips*** *(off-label)* [2]
- Palm thrips in ***protected aubergines*** *(off-label)*, ***protected courgettes*** *(off-label)*, ***protected cucumbers*** *(off-label)*, ***protected ornamentals*** *(off-label)*, ***protected peppers*** *(off-label)*, ***protected tomatoes*** *(off-label)* [3]
- Peach-potato aphid in ***leaf brassicas*** *(off-label - baby leaf production)*, ***protected herbs (see appendix 6)*** *(off-label)*, ***protected lamb's lettuce*** *(off-label)*, ***protected lettuce*** *(off-label)* [1]
- Pear midge in ***pears*** *(off-label)* [1, 3]
- Pollen beetle in ***mustard*** [2]; ***spring oilseed rape***, ***winter oilseed rape*** [2, 5]
- Raspberry beetle in ***protected blackberries*** *(off-label)*, ***protected raspberries*** *(off-label)* [3]
- Rosy apple aphid in ***apples*** [1, 3]
- South American leaf miner in ***protected aubergines*** *(off-label)*, ***protected courgettes*** *(off-label)*, ***protected cucumbers*** *(off-label)*, ***protected ornamentals*** *(off-label)*, ***protected peppers*** *(off-label)*, ***protected tomatoes*** *(off-label)* [3]
- Tarnished plant bug in ***blackberries*** *(off-label)*, ***protected blackberries*** *(off-label)*, ***protected raspberries*** *(off-label)*, ***protected strawberries*** *(off-label)*, ***raspberries*** *(off-label)*, ***rubus hybrids*** *(off-label)* [3]
- Thrips in ***protected aubergines*** *(off-label)*, ***protected courgettes*** *(off-label)*, ***protected cucumbers*** *(off-label)*, ***protected ornamentals*** *(off-label)*, ***protected peppers*** *(off-label)*, ***protected tomatoes*** *(off-label)* [1]
- Tobacco whitefly in ***protected aubergines*** *(off-label)*, ***protected courgettes*** *(off-label)*, ***protected cucumbers*** *(off-label)*, ***protected ornamentals*** *(off-label)*, ***protected peppers*** *(off-label)*, ***protected tomatoes*** *(off-label)* [3]
- Vine weevil in ***bedding plants***, ***hardy ornamental nursery stock***, ***ornamental plant production***, ***pot plants***, ***protected ornamentals*** [4]
- Western flower thrips in ***protected aubergines*** *(off-label)*, ***protected courgettes*** *(off-label)*, ***protected cucumbers*** *(off-label)*, ***protected ornamentals*** *(off-label)*, ***protected peppers*** *(off-label)*, ***protected tomatoes*** *(off-label)* [1, 3]
- Wheat-blossom midge in ***spring wheat***, ***winter wheat*** [2]
- Whitefly in ***bedding plants***, ***hardy ornamental nursery stock***, ***ornamental plant production***, ***pot plants***, ***protected ornamentals*** [4]; ***protected aubergines*** *(off-label)*, ***protected courgettes*** *(off-label)*, ***protected cucumbers*** *(off-label)*, ***protected ornamentals*** *(off-label)*, ***protected peppers*** *(off-label)*, ***protected tomatoes*** *(off-label)* [1]
- Willow aphid in ***carrots*** *(off-label)*, ***horseradish*** *(off-label)*, ***parsley root*** *(off-label)*, ***parsnips*** *(off-label)*, ***red beet*** *(off-label)*, ***salsify*** *(off-label)* [2]
- Woolly currant scale in ***bilberries*** *(off-label)*, ***blackcurrants*** *(off-label)*, ***blueberries*** *(off-label)*, ***cranberries*** *(off-label)*, ***gooseberries*** *(off-label)*, ***redcurrants*** *(off-label)*, ***whitecurrants*** *(off-label)* [3]

Specific Off-Label Approvals (SOLAs)

- ***all edible crops (outdoor)*** *20082825* [2]
- ***all non-edible crops (outdoor)*** *20082825* [2]
- ***almonds*** *20080471* [1]
- ***bilberries*** *20080466* [1]
- ***blackberries*** *20080475* [1]
- ***blackcurrants*** *20080466* [1]
- ***blueberries*** *20080466* [1]
- ***broccoli*** *20073237* [2]
- ***brussels sprouts*** *20073237* [2]

- ***cabbages*** *20073237* [2]
- ***calabrese*** *20073237* [2]
- ***carrots*** *20092424* [2]
- ***cauliflowers*** *20073237* [2]
- ***cherries*** *(under temporary protective rain covers) 20080469* [1]
- ***chestnuts*** *20080471* [1]
- ***chinese cabbage*** *20082266* [2]
- ***choi sum*** *20082266* [2]
- ***cob nuts*** *20080471* [1]
- ***collards*** *20082266* [2]
- ***courgettes*** *20081006* [3]
- ***cranberries*** *20080466* [1]
- ***gherkins*** *20081006* [3]
- ***gooseberries*** *20080466* [1]
- ***hazel nuts*** *20080471* [1]
- ***hops*** *20082831* [3]
- ***horseradish*** *20092424* [2]
- ***kale*** *20082266* [2]
- ***leaf brassicas*** *(baby leaf production) 20080470* [1]
- ***marrows*** *20081006* [3]
- ***mirabelles*** *(under temporary protective rain covers) 20080469* [1]
- ***ornamental plant production*** *20082825* [2], *20082831* [3]
- ***pak choi*** *20082266* [2]
- ***parsley root*** *20092424* [2]
- ***parsnips*** *20092424* [2]
- ***pears*** *20080464* [1], *20060324* [3]
- ***plums*** *20080468* [1]
- ***protected aubergines*** *20080474* [1]
- ***protected blackberries*** *20080467* [1], *20070534* [3]
- ***protected blueberry*** *20081332* [3]
- ***protected courgettes*** *20080474* [1]
- ***protected cucumbers*** *20080474* [1]
- ***protected forest nurseries*** *20082831* [3]
- ***protected herbs (see appendix 6)*** *20080472* [1]
- ***protected lamb's lettuce*** *20080472* [1]
- ***protected lettuce*** *20080472* [1]
- ***protected ornamentals*** *20080474* [1]
- ***protected peppers*** *20080474* [1]
- ***protected raspberries*** *20080467* [1], *20070534* [3]
- ***protected tomatoes*** *20080474* [1]
- ***raspberries*** *20080475* [1]
- ***red beet*** *20092424* [2]
- ***redcurrants*** *20080466* [1]
- ***rubus hybrids*** *20080475* [1]
- ***salsify*** *20092424* [2]
- ***soft fruit*** *20082831* [3]
- ***spring greens*** *20082266* [2]
- ***strawberries*** *20080465* [1]
- ***swedes*** *20073132* [2]
- ***tatsoi*** *20082266* [2]
- ***top fruit*** *20082831* [3]
- ***turnips*** *20073132* [2]
- ***walnuts*** *20080471* [1]
- ***whitecurrants*** *20080466* [1], [3]

Approval information

- Thiacloprid included in Annex I under EC Directive 91/414

Efficacy guidance

- Best results in apples obtained from a programme of sprays commencing pre-blossom at the first sign of aphids [3]

- For best pear midge control treat under warm conditions (mid-day) when the adults are flying (temperature higher than 12 °C). Applications on cool, damp days are unlikely to be very successful [3]
- Best results in field crops obtained from treatments when target pests reach threshold levels. For wheat blossom midge use pheromone or yellow sticky traps to monitor adult activity in crops at risk [2, 5]
- Treatment of ornamentals and protected ornamentals is by incorporation into peat-based growing media prior to sowing or planting, using suitable automated equipment [4]
- Ensure thorough mixing in the compost to achieve maximum control. Top dressing is ineffective [4]
- Unless transplants have been treated with a suitable pesticide prior to planting in treated compost full protection against target pests is not guaranteed [4]
- Minimise the possibility of development of resistance by alternating insecticides with different modes of action in the programme
- In dense canopies and on larger trees increase water volume to ensure full coverage

Restrictions

- Maximum number of treatments 2 per yr on apples, seed potatoes; 1 per yr on other field crops and for compost incorporation
- Consult processor before use on crops for processing [2, 5]
- Do not mix treated compost with any other bulky materials such as perlite or bark [4]
- Use treated compost as soon as possible, and within 4 wk of mixing [4]

Crop-specific information

- Latest use: up to and including flowering just complete (GS 69) for wheat; before sowing or planting for treated compost
- HI apples, potatoes 14 d;oilseed rape, mustard 30 d
- Test tolerance of ornamental species before large scale use [4]

Environmental safety

- Dangerous for the environment [2, 4, 5]
- Very toxic or toxic to aquatic organisms
- Risk to certain non-target insects or other arthropods. See directions for use [3]
- Broadcast air-assisted LERAP (30 m) [3]

Hazard classification and safety precautions

Hazard Harmful [1-5]; Dangerous for the environment [2, 4, 5]

Transport code 6.1 [1, 3]; 9 [4]

Packaging group III [1, 3, 4]

UN number 2902 [1, 3]; 3077 [4]; N/C [2, 5]

Risk phrases R20, R52 [1, 3]; R22a [1-3, 5]; R36, R38, R51 [2, 5]; R40, R53a [1-5]; R43 [1, 3, 4]; R50 [4]

Operator protection A [1-5]; C [2, 5]; D [4]; H [1, 3]; U05a [1-5]; U11, U19a [2, 5]; U14 [1, 3]; U16a [2]; U20b [1-3, 5]; U20c [4]

Environmental protection E13c, E22b [3]; E15a, E38 [4]; E15b [1, 2, 5]; E17b [1] (18 m); E17b [3] (30 m); E22c [1]; E34 [1-5]

Storage and disposal D01, D02, D09a [1-5]; D10b [1, 3]; D10c [2, 5]; D11a [4]; D12a [1, 2, 5]

Treated seed S04a [4]

Medical advice M03 [1-5]; M05a [1]; M05b [3]

438 thiamethoxam

A neonicotinoid insecticide for apples, pears, potatoes and beet crops
IRAC mode of action code: 4A

See also fludioxonil + metalaxyl-M + thiamethoxam

Products

1	Actara	Syngenta	25% w/w	WG	13728
2	Centric	Syngenta	25% w/w	WG	13954
3	Cruiser SB	Syngenta	600 g/l	FS	12958

Uses

- Aphids in ***apples***, ***hops*** *(off-label)*, ***pears***, ***soft fruit*** *(off-label)* [2]; ***potatoes***, ***potatoes grown for seed*** [1]
- Beet leaf miner in ***fodder beet*** *(seed treatment)*, ***sugar beet*** *(seed treatment)* [3]
- Flea beetle in ***fodder beet*** *(seed treatment)*, ***sugar beet*** *(seed treatment)* [3]
- Foliar disease control in ***ornamental plant production*** *(off-label)* [1]
- Green aphid in ***apples***, ***pears*** [2]
- Millipedes in ***fodder beet*** *(seed treatment)*, ***sugar beet*** *(seed treatment)* [3]
- Pear sucker in ***pears*** [2]
- Pygmy beetle in ***fodder beet*** *(seed treatment)*, ***sugar beet*** *(seed treatment)* [3]
- Rosy apple aphid in ***apples***, ***pears*** [2]
- Springtails in ***fodder beet*** *(seed treatment)*, ***sugar beet*** *(seed treatment)* [3]
- Symphylids in ***fodder beet*** *(seed treatment)*, ***sugar beet*** *(seed treatment)* [3]
- Western flower thrips in ***ornamental plant production*** *(off-label)* [2]
- Wireworm in ***fodder beet*** *(seed treatment - reduction)*, ***sugar beet*** *(seed treatment - reduction)* [3]
- Woolly aphid in ***apples***, ***pears*** [2]

Specific Off-Label Approvals (SOLAs)

- ***hops*** *20082832* [2]
- ***ornamental plant production*** *20082801* [1], *20082230* [2]
- ***soft fruit*** *20082832* [2]

Approval information

- Thiamethoxam included in Annex I under EC Directive 91/414
- Approval expiry 20 Oct 2010 [3]

Efficacy guidance

- Apply to sugar beet or fodder beet seed as part of the normal commercial pelleting process using special treatment machinery [3]
- Seed drills must be suitable for use with polymer-coated seeds. Standard drill settings should not need to be changed [3]
- Drill treated seed into a firm even seedbed. Poor seedbed quality or seedbed conditions may results in delayed emergence and poor establishment [3]
- Where very high populations of soil pests are present protection may be inadequate to achieve an optimum plant stand [3]
- Use in line with latest IRAG guidelines

Restrictions

- Maximum number of treatments 1 per seed batch [3]
- Avoid deep or shallow drilling which may adversely affect establishment and reduce the level of pest control [3]
- Do not use herbicides containing lenacil pre-emergence on treated crops [3]

Crop-specific information

- Latest use: before drilling for fodder beet, sugar beet [3]

Environmental safety

- Dangerous for the environment
- Very toxic to aquatic organisms

Hazard classification and safety precautions

Hazard Dangerous for the environment
UN number N/C
Risk phrases R50 [1, 3]; R51 [2]; R53a [1-3]
Operator protection A, D, H [3]; U05a, U07 [3]; U05b, U20c [1, 2]
Environmental protection E03, E34, E36a [3]; E12c, E12e [1, 2]; E15b, E38 [1-3]
Storage and disposal D01, D02, D09a, D12a [1-3]; D05, D11a, D14 [3]; D10c [1, 2]
Treated seed S02, S04b, S05, S06a, S06b, S08 [3]

439 thifensulfuron-methyl

A translocated sulfonylurea herbicide
HRAC mode of action code: B

See also *flupyrsulfuron-methyl + thifensulfuron-methyl*
fluroxypyr + thifensulfuron-methyl + tribenuron-methyl
metsulfuron-methyl + thifensulfuron-methyl

Products

Pinnacle	Headland	50% w/w	SG	12285

Uses

- Docks in ***permanent grassland*, *rotational grass***
- Green cover in ***land temporarily removed from production***

Approval information

- Thifensulfuron-methyl included in Annex I under EC Directive 91/414
- Accepted by BBPA for use on malting barley

Efficacy guidance

- Best results achieved from application to small emerged weeds when growing actively. Broad-leaved docks are susceptible during the rosette stage up to onset of stem extension
- Ensure good spray coverage and apply to dry foliage
- Susceptible weeds stop growing almost immediately but symptoms may not be visible for about 2 wk
- Product sold in twin pack with mecoprop-P to provide option for improved control of cleavers
- Only broad-leaved docks (*Rumex obtusifolius*) are controlled; curled docks (*Rumex crispus*) are resistant
- Docks with developing or mature seed heads should be topped and the regrowth treated later
- Established docks with large tap roots may require follow-up treatment
- High populations of docks in grassland will require further treatment in following yr
- Thifensulfuron-methyl is a member of the ALS-inhibitor group of herbicides and products should be used in a planned Resistance Management strategy. See Section 5 for more information

Restrictions

- Maximum number of treatments 1 per crop for cereals or one per yr for grassland and cereals. Must only be applied from 1 Feb in yr of harvest
- Do not treat new leys in year of sowing
- Do not treat where nutrient imbalances, drought, waterlogging, low temperatures, lime deficiency, pest or disease attack have reduced crop or sward vigour
- Do not roll or harrow within 7 d of spraying
- Do not graze grass crops within 7 d of spraying
- Specific restrictions apply to use in sequence or tank mixture with other sulfonylurea or ALS-inhibiting herbicides. See label for details
- Only one application of a sulfonylurea product may be applied per calendar yr to grassland and green cover on land temporarily removed from production

Crop-specific information

- Latest use: before 1 Aug on grass; flag leaf fully emerged stage (GS 39) on cereals
- On grass apply 7-10 d before grazing and do not graze for 7 d afterwards
- Product may cause a check to both sward and clover which is usually outgrown

Following crops guidance

- Only grass or cereals may be sown within 4 wk of application to grassland or setaside, or in the event of failure of any treated crop
- No restrictions apply after normal harvest of a treated cereal crop

Environmental safety

- Dangerous for the environment
- Very toxic to aquatic organisms
- Keep livestock out of treated areas for at least 7 d following treatment
- Take extreme care to avoid drift onto broad-leaved plants outside the target area or onto surface waters or ditches, or land intended for cropping

SEE SECTION 3 FOR PRODUCTS ALSO REGISTERED

- Spraying equipment should not be drained or flushed onto land planted, or to be planted, with trees or crops other than cereals and should be thoroughly cleansed after use - see label for instructions

Hazard classification and safety precautions

Hazard Dangerous for the environment
Transport code 9
Packaging group III
UN number 3077
Risk phrases R50, R53a
Operator protection U19a, U20b
Environmental protection E06a (7 d); E15a, E38
Storage and disposal D09a, D11a, D12a

440 thifensulfuron-methyl + tribenuron-methyl

A mixture of two sulfonylurea herbicides for cereals
HRAC mode of action code: B + B

See also tribenuron-methyl

Products

1	Calibre SX	DuPont	33.3:16.7% w/w	SG	12241
2	Inka SX	DuPont	25:25% w/w	SG	13601
3	Ratio SX	DuPont	40:10% w/w	SG	12601

Uses

- Annual dicotyledons in ***game cover*** *(off-label)*, ***ornamental plant production*** *(off-label)*, ***spring oats***, ***spring rye***, ***triticale***, ***winter oats***, ***winter rye*** [2]; ***spring barley***, ***spring wheat***, ***winter barley***, ***winter wheat*** [1-3]
- Charlock in ***spring barley***, ***spring wheat***, ***winter barley***, ***winter wheat*** [1-3]; ***spring oats***, ***spring rye***, ***triticale***, ***winter oats***, ***winter rye*** [2]
- Chickweed in ***spring barley***, ***spring wheat***, ***winter barley***, ***winter wheat*** [1-3]; ***spring oats***, ***spring rye***, ***triticale***, ***winter oats***, ***winter rye*** [2]
- Docks in ***spring barley***, ***spring oats***, ***spring rye***, ***spring wheat***, ***triticale***, ***winter barley***, ***winter oats***, ***winter rye***, ***winter wheat*** [2]
- Mayweeds in ***spring barley***, ***spring wheat***, ***winter barley***, ***winter wheat*** [1-3]; ***spring oats***, ***spring rye***, ***triticale***, ***winter oats***, ***winter rye*** [2]

Specific Off-Label Approvals (SOLAs)

- ***game cover*** *20082875* [2]
- ***ornamental plant production*** *20082875* [2]

Approval information

- Thifensulfuron-methyl and tribenuron-methyl included in Annex I under EC Directive 91/414
- Accepted by BBPA for use on malting barley

Efficacy guidance

- Apply when weeds are small and actively growing
- Apply after 1 Feb [1, 3]
- Apply after end of Feb in year of harvest [2]
- Ensure good spray cover of the weeds
- Susceptible weeds cease growth almost immediately after application and symptoms become evident 2 wk later
- Effectiveness reduced by rain within 4 h of treatment and in very dry conditions
- Various tank mixtures recommended to broaden weed control spectrum [1]
- Thifensulfuron-methyl and tribenuron-methyl are members of the ALS-inhibitor group of herbicides and products should be used in a planned Resistance Management strategy. See Section 5 for more information

Restrictions

- Maximum number of treatments 1 per crop
- Do not apply to cereals undersown with grass, clover or other legumes, or any other broad-leaved crop

- Do not apply within 7 d of rolling
- Specific restrictions apply to use in sequence or tank mixture with other sulfonylurea or ALS-inhibiting herbicides. See label for details
- Do not apply to any crop suffering from stress
- Consult contract agents before use on crops grown for seed

Crop-specific information

- Latest use: before flag leaf ligule first visible (GS 39) for all crops

Following crops guidance

- Only cereals, field beans or oilseed rape may be sown in the same calendar year as harvest of a treated crop
- In the event of failure of a treated crop sow only a cereal crop within 3 mth of product application and after ploughing and cultivating to at least 15 cm. After 3 mth field beans or oilseed rape may also be sown

Environmental safety

- Dangerous for the environment
- Very toxic to aquatic organisms
- Spraying equipment should not be drained or flushed onto land planted, or to be planted, with trees or crops other than cereals and should be thoroughly cleansed after use - see label for instructions
- Take particular care to avoid damage by drift onto broad-leaved plants outside the target area or onto surface waters or ditches
- LERAP Category B [2, 3]

Hazard classification and safety precautions

Hazard Irritant [1, 3]; Dangerous for the environment [1-3]
Transport code 9
Packaging group III
UN number 3077
Risk phrases R43 [1, 3]; R50, R53a [1-3]
Operator protection A, H [1, 3]; U05a, U08, U20a [1, 3]; U14 [3]; U20c [2]
Environmental protection E15a [1, 3]; E15b [2]; E16a [2, 3]; E38 [1-3]
Storage and disposal D01, D02 [1, 3]; D09a, D11a, D12a [1-3]

441 thiophanate-methyl

A thiophanate fungicide with protectant and curative activity
FRAC mode of action code: 1

See also iprodione + thiophanate-methyl

Products

1	Cercobin WG	Certis	70% w/w	WG	13854
2	Topsin WG	Certis	70% w/w	WG	13988

Uses

- Canker in ***apples*** *(off-label)*, ***crab apples*** *(off-label)*, ***pears*** *(off-label)*, ***quinces*** *(off-label)* [1]
- Disease control in ***hops*** *(off-label - do not harvest for human or animal consumption (including idling) within 12 months of treatment.)*, ***soft fruit*** *(off-label - any fruit harvested within 12 months of treatment must be destroyed.)* [1]
- Fusarium in ***durum wheat***, ***spring oilseed rape***, ***winter oilseed rape*** [2]; ***leeks*** *(off-label)* [1]; ***spring wheat***, ***triticale***, ***winter wheat*** [1, 2]
- Fusarium root rot in ***ornamental plant production*** *(off-label)* [1]
- Gloeosporium rot in ***apples*** *(off-label)*, ***crab apples*** *(off-label)*, ***pears*** *(off-label)*, ***quinces*** *(off-label)* [1]
- Grey mould in ***protected aubergines*** *(off-label)*, ***protected melons*** *(off-label)*, ***protected pumpkins*** *(off-label)*, ***protected squashes*** *(off-label)* [1]
- Leaf curl in ***ornamental plant production*** *(off-label)* [1]
- Mycotoxins in ***durum wheat*** *(reduction only)*, ***spring oilseed rape*** *(reduction only)*, ***winter oilseed rape*** *(reduction only)* [2]; ***spring wheat*** *(reduction only)*, ***triticale*** *(reduction only)*, ***winter wheat*** *(reduction only)* [1, 2]

SEE SECTION 3 FOR PRODUCTS ALSO REGISTERED

- Phoma in ***ornamental plant production*** *(off-label)* [1]
- Sclerotinia in ***protected aubergines*** *(off-label)*, ***protected melons*** *(off-label)*, ***protected pumpkins*** *(off-label)*, ***protected squashes*** *(off-label)* [1]
- Verticillium wilt in ***protected tomatoes*** *(off-label)*, ***strawberries*** *(off-label)* [1]

Specific Off-Label Approvals (SOLAs)

- ***apples*** *20081813* [1]
- ***crab apples*** *20081813* [1]
- ***hops*** *(do not harvest for human or animal consumption (including idling) within 12 months of treatment.) 20082833* [1]
- ***leeks*** *20081383* [1]
- ***ornamental plant production*** *20081384* [1]
- ***pears*** *20081813* [1]
- ***protected aubergines*** *20081382* [1]
- ***protected melons*** *20081382* [1]
- ***protected pumpkins*** *20081382* [1]
- ***protected squashes*** *20081382* [1]
- ***protected tomatoes*** *20091969* [1]
- ***quinces*** *20081813* [1]
- ***soft fruit*** *(any fruit harvested within 12 months of treatment must be destroyed.) 20082833* [1]
- ***strawberries*** *20081381* [1]

Approval information

- Thiophanate-methyl included in Annex I under EC Directive 91/414

Efficacy guidance

- To avoid the development of resistance, a maximum of 2 applications of any MBC product (thiophanate-methyl or carbendazim) are allowed in any one crop. Avoid using MBC fungicides alone.

Environmental safety

- Dangerous for the environment
- Very toxic to aquatic organisms

Hazard classification and safety precautions

Transport code 9
Packaging group III
UN number 3077
Operator protection A; U19a [2]
Environmental protection E15b, E16a, E16b [1, 2]; E38 [2]
Storage and disposal D01, D02, D12a [2]; D10b [1, 2]
Medical advice M05a [2]

442 thiram

A protectant dithiocarbamate fungicide
FRAC mode of action code: M3

See also *carboxin + thiram*
prochloraz + thiram
thiabendazole + thiram

Products

1	Agrichem Flowable Thiram	Agrichem	600 g/l	FS	10784
2	Thianosan DG	Unicrop	80% w/w	WG	13404
3	Thiraflo	Chemtura	480 g/l	FS	13338
4	Thyram Plus	Agrichem	600 g/l	FS	10785
5	Triptam	Certis	80% w/w	WG	14014

Uses

- Botrytis in ***apples***, ***ornamental plant production***, ***pears***, ***protected strawberries***, ***strawberries*** [5]; ***ornamental plant production*** *(except Hydrangea)*, ***tulips*** [2]
- Botrytis fruit rot in ***apples*** [2]
- Carnation rust in ***carnations*** [2]

- Chrysanthemum brown rust in ***chrysanthemums*** [2]
- Damping off in ***broad beans*** *(seed treatment)*, ***combining peas*** *(seed treatment)*, ***dwarf beans*** *(seed treatment)*, ***forage maize*** *(seed treatment)*, ***grass seed*** *(seed treatment)*, ***runner beans*** *(seed treatment)*, ***spring field beans*** *(seed treatment)*, ***spring oilseed rape*** *(seed treatment)*, ***vining peas*** *(seed treatment)*, ***winter field beans*** *(seed treatment)*, ***winter oilseed rape*** *(seed treatment)* [1, 3, 4]; ***bulb onions*** *(seed treatment)*, ***cabbages*** *(seed treatment)*, ***cauliflowers*** *(seed treatment)*, ***edible podded peas*** *(seed treatment)*, ***leeks*** *(seed treatment)*, ***radishes*** *(seed treatment)*, ***salad onions*** *(seed treatment)*, ***turnips*** *(seed treatment)* [1, 4]; ***cress*** *(off-label - seed treatment)*, ***endives*** *(off-label - seed treatment)*, ***frise*** *(off-label - seed treatment)*, ***garlic*** *(off-label - seed treatment)*, ***lamb's lettuce*** *(off-label - seed treatment)*, ***lettuce*** *(off-label - seed treatment)*, ***lupins*** *(off-label - seed treatment)*, ***poppies for morphine production*** *(off-label - seed treatment)*, ***radicchio*** *(off-label - seed treatment)*, ***rhubarb*** *(off-label - seed treatment)*, ***scarole*** *(off-label - seed treatment)*, ***shallots*** *(off-label - seed treatment)*, ***soya beans*** *(seed treatment - qualified minor use))*, ***swedes*** *(off-label - seed treatment)* [1]; ***soya beans*** *(seed treatment - qualified minor use)* [4]
- Gloeosporium in ***apples*** [2]
- Grey mould in ***chrysanthemums***, ***freesias***, ***protected strawberries***, ***strawberries*** [2]
- Pythium in ***poppies for morphine production*** *(off-label - seed treatment)* [1]
- Scab in ***apples***, ***pears*** [2]
- Seed-borne diseases in ***carrots*** *(seed soak)*, ***celery (outdoor)*** *(seed soak)*, ***fodder beet*** *(seed soak)*, ***mangels*** *(seed soak)*, ***parsley*** *(seed soak)*, ***red beet*** *(seed soak)*, ***sugar beet*** *(seed soak)* [1]; ***garlic*** *(off-label - Seed treatment)*, ***lupins*** *(off-label - Seed treatment)*, ***shallots*** *(off-label - Seed treatment)*, ***swedes*** *(off-label - Seed treatment)* [4]

Specific Off-Label Approvals (SOLAs)

- ***garlic*** *(seed treatment) 20052395* [1], *(Seed treatment) 20052391* [4]
- ***lupins*** *(seed treatment) 20052394* [1], *(Seed treatment) 20052389* [4]
- ***poppies for morphine production*** *(seed treatment) 20040681* [1]
- ***rhubarb*** *(seed treatment) 20052396* [1]
- ***shallots*** *(seed treatment) 20052395* [1], *(Seed treatment) 20052391* [4]
- ***swedes*** *(seed treatment) 20052397* [1], *(Seed treatment) 20052390* [4]

Approval information

- Thiram included in Annex I under EC Directive 91/414

Efficacy guidance

- Spray before onset of disease and repeat every 7-14 d. Spray interval varies with crop and disease. See label for details [2]
- Seed treatments may be applied through most types of seed treatment machinery, from automated continuous flow machines to smaller batch treating apparatus [1, 3, 4]
- Co-application of 175 ml water per 100 kg of seed likely to improve evenness of seed coverage [1, 3, 4]

Restrictions

- Maximum number of treatments 3 per crop for protected winter lettuce (thiram based products only; 2 per crop if sequence of thiram and other EBDC fungicides used) [2]; 2 per crop for protected summer lettuce [2]; 1 per batch of seed for seed treatments [1, 3, 4]
- Do not apply to hydrangeas [2]
- Notify processor before dusting or spraying crops for processing [2]
- Do not dip roots of forestry transplants [2]
- Do not spray when rain imminent [2]
- Do not treat seed of tomatoes, peppers or aubergines [1, 3, 4]
- Soya bean seed treatments restricted to crops that are to be harvested as a mature pulse crop only [1, 4]
- Treated seed should not be stored from one season to the next [1, 3, 4]

Crop-specific information

- Latest use: pre-drilling for seed treatments and seed soaks [1, 3, 4]; 21 d after planting out or 21 d before harvest, whichever is earlier, for protected winter lettuce; 14 d after planting out or 21 d before harvest, whichever is earlier, for spraying protected summer lettuce [2]
- HI 21 d for protected lettuce; 14 d for outdoor lettuce; 7 d for apples, pears, blackcurrants, raspberries, strawberries, tomatoes [2]
- Seed to be treated should be of satisfactory quality and moisture content [1, 3, 4]

SEE SECTION 3 FOR PRODUCTS ALSO REGISTERED

- Follow label instructions for treating small quantities of seed [1, 3, 4]
- For use on tulips, chrysanthemums and carnations add non-ionic wetter [2]

Environmental safety

- Dangerous for the environment [1, 3, 4]
- Very toxic to aquatic organisms [1, 3, 4]
- Do not use treated seed as food or feed [1, 3, 4]
- Treated seed harmful to game and wildlife [1, 3, 4]
- A red dye is available from manufacturer to colour treated seed, but not recommended as part of the seed steep. See label for details [1]

Hazard classification and safety precautions

Hazard Harmful, Dangerous for the environment
Transport code 9
Packaging group III
UN number 3077 [2, 5]; 3082 [1, 3, 4]
Risk phrases R20, R36, R38 [1, 4]; R22a, R48, R50, R53a [1-5]; R43 [1, 2, 4, 5]
Operator protection A [1-5]; P [5]; U02a, U11 [1, 4]; U05a, U20b [1-5]; U08 [1-4]; U09a [5]; U14, U19a [2, 5]
Environmental protection E13b [2, 5]; E15a, E34 [1, 4]; E38 [3]
Storage and disposal D01, D09a [1-5]; D02, D05 [1, 3, 4]; D10a [3]; D10c, D12b [1, 4]; D11a, D12a [2, 5]
Treated seed S01, S02, S03, S04b, S05, S06a, S07 [1, 3, 4]; S04a [1, 4]
Medical advice M03 [3]; M04a [1, 4]

443 tolclofos-methyl

A protectant nitroaniline fungicide for soil-borne diseases
FRAC mode of action code: 14

Products

1	Basilex	Scotts	50% w/w	WP	07494
2	Rizolex	Certis	10% w/w	DS	09673
3	Rizolex	Interfarm	10% w/w	DP	14204
4	Rizolex Flowable	Interfarm	500 g/l	FS	11399

Uses

- Black scurf and stem canker in ***potatoes*** *(tuber treatment)* [2, 3]; ***potatoes*** *(tuber treatment only to be used with automatic planters)* [4]
- Bottom rot in ***protected lettuce*** [1]
- Damping off in ***seedlings of ornamentals*** [1]
- Damping off and foot rot in ***radishes*** *(off-label)* [1]
- Damping off and wirestem in ***brussels sprouts***, ***cabbages***, ***calabrese***, ***cauliflowers***, ***chinese cabbage*** [1]
- Foot rot in ***ornamental plant production***, ***seedlings of ornamentals*** [1]
- Rhizoctonia in ***broccoli*** *(off-label)*, ***leaf brassicas*** *(off-label - baby leaf production only)*, ***protected celery*** *(off-label)*, ***protected radishes*** *(off-label)*, ***swedes*** *(with fleece or mesh covers)*, ***swedes*** *(without covers)*, ***turnips*** *(with fleece or mesh covers)*, ***turnips*** *(without covers)* [1]; ***potatoes*** *(off-label - chitted seed treatment)* [4]
- Root rot in ***ornamental plant production***, ***seedlings of ornamentals*** [1]

Specific Off-Label Approvals (SOLAs)

- ***broccoli*** *20063527* [1]
- ***leaf brassicas*** *(baby leaf production only) 20063528* [1]
- ***potatoes*** *(chitted seed treatment) 20041323* [4]
- ***protected celery*** *20011055* [1]
- ***protected radishes*** *20011054* [1]
- ***radishes*** *20020214* [1]

Approval information

- Tolclofos-methyl included in Annex I under EC Directive 91/414

- In 2006 CRD required that all products containing this active ingredient should carry the following warning in the main area of the container label: "Tolclofos-methyl is an anticholinesterase organophosphate. Handle with care"
- May be applied by misting equipment mounted over roller table. See label for details [2, 4]

Efficacy guidance

- Apply dust to seed potatoes during hopper loading [2]
- Apply flowable formulation to clean tubers with suitable misting equipment over a roller table. Spray as potatoes taken into store (first earlies) or as taken out of store (second earlies, maincrop, crops for seed) pre-chitting [4]
- Do not mix flowable formulation with any other product [4]
- To control Rhizoctonia in vegetables and ornamentals apply as drench before sowing, pricking out or planting [1]
- On established seedlings and pot plants apply as drench and rinse off foliage [1]

Restrictions

- Tolclofos-methyl is an atypical organophosphorus compound which has weak anticholinesterase activity. Do not use if under medical advice not to work with such compounds
- Maximum number of treatments 1 per batch for seed potatoes; 1 per crop for lettuce, brassicas, swedes, turnips; 1 at each stage of growth (ie sowing, pricking out, potting) to a maximum of 3, for ornamentals
- Only to be used with automatic planters [2]
- Not recommended for use on seed potatoes where hot water treatment used or to be used [2]
- Do not apply as overhead drench to vegetables or ornamentals when hot and sunny [1]
- Do not use on heathers [1]
- Must not be used via hand-held equipment
- Application to protected crops must only be made where the operator is outside the structure at the time of treatment

Crop-specific information

- Latest use: at planting of seed potatoes [2, 4]; before transplanting for lettuce, brassicas, ornamental specimens [1]; before 2 true lvs for swedes, turnips [1]

Environmental safety

- Dangerous for the environment [1, 2, 4]
- Very toxic to aquatic organisms
- LERAP Category B [1]
- Treated tubers to be used as seed only, not for food or feed [2]

Hazard classification and safety precautions

Hazard Irritant, Dangerous for the environment
Transport code 9
Packaging group III
UN number 3077 [1-3]; 3082 [4]
Risk phrases R36, R37, R38 [1-4]; R50 [2]; R51 [1, 3, 4]; R53a [1, 4]
Operator protection A, D, H [1-4]; C [4]; E [2-4]; J, M [2, 3]; U05a, U11 [2, 3]; U16a [4]; U19a, U20b [1-4]
Environmental protection E15a [2-4]; E16a [1]
Storage and disposal D01, D02, D12a [2, 3]; D08 [2-4]; D09a [1-4]; D10a [4]; D11a [1-3]
Medical advice M01 [1]; M04a [2-4]

444 tralkoxydim

A foliar applied oxime herbicide for grass weed control in cereals.
HRAC mode of action code: A

Products

1	Grasp 400 SC	Nufarm UK	400 g/l	SC	13594
2	Greencrop Gweedore	Greencrop	250 g/l	SC	09882
3	Standon Tralkoxydim	Standon	250 g/l	SC	09579
4	Strimma	Makhteshim	250 g/l	SC	12777

SEE SECTION 3 FOR PRODUCTS ALSO REGISTERED

Uses

- Awned canary grass in ***durum wheat, spring barley, spring wheat, triticale, winter barley, winter rye, winter wheat*** [1]
- Blackgrass in ***durum wheat, spring barley, spring wheat, triticale, winter barley, winter rye, winter wheat*** [1-4]; ***grass seed crops*** *(off-label)* [2, 3]
- Italian ryegrass in ***durum wheat, spring barley, spring wheat, triticale, winter barley, winter rye, winter wheat*** [1]
- Perennial ryegrass in ***durum wheat, spring barley, spring wheat, triticale, winter barley, winter rye, winter wheat*** [1]
- Wild oats in ***durum wheat, spring barley, spring wheat, triticale, winter barley, winter rye, winter wheat*** [1-4]; ***grass seed crops*** *(off-label)* [2, 3]
- Yorkshire fog in ***durum wheat, spring barley, spring wheat, triticale, winter barley, winter rye, winter wheat*** [1]

Specific Off-Label Approvals (SOLAs)

- ***grass seed crops*** *20063497* [2], *20063495* [3]

Approval information

- Accepted by BBPA for use on malting barley
- Tralkoxydim included in Annex I under EC Directive 91/414

Efficacy guidance

- Product leaf-absorbed and translocated rapidly to growing points. Best results achieved after completion of emergence of grass weeds when growing actively in competitive crops under warm humid conditions with adequate soil moisture
- Activity not dependent on soil type or condition. Weeds germinating after application will not be controlled
- Best control of wild oats obtained from 2 leaf to 1st node detectable stage of weeds, and of blackgrass up to 3 tillers
- Tralkoxydim is an ACCase inhibitor herbicide. To avoid the build up of resistance do not apply products containing an ACCase inhibitor herbicide more than twice to any crop. In addition do not use any product containing tralkoxydim in mixture or sequence with any other product containing the same ingredient
- Use these products as part of a resistance management strategy that includes cultural methods of control and does not use ACCase inhibitors as the sole chemical method of grass weed control
- Applying a second product containing an ACCase inhibitor to a crop will increase the risk of resistance development; only use a second ACCase inhibitor to control different weeds at a different timing
- Always follow WRAG guidelines for preventing and managing herbicide resistant weeds. See Section 5 for more information

Restrictions

- Maximum number of treatments 1 per crop. See label
- Authorised adjuvant must always be added. See label
- Do not use on oats
- Do not spray undersown crops or crops to be undersown
- Do not spray when foliage wet or covered in ice or crop otherwise under stress
- Do not spray if a protracted period of cold weather forecast
- Do not spray crops under stress from chemical treatment, grazing, pest attack, mineral deficiency or low fertility. Treatment of stressed crops may be followed by transient foliar discolouration
- Do not roll or harrow within 1 wk of spraying
- Restrictions apply to mixtures or sequences with phenoxy hormone or sulfonylurea herbicides. See label

Crop-specific information

- Latest use: before booting (GS 41)
- Apply to winter cereals from 2 leaves unfolded. If necessary winter cereals may be sprayed twice: once in autumn and once in spring
- Apply to spring cereals from end of tillering

Environmental safety

- Dangerous for the environment [4]

- Harmful to aquatic organisms [2-4]
- Take care to avoid drift onto neighbouring crops, especially oats

Hazard classification and safety precautions

Hazard Harmful, Dangerous for the environment [4]; Irritant [2, 3]
UN number N.C [3]; N/C [1, 2, 4]
Risk phrases R22a [4]; R36, R43 [2, 3]; R52, R53a [2-4]
Operator protection A, H [1-4]; C [2-4]; U02a, U04a, U08, U20b [1-4]; U05a [1-3]; U14 [2, 4]; U15 [4]; U19a [2]
Environmental protection E15a [2-4]; E15b [1]; E34 [3]; E38 [4]
Storage and disposal D01, D02, D05, D09a [1-4]; D10b [2, 3]; D10c [1, 4]; D12a [4]
Medical advice M03 [3]

445 tri-allate

A soil-acting thiocarbamate herbicide for grass weed control
HRAC mode of action code: N

Products

Avadex Excel 15G	Gowan	15% w/w	GR	12109

Uses

- Blackgrass in ***broad beans*** *(off-label)*, ***canary seed*** *(off-label - for wild bird seed production)*, ***combining peas***, ***durum wheat***, ***fodder beet***, ***lucerne***, ***lupins*** *(off-label)*, ***mangels***, ***red beet***, ***red clover***, ***sainfoin***, ***spring barley***, ***spring field beans***, ***sugar beet***, ***triticale***, ***vetches***, ***vining peas***, ***white clover***, ***winter barley***, ***winter field beans***, ***winter wheat***
- Meadow grasses in ***broad beans*** *(off-label)*, ***canary seed*** *(off-label - for wild bird seed production)*, ***combining peas***, ***durum wheat***, ***fodder beet***, ***lucerne***, ***lupins*** *(off-label)*, ***mangels***, ***red beet***, ***red clover***, ***sainfoin***, ***spring barley***, ***spring field beans***, ***sugar beet***, ***triticale***, ***vetches***, ***vining peas***, ***white clover***, ***winter barley***, ***winter field beans***, ***winter wheat***
- Wild oats in ***broad beans*** *(off-label)*, ***canary seed*** *(off-label - for wild bird seed production)*, ***combining peas***, ***durum wheat***, ***fodder beet***, ***lucerne***, ***lupins*** *(off-label)*, ***mangels***, ***red beet***, ***red clover***, ***sainfoin***, ***spring barley***, ***spring field beans***, ***sugar beet***, ***triticale***, ***vetches***, ***vining peas***, ***white clover***, ***winter barley***, ***winter field beans***, ***winter wheat***

Specific Off-Label Approvals (SOLAs)

- ***broad beans*** *20063531*
- ***canary seed*** *(for wild bird seed production) 20070867*
- ***lupins*** *20063530*

Approval information

- Approved for aerial application on wheat, barley, rye, triticale, field beans, peas, fodder beet, sugar beet, red beet, lucerne, sainfoin, vetches, clover. See Section 5 for more information
- Accepted by BBPA for use on malting barley

Efficacy guidance

- Incorporate or apply to surface pre-emergence (post-emergence application possible in winter cereals up to 2-leaf stage of wild oats)
- Do not use on soils with more than 10% organic matter
- Wild oats controlled up to 2-leaf stage
- If applied to dry soil, rainfall needed for full effectiveness, especially with granules on surface. Do not use if top 5-8 cm bone dry
- Do not apply with spinning disc granule applicator; see label for suitable types
- Do not apply to cloddy seedbeds
- Use sequential treatments to improve control of barren brome and annual dicotyledons (see label for details)

Restrictions

- Maximum number of treatments 1 per crop
- Consolidate loose, puffy seedbeds before drilling to avoid chemical contact with seed
- Drill cereals well below treated layer of soil (see label for safe drilling depths)
- Do not use on direct-drilled crops or undersow grasses into treated crops
- Do not sow oats or grasses within 1 yr of treatment

Crop-specific information

- Latest use: pre-drilling for beet crops; before crop emergence for field beans, spring barley, peas, forage legumes; before first node detectable stage (GS 31) for winter wheat, winter barley, durum wheat, triticale, winter rye

Environmental safety

- Irritating to eyes and skin
- May cause sensitization by skin contact
- Harmful to fish or other aquatic life. Do not contaminate surface waters or ditches with chemical or used container

Hazard classification and safety precautions

Hazard Irritant
Transport code 9
Packaging group III
UN number 3077
Risk phrases R36, R38, R43
Operator protection A, C, H; U02a, U05a, U20a
Environmental protection E13c
Storage and disposal D01, D02, D09a, D11a

446 triazoxide

A benzotriazine fungicide available only in mixtures
FRAC mode of action code: 35

See also *clothianidin + prothioconazole + tebuconazole + triazoxide*
prothioconazole + tebuconazole + triazoxide

447 tribenuron-methyl

A foliar acting sulfonylurea herbicide with some root activity for use in cereals
HRAC mode of action code: B

See also *flupyrsulfuron-methyl + tribenuron-methyl*
fluroxypyr + thifensulfuron-methyl + tribenuron-methyl
metsulfuron-methyl + tribenuron-methyl
thifensulfuron-methyl + tribenuron-methyl

Products

Thor	Nufarm UK	50% w/w	WG	14240

Uses

- Annual dicotyledons in ***spring barley***, ***spring oats***, ***spring wheat***, ***triticale***, ***winter barley***, ***winter oats***, ***winter rye***, ***winter wheat***

Approval information

- Tribenuron-methyl included in Annex I under EC Directive 91/414
- Accepted by BBPA for use on malting barley

Efficacy guidance

- Best control achieved when weeds small and actively growing
- Good spray cover must be achieved since larger weeds often become less susceptible
- Susceptible weeds cease growth almost immediately after treatment and symptoms can be seen in about 2 wk
- Weed control may be reduced when conditions very dry
- Tribenuron-methyl is a member of the ALS-inhibitor group of herbicides and products should be used in a planned Resistance Management strategy. See Section 5 for more information

Restrictions

- Maximum number of treatments 1 per crop
- Specific restrictions apply to use in sequence or tank mixture with other sulfonylurea or ALS-inhibiting herbicides. See label for details

- Do not apply to crops undersown with grass, clover or other broad-leaved crops
- Do not apply to any crop suffering stress from any cause or not actively growing
- Do not apply within 7 d of rolling

Crop-specific information
- Latest use: up to and including flag leaf ligule/collar just visible (GS 39)
- Apply in autumn or in spring from 3 leaf stage of crop

Following crops guidance
- Only cereals, field beans or oilseed rape may be sown in the same calendar yr as harvest of a treated crop
- In the event of crop failure sow only a cereal within 3 mth of application. After 3 mth field beans or oilseed rape may also be sown

Environmental safety
- Dangerous for the environment
- Very toxic to aquatic organisms
- Take extreme care to avoid drift onto broad-leaved plants outside the target area or onto surface waters or ditches, or land intended for cropping
- Spraying equipment should not be drained or flushed onto land planted, or to be planted, with trees or crops other than cereals and should be thoroughly cleansed after use - see label for instructions

Hazard classification and safety precautions

Hazard Irritant, Dangerous for the environment
Transport code 9
Packaging group III
UN number 3077
Risk phrases R43, R50, R53a, R70
Operator protection A, H; U05a, U08, U20b
Environmental protection E15a, E38
Storage and disposal D01, D02, D09a, D11a, D12a

448 triclopyr

A pyridinecarboxylic acid herbicide for perennial and woody weed control
HRAC mode of action code: O

See also *2,4-D + dicamba + triclopyr*
aminopyralid + triclopyr
clopyralid + fluroxypyr + triclopyr
clopyralid + triclopyr
fluroxypyr + triclopyr

Products

1	Timbrel	Dow	480 g/l	EC	05815
2	Woody	AgChem Access	480 g/l	EC	14163

Uses
- Brambles in ***forest***, ***land not intended to bear vegetation*** [1, 2]
- Broom in ***forest***, ***land not intended to bear vegetation*** [1, 2]
- Brush clearance in ***industrial sites*** [1, 2]
- Docks in ***forest***, ***land not intended to bear vegetation*** [1, 2]
- Gorse in ***forest***, ***land not intended to bear vegetation*** [1, 2]
- Perennial dicotyledons in ***asparagus*** *(off-label - directed treatment)* [1]; ***forest***, ***industrial sites***, ***land not intended to bear vegetation*** [1, 2]
- Perennial grasses in ***asparagus*** *(off-label - directed treatment)* [1]
- Rhododendrons in ***forest***, ***land not intended to bear vegetation*** [1, 2]
- Scrub clearance in ***industrial sites*** [1, 2]
- Stinging nettle in ***forest***, ***land not intended to bear vegetation*** [1, 2]
- Woody weeds in ***forest***, ***industrial sites***, ***land not intended to bear vegetation*** [1, 2]

Specific Off-Label Approvals (SOLAs)
- ***asparagus*** *(directed treatment) 20050530* [1]

SEE SECTION 3 FOR PRODUCTS ALSO REGISTERED

Approval information

- Triclopyr included in Annex I under EC Directive 91/414

Efficacy guidance

- Apply in grassland as spot treatment or overall foliage spray when weeds in active growth in spring or summer. Details of dose and timing vary with species. See label
- Apply to woody weeds as summer foliage, winter shoot, basal bark, cut stump or tree injection treatment
- Apply foliage spray in water when leaves fully expanded but not senescent
- Do not spray in drought, in very hot or cold conditions
- Control may be reduced if rain falls within 2 h of application
- Control of rhododendron can be variable. If higher than 1.8 m cut stump treatment recommended. A follow-up shoot treatment may be required
- See label for maximum concentrations when applying in oil, water or via watering can

Restrictions

- Maximum number of treatments 2 per yr
- Do not apply to grass leys less than 1 yr old
- Do not drill kale, swedes, turnips, grass or mixtures containing clover within 6 wk of treatment. Allow at least 6 wk before planting trees
- Not to be used on food crops
- Do not apply through hand held rotary atomisers
- Not to be applied in or near water

Crop-specific information

- Latest use: 6 wk before replanting; 7 d before grazing
- Uses on land not intended for cropping include grassland of no agricultural interest such as roadside verges, railway and motorway embankments
- Apply winter shoot, basal bark or cut stump sprays in paraffin or diesel oil. Dose and timing vary with species. See label for details
- Inject undiluted or 1:1 dilution into cuts spaced every 7.5 cm round trunk
- Clover will be killed or severely checked by application in grassland

Environmental safety

- Dangerous for the environment
- Very toxic to aquatic organisms
- LERAP Category B
- Do not allow spray to drift onto agricultural or horticultural crops, amenity plantings, gardens, ponds, lakes or water courses. Vapour drift may occur under hot conditions
- Keep livestock out of treated areas for at least 7 d and until foliage of any poisonous weeds such as buttercups or ragwort has died and become unpalatable

Hazard classification and safety precautions

Hazard Harmful, Dangerous for the environment
Transport code 9
Packaging group III
UN number 3082
Risk phrases R22a, R22b, R38, R43, R50, R53a
Operator protection A, C, H, M; U02a, U05a, U08, U14, U20b
Environmental protection E07a, E15a, E16a, E16b, E23, E34, E38
Consumer protection C01
Storage and disposal D01, D02, D09a, D10b, D12a
Medical advice M05b

449 trifloxystrobin

A protectant strobilurin fungicide for cereals and managed amenity turf
FRAC mode of action code: 11

See also *cyproconazole + trifloxystrobin*
fluoxastrobin + prothioconazole + trifloxystrobin
prothioconazole + trifloxystrobin
tebuconazole + trifloxystrobin

Products

1	Martinet	AgChem Access	500 g/l	SC	14614
2	Mascot Defender	Rigby Taylor	500 g/l	SG	14065
3	Micro-Turf	AgriGuard	50% w/w	WG	12988
4	Pan Aquarius	Pan Agriculture	50% w/w	WG	13018
5	Scorpio	Bayer Environ.	50% w/w	WG	12293
6	Swift SC	Bayer CropScience	500 g/l	SC	11227

Uses

- Brown rust in ***spring barley***, ***winter barley***, ***winter wheat*** [1, 6]
- Foliar disease control in ***durum wheat*** *(off-label)*, ***grass seed crops*** *(off-label)*, ***ornamental specimens*** *(off-label)*, ***spring rye*** *(off-label)*, ***triticale*** *(off-label)*, ***winter rye*** *(off-label)* [6]
- Fusarium patch in ***amenity grassland***, ***managed amenity turf*** [2-5]
- Glume blotch in ***winter wheat*** [1, 6]
- Net blotch in ***spring barley***, ***winter barley*** [1, 6]
- Red thread in ***amenity grassland***, ***managed amenity turf*** [2-5]
- Rhynchosporium in ***spring barley***, ***winter barley*** [1, 6]
- Septoria leaf blotch in ***winter wheat*** [1, 6]

Specific Off-Label Approvals (SOLAs)

- ***durum wheat*** *20061287* [6]
- ***grass seed crops*** *20061287* [6]
- ***ornamental specimens*** *20082882* [6]
- ***spring rye*** *20061287* [6]
- ***triticale*** *20061287* [6]
- ***winter rye*** *20061287* [6]

Approval information

- Trifloxystrobin included in Annex I under EC Directive 91/414
- Accepted by BBPA for use on malting barley

Efficacy guidance

- Should be used protectively before disease is established in crop. Further treatment may be necessary if disease attack prolonged
- Treat grass after cutting and do not mow for at least 48 h afterwards to allow adequate systemic movement [3-5]
- Trifloxystrobin is a member of the QoI cross resistance group. Product should be used preventatively and not relied on for its curative potential
- Use product as part of an Integrated Crop Management strategy incorporating other methods of control, including where appropriate other fungicides with a different mode of action. Do not apply more than two foliar applications of QoI containing products to any cereal crop
- There is a significant risk of widespread resistance occurring in *Septoria tritici* populations in UK. Failure to follow resistance management action may result in reduced levels of disease control
- On cereal crops product must always be used in mixture with another product, recommended for control of the same target disease, that contains a fungicide from a different cross resistance group and is applied at a dose that will give robust control
- Strains of barley powdery mildew resistant to QoIs are common in the UK

Restrictions

- Maximum number of treatments 2 per crop per yr
- Do not apply to turf during dought conditions or to frozen turf [3-5]

Crop-specific information

- HI barley, wheat 35 d [6]

SEE SECTION 3 FOR PRODUCTS ALSO REGISTERED

Environmental safety

- Dangerous for the environment
- Very toxic to aquatic organisms
- LERAP category B [3-5]

Hazard classification and safety precautions

Hazard Irritant [2-5]; Dangerous for the environment [1-6]
Transport code 9
Packaging group III
UN number 3077 [3-5]; 3082 [1, 2, 6]
Risk phrases R43 [2-5]; R50, R53a [1-6]
Operator protection A, H; U02a, U09a, U19a [1, 6]; U05a, U20b [1-6]; U11, U13, U14, U15 [2-5]
Environmental protection E13b [1, 6]; E15b, E16a, E16b, E34 [2-5]; E38 [1-6]
Consumer protection C02a [1, 6] (35 d)
Storage and disposal D01, D02, D09a, D10c, D12a
Medical advice M03 [2-5]

450 trifluralin

A soil-incorporated dinitroaniline herbicide for use in various crops; all approvals expired on 20 March 2009.
HRAC mode of action code: K1

451 triflusulfuron-methyl

A sulfonyl urea herbicide for beet crops
HRAC mode of action code: B

See also lenacil + triflusulfuron-methyl

Products

Debut	DuPont	50% w/w	WG	07804

Uses

- Annual dicotyledons in ***chicory*** *(off-label)*, ***fodder beet***, ***sugar beet***
- Charlock in ***red beet*** *(off-label)*
- Cleavers in ***red beet*** *(off-label)*
- Flixweed in ***red beet*** *(off-label)*
- Fool's parsley in ***red beet*** *(off-label)*
- Nipplewort in ***red beet*** *(off-label)*
- Ox-eye daisy in ***red beet*** *(off-label)*

Specific Off-Label Approvals (SOLAs)

- ***chicory*** *20050937*
- ***red beet*** *20023629*

Efficacy guidance

- Product should be used with a recommended adjuvant or a suitable herbicide tank-mix partner - see label for details
- Product acts by foliar action. Best results obtained from good spray cover of small actively growing weeds
- Susceptible weeds cease growth immediately and symptoms can be seen 5-10 d later
- Best results achieved from a programme of up to 4 treatments starting when first weeds have emerged with subsequent applications every 5-14 d when new weed flushes at cotyledon stage
- Weed spectrum can be broadened by tank mixture with other herbicides. See label for details
- Product may be applied overall or via band sprayer
- Triflusulfuron-methyl is a member of the ALS-inhibitor group of herbicides

Restrictions

- Maximum number of treatments 4 per crop
- Do not apply to any crop stressed by drought, water-logging, low temperatures, pest or disease attack, nutrient or lime deficiency

Crop-specific information

- Latest use: before crop leaves meet between rows
- HI 4 wk for red beet
- All varieties of sugar beet and fodder beet may be treated from early cotyledon stage until the leaves begin to meet between the rows

Following crops guidance

- Only winter cereals should follow a treated crop in the same calendar yr. Any crop may be sown in the next calendar yr
- After failure of a treated crop, sow only spring barley, linseed or sugar beet within 4 mth of spraying unless prohibited by tank-mix partner

Environmental safety

- Dangerous for the environment
- Very toxic to aquatic organisms
- Extremely dangerous to fish or other aquatic life. Do not contaminate surface waters or ditches with chemical or used container
- LERAP Category B
- Take extreme care to avoid drift onto broad-leaved plants outside the target area or onto surface waters or ditches, or land intended for cropping
- Spraying equipment should not be drained or flushed onto land planted, or to be planted, with trees or crops other than sugar beet and should be thoroughly cleansed after use - see label for instructions

Hazard classification and safety precautions

Hazard Dangerous for the environment
Transport code 9
Packaging group III
UN number 3077
Risk phrases R50, R53a
Operator protection A; U05a, U08, U19a, U20a
Environmental protection E13a, E15a, E16a, E16b, E38
Storage and disposal D01, D02, D05, D09a, D11a, D12a

452 trinexapac-ethyl

A novel cyclohexanecarboxylate plant growth regulator for cereals, turf and amenity grassland

Products

1	Clayton Truss	Clayton	250 g/l	EC	13129
2	Moddus	Syngenta	250 g/l	EC	08801
3	Pan Tepee	Pan Agriculture	250 g/l	EC	12279
4	Primo Maxx	Syngenta	121 g/l	SL	13374
5	Shrink	AgChem Access	250 g/l	EC	13762
6	Standon Pygmy	Standon	250 g/l	EC	13858
7	Stunt	AgriGuard	250 g/l	EC	12859
8	Tacet	Goldengrass	250 g/l	EC	14342
9	Tempo	Syngenta	250 g/l	EC	13799

Uses

- Growth regulation in ***durum wheat***, ***ryegrass seed crops***, ***spring barley***, ***spring oats***, ***spring rye***, ***triticale***, ***winter barley***, ***winter oats***, ***winter rye***, ***winter wheat*** [5, 6, 8]
- Growth retardation in ***amenity grassland***, ***managed amenity turf*** [4]
- Lodging control in ***canary seed*** *(off-label - for wild bird seed production)*, ***spring wheat*** *(off-label - cv A C Barrie only)* [2]; ***durum wheat***, ***ryegrass seed crops***, ***spring barley***, ***spring oats***, ***spring rye***, ***triticale***, ***winter barley***, ***winter oats***, ***winter rye***, ***winter wheat*** [1-3, 7, 9]

Specific Off-Label Approvals (SOLAs)

- ***canary seed*** *(for wild bird seed production) 20070869* [2]
- ***spring wheat*** *(cv A C Barrie only) 20042300* [2]

Approval information

- Trinexapac-ethyl included in Annex I under EC Directive 91/414
- Accepted by BBPA for use on malting barley

SEE SECTION 3 FOR PRODUCTS ALSO REGISTERED

Efficacy guidance

- Best results on cereals and ryegrass seed crops obtained from treatment from the leaf sheath erect stage [1-3, 7]
- Best results on turf achieved from application to actively growing weed free turf grass that is adequately fertilized and watered and is not under stress. Adequate soil moisture is essential [4]
- Turf should be dry and weed free before application [4]
- Environmental conditions, management and cultural practices that affect turf growth and vigour will influence effectiveness of treatment [4]
- Repeat treatments on turf up to the maximum approved dose may be made as soon as growth resumes [4]

Restrictions

- Maximum total dose equivalent to one full dose on cereals, ryegrass seed crops [1-3, 7]
- Maximum number of treatments on turf equivalent to five full dose treatments [4]
- Do not apply if rain or frost expected or if crop wet. Products are rainfast after 12 h
- Only use on crops at risk of lodging [1-3, 7]
- Do not apply within 12 h of mowing turf [4]
- Do not treat newly sown turf [4]
- Not to be used on food crops [4]
- Do not compost or mulch grass clippings [4]

Crop-specific information

- Latest use: before 2nd node detectable (GS 32) for oats, ryegrass seed crops; before 3rd node detectable (GS 33) for durum wheat, spring barley, triticale, rye; before flag leaf sheath extending (GS 41) for winter barley, winter wheat
- On wheat apply as single treatment between leaf sheath erect stage (GS 30) and flag leaf fully emerged (GS 39) [1-3, 7]
- On barley, rye, triticale and durum wheat apply as single treatment between leaf sheath erect stage (GS 30) and second node detectable (GS 32), or on winter barley at higher dose between flag leaf just visible (GS 37) and flag leaf fully emerged (GS 39) [1-3, 7]
- On oats and ryegrass seed crops apply between leaf sheath erect stage (GS 30) and first node detectable stage (GS 32) [1-3, 7]
- Treatment may cause ears of cereals to remain erect through to harvest [1-3, 7]
- Turf under stress when treated may show signs of damage [4]
- Any weed control in turf must be carried out before application of the growth regulator [4]

Environmental safety

- Dangerous for the environment [1-3, 7]
- Toxic to aquatic organisms [1-3, 7]

Hazard classification and safety precautions

Hazard Irritant, Dangerous for the environment [1-3, 5-9]
Transport code 9 [1-3, 5-9]
Packaging group III [1-3, 5-9]
UN number 3082 [1-3, 5-9]; N/C [4]
Risk phrases R43, R51, R53a [1-3, 5-9]
Operator protection A, C [1-9]; H [1, 3, 4, 7]; K [4]; U05a [1-9]; U15, U20c [1-3, 5-9]; U20b [4]
Environmental protection E15a [2, 3, 5-9]; E15b [1]; E34 [7]; E38 [2, 3, 5, 6, 8, 9]
Consumer protection C01 [4]
Storage and disposal D01, D02, D05 [1-9]; D09a [1-3, 5-9]; D10b [7]; D10c [1-6, 8, 9]; D12a [1-3, 5, 6, 8, 9]

453 triticonazole

A triazole fungicide available only in mixtures
FRAC mode of action code: 3

See also prochloraz + triticonazole

454 Verticillium lecanii

A fungal parasite of aphids and whitefly

Products

1 Mycotal	Koppert	16.1% w/w	WP	04782
2 Vertalec	Koppert	2.5% w/w	WP	04781

Uses

- Aphids in ***aubergines, protected beans, protected cayenne peppers*** *(off-label)*, ***protected chrysanthemums, protected cucumbers, protected edible podded peas*** *(off-label)*, ***protected herbs (see appendix 6)*** *(off-label)*, ***protected lettuce, protected ornamentals, protected peppers, protected roses, protected tomatoes*** [2]
- Whitefly in ***aubergines, protected beans, protected cayenne peppers*** *(off-label)*, ***protected chives*** *(off-label)*, ***protected cucumbers, protected edible podded peas*** *(off-label)*, ***protected herbs (see appendix 6)*** *(off-label)*, ***protected lettuce, protected ornamentals, protected parsley*** *(off-label)*, ***protected peppers, protected salad brassicas*** *(off-label)*, ***protected tomatoes*** [1]

Specific Off-Label Approvals (SOLAs)

- ***protected cayenne peppers*** *20070862* [1], *20063442* [2]
- ***protected chives*** *20070861* [1]
- ***protected edible podded peas*** *20070863* [1], *20063443* [2]
- ***protected herbs (see appendix 6)*** *20070861* [1], *20063444* [2]
- ***protected parsley*** *20070861* [1]
- ***protected salad brassicas*** *20070861* [1]

Approval information

- Verticillium included in Annex I under EC Directive 91/414

Efficacy guidance

- *Verticillium lecanii* is a pathogenic fungus that infects the target pests and destroys them
- Apply spore powder as spray as part of biological control programme keeping the spray liquid well agitated
- Pre-soak the product for 2-4 h before application to rehydrate the spores and assist in dispersion
- Treat before infestations build to high levels and repeat as directed on the label
- Spray during late afternoon and early evening directing spray onto underside of leaves and to growing points
- Best results require minimum 80% relative humidity and 18 °C within the crop canopy
- Product highly infective to many aphid species except the chrysanthemum aphid. Follow specific label directions for this pest [2]

Restrictions

- Never use in tank mixture
- Do not use a fungicide within 3 d of treatment. Pesticides containing captan, chlorothalonil, fenarimol, dichlofluanid, imazalil, maneb, prochloraz, quinomethionate, thiram or tolylfluanid may not be used on the same crop
- Keep in a refrigerated store at 2-6 °C but do not freeze

Environmental safety

- Products have negligible effects on commercially available natural predators or parasites but consult manufacturer before using with a particular biological control agent for the first time

Hazard classification and safety precautions

UN number N/C
Operator protection U19a, U20b
Environmental protection E15a
Storage and disposal D09a, D11a

455 warfarin

A coumarin anti-coagulant rodenticide

Products

1	Grey Squirrel Bait	Killgerm	0.02% w/w	RB	13020
2	Sakarat Ready-to-Use (Cut Wheat Base)	Killgerm	0.025% w/w	RB	H6807
3	Sakarat Ready-to-Use (Whole Wheat)	Killgerm	0.025% w/w	RB	H6808
4	Sakarat X	Killgerm	0.05% w/w	RB	H6809
5	Sewarin Extra	Killgerm	0.05% w/w	RB	H6810
6	Sewarin P	Killgerm	0.025% w/w	RB	H6811
7	Sewercide Cut Wheat Rat Bait	Killgerm	0.05% w/w	RB	H6805
8	Sewercide Whole Wheat Rat Bait	Killgerm	0.05% w/w	RB	H6806
9	Warfarin 0.5% Concentrate	B H & B	0.5% w/w	CB	H6815
10	Warfarin Ready Mixed Bait	B H & B	0.025% w/w	RB	H6816

Uses

- Grey squirrels in ***agricultural premises, forest, trees*** *(nuts)* [1]
- Rats in ***agricultural premises*** [2-10]

Approval information

- Warfarin included in Annex I under EC Directive 91/414

Efficacy guidance

- For rodent control place ready-to-use or prepared baits at many points wherever rats active. Out of doors shelter bait from weather [2-10]
- Inspect baits frequently and replace or top up as long as evidence of feeding. Do not underbait [2-10]
- Use grey squirrel baits when bark stripping is evident in farm forestry or forestry or where there is risk of damage to trees grown for their nuts [1]
- Use squirrel bait in specially constructed hoppers and inspect every 2-3 d. Place bait hoppers so as to prevent rainwater entry and replace as necessary [1]

Restrictions

- For use only by local authorities, professional operators providing a pest control service and persons occupying industrial, agricultural or horticultural premises
- For use only between 15 Mar and 15 Aug for tree protection [1]
- Squirrel bait hoppers must be clearly labelled "Poison" or carry the instruction "Tree protection, do not disturb" [1]
- Warfarin baits for grey squirrel control must only be used outdoors in specially constructed hoppers which comply with the Grey Squirrel Order 1973, in Scotland and specified counties of England and Wales. See label [1]
- The use of warfarin to control grey squirrels is illegal unless the provisions of the Grey Squirrels Order 1973 are observed [1]

Environmental safety

- Harmful to wildlife
- Prevent access to baits by children and animals, especially cats, dogs and pigs
- Rodent bodies must be searched for and burned or buried, not placed in refuse bins or rubbish tips. Remains of bait and containers must be removed after treatment and burned or buried
- Bait must not be used where food, feed or water could become contaminated
- If squirrel bait hoppers have obviously been disturbed by badgers or other animals, change the site or lift onto tables or platforms [1]
- Warfarin baits must nor be used outdoors where pine martens are know to occur naturally [1]
- Under the terms of the Wildlife and Countryside Act 1981 warfarin squirrel baits must not be used outdoors in England, Scotland or Wales where red squrrels are know to occur [1]

Hazard classification and safety precautions

UN number N/C [1-8, 10]
Operator protection A [1-8]; C, D, E, H [2-8]; U13 [1-10]; U20a [2-6]; U20b [1, 7-10]
Environmental protection E10b, E15a [1]

Storage and disposal D09a [1-10]; D10a [9, 10]; D11a [1-8]
Vertebrate/rodent control products V01a, V03a [4, 7-10]; V01b [1-3, 5, 6]; V02 [1-10]; V03b, V04b [2, 3, 5, 6]; V04a [1, 4, 7-10]
Medical advice M03 [2, 3, 5, 6]

456 zeta-cypermethrin

A contact and stomach acting pyrethroid insecticide
IRAC mode of action code: 3

Products

1	Angri	AgChem Access	100 g/l	EW	13730
2	Fury 10 EW	Belchim	100 g/l	EW	12248
3	Minuet EW	Belchim	100 g/l	EW	12304
4	Symphony	AgriGuard	100 g/l	EW	13021

Uses

- Aphids in ***broad beans*** *(off-label)*, ***durum wheat*** *(off-label)*, ***fodder beet*** *(off-label)*, ***grass seed crops*** *(off-label)*, ***lupins*** *(off-label)*, ***spring rye*** *(off-label)*, ***triticale*** *(off-label)*, ***winter rye*** *(off-label)* [2]; ***spring barley***, ***spring oats***, ***spring wheat***, ***winter barley***, ***winter oats***, ***winter wheat*** [1-4]
- Barley yellow dwarf virus vectors in ***spring barley***, ***spring wheat***, ***winter barley***, ***winter wheat*** [1-4]
- Cabbage seed weevil in ***spring oilseed rape***, ***winter oilseed rape*** [1-4]
- Cabbage stem flea beetle in ***spring oilseed rape***, ***winter oilseed rape*** [1-4]
- Cutworms in ***potatoes***, ***sugar beet*** [1-4]
- Flax flea beetle in ***linseed*** [1-4]
- Flea beetle in ***spring oilseed rape***, ***winter oilseed rape*** [1-4]
- Insect pests in ***broad beans*** *(off-label)*, ***durum wheat*** *(off-label)*, ***fodder beet*** *(off-label)*, ***grass seed crops*** *(off-label)*, ***lupins*** *(off-label)*, ***spring rye*** *(off-label)*, ***triticale*** *(off-label)*, ***winter rye*** *(off-label)* [2]
- Large flax flea beetle in ***linseed*** [1-4]
- Pea and bean weevil in ***combining peas***, ***spring field beans***, ***vining peas***, ***winter field beans*** [1-4]
- Pea aphid in ***combining peas***, ***vining peas*** [1-4]
- Pea moth in ***combining peas***, ***vining peas*** [1-4]
- Pod midge in ***spring oilseed rape***, ***winter oilseed rape*** [1-4]
- Pollen beetle in ***spring oilseed rape***, ***winter oilseed rape*** [1-4]
- Rape winter stem weevil in ***spring oilseed rape***, ***winter oilseed rape*** [1-4]

Specific Off-Label Approvals (SOLAs)

- ***broad beans*** *20063573* [2]
- ***durum wheat*** *20063570* [2]
- ***fodder beet*** *20063572* [2]
- ***grass seed crops*** *20063570* [2]
- ***lupins*** *20063571* [2]
- ***spring rye*** *20063570* [2]
- ***triticale*** *20063570* [2]
- ***winter rye*** *20063570* [2]

Approval information

- Zeta-cypermethrin included in Annex I under EC Directive 91/414
- Accepted by BBPA for use on malting barley

Efficacy guidance

- On winter cereals spray when aphids first found in the autumn for BYDV control. A second spray may be required on late drilled crops or in mild conditions
- For summer aphids on cereals spray when treatment threshold reached
- For listed pests in other crops spray when feeding damage first seen or when treatment threshold reached. Under high infestation pressure a second treatment may be necessary
- Best results for pod midge and seed weevil control in oilseed rape obtained from treatment after pod set but before 80% petal fall

- Pea moth treatments should be applied according to ADAS/PGRO warnings or when economic thresholds reached as indicated by pheromone traps
- Treatments for cutworms should be made at egg hatch and repeated no sooner than 10 d later

Restrictions

- Maximum number of treatments 2 per crop [4]
- Maximum total dose equivalent to two full dose treatments on all crops [2, 3]
- Consult processors before use on crops for processing

Crop-specific information

- Latest use: before 4 true leaves for linseed, before end of flowering for oilseed rape, cereals
- HI potatoes, field beans, peas 14 d; sugar beet 60 d

Environmental safety

- Dangerous for the environment
- Very toxic to aquatic organisms
- High risk to non-target insects or other arthropods. Do not spray within 6 m of the field boundary
- LERAP Category A

Hazard classification and safety precautions

Hazard Harmful, Dangerous for the environment
Transport code 6.1
Packaging group III
UN number 3352
Risk phrases R20, R22a, R43, R50, R53a
Operator protection A, C, H; U05a, U08, U14, U15, U19a, U20b
Environmental protection E15b, E16c, E16d, E22a, E34 [1-4]; E38 [1-3]
Storage and disposal D01, D02, D09a, D10b [1-4]; D05 [4]; D12a [1-3]
Medical advice M05a [1-3]

457 zoxamide

A substituted benzamide fungicide available only in mixtures
FRAC mode of action code: 22

See also mancozeb + zoxamide

SECTION 3
PRODUCTS ALSO REGISTERED

Products also Registered

Products listed in the table below have not been notified for inclusion in Section 2 of this edition of the *Guide*. However most have extant approval at least until 31 December 2013 unless an different expiry date is shown. These products may legally be stored and used in accordance with their label until their approval expires, but they may not still be available for purchase.

Product	Approval holder	MAPP No.	Expiry Date
2,4-D + triclopyr			
Genoxone ZX EC	Agriphar	13131	
abamectin			
Acaramilk	Rotam Agrochemical	14344	
Mectinide	AgriGuard	13953	
RouteOne Abamectin 18	Albaugh UK	13560	
acetamiprid			
Gazelle	Certis	12909	31 Dec 2014
acetic acid			
Natural Weed Spray NO. 1	Punya	12328	
acibenzolar-S-methyl			
Bion	Syngenta	09803	31 Oct 2011
alpha-cypermethrin			
Alert	BASF	13632	
Alpha C 6 ED	Techneat	13611	28 Feb 2015
Alpha C 6 ED	Techneat	11838	31 Aug 2010
Alphamex 100 EC	MAC	14071	
Fastac	BASF	13604	
Fastac Dry	BASF	10221	
aluminium ammonium sulphate			
Curb S	Sphere	13564	
Rezist	Barrettine	08576	
Rezist Liquid	Barrett	14643	
aluminium phosphide			
Degesch Fumigation Pellets	Rentokil	11436	
Detia Gas Ex-P	Igrox	09802	
Detia Gas-Ex-B	Detia Degesch	06927	
Detia Gas-Ex-T	Igrox	03792	
Luxan Talunex	Luxan	06563	31 Oct 2011
Phostoxin I	Rentokil	05694	
Talunex	Certis	14608	

Product	Approval holder	MAPP No.	Expiry Date
amidosulfuron			
Amidosulf	Goldengrass	14409	
Barclay Cleave	Barclay	11340	
Landgold Amidosulfuron	Landgold	09021	31 May 2012
Landgold Amidosulfuron	Teliton	12110	31 Mar 2013
Pursuit	Bayer CropScience	07333	
Squire	Bayer CropScience	08715	
Squire Ultra	Bayer CropScience	13125	
aminopyralid + fluroxypyr			
Halcyon	Dow	12749	29 Jul 2010
Halcyon	Dow	14709	29 Jul 2011
Mileway	Dow	14702	29 Jul 2011
Synero	Dow	14059	29 Jul 2010
Synero	Dow	14708	29 Jul 2011
amitrole			
Aminotriazole Technical	Nufarm UK	11137	
Weedazol Pro	Nufarm UK	11995	
asulam			
Cleancrop Asulam	United Agri	10465	
I T Asulam	I T Agro	10186	
Milentus Asulam	Milentus	12245	
Noble Asulam	Barclay	12939	
azoxystrobin			
5504	Syngenta	12351	31 Dec 2011
Alpha Azoxystrobin	Makhteshim	14400	31 Dec 2011
Barclay ZX	Barclay	11336	31 Dec 2011
Clayton Stobik	Clayton	09440	31 Dec 2011
Cleancrop Celeb	United Agri	13294	31 Dec 2011
Kingdom Turf	AgriGuard	13833	31 Dec 2011
Landgold Strobilurin 250	Teliton	12128	31 Dec 2011
Me2 Azoxystrobin	Me2	09654	31 Dec 2011
Ortiva	Syngenta	10542	31 Dec 2011
Priori	Syngenta	10543	31 Dec 2011
RouteOne Roxybin 25	Albaugh UK	13653	31 Dec 2011
Roxybin 25	RouteOne	13268	31 Dec 2011
azoxystrobin + chlorothalonil			
Amistar Opti	Syngenta	14582	31 Dec 2011
BTP	Syngenta	13400	31 Aug 2011
azoxystrobin + fenpropimorph			
RouteOne Roxypro FP	Albaugh UK	13634	

Product	Approval holder	MAPP No.	Expiry Date
azoxystrobin + propiconazole			
Headway	Syngenta	14396	12 Feb 2012
Bacillus thuringiensis			
DiPel DF	Fargro	11184	31 Aug 2012
Beauveria bassiana			
Naturalis-L	Belchim	14655	01 Apr 2013
benalaxyl + mancozeb			
Galben M	Sipcam	07220	30 Sep 2013
Galben M	Belchim	14661	
Intro Plus	Belchim	14666	
Tairel	Belchim	14659	
Tairel	Sipcam	07767	30 Sep 2013
bentazone			
Basagran	BASF	00188	31 Jul 2011
Bently	Chem-Wise	14210	31 Jul 2011
IT Bentazone	I T Agro	13132	31 Jul 2011
Landgold Bentazone SL	Teliton	13418	31 Jul 2011
RouteOne Benta 48	Albaugh UK	13662	31 Jul 2011
RouteOne Bentazone 48	Albaugh UK	13664	31 Jul 2011
bentazone + MCPB			
Pulsar	BASF	04002	
bentazone + pendimethalin			
Impuls	BASF	13372	31 Jul 2011
benzoic acid			
Menno Florades	Brinkman	12472	30 Apr 2012
Menno Florades	Fargro	13985	31 May 2014
beta-cyfluthrin			
Gandalf	Makhteshim	12865	
beta-cyfluthrin + imidacloprid			
Chinook	Bayer CropScience	13696	
bifenazate			
Inter Bifenazate 240 SC	Iticon	14543	30 Nov 2015
bifenox			
Cleancrop Diode	United Agri	14620	
bifenox + MCPA + mecoprop-P			
Quickfire	Headland Amenity	13053	
Sirocco	Makhteshim	13051	

Product	Approval holder	MAPP No.	Expiry Date
bifenthrin			
Gyro	Certis	13026	31 Jul 2010
UPL Bifenthrin	United Phosphorus	12653	31 Jul 2010
bitertanol + fuberidazole			
UK 743	Bayer CropScience	11315	28 Feb 2011
boscalid			
RouteOne Boscalid 50	Albaugh UK	13660	30 Sep 2013
boscalid + epoxiconazole			
Injun	Goldengrass	14305	
Maitre	Agro Trade	13878	
RouteOne Rocker	Albaugh UK	13642	
boscalid + pyraclostrobin			
RouteOne Bosca 25/12	Albaugh UK	13952	30 Sep 2013
bromoxynil			
Emblem	Nufarm UK	11559	31 Aug 2010
bromoxynil + diflufenican + ioxynil			
Capture	Bayer CropScience	09982	
bromoxynil + ioxynil			
Deloxil	Bayer CropScience	09987	31 Aug 2010
Mextrol Biox	Nufarm UK	09470	30 Nov 2010
Mextrol-Biox	Nufarm UK	14697	18 Aug 2013
Oxytril CM	Bayer CropScience	10005	30 Nov 2010
Stellox	Nufarm UK	11551	30 Nov 2010
Stellox	Nufarm UK	14696	18 Aug 2013
bromoxynil + ioxynil + mecoprop-P			
Astalavista	Nufarm UK	12335	31 Aug 2010
Swipe P	Nufarm UK	11955	31 Aug 2010
bromoxynil + ioxynil + triasulfuron			
Teal	Nufarm UK	11940	31 Aug 2010
bromoxynil + terbuthylazine			
Alpha Bromotril PT	Makhteshim	09435	
Cleancrop Amaize	United Agri	11990	
bromuconazole			
Tote	Interfarm	13780	03 May 2010
buprofezin			
Applaud	Certis	11532	30 Mar 2010
Candida Oleophila Strain 0			
NEXY 1	BioNext	13609	22 Jul 2011

EXPIRY DATE IS 31/12/2013 UNLESS OTHERWISE SHOWN

Product	Approval holder	MAPP No.	Expiry Date
captan			
Akotan 80 WG	Aako	14490	
Alpha Captan 50 WP	Makhteshim	04797	30 Sep 2012
PP Captan 80 WG	Calliope SAS	12435	
PP Captan 83	Calliope SAS	12330	
carbendazim			
Caste Off	Nufarm UK	12832	
Clayton Am-Carb	Clayton	11906	
Cleancrop Curve	United Agri	11774	
Mascot Systemic	Scotts	09132	
Mascot Systemic	Rigby Taylor	14488	
Nuturf Carbendazim	Nufarm UK	11469	
Turf Systemic Fungicide	Barclay	13291	
Turfclear	Scotts	07506	
carbetamide			
Carbetamex	Makhteshim	13045	
carfentrazone-ethyl			
Aurora 40 WG	Belchim	11614	30 Sep 2013
Harrier	Belchim	14164	30 Sep 2013
Platform	Belchim	11615	30 Sep 2013
carfentrazone-ethyl + mecoprop-P			
Jewel	Scotts	14327	09 Dec 2012
carfentrazone-ethyl + metsulfuron-methyl			
Ally Express	DuPont	08640	30 Jun 2011
chloridazon			
Cleancrop Akazon	United Agri	11903	31 Dec 2010
Gladiator DF	Tripart	06342	31 Dec 2010
Globazone 430 SC	Globachem	12878	31 Oct 2010
Portman Weedmaster	Agform	06018	31 Dec 2010
Pyramin FL	BASF	11628	
Sculptor	Sipcam	08836	31 Dec 2010
Starter Flowable	Truchem	03421	31 Dec 2010
Weedmaster SC	Agform	08793	31 Dec 2010
chloridazon + ethofumesate			
Gremlin	Sipcam	09468	
chloridazon + lenacil			
Advisor S	DuPont	10948	31 Mar 2010
Advizor	DuPont	06571	31 Mar 2010
chloridazon + metamitron			
Volcan Combi FL	Sipcam	11442	

SECTION 3

Product	Approval holder	MAPP No.	Expiry Date
chloridazon + propachlor			
Ashlade CP	Nufarm UK	06481	18 Mar 2010
chlormequat			
3C Chlormequat 720	BASF	13973	
3C Chlormequat 750	BASF	13984	
Adjust	Mandops	13148	30 Apr 2012
Agriguard 5C Chlormequat 460	AgriGuard	09851	
Agriguard Chlormequat 700	AgriGuard	09782	
AgriGuard Chlormequat 720	AgriGuard	14120	
AgriGuard Chlormequat 720	AgriGuard	14125	
AgriGuard Chlormequat 750	AgriGuard	14144	
AgriGuard Chlormequat 750	AgriGuard	14180	
Agriguard Chlormequat 760	AgriGuard	10290	
Alpha Chlormequat 460	Makhteshim	04804	
Alpha Pentagan	Makhteshim	04794	
Alpha Pentagan Extra	Makhteshim	04796	
Atlas 5C Quintacel	Nufarm UK	11130	
Barclay Holdup	Barclay	06799	
Barclay Holdup 600	Barclay	08794	
Barclay Holdup 600	Barclay	11373	
Barclay Holdup 640	Barclay	08795	
Barclay Liffey	Barclay	11366	
Barclay Liffey	Barclay	09856	
Barclay Lucan	Barclay	09855	
Barclay Lucan	Barclay	11367	
Barclay Take 5	Barclay	08524	
Barleyquat B	Mandops	13187	30 Apr 2012
Barleyquat B	Taminco	13965	
BASF 3C Chlormequat 600	BASF	04077	
BASF 3C Chlormequat 750	BASF	06878	
Belcocel	Taminco	11881	
Bettaquat B	Mandops	13192	30 Apr 2012
Chlormequat 46	Nufarm UK	11504	
Ciba Chlormequat 460	Ciba Specialty	09525	
Ciba Chlormequat 5C 460:320	Ciba Specialty	09527	
Ciba Chlormequat 730	Ciba Specialty	09526	
Clayton CCC 750	Clayton	07952	
Clayton Manquat	Clayton	09916	
Cleancrop Chlormequat 700	United Agri	10143	
Cropsafe 5C Chlormequat	Certis	11179	
Cropsafe 5C Chlormequat	Certis	07897	
Hyquat 70	Agrichem	03364	
K2	Mandops	13190	30 Apr 2012
Mandops Chlormequat 460	Taminco	13969	
Mandops Chlormequat 700	Mandops	06002	30 Apr 2012

Product	Approval holder	MAPP No.	Expiry Date
Manipulator	Mandops	13189	30 Apr 2012
Midget	Nufarm UK	12283	
MSS Mirquat	Mirfield	08166	
Portman Chlormequat 400	Agform	01523	
Portman Chlormequat 460	Agform	02549	
Portman Chlormequat 700	Agform	03465	
Portman Supaquat	Agform	03466	
Selon	Mandops	13826	30 Apr 2012
Stabilan 750	Nufarm UK	09303	
Supaquat 720	Agform	09381	
Tricol	Nufarm UK	12682	
Tripart 5C	Tripart	04726	
Tripart Brevis	Tripart	03754	
Tripart Brevis 2	Tripart	06612	
Tripart Chlormequat 460	Tripart	03685	
Uplift	United Phosphorus	07527	
chlormequat + 2-chloroethylphosphonic acid			
Barclay Banshee XL	Barclay	11339	
Sypex	BASF	04650	
Terpal C	BASF	07062	
chlormequat + 2-chloroethylphosphonic acid + imazaquin			
Satellite	BASF	10395	
2-chloroethylphosphonic acid			
Ethro 48	RouteOne	13333	31 Jul 2012
Milentus Ethephon	Milentus	12185	
RouteOne Ethro 48	Albaugh UK	13639	31 Jul 2012
2-chloroethylphosphonic acid + mepiquat chloride			
Barclay Banshee	Barclay	11343	
CleanCrop Fonic M	United Agri	09553	
Guilder	Nufarm UK	11894	28 Feb 2011
Mepiquat Plus	RouteOne	13297	
RouteOne Mepiquat Plus	Albaugh UK	13636	
Terpitz	Me2	09634	
chloropicrin			
Chloropicrin Fumigant	Dewco-Lloyd	04216	
Custo-Fume	Custodian	13272	
chlorothalonil			
Agrotech Chlorthalonil 500 SC	Agrotech Trading	13176	28 Feb 2011
Bravo 500	Syngenta	14548	28 Feb 2016
CleanCrop Rio	United Agri	13332	31 Aug 2011
Cropguard 2	Nufarm UK	13193	31 Aug 2011
Daconil Turf	Syngenta	13602	31 Aug 2011

Product	Approval holder	MAPP No.	Expiry Date
Jupital	Syngenta	14558	28 Feb 2016
Jupital	Syngenta	10528	31 Aug 2011
Landgold Chlorothalonil 50	Teliton	13473	28 Feb 2011
Mainstay	Quadrangle	05625	31 Aug 2011
Repulse	Certis	07641	30 Sep 2011
RouteOne Ronil 50	Albaugh UK	13637	28 Feb 2011
chlorothalonil + cyproconazole			
Bravo Xtra	Syngenta	11824	
Cleancrop Cyprothal	United Agri	09580	
chlorothalonil + cyproconazole + propiconazole			
Apache	Syngenta	14255	
UK 282	Syngenta	13204	30 Sep 2010
chlorothalonil + flutriafol			
Argon	Headland	12312	
Halo	Headland	11546	
chlorothalonil + metalaxyl-M			
Folio Gold	Syngenta	14368	30 Sep 2012
chlorothalonil + propamocarb hydrochloride			
Merlin	Bayer CropScience	07943	
Pan Wizard	Pan Agriculture	11953	
chlorothalonil + tebuconazole			
Timpani	Nufarm UK	14651	
chlorothalonil + tetraconazole			
Eminent Star	Isagro	10447	
chlorotoluron			
CTU Minrinse	Makhteshim	12458	31 Aug 2011
Lentipur 700	Nufarm UK	13526	31 Aug 2011
chlorpropham			
Aceto Chlorpropham 50M	Aceto	08929	31 Jul 2010
Aceto Chlorpropham 50M	Aceto	14134	07 Aug 2012
Aceto Sprout Nip	Aceto	14156	02 Sep 2012
Atlas CIPC 40	Atlas	07710	31 Jul 2010
BL500	Whyte Agrochemicals	00279	31 Jul 2010
BL500	Whyte Agrochemicals	14153	07 Aug 2012
Comrade	United Phosphorus	13566	31 Jul 2010
Gro-Stop 100	Certis	13742	31 Jul 2010
Gro-Stop 300 EC	Certis	13740	31 Jul 2010
Gro-Stop Basis	Certis	13739	31 Jul 2010
Gro-Stop Fog	Certis	13741	31 Jul 2010
Gro-Stop HN	Certis	13738	31 Jul 2010

Product	Approval holder	MAPP No.	Expiry Date
Gro-Stop Innovator	Certis	14147	31 Jan 2015
Gro-Stop Innovator	Certis	13743	31 Jul 2010
Jupiter 40 EC	United Phosphorus	12852	31 Jul 2010
Luxan Gro Stop 100	Luxan	12052	31 Jul 2010
Luxan Gro-Stop 300 EC	Luxan	08602	31 Jul 2010
Luxan Gro-Stop Basis	Luxan	08601	31 Jul 2010
Luxan Gro-Stop Fog	Luxan	09388	31 Jul 2010
Luxan Gro-Stop HN	Luxan	07689	31 Jul 2010
Luxan Gro-Stop Innovator	Luxan	12340	31 Jul 2010
MSS CIPC 30M	Nufarm UK	10064	31 Jul 2010
MSS CIPC 5 G	Whyte Agrochemicals	01402	31 Jul 2010
MSS CIPC 50 LF	Whyte Agrochemicals	03285	31 Jul 2010
MSS CIPC 50LF	United Phosphorus	14635	31 Jan 2015
MSS CIPC 50M	Whyte Agrochemicals	14151	07 Aug 2012
MSS CIPC 50M	Nufarm UK	11165	31 Jul 2010
MSS Sprout Nip	Whyte Agrochemicals	11804	31 Jul 2010
Pro-Long	Whyte Agrochemicals	12600	31 Jul 2010
Pro-Long	Whyte Agrochemicals	14152	07 Aug 2012
Sprout Nip	Aceto	11786	31 Jul 2010
Standon CIPC 300 HN	Standon	09187	31 Jul 2010
chlorpropham with cetrimide			
Croptex Pewter	Certis	11181	31 Jul 2010
chlorpyrifos			
Akofos 480 EC	Aako	14211	
Clayton Pontoon	Clayton	13219	30 Jun 2011
Lorsban WG	Landseer	11962	31 Mar 2013
Lorsban WG	Dow	10139	31 Mar 2013
Pirisect	AgriGuard	13371	30 Jun 2011
Pyrinex 48 EC	Makhteshim	13534	
Suscon Indigo	Fargro	09902	
chlorthal-dimethyl			
Dacthal W-75	AMVAC	10289	23 Mar 2011
RouteOne Chlorthal-D75	Albaugh UK	13649	23 Mar 2011
chlorthal-dimethyl + propachlor			
Decimate	Certis	13419	18 Mar 2010
cinidon-ethyl			
Lotus	BASF	09231	31 Oct 2011
clodinafop-propargyl			
Agrilens Clodinafop	Agrilens	14331	31 Jan 2012
Landgold Clodinafop	Teliton	13329	31 Jan 2012
RouteOne Roclod 24	RouteOne	13638	31 Jan 2012

Product	Approval holder	MAPP No.	Expiry Date
clodinafop-propargyl + diflufenican			
Amazon	Syngenta	08128	31 Dec 2010
Amazon	Bayer CropScience	10266	31 Dec 2010
clofentezine			
Acaristop 500 SC	Aako	14405	
clomazone			
IT Clomazone	Inter-Trade	14695	
clomazone + linuron			
Linzone	Belchim	14688	
clomazone + metazachlor			
Centium Plus	Belchim	14634	
clopyralid			
Cliophar	Agriphar	13360	
clopyralid + 2,4-D + MCPA			
Lonpar	Dow	08686	
clopyralid + florasulam + fluroxypyr			
Bofix FFC	Dow	13152	31 Oct 2012
Bofix FFC	Dow	14179	31 Dec 2011
Galaxy	Dow	13127	31 Oct 2012
clopyralid + fluroxypyr + MCPA			
Interfix	Iticon	14386	
clopyralid + fluroxypyr + triclopyr			
Charter	AgriGuard	13908	
clopyralid + picloram			
Euro Pyralid Extra	Euro	14681	
Prevail	Dow	13205	
clopyralid + propyzamide			
Matrikerb	Dow	10806	
copper hydroxide			
Spin Out	DuPont	12069	
copper silicate			
Socusil Slug Spray	Doff Portland	11817	
copper sulphate			
Cuproxat FL	Nufarm UK	13241	
cyazofamid + cymoxanil			
Ranman Super	Belchim	14024	

Product	Approval holder	MAPP No.	Expiry Date
cyazofamid + polyalkyleneoxide modheptamethylsiloxane			
Roazafod	RouteOne	13530	30 Jun 2013
RouteOne Roazafod	Albaugh UK	13666	30 Jun 2013
cycloxydim			
Stratos	BASF	13012	31 Jan 2011
Cydia pomonella GV			
Cyd-X Xtra	Certis	14397	
cymoxanil			
Cleancrop Covert	United Agri	13840	
Cymostraight 45	Belchim	14615	
Cymoxanil 45% WG	Globachem	13727	
Sipcam C50 WG	Sipcam	14650	
cymoxanil + mancozeb			
Cleancrop Xanilite	United Agri	10050	
Curzate M68	DuPont	08072	
Curzate M68 WSB	DuPont	08073	
Me2 Cymoxeb	Me2	09486	
Nautile DG	Nufarm UK	13416	
Solace WG	Nufarm UK	13729	31 Dec 2012
Standon Cymoxanil Extra	Standon	09442	
Zetanil	Sipcam	11993	
cypermethrin			
Afrisect 10	Agriphar	13159	31 Aug 2011
AgriGuard Cypermethrin 250EC	AgriGuard	14076	31 Aug 2011
Cyperkill 10	Agriphar	13157	31 Aug 2011
Cyperkill 25	Agriphar	13160	31 Aug 2011
Cyperkill 5	Agriphar	13162	31 Aug 2011
MCC 25 EC	Agriphar	13161	31 Aug 2011
Sherpa 100 EC	SBM	13570	28 Feb 2011
Syper 100	Goldengrass	14194	31 Aug 2011
Syper 250	Goldengrass	14460	31 Aug 2011
cyproconazole + picoxystrobin			
Furlong	DuPont	13818	
cyproconazole + propiconazole			
Alto Xtra	Syngenta	13058	31 Mar 2013
Menara	Syngenta	13035	28 Feb 2013
cyprodinil			
Barclay Amtrak	Barclay	11338	
Cleancrop Cyprodinil	United Agri	09668	
Standon Cyprodinil	Standon	09345	

SECTION 3

Product	Approval holder	MAPP No.	Expiry Date
2,4-D			
2,4-D Amine 500	Nufarm UK	14360	30 Sep 2012
Damine	Agriphar	13366	01 Oct 2012
Dioweed 50	United Phosphorus	13197	30 Sep 2012
Growell 2,4-D Amine	GroWell	13146	30 Sep 2012
Herboxone	Headland	13144	30 Apr 2012
Herboxone	Nufarm UK	14101	30 Apr 2012
Herboxone 60	Headland	14080	30 Sep 2012
Herboxone 60	Marks	13145	31 Aug 2012
Maton	Headland	13234	30 Nov 2010
2,4-D + dicamba			
Landscaper Pro Weed Control + Fertilizer	Scotts	14128	
Lawn Builder Plus Weed Control	Scotts	08499	
Lawn Spot Weeder	Bayer CropScience	13116	
Scotts Feed and Weed	Scotts	13076	
2,4-D + dicamba + dichlorprop-P			
NUB-041	Nufarm UK	14287	
2,4-D + dicamba + fluroxypyr			
Holster	SumiAgro Amenity	10593	31 Dec 2010
2,4-D + dicamba + MCPA + mecoprop-P			
Dicophar	Agriphar	13256	
Longbow	Bayer Environ.	14316	
2,4-D + dicamba + triclopyr			
Cleancrop Broadshot	United Agri	11664	
Greengard	SumiAgro Amenity	11715	
2,4-D + MCPA			
Agroxone Combi	Marks	11025	30 Sep 2012
Agroxone Combi	Nufarm UK	14094	
2,4-D + picloram			
Atladox HI	Nomix Enviro	05559	31 Dec 2010
Atladox Hi	Nomix Enviro	13867	31 Mar 2013
daminozide			
B-Nine SG	Chemtura	12698	31 Aug 2011
B-Nine SG	Chemtura	14434	28 Feb 2016
dazomet			
Assassin	AgriGuard	13737	
Basamid	Kanesho	12895	
2,4-DB			
Butoxone DB	Marks	13680	30 Sep 2012

EXPIRY DATE IS 31/12/2013 UNLESS OTHERWISE SHOWN

Product	Approval holder	MAPP No.	Expiry Date
Butoxone DB	Nufarm UK	14100	
DB Straight	United Phosphorus	13736	
2,4-DB + MCPA			
Butoxone DB Extra	Marks	12152	30 Sep 2012
Butoxone DB Extra	Nufarm UK	14095	
MSS 2,4-DB + MCPA	Mirfield	01392	
deltamethrin			
Agriguard Deltamethrin	AgriGuard	10770	
Agrotech Deltamethrin	Agrotech-Trading	12165	
Cleancrop Decathlon	United Agri	12834	
Delta-M 2.5 EC	MAC	14212	
K-Obiol EC 25	Bayer Environ.	13573	14 Jun 2011
K-Obiol ULV 6	Bayer Environ.	13572	14 Jun 2011
Landgold Deltaland	Teliton	12114	
Milentus Deltamethrin	Milentus	12219	
Pearl Micro	Bayer CropScience	08620	
Routeone Deltam 10	Albaugh UK	13615	
desmedipham + ethofumesate + phenmedipham			
Agrotech-Desmedipham Plus EC	Agrotech-Trading	12381	31 Aug 2010
Betanal Expert	Bayer CropScience	10592	31 Aug 2010
Landgold Deputy	Landgold	11975	31 Aug 2010
Landgold Deputy	Teliton	12115	31 Aug 2010
desmedipham + phenmedipham			
Betanal Carrera	Bayer CropScience	11029	31 Aug 2010
dicamba			
Cadence	Barclay	09578	
Cadence	Syngenta	08796	
Dicavel 480	Globachem	14063	31 Dec 2010
I T Dicamba	I T Agro	10976	31 Dec 2010
dicamba + dichlorprop-P + ferrous sulphate + MCPA			
Renovator 2	Scotts	11411	
dicamba + MCPA + mecoprop-P			
ALS Premier Selective Plus	Amenity Land	08940	
Banlene Super	Bayer CropScience	10053	31 Dec 2010
Grassland Herbicide	United Phosphorus	14623	
Mascot Super Selective Plus	Barclay	12839	
Mircam Super	Nufarm UK	11836	
Nomix Turf	Nomix Enviro	14375	
Nomix Turf Plus	Nomix Enviro	14377	
Premier Amenity Selective	Amenity Land	10098	
Tribute	Nomix Enviro	06921	31 Dec 2010

SECTION 3

Product	Approval holder	MAPP No.	Expiry Date
Tribute Plus	Nomix Enviro	09493	31 Dec 2010
Trireme	Nufarm UK	11524	
Tritox	Scotts	07764	
dicamba + mecoprop-P			
Camber	Headland	09901	
Foundation	Syngenta	08475	
Headland Swift	Headland	11945	
Optica Forte	Marks	11913	31 Dec 2012
Optica Forte	Nufarm UK	14334	
dichlobenil			
Casoron G	Miracle	07926	18 Mar 2010
Casoron G	Certis	11632	18 Mar 2010
Casoron G	Nomix Enviro	09023	18 Mar 2010
Casoron G	Chemtura	09022	18 Mar 2010
Casoron G4	Miracle	07927	18 Mar 2010
Casoron G4	Chemtura	09215	18 Mar 2010
Casoron G-SR	Miracle	07925	18 Mar 2010
Certis Dichlobenil Granules	Certis	13705	18 Mar 2010
Dichlobenil Granules	Certis	11871	18 Mar 2010
Embargo G	Barclay	12690	18 Mar 2010
Luxan Dichlobenil Granules	Luxan	09250	18 Mar 2010
Osorno	Nufarm UK	12486	18 Mar 2010
Resimate	Certis	13789	18 Mar 2010
Sierraron G	Scotts	09675	18 Mar 2010
Sierraron G	Scotts	09263	18 Mar 2010
Sierraron G4	Scotts	10491	18 Mar 2010
Standon Dichlobenil 6 G	Standon	08874	18 Mar 2010
Viking Granules	Certis	13883	18 Mar 2010
Viking Granules	Nomix Enviro	11859	18 Mar 2010
dichlorprop-P			
Optica DP	Marks	11067	30 Sep 2012
Optica DP	Nufarm UK	14097	
dichlorprop-P + ferrous sulphate + MCPA			
Vitagrow Granular Feed, Weed & Mosskiller	Sinclair	10971	
dichlorprop-P + MCPA			
Optica Duo	Marks	10343	30 Sep 2012
Optica Duo	Nufarm UK	14098	
dichlorprop-P + MCPA + mecoprop-P			
Optica Trio	Marks	09747	30 Sep 2012
Optica Trio	Nufarm UK	14099	

Product	Approval holder	MAPP No.	Expiry Date
diclofop-methyl + fenoxaprop-P-ethyl			
Corniche	Bayer CropScience	08947	
Tigress Ultra	Bayer CropScience	08946	
difenoconazole			
COM 302 09 F AL	Westland Horticulture	14021	
COM 302 8 F EC	Westland Horticulture	14020	31 Mar 2013
Fungus Attack 2 RTU	Westland Horticulture	14454	30 Sep 2013
Landgold Difenoconazole	Landgold	09964	31 May 2012
Plant Rescue Fungus Ready to Use	Westland Horticulture	14646	
diflubenzuron			
Dimilin 25-WP	Chemtura	08902	
diflufenican			
Alliance Pro	Nufarm UK	12434	31 Dec 2010
Alpha Diflufenican 500 SC	Makhteshim	13182	
CleanCrop Precinct	United Agri	13472	31 Dec 2010
Dazzler	AgriGuard	13620	31 Dec 2010
Dina 50	Globachem	13627	
IT Diflufenican	I T Agro	13044	31 Dec 2010
MAC-DIFLUFENICAN 500 SC	MAC	14109	
Pelican	Headland	12569	31 Dec 2010
RouteOne Flucan 50 UK	Albaugh UK	14130	
Solo D	Sipcam	14667	
Viola	Bayer CropScience	13321	
diflufenican + metsulfuron-methyl			
Pelican Delta	Headland	14312	
dimethoate			
Danadim	Headland	11550	
Dimethoate 40	BASF	13949	
Rogor L40	Interfarm	07611	
Sector	Cheminova	10492	
dimethomorph + mancozeb			
Saracen	BASF	12005	
diphenylamine			
No Scald DPA 31	Decco	14141	
diquat			
A1412A2	Syngenta	13440	31 Dec 2011
Cleancrop Diquat	United Agri	13348	31 Dec 2011
Diqua	Sharda	14032	20 May 2012
Diquanet	Hermoo	13413	31 Dec 2011
I.T. Diquat	I T Agro	13557	31 Dec 2011
Inter Diquat	I T Agro	14196	22 Sep 2012

Product	Approval holder	MAPP No.	Expiry Date
Quad S	Q-Chem	13578	28 Jun 2011
Standon Googly	Standon	12995	31 Dec 2011
Tuber Mission	AgriChem BV	14542	31 Dec 2011
dithianon			
Barclay Cluster	Barclay	11347	
RouteOne Dith-WG 70	Albaugh UK	13851	
dodeca-8,10-dienyl acetate			
Exosex CM	Exosect	12103	
dodine			
Barclay Dodex	Barclay	11351	
Syllit 400 SC	Chimac-Agriphar	11079	
Syllit 400 SC	Agriphar	13363	
epoxiconazole			
Agrotech-Epoxiconazole 125 SC	Agrotech-Trading	12382	
Bassoon	BASF	14402	
Epoxi 125	Goldengrass	14291	
Epro 125	RouteOne	13375	
Fathom	Headland	14267	
Landgold Epoxi	Makhteshim	14062	30 Nov 2012
Landgold Epoxiconazole	Teliton	12117	
Landgold Epoxiconazole	Landgold	09821	31 May 2012
Milentus Epoxiconazole	Milentus	12363	
RouteOne Epro 125	Albaugh UK	13640	
Spike SC	Globachem	14185	30 Apr 2011
Starburst	Chem-Wise	13981	
Strand	Headland	14261	
Supo	Makhteshim	13903	
epoxiconazole + fenpropimorph			
Barclay Riverdance	Barclay	11341	
Landgold Epoxiconazole FM	Landgold	08806	31 May 2012
Optiorb	AgriGuard	13426	
Tango Super	BASF	13283	
epoxiconazole + fenpropimorph + kresoxim-methyl			
Allegro Plus	BASF	12218	
Asana	BASF	11934	
BAS 493F	BASF	11748	
Cleancrop Chant	United Agri	11746	
Mastiff	BASF	11747	
epoxiconazole + fenpropimorph + pyraclostrobin			
Diamant	BASF	11557	
Diamant	BASF	13253	

Product	Approval holder	MAPP No.	Expiry Date
RouteOne Super 3	Albaugh UK	13721	
Tourmaline	Agro Trade	13934	
epoxiconazole + kresoxim-methyl			
Allegro	BASF	12220	
Barclay Avalon	Barclay	11337	
Cleancrop Kresoxazole	United Agri	09698	
Landgold Strobilurin KE	Landgold	09908	31 May 2012
epoxiconazole + kresoxim-methyl + pyraclostrobin			
Covershield	BASF	10900	
epoxiconazole + pyraclostrobin			
Euro	Me2	11511	
Ibex	BASF	12168	
esfenvalerate			
Sumi-Alpha	BASF	10401	31 May 2010
ethanol			
Ethy-Gen II	Banana-Rite	13412	
Restrain Fuel	Restrain	14520	
ethofumesate			
Agriguard Ethofumesate 200	AgriGuard	14129	31 Aug 2012
Barclay Keeper 500 SC	Barclay	13430	28 Feb 2013
ethofumesate + metamitron			
Galahad	Bayer CropScience	10727	
Goltix Plus	Aako	14461	
ethofumesate + metamitron + phenmedipham			
MAUK 540	Makhteshim	11545	
ethofumesate + phenmedipham			
Betosip Combi FL	Sipcam	14403	28 Feb 2013
Duo 400 SC	AgriGuard	14323	28 Feb 2013
Fenlander 2	Makhteshim	13124	31 Aug 2010
Magic Tandem	Bayer CropScience	13925	28 Feb 2013
Magic Tandem	Bayer CropScience	11814	31 Aug 2010
Powertwin	Makhteshim	12373	31 Aug 2010
Thunder	Makhteshim	12464	31 Aug 2010
Thunder	Makhteshim	14031	28 Feb 2013
Twin	Makhteshim	13114	31 Aug 2010
ethylene			
Biofresh Safestore	Biofresh	14579	

Product	Approval holder	MAPP No.	Expiry Date
famoxadone + flusilazole			
Caynil	DuPont	14641	
Mandolin	DuPont	14521	
fatty acids			
Finalsan Moss Control for Lawns Concentrate	Vitax	13141	
Finalsan Weed Killer RTU	Vitax	13143	
NEU1170 H	Growing Success	12971	
Organic Moss Control for Lawns Concentrate	Growing Success	13140	
Organic Weed Killer RTU	Growing Success	13142	
Safers Insecticidal Soap	Woodstream	07197	
fatty acids + maleic hydrazide			
Ultima	Neudorff	13347	
fenamidone + mancozeb			
Sonata	Bayer CropScience	11570	
fenazaquin			
Matador 200 SC	Margarita	11058	
fenbuconazole			
Indar 5 EW	Whelehan	09644	
fenbuconazole + propiconazole			
Graphic	Interfarm	10987	
fenhexamid			
Druid	AgriGuard	13901	31 May 2011
RouteOne Fenhex 50	Albaugh UK	13665	31 May 2011
fenoxaprop-P-ethyl			
Cheetah Energy	Bayer CropScience	13599	
fenpropidin			
Cleancrop Fulmar	United Agri	12033	
Instinct	Headland	14512	
Landgold Fenpropidin 750	Landgold	08973	31 May 2012
Mallard	Syngenta	08662	
Patrol	Syngenta	10531	
fenpropidin + prochloraz			
SL 552A	Syngenta	08673	31 Dec 2010
Sponsor	BASF	11902	31 Dec 2010
fenpropidin + prochloraz + tebuconazole			
Artemis	Makhteshim	14178	

Product	Approval holder	MAPP No.	Expiry Date
fenpropidin + propiconazole			
Prophet	Syngenta	08433	
fenpropidin + tebuconazole			
SL 556 500 EC	Syngenta	08449	31 Dec 2010
fenpropimorph			
Cleancrop Fenpro	United Agri	09885	
Cleancrop Fenpropimorph	United Agri	09445	
Keetak	BASF	06950	
Landgold Fenpropimorph 750	Landgold	10472	31 May 2012
Landgold Fenpropimorph 750	Teliton	12118	31 Mar 2013
Marnoch Phorm	Me2	11087	
Propimorf 750	Goldengrass	14410	
fenpropimorph + kresoxim-methyl			
Cleancrop Duster	United Agri	12306	
Landgold Strobilurin KF	Landgold	09196	31 May 2012
fenpropimorph + pyraclostrobin			
BAS 528 00f	BASF	11444	
Jemker	BASF	12225	
fenpropimorph + quinoxyfen			
Jackdraw	BASF	13769	
fenpyroximate			
NNI-850 5 SC	Certis	12658	
ferric phosphate			
Ferramol Max	Certis	14463	31 Oct 2011
Ferrox	Neudorff	14356	31 Oct 2011
Growing Success Advanced Slug Killer	Growing Success	12378	31 Oct 2011
NEU 1181 M	Neudorff	14355	31 Oct 2011
ferrous sulphate			
Aitken's Lawn Sand	Aitken	05253	
Greenmaster Mosskiller	Scotts	12197	
Greentec Mosskiller Pro	Headland Amenity	14645	
Landscaper Pro Moss Control + Fertilizer	Scotts	14317	
Maxicrop Moss Killer & Conditioner	Maxicrop	04635	
Moss Control Plus Lawn Fertilizer	Miracle	07912	
No More Moss Lawn Feed	Wolf	10754	
Pentagon Prestige Lawn Sand	Sinclair	10456	
Scotts Moss Control	Scotts	12928	

SECTION 3

Product	Approval holder	MAPP No.	Expiry Date
flazasulfuron			
Chikara	Nomix Enviro	13775	31 May 2014
flonicamid			
RouteOne Ski	Albaugh UK	13777	31 May 2014
florasulam			
Primus 25 SC	Dow	10175	30 Sep 2012
RouteOne Florasul 50	Albaugh UK	13668	30 Sep 2012
florasulam + fluroxypyr			
Image	Dow	13771	31 Dec 2011
Slalom	Dow	13772	31 Dec 2011
fluazifop-P-butyl			
Clayton Crowe	Clayton	12572	
Clean Crop Clifford	United Agri	13597	
Fluazifop +	RouteOne	13382	
Fusilade 250 EW	Syngenta	10525	
PP 007	Syngenta	10533	
RouteOne Fluazifop +	Albaugh UK	13641	
Wizzard	Syngenta	10539	
fluazinam			
Alpha Fluazinam 50SC	Makhteshim	13622	
Barclay Cobbler	Barclay	11348	
Blizzard	AgriGuard	12854	
Boyano	Belchim	14571	
Boyano	Hermoo	13849	28 Feb 2010
Fluazinam 500	Goldengrass	14411	
Ibiza 500	Belchim	14393	
Ibiza 500	Q-Chem	14002	28 Feb 2013
Landgold Fluazinam	Landgold	08060	31 May 2012
Landgold Fluazinam	Teliton	12119	31 Mar 2013
Legacy	ISK Biosciences	09966	31 Oct 2011
Legacy	Syngenta	13735	
Shirlan Programme	Syngenta	10574	
Standon Fluazinam 500	Standon	08670	
fluazinam + metalaxyl-M			
Fluazmet M	RouteOne	13519	
RouteOne Fluazmet M	Albaugh UK	13650	
fludioxonil			
Beret Gold	BASF	12656	
fludioxonil + tefluthrin			
Austral Plus	Bayer CropScience	13376	

EXPIRY DATE IS 31/12/2013 UNLESS OTHERWISE SHOWN

Product	Approval holder	MAPP No.	Expiry Date
flufenacet + isoxaflutole			
RouteOne Oxanet 481	Albaugh UK	13828	30 Sep 2013
fluoxastrobin			
Bayer UK 831	Bayer CropScience	12091	30 Apr 2013
fluoxastrobin + prothioconazole + tebuconazole			
Scenic	Bayer CropScience	12289	
flupyrsulfuron-methyl			
Bullion	DuPont	14058	30 Jun 2011
DP 459	DuPont	12996	30 Jun 2011
Oriel 50SX	DuPont	14640	30 Jun 2011
flupyrsulfuron-methyl + tribenuron-methyl			
Excalibur	DuPont	12047	28 Feb 2010
fluquinconazole			
Diablo	BASF	11688	
Jockey F	BASF	11690	30 Nov 2012
Sahara	Bayer CropScience	11905	
Triplex	BASF	12443	
fluquinconazole + prochloraz			
Baron	BASF	11686	
fluroxypyr			
Agrotech Fluroxypyr	Agrotech-Trading	12294	31 Dec 2011
Alpha Fluroxypyr 20EC	Makhteshim	13352	31 Dec 2011
Cascade	Chem-Wise	13939	31 Dec 2011
Cerastar	Cera Chem	13183	31 Dec 2011
Clean Crop Gallifrey	Inter-Trade	13040	08 May 2010
Clean Crop Gallifrey 200	Globachem	13813	31 Dec 2011
Floxy	Chimac-Agriphar	12984	31 Dec 2011
Floxy	Agriphar	13367	31 Dec 2011
Flurox 180	Stockton	13959	30 Dec 2011
Flurox 200	Stockton	13938	31 Dec 2011
Fluroxy 200	Goldengrass	14421	31 Dec 2011
Gala	Dow	12019	31 Dec 2011
GAL-GONE	Globachem	13821	31 Dec 2011
Hatchet 180EC	AgriGuard	14054	31 Dec 2011
Landgold Fluroxypyr	Teliton	13356	31 Dec 2011
Milentus Fluroxypyr	Milentus	12266	31 Dec 2011
NUB-010	Nufarm UK	13829	25 Nov 2011
Standon Homerun	Standon	13177	31 Dec 2011
Taipan	Cera Chem	14155	18 Aug 2012

EXPIRY DATE IS 31/12/2013 UNLESS OTHERWISE SHOWN

Product	Approval holder	MAPP No.	Expiry Date
flutolanil			
NNF-136	Nihon Nohyaku	13120	31 Dec 2012
NNF-136	Nihon	14302	
flutriafol			
Impact	Headland	12776	
fosetyl-aluminium			
Fosal	AgriGuard	13384	
Routeone Fosetyl 80	Albaugh UK	13625	
fuberidazole + imidacloprid + triadimenol			
Baytan Secur	Bayer CropScience	11253	28 Feb 2011
fuberidazole + triadimenol			
Baytan Flowable	Makhteshim	11714	
gibberellins			
Berelex	Interfarm	08903	31 Aug 2012
Regulex	Sumitomo	10147	
Regulex 10 SG	Sumitomo	13323	31 Aug 2012
glufosinate-ammonium			
Challenge	Bayer CropScience	07306	
Challenge 60	Fargro	08236	
glyphosate			
Accelerate	Headland	13390	30 Jun 2012
Acomac	CP AGRO	14254	30 Jun 2012
Acrion	Bayer Environ.	12677	30 Jun 2012
Amenity Glyphosate 360	Monsanto	13000	30 Jun 2012
Azural	Monsanto	14361	30 Jun 2012
Azural	Cardel	12666	30 Jun 2012
Barclay Barbarian	Barclay	12714	30 Jun 2012
Barclay Garryowen	Barclay	12715	30 Jun 2012
Cleancrop Hoedown	United Agri	12913	30 Jun 2012
Cleancrop Tungsten	United Agri	13049	30 Jun 2012
Clinic	Nufarm UK	12678	30 Jun 2012
Credit	Nufarm UK	13032	12 Jul 2010
Etna	AgriChem BV	14674	30 Jun 2012
Euro Glyfo 360	Euro	14691	30 Jun 2012
First Line	Linemark	12759	02 Feb 2010
Frontierland Glyphosate	Soyl	14638	31 Oct 2013
Frontsweep	Soyl	14700	30 Jun 2012
Gallup Hi- Aktiv Amenity	Barclay	12898	30 Jun 2012
Glister	Sinon EU	12990	30 Jun 2012
Glydate	Nufarm UK	12679	30 Jun 2012
Glyder 360	Agform	14006	30 Jun 2012

Product	Approval holder	MAPP No.	Expiry Date
Glyfos Gold	Nomix Enviro	10570	30 Jun 2012
Glymark	Nomix Enviro	12994	31 Jan 2012
Glymark	Nomix Enviro	13870	30 Jun 2012
Glyper	SBM	14383	30 Jun 2012
Glyper	Interfarm	13105	30 Jun 2012
Glyphosate 360	SumiAgro	12680	30 Jun 2012
Glyphosate 360	Monsanto	12669	30 Jun 2012
Glypoo	AgChem Access	13787	30 Jun 2012
Glyweed	Sabero	14444	30 Jun 2012
Habitat	Barclay	12717	30 Jun 2012
Hilite	Nomix Enviro	12922	31 Jan 2012
Initial Line	Linemark	12758	02 Feb 2010
KN 540	Monsanto	12009	30 Jun 2012
MON 79351	Monsanto	12860	30 Jun 2012
MON 79376	Monsanto	12664	30 Jun 2012
MON 79545	Monsanto	12663	30 Jun 2012
MON 79632	Monsanto	12582	30 Jun 2012
Nomix G	Nomix Enviro	13872	30 Jun 2012
Nomix G	Nomix Enviro	12921	31 Jan 2012
Nomix Nova	Nomix Enviro	12920	31 Jan 2012
Nomix Nova	Nomix Enviro	13873	30 Jun 2012
Nomix Revenge	Nomix Enviro	13874	30 Jun 2012
Nomix Revenge	Nomix Enviro	12993	31 Jan 2012
Nufosate	Nufarm UK	12699	30 Jun 2012
Onslaught	Procam	14564	30 Jun 2012
Pontil 360	Barclay	14322	30 Jun 2012
Pontil 360	Mastra	12733	30 Jun 2012
Preline	Linemark	12756	30 Jun 2012
Preline Plus	Linemark	12757	02 Feb 2010
Romany	Greencrop	12681	30 Jun 2012
Rosate 36	Albaugh UK	14459	30 Jun 2012
Rosate 36	Albaugh UK	14104	30 Jun 2012
Roundup	Monsanto	12645	30 Jun 2012
Roundup Amenity	Monsanto	12672	30 Jun 2012
Roundup Biactive 3G	Monsanto	13409	30 Jun 2012
Roundup Biactive Dry	Monsanto	12646	30 Jun 2012
Roundup Express	Monsanto	12526	30 Jun 2012
Roundup Gold	Monsanto	10975	30 Jun 2012
Roundup Greenscape	Monsanto	12731	30 Jun 2012
Roundup Pro-Green	Monsanto	11907	30 Jun 2012
Roundup Provide	Monsanto	12953	30 Jun 2012
Roundup Rail	Monsanto	12671	30 Jun 2012
Roundup Ultimate	Monsanto	12774	30 Jun 2012
Roundup Ultra ST	Monsanto	12732	30 Jun 2012
RouteOne Glyphosate 360	Albaugh UK	14133	30 Jun 2012

Product	Approval holder	MAPP No.	Expiry Date
RouteOne Glyphosate 360	Albaugh UK	14458	30 Jun 2012
Routeone Rosate 36	Albaugh UK	13889	30 Jun 2012
RouteOne Rosate 360	Albaugh UK	13659	30 Jun 2012
RouteOne Rosate 360	Albaugh UK	13661	30 Jun 2012
Scorpion	Cardel	12673	30 Jun 2012
Silvio	Rotam Agrochemical	14626	30 Jun 2012
Statis	AgriGuard	13079	30 Jun 2012
Statis 360	AgriGuard	14568	30 Jun 2012
Stirrup	Nomix Enviro	12923	31 Jan 2012
Task 360	Albaugh UK	14570	30 Jun 2012
Trustee Amenity	Barclay	12897	30 Jun 2012
Trustee Elite	Barclay	12662	31 Jan 2012
Tumbleweed Pro-Active	Scotts	12729	30 Jun 2012
Vival	Belchim	14550	30 Jun 2012
glyphosate + pyraflufen-ethyl			
Thunderbolt	Nichino	14102	20 Jul 2012
glyphosate + sulfosulfuron			
Nomix Blade	Nomix Enviro	13907	27 Jan 2012
imazalil			
I.T. Imazalil	Inter-Trade	14166	04 Jun 2011
Magnate 100SL	Makhteshim	14660	31 Dec 2011
Neozil 10 SL	Laboratorios Agrochem	13034	31 Mar 2010
Sphinx	BASF	11764	31 Dec 2011
imazalil + thiabendazole			
Storite Super	Syngenta	13135	05 Oct 2010
imazalil + triticonazole			
Robust	BASF	11807	30 Apr 2010
imazamox + pendimethalin			
Nirvana	BASF	13220	30 Jun 2013
imidacloprid			
Gaucho	Bayer CropScience	11281	
Gaucho FS	Bayer CropScience	11282	
Imidasect	Biologic HC	12899	
Imidasect 5GR	Fargro	14574	
Imidasect 60 FS	Bioscientific	14573	
Imidasect 60 FS	Bioscientific	14077	
Imidasect 70 WS	Biologic HC	14572	
Imidasect 70 WS	Biologic HC	13191	
Intercept 5GR	Scotts	14091	
Jive	Makhteshim	14514	
Mido 70% WDG	Sharda	14174	

Product	Approval holder	MAPP No.	Expiry Date
Neptune	Makhteshim	12992	
Picus 600 FS	Headland	14598	
imidacloprid + tebuconazole + triazoxide			
Raxil Secur	Bayer CropScience	11298	
4-indol-3-ylbutyric acid			
Seradix 1	Certis	10422	
Seradix 2	Certis	10423	
Seradix 3	Certis	10424	
iodosulfuron-methyl-sodium + mesosulfuron-methyl			
Hatra	Bayer CropScience	14524	
Horus	Bayer CropScience	14541	
Ocean	Goldengrass	14422	
Octavian	Bayer CropScience	14604	
RouteOne Seafarer	Albaugh UK	13654	
Teliton Ocean	Teliton	13163	31 Mar 2013
iodosulfuron-methyl-sodium + propoxycarbazone-sodium			
Caliban Duo	Headland	14283	31 Mar 2013
ipconazole			
Crusoe	Chemtura	13955	23 Apr 2012
iprodione			
Superturf	AgriGuard	14495	
isoxaben			
Flexidor	Dow	05121	
Flexidor 125	Dow	05104	
Gallery 125	Rigby Taylor	06889	
Knot Out	Vitax	05163	
lambda-cyhalothrin			
Barbarossa	Clayton	13510	03 Apr 2011
Clayton Zen	Clayton	13512	03 Apr 2011
CleanCrop Corsair	United Agri	14124	03 Apr 2011
Dalda 5	Globachem	13688	28 Jun 2011
IT Lambda	I T Agro	13451	03 Apr 2011
Komodo 10 EC	United Phosphorus	14703	31 Dec 2012
Major	Makhteshim	14527	31 Dec 2011
Markate 50	Agrovista	13529	28 Jun 2011
Stealth	Syngenta	14551	31 Dec 2011
laminarin			
Vacciplant	Goemar	13260	31 Mar 2015
lenacil			
Cleancrop Lenflow	United Agri	10059	31 Dec 2010

Product	Approval holder	MAPP No.	Expiry Date
Lenacil FL	United Phosphorus	12183	31 Dec 2010
Lenazar Flowable	Hermoo	11068	31 Dec 2010
Venzar 80 WP	DuPont	09981	
Volcano	United Phosphorus	12301	31 Dec 2010
linuron			
Aredios 45 SC	Novastar	14586	09 Oct 2012
Linurex 50 SC	Makhteshim	14652	
Linurex 50SC	Makhteshim	14268	14 Oct 2012
magnesium phosphide			
Detia Gas-Ex-B Forte	Detia Degesch	10661	
maleic hydrazide			
Cleancrop Malahide	United Agri	13629	22 Jul 2011
Fazor	Chemtura	13617	
Source II	Chiltern	13618	22 Jul 2011
mancozeb			
Agrizeb	Agriphar	13365	
Agrizeb	Chimac-Agriphar	10980	
Beacon	AgriGuard	13368	30 Jun 2011
Cleancrop Mancozeb	United Agri	11193	
Cleancrop Mandrake	United Agri	12500	
Dithane 945	Interfarm	12585	31 Jul 2011
Dithane 945	Dow	12545	31 Jul 2011
Dithane Dry Flowable Neotec	Dow	14631	13 Sep 2013
Dithane Dry Flowable Newtec	Dow	12564	30 Sep 2011
Dithane NT Dry Flowable	Interfarm	14704	13 Sep 2013
Karamate Dry Flo Neotec	Landseer	14632	14 Oct 2013
Laminator DG	Interfarm	11073	
Laminator WP	Interfarm	11071	
Manfil 80 WP	Indofil	12941	
Manfill 75 WG	Indofil	12419	
Manzate 75 WG	Headland	12070	
Micene 80	Sipcam	14622	02 Jul 2013
Micene DF	Sipcam	14705	13 Sep 2013
Penncozeb 80 WP	Cerexagri	14718	14 Oct 2013
Trimanzone	Intracrop	09584	
mancozeb + metalaxyl-M			
Fubol Gold WG	Syngenta	14605	30 Jul 2013
MetMan 680	RouteOne	13553	30 Jun 2011
RouteOne MetMan 680	Albaugh UK	13652	30 Jun 2011
mancozeb + propamocarb hydrochloride			
Tattoo	Bayer CropScience	07293	

Product	Approval holder	MAPP No.	Expiry Date
mancozeb + zoxamide			
Electis 75 WG	Dow	11013	
Roxam 75 WG	Dow	11017	
Unikat 75 WG	Gowan	11018	
Unikat 75WG	Gowan	14200	
maneb			
Trimangol 80	United Phosphorus	06871	
Trimangol WDG	United Phosphorus	06992	
MCPA			
Agrichem MCPA-50	Agrichem	04097	
Agricorn 500 II	Nufarm UK	09155	
Agritox Dry	Nufarm UK	10554	
Agroxone 40	Marks	10264	30 Sep 2012
Agroxone 40	Nufarm UK	14093	
Agroxone 75	Nufarm UK	14419	
Agroxone 75	Marks	09208	31 Mar 2013
Circium II	Nufarm UK	11801	
Dow MCPA Amine 50	Dow	14517	
Go-Low Power	DuPont	13812	30 Apr 2016
Growell MCPA DMA 500	GroWell	12989	
MCPA 25%	Nufarm UK	07998	
MCPA 500	Nufarm UK	08655	
Nufarm MCPA 750	Nufarm UK	11768	30 Apr 2011
Nufarm MCPA Amine 50	Nufarm UK	12046	31 May 2013
POL-MCPA 500 SL	Nufarm UK	14498	30 Apr 2011
POL-MCPA 500 SL	Zaklady	14528	30 Apr 2011
MCPA + MCPB			
Butoxone Plus	Marks	11080	
Trifolex-Tra	BASF	10396	
MCPA + mecoprop-P			
No More Weeds Lawn Feed	Wolf	10747	
Optica Combi	Marks	10118	30 Sep 2012
Optica Combi	Nufarm UK	14096	
MCPB			
Bellmac Straight	United Phosphorus	14448	30 Apr 2016
Butoxone	Headland	10501	
Tropotox	Nufarm UK	11141	
mecoprop-P			
Clenecorn Super	Nufarm UK	14628	31 May 2014
Optica	Headland	14373	31 May 2014

Product	Approval holder	MAPP No.	Expiry Date
mecoprop-P + metsulfuron-methyl			
Headland Neptune	Headland	10230	
mesotrione + terbuthylazine			
RouteOne Mesot	Albaugh UK	13576	
metalaxyl-M			
Apron XL	Syngenta	14654	30 Sep 2012
Fongarid Gold	Syngenta	12547	30 Sep 2012
metaldehyde			
Agosto	Makhteshim	14171	
Allure	Chiltern	12651	
Antares	De Sangosse	12998	
Appeal	Chiltern	12713	
Aristo	De Sangosse	11797	
Ascari	Makhteshim	13991	
Attract	Chiltern	12712	
Blue Prince	Doff Portland	13061	
Brits	Doff Portland	11792	
Brits M	De Sangosse	12956	28 Feb 2011
CDP H	De Sangosse	14432	
Certis Metaldehyde 5	Certis	13490	31 Mar 2013
Certis Metaldehyde 5	Certis	14457	
Certis Red 5	Certis	14479	
Certis Red 5	Certis	13491	30 Apr 2013
Chiltern Blues	Chiltern	12702	
Chiltern Blues	Chiltern	10071	
Chiltern Hardy	Chiltern	06948	
Chiltern Hardy	Chiltern	12647	
Chiltern Hundreds	Chiltern	12710	
Clartex	De Sangosse	12935	
Clean Crop Hyde	United Agri	12025	
Clean Crop Hyde	United Agri	12707	
Clean Crop Jekyll D 5	Doff Portland	13060	
Cleancrop Hyde 2	United Agri	13199	
Cleancrop Hyde 2	Makhteshim	14042	
Cleancrop Jekyll	Chiltern	12703	
Cleancrop Jekyll	United Agri	11724	
CleanCrop Jekyll 5	United Agri	14117	
CleanCrop Jekyll 5	United Agri	14534	
Cleancrop Jekyll D3	United Agri	14369	
Cleancrop Potent	United Agri	14228	
Cleancrop Tremolo	United Agri	14225	
Condor 3	Doff Portland	14324	
Duracoat Slug Killer	Doff Portland	13134	
Entice	De Sangosse	12933	28 Feb 2011

Product	Approval holder	MAPP No.	Expiry Date
Escar-Go 6	Chiltern	12649	
Escar-Go 6	Chiltern	06076	
ESP W	De Sangosse	14429	
Gusto 4	Makhteshim	14165	
Helimax	De Sangosse	13108	
Helimax W	De Sangosse	14424	
Keel Over	De Sangosse	13110	
Keel Over W	De Sangosse	14431	
Kimble	Chiltern	12907	
Limagri GR	SBM	13475	
Luxan 9363 Blue	Luxan	12391	31 Aug 2012
Luxan Metaldehyde 5	Luxan	12401	20 Feb 2010
Luxan Red 5	Luxan	12390	31 Aug 2012
Luxan Trigger 5	Luxan	12389	31 Aug 2012
Metadisque	Frunol Delicia	14665	
MetaPads	Frunol Delicia	14658	
Metarex Amba	De Sangosse	13245	
Metarex Green	De Sangosse	13069	
Metarex RG	De Sangosse	13070	
Mifaslug 5	Luxan	12392	31 Aug 2012
Molotov	Chiltern	08295	
Molto	Makhteshim	14170	
Optimol XL	De Sangosse	12937	28 Feb 2011
Pesta	De Sangosse	11796	
Prego 4	Makhteshim	14168	
Regel	De Sangosse	13115	
Regel Star	De Sangosse	13071	
Regel W	De Sangosse	14425	
Regiment	Certis	14338	
Rowent	Chiltern	12024	
Rowent	Chiltern	12709	
Sloggy	SBM	13373	
Slug Pellets	De Sangosse	12936	28 Feb 2011
Slugdown	Doff Portland	13063	
Slugdown 3	Unicrop	14630	
Slug-Lentils	Frunol Delicia	13613	
Stanza	Makhteshim	14169	
Steadfast	Doff Portland	13062	
Steadfast 3	Doff Portland	14367	
Super-Flor 6% Metaldehyde Slug Killer Mini Pellets	CMI	11789	
Tempt	Chiltern	14227	
Terminator	Chiltern	14229	
Tinto	Certis	14336	
Top Gun	Certis	14325	
Trigger 3	Certis	14304	

EXPIRY DATE IS 31/12/2013 UNLESS OTHERWISE SHOWN

Product	Approval holder	MAPP No.	Expiry Date
Trigger 5	Certis	14480	
Trigger 5	Certis	13493	30 Apr 2013
Trounce	Chiltern	14222	
Unicrop 6% Mini Slug Pellets	Unicrop	11795	
metamitron			
Alpha Metamitron	Makhteshim	11081	
Alpha Metamitron 70 SC	Makhteshim	12868	
Barclay Seismic	Barclay	11378	
Bettix 70 WG	United Phosphorus	11019	
Celmitron 70% WDG	AgriChem BV	14408	
Celmitron 70WDG	SD Agchem	13630	31 Mar 2013
Defiant WG	United Phosphorus	12300	
Fernpath Haptol Flo	AgriGuard	13549	
Homer	Makhteshim	11544	
Inter Metamitron WG	I T Agro	12388	28 Feb 2010
IT Metamitron SC	I T Agro	13059	
IT Metamitron WG	I T Agro	12467	
Landgold Metamitron	Landgold	06287	31 May 2012
Landgold Metamitron	Teliton	12123	
Lektan	Makhteshim	11543	
Maymat 70 WDG	Global Crop Care	12477	
MM 70	Nufarm UK	11582	
MM70	Gharda	09490	
Standon Metamitron	Standon	07885	
Volcan	Sipcam	09295	
metam-sodium			
Fumetham	Chemical Nutrition	10047	
Metham Sodium 400	United Phosphorus	08051	
Sistan	Unicrop	01957	
Sistan 38	Unicrop	08646	
Vapam	Willmot Pertwee	09194	
metazachlor			
Agrotech Metazachlor 500 SC	Agrotech-Trading	12454	
Barclay Metaza	Barclay	11359	
Booty	Me2	10659	
Butey	Chem-Wise	14176	
Butisan S	BASF	00357	
Clean Crop MTZ 500	United Agri	09222	
Fuego	Makhteshim	11177	
Fuego 50	Makhteshim	11473	
Gharda Bonanza 500 SC	Nufarm UK	13628	
Landgold Metazachlor 50	Landgold	09726	31 May 2012
Landgold Metazachlor 50	Teliton	12124	31 Mar 2013
Landgold Metazachlor 50 SC	Teliton	12133	31 Mar 2013

Product	Approval holder	MAPP No.	Expiry Date
Makila 500 SC	Novastar	14492	
Metarap 500 SC	Global Crop Care	12540	
Metaz 50	RouteOne	13467	
Metaz 500	Goldengrass	14412	
Metaz 500 SC	Goldengrass	14413	
Metazachlore GL 500	Globachem	12528	
Noble Metazachlor	Barclay	12892	
Rapsan 500 Sc	Nufarm UK	12365	28 Feb 2012
RouteOne Metaz 50	Albaugh UK	13648	
Triton SC	Nuvaros	14047	30 Jun 2013
metazachlor + quinmerac			
Katamaran	BASF	11732	
Rapsan TDI	Q-Chem	13920	31 Jul 2013
Syndicate	AgriGuard	12262	
metconazole			
Conzole	AgriGuard	13444	
methiocarb			
Draza Wetex	Bayer CropScience	11271	
Lupus	Bayer CropScience	11291	
methyl bromide			
Mebrom 100	Mebrom	04869	18 Mar 2010
Methyl Bromide 100%	Bromine & Chem.	01336	18 Mar 2010
methyl bromide with chloropicrin			
Methyl Bromide 98	Bromine & Chem.	01335	18 Mar 2010
1-methylcyclopropene			
SmartFresh SmartTabs	Landseer	12684	31 Mar 2016
metiram			
Polyram DF	BASF	08234	
metribuzin			
AgriGuard Metribuzin	AgriGuard	14084	31 Jul 2013
Buzin WG	Goldengrass	14415	
Citation 70	United Phosphorus	09370	
Landgold Metribuzin WG	Teliton	12468	31 Mar 2013
Lexone 2	DuPont	13804	
Lexone 2	DuPont	12051	30 Sep 2011
Milentus Metribuzin 70 WG	Milentus	12224	
Promor	Bayer CropScience	12502	
Python	Makhteshim	13802	
Python	Makhteshim	13802	
Rapture	AgriGuard	14536	
Shotput	Makhteshim	11960	30 Sep 2011

SECTION 3

EXPIRY DATE IS 31/12/2013 UNLESS OTHERWISE SHOWN

Product	Approval holder	MAPP No.	Expiry Date
metsulfuron-methyl			
Accurate	Headland	13224	28 Nov 2010
Alias	DuPont	13397	30 Jun 2011
Alias SX	DuPont	13398	30 Jun 2011
Ally SX	DuPont	12059	30 Jun 2011
Cleancrop Mondial	United Agri	13353	30 Jun 2011
Deft	Sipcam	13263	18 Dec 2010
Landgold Metsulfuron	Teliton	13421	30 Jun 2011
Lorate	DuPont	12743	30 Jun 2011
Metro 20	AgriGuard	13496	30 Jun 2011
Metro Solo	AgriGuard	14067	25 Apr 2011
Minx PX	Nufarm UK	13337	30 Jun 2011
Pike	Nufarm UK	12746	30 Jun 2011
Pike PX	Nufarm UK	13249	30 Jun 2011
Revenge 20	Agform	14449	30 Jun 2011
Revenge 20 SG	Agform	13633	30 Jun 2011
RouteOne Romet 20	Albaugh UK	13657	30 Jun 2011
Savvy	Rotam Agrochemical	14266	18 Dec 2010
Savvy Clear	Rotam Agrochemical	14474	30 Jun 2011
metsulfuron-methyl + thifensulfuron-methyl			
Accurate Extra	Headland	14167	23 Jul 2012
Choir	Nufarm UK	14081	30 Jun 2011
Ergon	Rotam Agrochemical	14382	30 Jun 2011
metsulfuron-methyl + tribenuron-methyl			
DP 911 WSB	DuPont	09867	31 Aug 2011
myclobutanil			
Robut 20	RouteOne	13437	
RouteOne Robut 20	Albaugh UK	13646	
Systhane 20 EW	Whelehan	09397	31 Mar 2013
Systhane 6 W	Dow	10808	31 Mar 2013
1-naphthylacetic acid			
Tipoff	Hatrim	12292	
napropamide			
Associate	FMC	13933	
Banweed	United Phosphorus	09376	
Devrinol	United Phosphorus	09375	
nicosulfuron			
Landgold Nicosulfuron	Teliton	13464	
Landgold Nicosulfuron	Goldengrass	13997	30 Nov 2012
Maizon	AgriGuard	13222	
RouteOne Strong 4	Albaugh UK	13830	
Samson	Syngenta	12141	

Product	Approval holder	MAPP No.	Expiry Date
nicotine			
No-Fid	Certis	11183	08 Jun 2010
No-FID	Certis	07959	08 Jun 2010
Stalwart	United Phosphorus	11877	08 Jun 2010
XL-All Nicotine 95%	Vitax	07402	08 Jun 2010
oxadiazon			
Festival	Bayer Environ.	14606	
Noble Oxadiazon	Barclay	12889	
paclobutrazol			
Carousel	Fine	13225	
Pirouette	Fine	13137	
penconazole			
RouteOne Pencon 10	Albaugh UK	13651	
pencycuron			
Me2 Penny	Me2	10369	
Monceren Flowable	Bayer CropScience	11425	
Tubercare 12.5 DS	AgriChem BV	12194	
pendimethalin			
Akolin 330 EC	Makhteshim	14005	31 Dec 2012
Akolin 330 EC	Aako	14282	
Bunker	Makhteshim	13816	18 Nov 2011
Claymore	BASF	13441	
CleanCrop Stomp	United Agri	13443	
Convey 330EC	AgriGuard	14535	
Routeone Penthal 330	Albaugh UK	13768	
Sovereign	BASF	13442	
pendimethalin + terbuthylazine			
Bullet XL	Makhteshim	14469	
Peniophora gigantea			
PG Suspension	Forest Research	11772	
phenmedipham			
Alpha Phenmedipham 320 SC	Makhteshim	12320	31 Aug 2010
Beetup Flo	United Phosphorus	14328	
Beetup Flo 160	United Phosphorus	10382	31 Aug 2010
Betamax	AgriGuard	12885	31 Aug 2010
Betasana SC	United Phosphorus	12353	28 Feb 2010
Crotale	Sipcam	11593	31 Aug 2010
Crotale	Sipcam	13909	28 Feb 2015
Dancer Flow	Sipcam	14395	28 Feb 2015
Galcon SC	Novastar	14607	
Pump	Makhteshim	14087	28 Feb 2015

Product	Approval holder	MAPP No.	Expiry Date
Pump	Makhteshim	12830	31 Aug 2010
Shrapnel	Makhteshim	13593	31 Aug 2010
Shrapnel	Makhteshim	14088	28 Feb 2015
picloram			
Pantheon	Pan Agriculture	13695	31 Dec 2010
RouteOne Loram 24	Albaugh UK	13951	
Tordon 22K	Dow	05083	
Tordon 22K	Dow	13869	31 Mar 2013
picolinafen			
AC 900001	BASF	10714	30 Sep 2012
picoxystrobin			
Acanto	DuPont	13043	
pinoxaden			
Rescue	Syngenta	14518	19 Sep 2011
Roaxe 100	RouteOne	13304	19 Sep 2011
RouteOne Roaxe 100	Albaugh UK	13655	19 Sep 2011
Symmetry	AgriGuard	13577	19 Sep 2011
pirimicarb			
Agrotech-Pirimicarb 50 WG	Agrotech-Trading	12269	
Cleancrop Miricide	United Agri	11776	
Milentus Pirimicarb	Milentus	12268	
RouteOne Primro 50 WG	Albaugh UK	13644	31 Jan 2012
pirimiphos-methyl			
Actellic Smoke Generator No. 10	Syngenta	10448	
prochloraz			
Alpha Prochloraz 40 EC	Makhteshim	11002	
Barclay Eyetak 40	Barclay	11352	
Cloraz 400	Goldengrass	14657	
Mirage 45 SC	Makhteshim	10894	
Octave	BASF	11741	
Piper	Nufarm UK	13793	
Prelude 20 LF	BASF	11904	
Sporgon 50 WP	Sylvan	08802	
Sportak 45 EW	BASF	11696	
Sportak 45 HF	BASF	11697	
prochloraz + propiconazole			
Mambo	Headland Amenity	13387	
prochloraz + tebuconazole			
Orius P	Makhteshim	12880	
System Turf	AgriGuard	14538	

Product	Approval holder	MAPP No.	Expiry Date
propachlor			
Alpha Propachlor 50 SC	Makhteshim	13585	18 Mar 2010
Alpha Propachlor 65 WP	Makhteshim	04807	18 Mar 2010
Brasson	Monsanto	10560	18 Mar 2010
Ramrod 20 Granular	Monsanto	10313	18 Mar 2010
Ramrod Flowable	Monsanto	10314	18 Mar 2010
Tripart Sentinel	Tripart	03250	18 Mar 2010
Tripart Sentinel 2	Tripart	05140	18 Mar 2010
propamocarb hydrochloride			
Previcur N	Bayer CropScience	08575	
propaquizafop			
Agil	Makhteshim	11048	31 Jul 2013
Agil	Makhteshim	14562	
Alpha Propaquizafop 100 EC	Makhteshim	13689	31 Jul 2013
Alpha Propaquizafop 100 EC	Makhteshim	14563	
Barclay Rebel	Barclay	11648	
Landgold PQF 100	Teliton	12127	31 Mar 2013
Landgold PQF 100	Landgold	10763	31 May 2012
Noble Propaquizafop	Barclay	12890	
Osprey	Makhteshim	13731	
PQF 100	Goldengrass	14414	
RouteOne Quizro 10	Albaugh UK	13643	
propyzamide			
Edge 400	Albaugh UK	14683	14 Sep 2013
Kerb Pro Granules	Barclay	14280	05 Oct 2013
Ponder	ChemSource	14693	31 Mar 2014
Precis	Dow	14202	31 Mar 2014
RouteOne Zamide Flo	Albaugh UK	14559	31 Mar 2014
Setanta Flo	AgriGuard	14678	13 Sep 2013
Solitaire	AgriGuard	14698	13 Sep 2013
Stroller	AgChem Access	14603	31 Mar 2014
Verdah 400	DAPT	14680	31 Mar 2014
Verge 400	Albaugh UK	14682	14 Sep 2013
Zamide 80 WG	Albaugh UK	14723	31 Mar 2014
Zamide Flo	Albaugh UK	14679	14 Sep 2013
prosulfocarb			
Arcade	Syngenta	12641	
RouteOne Rosulfocarb 80	Albaugh UK	13635	
prothioconazole			
Banguy De	PSI	13575	30 Apr 2013
Prothro 25	RouteOne	13392	30 Apr 2013
RouteOne Prothro 25	Albaugh UK	13658	31 Jan 2011

SECTION 3

EXPIRY DATE IS 31/12/2013 UNLESS OTHERWISE SHOWN

Product	Approval holder	MAPP No.	Expiry Date
prothioconazole + spiroxamine			
RouteOne Prothioxamine	Albaugh UK	13656	31 Jan 2011
Pseudomonas chlororaphis MA 342			
Cerall	Chemtura	14546	11 Jun 2013
pyraclostrobin			
BAS 500 06F	BASF	12338	31 May 2014
BASF Insignia	BASF	11900	31 May 2014
Comet	BASF	10875	31 May 2014
Platoon	BASF	12325	31 May 2014
pyraflufen-ethyl			
OS159	Ceres	12723	31 Oct 2011
pyrethrins			
Pyrethrum 5 EC	Gooddeed	12638	31 Oct 2010
pyrimethanil			
Pyrus 400 SC	Agriphar	13286	31 May 2012
RouteOne Pyrimet 40	Albaugh UK	13574	
Standon Pyrimethanil	Standon	10576	
pyroxsulam			
GF 1274	Dow	14089	05 Aug 2012
quinoxyfen			
Clean Crop QFN	United Agri	11966	31 Mar 2013
quizalofop-P-ethyl			
Sceptre	Bayer CropScience	08043	
quizalofop-P-tefuryl			
Rango	Certis	14039	
rimsulfuron			
Agrotech-Rimsulfuron 25 WG	Agrotech-Trading	12427	
RouteOne Empera 25	Albaugh UK	13823	31 Jan 2012
rotenone			
Devcol Liquid Derris	Nehra	06063	31 Oct 2011
Liquid Derris	Law Fertilisers	01213	31 Oct 2011
Liquid Derris	Law Fertilisers	13370	31 Oct 2011
S-metolachlor			
Dual Gold	Syngenta	14649	31 Mar 2015
sodium chlorate			
Atlacide Soluble Powder Weedkiller	Nomix Enviro	13863	10 May 2010
Aztec Complete Weedkiller	Aztec Chemicals	12485	10 May 2010

Product	Approval holder	MAPP No.	Expiry Date
Blagden Professional Sodium Chlorate Weedkiller	Blagden	12624	10 May 2010
Blitzweed	Micromix	12393	10 May 2010
Cooke's Professional Sodium Chlorate Weedkiller with Fire Depressant	Cooke's Chemicals	06796	10 May 2010
Deadfast Sodium Chlorate Weed Killer - Professional Use	Growing Success	12510	10 May 2010
Deosan Chlorate Weedkiller (Fire Suppressed)	JohnsonDiversey	08521	10 May 2010
Gem Sodium Chlorate Weedkiller	Metcalf	04276	10 May 2010
Growing Success Sodium Chlorate Weedkiller	Growing Success	10787	10 May 2010
Sodium Chlorate	Marlow Chemical	06294	10 May 2010
Strathclyde Sodium Chlorate Weedclear	Dremm	12015	10 May 2010
t.w.k	Apex	12453	10 May 2010
spiroxamine			
Torch	Bayer CropScience	11258	31 Dec 2011
Zenon	Bayer CropScience	11232	31 Dec 2011
spiroxamine + tebuconazole			
Draco	Bayer CropScience	11267	
sulfuryl fluoride			
Profume	Dow	12035	30 Apr 2013
sulphur			
Cosavet DF	Headland	11477	
Headland Venus	Headland	10611	
tau-fluvalinate			
Alpha Tau-Fluvalinate 240 EW	Makhteshim	13605	
tebuconazole			
Agrotech-Tebuconazole 250 EW	Agrotech-Trading	12523	
Aurigen	Syngenta	12043	
Barclay Busker	Barclay	11345	
Bayer UK226	Bayer CropScience	12412	
CHA 1640	Headland	14583	
Flail	Headland	12728	
I.T. Tebu	I T Agro	12882	31 Oct 2010
Manicure	Chem-Wise	13906	
Me2 Tebuconazole	Me2	09751	
Noble Tebuconazole	Barclay	12874	
Odin	Rotam Agrochemical	13468	
Raxil	Bayer CropScience	11295	
RouteOne Tebbro 25	RouteOne	13647	

SECTION 3

Product	Approval holder	MAPP No.	Expiry Date
Tebucur 250	Globachem	13975	
Toledo	Rotam Agrochemical	14036	
tebuconazole + triadimenol			
Veto F	Bayer CropScience	11317	30 Nov 2012
tebuconazole + triazoxide			
Raxil S	Bayer CropScience	11296	
tebuconazole + trifloxystrobin			
Eradicate	Pan Agriculture	13986	
Mascot Fission	Rigby Taylor	14114	
Mascot Indicate	Rigby Taylor	14057	
Pan Pandora	Pan Agriculture	14115	
tebufenpyrad			
Masai G	BASF	10224	
tefluthrin			
Force ST	Bayer CropScience	11671	
tepraloxydim			
Landgold Tepraloxydim	Teliton	13379	31 May 2015
Omara	Chem-Wise	13682	31 May 2015
RouteOne Tepralox 50	Albaugh UK	13667	31 May 2015
tetraconazole			
Domark	Isagro	13048	
Juggler	Sipcam	13077	
thifensulfuron-methyl			
Harmony SX	DuPont	12181	30 Jun 2012
Prospect SX	DuPont	12212	30 Jun 2012
thifensulfuron-methyl + tribenuron-methyl			
Calibre	DuPont	07795	31 Aug 2011
thiram			
Thianosan DG	Unicrop	13404	
tolclofos-methyl			
Rizolex 50 WP	Interfarm	13592	31 Oct 2012
Rizolex 50 WP	Interfarm	14217	
Rizolex Flowable	Interfarm	14207	
tralkoxydim			
Alpha Tralkoxydim 250 SC	Makhteshim	12884	
Landgold Tralkoxydim	Landgold	08604	31 May 2012
Landgold Tralkoxydim	Teliton	12130	

Product	Approval holder	MAPP No.	Expiry Date
triadimenol			
Bayfidan	Bayer CropScience	11417	
Hi-Shot	BASF	10397	
Spinnaker	BASF	10398	
tri-allate			
Avadex BW Granular	Gowan	12104	
tribenuron-methyl			
Quantum	DuPont	06270	
Quantum PX	DuPont	10843	31 Aug 2011
Quantum SX	DuPont	12239	
Triad	Headland	12751	
triclopyr			
Altix 240 EC	Agriphar	13362	
Cleancrop Triptic 48 EC	United Agri	11568	
Cleancrop Unival	United Agri	11388	
Garlon 4	Dow	05090	
Nomix Garlon 4	Nomix Enviro	13868	
Nomix Garlon 4	Nomix Enviro	12081	31 Jan 2012
trifloxystrobin			
Flint	Bayer CropScience	11259	30 Sep 2013
Pan Tees	Pan Agriculture	12894	30 Sep 2013
Twist	Bayer CropScience	11230	30 Sep 2013
Twist 500 SC	Bayer CropScience	11231	30 Sep 2013
Zest SC	Bayer CropScience	11233	30 Sep 2013
triflusulfuron-methyl			
Debut WSB	DuPont	07809	
Landgold TFS 50	Landgold	08941	31 May 2012
Standon Triflusulfuron	Standon	09487	
trinexapac-ethyl			
Cleancrop Alatrin	United Agri	11805	
Cutaway	Syngenta	14445	30 Apr 2017
Groslow	Pan Amenity	13694	30 Apr 2012
Landgold Tacet	Teliton	13036	31 Mar 2013
Peg	Goldengrass	14416	
RouteOne Rotrinex 25	RouteOne	13645	
Shortcut	Syngenta	13320	
Standon Pygmy	Standon	13001	
zeta-cypermethrin			
RouteOne Zeta 10	Albaugh UK	13608	

SECTION 4
ADJUVANTS

Adjuvants

Adjuvants are not themselves classed as pesticides and there is considerable misunderstanding over the extent to which they are legally controlled under the Food and Environment Protection Act. An adjuvant is a substance other than water which enhances the effectiveness of a pesticide with which it is mixed. Consent C(i)5 under the Control of Pesticides Regulations allows that an adjuvant can be used with a pesticide only if that adjuvant is authorised and on a list published from time to time in *The Pesticides Register*. An authorised adjuvant has an *adjuvant number* and may have specific requirements about the circumstances in which it may be used.

Adjuvant product labels must be consulted for full details of authorised use, but the table below provides a summary of the label information to indicate the area of use of the adjuvant. Label precautions refer to the keys given in Appendix 4, and may include warnings about products harmful or dangerous to fish. The table includes all adjuvants notified by suppliers as available in 2009.

Product	**Supplier**	**Adj. No.**	**Type**
Abacus	De Sangosse	A0543	vegetable oil
Contains	53.43% esterified rapeseed oil, 20% w/w alcohol ethoxylate and 9.0% w/w natural fatty acids		
Use with	All approved pesticides on all edible crops when used at half their recommended dose or less, and on all non-edible crops up their full recommended dose. Also at a maximum concentration of 0.1% with approved pesticides on listed crops up to specified growth stages		
Protective clothing	A, C, H		
Precautions	R36, U05a, U11, U14, U19a, U20b, E15a, E19b, E34, D01, D02, D05, D10a, D12a, H04		
Activator 90	De Sangosse	A0547	non-ionic surfactant/wetter
Contains	750 g/l alcohol ethoxylates and 150 g/l natural fatty acids		
Use with	All approved pesticides on all edible crops when used at half their recommended dose or less, and on all non-edible crops up their full recommended dose. Also at a maximum concentration of 0.1% with approved pesticides on listed crops up to specified growth stages		
Protective clothing	A, C, H		
Precautions	R36, R38, R53a, U02a, U04a, U05a, U10, U11, U20b, E15a, E19b, D01, D02, D05, D10a, H04		
Adigor	Syngenta	A0522	wetter
Contains	47% w/w methylated rapeseed oil		
Use with	Topik, Axial, Trazos, Amazon and Viscount on cereals in accordance with recommendations on the respective herbicide labels		
Protective clothing	A, C, H		
Precautions	R43, R51, R53a, U02a, U05a, U09a, U20b, E15b, E38, D01, D02, D05, D09a, D10c, D12a, H04, H11		

Product	Supplier	Adj. No.	Type
Admix-P	De Sangosse	A0301	wetter
Contains	80% w/w polyalkylene oxide modified heptamethyltrisiloxane and a maximum of 20 % w/w allyloxypolyethylene glycol methyl ether		
Use with	A wide range of pesticides applied as corm, tuber, onion and other bulb treatments in seed production and in seed potato treatment		
Protective clothing	A, C, H		
Precautions	R20, R21, R22a, R36, R43, R48, R51, R58, U11, U15, U19a, E15a, E19b, D01, D02, D05, D10a, D12a, H03, H11		
Agropen	Intracrop	A0227	vegetable oil
Contains	95% refined rapeseed oil		
Use with	Recommended rates of approved pesticides up to the growth stage indicated for specified crops, and with half or less than the recommended rate on these crops after the stated growth stages. Also for use with pesticides on grassland at half their recommended rate		
Protective clothing	A, C		
Precautions	U08, U20b, E13c, E37, D09a, D10b		
Amber	Interagro	A0367	vegetable oil
Contains	95% w/w methylated rapeseed oil		
Use with	Sugar beet herbicides, oilseed rape herbicides, cereal graminicides and a wide range of other pesticides that have a label recommendation for use with authorised adjuvant oils on specified crops. See label for details		
Protective clothing	A, C		
Precautions	U05a, U20b, E15a, D01, D02, D05, D09a, D10a		
Anoint	Intracrop	A0236	spreader/vegetable oil/wetter
Contains	95% rapeseed oil		
Use with	Pesticides that have a recommendation for the addition of a wetter/ spreader		
Protective clothing	A, C		
Precautions	R36, R38, U05a, U08, U19a, U20b, E13c, E34, D01, D02, D09a, D10a, H04		
Arma	Interagro	A0306	penetrant
Contains	500 g/l alkoxylated fatty amine + 500 g/l polyoxyethylene monolaurate		
Use with	Cereal growth regulators, cereal herbicides, cereal fungicides, oilseed rape fungicides and a wide range of other pesticides on specified crops		
Protective clothing	A, C		
Precautions	R51, R58, U05a, E15a, E34, E37, D01, D02, D05, D09a, D10a, H11		
Asu-Flex	Greenaway	A0677	spreader/sticker/wetter
Contains	10.0 % w/w rapeseed oil		
Use with	Asulox, Greencrop Found, I T Asulam, Inter Asulam and Spitfire		
Protective clothing	A, C		
Precautions	U11, U12, U15, U20b, E15a, E34, D02		

Product	Supplier	Adj. No.	Type
BackRow	Interagro	A0472	mineral oil
Contains	60% w/w refined paraffinic petroleum oil		
Use with	Pre-emergence herbicides. Refer to label or contact supplier for further details		
Protective clothing	A		
Precautions	R22b, R38, U02a, U05a, U08, U20b, E15a, D01, D02, D05, D09a, D10a, M05b, H03		
Ballista	Interagro	A0524	spreader/wetter
Contains	10% w/w alkoxylated triglycerides		
Use with	All approved pesticides and plant growth regulators for use in winter and spring cereals applied at full rate up to and including GS 52, and at half the approved rate thereafter. May be applied with approved pesticides for other specified crops at full rate up to specified growth stages and at half rate thereafter		
Protective clothing	A		
Precautions	U05a, U20b, E15a, E34, D01, D02, D09a		
Bandrift Plus	Ciba Specialty	xx	drift retardant
Contains	A non-ionic polyamide dispersed in oil		
Use with	A wide range of pesticides. See label for restrictions on use with wettable powders, suspension concentrates and water dispersible granules, and other usage limitations		
Protective clothing	A, C, M		
Precautions	U05a, U08, U20a, E13c, E34, D01, D02, D09a		
Banka	Interagro	A0245	spreader/wetter
Contains	29.2% w/w alkyl pyrrolidones		
Use with	Potato fungicides and a wide range of other pesticides on specified crops		
Protective clothing	A, C		
Precautions	R38, R41, R52, R58, U02a, U05a, U11, U19a, U20b, E15a, E34, E37, D01, D02, D05, D09a, H04		
Barclay Clinger	Barclay	A0198	sticker/wetter
Contains	96% w/v poly-1-p-menthene		
Use with	Barclay Gallup, Barclay Gallup Amenity and a wide range of approved fungicides and insecticides. See label for details		
Protective clothing	A		
Precautions	U19a, U20b, E13c, D09a, D10a		
Barramundi	Interagro	A0376	mineral oil
Contains	95% w/w mineral oil		
Use with	Sugar beet herbicides, oilseed rape herbicides, cereal graminicides and a wide range of other pesticides that have a label recommendation for use with authorised adjuvant oils on specified crops. Refer to label or contact supplier for further details		
Protective clothing	A		
Precautions	R22b, R38, U02a, U05a, U08, U20b, E15a, D01, D02, D05, D09a, D10a, M05b, H03		

Product	Supplier	Adj. No.	Type
Bazuka	Intracrop	A0241	spreader/wetter
Contains	800 g/l polyalkyleneoxide modified heptamethyltrisiloxane		
Use with	All fungicides applied to spring and winter sown wheat, spring and winter sown barley, spring and winter sown oats, triticale, rye and durum wheat where the pesticide label recommends the addition of wetters and spreaders		
Protective clothing	A, C		
Precautions	R21, R22a, R43, U02a, U05a, U08, U19a, U20b, E13c, E34, D01, D02, D09a, D10a, M03, H03, H04		
Binder	Nufarm UK	A0598	spreader/wetter
Contains	30% alkyl ethoxylate		
Use with	Recommended rates of approved pesticides up to the growth stage indicated for specified crops, and with half or less than the recommended rate on these crops after the stated growth stages		
Protective clothing	A		
Precautions	R41, R52, R53a, U02a, U05a, U08, U11, U13, U15, U20b, E34, E37, D01, D09a, M03, H04		
Bio Syl	Intracrop	A0385	spreader/sticker/wetter
Contains	1% w/w polyoxyethylen-alpha-methyl-omega-[3-(1,3,3,3-tetramethyl-3-trimethylsiloxy)-disiloxanyl] propylether and 32.67% w/w ethylene oxide condensate		
Use with	Recommended rates of approved pesticides on a range of specified fruit and vegetable crops and on managed amenity turf. See label for details		
Protective clothing	A, C		
Precautions	R36, R38, U05a, U08, U20c, E15a, D01, D02, D10a, H04		
Bioduo	Intracrop	A0606	wetter
Contains	85% alkyl polyglycol ether and fatty acids		
Use with	A wide range of pesticides used in grassland, agriculture and horticulture and with pesticides used in non-crop situations		
Protective clothing	A, C		
Precautions	R22a, R36, R38, U05a, U08, U20c, E13c, E34, D01, D02, D09a, D10a, M03, H03, H04, H08		
Biofilm	Intracrop	A0634	sticker/wetter
Contains	96% poly-1-p-menthene		
Use with	All approved fungicides and insecticides on edible crops and all approved formulations of glyphosate used pre-harvest on wheat, barley, oilseed rape, stubble, and in non-crop situations and grassland destruction		
Protective clothing	A, C		
Precautions	U19a, U20c, E13c, D09a, D11a		

Product	Supplier	Adj. No.	Type
BioPower	Bayer CropScience	A0617	wetter
Contains	6.7% w/w 3,6-dioxaeicosylsulphate sodium salt and 20.2% w/w 3,6-dioxaoctadecylsulphate sodium salt		
Use with	Atlantis and all other approved cereal herbicides		
Protective clothing	A, C		
Precautions	R36, R38, U02a, U05a, U08, U13, U19a, U20b, E13c, E34, D01, D02, D05, D09a, D10a, H04		
Biothene	Intracrop	A0633	sticker/wetter
Contains	96% poly-1-p-menthene		
Use with	All approved fungicides and insecticides on edible crops up to 30 d before harvest, and all approved formulations of glyphosate used pre-harvest on wheat, barley, oilseed rape, stubble, and in non-crop situations and grassland destruction. Must not be used in mixture with adjuvant oils or surfactants		
Precautions	U19a, U20c, E13c, D09a, D11a		
Bond	De Sangosse	A0556	extender/sticker/wetter
Contains	900 g/l synthetic latex solution and 101 g/l alcohol alkoxylate		
Use with	All approved potato blight fungicides. Also with all approved pesticides on all edible crops when used at half their recommended dose or less, and on all non-edible crops up their full recommended dose. Also at a maximum concentration of 0.14% with approved pesticides on listed crops up to specified growth stages		
Protective clothing	A, C, H		
Precautions	R36, R38, U11, U14, U16b, U19a, E15a, E19b, D01, D02, D05, D10a, D12a, H04		
Bonser	Nufarm UK	A0601	spreader/wetter
Contains	258 g/l ammonium sulphate, 15.0 g/l modified soya lecithin, 15.0 g/l propionic acid and 4.0 g/l polyoxyethylene (5-7EO) C10-C15 primary alcohol		
Use with	Credit DST (MAPP 13822) on specified crops, and with all approved pesticides on non-edible crops at up to their full dose, and with all approved pesticides on edible crops at up to half their approved dose		
Break-Thru S 240	PP Products	A0431	wetter
Contains	85% w/w polyalkylene oxide modified heptamethyl trisiloxane		
Use with	All approved pesticides which have a recommendation for use with a wetting agent applied to cereals, permanent grassland and rotational grass. Also with all approved pesticides at full rate on non-edible crops and on listed edible crops up to specified growth stages and on all edible crops when used at half or less of their approved full rate		
Protective clothing	A, C		
Precautions	R20, R21, R38, R41, R51, R53a, U11, U16b, U19a, E19b, D01, D02, D09a, D10a, M05a, H03, H11		

Product	Supplier	Adj. No.	Type
Broad-Flex	Greenaway	A0680	spreader/sticker/wetter
Contains	10.0 % w/w rapeseed oil		
Use with	Broad Sword or Green Guard		
Protective clothing	A, C		
Precautions	U11, U12, U15, U20b, E15a, E34, D02		
Byo-Flex	Greenaway	A0545	sticker/wetter
Contains	10.0 % w/w rapeseed oil		
Use with	GLY 490 (MAPP 12718)		
Protective clothing	A, C		
Precautions	U11, U12, U15, U20b, E34		
C-Cure	Interagro	A0467	mineral oil
Contains	60% w/w refined mineral oil		
Use with	Pre-emergence herbicides		
Protective clothing	A		
Precautions	R22b, R38, U02a, U05a, U08, U20b, E15a, D01, D02, D05, D09a, D10a, M05b, H03		
Ceres Platinum	Interagro	A0445	penetrant/spreader
Contains	500 g/l alkoxylated fatty amine + 500 g/l polyoxyethylene monolaurate		
Use with	Cereal growth regulators, cereal herbicides, cereal fungicides, oilseed rape fungicides and a wide range of other pesticides on specified crops		
Precautions	R51, R58, U05a, E15a, E34, D01, D02, D05, D09a, D10a, H11		
Cerround	Helena	A0510	adjuvant/vegetable oil
Contains	80% w/w methylated soybean oil and 15% w/w polyalkylene oxide modified heptamethyl trisiloxane		
Use with	Approved pesticides on non-edible crops, cereals and stubbles of all edible crops, grassland (destruction), and pesticides approved for use on growing edible crops when used at half recommended dose or less. On specified crops product may be used at a maximum spray concentration of 0.5% with approved pesticides at their full approved rate up to the growth stages shown in the label		
Protective clothing	A, H		
Precautions	R38, R41, U02a, U05a, U08, U11, U19a, U20b, E13c, E34, E37, D01, D02, D05, D09a, M03, H04		
Codacide Oil	Microcide	A0629	vegetable oil
Contains	95% emulsified vegetable oil		
Use with	All approved pesticides and tank mixes. See label for details		
Protective clothing	A, C		
Precautions	U20b, D09a, D10b		
Companion PCT12	Ciba Specialty	A0482	spreader/sticker/wetter
Contains	25% w/w polyacryalmide		
Use with	Approved herbicides on cereals and any pesticide on non-food crops. See label for restrictions on use with wettable powders, suspension concentrates and water dispersible granules, and other usage limitations		
Precautions	U05a, U08, E13c, D01, D02, D09a		

Product	Supplier	Adj. No.	Type
Compliment	United Agri	A0578	spreader/vegetable oil/wetter
Contains	75% mixed fatty acid esters of rapeseed oil		
Use with	All approved pesticides in non-edible crops, all approved pesticides on edible crops when used at half or less their recommended rate, morpholine or triazine fungicides on cereals, and with all pesticides on listed crops up to specified growth stages		
Protective clothing	A, C		
Precautions	R43, R52, R53a, U14, E38, D01, D05, D07, D08, H04		
Conjoin	AgriGuard	A0676	adjuvant
Contains	47% w/w methylated rapeseed oil		
Use with	Amazon, Symmetry, Topik, Traxos and Viscount		
Protective clothing	A, C, H		
Precautions	R43, R50, R53a, U02a, U05a, U08, U20b, E13c, E15b, D01, D02, D09a, D10c, H04, H11		
Contact Plus	Interagro	A0418	mineral oil
Contains	95% w/w mineral oil		
Use with	Sugar beet herbicides, oilseed rape herbicides, cereal graminicides and a wide range of other pesticides that have a label recommendation for use with authorised adjuvant oils on specified crops. Refer to label or contact supplier for further details		
Protective clothing	A, C		
Precautions	R22b, R38, U02a, U05a, U08, U20b, E15a, D01, D02, D09a, D10a, M05b, H03		
Cropspray 11-E	De Sangosse	A0537	adjuvant/mineral oil
Contains	99% highly refined paraffinic oil		
Use with	All approved pesticides on all edible crops when used at half their recommended dose or less, and on all non-edible crops up to their full recommended dose. Also with listed herbicides on a range of specified crops and with all pesticides on a range of specified crops up to specified growth stages.		
Protective clothing	A, C		
Precautions	R22a, U10, U16b, U19a, E15a, E19b, E34, E37, D01, D02, D05, D10a, D12a, M05b, H03		
Designer	De Sangosse	A0322	drift retardant/extender/sticker/wetter
Contains	8.44% w/w organosilicone wetter and 50% w/w synthetic latex		
Use with	A wide range of fungicides, insecticides and trace elements for cereals and specified agricultural and horticultural crops		
Protective clothing	A, C, H		
Precautions	R36, R38, R52, U02a, U11, U14, U15, U19a, E15a, E37, D01, D02, D05, D09a, D10a, H04		

SECTION 4

Product	Supplier	Adj. No.	Type
Desikote Max	Taminco	A0666	extender/sticker/wetter
Contains	400 g/l di-1-p-menthene		
Use with	Approved pesticides on all edible crops (except herbicides on peas) and as an anti-transpirant on vegetables before transplanting, evergreens, deciduous trees, shrubs, bushes and on turf		
Protective clothing	A		
Precautions	R38, R50, R53a, U05a, U14, U20b, E15a, E34, E37, D01, D09a, D10b, D12b, M03, H04, H11		
Diagor	AgChem Access	A0671	wetter
Contains	47% w/w methylated rapeseed oil		
Use with	Topik, Axial, Trazos, Amazon and Viscount on cereals in accordance with recommendations on the respective herbicide labels		
Protective clothing	A, C, H		
Precautions	R43, R51, R53a, U02a, U05a, U09a, U20b, E15b, E38, D01, D02, D05, D09a, D10c, D12a, H04, H11		
Drill	De Sangosse	A0544	adjuvant
Contains	63.34% w/w esterified rapeseed oil, 15% w/w alcohol ethoxylate and 7.5% w/w natural fatty acids		
Use with	All approved pesticides on all edible crops when used at half their recommended dose or less, and on all non-edible crops up to their full recommended dose. Also at a maximum concentration of 0.1% with approved pesticides on listed crops up to specified growth stages		
Protective clothing	A, C, H		
Precautions	R36, U05a, U11, U14, U19a, U20b, E15a, E19b, E34, D01, D02, D05, D10a, D12a, H04		
Elan	Intracrop	A0392	spreader/sticker/wetter
Contains	82% w/w polyoxyethylen-alpha-methyl-omega-disiloxanyl] propylether		
Use with	Recommended rates of approved pesticides up to the growth stage indicated for specified crops, and with half or less than the recommended rate on these crops after the stated growth stages. Also for use with recommended rates of herbicides on managed amenity turf		
Protective clothing	A, C, H		
Precautions	R36, R38, U05a, U08, U20c, E13c, E15a, E37, D01, D02, D09a, D10a, H04		
Emerald	Intracrop	A0636	anti-transpirant/extender
Contains	96% di-1-p-menthene		
Use with	Recommended rates of approved pesticides up to the growth stage indicated for specified crops, and with half or less than the recommended rate on these crops after the stated growth stages. Also for use alone on transplants, turf, fruit crops, glasshouse crops and Christmas trees. Must not be used in mixture with adjuvant oils or surfactants		
Precautions	U19a, U20c, E13c, E37, D09a, D11a		

Product	Supplier	Adj. No.	Type
Esterol	Interagro	A0330	vegetable oil
Contains	95% w/w methylated rapeseed oil		
Use with	Plant protection products that have a label recommendation for use with adjuvant oils. Refer to label or contact supplier for further details		
Precautions	U05a, U20b, E15a, D01, D02, D09a, D10a		
Euroagkem Pace	Intracrop	A0562	spreader/wetter
Contains	42% w/w propionic acid		
Use with	All approved formulations of chlormequat, all approved formulations of glyphosate, diquat, fenoxaprop-P-ethyl, tralkoxydim, clodinafop-propargyl, fluazifop-P-butyl, cycloxydim and propaquizafop at half or less than half the approved pesticide rate in edible crops, at full rate in non-edible crops.		
Protective clothing	A, C		
Precautions	U02a, U04a, U05a, U08, U10, U11, U13, U14, U15, U19a, U20b, E15a, D01, D02, D09a, D10a, M05a, H05		
Euroagkem Pen-e-trate	Intracrop	A0564	spreader/wetter
Contains	350 g/l propionic acid		
Use with	All approved pesticides which have a recommendation for use with a wetting agent; pesticides must be used at half approved dose rate or less on edible crops, up to full dose rate on non-edible crops		
Protective clothing	A, C		
Precautions	U05a, U08, U11, U14, U15, U19a, U20c, E13e, E15a, D01, D02, D09a, D10b, M03, H05		
Felix	Intracrop	A0178	spreader/wetter
Contains	60% ethylene oxide condensate		
Use with	Mecoprop, 2,4-D in cereals and amenity turf, and a range of grass weedkillers in agriculture. See label for details		
Protective clothing	A, C		
Precautions	R36, R38, U05a, U08, U20a, E15a, D01, D02, D09a, D10a, H04, H08		
Firebrand	Barclay	-	fertiliser/water conditioner
Contains	500 g/l ammonium sulphate		
Use with	glyphosate		
Frigate	Unicrop	A0325	sticker/wetter
Contains	800 g/l tallow amine ethoxylate		
Use with	Roundup formulations		
Protective clothing	A, C		
Precautions	R22a, R23, R41, R51, R58, R67, U05a, U11, U19a, E38, D01, D02, D12a, M04a, H02, H08, H11		

Product	Supplier	Adj. No.	Type
Gateway	United Agri	A0587	extender/sticker/wetter
Contains	73% w/v synthetic latex solution and 8.5% w/v polyether modified trisiloxane		
Use with	All approved pesticides in crops not destined for human or animal consumption, all approved pesticides on edible crops when used at half or less their recommended rate, and with all pesticides on listed crops up to specified growth stages		
Protective clothing	A, C, H		
Precautions	R41, R52, R53a, U11, U15, E37, E38, D01, D05, D07, D08, H04		
Gladiator	De Sangosse	A0542	penetrant/wetter
Contains	210 g/kg alkoxylated tallow amine, 380 g/kg alcohol ethoxylates, 75 g/l natural fatty acids and 210 g/kg polyalkylene glycol		
Use with	All approved pesticides on all edible crops when used at half their recommended dose or less, and on all non-edible crops up their full recommended dose. Also at a maximum concentration of 0.1% with approved pesticides on listed crops up to specified growth stages		
Protective clothing	A, C, H		
Precautions	R36, R38, R41, R52, R58, U05a, U10, U11, U19a, E15a, E19b, E34, E37, D01, D02, D05, D10a, D12a, M03, H04		
Gly-Flex	Greenaway	A0588	spreader/sticker/wetter
Contains	95% refined rapeseed oil		
Use with	GLY-490 (MAPP 12718) or any approved formulations of 490 g/l glyphosate		
Protective clothing	A, C		
Precautions	R36, U11, U12, U15, U20b, E13c, E34, E37, H04		
Gly-Plus A	Greenaway	A0377	spreader/sticker/wetter
Contains	95% refined rapeseed oil		
Use with	Any approved 360 g/l glyphosate		
Protective clothing	A, C		
Precautions	R36, U05a, U08, U11, U12, U15, U20b, E13c, E34, E37, E38, H04		
Green Gold	Intracrop	A0250	spreader/wetter
Contains	95% refined rapeseed oil		
Use with	All pesticides which have a recommendation for the addition of a wetter/spreader		
Precautions	U08, U20b, E13c, E34, D09a, D10b		
Greencrop Astra	Greencrop	A0417	adjuvant/penetrant/spreader
Contains	500 g/l alkoxylated fatty amine and 500 g/l polyoxylene monolaurate		
Use with	Approved pesticides on a range of specified arable crops, and in forestry, managed amenity turf, stubbles and non-crop areas and situations		
Precautions	U05a, E13c, E34, D01, D02, D05, D09a, D10b		

Product	Supplier	Adj. No.	Type
Greencrop Dolmen	Greencrop	A0419	adjuvant/spreader
Contains	64% w/w polyalkeneoxide modified heptamethyltrisiloxane		
Use with	Approved pesticides on a range of specified arable crops, and in forestry, managed amenity turf, stubbles and non-crop areas and situations		
Protective clothing	A, C, H		
Precautions	R21, R22a, R38, R41, R43, U02a, U05a, U08, U11, E13c, E34, D01, D02, D09a, D10b, M03, H03, H04		
Greencrop Fenit	Greencrop	A0602	wetter
Contains	47% methylated rapeseed oil		
Use with	Greencrop Helvick		
Precautions	R43, R50, R53a, U02a, U05a, U09a, U20b, E15b, D01, D02, D05, D09a, D10c, D12a, H04, H11		
Greencrop Rookie	Greencrop	A0515	wetter
Contains	20.2% w/w 3,6-dioxoctadecylsulphate sodium salt and 6.7% w/w 3,6-dioxaeicosylsulphate sodium salt		
Use with	All approved herbicides in cereals		
Protective clothing	A, C		
Precautions	R36, R38, R52, U02a, U05a, U08, U11, U13, U19a, E15a, D01, D02, D05, D09a, D10b, H04		
Grenadier	Intracrop	A0608	spreader/sticker/wetter
Contains	700 g/l polyoxyethylene (5-8 EO) C10-C15 primary alcohol and 150 g/l fatty acids		
Use with			
Grip	De Sangosse	A0558	extender/sticker/wetter
Contains	900 g/l synthetic latex solution and 101 g/l alcohol alkoxylate		
Use with	All approved potato blight fungicides. Also with all approved pesticides on all edible crops when used at half their recommended dose or less, and on all non-edible crops up their full recommended dose. Also at a maximum concentration of 0.14% with approved pesticides on listed crops up to specified growth stages		
Protective clothing	A, C, H		
Precautions	R36, R38, U11, U14, U16b, U19a, E15a, E19b, D01, D02, D05, D10a, D12a, H04		
Grounded	Helena	A0456	mineral oil
Contains	732 g/l refined paraffinic petroleum oil		
Use with	All approved pesticides on edible and non-edible crops when used at half their approved dose or less. Also with approved pesticides on specified crops, up to specified growth stages, at up to their full approved dose		
Protective clothing	A, C		
Precautions	R53a, U02a, U05a, U08, U20b, E15a, E34, E37, D01, D02, D05, D09a, D10c, H11		

Product	Supplier	Adj. No.	Type
Guide	De Sangosse	A0528	acidifier/drift retardant/penetrant
Contains	35% w/w lecithin, 35% w/w propionic acid and 9.39%w/w alcohol ethoxylate		
Use with	All approved pesticides on all edible crops when used at half their recommended dose or less, and on all non-edible crops up their full recommended dose. Also with morpholine fungicides on cereals and with pirimicarb on legumes when used at full dose; with iprodione on brassicas when used at 75% dose, and at a maximum concentration of 0.5% with all pesticides on listed crops at full dose up to specified growth stages		
Protective clothing	A, C, H		
Precautions	R36, R38, U11, U14, U15, U19a, E15a, E19b, D01, D02, D05, D10a, D12a, H04		
Headland Fortune	Headland	A0277	penetrant/spreader/vegetable oil/wetter
Contains	75% w/w mixed methylated fatty acid esters of seed oil and N-butanol		
Use with	Herbicides and fungicides in a wide range of crops. See label for details		
Protective clothing	A, C		
Precautions	R43, U02a, U05a, U14, U20a, E15a, E34, E38, D01, D02, D05, D09a, D10b, H04		
Headland Guard 2000	Headland	A0652	extender/sticker/wetter
Contains	10% w/w styrene/butadiene co-polymer		
Use with	All approved pesticides that recommend the use of a sticking, wetting or extending agent; also with all approved pesticides for use on crops not destined for human or animal consumption; also with all approved pesticides on beans, peas, edible podded peas, oilseed rape, linseed, sugar beet, cereals, maize, Brussels sprouts, potatoes, cauliflowers		
Precautions	U05a, U08, U19a, U20b, E13b, E34, E37, D01, D02, D05, D10b		
Headland Guard Pro	Headland Amenity	A0423	extender/sticker
Contains	52% synthetic latex solution		
Use with	All approved pesticides, micronutrients and sea-weed based plant growth stimulants in amenity grass, managed amenity turf and amenity vegetation		
Precautions	U05a, U08, U19a, U20b, E13b, E34, E37, D01, D02, D05, D10b		
Headland Inflo XL	Headland Amenity	A0329	wetter
Contains	85% w/w polyalkylene oxide modified heptamethyl siloxane		
Use with	Any pesticide for use on amenity grassland or managed amenity turf (except any product applied in or near water) where the use of a wetting, spreading and penetrating surfactant is recommended to improve foliar coverage, and with any approved pesticide on crops not intended for human or animal consumption		
Protective clothing	A, C		
Precautions	R21, R22a, R38, R41, R43, R51, R58, U02a, U05a, U08, U14, U19a, U20b, E15a, E34, E38, D01, D02, D05, D09a, D10b, M03, H03, H11		

Product	Supplier	Adj. No.	Type
Headland Intake	Headland	A0074	penetrant
Contains	450 g/l propionic acid		
Use with	All approved pesticides on any crop not intended for human or animal consumption, and with all approved pesticides on beans, peas, edible podded peas, oilseed rape, linseed, sugar beet, cereals (except triazole fungicides), maize, Brussels sprouts, potatoes, cauliflowers. See label for detailed advice on timing on these crops		
Protective clothing	A, C		
Precautions	R34, U02a, U05a, U10, U11, U14, U15, U19a, E34, D01, D02, D05, D09b, D10b, M04a, H05		
Headland Rhino	Headland	A0328	wetter
Contains	85% w/w polyalkylene oxide modified heptamethyl siloxane		
Use with	Any herbicide, systemic fungicide, systemic insecticide or plant growth regulator (except any product applied in or near water) where the use of a wetting, spreading and penetrating surfactant is recommended to improve foliar coverage		
Protective clothing	A, C		
Precautions	R21, R22a, R38, R41, R43, R51, R58, U02a, U05a, U08, U11, U14, U19a, U20b, E13c, E34, D01, D02, D05, D09a, D10b, M03, H03, H11		
Holster	Greenaway	A0682	spreader/sticker/wetter
Contains	95% refined rapeseed oil		
Use with	Any approved herbicide		
Protective clothing	A, H, M		
Precautions	U11, U12, U15, U20b, E15a, E34, D02		
Impala	Intracrop	A0607	spreader/sticker/wetter
Contains	700 g/l polyoxyethylene (5-8 EO) C10-C15 primary alcohol and 150 g/l fatty acids		
Use with			
Intracrop Agwet	Intracrop	A0616	sticker/wetter
Contains	900 g/l polyoxyethylene (5-8 EO) C10-C15 primary alcohol		
Use with			
Intracrop Agwet Gtx	Intracrop	A0646	sticker/wetter
Contains	500 g/l polyoxyethylene (5-8 EO) C10-C15 primary alcohol		
Use with			
Intracrop Archer	Intracrop	A0242	spreader/wetter
Contains	800 g/l polyalkyleneoxide modified heptamethyltrisiloxane		
Use with	All fungicides used in cereals where the addition of wetters/spreaders is recommended on the pesticide label		
Protective clothing	A, C		
Precautions	R21, R22a, R43, U02a, U05a, U08, U19a, U20b, E13c, E34, D01, D02, D09a, D10a, M03, H03, H04		

SECTION 4

Product	Supplier	Adj. No.	Type
Intracrop BLA	Intracrop	A0453	extender/sticker
Contains	41.5% w/w styrene butadiene co-polymer solution		
Use with	All potato blight fungicides. Also with recommended rates of approved pesticides in certain non-crop situations and up to the growth stage indicated for specified crops, and with half or less than the recommended rate on these crops after the stated growth stages. Also with pesticides in grassland at half or less their recommended rate		
Precautions	U08, U20b, E13c, E34, E37, D01, D05, D09a, D10a		
Intracrop Bla-Tex	Intracrop	A0656	sticker
Contains	22.0 % w/w styrene-butadiene co-polymer		
Use with			
Intracrop Dictate	Intracrop	A0673	adjuvant
Contains	91% w/w methylated rapeseed oil		
Use with	All approved pesticides on non-edible crops; all approved pesticides applied at half or less than half dose on edible crops		
Protective clothing	A, C		
Precautions	U19a, U20b, D05, D09a, D10b		
Intracrop Era	Intracrop	A0664	spreader/sticker/wetter
Contains	27% w/w polyoxyethylen- -methyl- -[3-(1,3,3,3-tetramethyl-3-trimethylsiloxy) disiloxanyl] propylether		
Use with	All approved pesticides at half or less than half the approved rate		
Protective clothing	A, C		
Precautions	U05a, U08, U20b, E13e, D01, D02, D09a, D10a		
Intracrop Green Oil	Intracrop	A0262	wetter
Contains	95% refined rapeseed oil		
Use with	Recommended rates of approved pesticides up to the growth stage indicated for specified crops, and with half or less than the recommended rate on these crops after the stated growth stages. Also for use with pesticides on grassland at half their recommended rate		
Precautions	U08, U20b, E13c, E37, D09a, D10b		
Intracrop Impetus	Intracrop	A0647	spreader/wetter
Contains	50.0 % w/w polyoxyethylene (5-8 EO) C10-C15 primary alcohol		
Use with			
Intracrop Iona	Intracrop	A0583	spreader/wetter
Contains	90.0 % w/w polyoxyethylene (3EO) C12-C15 primary alcohol		
Use with			
Intracrop Iona Low Foam	Intracrop	A0582	spreader/wetter
Contains	89.8 % w/w polyoxyethylene (3EO) C12-C15 primary alcohol		
Use with			

Product	Supplier	Adj. No.	Type
Intracrop Mica	Intracrop	A0663	spreader/sticker/wetter
Contains	27% w/w polyoxyethylen- -methyl- -[3-(1,3,3,3-tetramethyl-3-trimethylsiloxy) disiloxanyl] propylether		
Use with	All approved pesticides at half or less than half the approved rate		
Protective clothing	A, C		
Precautions	U05a, U08, U20c, E13e, D01, D02, D09a, D10a		
Intracrop Micotex	Intracrop	A0688	activator/spreader/wetter
Contains	25.0% w/w polyoxyethylene (12E))/polyoxypropylene (30EO) block copolymer		
Use with	All approved pesticides on non-edible crops and non crop production uses; All approved pesticides at half or less than half the approved pesticide rate on edible crops		
Protective clothing	A, C		
Precautions	R36, R38, U05b, U08, U20c, E15a, D01, D02, D09a, D10b, H04		
Intracrop Neotex	Intracrop	A0460	extender/sticker
Contains	42.5% w/w styrene butadiene co-polymer solution		
Use with	Recommended rates of approved pesticides up to specified growth stages of a wide range of agricultural arable and horticultural crops, and for non-crop uses. Use on edible crops beyond specified growth stages, and in grass, should only be with half recommended rates of the pesticide or less. See label for details of growth stage restrictions. In addition may be used with all potato blight fungicides at their recommended rates of use up to the latest recommended timing of the fungicide		
Precautions	U08, U20b, E13c, E34, E37, D01, D05, D09a, D10a		
Intracrop Novatex	Intracrop	A0566	extender/sticker
Contains	42.5% w/w styrene butadiene co-polymer solution		
Use with	Recommended rates of approved pesticides up to the growth stage indicated for specified crops, and with half or less than the recommended rate on these crops after the stated growth stages. Also for use with recommended rates of pesticides on grassland and specified non-crop situations		
Precautions	U08, U20b, E13c, E34, E37, D01, D09a, D10a		
Intracrop Perm-E8	Intracrop	A0565	spreader/wetter
Contains	42% w/w propionic acid		
Use with	All approved formulations of chlormequat, all approved formulations of glyphosate, diquat, fenoxaprop-P-ethyl, tralkoxydim, clodinafop-propargyl, fluazifop-P-butyl, cycloxydim and propaquizafop at half or less than half the approved pesticide rate in edible crops, at full rate in non-edible crops.		
Protective clothing	A, C		
Precautions	U02a, U04a, U05a, U08, U10, U11, U13, U14, U15, U19a, U20b, D01, D02, D09a, D10a, H05		

Product	Supplier	Adj. No.	Type
Intracrop Predict	Intracrop	A0503	adjuvant/vegetable oil
Contains	91% w/w methylated rapeseed oil		
Use with	All approved pesticides on non-edible crops, and all pesticides approved for use on growing edible crops when used at half recommended dose or less. On specified crops product may be used at a maximum spray concentration of 1% with approved pesticides at their full approved rate up to the growth stages shown in the label		
Precautions	U19a, U20b, D05, D09a, D10b		
Intracrop Protocol	Intracrop	A0687	spreader/wetter
Contains	25.0% w/w polyoxyethylene (12E))/polyoxypropylene (30EO) block copolymer		
Use with	All approved pesticides on all non-edible crops and non crop production; all approved pesticides at half or less than half the approved pesticide rate on all edible crops		
Protective clothing	A, C		
Precautions	R36, R38, U05a, U08, U20c, E15a, D01, D02, D09a, D10b, H04		
Intracrop Questor	Intracrop	A0495	activator/non-ionic surfactant/spreader
Contains	75% polyoxyethylene polypropoxypropanol		
Use with	All approved pesticides on non-edible crops and pesticides used in non-crop production, and all pesticides approved for use on growing edible crops when used at half recommended dose or less. On specified crops product may be used at a maximum spray concentration of 0.3% with approved pesticides at their full approved rate up to the growth stages shown in the label		
Protective clothing	A, C		
Precautions	R36, R38, U04a, U05a, U08, U19a, E13b, D01, D02, D09a, D10b, M03, M05a, H04		
Intracrop Rapeze	Intracrop	A0643	vegetable oil
Contains	842 g/l methylated rapeseed oil		
Use with			
Intracrop Rapide	Intracrop	A0563	penetrant/surfactant
Contains	40% propionic acid		
Use with	A wide range of pesticides and growth regulators, especially chlormequat		
Protective clothing	A, C		
Precautions	R34, R36, R38, U02a, U05a, U08, U19a, U20a, D01, D02, D09a, D10a, H04, H05		
Intracrop Rapide Beta	Intracrop	A0672	wetter
Contains	350 g/l propionic acid and 100 g/l polyoxyethylene (5-8 EO) C10-C15 primary alcohol		
Use with	All approved pesticides which have a recommendation for use with a wetting agent; pesticides must be used at half approved dose rate or less on edible crops, up to full dose rate on non-edible crops		
Protective clothing	A, C		
Precautions	U02a, U04a, U05a, U08, U10, U11, U13, U14, U15, U19a, U20b, E15a, D01, D02, D09a, D10a, M05a, H05		

Product	Supplier	Adj. No.	Type
Intracrop Retainer	Intracrop	A0508	adjuvant/vegetable oil
Contains	60% w/w methylated rapeseed oil		
Use with	All approved pesticides on non-edible crops, and all pesticides approved for use on growing edible crops when used at half recommended dose or less. On specified crops product may be used at a maximum spray concentration of 1% with approved pesticides at their full approved rate up to the growth stages shown in the label		
Precautions	U19a, U20b, D05, D09a, D10b		
Intracrop Rigger	Intracrop	A0572	adjuvant/vegetable oil
Contains	91.7% w/w methylated rapeseed oil		
Use with	All approved pesticides on non-edible crops, and all pesticides approved for use on growing edible crops when used at half recommended dose or less. On specified crops product may be used at a maximum spray concentration of 1.78% with approved pesticides at their full approved rate up to the growth stages shown in the label		
Precautions	U19a, U20b, D05, D09a, D10b		
Intracrop Salute	Intracrop	A0683	spreader/sticker/wetter
Contains	27% w/w polyoxyethylen-methyl- [3-(1,3,3,3-tetramethyl-3-trimethylsiloxy) disiloxanyl] propylether		
Use with	All approved pesticides on non-edible crops and non crop production uses; All approved pesticides at half or less than half the approved pesticide rate on edible crops		
Protective clothing	A, C		
Precautions	U05a, U08, U20c, E13e, D01, D02, D09a, D10a		
Intracrop Sapper	Intracrop	A0684	spreader/sticker/wetter
Contains	27% w/w polyoxyethylen-methyl-[3-(1,3,3,3-tetramethyl- 3-trimethylsiloxy) disiloxanyl] propylether		
Use with	All approved pesticides on non-edible crops and non crop production uses; All approved pesticides at half or less than half the approved pesticide rate on edible crops		
Protective clothing	A, C		
Precautions	U05a, U08, U20c, E13c, D01, D02, D09a, D10a		
Intracrop Saturn	Intracrop	A0494	activator/non-ionic surfactant/spreader
Contains	75% polyoxyethylene polypropoxypropanol		
Use with	All approved pesticides on non-edible crops and pesticides used in non-crop production, and all pesticides approved for use on growing edible crops when used at half recommended dose or less. On specified crops product may be used at a maximum spray concentration of 0.3% with approved pesticides at their full approved rate up to the growth stages shown in the label		
Protective clothing	A, C		
Precautions	R36, R38, U04a, U05a, U08, U19a, E13b, D01, D02, D09a, D10b, M03, M05a, H04		

Product	Supplier	Adj. No.	Type
Intracrop Signal XL	Intracrop	A0659	extender/sticker
Contains	62.5% liquid concentrate formulation containing 22% w/w latex solidsco-formulated with 20% alcohol ethoxylate		
Use with	All approved pesticides on non-edible crops and non crop production uses; All approved pesticides at half or less than half the approved pesticide rate on edible crops		
Protective clothing	A, C		
Precautions	U05a, U19a, U20b, E13c, E34, D01, D02, D09a, D10b, M04a, M05a		
Intracrop Spread-n-wet	Intracrop	A0624	spreader/wetter
Contains	900 g/l polyoxyethylene (5-8 EO) C10-C15 primary alcohol		
Use with			
Intracrop Sprinter	Intracrop	A0513	spreader/wetter
Contains	192 g/l primary alcohol ethoxylate		
Use with	Recommended rates of approved pesticides up to the growth stage indicated for specified crops, and with half or less than the recommended rate on these crops after the stated growth stages. Also with herbicides on managed amenity turf at recommended rates, and on grassland at half or less than recommended rates		
Protective clothing	A, C		
Precautions	R36, R38, R41, U05a, U08, U11, U14, U15, U20c, E15a, D01, D02, D09a, D10b, M03, H04		
Intracrop Status	Intracrop	A0506	adjuvant/vegetable oil
Contains	91% w/w methylated rapeseed oil		
Use with	All approved pesticides on non-edible crops, and all pesticides approved for use on growing edible crops when used at half recommended dose or less. On specified crops product may be used at a maximum spray concentration of 1.0% with approved pesticides at their full approved rate up to the growth stages shown in the label		
Precautions	U19a, U20b, D05, D09a, D10b		
Intracrop Stay-Put	Intracrop	A0507	vegetable oil
Contains	91% w/w methylated rapeseed oil		
Use with	Recommended rates of approved pesticides for non-crop uses and with recommended rates of approved pesticides up to the growth stage indicated for specified crops, and with half or less than the recommended rate on these crops after the stated growth stages		
Precautions	U19a, U20b, D05, D09a, D10b		
Intracrop Super Rapeze MSO	Intracrop	A0491	adjuvant/vegetable oil
Contains	91.7% w/w methylated rapeseed oil		
Use with	All approved pesticides on non-edible crops, and all pesticides approved for use on growing edible crops when used at half recommended dose or less. On specified crops product may be used at a maximum spray concentration of 1.78% with approved pesticides at their full approved rate up to the growth stages shown in the label		
Precautions	U19a, U20b, D05, D09a, D10b		

Product	Supplier	Adj. No.	Type
Intracrop Warrier	Intracrop	0514	spreader/wetter
Contains	192 g/l primary alcohol ethylene		
Use with	Recommended rates of approved pesticides up to the growth stage indicated for specified crops, and with half or less than the recommended rate on these crops after the stated growth stages. Also with herbicides on managed amenity turf at recommended rates, and on grassland at half or less than recommended rates		
Protective clothing	A, C		
Precautions	R36, R38, R41, U05a, U08, U11, U14, U15, U20c, E15a, D01, D02, D09a, D10b, M03, H04		
Intracrop Zenith	Intracrop	A0665	spreader/sticker/wetter
Contains	A liquid Containing 27% w/w polyoxyethylen- -methyl- -[3-(1,3,3,3-tetramethyl-3-trimethylsiloxy) disiloxanyl] propylether co-formulated with a betaine latex polymer emulsifier complex		
Use with	All approved pesticides at half or less than half the approved pesticide rate		
Protective clothing	A, C		
Precautions	R36, R38, R41, U05a, U08, U11, U20c, E13e, D01, D02, D09a, D10a, H04		
Jogral	Intracrop	A0226	cationic surfactant
Contains	800 g/l tallow amine ethoxylate		
Use with	Glyphosate		
Protective clothing	A, C		
Precautions	R22a, U08, U19a, U20a, E13b, E34, D01, D02, D06b, D10b, D11a, M03, H03, H08		
Kantor	Interagro	A0623	spreader/wetter
Contains	790 g/l alkoxylated triglycerides		
Use with	All approved pesticides		
Precautions	U05a, U20b, E15b, E19b, E34, D01, D02, D09a, D10a, D12a		
Katalyst	Interagro	A0450	penetrant/water conditioner
Contains	90% w/w alkoxylated fatty amine		
Use with	Glyphosate and a wide range of other pesticides on specified crops. Refer to label or contact supplier for further details		
Protective clothing	A, C		
Precautions	R22a, R36, R38, R50, R58, U02a, U05a, U08, U19a, U20b, E15a, E34, E37, D01, D02, D05, D09a, D10a, M03, H03, H11		

Product	Supplier	Adj. No.	Type
Kinetic	Helena	A0252	spreader/wetter
Contains	80% w/w polyoxypropylene-polyoxyethylene glycol and 20% w/w polyalkylene oxide modified heptamethyl trisiloxane		
Use with	Approved pesticides on non-edible crops and cereals and stubbles of all edible crops when used at full recommended dose, and with pesticides approved for use on growing edible crops when used at half recommended dose or less. On specified crops product may be used at a maximum spray concentration of 0.2% with approved pesticides at their full approved rate up to the growth stages shown in the label		
Protective clothing	A, H		
Precautions	R38, R41, U02a, U05a, U08, U19a, U20b, E13c, E34, E37, D01, D02, D05, D09a, D10c, M03, H04		
Klipper	Amega	A0260	spreader/wetter
Contains	600 g/l ethylene oxide condensate		
Use with	Approved salt formulations of mecoprop alone or in mixtures with 2,4-D on managed amenity turf		
Protective clothing	A, C		
Precautions	R36, R38, U05a, U08, U20c, E15a, D01, D02, D09a, D10a, H04, H08		
Leaf-Koat	Helena	A0511	spreader/wetter
Contains	80% w/w polyoxypropylene-polyoxyethylene glycol and 20% w/w polyalkylene oxide modified heptamethyl trisiloxane		
Use with	Approved pesticides on non-edible crops and cereals and stubbles of all edible crops when used at full recommended dose, and with pesticides approved for use on growing edible crops when used at half recommended dose or less. On specified crops product may be used at a maximum spray concentration of 0.2% with approved pesticides at their full approved rate up to the growth stages shown in the label		
Protective clothing	A, H		
Precautions	R38, R41, U02a, U05a, U08, U19a, U20b, E13c, E34, E37, D01, D02, D05, D09a, D10c, M03, H04		
Level	United Agri	A0580	extender/sticker
Contains	52% synthetic latex solution		
Use with	All approved pesticides in non-edible crops, all approved pesticides on edible crops when used at half or less their recommended rate, potato blight fungicides and with all pesticides on listed crops up to specified growth stages		
Precautions	E37, D01, D02, D05		

Product	Supplier	Adj. No.	Type
Li-700	De Sangosse	A0529	acidifier/drift retardant/penetrant
Contains	35% w/w lecithin, 35% w/w propionic acid and 9.39%w/w alcohol ethoxylate		
Use with	All approved pesticides on all edible crops when used at half their recommended dose or less, and on all non-edible crops up their full recommended dose. Also with morpholine fungicides on cereals and with pirimicarb on legumes when used at full dose; with iprodione on brassicas when used at 75% dose, and at a maximum concentration of 0.5% with all pesticides on listed crops at full dose up to specified growth stages		
Protective clothing	A, C, H		
Precautions	R36, R38, U11, U14, U15, U19a, E15a, E19b, D01, D02, D05, D10a, D12a, H04		
Libsorb	Ciba Specialty	A0438	spreader/wetter
Contains	alkyl alcohol ethoxylate		
Use with	Any spray for which additional wetter is approved and recommended		
Protective clothing	A, C		
Precautions	R36, R38, U05a, U08, U20a, E13c, D01, D02, D09a, D10a, H04		
Logic	Microcide	A0288	vegetable oil
Contains	95% emulsified vegetable oil		
Use with	All approved pesticides on edible and non-edible crops for ground or aerial application		
Protective clothing	A, C		
Precautions	U19a, U20b, D09a, D10b		
Logic Oil	Microcide	A0630	vegetable oil
Contains	95% emulsified vegetable oil		
Use with	All approved pesticides on edible and non-edible crops for ground or aerial application		
Protective clothing	A, C		
Precautions	U19a, U20b, D09a, D10b		
Low Down	Helena	A0459	mineral oil
Contains	732 g/l refined paraffinic petroleum oil		
Use with	All approved pesticides on edible and non-edible crops when used at half their approved dose or less. Also with approved pesticides on specified crops, up to specified growth stages, at up to their full approved dose		
Protective clothing	A, C		
Precautions	R53a, U02a, U05a, U08, U20b, E15a, E34, E37, D01, D02, D05, D09a, D10c, H11		
Mangard	Taminco	A0667	adjuvant/anti-transpirant
Contains	96% w/w di-1-p-menthene		
Use with	Glyphosate and a wide range of other herbicides, fungicides and insecticides. See label for details		
Precautions	U20c, E15a, D09a, D10a		

SECTION 4

Product	Supplier	Adj. No.	Type
Meco-Flex	Greenaway	A0678	spreader/sticker/wetter
Contains	10.0 % w/w rapeseed oil		
Use with	Re-Act, Headland Relay Depitox		
Protective clothing	A, C		
Precautions	U11, U12, U15, U20b, E15a, E34, D02		
Mediator Sun	Nufarm UK	A0512	wetter
Contains	655 g/l terpenic alcohols		
Use with	All approved pesticides and plant growth regulators on a wide range of specified crops at full dose up to specified growth stages, and at half dose at later growth stages		
Precautions	R36, R52, R53a, U11, U15, D01, M05a, H04		
Mixture B NF	Amega	A0570	non-ionic surfactant/spreader/wetter
Contains	42.5% polyoxyethylene (3EO) C12-C15 primary alcohol and 38.25 % w/w polyoxyethylene (7EO) C12-C15 primary alcohol		
Use with	All approved pesticides on non-edible crops at their full recommended rate. Also with Timbrel (MAPP 05815) and all approved formulations of glyphosate in non-crop situations. Also with a range of specified herbicides when used at less than half their recommended rate on forest and grassland, and with all approved pesticides at full rate on a wide range of specified edible crops up to specified growth stages		
Protective clothing	A, C		
Precautions	R21, R22a, R36, R38, U05a, U08, U20b, E13c, E34, E37, D01, D02, D09a, D10a, M03, H03, H04		
Nettle	Interagro	A0466	mineral oil
Contains	60% w/w mineral oil		
Use with	Cereal graminicides and a wide range of other pesticides that have a label recommendation for use with authorised adjuvant oils on specified crops. Refer to label or contact supplier for further details		
Protective clothing	A		
Precautions	R22b, R38, U02a, U05a, U08, U20b, E15a, D01, D02, D05, D09a, D10a, M05b, H03		
Newman's T-80	De Sangosse	A0192	spreader/wetter
Contains	78% w/w polyoxyethylene tallow amine		
Use with	Glyphosate		
Protective clothing	A, C, H		
Precautions	R22a, R37, R38, R41, R50, R58, R67, U05a, U11, U14, U19a, U20a, E15a, E19b, E34, E37, D01, D02, D05, D10a, D12a, M03, M04a, H03, H08, H11		

Product	Supplier	Adj. No.	Type
Nion	Amega	A0415	spreader/wetter
Contains	90% (900 g/l) ethylene oxide condensate		
Use with	Recommended rates of approved pesticides up to specified growth stages of a wide range of agricultural arable and horticultural crops. Use on edible crops beyond specified growth stages should only be with half recommended rates of the pesticide or less. See label for details of growth stage restrictions.		
Protective clothing	A, C		
Precautions	R36, R38, U05a, U08, U19a, U20b, E13c, E37, D01, D02, D09a, D10a, H04		
Nu Film P	Intracrop	A0635	sticker/wetter
Contains	96% poly-1-p-menthene		
Use with	Glyphosate and many other pesticides and growth regulators for which a protectant is recommended. Do not use in mixture with adjuvant oils or surfactants		
Precautions	U19a, U20b, E13c, D09a, D10a		
Nufarm Cropoil	Nufarm UK	A0447	mineral oil
Contains	99% highly refined mineral oil		
Use with	Approved pesticides for use on certain specified crops (cereals, combinable break crops, field and dwarf beans, peas, oilseed rape, brassicas, potatoes, carrots, parsnips, sugar beet, fodder beet, mangels, red beet, maize, sweetcorn, onions, leeks, horticultural crops, forestry, amenity, grassland)		
Protective clothing	A, C		
Precautions	R43, U19a, U20b, E13c, D05, D09a, D10b, H04		
Nufarm Cropoil Gold	Nufarm UK	A0446	adjuvant/vegetable oil
Contains	97.5% w/w methylated rapeseed oil		
Use with	A wide range of pesticides used in agriculture, horticulture, forestry and amenity situations. See label for details of crops and dose rates.		
Protective clothing	A, H		
Precautions	R43, U05a, U15, U19a, U20b, E13c, D01, D02, D05, D09a, D10b, H04		
Orlin-one	Intracrop	A0597	spreader/sticker/wetter
Contains	25.0 % w/w polyoxyethylene (12EO)/ polyoxypropylene (30EO) block copolymer		
Use with			
Output	Nufarm UK	A0644	mineral oil/surfactant
Contains	60% w/w mineral oil and 40% w/w surfactants		
Use with	Grasp		
Protective clothing	A, C		
Precautions	R36, R38, R52, U05a, U08, U10, U19a, U20b, E15a, E38, D01, D02, D09a, D10b, D12a, H04		

SECTION 4

Product	Supplier	Adj. No.	Type
Output	Syngenta	A0429	mineral oil/surfactant
Contains	60% mineral oil and 40% surfactants		
Use with	Grasp		
Protective clothing	A, C		
Precautions	R38, U05a, U08, U19a, U20a, E13c, D01, D02, D09a, D10b, H04		
Pan Oasis	Pan Agriculture	A0411	vegetable oil
Contains	95% w/w methylated rapeseed oil		
Use with	A wide range of pesticides that have a label recommendation for use with authorised adjuvant oils. Contact distributor for further details		
Protective clothing	A		
Precautions	R36, U02a, U05a, U08, U20b, E15a, D01, D02, D05, D09a, D10a, H04		
Pan Panorama	Pan Agriculture	A0412	mineral oil
Contains	95% w/w mineral oil		
Use with	A wide range of pesticides that have a label recommendation for use with adjuvant oils. Contact distributor for details		
Protective clothing	A		
Precautions	R22b, R38, U02a, U05a, U08, U20b, E13c, D01, D02, D05, D09a, D10a, M05b, H03		
Paramount	United Agri	0585	wetter
Contains	85% polyalkylene oxide modified heptamethyl siloxane		
Use with	Approved pesticides on cereals and grassland for which a wetter is recommended, all approved pesticides in non-edible crops, all approved pesticides on edible crops when used at half or less their recommended rate, and with all pesticides on listed crops up to specified growth stages		
Protective clothing	A, C		
Precautions	R21, R22a, R38, R41, R51, R53a, U11, U14, U16b, E37, E38, D01, D05, D07, D08, H03, H11		
Phase II	De Sangosse	A0622	vegetable oil
Contains	95% w/w esterified rapeseed oil		
Use with	Pesticides approved for use in sugar beet, oilseed rape, cereals (for grass weed control), and other specified agricultural and horticultural crops		
Protective clothing	A, C, H		
Precautions	U19a, U20b, E15a, E19b, E34, D01, D02, D05, D10a, D12a		
Pik-Flex	Greenaway	A0681	spreader/sticker/wetter
Contains	10.0 % w/w rapeseed oil		
Use with	Tordon 22K or Tordon 101		
Protective clothing	A, C		
Precautions	U11, U12, U15, U20b, E15a, E34, D02		

Product	Supplier	Adj. No.	Type
Scout	De Sangosse	A0533	acidifier/drift retardant/penetrant
Contains	35% w/w lecithin, 35% w/w propionic acid and 9.39%w/w alcohol ethoxylate		
Use with	All approved pesticides on all edible crops when used at half their recommended dose or less, and on all non-edible crops up their full recommended dose. Also with morpholine fungicides on cereals and with pirimicarb on legumes when used at full dose; with iprodione on brassicas when used at 75% dose, and at a maximum concentration of 0.5% with all pesticides on listed crops at full dose up to specified growth stages		
Protective clothing	A, C, H		
Precautions	R36, R38, U11, U14, U15, U19a, E15a, E19b, D01, D02, D05, D10a, D12a, H04		
Siltex	Intracrop	A0398	spreader/sticker/wetter
Contains	27% w/w polyoxyethylen-alpha-methyl-omega-disiloxanyl] propylether		
Use with	Approved pesticides used at full dose on a range of specified fruit and vegetable crops, and on managed amenity turf		
Protective clothing	A, C		
Precautions	U05a, U08, U20c, E13c, E37, D01, D02, D09a, D10a		
Silwet L-77	De Sangosse	A0640	drift retardant/spreader/wetter
Contains	80% w/w polyalkylene oxide modified heptamethyltrisiloxane and a maximum of 20 % w/w allyloxypolyethylene glycol methyl ether		
Use with	All approved fungicides on winter and spring sown cereals; all approved pesticides applied at 50% or less of their full approved dose. A wide range of other uses. See label or contact supplier for details		
Protective clothing	A, C, H		
Precautions	R20, R21, R22a, R36, R43, R48, R51, R58, U11, U15, U19a, E15a, E19b, D01, D02, D05, D10a, D12a, H03, H11		
Slippa	Interagro	A0206	spreader/wetter
Contains	655 g/l polyalkyleneoxide modified heptamethyltrisiloxane		
Use with	Cereal fungicides and a wide range of other pesticides and trace elements on specified crops		
Protective clothing	A, C, H		
Precautions	R20, R38, R41, R43, R48, R51, R58, U02a, U05a, U08, U11, U19a, U20a, E15a, E34, E37, D01, D02, D05, D09a, D10a, M03, H03, H11		
SM-99	De Sangosse	A0534	mineral oil
Contains	99% highly refined paraffinic oil		
Use with	All pesticides on all crops when used up to 50% of maximum approved dose for that use (mixtures with Roundup must only be used for treatment of stubbles). Also with all approved pesticides at full dose on listed crops (see label). Also with listed herbicides on specified crops		
Protective clothing	A, C		
Precautions	R22a, U10, U16b, U19a, E15a, E19b, E34, E37, D01, D02, D05, D10a, D12a, M05b, H03		

SECTION 4

Product	Supplier	Adj. No.	Type
Solar	Intracrop	A0225	activator/non-ionic surfactant/spreader
Contains	75% polypropoxypropanol		
Use with	Foliar applied plant growth regulators		
Protective clothing	A, C		
Precautions	R36, R38, U04a, U05a, U08, U20a, E13c, E34, D01, D02, D09a, D10b, D11a, M03, H04		
Spartan	Interagro	A0375	penetrant/water conditioner
Contains	500 g/l alkoxylated fatty amine + 400 g/l polyoxyethylene monolaurate		
Use with	Cereal growth regulators, cereal herbicides, oilseed rape fungicides and a wide range of other pesticides on specified crops. Refer to label or contact supplier for further details		
Protective clothing	A, C		
Precautions	R22a, R36, R38, R51, R58, U02a, U05a, U19a, U20b, E15a, E34, E37, D01, D02, D05, D09a, D10a, M03, H03, H11		
Sprayfast	Taminco	A0668	extender/sticker/wetter
Contains	334 g/l di-1-p-menthene		
Use with	Glyphosate and other pesticides, growth regulators or nutrients for which a coating agent is approved and recommended		
Precautions	U20b, E13c, D09a, D10a		
Spray-fix	De Sangosse	A0559	extender/sticker/wetter
Contains	900 g/l synthetic latex solution and 101 g/l alcohol alkoxylate		
Use with	All approved potato blight fungicides. Also with all approved pesticides on all edible crops when used at half their recommended dose or less, and on all non-edible crops up their full recommended dose. Also at a maximum concentration of 0.14% with approved pesticides on listed crops up to specified growth stages		
Protective clothing	A, C, H		
Precautions	R36, R38, U11, U14, U16b, U19a, E15a, E19b, D01, D02, D05, D10a, D12a, H04		
Spraygard	Taminco	A0669	extender/sticker/wetter
Contains	400 g/l di-1-p-menthene		
Use with	Approved pesticides on all edible crops (except herbicides on peas) and as an anti-transpirant on vegetables before transplanting, evergreens, deciduous trees, shrubs, bushes and on turf		
Protective clothing	A		
Precautions	R38, R50, R53a, U05a, U14, U20b, E15a, E34, E37, D01, D09a, D10b, D12b, M03, H04, H11		

Product	Supplier	Adj. No.	Type
Spraymac	De Sangosse	A0549	acidifier/non-ionic surfactant
Contains	350 g/l propionic acid and 100 g/l alcohol ethoxylate		
Use with	All approved pesticides on all edible crops when used at half their recommended dose or less, and on all non-edible crops up their full recommended dose. Also at a maximum concentration of 0.5% with approved pesticides on listed crops up to specified growth stages		
Protective clothing	A, C, H		
Precautions	R34, U04a, U10, U11, U14, U19a, U20b, E15a, E19b, E34, D01, D02, D05, D10a, D12a, M04a, H05		
Stamina	Interagro	A0202	penetrant
Contains	100% w/w alkoxylated fatty amine		
Use with	Glyphosate and a wide range of other pesticides on specified crops		
Protective clothing	A, C		
Precautions	R22a, R38, R50, R58, U02a, U05a, U20a, E15a, E34, D01, D02, D05, D09a, D10a, M03, H03, H11		
Standon Shiva	Standon	A0519	wetter
Contains	20.2 % w/w sodium salt of 3,6-dioxaoctadecylsulphate and 6.7% w/w sodium salt of 3,6-dioxaeicosylsulphate		
Use with	Approved cereal herbicides		
Protective clothing	A, C		
Precautions	R36, R38, R52, U02a, U05a, U08, U13, U19a, E15a, D01, D02, D05, D09a, D10b, H04		
Stika	De Sangosse	A0557	extender/sticker/wetter
Contains	450 g/l synthetic latex solution and 101 g/l alcohol alkoxylate		
Use with	All approved potato blight fungicides. Also with all approved pesticides on all edible crops when used at half their recommended dose or less, and on all non-edible crops up their full recommended dose. Also at a maximum concentration of 0.14% with approved pesticides on listed crops up to specified growth stages		
Protective clothing	A, C, H		
Precautions	R36, R38, U11, U14, U16b, U19a, E15a, E19b, D01, D02, D05, D10a, D12a, H04		
Super Nova	Interagro	A0364	penetrant
Contains	500 g/l alkoxylated fatty amine + 500 g/l polyoxyethylene monolaurate		
Use with	Cereal growth regulators, cereal herbicides, cereal fungicides, oilseed rape fungicides and a wide range of other pesticides on specified crops		
Precautions	R51, R58, U05a, E15a, E34, D01, D02, D05, D09a, D10a, H11		
Sward	Amega	A0215	spreader/wetter
Contains	15.2% w/w polyalkylene oxide modified heptomethyltrisiloxane		
Use with	Specified pesticides in amenity situations		
Protective clothing	A, C		
Precautions	R41, R43, U05a, U08, U20c, E13c, D01, D02, D09a, D10a, H04		

SECTION 4

Product	Supplier	Adj. No.	Type
Talzene	Intracrop	A0233	wetter
Contains	800 g/l polyoxyethylene tallow amine		
Use with	Herbicides and dessicants up to their latest recommended time of application in agriculture, horticulture, amenity and forestry		
Protective clothing	A, C		
Precautions	R22a, R36, R38, R41, R48, U05a, U08, U11, U13, U19a, U20b, E13c, E34, D01, D02, D09a, D10a, M03, H03		
Teliton Ocean Power	Teliton	A0615	wetter
Contains	20.2% w/w 3,6-dioxoctadecylsulphate sodium salt and 6.7% w/w 3,6-dioxaeicosylsulphate sodium salt		
Use with	All approved herbicides for use on cereals		
Protective clothing	A, C		
Precautions	R36, R38, R52, U02a, U05a, U08, U13, U19a, E15a, D01, D02, D09a, D10a, H04		
TM 1088	Intracrop	A0675	wetter
Contains	750 g/l polyoxyethylene, polypropoxypropanol		
Use with	All approved pesticides on non-edible crops and non crop production uses; All approved pesticides at half or less than half the approved pesticide rate on edible crops		
Protective clothing	A, C		
Precautions	R36, R38, U04a, U05a, U08, U19a, U20b, E13b, E34, D01, D02, D09a, D10a, M03, M05a, H04		
Toil	Interagro	A0248	vegetable oil
Contains	95% w/w methylated rapeseed oil		
Use with	Sugar beet herbicides, oilseed rape herbicides, cereal graminicides and a wide range of other pesticides that have a label recommendation for use with authorised adjuvant oils on specified crops. Refer to label or contact supplier for further details		
Precautions	U05a, U20b, E15a, D01, D02, D05, D09a, D10a		
Torpedo-II	De Sangosse	A0541	penetrant/wetter
Contains	210 g/kg alkoxylated tallow amine, 380 g/kg alcohol ethoxylates, 75 g/l natural fatty acids and 210 g/kg polyalkylene glycol		
Use with	All approved pesticides on all edible crops when used at half their recommended dose or less, and on all non-edible crops up their full recommended dose. Also at a maximum concentration of 0.1% with approved pesticides on listed crops up to specified growth stages		
Protective clothing	A, C, H		
Precautions	R36, R38, R41, R52, R58, U05a, U10, U11, U19a, E15a, E19b, E34, E37, D01, D02, D05, D10a, D12a, M03, H04		

Product	Supplier	Adj. No.	Type
Transact	United Agri	A0584	penetrant
Contains	40% w/w propionic acid		
Use with	Plant growth regulators in cereals, all approved pesticides in non-edible crops, all approved pesticides on edible crops when used at half or less their recommended rate, and with all pesticides on listed crops up to specified growth stages		
Protective clothing	A, C		
Precautions	R34, U10, U11, U14, U15, U19a, D01, D05, D09b, M04a, H05		
Transcend	Helena	A0333	adjuvant/vegetable oil
Contains	80% w/w methylated soybean oil and 15% w/w polyalkylene oxide modified heptamethyl trisiloxane		
Use with	Approved pesticides on non-edible crops, cereals and stubbles of all edible crops, grassland (destruction), and pesticides approved for use on growing edible crops when used at half recommended dose or less. On specified crops product may be used at a maximum spray concentration of 0.5% with approved pesticides at their full approved rate up to the growth stages shown in the label		
Protective clothing	A, H		
Precautions	R38, R41, U02a, U05a, U08, U11, U19a, U20b, E13c, E34, E37, D01, D02, D05, D09a, M03, H04		
Try-Flex	Greenaway	A0489	sticker/wetter
Contains	95% refined rapeseed oil		
Use with	Gly-480 (MAPP 12718) and Freeway (MAPP 11129) on natural surfaces not intended to bear vegetation, permeable surfaces overlying soil, hard surfaces overlying soil and forest		
Protective clothing	A, C, D, H, M		
Precautions	U08, U20b, E37, C11, D05, D10a		
Validate	De Sangosse	A0500	adjuvant/wetter
Contains	50% w/w lecithin, 25% w/w esterified vegetable oil and 25% w/w alcohol ethoxylate		
Use with	Approved pesticides on non-edible crops where the addition of a wetter/spreader or adjuvant oil is recommended on the pesticide label, and pesticides approved for use on growing edible crops when used at half recommended dose or less. On specified crops product may be used at a maximum spray concentration of 0.5% with approved pesticides at their full approved rate up to the growth stages shown in the label		
Protective clothing	A, C, H		
Precautions	R51, R53a, U05a, U11, U14, U19a, U20b, E15a, E19b, E34, D01, D02, D10a, D12a, H11		
Wetcit	Plant Solutions	A0586	penetrant/wetter
Contains	8.15% alcohol ethoxylate		
Use with	All approved pesticides on all edible crops when used at half their recommended dose or less, and on all non-edible crops up their full recommended dose. Also at a maximum concentration of 0.25% with approved pesticides on listed crops up to specified growth stages at full rate, and at half rate thereafter		
Precautions	R22a, R38, R41, U08, U11, U19a, U20b, E13c, E15a, E34, E37, D05, D08, D09a, D10b, H03		

SECTION 4

Product	Supplier	Adj. No.	Type
X-Wet	De Sangosse	A0548	non-ionic surfactant/wetter
Contains	750 g/l alcohol ethoxylates and 150 g/l natural fatty acids		
Use with	All approved pesticides on all edible crops when used at half their recommended dose or less, and on all non-edible crops up their full recommended dose. Also at a maximum concentration of 0.1% with approved pesticides on listed crops up to specified growth stages		
Protective clothing	A, C, H		
Precautions	R36, R38, R53a, U02a, U04a, U05a, U10, U11, U20b, E15a, E19b, D01, D02, D05, D10a, H04		
Zarado	De Sangosse	A0516	vegetable oil
Contains	70% w/w esterified rapeseed oil		
Use with	All approved pesticides on edible and non-edible crops when used at half their approved dose or less. Also with approved pesticides on specified crops, up to specified growth stages, at up to their full approved dose		
Protective clothing	A, C, H		
Precautions	R43, R52, U05a, U13, U19a, U20b, E15a, E19b, E34, D01, D02, D05, D10a, M05a, H04		
Zeal	Interagro	0685	water conditioner/wetter
Contains	60% w/w alkoxylated fatty amine		
Use with	Trace elements such as calcium, copper, manganese and sulphur and with macronutrients such as phosphates		
Protective clothing	A, C, H		
Precautions	R22a, R36, R38, R51, R53a, U05a, U09a, U11, U14, U15, U19a, E15b, E34, D01, D02, D09a, D10a, M03, H03, H11		
Zigzag	Headland	A0692	extender/sticker/wetter
Contains	36.9% w/v styrene/butadiene co-polymer and 6.375% w/v polyether-modified trisiloxane		
Use with	Approved glyphosate formulations on cereal crops; All approved pesticides applied at half-rate or less; All approved pesticides for use on crops not destined for human or animal consumption; With all approved pesticides on edible crops up to the latest growth stage specified on the label.		
Protective clothing	C, H		
Precautions	U11, U15, E38, D01		
Zinzan	Certis	A0600	extender/spreader/wetter
Contains	70% 1,2 bis (2-ethylhexyloxycarbonyl) ethanesulponate		
Use with	All approved pesticides on non-edible crops, and all pesticides approved for use on growing edible crops when used at half recommended dose or less. On specified edible crops product may be used at a maximum spray concentration of 0.075% with approved pesticides at their full approved rate up to the growth stages shown in the label		
Protective clothing	A, C, P		
Precautions	R38, R41, U11, U14, U20c, D01, D09a, D10a, H04		

SECTION 5
USEFUL INFORMATION

Pesticide Legislation

Anyone who advertises, sells, supplies, stores or uses a pesticide is bound by legislation, including those who use pesticides in their own homes, gardens or allotments. There are numerous UK statutory controls, but the major legal instruments are outlined below.

The Food and Environment Protection Act 1985 (FEPA) and Control of Pesticides Regulations 1986 (COPR)

FEPA introduced statutory powers to control pesticides with the aims of protecting human beings, creatures and plants, safeguarding the environment, ensuring safe, effective and humane methods of controlling pests, and making pesticide information available to the public. Control of pesticides is achieved by COPR. These Regulations lay down the Approvals required before any pesticide may be sold, stored, supplied, advertised or used, subject to conditions which are contained in a series of Schedules. Schedule 1 relates to advertisement; Schedule 2 to sale, supply and storage; Schedule 3 to use; and Schedule 4 to aerial application of pesticides. The Schedules may be changed at any time, following Parliamentary approval. Details are given on the websites of the Chemicals Regulation Directorate (CRD) and the Health and Safety Executive (HSE).

The controls currently in force include the following.

- Only approved products may be sold, supplied, stored, advertised or used.
- No advertisement may contain any claim for safety beyond that which is permitted in the approved label text.
- Only products specifically approved for the purpose may be applied from the air.
- A recognised Storeman's Certificate of Competence is required by anyone who stores for sale or supply pesticides approved for agricultural use.
- A recognised Certificate of Competence is required by anyone who gives advice when selling or supplying pesticides approved for agricultural use.
- Users of pesticides must comply with the Conditions of Approval relating to use.
- A recognised Certificate of Competence is required for all contractors and persons born after 31 December 1964 applying pesticides approved for agricultural use (unless working under direct supervision of a certificate holder). Proposals are now in place that every user of agricultural pesticides will have to hold a Certificate of Competence, regardless of age or supervision. A suitable transition period will allow this requirement to be enacted.
- Only those adjuvants authorised by CRD may be used.
- Regarding tank-mixes, 'no person shall combine or mix for use two or more pesticides which are anti-cholinesterase compounds unless the approved label of at least one of the pesticide products states that the mixture may be made; and no person shall combine or mix for use two or more pesticides if all the conditions of the Approval relating to this use cannot be complied with'.

Plant Protection Products Regulation

European Council Directive 91/414/EEC, drafted to harmonise national arrangements for the authorisation of plant protection products within the European Community, is currently being upgraded to a full 'Regulation', which means that it becomes mandatory rather than advisory. It first became effective in the UK on 25 July 1993 and the new proposals will come into force in April 2011. The new rules switch from risk management to hazard criteria, and growers will be encouraged to adopt lower-risk practices and products. The aim of the new legislation is to minimise the risk to health and the environment by reducing the levels of harmful pesticides through substitution with safer alternatives, banning or severely limiting the use of the more dangerous pesticides, and encouraging the adoption of best practice for authorised products.

Some of the changes to be introduced include a ban on aerial spraying, and measures to protect water, public spaces and special conservation areas. National Action Plans will be set to establish indicators of risk and impact reduction, and to encourage the development of integrated pest management and alternative techniques to reduce the perceived dependency on pesticides. To simplify pesticide approval, national approvals will be replaced by zonal approvals, but with national veto if a product is deemed to be unsafe to use in a particular country. Much of the fine detail has yet to be established, but this should be achieved by the introduction date of April 2011.

Dangerous Preparations Directive (1999/45/EC)

The Dangerous Preparations Directive (DPD) came into force in UK for pesticide and biocidal products on 30 July 2004. Its aim is to achieve a uniform approach across all Member States to the classification, packaging and labelling of most dangerous preparations, including crop protection products. The Directive is implemented in UK under the Chemicals (Hazard Information and Packaging Supply) Regulations 2002, often referred to under the acronym CHIP3. In most cases the Regulations have led to additional hazard symbols, and associated risk and safety phrases relating to environmental and health hazards, appearing on the label. All products affected by the DPD entering the supply chain from the implementation date above must be so labelled. The new environmental hazard classifications and risk phrases are included.

Under the DPD, the labels of all plant protection products now state 'To avoid risks to man and the environment, comply with the instructions for use'. As the instruction applies to all products, it has not been repeated in each fact sheet.

The Review Programme

The process of reviewing existing active ingredients, as required by the Plant Protection Products Directive 91/414, is nearing completion. The programme is designed to ensure all available plant protection products are supported by up-to-date information on safety and efficacy. Approximately 850 substances on the EU market are being handled in four stages but, because of the cost of providing this information, many are not being supported and their approvals are being revoked. Over half the original list has been lost because of this, and others are being revoked following scientific risk assessments. More are likely to be lost as Europe changes from risk management to hazard criteria and the Annex 1 Review programme (AIR) is embarked upon.

New approvals will be granted for 10 years, with subsequent approvals being for 15 years. This will give 10 years' protection to the data generated for approval and so will protect new actives from generic registrations using that data for 10 years. However, mutual recognition within (and between) zones will be encouraged, and as parallel approvals help to bring down the cost of products to the end user, they will also be encouraged.

Control of Substances Hazardous to Health Regulations 1988 (COSHH)

The COSHH regulations, which came into force on 1 October 1989, were made under the Health and Safety at Work Act 1974, and are also important as a means of regulating the use of pesticides. The regulations cover virtually all substances hazardous to health, including those pesticides classed as Very toxic, Toxic, Harmful, Irritant or Corrosive, other chemicals used in farming or industry, and substances with occupational exposure limits. They also cover harmful micro-organisms, dusts and any other material, mixture, or compound used at work which can harm people's health.

The original Regulations, together with all subsequent amendments, have been consolidated into a single set of regulations: The Control of Substances Hazardous to Health Regulations 1994 (COSHH 1994).

The basic principle underlying the COSHH regulations is that the risks associated with the use of any substance hazardous to health must be assessed before it is used, and the appropriate measures taken to control the risk. The emphasis is changed from that pertaining under the Poisonous Substances in Agriculture Regulations 1984 (now repealed) – whereby the principal method of ensuring safety was the use of protective clothing – to the prevention or control of exposure to hazardous substances by a combination of measures. In order of preference the measures should be:

(a) substitution with a less hazardous chemical or product

(b) technical or engineering controls (e.g. the use of closed handling systems, etc.)

(c) operational controls (e.g. operators located in cabs fitted with air-filtration systems, etc.)

(d) use of personal protective equipment (PPE), which includes protective clothing.

Consideration must be given as to whether it is necessary to use a pesticide at all in a given situation and, if so, the product posing the least risk to humans, animals and the environment must be selected. Where other measures do not provide adequate control of exposure and the use of PPE is necessary, the items stipulated on the product label must be used as a minimum. It is essential that equipment is properly maintained and the correct procedures adopted. Where necessary, the exposure of workers must be monitored, health checks carried out, and employees instructed and trained in precautionary techniques. Adequate records of all operations involving pesticide application must be made and retained for at least 3 years.

Certificates of Competence – the roles of BASIS and NPTC

COPR, COSHH and other legislation places certain obligations on those who handle and use pesticides. Minimum standards are laid down for the transport, storage and use of pesticides, and the law requires those who act as storekeepers, sellers and advisors to hold recognised Certificates of Competence.

BASIS

BASIS is an independent Registration Scheme for the pesticide industry. It is responsible for organising training courses and examinations to enable such staff to obtain a Certificate of Competence.

In addition, BASIS undertakes annual assessment of pesticide supply stores, enabling distributors, contractors and seedsmen to meet their obligations under the Code of Practice for Suppliers of Pesticides. Further information can be obtained from BASIS.

Certificates of Competence

Storage

- BASIS Certificate of Competence in the Storage and Handling of Crop Protection Products

Sale and supply

- BASIS Certificate in Crop Protection (Agriculture)
- BASIS Certificate in Crop Protection (Commercial Horticulture)
- BASIS Certificate in Crop Protection (Amenity Horticulture)
- BASIS Certificate in Crop Protection (Forestry)
- BASIS Certificate in Crop Protection (Seed Treatment)
- BASIS Certificate in Crop Protection (Seed Sellers)
- BASIS Certificate in Crop Protection (Field Vegetables)
- BASIS Certificate in Crop Protection (Potatoes)
- BASIS Certificate in Crop Protection (Aquatic)
- BASIS Certificate in Crop Protection (Indoor Landscaping)
- BASIS Certificate in Crop Protection (Grassland and Forage Crops)
- BASIS/LEAF ICM Certificate
- BASIS Advanced Certificate
- FACTS (Fertiliser Advisers Certification and Training Scheme) Certificate

NPTC

Certain spray operators also require certificates of competence under the Control of Pesticides Regulations. NPTC's Pesticides Award is a recognised Certificate of Competence under COPR and it is aimed principally at those people who use pesticide products approved for use in agriculture, horticulture (including amenity horticulture) and forestry. All contractors, and anyone else born after 31 December 1964, must possess a certificate if they spray such products unless they are working under the direct supervision of a certificate holder.

Because the required spraying skills vary widely among the uses listed above, candidates are assessed under one (or more) modules that are most appropriate for their professional work. Assessments are carried out by an approved NPTC or Scottish Skills Testing Service Assessor. All candidates must first complete a foundation module (PA1), for which a certificate is not issued, before taking one of the specialist modules. Certificate holders who change their work so that a different specialist module becomes more appropriate may need to obtain a new certificate under the new module.

Holders are required to produce on demand their Certificate of Competence for inspection to any authorised person. Further information can be obtained from NPTC.

Certificates of Competence for spray operators

- PA1 Foundation Module
- PA2 Ground Crop Sprayers – mounted or trailed
- PA3 Broadcast Air Blast Sprayer. Variable Geometry Boom Air Assisted Sprayer
- PA4 Granule Applicator – Mounted or Trailed
- PA5 Boat Mounted Applicators
- PA6 Hand Held Applicators
- PA7 Aerial Application
- PA8 Mixer/Loader
- PA9 Fogging, Misting and Smokes
- PA10 Dipping Bulbs, Corms, Plant Material or Containers
- PA11 Seed treating Equipment
- PA12 Application of Pesticides to Material as a Continuous or Batch Process

Maximum Residue Levels

A small number of pesticides are liable to leave residues in foodstuffs, even when used correctly. Where residues can occur, statutory limits, known as maximum residue levels (MRLs), have been established. MRLs provide a check that products have been used as directed; **they are not safety limits**. However, they do take account of consumer safety because they are set at levels that ensure normal dietary intake of residues presents no risk to health. Wide safety margins are built in, and eating food containing residues above the MRL does not automatically imply a risk to health. Nevertheless, it is an offence to put into circulation any produce where the MRL is exceeded.

The surrounding legislation is complex. MRLs may be specified by several different bodies. The UK has set statutory MRLs since 1988. The European Union intends eventually to introduce MRLs for all pesticide/commodity combinations. These are being introduced initially by a series of priority lists, but will subsequently be covered by the review programme under Directive 91/414. However, in cases where no information is available, EU Directive 2000/42/EC requires many MRLs to be set at the limit of determination (LOD). As a result, certain approvals are being withdrawn where such use would leave residues above the MRL set in the Directive.

MRLs apply to imported as well as home-produced foodstuffs. Details of those that have been set have been published in *The Pesticides (Maximum Residue Levels in Crops, Food and Feeding Stuffs) Regulations 1994*, and successive amendments to these Regulations. These Statutory Instruments are available from The Stationery Office (www.thestationeryoffice.com).

MRLs are set for many chemicals not currently marketed in Britain. Because of this and the ever-changing information on MRLs, direct access is provided to the comprehensive MRL databases on the Chemicals Regulation Directorate website (www.pesticides.gov.uk). This online database sets out in table form the levels specified by UK Regulations, EC Directives, and the Codex Alimentarius for each commodity.

Approval (On-label and Off-label)

Only officially approved pesticides can be marketed and used in the UK. Approvals are granted by UK Government Ministers in response to applications that are supported by satisfactory data on safety, efficacy and, where relevant, humaneness. The Chemicals Regulation Directorate (CRD, www.pesticides.gov.uk) comes under the Health and Safety Executive (HSE, www.hse.gov.uk), and is the UK Government Agency for regulating agricultural pesticides and plant protection products. The HSE currently fulfils the same role for other pesticides, with the two organisations in the process of being merged. The main focus of the regulatory process in both bodies is the protection of human health and the environment.

Statutory Conditions of Use

Approvals are normally granted only in relation to individual products and for specified uses. It is an offence to use non-approved products or to use approved products in a manner that does not comply with the statutory conditions of use, except where the crop or situation is the subject of an off-label extension of use (see below).

Statutory conditions have been laid down for the use of individual products and may include:

- field of use (e.g. agriculture, horticulture etc.)
- crop or situations for which treatment is permitted
- maximum individual dose
- maximum number of treatments or maximum total dose
- maximum area or quantity which may be treated
- latest time of application or harvest interval
- operator protection or training requirements
- environmental protection
- any other specific restrictions relating to particular pesticides.

Products must display these statutory conditions in a boxed area on the label entitled 'Important Information', or words to that effect. At the bottom of the boxed area must be shown a bold text statement: **'Read the label before use. Using this product in a manner that is inconsistent with the label may be an offence. Follow the Code of Practice for Using Plant Protection Products'**. This requirement came into effect in October 2006, and replaced the previous 'Statutory Box'.

Types of Approval

There are three categories of approval that may be granted, under the Control of Pesticides Regulations, to products containing active ingredients that were already on the market by 25 July 1993.

- **Full** – granted for an unstipulated period.
- **Provisional** – granted for a stipulated period.
- **Experimental Permit** – granted for the purposes of testing and developing new products, formulations or uses. Products with only an Experimental Permit may not be advertised or sold and do not appear in this *Guide*.

Products containing new active substances, and those containing older ingredients once they have been listed in Annex I to Directive 91/414 (see The Plant Protection Products Regulation), will be granted approval under the Plant Protection Products Regulations. Again, there are three categories, similar to those listed above.

- **Standard approval** – granted for a period not exceeding 10 years. Only applicable to products for which active substances are listed in Annex I.
- **Provisional approval** – granted for a period not exceeding three years, but renewable. Applicable where Annex I listing of the active substance is awaited.
- **Approval for research and development** – to enable field experiments or tests to be carried out with active substances that are not otherwise approved.

The official lists of approved products, including all the above categories except those for experimental purposes, are shown on the websites of CRD and HSE. Details of new approvals, amendments to existing approvals and off-label approvals are published regularly on those websites.

Withdrawal of Approval

Product approvals may be reviewed, amended, suspended or revoked at any time. Revocation may occur for various reasons, such as commercial withdrawal, or failure by the approval holder to meet data requirements.

From September 2007, where an approval is being revoked for purely administrative, 'housekeeping' reasons, the existing approval will be revoked and in its place will be issued:

- an approval for advertisement, sale and supply by any person for 24 months, and
- an approval for storage and use by any person for 48 months.

Where an approval is replaced by a newer approval, such that there are no safety concerns with the previous approval, but the newer updated approval is more appropriate in terms of reflecting the latest regulatory standard, the existing approval will be revoked and in its place will be issued:

- an approval for advertisement, sale and supply by any person for 12 months, and
- an approval for storage and use by any person for 24 months.

Where there is a need for tighter control of the withdrawal of the product from the supply chain, for example failure to meet data submission deadlines, the current timelines will be retained:

- immediate revocation for advertisement, sale or supply for the approval holder
- approval for 6 months for advertisement, sale and supply by 'others', and
- approval for storage and use by any person for 18 months.

Immediate revocation and product withdrawal remain an option where serious concerns are identified, with immediate revocation of all approvals and approval for storage only by anyone for 3 months, to allow for disposal of product.

The expiry date shown in the product fact sheet is the final date of legal use of the product.

Off-label Extension of Use

Products may legally be used in a manner not covered by the printed label in several ways:

- In accordance with a specific off-label approval (SOLA). SOLAs are uses for which individuals or organisations other than the manufacturers have sought approval. The Notices of Approval are published by CRD and are widely available from ADAS or NFU offices. Users of SOLAs must first obtain a copy of the relevant Notice of Approval and comply strictly with the conditions laid down therein. Users of this Guide will find details of extant SOLAs and direct links to the CRD website, where SOLA notices can also be accessed.
- In tank mixture with other approved pesticides in accordance with Consent C(i) made under FEPA. Full details of Consent C(i) are given in Annex A of Guide to Pesticides on the PSD website, but there are two essential requirements for tank mixes. First, all the conditions of approval of all the components of a mixture must be complied with. Second, no person may mix or combine pesticides that are cholinesterase compounds unless allowed by the label of at least one of the pesticides in the mixture.
- In conjunction with authorised adjuvants.
- In reduced spray volume under certain conditions.
- In the use of certain herbicides on specified set-aside areas subject to restrictions, which differ between Scotland and the rest of UK.
- By mutual recognition of a use fully approved in another Member State of the European Union and authorised by CRD.

Although approved, off-label uses are not endorsed by manufacturers and such treatments are made entirely at the risk of the user.

Using Crop Protection Chemicals

Use of Herbicides In or Near Water

Products in this *Guide* approved for use in or near water are listed in Table 5.2. Before use of any product in or near water, the appropriate water regulatory body (Environment Agency/Local Rivers Purification Authority; or in Scotland the Scottish Environment Protection Agency) must be consulted. Guidance and definitions of the situation covered by approved labels are given in the Defra publication *Guidelines for the Use of Herbicides on Weeds in or near Watercourses and Lakes.* Always read the label before use.

Table 5.2 Products approved for use in or near water

Chemical	Product
2,4-D	Depitox
dichlobenil	Casoron G, Dicore, Midstream GSR, Scotts Dichlo G Macro, Scotts Dichlo G Micro
glyphosate	Asteroid, Barclay Gallup 360, Barclay Gallup Amenity, Barclay Gallup Biograde Amenity, Barclay Gallup Hi-Aktiv, Buggy XTG, Discman Biograde, Dow Agrosciences Glyphosate 360, Envision, Glyfo TDI, Glyfos Dakar, Glyfos Dakar Pro, Glyfos Proactive, Glypho-Rapid 450, Greenaway Gly-490, Kernel, Manifest, Roundup Biactive, Roundup Pro Biactive, Roundup ProBiactive 450, Tangent

Use of Pesticides in Forestry

Table 5.3 Products in this *Guide* approved for use in forestry

Chemical	Product	Use
2,4-D + dicamba + triclopyr	Broadsword, Nu-Shot	Herbicide
4-indol-3-ylbutyric acid + 1-naphthylacetic acid	Synergol	Growth regulator/stimulator/sprout suppressant
aluminium ammonium sulphate	Curb Crop Spray Powder, Guardsman, Liquid Curb Crop Spray	Animal deterrent/repellent
asulam	Asulox, Formule 1, Greencrop Frond	Herbicide
chlorpyrifos	Agriguard Chlorpyrifos, Dursban WG, Equity, Govern, Parapet	Acaricide, Insecticide
cycloxydim	Laser	Herbicide
cypermethrin	Forester	Insecticide
diflubenzuron	Dimilin Flo	Insecticide
ferric phosphate	Sluxx	Molluscicide
fluazifop-P-butyl	Clayton Maximus, Fusilade Max, Greencrop Bantry	Herbicide
glufosinate-ammonium	Harvest, Kaspar, Kibosh	Herbicide
glyphosate	Amega Duo, Barclay Gallup 360, Barclay Gallup Amenity, Barclay Gallup Biograde Amenity, Barclay Gallup Hi-Aktiv, Buggy XTG, Charger, Charger B, Charger C, Clinic Ace, Credit DST, Discman Biograde, Dow Agrosciences Glyphosate 360, Envision, Glyfo TDI, Glyfos, Glyfos Dakar Pro, Glyfos Proactive, Glyfos Supreme, Glyfosat 36, Glyphogan, Glypho-Rapid 450, Greenaway Gly-490, Hilite, Kernel, Manifest, Nomix Conqueror, Nufosate Ace, Roundup Metro, Roundup Pro Biactive, Roundup ProBiactive 450, Slingshot, Stacato, Stirrup, Tangent, Typhoon 360	Herbicide
isoxaben	Agriguard Isoxaben, Flexidor 125	Herbicide

Chemical	Product	Use
metazachlor	Alpha Metazachlor 50 SC, Clayton Buzz, Clayton Metazachlor, Clayton Metazachlor 50 SC, Greencrop Monogram, Marksman, Mezzanine, Rapsan 500 SC, Standon Metazachlor 500, Sultan 50 SC	Herbicide
napropamide	MAC-Napropamide 450 SC	Herbicide
propaquizafop	Bulldog, Clayton Orleans, Cleancrop GYR, Cleancrop GYR 2, Falcon, Greencrop Satchmo, MAC-Propaquizafop 100 EC, Raptor, Shogun, Standon Propaquizafop, Standon Zing PQF, Zealot	Herbicide
propyzamide	Engage, Kerb 50 W, Kerb Granules, Menace 80 EDF, Pizza 400 SC, Quaver Flo, Setanta 50 WP	Herbicide
pyrethrins	Spruzit	Insecticide
triclopyr	Timbrel, Woody	Herbicide

Pesticides Used as Seed Treatments

Information on the target pests for these products can be found in the relevant pesticide profile in Section 2.

Table 5.4 Products used as seed treatments (including treatments on seed potatoes)

Chemical	Product	Formulation	Crop(s)
beta-cyfluthrin + clothianidin	Modesto	FS	Poppies for morphine production, Winter oilseed rape
	Poncho Beta	FS	Fodder beet, Sugar beet
beta-cyfluthrin + imidacloprid	Chinook Blue	LS	Evening primrose, Honesty, Linseed, Mustard, Spring oilseed rape, Winter oilseed rape
	Chinook Colourless	LS	Evening primrose, Honesty, Linseed, Mustard, Winter oilseed rape
carboxin + thiram	Anchor	FS	Spring barley, Spring oats, Spring rye, Spring wheat, Triticale, Winter barley, Winter oats, Winter rye, Winter wheat
clothianidin	Deter	FS	Durum wheat, Triticale, Winter barley, Winter oats, Winter rye, Winter wheat
clothianidin + prothioconazole	Redigo Deter	FS	Durum wheat, Triticale, Winter barley, Winter oats, Winter rye, Winter wheat
clothianidin + prothioconazole + tebuconazole + triazoxide	Raxil Deter	FS	Winter barley
cymoxanil + fludioxonil + metalaxyl-M	Wakil XL	WS	Broad beans, Carrots, Chard, Chives, Combining peas, Herbs (see appendix 6), Leaf spinach, Lupins, Parsley, Parsnips, Poppies for morphine production, Red beet, Salad brassicas, Spinach beet, Spring field beans, Vining peas, Winter field beans

Chemical	Product	Formulation	Crop(s)
difenoconazole + fludioxonil	Celest Extra	FS	Winter oats, Winter rye, Winter wheat
fludioxonil	Beret Gold	FS	Durum wheat, Spring barley, Spring oats, Spring rye, Spring wheat, Triticale, Winter barley, Winter oats, Winter rye, Winter wheat
fludioxonil + flutriafol	Beret Multi	FS	Spring barley, Spring oats, Spring wheat, Winter barley, Winter oats, Winter wheat
fludioxonil + metalaxyl M	Maxim XL	FS	Forage maize
fludioxonil + metalaxyl-M + thiamethoxam	Cruiser OSR	FS	Fodder rape, Mustard, Spring oilseed rape, Winter oilseed rape
fludioxonil + tefluthrin	Austral Plus	FS	Spring barley, Spring oats, Spring wheat, Triticale seed crop, Winter barley, Winter oats, Winter wheat
fluoxastrobin + prothioconazole	Redigo Twin	FS	Durum wheat, Spring rye, Triticale, Winter oats, Winter rye, Winter wheat
	Redigo Twin TXC	FS	Durum wheat, Spring oats, Spring rye, Spring wheat, Triticale, Winter oats, Winter rye, Winter wheat
fluquinconazole	Galmano	FS	Durum wheat, Triticale, Winter wheat
	Jockey Solo	FS	Winter barley, Winter wheat
fluquinconazole + prochloraz	Epona	FS	Winter wheat
	Galmano Plus	FS	Durum wheat, Triticale, Winter wheat
	Jockey	FS	Winter barley, Winter wheat
flutolanil	Rhino	FS	Potatoes, Seed potatoes

Chemical	Product	Formulation	Crop(s)
	Rhino DS	DS	Potatoes
fuberidazole + imidacloprid + triadimenol	Tripod Plus	LS	Winter barley, Winter oats, Winter wheat
fuberidazole + triadimenol	Tripod	FS	Spring barley, Spring oats, Spring rye, Spring wheat, Triticale, Winter barley, Winter oats, Winter rye, Winter wheat
guazatine	Panoctine	LS	Spring barley, Spring oats, Spring wheat, Winter barley, Winter oats, Winter wheat
	Ravine	LS	Spring barley, Spring oats, Spring wheat, Winter barley, Winter oats, Winter wheat
guazatine + imazalil	Panoctine Plus	LS	Spring barley, Spring oats, Winter barley, Winter oats
	Ravine Plus	LS	Spring barley, Spring oats, Winter barley, Winter oats
imazalil	Fungazil 50LS	LS	Spring barley, Winter barley
	IMA 100 SL	LS	Seed potatoes, Ware potatoes
imazalil + pencycuron	Monceren IM	DS	Potatoes
imidacloprid	Nuprid 600FS	FS	Fodder beet, Sugar beet
pencycuron	Agriguard Pencycuron	DS	Potatoes
	Monceren DS	DS	Potatoes
	Pency Flowable	FS	Potatoes
	Standon Pencycuron DP	DS	Potatoes
physical pest control	Silico-Sec	DS	Stored grain

Chemical	Product	Formulation	Crop(s)
prochloraz	Prelude 20LF	LS	Flax, Hemp for oilseed production, Linseed
prochloraz + thiram	Agrichem Hy-Pro Duet	FS	Winter oilseed rape
prochloraz + triticonazole	Kinto	FS	Durum wheat, Triticale, Winter barley, Winter oats, Winter rye, Winter wheat
prothioconazole	Redigo	FS	Durum wheat, Spring oats, Spring rye, Spring wheat, Triticale, Winter barley, Winter oats, Winter rye, Winter wheat
prothioconazole + tebuconazole + triazoxide	Raxil Pro	FS	Spring barley, Winter barley
silthiofam	Latitude	FS	Durum wheat, Spring rye, Spring wheat, Triticale, Winter barley, Winter rye, Winter wheat
tebuconazole	Gizmo	FS	Spring barley, Winter barley
thiamethoxam	Cruiser SB	FS	Fodder beet, Sugar beet
thiram	Agrichem Flowable Thiram	FS	Broad beans, Bulb onions, Cabbages, Carrots, Cauliflowers, Celery (outdoor), Combining peas, Cress, Dwarf beans, Edible podded peas, Endives, Fodder beet, Forage maize, Frise, Garlic, Grass seed, Lamb's lettuce, Leeks, Lettuce, Lupins, Mangels, Parsley, Poppies for morphine production, Radicchio, Radishes, Red beet, Rhubarb, Runner beans, Salad onions, Scarole, Shallots, Soya beans, Spring field beans, Spring oilseed rape, Sugar beet, Swedes, Turnips, Vining peas, Winter field beans, Winter oilseed rape

Chemical	Product	Formulation	Crop(s)
	Thiraflo	FS	Broad beans, Combining peas, Dwarf beans, Forage maize, Grass seed, Runner beans, Spring field beans, Spring oilseed rape, Vining peas, Winter field beans, Winter oilseed rape
	Thyram Plus	FS	Broad beans, Bulb onions, Cabbages, Cauliflowers, Combining peas, Dwarf beans, Edible podded peas, Forage maize, Garlic, Grass seed, Leeks, Lupins, Radishes, Runner beans, Salad onions, Shallots, Soya beans, Spring field beans, Spring oilseed rape, Swedes, Turnips, Vining peas, Winter field beans, Winter oilseed rape
tolclofos-methyl	Rizolex	DS	Potatoes
	Rizolex Flowable	FS	Potatoes

Aerial Application of Pesticides

Only those products specifically approved for aerial application may be so applied, and they may only be applied to specific crops or for specified uses. A complete list of products approved for application from the air can be accessed in Guide to Pesticides on the website of the Pesticides Safety Directorate (PSD, www.pesticides.gov.uk). The list in Table 5.5 is taken mainly from this source, with updating from issues of *The Pesticides Monitor* and product labels.

It is emphasised that the list is for guidance only – reference must be made to the product labels for detailed conditions of use, which must be complied with. The list does not include those products which have been granted restricted aerial application approval limiting the area which may be treated.

Detailed rules are imposed on aerial application regarding prior notification of the Nature Conservancy, water authorities, bee keepers, Environmental Health Officers, neighbours, hospitals, schools, etc. and the conditions under which application may be made. The full conditions are available from Defra and must be consulted before any aerial application is made.

Table 5.5 Products approved for aerial application

Chemical	Product	Crops
2-chloroethylphosphonic acid	Agriguard Cerusite	Winter barley
	Agrotech Ethephon	Winter barley
	Cerone	Winter barley
	Ethefon 480	Winter barley
	Pan Stiffen	Winter barley
asulam	Agrotech Asulam	Amenity grassland, Forest, Permanent grassland, Rotational grass
	Asulox	Amenity grassland, Forest, Permanent grassland, Rotational grass
	Brack-N	Amenity grassland, Forest, Permanent grassland
	Greencrop Frond	Amenity grassland, Forest, Permanent grassland
benalaxyl + mancozeb	Intro Plus	Potatoes
chlormequat	Agriguard Chlormequat 700	Spring oats, Spring wheat, Winter oats, Winter wheat
	Agriguard Chlormequat 720	Spring oats, Spring wheat, Winter oats, Winter wheat
	Agrovista 3 See	Spring oats, Spring wheat, Winter oats, Winter wheat
	Barclay Holdup	Spring oats, Spring wheat, Winter barley, Winter oats, Winter wheat

Chemical	Product	Crops
	Barclay Holdup 640	Spring oats, Spring wheat, Winter barley, Winter oats, Winter wheat
	Barclay Take 5	Spring oats, Spring wheat, Triticale, Winter barley, Winter oats, Winter rye, Winter wheat
	BASF 3C Chlormequat 720	Spring oats, Spring rye, Spring wheat, Triticale, Winter barley, Winter oats, Winter rye, Winter wheat
	Greencrop Carna	Spring oats, Spring wheat, Winter barley, Winter oats, Winter wheat
	Hive	Winter wheat
	Mirquat	Spring oats, Spring wheat, Winter oats, Winter wheat
	New 5C Cycocel	Spring oats, Spring rye, Spring wheat, Triticale, Winter barley, Winter oats, Winter rye, Winter wheat
	New 5C Quintacel	Spring oats, Spring rye, Spring wheat, Triticale, Winter barley, Winter oats, Winter rye, Winter wheat
	Sigma PCT	Spring wheat, Winter barley, Winter wheat
	Terbine	Spring oats, Spring wheat, Winter oats, Winter wheat
chlorothalonil	Agriguard Chlorothalonil	Potatoes
	Bravo 500	Potatoes
	Busa	Potatoes
	Cheer 500	Potatoes
	CTL 500	Potatoes
	Greencrop Orchid 2	Potatoes
	Greencrop Orchid B	Potatoes
	Joules	Potatoes
	Juliet	Potatoes

Chemical	Product	Crops
	Landgold Chlorothalonil 50	Potatoes
	Repulse	Potatoes
	Sanspor	Potatoes
	Sonar	Potatoes
chlorotoluron	Lentipur CL 500	Durum wheat, Triticale, Winter barley, Winter wheat
copper oxychloride	Cuprokylt	Potatoes
diflubenzuron	Dimilin Flo	Forest
dimethoate	BASF Dimethoate 40	Spring rye, Spring wheat, Sugar beet, Sugar beet seed crops, Triticale, Winter rye, Winter wheat
	Danadim Progress	Spring rye, Spring wheat, Sugar beet, Triticale, Winter rye, Winter wheat
mancozeb	Dithane NT Dry Flowable	Potatoes
	Laminator FL	Potatoes
	Malvi	Potatoes
	Micene 80	Potatoes
	Micene DF	Potatoes
	Penncozeb WDG	Potatoes
	Quell Flo	Potatoes
metaldehyde	CDP Minis	All edible crops (outdoor), All non-edible crops (outdoor), Natural surfaces not intended to bear vegetation
	Condor 5	All edible crops (outdoor), All non-edible crops (outdoor), Natural surfaces not intended to bear vegetation
	Dixie 6	All edible crops (outdoor), All non-edible crops (outdoor), Natural surfaces not intended to bear vegetation

Chemical	Product	Crops
	Doff Horticultural Slug Killer Blue Mini Pellets	All edible crops (outdoor), All non-edible crops (outdoor), Cultivated land/soil
	ESP	All edible crops (outdoor), All non-edible crops (outdoor)
	Lynx S	All edible crops (outdoor), All non-edible crops (outdoor)
	Optimol	All edible crops (outdoor), All non-edible crops (outdoor), Natural surfaces not intended to bear vegetation
	Pesta M	All edible crops (outdoor), All non-edible crops (outdoor)
	Polymet 5	All edible crops (outdoor), All non-edible crops (outdoor), Natural surfaces not intended to bear vegetation
	Super 5	All edible crops (outdoor), All non-edible crops (outdoor), Natural surfaces not intended to bear vegetation
pirimicarb	Aphox	Durum wheat, Spring barley, Spring oats, Spring rye, Spring wheat, Triticale, Winter barley, Winter oats, Winter rye, Winter wheat
	Arena	Spring barley, Spring oats, Spring wheat, Winter barley, Winter oats, Winter wheat
	Phantom	Durum wheat, Spring barley, Spring oats, Spring rye, Spring wheat, Triticale, Winter barley, Winter oats, Winter rye, Winter wheat
	Pirimate	Spring barley, Spring oats, Spring wheat, Winter barley, Winter oats, Winter wheat
	Pirimicarb 50	Durum wheat, Spring barley, Spring oats, Spring rye, Spring wheat, Triticale, Winter barley, Winter oats, Winter rye, Winter wheat
	Reynard	Durum wheat, Spring barley, Spring oats, Spring rye, Spring wheat, Triticale, Winter barley, Winter oats, Winter rye, Winter wheat

Chemical	Product	Crops
	Standon Pirimicarb 50	Durum wheat, Spring barley, Spring oats, Spring rye, Spring wheat, Triticale, Winter barley, Winter oats, Winter rye, Winter wheat
tri-allate	Avadex Excel 15G	Combining peas, Fodder beet, Lucerne, Red beet, Red clover, Sainfoin, Spring barley, Spring field beans, Sugar beet, Triticale, Vetches, Vining peas, White clover, Winter barley, Winter field beans, Winter wheat

Resistance Management

Pest species are, by definition, adaptable organisms. The development of resistance to some crop protection chemicals is just one example of this adaptability. Repeated use of products with the same mode of action will clearly favour those individuals in the pest population able to tolerate the treatment. This leads to a situation where the tolerant (or resistant) individuals can dominate the population and the product becomes ineffective. In general, the more rapidly the pest species reproduces and the more mobile it is, the faster the emergence of resistant populations, although some weeds seem able to evolve resistance more quickly than would be expected. In the UK, key independent research organisations, chemical manufacturers and other organisations have collaborated to share knowledge and expertise on resistance issues through three action groups. Participants include ADAS, the Chemicals Regulation Directorate, universities, colleges and the Home Grown Cereals Authority. The groups have a common aim of monitoring resistance in UK and devising and publishing management strategies designed to combat it where it occurs.

- **Weed Resistance Action Group (WRAG)**, formed in 1989 (Secretary: Dr Stephen Moss, Rothamsted Research, Harpenden, Herts AL5 2JQ *Tel: 01582 7631330 ext. 2521*).
- **Fungicide Resistance Action Group (FRAG)**, formed in 1995 (Secretary: Mr Oliver Macdonald, Chemicals Regulation Directorate, Mallard House, King's Pool, 3 Peasholme Green, York YO1 2PX *Tel: 01904 455864*).
- **Insecticide Resistance Action Group (IRAG)**, formed in 1997 (Secretary: Dr Stephen Foster, Rothamsted Research, Harpenden, Herts AL5 2JQ *Tel: 01582 763133 ext. 2324*).

The above groups publish detailed advice on resistance management relevant for each sector and, in some cases, specific to a pest problem. This information, together with further details about the function of each group, can be obtained from the Chemicals Regulation Directorate website (www.pesticides.gov.uk).

The speed at which resistance appears depends on the mode of action of the crop protection chemicals, as well as the manner in which they are used. Resistance among insects and fungal diseases has been evident for much longer than weed resistance to herbicides, but examples in all three categories are now widespread and increasing. This has created a need for agreement on the advice given for the use of crop protection chemicals in order to reduce the likelihood of the development of resistance and to avoid the loss of potentially valuable products in the chemical armoury. Mixing or alternating modes of action is one of the guiding principles of resistance management. To assist appropriate product choices, mode of action codes, published by the international resistance action committees (see below), are shown in the respective active ingredient fact sheets. The product label and/or a professional advisor should always be consulted before making decisions. The general guidelines for resistance management are similar for all three problem areas.

Preparation in advance

- Be aware of factors that favour the development of resistance, such as repeated annual use of the same product, and assess the risk.
- Plan ahead and aim to integrate all possible means of control.
- Use cultural measures, such as rotations, stubble hygiene, variety selection and, for fungicides, removal of primary inoculum sources, to reduce reliance on chemical control.
- Monitor crops regularly.
- Keep aware of local resistance problems.
- Monitor effectiveness of actions taken and take professional advice, especially in cases of unexplained poor control.

Using crop protection products

- Optimise product efficacy by using them as directed, at the right time, in good condition.
- Treat pest problems early.
- Mix or alternate chemicals with different modes of action.
- Avoid repeated applications of very low doses.
- Keep accurate field records.

Label guidance depends on the appropriate strategy for the product. Most frequently it consists of a warning of the possibility of poor performance due to resistance, and a restriction on the number of treatments that should be applied in order to minimise the development of resistance. This information is summarised in the profiles in the active ingredient fact sheets, but detailed guidance must always be obtained by reading the label itself before use.

International Action Committees

Resistance to crop protection products is an international problem. Agrochemical industry collaboration on a global scale is via three action committees whose aims are to support a coordinated industry approach to the management of resistance worldwide. In particular they produce lists of crop protection chemicals classified according to their mode of action. These lists and other information can be obtained from the respective websites.

Herbicide Resistance Action Committee – www.hracglobal.com

Fungicide Resistance Action Committee – www.frac.info/frac/index.htm

Insecticide Resistance Action Committee – www.irac-online.org

Poisons and Poisoning

Chemicals Subject to the Poison Law

Certain products are subject to the provisions of the Poisons Act 1972, the Poisons List order 1982 and the Poisons Rules 1982 (copies obtainable from The Stationery Office, www.tso.co.uk). These Rules include general and specific provisions for the storage and sale and supply of listed non-medicine poisons. Full details can be accessed in Annex C of the Guide to Pesticides on the Chemicals Regulation Directorate (CRD) website. The nature of the formulation and the concentration of the active ingredient allow some products to be exempted from the Rules, while others with the same active ingredient are included (see below). The chemicals approved for use in the UK are specified under Parts I and II of the Poisons List as follows.

Part I Poisons (sale restricted to registered retail pharmacists and to registered non-pharmacy businesses provided sales do not take place on retail premises):

- aluminium phosphide
- methyl bromide
- chloropicrin
- sodium cyanide
- magnesium phosphide
- strychnine.

Part II Poisons (sale restricted to registered retail pharmacists and listed sellers registered with a local authority):

- alphachloralose
- fentin acetate
- phorate (a)
- ammonium bifluoride
- fentin hydroxide
- sodium fluoride, carbofuran (a)
- formaldehyde
- sulphuric acid
- chlorfenvinphos (a,b)
- mephosfolan
- thiometon
- demeton-s-methyl
- methomyl (e)
- triazaphos (a)
- dichlorvos (a,d)
- nicotine (c)
- zinc phosphide
- disulfuton (a)
- oxamyl (a).

Notes

(a) Granular formulations which do not contain more than 12% w/w of this, or a combination of similarly flagged poisons, are exempt

(b) Treatments on seeds are exempt

(c) Formulations containing not more than 7.5% of nicotine are exempt

(d) Preparations in aerosol dispensers containing not more than 1% w/w active ingredient are exempt. Materials impregnated with dichlorvos for slow release are exempt

(e) Solid substances containing not more than 1% w/w a.i. are exempt.

Occupational Exposure Limits

A fundamental requirement of the COSHH Regulations is that exposure of employees to substances hazardous to health should be prevented or adequately controlled. Exposure by inhalation is usually the main hazard, and in order to measure the adequacy of control of exposure by this route various substances have been assigned occupational exposure limits.

There are two types of occupational exposure limits defined under COSHH: Occupational Exposure Standards (OES) and Maximum Exposure Limits (MEL). The key difference is that an OES is set at a level at which there is no indication of risk to health; for a MEL a residual risk may exist and the level takes socio-economic factors into account. In practice, MELs have been most often allocated to carcinogens and to other substances for which no threshold of effect can be identified and for which there is no doubt about the seriousness of the effects of exposure.

OESs and MELs are set on the recommendation of the Advisory Committee on Toxic Substances (ACTS). Full details are published by HSE in *EH 40/95 Occupational Exposure Limits 1995*, ISBN 0 7176 0876 X.

As far as pesticides are concerned, OESs and MELs have been set for relatively few active ingredients. This is because pesticide products usually contain other substances in their formulation, including solvents, which may have their own OES/MEL. In practice inhalation of solvent may be at least, or more, important than that of the active ingredient. These factors are taken into account by the regulators when approving a pesticide product under the Control of Pesticide Regulations. This indicates one of the reasons why a change of pesticide formulation usually necessitates a new approval assessment under COPR.

First Aid Measures

If pesticides are handled in accordance with the required safety precautions, as given on the container label, poisoning should not occur. It is difficult, however, to guard completely against the occasional accidental exposure. Thus, if a person handling, or exposed to, pesticides becomes ill, it is a wise precaution to apply first aid measures appropriate to pesticide poisoning even though the cause of illness may eventually prove to have been quite different. An employer has a legal duty to make adequate first aid provision for employees. Regular pesticide users should consider appointing a trained first aider even if numbers of employees are not large, since there is a specific hazard.

The first essential in a case of suspected poisoning is for the person involved to stop work, to be moved away from any area of possible contamination and for a doctor to be called at once. If no doctor is available the patient should be taken to hospital as quickly as possible. In either event it is most important that the name of the chemical being used should be recorded and preferably the whole product label or leaflet should be shown to the doctor or hospital concerned.

Some pesticides, which are unlikely to cause poisoning in normal use, are extremely toxic if swallowed accidentally or deliberately. In such cases get the patient to hospital as quickly as possible, with all the information you have. Some labels now include Material Safety Data Sheets and these contain valuable information for both the first aider and medical staff.

General Measures

Measures appropriate in all cases of suspected poisoning include the following.

- Remove any protective or other contaminated clothing (taking care to avoid personal contamination).
- Wash any contaminated areas carefully with water or with soap and water if available.
- In cases of eye contamination, flush with plenty of clean water for at least 15 minutes.
- Lay the patient down, keep at rest and under shelter. Cover with one clean blanket or coat, etc. Avoid overheating.
- Monitor level of consciousness, breathing and pulse rate.
- If consciousness is lost, place the casualty in the recovery position (on his/her side with head down and tongue forward to prevent inhalation of vomit).

Reporting of Pesticide Poisoning

Any cases of poisoning by pesticides must be reported without delay to an HM Agricultural Inspector of the Health and Safety Executive. In addition any cases of poisoning by substances named in schedule 2 of The Reporting of Injuries, Diseases and Dangerous Occurrences Regulations 1985, must also be reported to HM Agricultural Inspectorate (this includes organophosphorus chemicals, mercury and some fumigants).

Cases of pesticide poisoning should also be reported to the manufacturer concerned.

Additional Information

General advice on the safe use of pesticides is given in a range of Health and Safety Executive leaflets available from HSE Books. The major agrochemical companies are able to provide authoritative medical advice about their own pesticide products. Useful information is now available from the Material Safety Data Sheet available from the manufacturer or the manufacturer's website.

A useful booklet, *Guidelines for Emergency Measures in Cases of Pesticide Poisoning,* is available from CropLife International.

New arrangements for the provision of information to doctors about poisons and the management of poisonings have been introduced as part of the modernisation of the National Poisons Information Service (NPIS). The Service provides a year-round, 24-hour-a-day service for healthcare staff on the diagnosis, treatment and management of patients who may have been poisoned. The new arrangements are aimed at moving away from the telephone as the first point of contact for poisons information, to the use by doctors of an online database, supported by a second-tier, consultant-led information service for more complex clinical advice. NPIS no longer provides direct information on poisoning to members of the public. **Anyone suspecting poisoning by pesticides should seek professional medical help immediately via their GP or NHS Direct (www.nhsdirect.nhs.uk) or NHS 24 (www.nhs24.com).**

Environmental Protection

Protection of Bees

Honey bees

Honey bees are a source of income for their owners and important to farmers and growers as pollinators of their crops. It is irresponsible and unnecessary to use pesticides in such a way that may endanger them. Pesticides vary in their toxicity to bees, but those that present a special hazard carry a specific warning in the precautions section of the label. They are indicated in this *Guide* in the hazard classification and safety precautions section of the pesticide profile.

Product labels indicate the necessary environmental precautions to take, but where use of an insecticide on a flowering crop is contemplated the British Beekeepers Association have produced the following guidelines for growers:

- Target insect pests with the most appropriate product.
- Choose a product that will cause minimal harm to beneficial species.
- Follow the manufacturer's instructions carefully.
- Inspect and monitor crops regularly.
- Avoid spraying crops in flower or where bees are actively foraging.
- Keep down flowering weeds.
- Spray late in the day, in still conditions.
- Avoid excessive spray volume and run-off.
- Adjust sprayer pressure to reduce production of fine droplets and drift.
- Give local beekeepers as much warning of your intention as possible.

Wild bees

Wild bees also play an important role. Bumblebees are useful pollinators of spring flowering crops and fruit trees because they forage in cool, dull weather when honey bees are inactive. They play a particularly important part in pollinating field beans, red and white clover, lucerne and borage. Bumblebees nest and overwinter in field margins and woodland edges. Avoidance of direct or indirect spray contamination of these areas, in addition to the creation of hedgerows and field margins, and late cutting or grazing of meadows, all help the survival of these valuable insects.

The Campaign Against Illegal Poisoning of Wildlife

The Campaign Against Illegal Poisoning of Wildlife, aimed at protecting some of Britain's rarest birds of prey and wildlife whilst also safeguarding domestic animals, was launched in March 1991 by the (then) Ministry of Agriculture, Fisheries and Food and the Department of the Environment, Transport and the Regions. The main objective is to deter those who may be considering using pesticides illegally.

The Campaign is supported by a range of organisations associated with animal welfare, nature preservation, field sports and gamekeeping including the RSPB, English Nature, the Countryside Alliance and the Game Conservancy Trust.

The three objectives are:

- To advise farmers, gamekeepers and other land managers on legal ways of controlling pests;
- To advise the public on how to report illegal poisoning incidents and to respect the need for legal alternatives;
- To investigate incidents and prosecute offenders.

A freephone number (0800 321 600) is available to make it easier for the public to report incidents, and numerous leaflets, posters, postcards and stickers have been created to publicise the existence of the Campaign. A video has also been produced to illustrate the many talks, demonstrations and exhibitions.

The Campaign arose from the results of the Wildlife Incident Investigation Scheme for the investigation of possible cases of illegal poisoning. Under this scheme, all reported incidents are considered and thoroughly investigated where appropriate. Enforcement action is taken wherever sufficient evidence of an offence can be obtained and numerous prosecutions have been made since the start of the Campaign.

Further information about the Campaign is available from: Chemicals Regulation Directorate, Research Co-ordination and Environmental Policy Branch, Room 317, Mallard House, 3 Peasholme Green, York YO1 7PX, or from the website: www.caip-uk.info/default.aspx. E-mail: wiis@hse.gsi.gov.uk

Water Quality

Even when diluted, some pesticides are potentially dangerous to fish and other aquatic life. Not only this, but many watercourses and groundwaters are sources of drinking water, and it requires only a tiny amount of contamination to breach the stringent European Union water quality standards. The EEC Drinking Water Directive sets a maximum admissible level in drinking water for any pesticide, regardless of its toxicity, at 1 part in 10,000 million. As little as 250 grams could be enough to cause the daily supply to a city the size of London to exceed the permitted levels (although it would be very unlikely to present a health hazard to consumers).

The protection of groundwater quality is therefore vital. The Food and Environment Protection Act 1985 (FEPA) places a special obligation on users of pesticides to "safeguard the environment and in particular avoid the pollution of water". Under the Water Resources Act 1991 it is an offence to pollute any controlled waters (watercourses or groundwater), either deliberately or accidentally. Protection of controlled waters from pollution is the responsibility of the Environment Agency (in England and Wales) and the Scottish Environmental Protection Agency. Addresses for both organisations can be found under Contacts.

Users of pesticides therefore have a duty to adopt responsible working practices and, unless they are applying herbicides in or near water, to prevent them getting into water. Guidance on how to achieve this is given in the Defra *Code of Good Agricultural Practice for the Protection of Water.*

The duty of care covers not only the way in which a pesticide is sprayed, but also its storage, preparation and disposal of surplus, sprayer washings and the container. Products that are a major hazard to fish, other aquatic life or aquatic higher plants carry one of several specific label precautions in their fact sheet, depending on the assessed hazard level.

Where to get information

For general enquiries telephone 0645 333 111 (Environment Agency) or 01786 457 700 (SEPA). Both Agencies operate a 24-hour emergency hotline for reporting all environmental incidents relating to air, land and water:

0800 80 70 60 (for incidents in England and Wales)

0345 73 72 71 (for incidents in Scotland)

In addition, printed literature concerning the protection of water is available from the Environment Agency and the Crop Protection Association (see Key References).

Protecting surface waters

Surface waters are particularly vulnerable to contamination. One of the best ways of preventing those pesticides that carry the greatest risk to aquatic wildlife from reaching surface waters is to prohibit their application within a boundary adjacent to the water. Such areas are known as no-spray, or buffer, zones. Certain products are restricted in this way and have a legally binding label precaution to make sure the potential exposure of aquatic organisms to pesticides that might harm them is minimised.

Before 1999 the protected zones were measured from the edge of the water. The distances were 2 metres for hand-held or knapsack sprayers, 6 metres for ground crop sprayers, and a variable distance (but often 18 metres) for broadcast air-assisted applications, such as in orchards. The introduction of LERAPs (see below) has changed the method of measuring buffer zones.

'Surface water' includes lakes, ponds, reservoirs, streams, rivers and watercourses (natural or artificial). It also includes temporarily or seasonally dry ditches, which have the potential to carry water at different times of the year. Buffer zone restrictions do not necessarily apply to all products containing the same active ingredient. Those in formulations that are not likely to contaminate surface water through spray drift do not pose the same risk to aquatic life and are not subject to the restrictions.

Local Environmental Risk Assessment for Pesticides (LERAPs)

Local Environment Risk Assessments for Pesticides (LERAPs) were introduced in March 1999, and revised guidelines were issued in January 2002. They give users of most products currently subject to a buffer zone restriction the option of continuing to comply with the existing buffer zone restriction (using the new method of measurement), or carrying out a LERAP and possibly reducing the size of the buffer zone as a result. In either case, there is a new legal obligation for the user to record his decision, including the results of the LERAP.

The scheme has changed the method of measuring the buffer zone. Previously the zone was measured from the edge of the water, but it is now the distance from the top of the bank of the watercourse to the edge of the spray area.

The LERAP provides a mechanism for taking into account other factors that may reduce the risk, such as dose reduction, the use of low drift nozzles, and whether the watercourse is dry or flowing. The previous arrangements applied the same restriction regardless of whether there was actually water present. Now there is a standard zone of 1 m from the top of a dry ditch bank.

Other factors to include in a LERAP that may allow a reduction in the buffer zone are:

- The size of the watercourse, because the wider it is, the greater the dilution factor, and the lower the risk of serious pollution.
- The dose applied. The lower the dose, the less is the risk.
- The application equipment. Sprayers and nozzles are star-rated according to their ability to reduce spray drift fallout. Equipment offering the greatest reductions achieves the highest rating of three stars. The scheme was originally restricted to ground crop sprayers; new, more flexible rules introduced in February 2002 included broadcast air-assisted orchard and hop sprayers.
- Other changes introduced in 2002 allow the reduction of a buffer zone if there is an appropriate living windbreak between the sprayed area and a watercourse.

Not all products that had a label buffer zone restriction are included in the LERAP scheme. The option to reduce the buffer zone does not to apply to organophosphorus or synthetic pyrethroid insecticides. This group are classified as Category A products. All other products that had a label buffer zone restriction are classified as Category B. In addition some products have a 'Broadcast air-assisted LERAP' classification. The wording of the buffer zone label precautions has been amended for all products to take account of the new method of measurement, and whether or not the particular product qualifies for inclusion in the LERAP scheme.

Products in this *Guide* that are in Category A or B, or have a broadcast air-assisted LERAP, are identified with an appropriate icon on the product fact sheet. Updates to the list are published regularly by CRD and details can be obtained from their website at www.pesticides.gov.uk.

The introduction of LERAPs was an important step forward because it demonstrated a willingness to reduce the impact of regulation on users of pesticides by allowing flexibility where local conditions make it safe to do so. This places a legal responsibility on the user to ensure the risk assessment is done either by himself or by the spray operator or by a professional consultant or advisor. It is compulsory to record the LERAP and make it available for inspection by enforcement authorities.

More details of the LERAP arrangements and guidance on how to carry out assessments are published in the Ministry booklets PB5621 *Local Environmental Risk Assessment for Pesticides – Horizontal Boom Sprayers*, and booklet PB6533 *Local Environment Risk Assessment for Pesticides – Broadcast Air-Assisted Sprayers*. Additional booklets, PB2088 *Keeping Pesticides Out of Water*, and PB3160 *Is Your Sprayer Fit for Work* give general practical guidance.

Groundwater Regulations

Groundwater Regulations were introduced in 1999 to complete the implementation in UK of the EU Groundwater Directive (Protection of Groundwater Against Pollution Caused by Certain Dangerous Substances – 80/68/EEC). These Regulations help prevent the pollution of groundwater by controlling discharges or disposal of certain substances, including all pesticides.

Groundwater is defined under the Regulations as any water contained in the ground below the water table. Pesticides must not enter groundwater unless it is deemed by the appropriate Agency to be permanently unsuitable for other uses. The Agricultural Waste Regulations were introduced in May 2006 and apply to the disposal of all farm wastes, including pesticides. With certain exemptions farmers are required to obtain a waste management licence for most waste disposal activities.

As far as pesticides are concerned the new Regulations made little difference to existing controls, except for the disposal of empty containers and the disposal of sprayer washings and rinsings. It remains an offence to dispose of pesticides onto land without official authorisation. Normal use of a pesticide in accordance with product approval does not require authorisation. This includes spraying the washings and rinsings back on the crop provided that, in so doing, the maximum approved dose for that product on that crop is not exceeded. However, those wishing to use a lined biobed for this purpose need to obtain an exemption from the Agency.

In practice the best advice to farmers and growers is to plan to use all diluted spray within the crop and to dispose of all washings via the same route making sure that they stay within the conditions of approval of the product. The enforcing agencies for these Regulations are the Environment Agency (in England and Wales) and the Scottish Environmental Protection Agency. The Agencies can serve notice at any time to modify the conditions of an authorisation where necessary to prevent pollution of groundwater.

Integrated Farm Management (IFM)

Integrated farm management is a method of farming that balances the requirements of running a profitable farming business with the adoption of responsible and sensitive environmental management. It is a whole-farm, long-term strategy that combines the best of modern technology with some basic principles of good farming practice. It is a realistic way forward that addresses the justifiable concerns of the environmental impact of modern farming practices, at the same time as ensuring that the industry remains viable and continues to provide wholesome, affordable food. IFM embraces arable and livestock management. The phrase 'integrated crop management' is sometimes used where a farm is wholly arable.

Pest control is essential in any management system. Much can be done to minimise the incidence and impact of pests, but their presence is almost inevitable. IFM ensures that, where a pest problem needs to be contained, the action taken is the best combination of all available options. Pesticides are one of these options, and form an essential, but by no means exclusive, part of pest control strategy.

Where chemicals are to be used, the choice of product should be made not only with the pest problem in mind, but with an awareness of the environmental and social risks that might accompany its use. The aim should be to use as much as necessary, but as little as possible. A major part of the approval process aims to safeguard the environment, so that any approved product, when used as directed, will not cause long-term harm to wildlife or the environment. This is achieved by specifying on the label detailed rules for the way in which a product may be used.

The skill in implementing an IFM pest control strategy is the decision on how easily these rules may be complied with in the particular situation where use is contemplated. Although many chemical options may be available, some are likely to be more suitable than others. This *Guide* lists, under the heading **Special precautions/Environmental safety**, the key label precautions that need to be considered in this context, although the actual product label must always be read before a product is used.

SECTION 6
APPENDICES

Appendix 1
Suppliers of Pesticides and Adjuvants

AgChem Access: AgChemAccess Limited
Pure House,
64-66 Westwick Street
Norwich
Norfolk
NR2 4SZ
UK
Tel: 01603 624413
Fax: 01759 371971
Email: martin@agchemaccess.co.uk
Web: www.agchemaccess.com

Agform: Agform Ltd
Maidstone Heath
Blundell Lane
Burlesdon
Southampton
SO31 1AA
Tel: 023 8040 7831
Fax: 023 8040 7198
Email: info@agform.com

Agrichem: Agrichem (International) Ltd
Industrial Estate
Station Road
Whittlesey
Cambs.
PE7 2EY
Tel: (01733) 204019
Fax: (01733) 204162
Email: alison.williamson@agrichem.co.uk
Web: www.agrichem.co.uk

AgriChem BV: AgriChem BV
Koopvaardijweg 9
4906 CV Oosterhout
Postbus 295
4900 AG Osterhout
Netherlands
Tel: 0031 1624 31931
Fax: 0031 1624 56797
Email: info@agrichem.com
Web: www.agrichem.com

AgriGuard: AgriGuard Ltd
Unit 1
Broomfield Business Park
Malahide
Co. Dublin
Ireland
Tel: (+353) 1 846 2044
Fax: (+353) 1 846 2489
Email: info@agriguard.ie
Web: www.agriguard.ie

Agropharm: Agropharm Limited
Buckingham Place
Church Road
Penn
High Wycombe
Bucks.
HP10 8LN
Tel: (01494) 816575
Fax: (01494) 816578
Email: sales@agropharm.co.uk
Web: www.agropharm.co.uk

Agrovista: Agrovista UK Ltd
Cambridge House
Nottingham Road
Stapleford
Nottingham
NG9 8AB
Tel: (0115) 939 0202
Fax: (0115) 921 8498
Email: enquiries@agrovista.co.uk
Web: www.agrovista.co.uk

Amega: Amega Sciences
Lanchester Way
Royal Oak Industrial Estate
Daventry
Northants.
NN11 5PH
Tel: (01327) 704444
Fax: (01327) 71154
Email: admin@amega-sciences.com
Web: www.amega-sciences.com

Antec Biosentry: Antec Biosentry
DuPont Animal Health Solutions
Windham Road
Chilton Industrial Estate
Sudbury
Suffolk
CO10 2XD
Tel: (01787) 377305
Fax: (01787) 310846
Email: biosecurity@antecint.com
Web: www.antecint.com

B H & B: Battle Hayward & Bower Ltd
Victoria Chemical Works
Crofton Drive
Allenby Road Industrial Estate
Lincoln
LN3 4NP
Tel: (01522) 529206
Fax: (01522) 538960

Barclay: Barclay Chemicals Manufacturing Ltd
Damastown Way
Damastown Industrial Park
Mulhuddart
Dublin 15
Ireland
Tel: (+353) 1 811 2900
Fax: (+353) 1 822 4678
Email: margaret@barclay.ie
Web: www.barclay.ie

Barrier: Barrier BioTech Ltd
36/37 Haverscroft Industrial Estate
New Road
Attleborough
Norfolk
NR17 1YE
Tel: (01953) 456363
Fax: (01953) 455594
Email: sales@barrier-biotech.com
Web: www.barrier-biotech.com

BASF: BASF plc
Agricultural Divison
PO Box 4, Earl Road
Cheadle Hulme
Cheshire
SK8 6QG
Tel: (0845) 602 2553
Fax: (0161) 485 2229
Web: www.agricentre.co.uk

Bayer CropScience:
Bayer CropScience Limited
230 Cambridge Science Park
Milton Road
Cambridge
CB4 0WB
Tel: (01223) 226500
Fax: (01223) 426240
Web: www.bayercropscience.co.uk

Bayer Environ.: Bayer Environmental Science
230 Cambridge Science Park
Milton Road
Cambridge
CB4 0WB
Tel: (01223) 226680
Fax: (01223) 226635
Email:
claire.hazell@bayercropscience.com
Web: www.bayer-escience.co.uk

Belchim: Belchim Crop Protection Ltd
Unit 1b, Fenice Court
Eaton Socon
St Neots
Cambs.
PE19 8EP
Tel: (01480) 403333
Fax: (01480) 403444
Email: info@belchim.com
Web: www.belchim.com

Biofresh: Biofresh Ltd
Meadowcroft Lane
Ripponden
West Yorks
HX6 4AJ
UK
Tel: 01422 824 586
Email: info@bio-fresh.com

Certis: Certis
1b Mills Way
Boscombe Down Business Park
Amesbury
Wilts.
SP4 7RX
Tel: (01980) 676500
Fax: (01980) 626555
Email: certis@certiseurope.co.uk
Web: www.certiseurope.co.uk

ChemSource: ChemSource Ltd
5 Jupiter House
Calleva Park
Aldermaston
Reading
Berkshire
RG7 4NN
UK
Tel: 0845 459 4039
Email: emma@chemsource.co.uk

Chemtura: Chemtura Europe Ltd
Kennet House
4 Langley Quay
Slough
Berks.
SL3 6EH
Tel: (01753) 603000
Fax: (01753) 603077
Web: www.chemtura.com

Chiltern: Chiltern Farm Chemicals Ltd
East Mellwaters
Stainmore
Bowes
Barnard Castle
Co. Durham
DL12 9RH
Tel: (01833) 628282
Fax: (01833) 628020
Web: www.chilternfarm.com

Clayton: Clayton Plant Protection Ltd
Bracetown Business Park
Clonee
Co. Meath
Ireland
Tel: (+353) 1 821 0127
Fax: (+353) 81 841 1084
Email: info@cppltd.eu
Web: www.cppltd.eu

DAPT: DAPT Agrochemicals Ltd
14 Monks Walk
Southfleet
Gravesend
Kent
DA13 9NZ
Tel: (01474) 834448
Fax: (01474) 834449
Email: rkjltd@supanet.com

De Sangosse: De Sangosse Ltd
Hillside Mill
Quarry Lane
Swaffham Bulbeck
Cambridge
CB25 0LU
Tel: (01223) 811215
Fax: (01223) 810020
Email: info@desangosse.co.uk
Web: www.desangosse.co.uk

Doff Portland: Doff Portland Ltd
Aerial Way
Hucknall
Nottingham
NG15 6DW
Tel: (0115) 963 2842
Fax: (0115) 963 8657
Email: info@doff.co.uk
Web: www.doff.co.uk

Dow: Dow AgroSciences Ltd
Latchmore Court
Brand Street
Hitchin
Herts.
SG5 1NH
Tel: (01462) 457272
Fax: (01462) 426605
Email: fhihotl@dow.com
Web: www.dowagro.com/uk

DuPont: DuPont (UK) Ltd
Crop Protection Products Department
Wedgwood Way
Stevenage
Herts.
SG1 4QN
Tel: (01438) 734450
Fax: (01438) 734452
Web: www.gbr.ag.dupont.com

Elliott: Thomas Elliott Fertilisers
Bencewell Granary
Oakley Road
Bromley
Kent
BR2 8HG
Tel: (0208) 462 6622
Fax: (0208) 462 5599

European Ag: European Agrichemicals Ltd
Unit NR2, Hurst House
City Road
Radnage
Bucks
HP14 4DW
UK
Tel: 01494 483728
Email: Radnage

Fargro: Fargro Ltd
Toddington Lane
Littlehampton
Sussex
BN17 7PP
Tel: (01903) 721591
Fax: (01903) 730737
Email: hortec@fargro.co.uk
Web: www.fargro.co.uk

Fine: Fine Agrochemicals Ltd
Hill End House
Whittington
Worcester
WR5 2RQ
Tel: (01905) 361800
Fax: (01905) 361810
Email: enquire@fine.eu
Web: www.fine.eu

Frontier: Frontier Agriculture Ltd
Fleet Road Industrial Estate
Holbeach
Spalding
Lincs.
PE12 8LY
Tel: (01406) 421405
Email: les_sykes@bankscargill.co.uk
Web: www.bankscargill.co.uk

Globachem: Globachem NV
Leeuwerweg 138
BE-3803 Sint Truiden
Belgium
Tel: (+32) 11 785 717
Fax: (+32) 11 681 565
Email: globachem@globachem.com
Web: www.globachem.com

Goldengrass: Goldengrass Ltd
PO Box 280
Ware
SG12 8XY
Tel: 01920 486253
Fax: 01920 485123
Email: contact@goldengrass.uk.com

Gowan: Gowan International
51 Harlow Moor Drive
Harrogate
N. Yorkshire
HG2 0LE
Tel: (01432) 524500
Fax: (01432) 524500

Greenaway: Greenaway Amenity Ltd
7 Browntoft Lane
Donington
Spalding
Lincs.
PE11 4TQ
Tel: (01775) 821031
Fax: (01775) 821034
Email: greenawayamenity@aol.com
Web: www.greenawaycda.com

Greencrop: Greencrop Technology Ltd
See Clayton Plant Protection Ltd

Headland: Headland Agrochemicals Ltd
Rectors Lane
Pentre
Deeside
Flintshire
CH5 2DH
Tel: (01244) 537370
Fax: (01244) 532097
Email: enquiry@headlandgroup.com
Web: www.headland-ag.co.uk

Headland Amenity:
Headland Amenity Limited
1010 Cambourne Business Park
Cambourne
Cambs.
CB3 6DP
Tel: (01223) 597834
Fax: (01223) 598052
Email: info@headlandamenity.com
Web: www.headlandamenity.com

Helena: Helena Chemical Company
Cambridge House
Nottingham Road
Stapleford
Nottingham
NG9 8AB
Tel: (0115) 939 0202
Fax: (0115) 939 8031

Hermoo: Hermoo (UK)
21 Victoria Road
Wargrave
Berks.
RG10 8AD
Tel: (0118) 940 4264
Fax: (0118) 940 4264
Email: johnhudson23@aol.com
Web: www.hermoo.com

Interagro: Interagro (UK) Ltd
230 Avenue West,
Skyline 120,
Great Notley
Braintree
Essex
CM77 7AA
Tel: (01376) 552703
Fax: (01376) 567608
Email: info@interagro.co.uk
Web: www.interagro.co.uk

Interfarm: Interfarm (UK) Ltd
Kinghams's Place
36 Newgate Street
Doddington
Cambs.
PE15 0SR
Tel: (01354) 741414
Fax: (01354) 741004
Email: technical@interfarm.co.uk
Web: www.interfarm.co.uk

Intracrop: Intracrop
Little Hay
Broadwell
Lechlade
Glos.
GL7 3QS
Tel: (01367) 860255
Fax: (01926) 634798
Email: catherine@a-o-c.co.uk

Irish Drugs: Irish Drugs Ltd
Burnfoot
Lifford
Co. Donegal
Ireland
Tel: (+353) 74 9368104
Fax: (+353) 74 9368311
Email: jgmcivor@eircom.net
Web: www.idl-home.com

K & S Fumigation:
K & S Fumigation Services Ltd
Asparagus Farm
Court Lodge Road
Appledore
Ashford
Kent
TN26 2DH
Tel: (01233) 758252
Fax: (01233) 758343
Email: david@kstreatments.co.uk

Killgerm: Killgerm Chemicals Ltd
115 Wakefield Road
Flushdyke
Ossett
W. Yorks.
WF5 9AR
Tel: (01924) 268400
Fax: (01924) 264757
Email: info@killgerm.com
Web: www.killgerm.com

Koppert: Koppert (UK) Ltd
Unit 8
53 Hollands Road
Haverhill
Suffolk
CB9 8PJ
Tel: (01440) 704488
Fax: (01440) 704487
Email: info@koppert.co.uk
Web: www.koppert.com

Landseer: Landseer Ltd
Lodge Farm
Goat Hall Lane
Galleywood
Chelmsford
Essex
CM2 8PH
Tel: (01245) 357109
Fax: (01245) 494165
Web: www.lanfruit.co.uk

Law Fertilisers: Law Fertilisers Ltd
Eastwood End
Wimblington
Cambs.
PE15 0QJ
Tel: (01354) 740740
Fax: (01354) 740720
Email: sales@lawfertilisers.co.uk
Web: www.lawfertilisers.co.uk

Luxan: Luxan (UK) Ltd
Crown Business Park
Old Dalby
Leics.
LE14 3NQ
Tel: (01664) 820052
Fax: (01664) 820216
Email: enquiry@luxan.co.uk
Web: www.luxan.co.uk

Makhteshim: Makhteshim-Agan (UK) Ltd
Unit 16
Thatcham Business Village
Colthrop Way
Thatcham
Berks.
RG19 4LW
Tel: (01635) 860555
Fax: (01635) 861555
Email: admin@mauk.co.uk
Web: www.mauk.co.uk

Mandops: Mandops (UK) Ltd
36 Leigh Road
Eastleigh
Hants.
SO50 9DT
Tel: (023) 8064 1826
Fax: (023) 8062 9106
Email: enquiries@mandops.co.uk
Web: www.mandops.co.uk

Microcide: Microcide Ltd
Shepherds Grove
Stanton
Bury St. Edmunds
Suffolk
IP31 2AR
Tel: (01359) 251077
Fax: (01359) 251545
Email: microcide@microcide co.uk
Web: www.microcide.co.uk

Monsanto: Monsanto (UK) Ltd
PO Box 663
Cambourne
Cambridge
CB1 0LD
UK
Tel: (01954) 717575
Fax: (01954) 717579
Email: technical.helpline.uk@monsanto.com
Web: www.monsanto-ag.co.uk

Nissan: Nissan Chemical Europe Sarl
Parc d'Affaires de Crecy
2 Rue Claude Chappe
69371 St Didier au Mont d'Or
France
Tel: (+33) 4 376 44020
Fax: (+33) 4 376 46874

Nomix Enviro: Nomix Enviro, A Division of Frontier Agriculture Ltd
Portland Building
The Grain Silos
Weyhill Road
Andover
Hampshire
SP10 3NT
UK
Tel: (0117) 957 4574
Fax: (0117) 956 3461
Email: info@nomix.co.uk
Web: www.nomix.co.uk

Nufarm UK: Nufarm UK Ltd
Wyke Lane
Wyke
West Yorkshire
BD12 9EJ
UK
Tel: 01274 69 1234
Fax: 01274 69 1176
Email: infouk@uk.nufarm.com
Web: www.nufarm.co.uk

Omex: Omex Agriculture Ltd
Bardney Airfield
Tupholme
Lincoln
LN3 5TP
Tel: (01526) 396000
Fax: (01526) 396001
Email: enquire@omex.com
Web: www.omex.co.uk

Pan Agriculture: Pan Agriculture Ltd
8 Cromwell Mews
Station Road
St Ives
Huntingdon
Cambs.
PE27 5HJ
Tel: (01480) 467790
Fax: (01480) 467041
Email: info@panagriculture.co.uk

Pan Amenity: Pan Amenity Ltd
8 Cromwell Mews
Station Road
St Ives
Cambs.
PE27 5HJ
UK
Tel: 01480 467790
Fax: 01480 467041

Plant Solutions: Plant Solutions
Pyports
Downside Bridge Road
Cobham
Surrey
KT11 3EH
Tel: (01932) 576699
Fax: (01932) 868973
Email: sales@plantsolutionsltd.com
Web: www.plantsolutionsltd.com

PP Products: PP Products
Longmynd
Tunstead Road
Hoveton
Norwich
NR12 8QN
Tel: (01603) 784367
Fax: (01603) 784367
Email: mail@ppproducts.net

Q-Chem: Q-CHEM nv
Leeuwerweg 138
BE-3803
Belgium
Tel: (+32) 11 785 717
Fax: (+32) 11 681 565
Email: q-chem@skynet.be

Rentokil: Rentokil Initial plc
7-8 Foundry Court,
Foundry Lane
Horsham
W. Sussex
RH13 5PY
Tel: (01403) 214112
Fax: (01403) 214101
Email: anita.lockyer@rentokil-initial.com
Web: www.rentokil.com

Rhizopon: Rhizopon UK Ltd
Croda Rosa
12 Bixley Road
Ipswich
Suffolk
IP3 8PL
Tel: (01473) 712666
Fax: (01473) 712666

Rigby Taylor: Rigby Taylor Ltd
Rigby Taylor House
Crown Lane
Horwich
Bolton
Lancs.
BL6 5HP
Tel: (01204) 677777
Fax: (01204) 677765
Email: info@rigbytaylor.com
Web: www.rigbytaylor.com

Scotts: The Scotts Company (UK) Ltd
Paper Mill Lane
Bramford
Ipswich
Suffolk
IP8 4BZ
Tel: (01473) 830492
Fax: (01473) 830814
Web: www.scottsprofessional.co.uk

Sherriff Amenity: Sherriff Amenity Services
The Pines
Fordham Road
Newmarket
Suffolk
CB8 7LG
Tel: (01638) 721888
Fax: (01638) 721815
Web: www.sherriffamenity.com

Sinclair: William Sinclair Horticulture Ltd
Firth Road
Lincoln
LN6 7AH
Tel: (01522) 537561
Fax: (01522) 513609
Web: www.william-sinclair.co.uk

Sipcam: Sipcam UK Ltd
3 The Barn
27 Kneesworth Street
Royston
Herts.
SG8 5AB
Tel: (01763) 212100
Fax: (01763) 212101
Email: paul@sipcamuk.co.uk

Solufeed: Solufeed Ltd
Highground Orchards, Highground Lane
Barnham
Bognor Regis
West Sussex
PO22 0BT
Tel: 01243 554090
Fax: 01243 554568
Email: enquiries@solufeed.com

Sorex: Sorex Ltd
St Michael's Industrial Estate
Widnes
Cheshire
WA8 8TJ
Tel: (0151) 420 7151
Fax: (0151) 495 1163
Email: enquiries@sorex.com
Web: www.sorex.com

Sphere: Sphere Laboratories (London) Ltd
c/o Mainswood
Putley Common
Ledbury
Herefordshire
HR8 2RF
Tel: (07974) 732026
Fax: (01531) 670517
Email: homesfg@aol.com

Standon: Standon Chemicals Ltd
48 Grosvenor Square
London
W1K 2HT
Tel: (020) 7493 8648
Fax: (020) 7493 4219

Sumitomo:
Sumitomo Chemical Agro Europe SA
Horatio House
77-85 Fulham Palace Road
London
W6 8JA
Tel: 0208 600 7700
Fax: 0208 600 7717

Summit Agro: Summit Agro Europe Limited
Vintners' Place
68 Upper Thames Street
London
EC4V 3BJ
Tel: (020) 7246 3697
Fax: (020) 7246 3799
Email: summit@summit-agro.com

Syngenta:
Syngenta Crop Protection UK Limited
CPC4
Capital Park
Fulbourn
Cambridge
CB21 5XE
Tel: 01223 883400
Fax: (01223) 493700
Web: www.syngenta-crop.co.uk

Syngenta Bioline: Syngenta Bioline
Telstar Nursery
Holland Road
Little Clacton
Essex
CO16 9QG
Tel: (01255) 863200
Fax: (01255) 863206
Email: syngenta.bioline@syngenta.com
Web: www.syngenta-bioline.co.uk

Truchem: Truchem Ltd
The Knoll
The Cross
Horsley
Stroud
Gloucestershire
GL6 0PR
Tel: (01453) 833293
Fax: (01453) 833293
Email: nichollstruchem@aol.com

Unicrop: Universal Crop Protection Ltd
Park House
Maidenhead Road
Cookham
Berks.
SL6 9DS
Tel: (01628) 526083
Fax: (01628) 810457
Email: enquiries@unicrop.com

United Agri: United Agri Products Ltd
The Crossways
Alconbury Hill
Huntingdon
PE28 4JH
Tel: (01480) 418000
Fax: (01480) 418010

United Phosphorus: United Phosphorus Ltd
Chadwick House
Birchwood Park
Warrington
Cheshire
WA3 6AE
Tel: (01925) 819999
Fax: (01925) 817425
Email: lpinto@uniphos.com
Web: www.upleurope.com

Vitax: Vitax Ltd
Owen Street
Coalville
Leicester
LE67 3DE
Tel: (01530) 510060
Fax: (01530) 510299
Email: info@vitax.co.uk
Web: www.vitax.co.uk

Whyte Agrochemicals:
Whyte Agrochemicals Ltd
Marlborough House
298 Regents Park Road
Finchley
London
N3 2UA
Tel: (020) 8346 5946
Fax: (020) 8349 4589
Web: www.whytechemicals.co.uk

Appendix 2
Useful Contacts

Agriculture Industries Confederation Ltd
Confederation House
East of England Showground
Peterborough PE2 6XE
Tel: (01733) 385236
Fax: (01733) 385270
Web: www.agindustries.org.uk

BASIS Ltd
Bank Chambers
34 St John Street
Ashbourne
Derbyshire DE6 1GH
Tel: (01335) 343945/346138
Fax: (01335) 346488
Web: www.basis-reg.co.uk

British Beekeepers' Association
National Agricultural Centre
Stoneleigh
Kenilworth
Warwickshire CV8 2LZ
Tel: (024) 7669 6679
Web: www.bbka.org.uk

British Beer & Pub Association
Market Towers
1 Nine Elms Lane
London SW8 5NQ
Tel: (0207) 627 9191
Fax: (0207) 627 9123

British Crop Production Council (BCPC)
7 Omni Business Centre
Omega Park
Alton
Hampshire GU34 2QD
Tel: (01420) 593200
Fax: (01420) 593209
Web: www.bcpc.org

British Pest Control Association
1 Gleneagles House
Vernon Gate, South Street
Derby DE1 1UP
Tel: (01332) 294288
Fax: (01332) 295904
Web: www.bpca.org.uk

Crop Protection Association Ltd
2 Swan Court
Cygnet Park
Hampton
Peterborough PE7 8GX
Tel: (01733) 355370
Fax: (01733) 355371
Web: www.cropprotection.org.uk

CropLife International
Avenue Louise 143
B-1050 Brussels
Belgium
Tel: (+32) 2 542 0410
Fax: (+32) 2 542 0419
Web: www.croplife.org

Department of Agriculture and Rural Development (Northern Ireland)
Pesticides Section
Dundonald House
Upper Newtownards Road
Belfast BT4 3SB
Tel: (028) 9052 4704
Fax: (028) 9052 4059
Web: www.dardni.gov.uk

Department of Environment, Food and Rural Affairs (Defra)
Nobel House
17 Smith Square
London SW1P 3JR
Tel: (020) 7238 6000
Fax: (020) 7238 6591
Web: www.defra.gov.uk

Environment Agency
Rio House, Waterside Drive
Aztec West
Almondsbury
Bristol BS12 4UD
Tel: (01454) 624400
Fax: (01454) 624409
Web: www.environment-agency.gov.uk

European Crop Protection Association (ECPA)
Avenue E van Nieuwenhuyse 6
B-1160 Brussels
Belgium
Tel: (+32) 2 663 1550
Fax: (+32) 2 663 1560
Web: www.ecpa.be

Farmers' Union of Wales
Llys Amaeth
Queen's Square
Aberystwyth
Dyfed SY23 2EA
Tel: (01970) 612755
Fax: (01970)612755
Web: www.fuw.org.uk

Forestry Commission
231 Corstorphine Road
Edinburgh EH12 7AT
Tel: (0131) 334 0303
Fax: (0131) 334 3047
Web: www.forestry.gov.uk

Health and Safety Executive
Information Services
Room 318 Daniel House
Stanley Precinct, Bootle
Merseyside L20 3TW
Tel: (0151) 951 3191
Web: www.hse.gov.uk

Health and Safety Executive
Biocides & Pesticides Assessment Unit
Room 123 Magdalen House, Bootle
Merseyside L20 3QZ
Tel: (0151) 951 3535
Fax: (0151) 951 3317

Health and Safety Executive – Books
PO Box 1999
Sudbury
Suffolk CO10 2WA
Tel: (01787) 881165
Fax: (01787) 313995

Horticulture Development Council
Bradbourne House
East Malling
Kent ME19 6DZ
Tel: (01732) 848383
Fax: (01732) 848498
Web: www.hdc.org.uk

Lantra (previously ATB-LandBase)
National Agricultural Centre
Stoneleigh
Kenilworth
Warwickshire CV8 2LG
Tel: (024) 7669 6996
Fax: (024) 7669 6732
Web: www.lantra.co.uk

National Association of Agricultural Contractors (NAAC)
Samuelson House, Paxton Road
Orton Centre, Peterborough
Cambs PE2 5LT
Tel: (01733) 362920
Fax: (01733) 362921
Web: www.naac.co.uk

National Chemicals Emergency Centre
Building 329H, Harwell
Didcot
Oxon, OX11 0QJ
Tel: (0870) 190 6621
Fax: (0870) 190 6620
Web: http://the-ncec.com/

National Farmers' Union
Agriculture House
Stoneleigh Park
Stoneleigh
Warwickshire CV8 2TZ
Tel: (024) 7685 0662
Fax: (024) 7685 8501
Web: www.nfu.org.uk

National Poisons Information Service (Birmingham Unit)
City Hospital
Birmingham B18 7QH
Tel: (0121) 507 4123
Fax: (0121) 507 5580
Web: www.npis.org

National Turfgrass Council
Hunter's Lodge
Dr Brown's Road
Minchinhampton
Glos. GL6 9BT
Tel: (01453) 883588
Fax: (01453) 731449

NPTC
National Agricultural Centre
Stoneleigh
Kenilworth
Warwickshire CV8 2LG
Tel: (024) 7669 6553
Fax: (020) 7331 7313
Web: www.nptc.org.uk

Chemicals Regulation Directorate
Mallard House
King's Pool
3 Peasholme Green
York YO1 2PX
Tel: (01904) 640500
Fax: (01904) 455733
Web: www.pesticides.gov.uk

Processors and Growers Research Organisation
The Research Station
Great North Road, Thornhaugh
Peterborough, Cambs. PE8 6HJ
Tel: (01780) 782585
Fax: (01780) 783993
Web: www.pgro.co.uk

Scottish Beekeepers' Association
North Trinity House
114 Trinity Road
Edinburgh EH5 3JZ
Tel: (0131) 552 5341
Web: www.scottishbeekeepers.org.uk

Scottish Environment Protection Agency (SEPA)
Erskine Court
The Castle Business Park
Stirling FK9 4TR
Tel: (01786) 457 700
Fax: (01786) 446 885
Web: www.sepa.org.uk

TSO (The Stationery Office)
Publications Centre
PO Box 276
London SW8 5DT
Tel: (020) 7873 9090 (orders)
(020) 7873 0011 (enquiries)
Fax: (020) 7873 8200
Web: www.thestationeryoffice.com

Ulster Beekeepers' Association
57 Liberty Road
Carrickfergus
Co. Antrim BT38 9DJ
Tel: (01960) 362998
Web: www.ubka.org

Welsh Beekeepers' Association
Trem y Clawdd
Fron Isaf, Chirk
Wrexham
Clwyd LL14 5AH
Tel/Fax: (01691) 773300

Appendix 3
Keys to Crop and Weed Growth Stages

Decimal Code for the Growth Stages of Cereals

Illustrations of these growth stages can be found in the reference indicated below and in some company product manuals.

0 Germination

- 00 Dryseed
- 03 Imbibition complete
- 05 Radicle emerged from caryopsis
- 07 Coleoptile emerged from caryopsis
- 09 Leaf at coleoptile tip

1 Seedling growth

- 10 First leaf through coleoptile
- 11 First leaf unfolded
- 12 2 leaves unfolded
- 13 3 leaves unfolded
- 14 4 leaves unfolded
- 15 5 leaves unfolded
- 16 6 leaves unfolded
- 17 7 leaves unfolded
- 18 8 leaves unfolded
- 19 9 or more leaves unfolded

2 Tillering

- 20 Main shoot only
- 21 Main shoot and 1 tiller
- 22 Main shoot and 2 tillers
- 23 Main shoot and 3 tillers
- 24 Main shoot and 4 tillers
- 25 Main shoot and 5 tillers
- 26 Main shoot and 6 tillers
- 27 Main shoot and 7 tillers
- 28 Main shoot and 8 tillers
- 29 Main shoot and 9 or more tillers

3 Stem elongation

- 30 Ear at 1 cm
- 31 1st node detectable
- 32 2nd node detectable
- 33 3rd node detectable
- 34 4th node detectable
- 35 5th node detectable
- 36 6th node detectable
- 37 Flag leaf just visible
- 39 Flag leaf ligule/collar just visible

4 Booting

- 41 Flag leaf sheath extending
- 43 Boots just visibly swollen
- 45 Boots swollen
- 47 Flag leaf sheath opening
- 49 First awns visible

5 Inflorescence

- 51 First spikelet of inflorescence just visible
- 52 1/4 of inflorescence emerged
- 55 1/2 of inflorescence emerged
- 57 3/4 of inflorescence emerged
- 59 Emergence of inflorescence completed

6 Anthesis

- 60
- 61 } Beginning of anthesis
- 64
- 65 } Anthesis half way
- 68
- 69 } Anthesis complete

7 Milk development

- 71 Caryopsis watery ripe
- 73 Early milk
- 75 Medium milk
- 77 Late milk

8 Dough development

- 83 Early dough
- 85 Soft dough
- 87 Hard dough

9 Ripening

- 91 Caryopsis hard (difficult to divide by thumb-nail)
- 92 Caryopsis hard (can no longer be dented by thumb-nail)
- 93 Caryopsis loosening in daytime

(From Tottman, 1987. *Annals of Applied Biology*, **110**, 441–454)

Stages in Development of Oilseed Rape

Illustrations of these growth stages can be found in the reference indicated below and in some company product manuals.

0 Germination and emergence

1 Leaf production

1,0 Both cotyledons unfolded and green
1,1 First true leaf
1,2 Second true leaf
1,3 Third true leaf
1,4 Fourth true leaf
1,5 Fifth true leaf
1,10 About tenth true leaf
1,15 About fifteenth true leaf

2 Stem extension

2,0 No internodes ('rosette')
2,5 About five internodes

3 Flower bud development

3,0 Only leaf buds present
3,1 Flower buds present but enclosed by leaves
3,3 Flower buds visible from above ('green bud')
3,5 Flower buds raised above leaves
3,6 First flower stalks extending
3,7 First flower buds yellow ('yellow bud')

4 Flowering

4,0 First flower opened
4,1 10% all buds opened
4,3 30% all buds opened
4,5 50% all buds opened

5 Pod development

5,3 30% potential pods
5,5 50% potential pods
5,7 70% potential pods
5,9 All potential pods

6 Seed development

6,1 Seeds expanding
6,2 Most seeds translucent but full size
6,3 Most seeds green
6,4 Most seeds green-brown mottled
6,5 Most seeds brown
6,6 Most seeds dark brown
6,7 Most seeds black but soft
6,8 Most seeds black and hard
6,9 All seeds black and hard

7 Leaf senescence

8 Stem senescence

8,1 Most stem green
8,5 Half stem green
8,9 Little stem green

9 Pod senescence

9,1 Most pods green
9,5 Half pods green
9.9 Few pods green

(From Sylvester-Bradley, 1985. *Aspects of Applied Biology*, **10**, 395–400)

Stages in Development of Peas

Illustrations of these growth stages can be found in the reference indicated below and in some company product manuals.

0 Germination and emergence

- 000 Dry seed
- 001 Imbibed seed
- 002 Radicle apparent
- 003 Plumule and radicle apparent
- 004 Emergence

1 Vegetative stage

- 101 First node (leaf with one pair leaflets, no tendril)
- 102 Second node (leaf with one pair leaflets, simple tendril)
- 103 Third node (leaf with one pair leaflets, complex tendril)
- •
- •
- l0x X nodes (leaf with more than one pair leaflets, complex tendril)
- •
- •
- 10n Last recorded node

2 Reproductive stage (main stem)

- 201 Enclosed buds
- 202 Visible buds
- 203 First open flower
- 204 Pod set (small immature pod)
- 205 Flat pod
- 206 Pod swell (seeds small, immature)
- 207 Podfill
- 208 Pod green, wrinkled
- 209 Pod yellow, wrinkled (seeds rubbery)
- 210 Dry seed

3 Senescence stage

- 301 Desiccant application stage. Lower pods dry and brown, middle yellow, upper green. Overall moisture content of seed less than 45%
- 302 Pre-harvest stage. Lower and middle pods dry and brown, upper yellow. Overall moisture content of seed less than 30%
- 303 Dry harvest stage. All pods dry and brown, seed dry

(From Knott, 1987. *Annals of Applied Biology*, **111**, 233–244)

Stages in Development of Faba Beans

Illustrations of these growth stages can be found in the reference indicated below and in some company product manuals.

0 Germination and emergence

000 Dry seed
001 Imbibed seed
002 Radicle apparent
003 Plumule and radicle apparent
004 Emergence
005 First leaf unfolding
006 First leaf unfolded

1 Vegetative stage

101 First node
102 Second node
103 Third node
•
•
l0x X nodes
•
•
10n N, last recorded node

2 Reproductive stage (main stem)

201 Flower buds visible
203 First open flowers
204 First pod set
205 Pods fully formed, green
207 Pod fill, pods green
209 Seed rubbery, pods pliable, turning black
210 Seed dry and hard, pods dry and black

3 Pod senescence

301 10% pods dry and black
•
•
305 50% pods dry and black
•
•
308 80% pods dry and black, some upper pods green
309 90% pods dry and black, most seed dry. Desiccation stage.
310 All pods dry and black, seed hard. Pre-harvest (glyphosate application stage)

4 Stem senescence

401 10% stem brown/black
•
•
405 50% stem brown/black
•
•
409 90% stem brown/black
410 All stems brown/black. All pods dry and black, seed hard.

(From Knott, 1990. *Annals of Applied Biology*, **116**, 391–404)

SECTION 6

Stages in Development of Potato

Illustrations of these growth stages can be found in the reference indicated below and in some company product manuals.

0 Seed germination and seedling emergence

000 Dry seed
001 Imbibed seed
002 Radicle apparent
003 Elongation of hypocotyl
004 Seedling emergence
005 Cotyledons unfolded

1 Tuber dormancy

100 Innate dormancy (no sprout development under favourable conditions)
150 Enforced dormancy (sprout development inhibited by environmental conditions)

2 Tuber sprouting

200 Dormancy break, sprout development visible
21x Sprout with 1 node
22x Sprout with 2 nodes
•
•
29x Sprout with 9 nodes
21x(2) Second generation sprout with 1 node
22x(2) Second generation sprout with 2 nodes
•
•
29x(2) Second generation sprout with 9 nodes

Where x = 1, sprout <2 mm;
2, 2-5 mm; 3, 5-20 mm;
4, 20-30 mm; 5, 50-100 mm;
6, 100-150 mm long

3 Emergence and shoot expansion

300 Main stem emergence
301 Node 1
302 Node 2
•
•
319 Node 19
Second order branch
321 Node 1
•
•
Nth order branch
3N1 Node 1
•
•
3N9 Node 9

4 Flowering

Primary flower
400 No flowers
410 Appearance of flower bud
420 Flower unopen
430 Flower open
440 Flower closed
450 Berry swelling
460 Mature berry
Second order flowers
410(2) Appearance of flower bud
420(2) Flower unopen
430(2) Flower open
440(2) Flower closed
450(2) Berry swelling
460(2) Mature berry

5 Tuber development

500 No stolons
510 Stolon initials
520 Stolon elongation
530 Tuber initiation
540 Tuber bulking (>10 mm diam)
550 Skin set
560 Stolon development

6 Senescence

600 Onset of yellowing
650 Half leaves yellow
670 Yellowing of stems
690 Completely dead

(From Jefferies & Lawson, 1991. *Annals of Applied Biology*, **119**, 387–389)

Stages in Development of Linseed

Illustrations of these growth stages can be found in the reference indicated below and in some company product manuals.

0 Germination and emergence

- 00 Dry seed
- 01 Imbibed seed
- 02 Radicle apparent
- 04 Hypocotyl extending
- 05 Emergence
- 07 Cotyledon unfolding from seed case
- 09 Cotyledons unfolded and fully expanded

1 Vegetative stage (of main stem)

- 10 True leaves visible
- 12 First pair of true leaves fully expanded
- 13 Third pair of true leaves fully expanded
- 1*n* *n* leaf fully expanded

2 Basal branching

- 21 One branch
- 22 Two branches
- 23 Three branches
- 2*n* *n* branches

3 Flower bud development (on main stem)

- 31 Enclosed bud visible in leaf axils
- 33 Bud extending from axil
- 35 Corymb formed
- 37 Buds enclosed but petals visible
- 39 First flower open

4 Flowering (whole plant)

- 41 10% of flowers open
- 43 30% of flowers open
- 45 50% of flowers open
- 49 End of flowering

5 Capsule formation (whole plant)

- 51 10% of capsules formed
- 53 30% Of capsules formed
- 55 50% of capsules formed
- 59 End of capsule formation

6 Capsule senescence (on most advanced plant)

- 61 Capsules expanding
- 63 Capsules green and full size
- 65 Capsules turning yellow
- 67 Capsules all yellow brown but soft
- 69 Capsules brown, dry and senesced

7 Stem senescence (whole plant)

- 71 Stems mostly green below panicle
- 73 Most stems 30% brown
- 75 Most stems 50% brown
- 77 Stems 75% brown
- 79 Stems completely brown

8 Stems rotting (retting)

- 81 Outer tissue rotting
- 85 Vascular tissue easily removed
- 89 Stems completely collapsed

9 Seed development (whole plant)

- 91 Seeds expanding
- 92 Seeds white but full size
- 93 Most seeds turning ivory yellow
- 94 Most seeds turning brown
- 95 All seeds brown and hard
- 98 Some seeds shed from capsule
- 99 Most seeds shed from capsule

(From Freer, 1991. *Aspects of Applied Biology*, **28**, 33–40)

SECTION 6

Stages in Development of Annual Grass Weeds

Illustrations of these growth stages can be found in the reference indicated below and in some company product manuals.

0 Germination and emergence

- 00 Dry seed
- 01 Start of imbibition
- 03 Imbibition complete
- 05 Radicle emerged from caryopsis
- 07 Coleoptile emerged from caryopsis
- 09 Leaf just at coleoptile tip

1 Seedling growth

- 10 First leaf through coleoptile
- 11 First leaf unfolded
- 12 2 leaves unfolded
- 13 3 leaves unfolded
- 14 4 leaves unfolded
- 15 5 leaves unfolded
- 16 6 leaves unfolded
- 17 7 leaves unfolded
- 18 8 leaves unfolded
- 19 9 or more leaves unfolded

2 Tillering

- 20 Main shoot only
- 21 Main shoot and 1 tiller
- 22 Main shoot and 2 tillers
- 23 Main shoot and 3 tillers
- 24 Main shoot and 4 tillers
- 25 Main shoot and 5 tillers
- 26 Main shoot and 6 tillers
- 27 Main shoot and 7 tillers
- 28 Main shoot and 8 tillers
- 29 Main shoot and 9 or more tillers

3 Stem elongation

- 31 First node detectable
- 32 2nd node detectable
- 33 3rd node detectable
- 34 4th node detectable
- 35 5th node detectable
- 36 6th node detectable
- 37 Flag leaf just visible
- 39 Flag leaf ligule just visible

4 Booting

- 41 Flag leaf sheath extending
- 43 Boots just visibly swollen
- 45 Boots swollen
- 47 Flag leaf sheath opening
- 49 First awns visible

5 Inflorescence emergence

- 51 First spikelet of inflorescence just visible
- 53 1/4 of inflorescence emerged
- 55 1/2 of inflorescence emerged
- 57 3/4 of inflorescence emerged
- 59 Emergence of inflorescence completed

6 Anthesis

- 61 Beginning of anthesis
- 65 Anthesis half-way
- 69 Anthesis complete

(From Lawson & Read, 1992. *Annals of Applied Biology*, **12**, 211–214)

Growth Stages of Annual Broad-leaved Weeds

Preferred Descriptive Phrases

Illustrations of these growth stages can be found in the reference indicated below and in some company product manuals.

Pre-emergence
Early cotyledons
Expanded cotyledons
One expanded true leaf
Two expanded true leaves
Four expanded true leaves
Six expanded true leaves
Plants up to 25 mm across/high
Plants up to 50 mm across/high
Plants up to 100 mm across/high
Plants up to 150 mm across/high
Plants up to 250 mm across/high
Flower buds visible
Plant flowering
Plant senescent

(From Lutman & Tucker, 1987. *Annals of Applied Biology*, **110**, 683–687)

Appendix 4
Key to Hazard Classifications and Safety Precautions

Every product label contains information to warn users of the risks from using the product, together with precautions that must be followed in order to minimise the risks. A hazard classification (if any) and symbol is shown with associated risk phrases, followed by a series of safety precautions. These are represented in the pesticide profiles in Section 2 by code letters and numbers under the heading **Hazard classification and safety precautions**.

The codes are defined below, under the same sub-headings as they appear in the pesticide profiles.

Where a product label specifies the use of personal protective equipment (PPE), the requirements are listed under the sub-heading **Operator protection**, using letter codes to denote the protective items, according to the list below. Often PPE requirements are different for specified operations, e.g. handling the concentrate, cleaning equipment etc., but it is not possible to list them separately. The lists of PPE are therefore an indication of what the user may need to have available to use the product in different ways. **When making a COSHH assessment it is therefore essential that the product label is consulted for information on the particular use that is being assessed**.

Where the generalised wording includes a phrase such as '... for *xx* days', the specific requirement for each pesticide is shown in brackets after the code.

Hazard

H01	Very toxic
H02	Toxic
H03	Harmful
H04	Irritant
H05	Corrosive
H06	Extremely flammable
H07	Highly flammable
H08	Flammable
H09	Oxidising agent
H10	Explosive
H11	Dangerous for the environment

Risk phrases

R08	Contact with combustible material may cause fire
R09	Explosive when mixed with combustible material
R16	Explosive when mixed with oxidising substances
R20	Harmful by inhalation
R21	Harmful in contact with skin
R22a	Harmful if swallowed
R22b	May cause lung damage if swallowed
R23	Toxic by inhalation
R24	Toxic in contact with skin
R25	Toxic if swallowed
R25b	Toxic; Danger of serious damage to health by prolonged exposure if swallowed
R26	Very toxic by inhalation
R27	Very toxic in contact with skin
R28	Very toxic if swallowed
R34	Causes burns
R35	Causes severe burns
R36	Irritating to eyes

R37	Irritating to respiratory system
R38	Irritating to skin
R39	Danger of very serious irreversible effects
R40	Limited evidence of a carcinogenic effect
R41	Risk of serious damage to eyes
R42	May cause sensitization by inhalation
R43	May cause sensitization by skin contact
R45	May cause cancer
R46	May cause heritable genetic damage
R48	Danger of serious damage to health by prolonged exposure
R50	Very toxic to aquatic organisms
R51	Toxic to aquatic organisms
R52	Harmful to aquatic organisms
R53a	May cause long-term adverse effects in the aquatic environment
R53b	Dangerous to aquatic organisms
R54	Toxic to flora
R55	Toxic to fauna
R56	Toxic to soil organisms
R57	Toxic to bees
R58	May cause long-term adverse effects in the environment
R60	May impair fertility
R61	May cause harm to the unborn child
R62	Possible risk of impaired fertility
R63	Possible risk of harm to the unborn child
R64	May cause harm to breast-fed babies
R66	Repeated exposure may cause skin dryness or cracking
R67	Vapours may cause drowsiness and dizziness
R68	Possible risk of irreversible effects
R69	Danger of serious damage to health by prolonged oral exposure.
R70	May produce an allergic reaction

Operator protection

A	Suitable protective gloves (the product label should be consulted for any specific requirements about the material of which the gloves should be made)
B	Rubber gauntlet gloves
C	Face-shield
D	Approved respiratory protective equipment
E	Goggles
F	Dust mask
G	Full face-piece respirator
H	Coverall
J	Hood
K	Apron/Rubber apron
L	Waterproof coat
M	Rubber boots
N	Waterproof jacket and trousers
P	Suitable protective clothing
U01	To be used only by operators instructed or trained in the use of chemical/product/type of produce and familiar with the precautionary measures to be observed
U02a	Wash all protective clothing thoroughly after use, especially the inside of gloves
U02b	Avoid excessive contamination of coveralls and launder regularly
U02c	Remove and wash contaminated gloves immediately
U03	Wash splashes off gloves immediately
U04a	Take off immediately all contaminated clothing
U04b	Take off immediately all contaminated clothing and wash underlying skin. Wash clothes before re-use
U04c	Wash clothes before re-use
U05a	When using do not eat, drink or smoke
U05b	When using do not eat, drink, smoke or use naked lights
U06	Handle with care and mix only in a closed container
U07	Open the container only as directed (returnable containers only)
U08	Wash concentrate/dust from skin or eyes immediately

U09a Wash any contamination/splashes/dust/powder/concentrate from skin or eyes immediately
U09b Wash any contamination/splashes/dust/powder/concentrate from eyes immediately
U10 After contact with skin or eyes wash immediately with plenty of water
U11 In case of contact with eyes rinse immediately with plenty of water and seek medical advice
U12 In case of contact with skin rinse immediately with plenty of water and seek medical advice
U13 Avoid contact by mouth
U14 Avoid contact with skin
U15 Avoid contact with eyes
U16a Ensure adequate ventilation in confined spaces
U16b Use in a well ventilated area
U18 Extinguish all naked flames, including pilot lights, when applying the fumigant/dust/liquid/product
U19a Do not breathe dust/fog/fumes/gas/smoke/spray mist/vapour. Avoid working in spray mist
U19b Do not work in confined spaces or enter spaces in which high concentrations of vapour are present. Where this precaution cannot be observed distance breathing or self-contained breathing apparatus must be worn, and the work should be done by trained operators
U19c In case of insufficient ventilation, wear suitable respiratory equipment
U20a Wash hands and exposed skin before eating, drinking or smoking and after work
U20b Wash hands and exposed skin before eating and drinking and after work
U20c Wash hands before eating and drinking and after work
U20d Wash hands after use
U20e Wash hands and exposed skin after cleaning and re-calibrating equipment
U21 Before entering treated crops, cover exposed skin areas, particularly arms and legs
U22a Do not touch sachet with wet hands or gloves/Do not touch water soluble bag directly
U22b Protect sachets from rain or water
U23a Do not apply by knapsack sprayer/hand-held equipment
U23b Do not apply through hand held rotary atomisers
U23c Do not apply via tractor mounted horizontal boom sprayers
U24 Do not handle grain unnecessarily
U25 Open the container only as directed

Environmental protection

E02a Keep unprotected persons/animals out of treated/fumigation areas for at least xx hours/days
E02b Prevent access by livestock, pets and other non-target mammals and birds to buildings under fumigation and ventilation
E02c Vacate treatment areas before application
E02d Exclude all persons and animals during treatment
E03 Label treated seed with the appropriate precautions, using the printed sacks, labels or bag tags supplied
E05a Do not apply directly to livestock/poultry
E05b Keep poultry out of treated areas for at least xx days/weeks
E05c Do not apply directly to animals
E06a Keep livestock out of treated areas for at least xx days/weeks after treatment
E06b Dangerous to livestock. Keep all livestock out of treated areas/away from treated water for at least xx days/weeks. Bury or remove spillages
E06c Harmful to livestock. Keep all livestock out of treated areas/away from treated water for at least xx days/weeks. Bury or remove spillages
E06d Exclude livestock from treated fields. Livestock may not graze or be fed treated forage nor may it be used for hay silage or bedding
E07a Keep livestock out of treated areas for up to two weeks following treatment and until poisonous weeds, such as ragwort, have died down and become unpalatable
E07b Dangerous to livestock. Keep livestock out of treated areas/away from treated water for at least xx weeks and until foliage of any poisonous weeds, such as ragwort, has died and become unpalatable
E07c Harmful to livestock. Keep livestock out of treated areas/away from treated water for at least xx days/weeks and until foliage of any poisonous weeds such as ragwort has died and become unpalatable

E07d	Keep livestock out of treated areas for up to 4-6 weeks following treatment and until poisonous weeds, such as ragwort, have died down and become unpalatable
E08a	Do not feed treated straw or haulm to livestock within xx days/weeks of spraying
E08b	Do not use on grassland if the crop is to be used as animal feed or bedding
E09	Do not use straw or haulm from treated crops as animal feed or bedding for at least xx days after last application
E10a	Dangerous to game, wild birds and animals
E10b	Harmful to game, wild birds and animals
E10c	Dangerous to game, wild birds and animals. All spillages must be buried or removed
E11	Paraquat can be harmful to hares; spray stubbles early in the day
E12a	High risk to bees
E12b	Extremely dangerous to bees
E12c	Dangerous to bees
E12d	Harmful to bees
E12e	Do not apply to crops in flower or to those in which bees are actively foraging. Do not apply when flowering weeds are present
E12f	Do not apply to crops in flower, or to those in which bees are actively foraging, except as directed on [crop]. Do not apply when flowering weeds are present
E12g	Apply away from bees
E13a	Extremely dangerous to fish or other aquatic life. Do not contaminate surface waters or ditches with chemical or used container
E13b	Dangerous to fish or other aquatic life. Do not contaminate surface waters or ditches with chemical or used container
E13c	Harmful to fish or other aquatic life. Do not contaminate surface waters or ditches with chemical or used container
E13d	Apply away from fish
E13e	Harmful to fish or other aquatic life. The maximum concentration of active inggredient in treated water must not exceed XX ppm or such lower concentration as the appropriate water regulatory body may require.
E14a	Extremely dangerous to aquatic higher plants. Do not contaminate surface waters or ditches with chemical or used container
E14b	Dangerous to aquatic higher plants. Do not contaminate surface waters or ditches with chemical or used container
E15a	Do not contaminate surface waters or ditches with chemical or used container
E15b	Do not contaminate water with product or its container. Do not clean application equipment near surface water. Avoid contamination via drains from farmyards or roads
E15c	To protect groundwater, do not apply to grass leys less than 1 year old
E16a	Do not allow direct spray from horizontal boom sprayers to fall within 5 m of the top of the bank of a static or flowing waterbody, unless a Local Environment Risk Assessment for Pesticides (LERAP) permits a narrower buffer zone, or within 1 m of the top of a ditch which is dry at the time of application. Aim spray away from water
E16b	Do not allow direct spray from hand-held sprayers to fall within 1 m of the top of the bank of a static or flowing waterbody. Aim spray away from water
E16c	Do not allow direct spray from horizontal boom sprayers to fall within 5 m of the top of the bank of a static or flowing waterbody, or within 1m of the top of a ditch which is dry at the time of application. Aim spray away from water. This product is not eligible for buffer zone reduction under the LERAP horizontal boom sprayers scheme.
E16d	Do not allow direct spray from hand-held sprayers to fall within 1 m of the top of the bank of a static or flowing waterbody. Aim spray away from water. This product is not eligible for buffer zone reduction under the LERAP horizontal boom sprayers scheme scheme.
E16e	Do not allow direct spray from horizontal boom sprayers to fall within 5 m of the top of the bank of a static or flowing water body or within 1 m from the top of any ditch which is dry at the time of application. Spray from hand held sprayers must not in any case be allowed to fall within 1 m of the top of the bank of a static or flowing water body. Always direct spray away from water. The LERAP scheme does not extend to adjuvants. This product is therefore not eligible for a reduced buffer zone under the LERAP scheme.
E16f	Do not allow direct spray/granule applications from vehicle mounted/drawn hydraulic sprayers/applicators to fall within 6 m of surface waters or ditches/Do not allow direct spray/granule applications from hand-held sprayers/applicators to fall within 2 m of surface waters or ditches. Direct spray/applications away from water
E16g	Do not allow direct spray from train sprayers to fall within 5 m of the top of the bank of any static or flowing waterbody.

E17a	Do not allow direct spray from broadcast air-assisted sprayers to fall within xx m of surface waters or ditches. Direct spray away from water
E17b	Do not allow direct spray from broadcast air-assisted sprayers to fall within xx m of the top of the bank of a static or flowing waterbody, unless a Local Environmental Risk Assessment for Pesticides (LERAP) permits a narrower buffer zone, or within 5 m of the top of a ditch which is dry at the time of application. Aim spray away from water
E18	Do not spray from the air within 250 m horizontal distance of surface waters or ditches
E19a	Do not dump surplus herbicide in water or ditch bottoms
E19b	Do not empty into drains
E20	Prevent any surface run-off from entering storm drains
E21	Do not use treated water for irrigation purposes within xx days/weeks of treatment
E22a	High risk to non-target insects or other arthropods. Do not spray within 6 m of the field boundary
E22b	Risk to certain non-target insects or other arthropods. For advice on risk management and use in Integrated Pest Management (IPM) see directions for use
E22c	Risk to non-target insects or other arthropods
E23	Avoid damage by drift onto susceptible crops or water courses
E34	Do not re-use container for any purpose/Do not re-use container for any other purpose
E35	Do not burn this container
E36a	Do not rinse out container (returnable containers only)
E36b	Do not open or rinse out container (returnable containers only)
E37	Do not use with any pesticide which is to be applied in or near water
E38	Use appropriate containment to avoid environmental contamination
E39	Extreme care must be taken to avoid spray drift onto non-crop plants outside the target area
E40	To protect groundwater/soil organisms the maximum total dose of this or other products containing ethofumesate MUST NOT exceed 1.0 kg ethofumesate per hectare in any three year period
E41	Hay for silage must not be cut from treated crops for at least 21 days after treatment

Consumer protection

C01	Do not use on food crops
C02a	Do not harvest for human or animal consumption for at least xx days/weeks after last application
C02b	Do not remove from store for sale or processing for at least 21 days after application
C02c	Do not remove from store for sale or processing for at least 2 days after application
C04	Do not apply to surfaces on which food/feed is stored, prepared or eaten
C05	Remove/cover all foodstuffs before application
C06	Remove exposed milk before application
C07	Collect eggs before application
C08	Protect food preparing equipment and eating utensils from contamination during application
C09	Cover water storage tanks before application
C10	Protect exposed water/feed/milk machinery/milk containers from contamination
C11	Remove all pets/livestock/fish tanks before treatment/spraying
C12	Ventilate treated areas thoroughly when smoke has cleared/Ventilate treated rooms thoroughly before occupying

Storage and disposal

D01	Keep out of reach of children
D02	Keep away from food, drink and animal feeding-stuffs
D03	Store away from seeds, fertilizers, fungicides and insecticides
D04	Store well away from corms, bulbs, tubers and seeds
D05	Protect from frost
D06a	Store away from heat
D06b	Do not store near heat or open flame
D06c	Do not store in direct sunlight
D06d	Do not store above 30/35 C
D07	Store under cool, dry conditions
D08	Store in a safe, dry, frost-free place designated as an agrochemical store
D09a	Keep in original container, tightly closed, in a safe place
D09b	Keep in original container, tightly closed, in a safe place, under lock and key

D09c Store unused sachets in a safe place. Do not store half-used sachets
D10a Wash out container thoroughly and dispose of safely
D10b Wash out container thoroughly, empty washings into spray tank and dispose of safely
D10c Rinse container thoroughly by using an integrated pressure rinsing device or manually rinsing three times. Add washings to sprayer at time of filling and dispose of container safely
D10d Do not rinse out container
D11a Empty container completely and dispose of safely/Dispose of used generator safely
D11b Empty container completely and dispose of it in the specified manner
D12a This material (and its container) must be disposed of in a safe way
D12b This material and its container must be disposed of as hazardous waste
D13 Treat used container as if it contained pesticide
D14 Return empty container as instructed by supplier (returnable containers only)
D15 Store container in purpose built chemical store until returned to supplier for refilling (returnable containers only)
D16 Do not store below 4 degrees Centigrade
D17 Use immediately on removal of foil
D18 Place the tablets whole into the spray tank - do not break or crumble the tablets

Treated Seed

S01 Do not handle treated seed unnecessarily
S02 Do not use treated seed as food or feed
S03 Keep treated seed secure from people, domestic stock/pets and wildlife at all times during storage and use
S04a Bury or remove spillages
S04b Harmful to birds/game and wildlife. Treated seed should not be left on the soil surface. Bury or remove spillages
S04c Dangerous to birds/game and wildlife. Treated seed should not be left on the soil surface. Bury or remove spillages
S04d To protect birds/wild animals, treated seed should not be left on the soil surface. Bury or remove spillages
S05 Do not reuse sacks or containers that have been used for treated seed for food or feed
S06a Wash hands and exposed skin before meals and after work
S06b Wash hands and exposed skin after cleaning and re-calibrating equipment
S07 Do not apply treated seed from the air
S08 Treated seed should not be broadcast
S09 Label treated seed with the appropriate precautions using printed sacks, labels or bag tags supplied

Vertebrate/Rodent control products

V01a Prevent access to baits/powder by children, birds and other animals, particularly cats, dogs, pigs and poultry
V01b Prevent access to bait/gel/dust by children, birds and non-target animals, particularly dogs, cats, pigs, poultry
V02 Do not prepare/use/lay baits/dust/spray where food/feed/water could become contaminated
V03a Remove all remains of bait, tracking powder or bait containers after use and burn or bury
V03b Remove all remains of bait and bait containers/exposed dust/after treatment (except where used in sewers) and dispose of safely (e.g. burn/bury). Do not dispose of in refuse sacks or on open rubbish tips.
V04a Search for and burn or bury all rodent bodies. Do not place in refuse bins or on rubbish tips
V04b Search for rodent bodies (except where used in sewers) and dispose of safely (e.g. burn/bury). Do not dispose of in refuse sacks or on open rubbish tips
V04c Dispose of safely any rodent bodies and remains of bait and bait containers that are recovered after treatment (e.g. burn/bury). Do not dispose of in refuse sacks or on open rubbish tips
V05 Use bait containers clearly marked POISON at all surface baiting points

Medical advice

M01 This product contains an anticholinesterase organophosphorus compound. DO NOT USE if under medical advice NOT to work with such compounds

M02 This product contains an anticholinesterase carbamate compound. DO NOT USE if under medical advice NOT to work with such compounds

M03 If you feel unwell, seek medical advice immediately (show the label where possible)

M04a In case of accident or if you feel unwell, seek medical advice immediately (show the label where possible)

M04b In case of accident by inhalation, remove casualty to fresh air and keep at rest

M05a If swallowed, seek medical advice immediately and show this container or label

M05b If swallowed, do not induce vomiting: seek medical advice immediately and show this container or label

M05c If swallowed induce vomiting if not already occurring and take patient to hospital immediately

M06 This product contains an anticholinesterase carbamoyl triazole compound. DO NOT USE if under medical advice NOT to work with such compounds

Appendix 5
Key to Abbreviations and Acronyms

The abbreviations of formulation types in the following list are used in Section 2 (Pesticide Profiles) and are derived from the *Catalogue of Pesticide Formulation Types and International Coding System* (CropLife International Technical Monograph 2, 5th edn, March 2002).

1 Formulation Types

AB	Grain bait
AE	Aerosol generator
AL	Other liquids to be applied undiluted
AP	Any other powder
BB	Block bait
BR	Briquette
CB	Bait concentrate
CC	Capsule suspension in a suspension concentrate
CF	Capsule suspension for seed treatment
CG	Encapsulated granule (controlled release)
CL	Contact liquid or gel (for direct application)
CP	Contact powder (for direct application)
CR	Crystals
CS	Capsule suspension
DC	Dispersible concentrate
DP	Dustable powder
DS	Powder for dry seed treatment
EC	Emulsifiable concentrate
EG	Emulsifiable granule
EO	Water in oil emulsion
ES	Emulsion for seed treatment
EW	Oil in water emulsion
FG	Fine granules
FP	Smoke cartridge
FS	Flowable concentrate for seed treatment
FT	Smoke tablet
FU	Smoke generator
FW	Smoke pellets
GA	Gas
GB	Granular bait
GE	Gas-generating product
GG	Macrogranules
GL	Emulsifiable gel
GP	Flo-dust (for pneumatic application)
GR	Granules
GS	Grease
GW	Water soluble gel
HN	Hot fogging concentrate
KK	Combi-pack (solid/liquid)
KL	Combi-pack (liquid/liquid)
KN	Cold-fogging concentrate
KP	Combi-pack (solid/solid)
LA	Lacquer
LI	Liquid, unspecified
LS	Solution for seed treatment
ME	Microemulsion
MG	Microgranules
MS	Microemulsion for seed treatment
OD	Oil dispersion
OL	Oil miscible liquid
PA	Paste

PC	Gel or paste concentrate
PS	Seed coated with a pesticide
PT	Pellet
RB	Ready-to-use bait
RH	Ready-to-use spray in hand-operated sprayer
SA	Sand
SC	Suspension concentrate (= flowable)
SE	Suspo-emulsion
SG	Water soluble granules
SL	Soluble concentrate
SP	Water soluble powder
SS	Water soluble powder for seed treatment
ST	Water soluble tablet
SU	Ultra low-volume suspension
TB	Tablets
TC	Technical material
TP	Tracking powder
UL	Ultra low-volume liquid
VP	Vapour releasing product
WB	Water soluble bags
WG	Water dispersible granules
WP	Wettable powder
WS	Water dispersible powder for slurry treatment of seed
WT	Water dispersible tablet
XX	Other formulations
ZZ	Not Applicable

2 Other Abbreviations and Acronyms

ACP	Advisory Committee on Pesticides
ACTS	Advisory Committee on Toxic Substances
ADAS	Agricultural Development and Advisory Service
a.i.	active ingredient
AIC	Agriculture Industries Confederation
BBPA	British Beer and Pub Association
CDA	Controlled droplet application
CPA	Crop Protection Association
cm	centimetre(s)
COPR	Control of Pesticides Regulations 1986
COSHH	Control of Substances Hazardous to Health Regulations
CRD	Chemicals Regulation Directorate
d	day(s)
Defra	Department for Environment, Food and Rural Affairs
EA	Environment Agency
EBDC	ethylene-bis-dithiocarbamate fungicide
FEPA	Food and Environment Protection Act 1985
g	gram(s)
GCPF	Global Crop Protection Federation (now CropLife International)
GS	growth stage (unless in formulation column)
h	hour(s)
ha	hectare(s)
HBN	hydroxybenzonitrile herbicide
HI	harvest interval
HSE	Health and Safety Executive
ICM	integrated crop management
IPM	integrated pest management
kg	kilogram(s)
l	litre(s)
LERAP	Local Environmental Risk Assessments for Pesticides
m	metre(s)
MAFF	Ministry of Agriculture, Fisheries and Food (now Defra)
MBC	methyl benzimidazole carbamate fungicide
MEL	maximum exposure limit

min	minute(s)
mm	millimetre(s)
MRL	maximum residue level
mth	month(s)
NA	Notice of Approval
NFU	National Farmers' Union
OES	Occupational Exposure Standard
OLA	off-label approval
PPE	personal protective equipment
PPPR	Plant Protection Products Regulations
SOLA	specific off-label approval
ULV	ultra-low volume
VI	Voluntary Initiative
w/v	weight/volume
w/w	weight/weight
wk	week(s)
yr	year(s)

3 UN Number details

UN No.	Substance	EAC	APP	Hazards Class	Sub risks	HIN
1692	**Strychnine or Strychnine salts**	**2X**		6.1		66
1760	**Corrosive Liquid, N.O.S.,** Packing groups II & III	**2X**	B	8		88
1830	**Sulphuric acid,** with more than 51% acid	**2P**		8		80
1993	**Flammable Liquid, N.O.S.,** Packing group III	**•3Y**		3		30
2011	**Magnesium Phosphide**	**4W**[1]		4.3	6.1	
2588	**Pesticide, Solid, TOXIC, N.O.S.**	**2X**		6.1		66/60
2757	**Carbamate Pesticide, Solid, Toxic**	**2X**		6.1		66/60
2783	**Organophosphorus Pesticide, Solid, TOXIC**	**2X**		6.1		66/60
2902	**Pesticide, Liquid, TOXIC, N.O.S.,** Packing Groups I, II & III	**2X**	B	6.1		66/60
3077	**Environmentally hazardous substance, Solid, N.O.S.**	**2Z**		9		90
3082	**Environmentally hazardous substance, Solid, N.O.S.**	**•3Z**		9		90
3265	**Corrosive Liquid, Acidic, Organic, N.O.S.,** Packing group 1, II or III	**2X**	B	8		80/88
3351	**Pyrethroid Pesticide, Liquid, Toxic, Flammable,** Flash Point 23 °C or more	**•3W**	A(fl)	6.1	3	663/63
3352	**Pyrethroid Pesticide, Liquid, TOXIC,** Packing groups I, II & III	**2X**	B	6.1		66/60

Ref: Dangerous Goods Emergency Action Code List 2009 (TSO).
[1] Not applicable to the carriage of dangerous goods under RID or ADR.

Appendix 6
Definitions

The descriptions used in this *Guide* for the crops or situations in which products are approved for use are those used on the approved product labels. These are now standardised in a Crop Hierarchy published by the Pesticides Safety Directorate in which definitions are given. To assist users of this *Guide* the definitions of some of the terminology where misunderstandings can occur are reproduced below.

Rotational grass: Short-term grass crops grown on land that is likely to be growing different crops in future years (*e.g. short-term intensively managed leys for one to three years that may include clover*)

Permanent grassland: Grazed areas that are intended to be permanent in nature (*e.g. permanent pasture and moorland that can be grazed*).

Ornamental Plant Production: All ornamental plants that are grown for sale or are produced for replanting into their final growing position (*e.g. flowers, house plants, nursery stock, bulbs grown in containers or in the ground*).

Managed Amenity Turf: Areas of frequently mown, intensively managed, turf that is not intended to flower and set seed. It includes areas that may be for intensive public use (*e.g. all types of sports turf*).

Amenity Grassland: Areas of semi-natural or planted grassland subject to minimal management. It includes areas that may be accessed by the public (*e.g. railway and motorway embankments, airfields, and grassland nature reserves*). These areas may be managed for their botanical interest, and the relevant authority should be contacted before using pesticides in such locations.

Amenity Vegetation: Areas of semi-natural or ornamental vegetation, including trees, or bare soil around ornamental plants, or soil intended for ornamental planting. It includes areas to which the public have access. It does NOT include hedgerows around arable fields.

Natural surfaces not intended to bear vegetation: Areas of soil or natural outcroppings of rock that are not intended to bear vegetation, including areas such as sterile strips around fields. It may include areas to which the public have access. It does not include the land between rows of crops.

Hard surfaces: Man-made impermeable surfaces that are not intended to bear vegetation (*e.g. pavements, tennis courts, industrial areas, railway ballast*).

Permeable surfaces overlying soil: Any man-made permeable surface (excluding railway ballast) such as gravel that overlies soil and is not intended to bear vegetation

Green Cover on Land Temporarily Removed from Production: Includes fields covered by natural regeneration or by a planted green cover crop that will not be harvested (*e.g. green cover on setaside*). It does NOT include industrial crops.

Forest Nursery: Areas where young trees are raised outside for subsequent forest planting.

Forest: Groups of trees being grown in their final positions. Covers all woodland grown for whatever objective, including commercial timber production, amenity and recreation, conservation and landscaping, ancient traditional coppice and farm forestry, and trees from natural regeneration, colonisation or coppicing. Also includes restocking of established woodlands and new planting on both improved and unimproved land.

Farm forestry: Groups of trees established on arable land or improved grassland including those planted for short rotation coppicing. It includes mature hedgerows around arable fields.

Indoors (for rodenticide use): Situations where the bait is placed within a building or other enclosed structure, and where the target is living or feeding predominantly within that building or structure.

Herbs: Reference to Herbs or Protected Herbs when used in Section 2 may include any or all of the following. The particular label or SOLA Notice will indicate which species are included in the approval.

Agastache spp.
Angelica
Applemint
Balm
Basil
Bay
Borage (except when grown for oilseed)
Camomile
Caraway
Catnip
Chervil
Clary
Clary sage
Coriander
Curry plant
Dill
Dragonhead
English chamomile
Fennel
Fenugreek
Feverfew
French lavender
Gingermint
Hyssop
Korean mint
Land cress
Lavandin
Lavender
Lemon balm
Lemon peppermint
Lemon thyme
Lemon verbena
Lovage
Marigold
Marjoram
Mint
Mother of thyme
Nasturtium
Nettle
Oregano
Origanum heracleoticum
Parsley root
Peppermint
Pineapplemint
Rocket
Rosemary
Rue
Sage
Salad burnet
Savory
Sorrel
Spearmint
Spike lavender
Tarragon
Thyme
Thymus camphoratus
Violet
Winter savory
Woodruff

Herbs for Medicinal Uses: Reference to Herbs for Medicinal Uses when used in Section 2 may include any or all of the following. The particular label or SOLA Notice will indicate which species are included in the approval

Black cohosh
Burdock
Dandelion
Echinacea
Ginseng
Goldenseal
Liquorice
Nettle
Valerian

Appendix 7
References

The information given in *The UK Pesticide Guide* provides some of the answers needed to assess health risks, including the hazard classification and the level of operator protection required. However, the *Guide* cannot provide all the details needed for a complete hazard assessment, which must be based on the product label itself and, where necessary, the Health and Safety Data Sheet and other official literature.

Detailed guidance on how to comply with the Regulations is available from several sources.

Pesticides: Code of Practice

Code of Practice for Using Plant Protection Products, 2006 (Defra publication PB 11090)

Known as the 'Green Code', this revised Code of Practice, which came into effect on 15 December 2005, replaced all previous editions of the *Code of Practice for the Safe Use of Pesticides on Farms and Holdings*. It also replaced the voluntary code of practice for the use of pesticides in amenity and industrial areas (the 'Orange Code'). The code explains how plant protection products can be used safely and so meet the legal requirements of the Control of Pesticides Regulations 1986 (COPR) (ISBN 0-110675-10-X), and the Control of Substances Hazardous to Health Regulations 2002 (COSHH) (ISBN 0-110429-19-2). The principal source of information for making a COSHH assessment is the approved product label. In most cases the label provides all the necessary information, but in certain circumstances other sources must be consulted, and these are listed in the Code of Practice.

The code, which applies to England and Wales only, is available in CD-ROM format and electronically from the PSD and Defra websites. A Welsh language version is available in printed and electronic formats. A Scottish version (approved by the Scottish Parliament) is being produced in printed and electronic formats. Northern Ireland will produce its own version of the code in due course.

Other Codes of Practice

Code of Practice for Suppliers of Pesticides to Agriculture, Horticulture and Forestry (the 'Yellow Code') (Defra Booklet PB 0091)

Code of Good Agricultural Practice for the Protection of Soil (Defra Booklet PB 0617)

Code of Good Agricultural Practice for the Protection of Water (Defra Booklet PB 0587)

Code of Good Agricultural Practice for the Protection of Air (Defra Booklet PB 0618)

Approved Code of Practice for the Control of Substances Hazardous to Health in Fumigation Operations. Health and Safety Commission (ISBN 0-717611-95-7)

Code of Best Practice: Safe use of Sulphuric Acid as an Agricultural Desiccant. National Association of Agricultural Contractors, 2002. (Also available at www.naac.co.uk/?Codes/acidcode.asp)

Safe Use of Pesticides for Non-agricultural Purposes. HSE Approved Code of Practice. HSE L21 (ISBN 0-717624-88-9)

Other Guidance and Practical Advice

HSE (by mail order from HSE Books – See Appendix 2)

COSHH – A brief guide to the Regulations 2003 (INDG136)

A Step by Step Guide to COSHH Assessment, 2004 (HSG97) (ISBN 0-717627-85-3)

Defra (from The Stationery Office – see Appendix 2)

Local Environment Risk Assessments for Pesticides (LERAP): Horizontal Boom Sprayers.

Local Environment Risk Assessments for Pesticides (LERAP): Broadcast Air-assisted Sprayers.

Crop Protection Association and the Voluntary Initiative (see Appendix 2)

Every Drop Counts: Keeping Water Clean

Best Practice Guides. A range of leaflets giving guidance on best practice when dealing with pesticides before, during and after application.

H2OK? Best Practice Advice and Decision Trees - July 2009/10

BCPC (British Crop Production Council – see Appendix 2)

The Pesticide Manual (15th edition) (ISBN 978-1-901396-18-8)

The e-Pesticide Manual (version 5.0) (ISBN 978-1-901396-19-5)

The Manual of Biocontrol Agents (4th edition) (ISBN 978-1-901396-17-1)

IdentiPest PC CD-ROM (ISBN 1-901396-05-3)

Garden Detective PC CD-ROM (ISBN 1-901396-32-0)

Small Scale Spraying (ISBN 1-901396-07-X)

Field Scale Spraying (ISBN 1-901396-08-8)

Using Pesticides (ISBN 1-901396-10-X)

Safety Equipment Handbook (ISBN 1-901396-06-1)

The Environment Agency (see Appendix 2)

Best Farming Practices: Profiting from a Good Environment

Use of Herbicides in or Near Water

SECTION 7
INDEX

Index of Proprietary Names of Products

The references are to entry numbers, not to pages. Adjuvant names are referred to as 'Adj' and are listed separately in Section 4

REFERENCES ARE TO ENTRY NUMBERS NOT PAGES

REFERENCES ARE TO ENTRY NUMBERS NOT PAGES

REFERENCES ARE TO ENTRY NUMBERS NOT PAGES

REFERENCES ARE TO ENTRY NUMBERS NOT PAGES

REFERENCES ARE TO ENTRY NUMBERS NOT PAGES

THE UK PESTICIDE GUIDE 2010

RE-ORDERS

☐ Please send me ____ more copies of *The UK Pesticide Guide 2010* at £43.95 each

Postage: single copy £2.95; orders up to £150, £5.95; orders over £150, no charge.

Outside the UK please add £10 per order for delivery.

Name ______________________ Position ______________________

Institution ______________________ Department ______________________

Address ____

City ____________ Region ____________ Postcode ________ Country ____________

Tel ____________ Fax ____________ E-mail ____________

EU countries except UK – VAT No: ______________________

Payment (pre-payment is required):

☐ I enclose a cheque/draft for £________ payable to BCPC. Please send me a receipt.

☐ I wish to pay by credit card: ☐ Visa ☐ Mastercard ☐ Amex ☐ Switch

Please charge to my card £________ and send me a receipt. Name of issuing bank ______________

Card no. ☐☐☐☐ ☐☐☐☐ ☐☐☐☐ ☐☐☐☐

Expiry date ☐☐/☐☐ Security code ☐☐☐ Switch cards only: Start date ☐☐/☐☐ Issue no. ☐

Signature ______________________ Date ____________

Name and address of cardholder if different from above: ______________________

____

BCPC

Please photocopy and return to:
BCPC Publications Sales, 7 Omni Business Centre, Omega Park, Alton, Hants GU34 2QD, UK
Tel: 01420 593 200, Fax: 01420 593 209, Email: publications@bcpc.org, Web: www.bcpc.org

FUTURE EDITIONS

☐ I wish to take out an annual order for ____ copies of each new edition of *The UK Pesticide Guide*

☐ Please send me advance price details for the 2011 edition of *The UK Pesticide Guide* when available

Order by phone: 01420 593 200 or online: www.bcpc.org/bookshop

Bookshop orders to:

www.cabi.org

Customer Services, CAB International, Nosworthy Way,
Wallingford, Oxfordshire OX10 8DE, UK
Tel: 01491 832111, Fax: 01491 829292,
Email: enquiries@cabi.org, Web: www.cabi.org

Bulk discount:

100+ copies	30%
50–99	25%
10–49	15%
List price £43.95	

Thank you for your order